T0140059

Lecture Notes in Artificial Intelligence 13343

Subseries of Lecture Notes in Computer Science

Series Editors

Randy Goebel
University of Alberta, Edmonton, Canada

Wolfgang Wahlster
DFKI, Berlin, Germany

Zhi-Hua Zhou
Nanjing University, Nanjing, China

Founding Editor

Jörg Siekmann
DFKI and Saarland University, Saarbrücken, Germany

More information about this subseries at https://link.springer.com/bookseries/1244

Hamido Fujita · Philippe Fournier-Viger ·
Moonis Ali · Yinglin Wang (Eds.)

Advances and Trends in Artificial Intelligence

Theory and Practices in Artificial Intelligence

35th International Conference
on Industrial, Engineering and Other Applications
of Applied Intelligent Systems, IEA/AIE 2022
Kitakyushu, Japan, July 19–22, 2022
Proceedings

 Springer

Editors
Hamido Fujita (ID)
i-SOMET, Inc.
Morioka-shi, Iwate, Japan

Moonis Ali
Texas State University
San Marcos, TX, USA

Philippe Fournier-Viger (ID)
College of Computer Science and Software
Engineering
Shenzhen University
Shenzhen, Guangdong, China

Yinglin Wang
Shanghai University of Finance
and Economics
Shanghai, China

ISSN 0302-9743   ISSN 1611-3349 (electronic)
Lecture Notes in Artificial Intelligence
ISBN 978-3-031-08529-1   ISBN 978-3-031-08530-7 (eBook)
https://doi.org/10.1007/978-3-031-08530-7

LNCS Sublibrary: SL7 – Artificial Intelligence

This Springer imprint is published by the registered company Springer Nature Switzerland AG
The registered company address is: Gewerbestrasse 11, 6330 Cham, Switzerland

Preface

In the last few decades, there have been major societal transformations due to the ever-increasing usage of computing devices. Impacts can be observed in all fields including science, governance, healthcare, industry, and the lives of individuals. Computers can calculate faster, store more data, and are smaller, while also being cheaper. Improved and specialized computing architectures have also been developed such as GPUs and FPGAs. Besides, distributed computing and storage platforms have become common to process very large databases. Thanks to technological advances and also several theoretical breakthroughs, researchers and practitioners have pushed back the limits of artificial intelligence to build more effective intelligent systems to solve real-world complex problems. Moreover, innovative applications of artificial intelligence are continuously being proposed.

This volume contains the proceedings of the 35th edition of the International Conference on Industrial, Engineering, and other Applications of Applied Intelligent Systems (IEA AIE 2022), which was during July 19–22, 2022, in Kitakyushu, Japan. IEA AIE is a yearly conference that focuses on applications of applied intelligent systems to solve real-life problems in all areas including business and finance, science, engineering, industry, cyberspace, bioinformatics, automation, robotics, medicine and biomedicine, and human-machine interactions. IEA AIE 2022 was organized in cooperation with the ACM Special Interest Group on Artificial Intelligence (SIGAI). This year, 127 submissions were received. Each paper was evaluated using double-blind peer review by at least three reviewers from an international Program Committee consisting of 74 members from 23 countries. Based on the evaluation, a total of 67 papers were selected as full papers and 11 as short papers, which are presented in this book. We would like to thank all the reviewers for the time spent on writing detailed and constructive comments for the authors, and to the latter for the proposal of many high-quality papers.

In the program of IEA AIE 2022, five special sessions were organized: Collective Intelligence in Social Media (CISM 2022), Intelligent Knowledge Engineering in Decision Making Systems (IKEDS 2022), Intelligent Systems and e-Applications (ISeA 2022), Multi-Agent Systems and Metaheuristics for Complex Problems (MASMCP 2022), and Spatiotemporal Big Data Analytics (SBDA 2022). In addition, two keynote talks were given by two distinguished researchers, one by Sebastian Ventura from the University of Cordoba (Spain) and the other by Tao Wu from the Shanghai University of Medicine and Health Sciences (China). We would like to thank everyone who has contributed to the success of this year's edition of IEA AIE, that is the authors, reviewers, keynote speakers, Program Committee members, and organizers.

May 2021

Hamido Fujita
Philippe Fournier-Viger
Moonis Ali
Yinglin Wang

Organization

Honorary Chair

Tao Wu — Shanghai University of Medicine and Health Sciences, China

General Chairs

Hamido Fujita — Iwate Prefectural University, Japan
Moonis Ali — Texas State University, USA

Organizing Chair

Jun Sasaki — i-SOMET Inc., Japan

Program Committee Chairs

Philippe Fournier-Viger — Shenzhen University, China
Yinglin Wang — Shanghai University of Finance and Economics, China

Special Session Chairs

Ali Selamat — Universiti Teknologi Malaysia, Malaysia
Xing Wu — Shanghai University, China
Jerry Chun-Wei Lin — Western Norway University of Applied Sciences, Norway
Ngoc Thanh Nguyen — Wroclaw University of Technology, Poland

Program Committee

Moulay A. Akhloufi — Université de Moncton, Canada
Azri Azmi — Universiti Teknologi Malaysia, Malaysia
Hafewa Bargaoui — Institut Supérieur de Gestion de Tunis, Tunisia
Olfa Belkahla Driss — Ecole Supérieure de Commerce de Tunis, Tunisia
Ladjel Bellatreche — LIAS/ISAE-ENSMA, France
Zalan Bodo — Babes-Bolyai University, Romania
Zaki Brahmi — ISITCOM, Tunisia
Francisco J. Cabrerizo — University of Granada, Spain

Alberto Cano	Virginia Commonwealth University, USA
Andrew Tzer-Yeu Chen	University of Auckland, New Zealand
Chun-Hao Chen	National Taipei University of Technology, Taiwan
Shyi-Ming Chen	National Taiwan University of Science and Technology, Taiwan
Tai Dinh	JAIST, Japan
Youcef Djenouri	Southern Denmark University, Denmark
Alexander Ferrein	Aachen University of Applied Science, Germany
Philippe Fournier-Viger	Shenzhen University, China
Hamido Fujita	Iwate Prefectural University, Japan
Abdennaceur Ghandri	Higher Institute of Management of Gabes, Tunisia
Sergei Gorlatch	Muenster University, Germany
Deepak Gupta	NIT Arunachal Pradesh, India
Tzung-Pei Hong	National University of Kaohsiung, Taiwan
Ko-Wei Huang	National Kaohsiung University of Science and Technology, Taiwan
Miroslav Hudec	University of Economics in Bratislava, Slovakia
Dosam Hwang	Yeungnam University, South Korea
Marcin Jodłowiec	Wroclaw University of Science and Technology, Poland
Fadoua Khennou	Université de Moncton, Canada
Yun Sing Koh	University of Auckland, New Zealand
Adrianna Kozierkiewicz	Wroclaw University of Science and Technology, Poland
Marek Krótkiewicz	Wroclaw University of Science and Technology, Poland
Masaki Kurematsu	Iwate Prefectural University, Japan
Thomas Lacombe	University of Auckland, New Zealand
Shih Hsiung Lee	National Kaohsiung University of Applied Sciences, Taiwan
Arkadiusz Liber	Wroclaw University of Science and Technology, Poland
Jerry Chun-Wei Lin	Western Norway University of Applied Sciences, Norway
Wen-Yang Lin	National University of Kaohsiung, Taiwan
Yu-Chen Lin	Feng Chia University, Taiwan
Frederick Maier	University of Georgia, USA
Wolfgang Mayer	University of South Australia, Australia
Masurah Mohamad	Universiti Teknologi Malaysia, Malaysia
M. Rashedur Rahman	North South University, Bangladesh
Yasser Mohammed	Assiut University, Egypt
Tauheed Khan Mohd	Augustana College, USA
Anirban Mondal	Ashoka University, India

Contents

Industrial Applications

Comparative Study of Methods for the Real-Time Detection of Dynamic
Bottlenecks in Serial Production Lines 3
Nikolai West, Jörn Schwenken, and Jochen Deuse

Ultra-short-Term Load Forecasting Model Based on VMD and TGCN-GRU ... 15
*Meirong Ding, Hang Zhang, Biqing Zeng, Gaoyan Cai, Yuan Chai,
and Wensheng Gan*

Learning to Match Product Codes 29
Ying Excell and Sebastian Link

ResUnet: A Fully Convolutional Network for Speech Enhancement
in Industrial Robots 42
Yangyi Pu and Hongyang Yu

Surface Defect Detection and Classification Based on Fusing Multiple
Computer Vision Techniques 51
*Min Zhu, Bingqing Shen, Yan Sun, Chongyu Wang, Guoxin Hou,
Zhijie Yan, and Hongming Cai*

Development of a Multiagent Based Order Picking Simulator
for Optimizing Operations in a Logistics Warehouse 63
Takuto Sakuma, Minami Watanabe, Koya Ihara, and Shohei Kato

Health Informatics

Predicting Infection Area of Dengue Fever for Next Week Through
Multiple Factors 77
*Cong-Han Zheng, Ping-Yu Hsu, Ming-Shien Cheng, Ni Xu,
and Yu-Chun Chen*

Hospital Readmission Prediction via Personalized Feature Learning
and Embedding: A Novel Deep Learning Framework 89
Yuxi Liu and Shaowen Qin

Intelligent Medical Interactive Educational System for Cardiovascular
Disease 101
Sheng-Shan Chen, Hou-Tsan Lee, Tun-Wen Pai, and Chao-Hung Wang

Evolutionary Optimization for CNN Compression Using Thoracic X-Ray
Image Classification 112
    Hassen Louati, Slim Bechikh, Ali Louati, Abdulaziz Aldaej,
    and Lamjed Ben Said

An Oriented Attention Model for Infectious Disease Cases Prediction 124
    Peisong Zhang, Zhijin Wang, Guoqing Chao, Yaohui Huang,
    and Jingwen Yan

The Differential Gene Detecting Method for Identifying Leukemia Patients 137
    Mingzhao Wang, Weiliang Jiang, and Juanying Xie

Epidemic Modeling of the Spatiotemporal Spread of COVID-19
over an Intercity Population Mobility Network 147
    Yuxi Liu, Shaowen Qin, and Zhenhao Zhang

Skin Cancer Classification Using Different Backbones of Convolutional
Neural Networks 160
    Anh T. Huynh, Van-Dung Hoang, Sang Vu, Trong T. Le,
    and Hien D. Nguyen

Cardiovascular Disease Detection on X-Ray Images with Transfer Learning ... 173
    Nguyen Van-Binh and Nguyen Thai-Nghe

Causal Reasoning Methods in Medical Domain: A Review 184
    Xing Wu, Jingwen Li, Quan Qian, Yue Liu, and Yike Guo

Optimization

Enhancing a Multi-population Optimisation Approach with a Dynamic
Transformation Scheme 199
    Shengqi Dai, Vincent W. L. Tam, Zhenglong Li, and L. K. Yeung

A Model Driven Approach to Transform Business Vision-Oriented
Decision-Making Requirement into Solution-Oriented Optimization Model 211
    Liwen Zhang, Hervé Pingaud, Elyes Lamine, Franck Fontanili,
    Christophe Bortolaso, and Mustapha Derras

A Hybrid Approach Based on Genetic Algorithm with Ranking
Aggregation for Feature Selection 226
    Bui Quoc Trung, Le Minh Duc, and Bui Thi Mai Anh

A Novel Type-Based Genetic Algorithm for Extractive Summarization 240
    Bui Thi Mai Anh, Nguyen Thi Thu Trang, and Tran Thi Dinh

Dragonfly Algorithm for Multi-target Search Problem in Swarm Robotic
with Dynamic Environment Size 253
  Mohd Ghazali Mohd Hamami and Zool H. Ismail

Video and Image Processing

Improved Processing of Ultrasound Tongue Videos by Combining
ConvLSTM and 3D Convolutional Networks 265
  Amin Honarmandi Shandiz and László Tóth

Improvement of Text Image Super-Resolution Benefiting Multi-task
Learning 275
  Kosuke Honda, Hamido Fujita, and Masaki Kurematsu

Question Difficulty Estimation with Directional Modality Association
in Video Question Answering 287
  Bong-Min Kim and Seong-Bae Park

Natural Language Processing

Improving Neural Machine Translation by Efficiently Incorporating
Syntactic Templates 303
  Phuong Nguyen, Tung Le, Thanh-Le Ha, Thai Dang, Khanh Tran,
  Kim Anh Nguyen, and Nguyen Le Minh

Forensic Analysis of Text and Messages in Smartphones by a Unification
Rosetta Stone Procedure 315
  Claudio Tomazzoli, Simone Scannapieco, and Matteo Cristani

Relation-Level Vector Representation for Relation Extraction
and Classification on Specialized Data 327
  Camille Gosset, Mokhtar Boumedyen Billami, Mathieu Lafourcade,
  Christophe Bortolaso, and Mustapha Derras

SAKE: A Graph-Based Keyphrase Extraction Method Using Self-attention 339
  Ping Zhu, Chuanyang Gong, and Zhihua Wei

Synonym Prediction for Vietnamese Occupational Skills 351
  Hai-Nam Cao, Duc-Thai Do, Viet-Trung Tran, Tuan-Dung Cao,
  and Young-In Song

A Survey of Pretrained Embeddings for Japanese Legal Representation 363
  Ha-Thanh Nguyen, Le-Minh Nguyen, and Ken Satoh

Machine Reading Comprehension Model for Low-Resource Languages
and Experimenting on Vietnamese 370
   Bach Hoang Tien Nguyen, Dung Manh Nguyen,
   and Trang Thi Thu Nguyen

Inducing a Malay Lexicon from an Unlabelled Dataset Using Word
Embeddings 382
   Ian H. J. Ho, Hui-Ngo Goh, and Yi-Fei Tan

Agent and Group-Based Systems

Agent-Based Intermodal Behavior for Urban Toll 397
   Azise Oumar Diallo, Guillaume Lozenguez, Arnaud Doniec,
   and René Mandiau

Entropy Based Approach to Measuring Consensus in Group
Decision-Making Problems 409
   J. M. Tapia, F. Chiclana, M. J. del Moral, and E. Herrera–Viedma

Adaptation of HMIs According to Users' Feelings Based on Multi-agent
Systems 416
   Alia Maaloul, Houssem Eddine Nouri, Zied Trifa, and Olfa Belkahla Driss

Pattern Recognition

A Generalized Inverted Dirichlet Predictive Model for Activity
Recognition Using Small Training Data 431
   Jiaxun Guo, Manar Amayri, Wentao Fan, and Nizar Bouguila

Deepfake Detection Using CNN Trained on Eye Region 443
   David Johnson, Tony Gwyn, Letu Qingge, and Kaushik Roy

Face Authentication from Masked Face Images Using Deep Learning
on Periocular Biometrics 452
   Jeffrey J. Hernandez V., Rodney Dejournett, Udayasri Nannuri,
   Tony Gwyn, Xiaohong Yuan, and Kaushik Roy

An Optimization Algorithm for Extractive Multi-document Summarization
Based on Association of Sentences 460
   Chun-Hao Chen, Yi-Chen Yang, and Jerry Chun-Wei Lin

A Spatiotemporal Image Fusion Method for Predicting High-Resolution
Satellite Images 470
   Vipul Chhabra, R. Uday Kiran, Juan Xiao, P. Krishna Reddy,
   and Ram Avtar

Security

WHTE: Weighted Hoeffding Tree Ensemble for Network Attack Detection
at Fog-IoMT 485
*Shilan S. Hameed, Ali Selamat, Liza Abdul Latiff, Shukor A. Razak,
and Ondrej Krejcar*

An Improved Ensemble Deep Learning Model Based on CNN
for Malicious Website Detection 497
Nguyet Quang Do, Ali Selamat, Kok Cheng Lim, and Ondrej Krejcar

Intrusion-Based Attack Detection Using Machine Learning Techniques
for Connected Autonomous Vehicle 505
*Mansi Bhavsar, Kaushik Roy, Zhipeng Liu, John Kelly,
and Balakrishna Gokaraju*

Detection of Anti-forensics and Malware Applications in Volatile Memory
Acquisition 516
*Chandlor Ratcliffe, Biodoumoye George Bokolo, Damilola Oladimeji,
and Bing Zhou*

Malware Classification Based on Graph Convolutional Neural Networks
and Static Call Graph Features 528
Attila Mester and Zalán Bodó

Modelling and Diagnosis

The Java2CSP Debugging Tool Utilizing Constraint Solving
and Model-Based Diagnosis Principles 543
Franz Wotawa and Vlad Andrei Dumitru

Formal Modelling and Security Analysis of Inter-Operable Systems 555
Abdelhakim Baouya, Samir Ouchani, and Saddek Bensalem

Social Network Analysis

Content-Context-Based Graph Convolutional Network for Fake News
Detection 571
Huyen Trang Phan, Ngoc Thanh Nguyen, and Dosam Hwang

Multi-class Sentiment Classification for Customers' Reviews 583
*Cuong T. V. Nguyen, Anh M. Tran, Thao Nguyen, Trung T. Nguyen,
and Binh T. Nguyen*

Transportation and Urban Applications

MM-AQI: A Novel Framework to Understand the Associations Between
Urban Traffic, Visual Pollution, and Air Pollution 597
 Kazuki Tejima, Minh-Son Dao, and Koji Zettsu

Two-Stage Traffic Clustering Based on HNSW 609
 Xu Zhang, Xinzheng Niu, Philippe Fournier-Viger, and Bing Wang

Explainable Online Lane Change Predictions on a Digital Twin
with a Layer Normalized LSTM and Layer-wise Relevance Propagation 621
 Christoph Wehner, Francis Powlesland, Bashar Altakrouri,
 and Ute Schmid

An Agenda on the Employment of AI Technologies in Port Areas:
The TEBETS Project 633
 Adorni Emanuele, Rozhok Anastasiia, Revetria Roberto,
 and Suchev Sergey

Modelling and Solving the Green Share-a-Ride Problem 648
 Elhem Elkout and Olfa Belkahla Driss

Machine Learning Techniques to Predict Real Time Thermal Comfort,
Preference, Acceptability, and Sensation for Automation of HVAC
Temperature 659
 Yaa T. Acquaah, Balakrishna Gokaraju, Raymond C. Tesiero III,
 and Kaushik Roy

Neural Networks

Serially Disentangled Learning for Multi-Layered Neural Networks 669
 Ryotaro Kamimura and Ryozo Kitajima

Detecting Use Case Scenarios in Requirements Artifacts: A Deep Learning
Approach 682
 Munima Jahan, Zahra Shakeri Hossein Abad, and Behrouz Far

Hybrid Deep Neural Networks for Industrial Text Scoring 695
 Sidharrth Nagappan, Hui-Ngo Goh, and Amy Hui-Lan Lim

Benchmarking Training Methodologies for Dense Neural Networks 707
 Isaac Tonkin, Geoff Harris, and Volodymyr Novykov

Proposing Novel High-Performance Compounds by Nested VAEs Trained
Independently on Different Datasets 714
  Yoshihiro Osakabe and Akinori Asahara

Clustering

Monotonic Constrained Clustering: A First Approach 725
  Germán González-Almagro, Pablo Sánchez Bermejo, Juan Luis Suarez,
  José-Ramón Cano, and Salvador García

Extractive Text Summarization on Large-scale Dataset Using K-Means
Clustering 737
  Ti-Hon Nguyen and Thanh-Nghi Do

Multi-Granular Large Scale Group Decision-Making Method with a New
Consensus Measure Based on Clustering of Alternatives in Modifiable
Scenarios 747
  José Ramón Trillo, Ignacio Javier Pérez, Enrique Herrera-Viedma,
  Juan Antonio Morente-Molinera, and Francisco Javier Cabrerizo

Optimal User Categorization from a Hierarchical Clustering Tree
for Recommendation 759
  Wei Song and Siqi Liu

Classification

A Preliminary Approach for using Metric Learning in Monotonic
Classification 773
  Juan Luis Suárez, Germán González-Almagro, Salvador García,
  and Francisco Herrera

Deep Learning Architectures Extended from Transfer Learning
for Classification of Rice Leaf Diseases 785
  Hai Thanh Nguyen, Quyen Thuc Quach, Chi Le Hoang Tran,
  and Huong Hoang Luong

Height Estimation for Abrasive Grain of Synthetic Diamonds
on Microscope Images by Conditional Adversarial Networks 797
  Joe Brinton, Shota Oki, Xin Yang, and Maiko Shigeno

Pattern Mining and Tsetlin Machines

Fast Weighted Sequential Pattern Mining 807
  Zhenqiang Ye, Ziyang Li, Weibin Guo, Wensheng Gan, Shicheng Wan,
  and Jiahui Chen

Parallel High Utility Itemset Mining 819
Gaojuan Fan, Huaiyuan Xiao, Chongsheng Zhang, George Almpanidis,
Philippe Fournier-Viger, and Hamido Fujita

Towards Efficient Discovery of Stable Periodic Patterns in Big Columnar
Temporal Databases 831
Hong N. Dao, Penugonda Ravikumar, P. Likitha,
Bathala Venus Vikranth Raj, R. Uday Kiran, Yutaka Watanobe,
and Incheon Paik

Cyclostationary Random Number Sequences for the Tsetlin Machine 844
Svein Anders Tunheim, Rohan Kumar Yadav, Lei Jiao, Rishad Shafik,
and Ole-Christoffer Granmo

Logics and Ontologies

Evolution of Prioritized $\mathcal{EL}$ Ontologies 859
Rim Mohamed, Zied Loukil, Faiez Gargouri, and Zied Bouraoui

A Comparison of Resource Data Framework and Inductive Logic
Programing for Ontology Development 871
Durgesh Nandini

MDNCaching: A Strategy to Generate Quality Negatives for Knowledge
Graph Embedding 877
Tiroshan Madushanka and Ryutaro Ichise

Robotics, Games and Consumer Applications

Application of a Limit Theorem to the Construction of Japanese Crossword
Puzzles 891
Volodymyr Novykov, Geoff Harris, and Isaac Tonkin

Non Immersive Virtual Laboratory Applied to Robotics Arms 898
Daniela A. Bastidas, Luis F. Recalde, Patricia N. Constante,
Victor H. Andaluz, Dayana E. Gallegos, and José Varela-Aldás

An Improved Subject-Independent Stress Detection Model Applied
to Consumer-grade Wearable Devices 907
Van-Tu Ninh, Manh-Duy Nguyen, Sinéad Smyth, Minh-Triet Tran,
Graham Healy, Binh T. Nguyen, and Cathal Gurrin

WDTourism: A Personalized Tourism Recommendation System Based
on Semantic Web 920
   Kaiyu Dai, Pengfei Ji, Xiaorui Zuo, and Daixin Dai

Author Index 935

Industrial Applications

Comparative Study of Methods for the Real-Time Detection of Dynamic Bottlenecks in Serial Production Lines

Nikolai West[1]([✉]) [iD], Jörn Schwenken[1] [iD], and Jochen Deuse[1,2] [iD]

[1] Institute of Production Systems, Technical University Dortmund, Leonhard-Euler-Str. 5, 44227 Dortmund, Germany
nikolai.west@tu-dortmund.de
[2] Centre for Advanced Manufacturing, University of Technology Sydney, 11 Broadway, Ultimo, NSW 2007, Australia

Abstract. Capacity-limiting bottlenecks in manufacturing systems form the ideal starting point for measures of improvement. However, the inherent variability of modern systems leads to dynamic bottleneck behavior, causing them to shift between stations. Numerous methods for the detection of shifting bottlenecks exist in literature. In this paper, we present and compare three methods: Bottleneck Walk (BNW), Active Period Method (APM), and an adaptation of Interdeparture Time Variances (ITV). The comparative study deploys the methods in a serial production line with seven stations and eight buffers. We vary the individual locations of the bottlenecks by adding more process time. To compare the methods, we determine the overall average ratio of agreement between the three detection methods. APM and ITV have the highest agreement at an average of 80.10%. Pairings with BNW achieve significantly lower rates of agreement, with 56.33% for ITV, and 62.03%% when compared to the APM.

Keywords: Bottleneck analysis · Bottleneck detection · Material flow simulation

1 Introduction

According to the *Theory of Constraints* (TOC), the performance of a material flow system or a manufacturing system is inevitably limited by just one station, which the TOC calls a *bottleneck* [1]. Like the weakest link in a chain, such a bottleneck limits the output of the entire system. Focusing all improvement efforts on these links of the chain constitutes the most efficient approach to increasing the overall performance of the system [2]. Since optimization measures of non-bottleneck stations have no impact on the systemic output, only improvements at the bottleneck lead to a quantifiable enhancement. Despite 'bottleneck' being a commonly used term in general linguistics, there is no accepted definition in scientific literature [3]. The comparative study in this paper shows that different metrics for bottleneck identification also led to varying bottleneck definitions adopted by the respective authors. Therefore, we follow a metric-independent definition and refer to a bottleneck in a general fashion as "the resource that restricts a systemic output up to a specific limitation" [4, 5].

© Springer Nature Switzerland AG 2022
H. Fujita et al. (Eds.): IEA/AIE 2022, LNAI 13343, pp. 3–14, 2022.
https://doi.org/10.1007/978-3-031-08530-7_1

1.1 On the Dynamic Nature of Bottlenecks

As mentioned before, a bottleneck's location can change over time due to variability. Variability refers to a fluctuation of product and process variables and affects all real-world manufacturing systems. Thus, literature classifies a bottleneck according to its behavior in being *static* or *dynamic* [5–7]. Static bottlenecks have a fixed position in the value stream. They only affect one station during the entire observation period. Their occurrence can often be traced back to design flaws which cause a singular restriction of the material flow. In contrast, the place of occurrence of dynamic bottlenecks is variable. Such shifts occur either due to random events or due to gradually changing conditions in the manufacturing system [8].

Despite this distinction, in practice often one station is referred to as main bottleneck. However, static assumptions only apply to very simple systems and are rarely observed in practice [7]. In this paper, we therefore examine methods for identifying dynamic bottlenecks. Still, the methods are also suited to detect static bottlenecks.

1.2 The Need for Real-Time Bottleneck Detection

Instead of pointwise measurements with manual efforts, dynamic behavior requires continuous recording and evaluation to allow targeted measures of optimization [6]. The primary goal is to detect the bottleneck that is currently affecting the system. Similar to the way bottlenecks can be differentiated according to their behavior, detection methods can be divided into the two groups of *Average Value Methods* (AVM) and *Momentary Value Methods* (MVM) [4, 9]. AVM use defined periods as basis for the analysis. They determine bottlenecks using the average values of different production metrics. AVM for bottleneck detection are based on a variety of metrics, such as the *Overall Throughput Effectiveness* [10], the *degree of Utilization* [11], or *Interdeparture Times Variances* [12]. These methods are less suitable to handle dynamic systems and tend to achieve relatively low detection confidence. In contrast, MVM use a singular observation. They enable an identification of shifting bottlenecks in dynamic systems [13]. Examples for MVM methods are the *Arrow Method* [14, 15], the *Active Period Method* [7], or the *Turning Point Method* [16, 17]. This selection does not claim to be comprehensive, as a multitude of methods and method variants exists. We refer to [4, 12, 18] for an extensive review of such methods.

In summary, *Bottleneck Detection*, as the first phase of a holistic bottleneck analysis [4], aims to detect shifting bottlenecks. Therefore, it is essential to first detect the momentary bottleneck before calculating averages of the overall systemic impact. Any method that uses averages to detect the bottlenecks is likely to introduce errors in the detection of shifting bottlenecks [13]. For the study in Sec. 4, we use two MVM approaches and adapt an AVM approach for momentary bottleneck detection.

The remainder of this paper is divided into four parts. First, we briefly describe a selection of three methods for real-time bottleneck detection that use different metrics. Next, we introduce the design of the case study, followed by an overview of the results for a serial production line. Finally, we discuss and compare the results of the study, and lastly provide a brief outlook with recommendations for future research.

2 Related Work on Bottleneck Detection

According to the TOC, any measure of improvement first requires a determination of the position of the bottleneck. As mentioned before, there is a large number of potential methods for detecting bottlenecks. We do not aim for a comparison of all available methods, but again refer to the respective literature instead [4, 12, 13, 18, 19].

Through our selection of the following three detection methods, we promote the utilization of different key figures from the manufacturing system. The Bottleneck Walk is focusses primarily on examining the levels of the production buffers. The Active Period method uses machine states to identify bottlenecks. Lastly, Interdeparture Time Variance detection evaluates of the process times of the workstations [4].

2.1 Detection Using Bottleneck Walk with Buffer Levels

The *Bottleneck Walk* (BNW) is a method for the identification of dynamic bottlenecks in serial lines with finite buffers [20]. BNW is closely related to the *Arrow Method* [14]. It serves as a hands-on method for bottleneck detection that requires practitioners to take a tour through the manufacturing line. While walking along the line, said observer writes down inventory levels and process states of the line into a defined data sheet. These states are recorded according to the observations, while distinguishing the three states *processing*, *breakdown* and *waiting*. For waiting, BNW requires a second subdivision: either a station can be blocked due to a full subsequent buffer, or it can be starving due to an empty preceding buffer. In addition, BNW considers the level of the buffers, where a clearly defined buffer can be filled between 0% and 100% with regard to its maximum capacity. If the currently observed buffer level is lower than one third of the maximum capacity, the bottleneck is located upstream. If the level is higher than two thirds of the capacity, the BNW pinpoints the station downstream. BNW uses an arrow-based system to determine a bottleneck station. A station with arrows pointing to it from both sides represents the bottleneck [20].

In this usage mode, BNW requires a manual check of the system state to be performed several times a day. For a data-driven application in flexible manufacturing systems, the methodology was therefore adapted for real-time monitoring [5]. Focusing on virtual buffer levels and neglecting the station states, allows making equivalent statements about the bottleneck station. Given a sufficient availability of data, this adoption of the BNW allows determining a bottleneck at any time.

2.2 Detection Using Active Period Method with Machine States

According to the *Active Period Method* (APM), a bottleneck is the station within the material flow system that has been working the longest without any interruptions. APM relies on a similar understanding of machine states as BNW. Workstations are considered *active* when they are processing products as defined by their production program, or when they are otherwise busy due to *repair*, *setup* or *maintenance* operations. In contrast, a station is called *inactive* if it is waiting due to buffer-related starvation or blockage [6, 7]. APM considers a station *blocked* if the downstream buffer is filled to the maximum. Then it is unable to transfer another part to the following stations. For the opposite case, if the upstream buffer is empty and it cannot supply another part to the next station, the station is considered *starved* [14]. While we simply refer to station states in this paper, APM allows a grouping of states for different entities of a production system. For example, for a processing machine, the states 'working', 'in repair', 'changing tools' or 'being serviced' are all considered active states. For a factory worker, 'working' or 'being on scheduled break' are active states, while 'planned or unplanned waiting' are considered inactive states. Regardless of the observed entity, the longest active operating period then marks the bottleneck [7].

APM also includes the shifting state of a bottleneck by determining whether a station is the sole bottleneck or a shifting bottleneck. Shifting states occur at the overlap of the current and the subsequent bottleneck periods. An accurate distinction between sole bottleneck and shifting bottleneck requires a retrospective view of the states. Several active stations at an observation time simply lead to an identification of shifting bottlenecks. Despite this disputable limitation, APM allows near real-time determination of bottleneck stations based on current conditions of said stations.

2.3 Detection Using Interdeparture Time Variance with Process Times

The third method utilizes *Interdeparture Time Variances* (ITV) to detect dynamic bottlenecks. As a method, ITV is based on the assumption that the bottleneck of a manufacturing system has the lowest variance in the interdeparture times of all products, i.e. the lowest ITV [12]. The method relies on a chain of considerations: If a machine in the manufacturing system requires a longer *effective processing time* to complete products, this machine tends to be more utilized. Higher utilization is associated with a higher utilization rate, which leads to a longer queue in front of the machine, i.e. a higher buffer level. Analogously, this can lead to a less highly filled buffer behind the considered machine. This buffer behavior leads to less frequent *starvation* or *blocking* of the machine, and the proportion of idle states is correspondingly lower. Similarly, the proportion of active machine states is significantly higher, which in turn leads to other machines starving downstream and blocking upstream. Due to these recurring interruptions, the variance of the times between the completion of products, i.e. the ITV, increases for the other machines. A bottleneck can then be unambiguously determined as the machine that achieves the lowest ITV of the entire system [12]. In addition to this logical-argumentative reasoning, the assumption can also be described in mathematical terms using the *linking equation*, in accordance with [11]. The calculations are further elaborated by [12] with reference to [11].

For ITV calculation, the process time must be determined as the difference to the time stamp of the next product, starting from a station-specific time stamp per station. The variance of these distances is then calculated for an aggregation interval. When selecting the interval length, it is important to select a sufficiently large and representative set of observation times. At the same time, a loss of information due to excessive aggregation must be prevented. To enable real-time detection like a MVM, the variance can also be determined at any time, using a sliding window with a defined length. This allows a bottleneck detection that utilizes the current system state.

3   Design of the Comparative Study for Bottleneck Detection

To compare the selected bottleneck detection methods, we build a serial production line with seven workstations M_i and eight buffers B_i as shown in Fig. 1. All stations are connected through a buffer, with the first buffer B_0 and the last buffer B_7 forming the system boundaries. For the boundaries, we assume infinite supply and demand. We set a maximum capacity of all buffers BC_{Bi} to 5 units. Each station has a defined process time pt_{Mi}, although we introduce variability into the system by means of an appropriate distribution function. The distribution function used to determine the individual process times is shown in Fig. 2. The right-skewed distribution corresponds to a manufacturing system with occasional equipment downtimes.

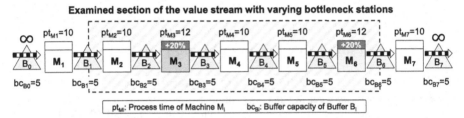

Examined section of the value stream with varying bottleneck stations

| pt_{Mi}: Process time of Machine M_i | bc_{Bi}: Buffer capacity of Buffer B_i |

Fig. 1. Layout of the serial production line with exemplary 20%-bottlenecks at M3 and M6

Since we need to apply the detection methods in systems with dynamic bottlenecks, we intentionally induce bottleneck states by applying a percentage factor to the process times of two selected stations. Figure 1, for example, shows a system configuration in which stations M_3 and M_6 have 20% higher process times. We show the corresponding distribution of process times in red in Fig. 2. For unchanged stations with an initial process time of 10, the average process time after variability adjustment is *17.94* units. Similarly, for modified stations with an initial process time of 12, the average time after variability penalty is *21.53*. Due to unlimited boundaries, M_1 does not starve while M_7 is never blocked. To compare different bottleneck situations, we use all bottleneck combinations, for M_2 to M_6, including singular bottleneck states with just one affected station. This results in 15 unique combinations and 25 in total.

For a better understanding of the studies result, we show a single example of our simulation results in Fig. 3, where M_3 and M_6 are again set to 20%-bottlenecks. In all following visualizations, the dashed line marks the bottleneck stations of the scenario. We run all simulations for 20,000 units, preceded by 5,000 units for initial settling of the system. The Y-axis shows the bottleneck station identified by the respective detection method, with the detection shown using the blue markers. In Fig. 3, for example, a shifting bottleneck occurs after about 9,500 units, the bottleneck at M_3 shifts to M_6. Moreover, since a detection is made for each point in time, unambiguous visualization is not always possible. In the example, after 14,500 time units, the bottleneck repeatedly alternates between station M_3 and M_6 for an interval of approx. 1,000 units.

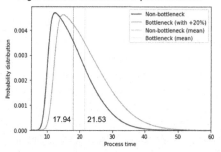
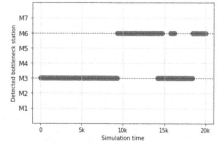

Fig. 2. Distribution of process times for bottleneck and none-bottleneck stations

Fig. 3. Exemplary structure of the detection results for a 20%-BN at M_3 and M_6

4   Detection Results using BNW, APM and ITV

All visualizations in Sec. 4 use a 20% factor for the bottleneck stations. The location of the such stations is varied between M_3 and M_6, leading to 25 combinations for each detection method. To compare the results, we use matrix plots that contain all 25 combinations. The diagonal of the matrix shows simulations with just one bottleneck. All other fields show simulations with two bottlenecks. Since we use the same random seed to determine the process time variability, the result matrix is symmetrical. As such, the same result is obtained for a M_2-M_4 system as for a M_4-M_2 system. The visualization logic of each tile in the matrix follows the format presented in Fig. 3.

4.1   Bottleneck Detection with Bottleneck Walk

Figure 4 shows the results of bottleneck detection using the BNW. As described above, the matrix is symmetrical. In the cases on the diagonal, in which only one station has an increased process time, the BNW identifies this station as a bottleneck in most cases. In the other cases, the stations marked with dashed lines, which are the stations with an increased process time, are alternatingly identified as bottlenecks. In addition to the apparently correctly identified bottleneck situations, all plots show a scattering of selectively identified bottlenecks at all stations, ranging from M_1 to M_7.

4.2   Bottleneck Detection Using the Active Period Method

Next, we identified bottlenecks using the APM. For the sake of clarity, we limit the use of the APM to an identification of sole bottlenecks. In situations with a shifting bottleneck, we refer to the station with the longest active period as bottleneck. Only after an interruption of this station, the other shifting bottleneck takes over the label of sole bottleneck. The results of this investigation are shown in Fig. 5. Again, clearly identified bottlenecks are evident in the five scenarios in the diagonal of the matrix. In the remaining scenarios, changing bottlenecks between the expected stations can be

Study results for bottleneck detection using the Bottleneck Walk with 20%-BN

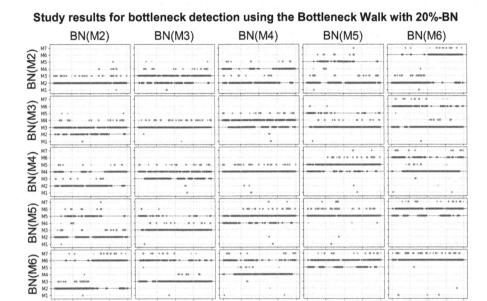

Fig. 4. Matrix plot of all bottleneck combinations (20%), using the Bottleneck Walk

Study results for bottleneck detection using the Act. Period Method with 20%-BN

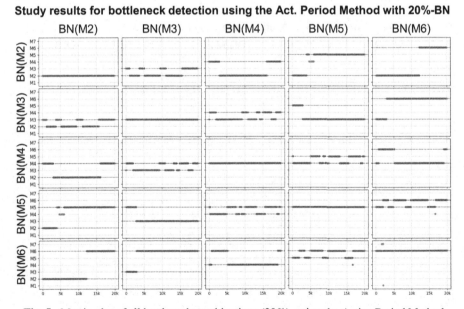

Fig. 5. Matrix plot of all bottleneck combinations (20%), using the Active Period Method

identified. It is noticeable that shifts in the bottleneck stations occur more frequently if the stations are located directly after one another or are only a short distance apart in the value stream. Although there are also isolated pointwise identified bottleneck stations, these occur much less frequently than previously for BNW. Overall, the results show a clear image of the expected system behavior in all scenarios.

4.3   Bottleneck Detection Using Interdeparture Time Variances

Figure 6 shows the bottleneck stations determined using ITV. Since ITV in the original proposal is an AVM, we adapted the method for momentary bottleneck detection. As already mentioned in Sec. 3, we use a sliding window to calculate time-dependent variances for each point of time in the simulations. We use 5,000 as the length of the variance interval, which corresponds to the period of the initial settling of the system. Then, for each point of time, the ITV within this sliding window is used, while the lowest ITV marks the respective bottleneck station. Once again, the individual bottlenecks are reliably identified, with only a minor anomaly in the M5-M5 scenario. In the other scenarios, the influenced stations are again identified as the main bottlenecks, with shifting states occurring here as well. Additionally, it is noticeable that the frequency of the shifting is not as strongly influenced by the distances between the stations in the value stream as seen for the APM detection. The number of individual identifications is slightly higher than it was for APM, but significantly lower than for BNW. In summary, the results of the ITV method show a slightly less clear result than the APM, but are considerably more uniform than the detection of BNW.

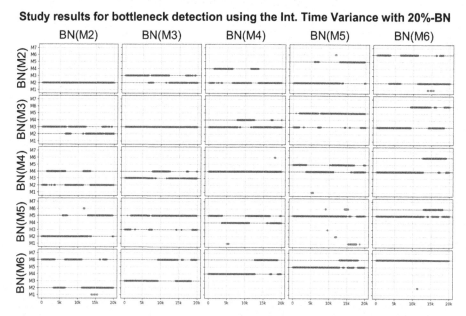

Fig. 6. Matrix plot of all bottleneck combinations (20%), using Interdeparture Time Variances

5 Comparison

We first address the comparison with the 20% bottlenecks as shown before. Since the process time increase has a significant effect on the system behavior, we also compare the results for 10% to 100% increases, focusing on the agreement of the methods.

5.1 Comparison of 20%-Bottleneck Results

With the help of the BNW, it is relatively simple to recognize bottleneck situations in the system and to detect bottlenecks accordingly. However, alternating bottlenecks, between two stations with increased process times, occur much more frequently, when compared to APM and ITV. Furthermore, only the visualization of the BNW shows an extensive scattering of selectively identified bottlenecks. These variations occurred most likely due to the ratio of average process time and maximum buffer capacity. With a capacity limit of 5 units assumed in the case study, the lower and upper decision limits are close to each other. On this scale, one third respectively two third, are unreliable capacity limits due to the rule-of-thumbs assumptions. It is possible that BNW is more suitable for bottleneck detection in systems of a different nature with a change in this ratio. However, in the setup with a 20% increase in process time at bottleneck stations, the BNW performs the worst.

APM has proven to be a more robust method for bottleneck detection. It shows almost no punctual deviations and designates mainly the influenced stations as bottlenecks. However, this observation has to be quantified to some degree. By focusing on continuous active periods, it is in the nature of APM that longer periods of utilized stations lead to a continuous bottleneck identification. The method is thus partially protected from short-term and variability-induced deviations. It may even be argued that APM is a memory-based method due to this implicit use of past systems, effectively becoming a hybrid of AVM and MVM. Regarding the decreasing frequencies of bottleneck shifting for more widely separated stations in the value stream, this is likely due to the increasing overall workload between stations. With only two stations with higher process times, the stations in between act like a shared buffer, making starvation or blocking conditions less likely to occur. Summarizing, the APM is proving effective in identifying shifting bottlenecks.

Like the APM, the modified ITV has also been shown to be well suited for bottleneck detection. The advantage of using ITV is that bottlenecks can be identified selectively, especially in longer value streams. The phenomenon of a split buffer at high workloads between two stations is less likely. A disadvantage is that immediate bottlenecks are subject to a slightly higher delay due to the average-based determination of the variance. In addition, the length of the selected variance interval has a significant impact on the quality of the detection. The determination of a suitable interval length is highly application-dependent. In order not to exceed the scope of this work, we had to refrain from an additional variation of this parameter.

We provide the simulation model, the generated data and all variations of bottleneck process time here: github.com/nikolaiwest/2021-bottleneck-detection-ieaaie.

5.2 Results for Varying Bottleneck Process Times (10% to 100%)

To determine the correctness of a detected bottleneck, a way to identify it according to a single source of truth is required. Such a truth can then be used to crosscheck the degree of correct identification. However, since such measures always depend on the selected bottleneck detection method, we have no way to determine an absolute truth regarding the bottleneck station. Instead, we have to compare the statements of the three previously used methods BNW, APM and ITV. All agreement ratios therefore consist of pairings of methods. In each case, we determine the average ratio of agreement from the 25 scenarios. This calculation is performed for bottleneck stations of varying increase of process times. We compare an extra of 10 to 100% of the basic process time. Table 1 summarizes the results.

Table 1. Average ratio of agreement between two detection methods for different additions to the stations' bottleneck process times

BN%		BNW - APM	BNW - ITV	APM - ITV
	+10%	49.91%	39.26%	61.63%
	+20%	52.87%	47.28%	72.74%
	+30%	58.29%	52.12%	81.15%
	+40%	63.52%	58.30%	77.53%
	+50%	68.01%	61.45%	81.91%
	+60%	60.78%	55.82%	85.38%
	+70%	64.17%	57.90%	84.46%
	+80%	64.30%	62.22%	77.37%
	+90%	69.15%	64.22%	90.55%
	+100%	69.27%	64.75%	88.30%
Average		62.03%	56.33%	80.10%

The ratios of agreement are given as percentages, with 100% corresponding to a full agreement and 0% to no agreement. BNW and ITV have the lowest degree of agreement at 39.26% with a bottleneck increase of 10%, and an average of 56.33%, APM and ITV have the highest agreement at 90.55% with a 90% bottleneck increase, while reaching an average of 80.10%. Through this representation, a clear ranking of the method pairs becomes apparent. ITV and APM always have a higher agreement ratio than BNW and APM, whereas the ratio for BNW and APM is always above BNW and ITV. These deviations are probably due to the sporadic deviations of the BNW. Although the method recognizes the influenced stations as bottlenecks in many cases, it is also frequently wrong due to the sporadically recognized bottlenecks.

Furthermore, all pairings show a tendency of an increased ratio of agreement with more additional process time for bottlenecks. This increase can be attributed to a tendency towards static bottlenecks in the front sections of the value stream. Due to the increasing differences in process times on bottlenecks, the duration and frequency of early bottlenecks increases. Consequently, the detection of such bottlenecks becomes easier for all three methods, leading to an increasing the rate of agreement.

6 Conclusion

From the results, we deduce that APM is well suited for detecting dynamic bottlenecks when the workload between possibly affected stations is not too extensive. A quantification of this length depends on the systems' properties. The use of ITV is particularly suitable when bottlenecks are sought in longer serial lines. It detects short-term alterations in process times and is thereby able to identify sporadic bottlenecks on a miniscule time scale. Although the BNW has proven to be a practical method, it is highly dependent on the maximum buffer capacity. This led to poorer detection results, but results may differ in systems set up differently. Summarizing, since even this comparison does not show a strictly dominant method, we recommend a comparative use of several detection methods, if possible. Depending on data availability, the methods open up different approaches for detecting dynamic bottlenecks.

In summary, the detection of bottlenecks is readily solvable with the methods presented. However, a mere detection of bottlenecks does not yet enable monetary savings. For future research, two promising possibilities arise within this area. On the one hand, the derivation of suitable measures to resolve detected bottlenecks has to be improved. With a suitable response to bottlenecks, their time of occurrence and their severity can be reduced, leading to an overall increase in systemic throughput. We recommend the development of a catalog of measures for dealing with identified bottlenecks. On the other hand, bottlenecks have to be predicted before they actually occur. Similar to fire-fighting strategies in maintenance, to this date, bottleneck detection only enables reactive measures. Appropriate bottleneck forecasts allow measures to be taken to widen the bottlenecks even before availability losses occur. Although some approaches to forecasting bottlenecks already exist, no generally applicable procedure or method is yet established in practical applications.

Acknowledgement. The work on this paper is a part of the project *'Prediction of dynamic bottlenecks in directed material flow systems using machine learning methods'* (PrEPFlow, 21595), which is funded by the German Federal Ministry of Economics and Technology (BMWi), through the Working Group of Industrial Research Associations (AIF).

References

1. Goldratt, E.M., Cox, J.F.: The Goal: Excellence in Manufacturing. North River Press, Croton-on-Hudson, NY (1984)
2. Gupta, A., Bhardwaj, A., Kanda, A.: Fundamental concepts of theory of constraints: An emerging philosophy. Int. J. o. Social, Behavioral, Educational, Economic, Business and Industrial Eng. 4(10), 2089–2095 (2010)
3. Lawrence, S.R., Buss, A.H.: Shifting production bottlenecks: causes, cures and conundrums. Prod. Oper. Manag. 3(1), 21–37 (1994)
4. West, N., Syberg, M., Deuse, J.: A holistic methodology for successive bottleneck analysis in dynamic value streams of manufacturing companies. In: Andersen, A.-L., et al. (eds.) Lecture Notes in Mechanical Engineering, Towards Sustainable Customization: Bridging Smart Products and Manufacturing Systems, pp. 612–619. Springer International Publishing, Cham (2022)

5. Klenner, F., Lenze, D., Schwarzer, S., Deuse, J., Friedrich, T.: Smart Data Analytics for the identification of dynamic bottlenecks in flexible manufacturing systems (ger.). *at*, **64**(7), 540–554 (2016)
6. Roser, C., Nakano, M., Tanaka, M.: Detecting shifting bottlenecks. In: International Symposium on Scheduling, pp. 59–62 (2002)
7. Roser, C., Nakano, M., Tanaka, M.: Shifting bottleneck detection. In: Proceedings of the Winter Simulation Conference, pp. 1079–1086 (2002)
8. Wang, Y., Zhao, Q., Zheng, D.: Bottlenecks in production networks: an overview. JSSSE **14**(3), 347–363 (2005)
9. Deuse, J., Lenze, D., Klenner, F., Friedrich, T.: Manufacturing data analytics for the identification of dynamic bottlenecks in production systems with high value-adding variability (ger.). Megatrend Digitalisierung, Berlin: GITO mbH Verlag **1**, 11–26 (2016)
10. Muthiah, K.M.N., Huang, S.: Overall throughput effectiveness (OTE) metric for factory-level performance monitoring and bottleneck detection. IJPR **45**(20), 4753–4769 (2007)
11. Hopp, W.J., Spearman, M.L.: Factory Physics, 3rd edn. Waveland Press Inc., Long Grove, IL (2008)
12. Betterton, C.E., Silver, S.J.: Detecting bottlenecks in serial production lines: a focus on interdeparture time variance. IJPR **50**(15), 4158–4174 (2012)
13. Roser, C., Nakano, M.: A quantitative comparison of bottleneck detection methods in manufacturing systems with particular consideration for shifting bottlenecks. Int. C. o. Adv. in Prod. Manage. **1**, 273–281 (2015)
14. Kuo, C.-T., Lim, J.-T., Meerkov, S.M.: Bottlenecks in serial production lines: a system-theoretiv approach. Mathematic Problems Eng. **2**(3), 233–276 (1996)
15. Biller, S., Li, J., Marin, S.P., Meerkov, S.M., Zhang, L.: Bottlenecks in production lines with rework: a systems approach. In: Proceedings of the 17th World Congress (IFAC, Seoul), **17**(1), pp. 1488–14893 (2008)
16. Li, L.: Bottleneck detection of complex manufacturing systems using a data-driven method. IJPR **47**(24), 6929–6940 (2009)
17. Li, L., Chang, Q., Ni, J.: Data driven bottleneck detection of manufacturing systems. IJPR **47**(18), 5019–5036 (2009)
18. Yu, C., Matta, A.: A statistical framework of data-driven bottleneck identification in manufacturing systems. IJPR **54**(21), 6317–6332 (2016)
19. Zhao, D., Tian, X., Geng, J.: A bottleneck detection algorithm for complex product assembly line based on maximum operation capacity. Mathematic Problems Eng. **3**(1), 1–9 (2014)
20. Roser, C., Lorentzen, K., Deuse, J.: Reliable shop floor bottleneck detection for flow lines through process and inventory observations. Procedia CRIP **19**(1), 63–69 (2014)

Ultra-short-Term Load Forecasting Model Based on VMD and TGCN-GRU

Meirong Ding[1], Hang Zhang[1,2], Biqing Zeng[1(✉)], Gaoyan Cai[2], Yuan Chai[3], and Wensheng Gan[4]

[1] School of Software, South China Normal University, Foshan 528225, China
2020023879@m.scnu.edu.cn
[2] Hodi Technology Co. Ltd., Foshan 528299, China
[3] Manning Selvage & Lee Public Relations Consultancy Beijing Co., LTD. Shanghai Branch, Guangzhou 20042, China
[4] College of Cyber Security, Jinan University, Guangzhou 510632, China

Abstract. Load forecasting is to use historical load information to estimate the load demand for a period of time in the future. At present, mode decomposition algorithm is often used in the field to improve the forecasting accuracy. However, mode decomposition will cause the accumulation of errors, in order to solve this problem, this paper analyzes the relationship between the original load and the Intrinsic Mode Functions (IMFs), and constructs an ultra-short-term load forecasting algorithm based on Variational Mode decomposition (VMD) and TGCN-GRU (Temporal Graph Convolution-Gated Recurrent Unit). Firstly, the model uses VMD to decompose the original load into multiple relatively stable IMFs. Then it inputs the original load, external factors that affect the load, and all IMFs into TGCN, and uses TGCN to extract the spatial relationships of each graph node and the timing characteristics of each graph node itself. Finally, these spatiotemporal features are input into the GRU unit for prediction. The model not only rationally combines all IMFs into a whole and to solve the problem of error accumulation, but also fully analyzes the interrelationship among the IMFs, the original load and the external factors affecting the load. We conducted comparison experiments using real load data and other load forecasting models, and the experiment results indicate, that the overall accuracy of this model is superior to the model compared.

Keywords: Ultra-short-term load forecasting · Variational mode decomposition · Temporal graph convolution network · Gated recurrent unit

1 Introduction

With the increasing awareness of environmental protection in human society, various renewable energy sources are beginning to be connected to the smart grid through microgrids and other means [1]. However, renewable energy is generally

© Springer Nature Switzerland AG 2022
H. Fujita et al. (Eds.): IEA/AIE 2022, LNAI 13343, pp. 15–28, 2022.
https://doi.org/10.1007/978-3-031-08530-7_2

unstable in generating electricity. To maintain a security and steady operation of the microgrid, it is essential to predict users' electricity consumption habits more accurately in a very short time, so that one can develop appropriate transport and distribution plans to improve the use of renewable energy improving.

Traditional load predictive models are based on statistical statistics, such as Multiple Linear Regression (MLR) model [2], Autoregression-based Moving Average model [3], Gray model [4], etc. Due to the poor robustness of load forecasting models based on statistical, they are generally used for linear data with low volatility. The machine learning (ML) model has a strong non-linear mapping capability, and it better able to deal with the sequence data of high complexity and strong volatility. ML models that are currently usually used in load forecasting includes kernel models [5], decision trees [6], support vector machines and their variants [7]. Although machine learning models have many advantages over statistical models, they are still sensitive to abnormal points and are difficult to process large amounts of data [8]. With the rapid development of chip technology and deep learning (DL), the Gated Recurrent Unit (GRU) and Long and Short-term Memory Network (LSTM) have attracted much academic interest due to their ability to preserve temporal information in data and prevent gradient explosion and disappearance. As the literature [9] used LSTM to forecast the load demand of residential customers and showed that LSTM outperformed other traditional load forecasting models. The literature [10] and [11] first used Convolutional Neural Network (CNN) to extract features in the sequence, then used LSTM or GRU to predict the load. The literature [12] first used BiLSTM to extract the timing information in the data, then used Attention Mechanism to focus on more important features to reduce the difficulty of forecasting.

Despite the significant merits of various predictive models based on DL, the performance of predictive models is still constrained by the quality of the data, whereas a user's power consumption is often affected by weather, temperature and various human factors, which makes the load data extremely volatile, increasing the difficulty of forecasting [13]. In order to minimize the effect of external factors as much as possible, mode decomposition technology is often used to pre-process the original data [13–16]. In literature [16], the IMFs decomposed by Ensemble Empirical Mode Decomposition (EEMD) are divided into high frequency and low frequency components, and then LSTM and MLR are used for prediction respectively. In literature [13] used the clustering method to divide the IMFs decomposed by Empirical Mode Decomposition (EMD) into different clusters, added the IMFs in the same cluster, then used CNN-LSTM for prediction. Although the use of mode decomposition techniques can reduce the difficulty of prediction, it requires the creation of multiple models to predict different IMFs or clusters, and the prediction of each IMFs or cluster is difficult to be completely correct, and this will inevitably result in prediction errors, and at the same time it is difficult to confirm whether these errors will cancel each other out, so it leads to the problem of error accumulation.

In order to solve the above issues, a predictive model based on VMD and TGCN-GRU(VTGG) is proposed. First, the original load is decomposed into multiple IMFs by the VMD, reducing the non-stationarity and volatility of the load data. Moreover, in order to accelerate the information transfer between each graph node, the original load is introduced as a connection node to connect all graph nodes. In addition, factors such as temperature is used as one of the inputs for analysing the effect of external factors on the load forecasting. Then input IMFs, original load, and external factors as a graph structure into TGCN, using the TGCN to analyze the spatial connection of each graph node and the timing information of each graph node itself. Finally, the features extracted by TGCN are fed into the GRU, which uses the GRU to analyze the temporal information in the features to achieve the prediction of the data.

The main work of this paper is as follows:

(1) An Ultra-short-term load forecasting model combining mode decomposition and graph neural networks is proposed. The IMFs decomposed by VMD are used as graph nodes to form a unified graph structure, which solves the error accumulation caused by the mode decomposition. At the same time, the graph neural network is used to analyze and to extract the global connection between each component, which effectively improves the prediction accuracy.
(2) Using real load data to test the VTGG model, the results of experimental analysis and comparison with the existing benchmark model and combination model prove the superiority of the VTGG model in ultra-short-term load forecasting.

2 Methodology

2.1 Variational Mode Decomposition

Mode decomposition techniques can decompose the original load into mode data and residual data at different frequencies. Because VMD can overcome the endpoint effects and the mode mixing that exists in EMD, it can decompose non-stationary time series into multiple different frequency scales and relatively stable IMFs and residual [17], so VMD was used to pre-process the power load data in the VTGG model.

Because VMD must ensure that the decomposed sub-sequences with limited bandwidth, and minimize the sum of the estimated bandwidth of each subsequence. Additionally, the sum of all subsequences must be equal to the original signal [17], so the following constraints are set:

$$\begin{cases} \min_{\{u_k\},\{\omega_k\}} \{\sum_k \left\| \partial_t[(\delta(t) + j/\pi t) \times u_k(t)]e^{-j\omega_k t}\right\|_2^2\} \\ s.t. \sum_{k=1}^{K} u_k = f \end{cases} \tag{1}$$

where K is the number of IMFs, u_k and ω_k correspond to the Kth IMF and the center frequency respectively, $\delta(t)$ represents the Dirac function, and f represents the original signal.

In order to solve the Eq. (1), the Lagrange multiplication operator λ can be introduced to convert the constrained problem into a non-constrained problem.

The Eq. (2) can be obtained:

$$L(\{u_k\}, \{\omega_k\}, \lambda) = \alpha \sum_k \left\| \partial_t [(\delta(t) + j/\pi t) \times u_k(t)] e^{-j\omega_k t} \right\|_2^2 + \tag{2}$$

$$\left\| \sum_{k=1}^{K} u_k - f \right\|_2^2 + \left\langle \lambda(t), f(t) - \sum_k u_k(t) \right\rangle$$

where α is the secondary penalty factor, which is used to reduce the interference of Gaussian noise. Then use the alternate direction multiplier iterative algorithm to solve the above equation [17].

2.2   Temporal Graph Convolution Network

Graph Convolutional Network. Through mode decomposition technology, the original load can be decomposed into multiple IMFs of different frequencies. There are certain connections between different IMFs and between each independent IMF and the original load. In addition, in order to predict the future power load, it is also necessary to comprehensively consider the local real-time temperature, weather and other features closely related to the power load. These factors together constitute a complex graph structure. Traditional CNNs, while better at extracting local features, are only applicable to Euclidean space, making it difficult to handle this situation appropriately. Compared to traditional CNN, Graph Convolutional Network (GCN) has significant advantages over traditional CNNs for processing such unstructured data [18]. GCN builds a filter in the Fourier domain, and each graph node extracts the spatial features of the first-order neighbouring nodes through this filter, and then builds the GCN model by superimposing multiple similar convolutional layers [18–20], As shown in Fig. 1, it can be expressed as:

$$H^{(l+1)} = \sigma(D^{-1/2} \hat{A} D^{-1/2} H^{(l)} W^{(l)}) \tag{3}$$

where $\hat{A} = A + E$, A represents the adjacency matrix; E represents the identity matrix; $D = \mathrm{diag}(\sum_j \hat{A}_{ij})$ represents the degree matrix; $H(l)$ represents the input of the lth layer; $W(l)$ represents the weight of the lth layer; $\sigma()$ represents the Sigmoid activation function.

Gated Recurrent Unit. GRU mainly has two gating structures, namely update gate, and reset gate. The update gate mainly saves the state information of the previous unit through certain rules. The reset gate is used to decide whether to combine the information of the previous unit with the current unit [21], as shown in Eq. (4) to (7):

$$z_t = \sigma(W_z[x_t, h_{t-1}] + b_z) \tag{4}$$

$$r_t = \sigma(W_r[x_t, h_{t-1}] + b_r) \tag{5}$$

$$c_t = tanh(W_c[x_t, r_t * h_{t-1}] + b_c) \tag{6}$$

$$h_t = (1 - z_t) * c_t + z_t * h_{t-1} \tag{7}$$

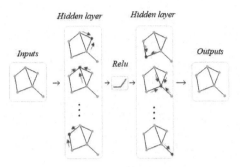

Fig. 1. Graph convolutional network

where h_{t-1} and h_t represent the output of the previous unit and the current unit respectively; x_t represents the input of the current unit, $[x_t, h_{t-1}]$ represents the splicing of h_{t-1} and x_t; z_t and r_t represent the output of the update gate and of the reset gate respectively, c_t represents the combination of the previous unit output and the current input; W_z, W_r, W_c and b_z, b_r, b_c represent the weight and bias respectively; $tanh()$ represents the hyperbolic tangent activation function, $*$ represents Multiply matrices.

Temporal Graph Convolution Network. Fusion of GCN and RNN, while extracting the connection of each node in the graph structure data, extract its own timing features [18–20]. The basic structure of TGCN is like GRU, but graph convolution is used on top of each gating mechanism, as shown in Fig. 2 and Eq. (8) to (11):

$$z_t = \sigma(W_z gc(A, [x_t, h_{t-1}]) + b_z) \tag{8}$$

$$r_t = \sigma(W_r gc(A, [x_t, h_{t-1}]) + b_r) \tag{9}$$

$$c_t = tanh(W_c gc(A, [x_t, r_t * h_{t-1}]) + b_c) \tag{10}$$

$$h_t = (1 - z_t) * c_t + z_t * h_{t-1} \tag{11}$$

where $gc()$ represents the graph convolution operation.

2.3 VTGG Model

This paper combines mode decomposition technology with graph neural network, proposes a VMD-based TGCN-GRU ultra-short-term load forecasting model. The model uses VMD to analyze and process the original load decomposition to reduce the effect of external factors on the load and improves the accuracy of load forecasting. Then, each IMF is used as a graph node to form a whole and input into the TGCN-GRU graph neural network to solve the problem of error accumulation.

Because ultra-short-term load data are influenced by a combination of factors, they are highly volatile and unstable, the VTGG model uses VMD techniques to

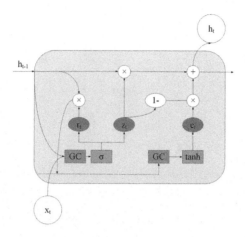

Fig. 2. TGCN structure

process the raw load data and decompose multiple relatively stable and regular components at different frequencies. If one builds a model for each IMF to make predictions, and then sum the prediction of each model, not only does one need to build multiple modes, increasing computational consumption, but also the prediction of each component may not be completely accurate, and whether the prediction errors of each IMF can cancel each other out would always remain a question.

Therefore, in order to tackle the above problems, the VTGG model uses each IMF as a graph node to construct a graph structure. It only needs to build one graph neural network to process all IMFs, and it can effectively assist each IMF in extracting relevant information about other graph nodes, thereby improving the accuracy of prediction.

The construction of the graph adjacency matrix A is mainly by calculating the Pearson coefficient r between every two graph nodes. The calculation method of the Pearson coefficient is shown in Eq. (12):

$$r(X,Y) = \frac{Cov(X,Y)}{\sqrt{Var[X]Var[Y]}} \tag{12}$$

where $Cov(X,Y)$ represents the covariance of X and Y, and $Var[X]$ represents the variance of X. When the absolute value of the Pearson coefficient r is greater than a preset threshold θ, it is believed that there is a stronger relationship between the two, and that the two graph nodes are connected. Because the IMFs decomposed by the mode decomposition technology are at different frequencies, the Pearson coefficient r between them is relatively low, it may not be possible to construct a fully connected graph and prevent the graph neural network from extracting the connections between individual IMFs. To solve this problem, this paper introduces the original load sequence that is used as a connectivity node to connect all other graph nodes to enhance information transfer between graph nodes. This strengthens the information transfer between each graph node, making the graph neural network easier to analyze and extract the spatial connection of each graph node. In addition, in order to analyze the effect of external factors on the accuracy of load forecasting, external features such as temperature are also taken as a graph node as input.

The VTGG prediction model is shown in Fig. 3. When the input time series is too long, it will cause frequent adjustment of the weights in the recurrent unit during one iteration process, substantially increasing the computational consumption, also making it difficult to preserve the information of the historical, thus decrease the accuracy of the load prediction. Therefore, before the data is input to the VTGG model, the "folding" method is used to splice the m time series data points as an input matrix to reduce the length of the time series input. This graph neural network is composed of a TGCN and a GRU layer. TGCN can not only extract the spatial features between each graph node, obtain the connection between the data, but can also obtain the changing laws and trends of each graph node itself by extracting the time series features in the sequence data. Generally, it is more effective to use a two-layer RNN to analyze the time series sequence. Due to the excessive number of TGCN parameters and the fact that TGCN has performed two graph convolution operations on the input data, in order to reduce the amount of calculation and prevent overfitting of the model, a GRU with a relatively simpler structure and without graph convolution operations is used to analyze the features extracted by TGCN to achieve load prediction.

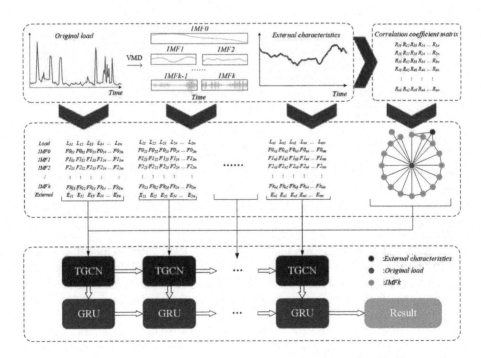

Fig. 3. Illustration of the fusion mode (IMF0 represents the residual)

3   Experiments and Discussions

3.1   Data

The data set selected in this paper comes from the University of Mons in Belgium [22], which is the load data of a Belgian power user from 7:00 on January 11th to 18:00 on May 27th, 2016, the time interval is 10 min. In addition to the load data, the data set also contains temperature, a feature that is closely related to the load. Because load data has strong daily and weekly periodicity, this study selects the data of the previous week, that is, 1008 data before the forecast point as input.

3.2   Evaluation Method

This paper selects the Mean Absolute Percentage Error (MAPE), the Mean Absolute Error (MAE), and the Root Mean Square Error (RMSE) as the evaluation indicators, as shown in Eq. (13) to (15):

$$MAPE = \frac{1}{N} \sum_{t=0}^{N} \frac{|\hat{y}_i - y_i|}{y_i} \times 100\% \tag{13}$$

$$MAE = \frac{1}{N} \sum_{t=0}^{N} |\hat{y}_i - y_i| \tag{14}$$

$$RMSE = \sqrt{\frac{1}{N} \sum_{t=0}^{N} (\hat{y}_i - y_i)^2} \tag{15}$$

where N is the length of the prediction data set, $\hat{y}_i$ and y_i represent the predicted value and true value at time i, respectively.

3.3   Contrast Experimental Model

In order to verify the performance of the VTGG model proposed in this paper, in addition to selecting a single prediction algorithm such as MLR, BPNN and LSTM as the comparison model, this paper also selects a variety of combined prediction models.

(1) Based on the attention mechanism of CNN-GRU [23]: The model first uses CNN-GRU to extract the time series features in the load data, then uses the attention mechanism to focus on more important information, and finally inputs the features reconstructed by the attention mechanism into the fully connected neural network for prediction.
(2) TGCN model [18]: The model uses TGCN to extract the spatiotemporal features between each graph node, and then uses a fully connected neural network to make predictions.

(3) EEMD-MLR-LSTM model (EML) [16]: The model first divides the IMFs decomposed by EEMD into high-frequency and low-frequency components, and then uses LSTM and MLR respectively for prediction. Finally, the prediction results of each IMFs are integrated to obtain the final prediction result.

(4) CEMD-CNN-LSTM model (CECL) [13]: Using the same test set as in this article. The model first uses EMD to decompose the original data into multiple IMFs. Then uses the K-means clustering method to analyze the IMF, and adds the data clustered into the same category. Finally, for every type of data, a CNN-LSTM model is established for prediction, and the predictions results of various types of data are integrated.

3.4 Experimental Environment and Parameter Settings

All experiments in this article are done using Python 3.6, and the DL framework used is TensorFlow 2.0. The experimental hardware platform uses Intel Core i5-7300HQ CPU and uses NVIDIA GeForce GTX 1050 for acceleration.

The hyperparameters of the model proposed in this paper are as follows: the K and α distributions of VMD are 17 and 2000, the number of neurons of TGCN and GRU are 96 and 1, respectively. The initial learning rate is 0.000115, the learning rate decay is 0.9, and the batch size is 64, the number of iterations is 3000. The threshold θ is 0.1, the number of "folding" m is 144.

3.5 Experimental Results

This paper uses VMD technology to decompose the original load into multiple IMFs. It can be seen from Fig. 4 that the decomposed IMFs at different frequencies are relatively stable, and there is no mode mixing phenomenon.

Comparison of Experimental Results. As shown in Table 1, compared with most of the comparison models selected in this experiment, the overall precision of prediction of the VTGG model is relatively better. From the results in Table 1, the following conclusions can be drawn:

Table 1. Comparison results of different models

Model	MAPE (%)	RMSE(W)	MAE(W)
MLR	33.073	77.43	43.391
BPNN	36.753	83.51	48.251
LSTM	28.323	77.94	41.094
Ref. [23]	28.777	72.018	38.782
TGCN [18]	28.256	76.289	39.779
EML [16]	21.631	42.887	25.954
CECL [13]	**3.821**	5.514	4.357
VTGG	4.788	**5.463**	**4.102**

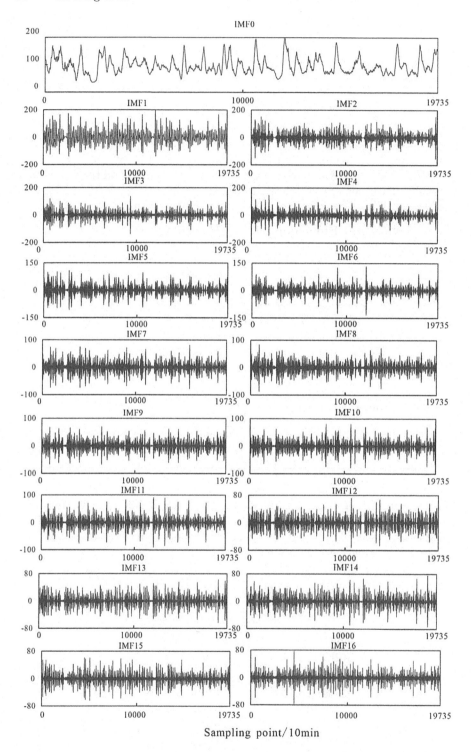

Fig. 4. VMD decomposition results

(1) Compared with MLR, BPNN and LSTM, the MAPE of VTGG dropped by 85.52%, 86.97% and 83.10%, respectively; the RMSE dropped by 92.94%, 93.46% and 93.00%, respectively; the MAE dropped by 90.55%, 91.50% and 90.02%, respectively. It shows that the precision of prediction of the VTGG is significantly better than that of the single prediction model.

(2) The overall accuracy of the ordinary combined model (Based on the attention mechanism of CNN-GRU, and TGCN) is relatively better than that of MLR, BPNN and LSTM, but the improved accuracy is limited. The overall accuracy of the combined prediction models (EML, CECL, and VTGG) based on mode decomposition is significantly better than other prediction models. This shows that the accuracy of prediction is obviously constrained by data quality, and mode decomposition technology can decompose load data with high complexity and strong volatility into relatively stable and more regular components, effectively improving the predictability of data, this therefore greatly reduces the difficulty and error of load forecasting. Moreover, the experimental results in Table 1 show that compared with other models, the prediction results of the combined forecasting model based on mode decomposition are more in line with the real variation pattern and distribution of the load than other models.

(3) Compared with EML and CECL, the RMSE of the VTGG model decreased by 87.26% and 9.25%, respectively, and the MAE decreased by 84.2% and 5.85%, respectively. Among them, the prediction result of the EML model is relatively the worst. This is because EML only analyzes the nature of each IMF itself, and divides the components into high-frequency and low-frequency components according to the frequency of each IMF, and it did not consider the relationship between IMFs. Although CECL analyzes the commonalities and associations between different IMFs through clustering, it only considers the local connection between each IMF, and the problem of error accumulation caused by mode decomposition remains not solved. Compared with the above two models, the VTGG model regards each IMF as a graph node and combines all IMFs into a whole, solving the problem of error accumulation, while also using the graph neural network to fully analyse the global connection between each IMF and extract the spatial features between each IMF, which plays a positive role in improving the accuracy of load prediction, and its RMSE and MAE both reach the optimum.

Effect of Original Load. In order to verify the role of the original load data as a graph node input into the TGCN model, the VTGG-O model is used for comparison experiments in the research. The VTGG-O model is a variant of the VTGG model. The difference from the VTGG model is that when the TGCN layer is input, the original load data is not used as the graph node, and the rest is unchanged. Therefore, through the comparison of these two experiments, the effect of the original load data on the accuracy of the model can be verified. From the experimental results in Table 2, it can be observed that the MAPE, RMSE and MAE of the VTGG model are reduced by 6.90%, 8.06% and 9.05%, respectively, compared with the VTGG-O model. As shown in the graph node in Fig. 3, the original load data acts as a link in this input. The original load is directly connected to other IMFs, which can help each IMF obtain the information of other graph nodes faster, so that the graph neural network is easier to extract the interconnection of each graph node.

Table 2. The influence of the original load and external factors

Model	MAPE (%)	RMSE(W)	MAE(W)
VTGG-O	5.143	5.942	4.51
VTGG-E	5.028	5.779	4.393
VTGG	**4.788**	**5.463**	**4.102**

The Influence of External Factors. External factors (such as temperature) are closely related to the electrical load. In order to verify this, the VTGG-E model is used in the study. Like the VTGG-O model, the difference between VTGG-E and VTGG is that external features are not used as input for a node of the graph, and the rest remain unchanged. Through the comparison of these two models, the influence of the input of external factors on the accuracy of the model can be analyzed. It can be found from Table 2 that the MAPE, RMSE, and MAE of the VTGG model are reduced by 4.77%, 5.47% and 6.62%, respectively, compared with the VTGG-E model. This shows that external factors, such as temperature, have a strong influence on the power load. By analyzing the relationship between external factors and load data, the forecast accuracy can be effectively improved.

4 Conclusion

In order to solve the problem of error accumulation caused by mode decomposition, this study proposes an ultra-short-term load forecasting model that combines mode decomposition technology and graph neural networks. This model treats each IMF decomposed by VMD as graph nodes, and the original load and external factors, then use TGCN-GRU to analyze the connection between each graph node and its own properties, solving the error accumulation problem caused by mode decomposition, and has achieved the prediction of ultra-short-term load. The model proposed in this paper was used to conduct experiments and comparative analysis of experimental results on a real load data set the experimental results prove that the accuracy of the model is significantly better than the single forecasting model, and the overall accuracy is better than that of the combined load forecasting model compared.

References

1. Kong, X., Li, C., Wang, C.: Short-term electrical load forecasting based on error correction using dynamic mode decomposition. Appl. Energy **261**, 114368 (2020)
2. Saber, A.Y., Alam, A.K.M.R.: Short term load forecasting using multiple linear regression for big data. In: IEEE Symposium Series on Computational Intelligence, pp. 1–6. Honolulu (2016)
3. Alberg, D., Mark, L.: Short-term load forecasting in smart meters with sliding window-based ARIMA algorithms. Vietnam J. Comput. Sci. **261**(3), 241–249 (2019)
4. Guefano, S., Tamba, J.G., Azong, T.E.W.: Forecast of electricity consumption in the Cameroonian residential sector by Grey and vector autoregressive models. Energy **214**, 118791 (2021)
5. Fan, G.F., Peng, L.L., Hong, W.C.: Short term load forecasting based on phase space reconstruction algorithm and bi-square kernel regression model. Appl. Energy **224**, 13–33 (2018)
6. Liu, S., Cui, Y., Ma, Y et al.:Short-term load forecasting based on GBDT combinatorial optimization. In: 2nd IEEE Conference on Energy Internet and Energy System Integration, pp. 1–5. Beijing (2018)
7. Kaytez, F.: A hybrid approach based on autoregressive integrated moving average and least-square support vector machine for long-term forecasting of net electricity consumption. Energy **197**, 118791 (2021)
8. Zhang., J, Xu. C.C., Zhang. Z: Electric load forecasting in smart grids using long-short-term-memory based recurrent neural network. In: The 51st Annual Conference on Information Sciences and Systems, pp. 1–6. Baltimore (2017)
9. Kong, W., Dong, Z.Y., Jia, Y.: Short-term residential load forecasting based on LSTM recurrent neural network. IEEE Trans. Smart Grid. Energy **10**(1), 841–851 (2017)
10. Kin, T.Y., Cho, S.B.: Predicting residential energy consumption using CNN-LSTM neural networks. Energy **182**, 72–81 (2019)
11. Sajjad, M., Khan, Z.A., Ullah, A.: A novel CNN-GRU-based hybrid approach for short-term residential load forecasting. IEEE Access. **8**, 143759–143768 (2020)
12. Wang, S., Wang, X., Wang, S.: Bi-directional long short-term memory method based on attention mechanism and rolling update for short-term load forecasting. Int. J. Electr. Power Energy Syst. **109**, 470–479 (2019)
13. Liu, Y.H., Zhao, Q.: Ultra-short-term power load forecasting method based on cluster empirical mode decomposition of CNN-LSTM. Power Syst. Technol. **45**(11), 4444–4451 (2021)
14. Tayab, U.B., Zia, A., Yang, F.: Short-term load forecasting for microgrid energy management system using hybrid HHO-FNN model with best-basis stationary wavelet packet transform. Energy **203**, 117857 (2020)
15. Liang, Y., Niu, D., Hong, W.C.: Short term load forecasting based on feature extraction and improved general regression neural network model. Energy **166**, 653–663 (2019)
16. Li, J., Deng, D., Zhao, J.: A novel hybrid short-term load forecasting method of smart grid using MLR and LSTM neural network. IEEE Trans. Ind. Informat. **17**(4), 2443–2452 (2020)
17. Han, L., Zhang, R., Wang, X.: Multi-step wind power forecast based on VMD-LSTM. IET Renew. Power Gener. **13**(10), 1690–1700 (2019)

18. Zhao, L., Zhao, Y., Zhao, C.: T-GCN: a temporal graph convolutional network for traffic prediction. IEEE Trans. Intell. Transp. Syst. **21**(9), 3848–3858 (2019)
19. Zhu, J., Wang, Q., Tao, C.: AST-GCN: attribute-augmented spatiotemporal graph convolutional network for traffic forecasting. IEEE Access. **9**, 35973–35983 (2021)
20. Hou, X., Wang, K., Zhong, C.: ST-Trader: a spatial-temporal deep neural network for modeling stock market movement. IEEE/CAA J. Automatica Sinica **8**(5), 1015–1024 (2021)
21. Jung, S., Moon, J., Park, S.: An attention-based multilayer GRU model for multistep-ahead short-term load forecasting. Sensors **21**(5), 1639 (2021)
22. Luis, C. https://archive-beta.ics.uci.edu/ml/datasets/appliances+energy+prediction. Accessed 12 Jan 2021
23. Zhao, B., Wang, Z.P., Ji, W.J.: A short-term power load forecasting method based on attention mechanism of CNN-GRU. Power Syst. Technol. **43**(12), 4370–4376 (2019)

Learning to Match Product Codes

Ying Excell and Sebastian Link$^{(\boxtimes)}$ (iD)

School of Computer Science, University of Auckland, Auckland, New Zealand
yexc128@aucklanduni.ac.nz, s.link@auckland.ac.nz

Abstract. Most businesses need to manually match their codes of products to the codes that suppliers use for the same products. Our industry-based project investigated two techniques for learning such matches automatically. The first approach uses synonyms when preprocessing data before applying approximate string matching. We found that trigram cosine distance matching outperforms the other six popular matching methods we evaluated. The second approach couples approximate string matching with deep learning. Here, the Siamese Manhattan biLSTM method has higher accuracy and lower run time compared to multiple LSTM 1D CNN. Suggesting the top three candidates to a domain expert leads to a near perfect accuracy with good turnaround time in our real-world business context.

Keywords: Approximate string matching · Cosine distance · Deep learning · NLP · Product code matching

1 Introduction

Businesses generate a lot of data by day-to-day operations. Broadly, we distinguish between business and product data [1]. Business data include accounting and sales data. These are analyzed using business reports. Even for the same product, data such as product names, bar codes, and dimensions, vary for each company. The purchase orders and sales contracts that companies receive daily include product details of their customers or suppliers. A lot of resources are spent on matching products across stakeholders of the supply chain.

Our company, which we keep anonymous in our submission, provides a cloud-based platform for suppliers, distributors, and retailers to share their product data. After collecting product data from the suppliers, the data is cleansed and linked with product data of the retailers. Retailers continuously receive new data in their ERP, POS and PIM systems. By acting as a central authority for matching product codes, our company reduces product mappings significantly. Indeed, without such a central authority, we would require product matches for each combination of s suppliers and c customers, that is, $s * c$ different matchings. With a central authority, we only require one product matching between our company and every supplier and retailer, that is, $s + c$ different matchings [2]. Benefits for all stakeholders include shorter order lead times, higher accuracy,

© Springer Nature Switzerland AG 2022
H. Fujita et al. (Eds.): IEA/AIE 2022, LNAI 13343, pp. 29–41, 2022.
https://doi.org/10.1007/978-3-031-08530-7_3

lower inventory cost, simpler control of stock levels, up-to-date information about promotions, price changes, product availability, as well as higher customer service level, and reduction in labour costs [3].

Main Contribution and Impact. We developed two off-line methods for learning matches between product codes for our industry partner. In practical terms we recommend our automated matches to each product vendor, who will choose to accept or reject our recommendations. Their choices will guide future recommendations to further improve the accuracy of our automated matches. Our work applies academic research to a real industry problem, enabling suppliers, distributors and retailers to benefit from automating manual work.

Organization. We review our use of previous work in Sect. 2, before commenting on how we preprocessed our data in Sect. 3. Section 4 summarizes the popular string matching functions we applied, and Sect. 5 summarizes the deep learning methods we employed. Our experiments and results are detailed in Sect. 7 before we conclude and outline future work in Sect. 8.

2   Related Work

Learning how to match product codes can be viewed as the problem of measuring the similarity of short texts. Various approaches have been used to measure the similarity of strings, sentences and documents. These approaches are based on different principles. String-based similarity determines the similarity between words by the similarity of string sequences and character composition. It can be divided into two aspects, character-based similarity (such as Levenshtein Distance [5] and Jaro Distance [6]) and term-based similarity (such as Manhattan distance [7]). Corpus-based similarity attempts to identify semantic similarities between words using information that is specifically derived from large corpora. Moreover, knowledge-based similarity measures the semantic similarity of two words by using information derived from semantic networks [8]. On the other hand, deep learning methods play recently an increasingly important role in natural language processing. By using pre-trained word embeddings, recurrent neural network models and 1D convolution neural network models achieve high performance in text representation. We have applied and combined state-of-the-art solutions from both areas to derive the best possible solution to our problem.

3   Data Wrangling

The customers of our industry partner are from various industries with a wide range of products. Their data includes product codes, store codes, global trade item numbers, product names, descriptions, and images of the products. We focus on learning matches between different names used for the same product.

Typical product names are made up of brand name, product features (such as colour, material), product description, and numbers (such as size, model). Some product names only provide information about two or three parts. Table 1

Table 1. Examples of product names

Product name	Brand name	Features	Product	Numbers
Samsung Galaxy S9 Unlocked - 64gb - Black	Samsung	Galaxy, Unlocked, Midnight Black	- (phone)	S9, 64gb
TimberTech Terrain Composite Decking Grooved Board 136 mm × 24 mm × 4.8 m Silver Maple	Timber Tech	Terrain Composite, Grooved, Silver Maple	Decking Board	136 mm × 24 mm × 4.8 m
Bloom Braced Garden Rake Green	Bloom	Green	Garden Rake	–

Table 2. Detail of data sets

Data sets	Number of products in ssupplier sets	Number of products in retailer sets	Number of good matches
Data1	4230	776	312
Data2	8041	4938	4938

shows examples of product names from large retailer websites. The first product name does not contain the product part, but humans easily recognize it as a phone. The third product name does not contain any numerical information since measurements are irrelevant for this product.

There are some practical guidelines around automating the search for matches. These formalize human common sense for computers. (1) Product names with different brands are different products. For example, "Comvita Manuka Honey UMF 10+ 250g" and "Go Health Manuka Honey UMF 10+ 250g" are different because of the different brands "Comvita" and "Go". (2) Product names with different features are different products. For instance, "Apple iPhone X 64GB Silver" and "Apple iPhone X 64GB Black" are different. (3) Suppliers may supply bulk products to the distributors who repack them into products with different names. (4) Product names with different numbers are different. For example, products with the same type of timber but different lengths are different. (5) The same product can be described in different ways, such as "day and night pills" and "am and pm tablets" refer to the same product, and "Stainless T316" is the same material as "Stainless A4". (6) Abbreviations are common practice that introduce problems, for example GH stands for the brand Good Health.

Our first method applies approximate string matching to data standardized by the use of synonym lists. Our second method applies approximate string matching to training data selected by stakeholders. Deep learning is then used to recommend matches for the remaining data.

In addressing the importance of numbers in product names, we collect two types of raw data sets: one focused on text (called Data1 in Table 2) and one including numbers (such as measurements) in product names (called Data2 in Table 2). Data1 and Data2 both contain two files: one from the suppliers and one from the retailers. In Data1, there are 312 matches between supplier and retailer data, and there are 8041 matches in Data2. For each file, we extract the product ID and product name columns. Each product has a unique ID, so the product ID is the foreign key that references the unique original data element.

The product names are typically created manually by stakeholders, or generated from existing products in the system. The raw data we collect suffers from incompleteness, inconsistency or other errors. Therefore, we pre-process the input data to resolve common issues. Below is a summary of our pre-processing steps to avoid false negatives. (1) Change all characters to lowercase. (2) Remove extra white spaces. (3) Remove punctuation. (4) Replace symbols with words. For example, replace "&" with "and" and replace "%" with "percent". (5) Change numbers to words. For instance, change "3" to the word "three" and change "123" to words "one hundred twenty three". This is to add weights to the numbers. For two product names with only one different number, the character-based similarity algorithms would incorrectly return a high similarity score. Using words to represent numbers will decrease the similarity score in this case. (6) Replace abbreviations and synonyms from a given list. The list has two columns, the first column contains the abbreviations or words in the product names, and the second column contains the corresponding full names of the abbreviations or the synonyms. This list is maintained manually since the abbreviations and words are very specific.

4   Approximate String Matching

Approximate string matching returns strings with the smallest distance to an input string. We regard product names as input strings, and apply seven popular distance functions to them. More details are in the cited papers.

Optimal String Alignment Distance (OSA). This extends the Levenshtein distance. To turn one string into another, it not only counts the number of insertions, deletions and substitutions, but also considers the number of transpositions of adjacent characters [9]. For example, $d_{osa}(ba, acb) = 3$ because we need to delete b from ba, insert c after a, and insert b after c.

Full Damerau-Levenshtein Distance (dl). This function [10] modifies the OSA distance by replacing simple transpositions with minimizing all possible swaps between the current character and all unprocessed characters [14].

Longest Common Substring Distance (LCS). The longest common substring(LCS) of strings s and t is the longest sequence obtained by respectively taking pairs of the same characters from s and t in left-to-right order. The longest common substring distance is the sum of the number of unpaired characters in strings s and t. It can also be considered as the number of deletions and insertions required to convert one string to another [11]. For example, the LSC of the words "leela" and "leia" is "lea", that is, $d_{lsc}(\text{leela}, \text{leia}) = 2 + 1 = 3$. Transforming the word "leela" to "leia" requires three steps, so $d_{lsc}(\text{leela}, \text{leia}) = 3$.

q-gram Distance. A q-gram is a string of q continuous characters. The q-grams of a string are obtained by sliding a q-character window from the beginning of the string and moving one character to the right each time until reaching the end of the string. For example, the di-grams of "happy" are "ha", "ap", "pp" and "py". The q-gram distance of two strings s and t are the number of the q-grams which only occur in s or t but not in both s and t [9].

Cosine Distance Between q-gram Profiles. We can generate vectors $v(s;q)$ and $v(t;q)$ based on q-grams of strings s and t. We also can use the cosine distance of $v(s;q)$ and $v(t;q)$ to represent the distance of strings s and t [12], which calculates the cosine of the angle between the two vectors [14].

Jaccard Distance Between q-gram Profiles. This distance is based on the idea of intersection over union. It simply lists the unique q-grams of two strings and finds the common q-grams appearing in both strings [12].

Jaro-Winkler Distance. This distance is an extension of Jaro distance [6]. It adds an extra penalty for mismatches in the first four characters [13].

The optimal string alignment distance, full Damerau-Levenshtein and longest common substring distance methods have a run time in $\mathcal{O}(|s||t|)$. The q-gram based methods store q-grams in a binary tree, so they use $\mathcal{O}(|Q(s;q)|)$ memory and $\mathcal{O}[(|s| - q - 1)\log|Q(s;q)|]$ time. The Jaro-Winkler distance takes $\mathcal{O}(|s||t|)$ time and $\mathcal{O}(\max\{|s|, |t|\})$ memory [14].

5 Deep Learning

We treat product names as sentences that follow a user-specific grammar, and use deep neural networks to learn how people describe their products in order to predict the similarity score of two products.

Word2Vec. Word2vec is a set of two-layer neural networks which are trained on a text corpus and produce a vector space with several hundred dimensions. Each unique word in the corpus is transformed into a vector. Words with common contexts in the corpus are located close to each other [15]. In our work, we treat Data1 and Data2 as two different corpora. Each vector in each corpus has 300 dimensions. Firstly, each unique word in a corpus is assigned an index. The sentence vector consists of the words' indices in the order they appear. The length of each vector is the length of the longest sentence, with prefix zero paddings to enforce the same length. Then, we use Word2vec to produce 300-dimensional word embeddings as the model input.

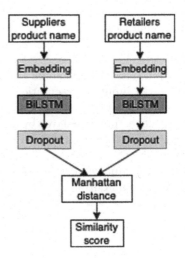

Fig. 1. Siamese manhattan

Siamese Manhattan BiLSTM. Siamese networks consist of two or more identical sub-networks [16]. In our model, the identical sub-network is from the embedding layer to the last LSTM layer. As shown in Fig. 1 our model contains one word embedding layer with two inputs, two bidirectional LSTM layers mixed with two 50% dropout layers, one layer calculates the Manhattan distance of the two vectors from the previous layer and an output layer. LSTMs can "understand" the semantics of a sentence by "analyzing" the words over time. They update their state by "remembering" important and "forgetting" unimportant information [18]. The bidirectional LSTMs not only train normal LSTMs but also contain a reversed copy of the input sequence. This provides additional context to improve the model [17]. The dropout layer prevents overfitting.

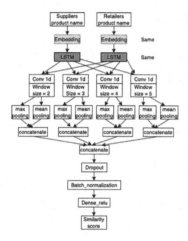

Fig. 2. Multiple LSTM 1D CNN

Multiple LSTM 1D CNN. Multiple LSTM 1D CNN is a very complex model, illustrated in Fig. 2. It starts with two copies of the same embedding layer for each input, followed by two LSTM layers. Then, the two LSTM layers are merged four times in parallel in four 1D convolution layers with sliding window size two, three, four and five, respectively. Each 1D convolution layer connects to one max pooling layer and one mean pooling layer, which are all combined in one concatenated layer at the end. Then, the concatenated layer is connected to several dropout layers, batch normalization layers and relu layers. The final output is a similarity score. The 1D convolution layer creates a sliding window that is convolved in one dimension with the input layer to produce an output tensor [19].

The max (mean) pooling layer has a filter moving across non-overlapping subregions to pick the maximum (mean) value of the subregions in order to downsample the input representation, reduce its dimensions and make assumptions of the features in the subregions. The batch normalization layer normalizes the output of the previous layer by subtracting the batch mean and dividing by the batch standard deviation. This layer increases stability and reduces the covariance shift. The relu layer is short for the rectified linear unit, it outputs the input if the input is positive, otherwise, it outputs zero. This layer is often used before the final output.

6   System Structure Design

We designed two systems for the product data matching (Fig. 3). In the first method, the input data is pre-processed by converting the upper cases to lower cases, removing the extra white spaces and punctuation, replacing symbols with words and replacing numbers with words. Also, based on synonyms lists, we normalized the abbreviations and terminologies to their synonyms. Then, we apply the best approximate string matching approach to find three suppliers' product names to match each retailers' product name. These three suppliers' product names have the highest similarity score from the retailers' product name. The user can choose the real matches of their product names. The second approach also applies the general pre-processing at the beginning. Without using synonyms lists, matching is done using approximate string matching.

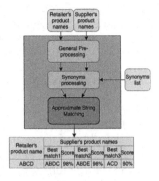

Fig. 3. Synonym+Matching

Then, a "best three matches" list is provided to the user. The user can tick the actual matches of their product names or choose not to tick any of them. This will generate a training data set to train the deep learning model. All the data without real matches (not chosen by the user) will get "three best matches" again based on the predictions of the neural networks. This also forms a "select-train-predict-select" cycle.

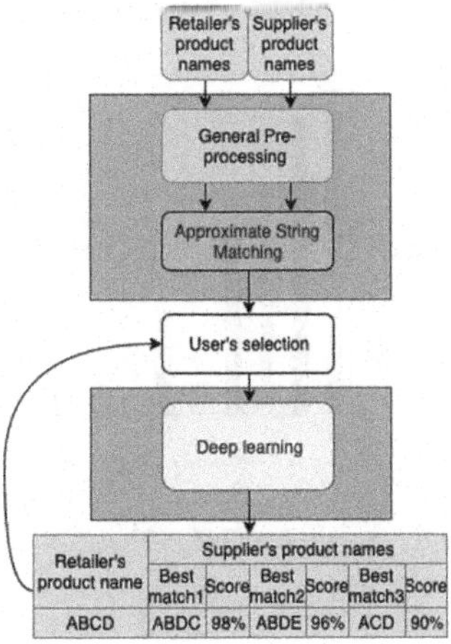

7   Experiments and Results

In this section, we design and conduct some experiments to compare the correct matching rates and run times of the approximate string matching methods on Data1 and Data2. We used (1) different parameters during approximate string matching, (2) applied and did not apply data pre-processing, and (3) change numbers to words or did not. In addition, there is a comparison of the efficiency and accuracy between the Siamese manhattan biLSTM and multiple LSTM 1D CNN methods on Data1 and Data2. All experiments were conducted on a machine with 2.7 GHz Dual-Core Intel Core i5 CPU with 8 GB RAM.

Fig. 4. Matching+Learning

7.1   Exploratory Data Analysis

Figure 5 shows the lengths of the product names in Data1 and Data2. In Data1, most product names are about 3 to 7 words long, while the product names in

Data2 have a length between 5 to 14 words. Figure 6 suggests that the number of frequently used words is similar for Data1 and Data2. A few words are used very often in the product names for both Data1 and Data2. These could be the most popular products. The occurrence rate of most words is low. Hence, product names are identifiable by a small number of different words.

7.2   Comparison of Approximate String Matching Methods

We pre-processed Data1 and Data2 by changing all the characters to lower case, removing punctuation and extra spaces, replacing symbols by words and replacing abbreviations and terminologies by their synonyms. For Data1 and Data2, we kept one data set with the original numbers and changed another data set with numbers to their corresponding words. This is to test if converting numbers to words can increase the correct matching rate. Then, we compared the seven approximate string matching methods on these data sets. We used $w_1 = w_2 = w_3 = w_4 = 1$ for the optimal string alignment distance and full Damerau-Levenshtein distance since we assigned deletions, insertions, substitutions and transpositions the same importance in the algorithms. For the three

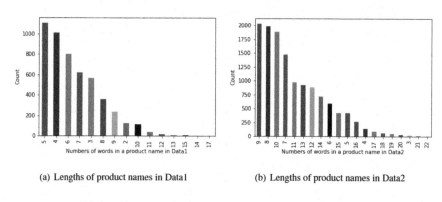

(a) Lengths of product names in Data1   (b) Lengths of product names in Data2

Fig. 5. Length of product names in Data1 and Data2

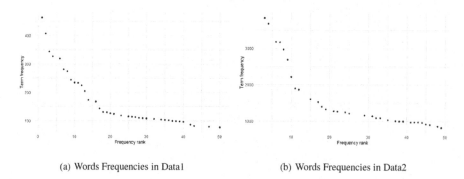

(a) Words Frequencies in Data1   (b) Words Frequencies in Data2

Fig. 6. Words Frequencies in Data1 and Data2

q-gram based methods, we chose $q = 1, 2, 3, 4$ to test the performance. In addition, we tested the Jaro-Winkler distance with $p = 0$, 0.1 and 0.2. We run each method ten times to get its average run time.

Table 3 and Table 4 show how the accuracy increases when suggesting the best three matches. The top match contributes the most to the accuracy (about 60% to 90%). However, the second and third ranked match contribute a lot, especially when the amount of data is large.

Table 3 shows that changing numbers to words slightly decreases the correct matching rate for Data1. Table 4 implies that replacing numbers with words can increase the correct matching rates for q-gram based cosine distance and Jaccard distance when $q = 2, 3$ or 4. For each of the string matching methods, the average run times are similar for the data sets with numbers, and with words that represent numbers.

Table 3. Correct matching rate and average run time in data1

Methods		Nunbers as Numbers					Numbers as words				
		Match1	Match2	Match3	Total	Avg time(s)	Match1	Match2	Match3	Total	Avg time(s)
OSA		73.72%	6.09%	2.88%	82.69%	13.39	66.99%	4.17%	2.88%	74.04%	16.68
DL		73.72%	6.09%	2.88%	82.69%	25.02	66.99%	4.17%	2.88%	74.04%	33.98
LCS		81.73%	6.41%	1.60%	89.74%	9.93	77.88%	5.45%	4.17%	87.50%	12.10
q-gram	$q=1$	69.87%	4.81%	1.28%	75.96%	3.73	52.56%	5.45%	2.24%	60.26%	3.84
	$q=2$	87.50%	5.77%	0.96%	94.23%	9.64	85.90%	6.41%	1.28%	93.59%	9.22
	$q=3$	86.54%	5.77%	1.28%	93.59%	27.37	86.54%	4.81%	0.96%	92.31%	26.14
	$q=4$	82.05%	8.01%	1.60%	91.67%	46.21	82.37%	6.73%	1.28%	90.38%	48.54
Cosine	$q=1$	75.64%	5.45%	2.88%	83.97%	3.61	60.58%	7.05%	0.96%	68.59%	3.91
	$q=2$	92.63%	4.17%	1.28%	98.08%	9.91	88.78%	4.81%	1.60%	95.19%	9.91
	$q=3$	94.55%	2.24%	1.60%	98.40%	30.55	91.03%	4.17%	0.32%	95.51%	28.26
	$q=4$	92.63%	5.13%	0.96%	98.72%	50.80	90.38%	5.77%	0.32%	96.47%	49.26
Jaccard	$q=1$	48.08%	0.00%	4.81%	52.88%	3.93	28.53%	10.26%	4.81%	43.59%	3.92
	$q=2$	93.59%	2.88%	0.64%	97.12%	11.38	89.74%	3.21%	0.64%	93.59%	11.53
	$q=3$	93.91%	3.21%	0.64%	97.76%	30.06	91.35%	3.21%	0.64%	95.19%	29.75
	$q=4$	92.63%	4.49%	0.96%	98.08%	47.61	90.06%	3.85%	1.60%	95.51%	48.62
JW	$p=0$	80.77%	5.77%	2.88%	89.42%	2.86	79.17%	6.09%	0.96%	86.22%	3.33
	$p=0.1$	83.33%	7.05%	1.92%	92.31%	2.81	82.05%	7.05%	1.28%	90.38%	3.35
	$p=0.2$	83.01%	7.37%	1.92%	92.31%	2.82	81.73%	7.37%	1.28%	90.38%	3.26

In general, all methods perform well for Data1, with q-gram based methods providing slight advantages. For Data2, q-gram based methods outperform other approaches when $q = 2, 3$ or 4. However, the average run time doubles whenever q increases by one. The Jaro-Winkler distance has the lowest average run time. We did not consider the memory usage here as our data sets are fairly small.

We also compared the correct matching rates with and without data pre-processing on OSA, cosine with $q = 3$, and JW with $p = 0.1$. These represent the distance based, q-gram based, and Jaro-Winkler methods. Figure 7 shows that data pre-processing increases the correct matching rate.

In conclusion, data pre-processing increase the accuracy signficantly. Considering the top three matches for each product name increases accuracy further. The trigram cosine distance function that replaces numbers by words performs well and stable with a median run time.

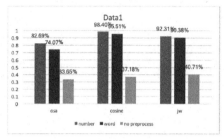 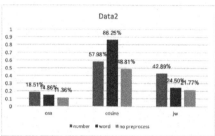

(a) Comparison of Correct Matching Rates - Data1   (b) Comparison of Correct Matching Rates - Data2

Fig. 7. Comparison of correct matching rates with and without data pre-processing

Table 4. Correct matching rate and average run time in Data2

Methods		Nunbers as Numbers					Numbers as words				
		Match1	Match2	Match3	Total	Avg time(s)	Match1	Match2	Match3	Total	Avg time(s)
OSA		12.47%	3.65%	2.39%	18.51%	222.19	10.98%	2.47%	1.42%	14.86%	735.14
DL		12.47%	3.60%	2.43%	18.51%	448.62	10.92%	2.51%	1.44%	14.86%	1236.94
LCS		20.49%	5.41%	2.31%	28.21%	214.12	16.46%	3.67%	1.96%	22.09%	570.32
q_gram	q = 1	16.06%	5.37%	3.18%	24.61%	62.34	10.09%	2.90%	1.68%	14.66%	70.77
	q = 2	26.39%	8.75%	5.97%	41.11%	130.43	23.82%	6.08%	3.73%	33.62%	146.89
	q = 3	17.11%	6.50%	4.64%	28.25%	308.46	24.61%	5.97%	3.26%	33.84%	337.77
	q = 4	10.35%	4.35%	3.95%	18.65%	559.35	20.58%	5.22%	0.00%	25.80%	601.66
Cosine	q = 1	23.84%	8.49%	5.39%	37.71%	75.39	14.07%	5.83%	3.93%	23.84%	83.26
	q = 2	43.05%	13.63%	6.50%	63.18%	143.45	48.99%	15.53%	7.41%	71.93%	162.11
	q = 3	36.01%	13.93%	8.04%	57.98%	342.93	59.52%	17.92%	8.81%	86.25%	311.11
	q = 4	27.76%	13.59%	7.80%	49.15%	647.29	56.44%	16.42%	7.65%	80.52%	604.65
Jaccard	q = 1	13.69%	6.34%	3.83%	23.86%	66.41	3.32%	2.05%	1.17%	6.54%	79.87
	q = 2	44.37%	14.74%	6.82%	65.94%	146.18	37.59%	12.96%	7.78%	58.32%	162.37
	q = 3	34.59%	13.93%	8.20%	56.72%	377.76	50.65%	15.07%	8.08%	73.80%	355.79
	q = 4	26.79%	13.16%	7.47%	47.43%	622.17	48.87%	14.99%	8.40%	72.26%	618.14
JW	p = 0	27.16%	7.15%	3.62%	37.93%	42.15	15.31%	6.40%	3.79%	25.50%	119.11
	p = 0.1	30.86%	8.00%	4.03%	42.89%	50.17	14.70%	6.08%	3.73%	24.50%	119.17
	p = 0.2	28.53%	7.13%	3.50%	39.17%	49.30	14.14%	5.85%	3.54%	23.53%	119.20

7.3   Comparison of Deep Learning Methods

Before applying the deep learning methods, we pre-processed Data1 and Data2 by changing all the characters to lower case, removing punctuation and extra spaces, replacing symbols by words and converting numbers to words. Then, we used the trigram cosine distance function to generate the training data sets. As before, we consider the top three matches for each product name. Every correct matching pair is marked with 1, and every incorrect matching pair is marked with 0. For Data1, we have 278 correct matches and 2050 incorrect matches in the training set and 34 pairs of product names from both supplier and retailer sets as the test set. For Data2, there are 3064 correct matching pairs and 11750 incorrect matching pairs in the training data, and 1875 pairs of product names as test data. Based on Data1 and Data2, we built two independent corpora for word and character embeddings. We tested the character and word embeddings for both Siamese Manhattan BiLSTM and multiple LSTM 1D CNN.

For the Siamese Manhattan BiLSTM method, we used the length of the longest product name as the length of every vector. Each input has two BiLSTM layers, with 128 neurons in the first BiLSTM layer and 32 neurons in the second BiLSTM layer. Each BiLSTM layer is followed by a 50% dropout layer. The multiple LSTM 1D CNN method takes vectors with a length of 25 as inputs. We trained the model with two connected 256 nodes LSTM layers for each of the two inputs. Then, the two LSTM layers were merged in four 1D convolution layers with 64 filters and sliding window size two, three, four and five, respectively. The pool size of maximum pooling and mean pooling layers is 25. After concatenating all the pooling layers into one layer, we used three dropout layers with 30% dropout rate, mixed with three batch normalization layers and two relu layers (with 512 and 256 nodes, respectively). We used binary cross-entropy as the loss function and performed 10-fold cross-validation on each model, with each validation being trained in 10 epochs.

Table 5 shows that both methods take much longer time to train on 10 epochs compared to the approximate string matching method, especially when the data set is large. The Siamese Manhattan BiLSTM method with character embedding performs the best, with high accuracy and low training time. Increasing the number of training epochs may increase accuracy. However, this will also increase the training time, which may be unrealistic.

Table 5. Comparison of correct matching rate and average run time

Methods	Embeddings	Data1		Data2	
		Accuracy	Avg time(s)	Accuracy	Avg time(s)
Siamese	Characters	67.65%	450.51	65.80%	4439.25
Manhattan BiLSTM	Words	88.24%	561.23	53.18%	6126.68
Multiple LSTM	Characters	25.88%	1059.21	56.51%	5572.73
1D CNN	Words	31.76%	1186.93	38.81%	7356.71

8 Conclusion and Future Work

We designed two useful and high-performing approaches to match product names in retailer and supplier data. The first method applies specific synonym lists and approximate string matching. The second method combines approximate string matching with deep neural networks. Both methods benefit from data pre-processing such as converting upper to lower cases, removing extra white spaces and punctuation, and replacing symbols and numbers by words. Our experiments suggest that the trigram cosine distance performs well and stable. Our second method saves any manual resources that go into the design and maintenance of the synonym list. Users are provided with a list of matches using the trigram cosine distance. The machine then learns and makes predictions based on which matches our users select. The Siamese Manhattan BiLSTM method outperforms the multiple LSTM 1D CNN method in terms of accuracy and efficiency. Our methods overcome the real-world challenge of detecting the same product using different descriptions. By adding weights to numbers in the description of the product names, we increase the matching rate using trigram cosine distance. Providing users with the top-three matches instead of only the top match increases accuracy by at least 20% and at most 40%. Our methods are applicable to different product data sets in different industries.

Future work is to build a synonym list that will be stored and utilized by a database. This introduces manual resources such as deep knowledge about the products. For the deep learning methods, building a large product corpus may help reduce the manual work of users. Another challenge is to scale the efficiency of our methods to large data sets. A well-designed database search algorithm may overcome this problem. Our final aim is to implement the most appropriate method online on the platform that our industry partner utilizes.

References

1. Yoo, S., Kim, Y.: Web-based knowledge management for sharing product data in virtual enterprises. Int. J. Prod. Econ. **75**, 173–183 (2002)
2. Fensel, D., et al.: Product data integration in B2B e-commerce. IEEE Intell. Syst. **16**, 54–59 (2001)
3. Hansen, J., Hill, N.: Control and audit of electronic data interchange. MIS Q. **13**, 403 (1989)
4. Vesta Central. https://vesta-central.com/. Accessed 01 Apr 2021
5. Vladimir, L.: Binary codes capable of correcting deletions, insertions, and reversals. Soviet Phys. Doklady **10**(8), 707–710 (1966)
6. Jaro, M.: Probabilistic linkage of large public health data files. Stat. Med. **14**, 491–498 (1995)
7. Dice, L.: Measures of the amount of ecologic association between species. Ecology **26**, 297–302 (1945)
8. Mihalcea, R., Corley, C., Strapparava, C.: Corpus-based and knowledge-based measures of text semantic similarity. In: AAAI, vol. 6, pp. 775–780 (2006)
9. Boytsov, L.: Indexing methods for approximate dictionary searching. J. Exp. Algorithmics **16**, 11 (2011)
10. Wagner, R., Lowrance, R.: An extension of the string-to-string correction problem. J. ACM **22**, 177–183 (1975)
11. Saul, B.N., Christian, D.W.: A general method applicable to the search for similarities in the amino acid sequence of two proteins. J. Mol. Biol. **48**, 443–453 (1970)

12. Ukkonen, E.: Approximate string-matching with q-grams and maximal matches. Theoret. Comput. Sci. **92**, 191–211 (1992)
13. Winkler, W.: String comparator metrics and enhanced decision rules in the fellegi-sunter model of record linkage. In: Proceedings of the Section on Survey Research Methods, pp: 354–359. American Statistical Association (1990)
14. Loo, M.: The stringdist package for approximate string matching. R J. **6**, 111 (2014)
15. Mikolov, T., Chen, K., Corrado, G., Dean, J.: Efficient estimation of word representations in vector space. In: 1st International Conference on Learning Representations, ICLR 2013, Scottsdale, Arizona, USA, May 2–4, 2013, Workshop Track Proceedings (2013)
16. Jonas, M., Aditya, T.: Siamese recurrent architectures for learning sentence similarity. In: Thirtieth AAAI Conference on Artificial Intelligence (2016)
17. Schuster, M., Paliwal, K.: Bidirectional recurrent neural networks. IEEE Trans. Signal Process. **45**, 2673–2681 (1997)
18. Pontes, l.L., Huet, S., Linhares, A.C., Torres-Moreno, J.-M.: Predicting the semantic textual similarity with siamese CNN and LSTM. In: CORIA-TALN-RJC 2018, Rennes, France, May 14–18, pp. 311–320 (2018)
19. Kim, Y.: Convolutional neural networks for sentence classification. In: Proceedings of the 2014 Conference on Empirical Methods in Natural Language Processing, EMNLP 2014, October 25–29, 2014, Doha, Qatar, A meeting of SIGDAT, a Special Interest Group of the ACL, pp. 1746–1751 (2014)

ResUnet: A Fully Convolutional Network for Speech Enhancement in Industrial Robots

Yangyi Pu and Hongyang Yu[✉]

University of Electronic Science and Technology of China, Chengdu, China
541847454@qq.com

Abstract. When conducting voice interactions, it is found that loud noises have terrible effects on the recognition of industrial robots. Some industrial machines like motors and fans, tend to produce loud noises when working, placing great obstacles to the voice interactions. Therefore, an end-to-end full convolutional network model, called Res-Unet, is proposed to solve this problem. The difference between this network and other conventional ones is the accession of residual networks to the encoder and the decoder. It has increased the convergence and complexity of the network and improved the expression ability of the network. In experiments on the decrease of noises caused by industrial machines, it is proved that the performance of Res-Unet was better than other networks, in respect of speech enhancement.

Keywords: Fully convolutional network · Industrial noise · Speech enhancement

1 Instruction

With the rapid development of industrial machines and the technology of speech recognition, industrial robots with voice control functions have been brought to world. However, the noises produced by working industrial robots have a certain impact on the command of speech recognition. For example, the main task of the automatic plastering and troweling robot is to spread plaster. Most of the noises come from the sound of the standby fan and the working motor. To tackle this problem, the method for speech enhancement is adopted to eliminate the influence of noise on speech recognition.

In the field of speech enhancement, classic speech enhancement techniques mainly include spectral subtraction, Wiener filtering, statistical model-based methods, and subspace-based methods [1]. The development of deep learning and acoustic modeling has made some contributions to the solution of speech enhancement in complex environments. According to the different goals of network learning, neural network-based speech enhancement is mainly divided into methods based on time-frequency masking and methods based on feature mapping. The time-frequency masking method, learning the relationship between

H. Fujita et al. (Eds.): IEA/AIE 2022, LNAI 13343, pp. 42–50, 2022.
https://doi.org/10.1007/978-3-031-08530-7_4

pure speech and noise, applies the obtained time-frequency masking estimation to the noisy speech, and synthesizes the time-domain waveform of the enhanced speech with the help of the inverse transform technology. Wang et al. [2] introduced Deep neural network (DNN) into the field of speech separation and noise reduction, and estimated the ideal binary mask (IBM) by the feed-forward DNN; subsequently, Narayanan et al. [3] proposed to estimation of the ideal ratio mask (IRM) in the Mel spectrum domain, which improves the robustness of speech recognition to a certain extent; Park et al. [4] put forward a redundant convolutional coding and decoding network, which optimizes the training process by deleting the pooling layer and adding jump connections and applies Convolutional Neural Network (CNN) to the spectrum mapping. Those two methods require the conversion of the time-domain waveform into the time-frequency domain and the process of amplitude spectrum or power spectrum of the signal, while the phase information has always been ignored. Afterwards, Jansson et al. [5] presented the idea to use U-Net network for speech enhancement and apply it to the separation of the accompaniment and the human voice.

In this paper, an end-to-end speech enhancement model, based on Res-Unet and with the spectrum as the input information, is proposed and applied to the complex working environment of industrial robots.

The rest of this paper is organized as follows. In Sect. 2, detailed information on the relevant methods is offered. In Sect. 3, an introduction to the Res-Unet architecture is made. In Sect. 4, the experimental methods are specified, and in Sect. 5 the experimental results are presented. In the end, the author finishes the part of the conclusion in Sect. 6.

2   Related Work

2.1   U-Net

U-Net [7] is one of the earliest algorithms with the use of fully convolutional networks to make semantic segmentation. The symmetric U-shaped structure that contains compression paths and expansion paths were very innovative at the time, and to some extent, it has affected the design of some segmentation networks. The network is in a shape of U, so it is called U-Net. There has been the application of it in audio to separate the human voice and the accompaniment [6], which has achieved relatively good results. Therefore, this author decides to improve the network on the basis of U-Net, so as to realize the voice enhancement on the industrial robot data set. The U-Net network structure is shown in Fig. 1.

It consists of three parts, the encoder, decoder, and skip connections. The encoder is a series of downsampling operations, composed of convolution and Max Pooling. This part is also called the compression path, made up of 4 blocks. Each block uses 3 effective convolutions and 1 Max Pooling downsampling. After each operation of downsampling, the number of Feature Maps doubles. Then, the network extracts local and structural features in the encoder. What's more, the decoder is called expansive path, which has 4 blocks as well. Before each

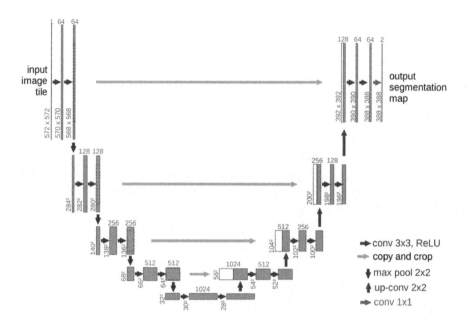

Fig. 1. U-Net structure, the left side is the encoder, the right side is the decoder, the middle is skip-connection.

convolution, the size of the Feature Map is doubled through upsampling, and the number is halved (the last layer is slightly different).

2.2   ResNet

ResNet [10] address the degradation of deep neural networks. As we know, by gradually adding layers to the shallow network, the performance of the model on the training set and the test set will get better, because the model has higher complexity and stronger expression ability. However, the degradation refers to the rapid degradation of network performance as more layers are added to the network. ResNet adjusted the model structure and called the stacked layers a block. For a block, its fitting function is $F(x)$. If the expected latent mapping is $H(x)$, then $F(x)$ will not directly learn the latent mapping, but learn the residual $H(x) - x$, that is, $F(x) = H(x) - x$. So the original forward path would be $F(x) + x$, and $F(x) + x$ would fit $H(x)$. Compared to learning $F(x)$ as an identity mapping, it is easier to learn $F(x)$ as 0, which you can easily do with an L2 regular. Thus, for redundant blocks, only $F(x)$ goes to 0 to obtain the identity mapping, with no loss of performance.

2.3   Huber Loss Function

Like MAE and MSE, the Huber loss function [11] is also a loss function used for regression problems, which combines the advantages of both MAE and MSE.

When the error value is large, the absolute value is used as the loss function, and when the error is small, the square value is used as the loss function. The size of the square error depends on the hyperparameter δ (delta):

$$L_\delta(y, f(x)) = \begin{cases} \frac{1}{2}(y - f(x))^2 & for \ |y - f(x)| \leq \delta \\ \delta|y - f(x)| - \frac{1}{2}\delta^2 & otherwise \end{cases} \tag{1}$$

Y is the predicted value, $f(x)$ is the true value, $y - f(x)$ represents the error, When the error approaches 0, Huber loss approaches MSE and when the error is a large number, Huber loss approaches MAE. This loss function turns out to be better convergence than MAE and MSE.

3   The Proposed Method

3.1   Overview of the Proposed Method

Among time-frequency decompositions, spectrograms have been proved to be a useful representation for audio processing [13]. They consist of 2D images of sequences of Short-Time Fourier Transform (STFT), with time and frequency as axes, and brightness as the strength of a frequency component at each time frame. Hence, the CNN architecture of images can be adopted and directly applied to sound. In this paper, magnitude spectrograms will be utilized as a representation of sound to predict the noise model and finally achieve the decrease of noise by using the original audio.

Firstly, the speech signal is divided into equal-length vectors with a fixed frame length, then short-time Fourier transform (STFT) is applied to each frame speech signal to compute the short-time feature vectors, which represents the magnitude spectrum.

To begin with, a fully convolutional network is put forward, which is the Residual U-Net (Res-Unet) network. It is a combination of U-Net and ResNet, with the Huber function as the loss function. Compared with U-Net, Res-Unet has a deeper network structure and more trained parameters than U-Net. Besides, its performance in speech enhancement has been significantly improved, though the training time is relatively longer.

3.2   Structure of Res-Unet

Res-Unet is made up of two parts: encoder and decoder. The encoder of it is similar to that in U-Net. However, the output results of each layer in Res-Unet are normalized first and then activated by the activation function. Each upsampling contains two 3×3 convolutional layers, a 1×1 "shortcut" and a 2×2 pooling layer. Each downsampling changes the image size to $1/2$ of the original size and doubles the number of convolution kernels. The decoder is similar to that in U-Net. Each upsampling contains two 3×3 convolutional layers and a 1×1 "shortcut", and the corresponding results in the encoder need to be combined before each upsampling. Similar to the encoder, the output results of each layer

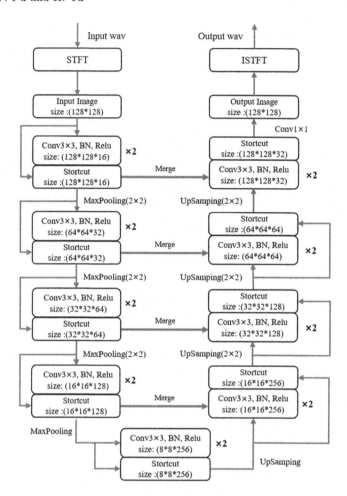

Fig. 2. The structure of Res-Unet

in the decoder need to be standardized and then activated by the activation function. Finally, a 1×1 convolutional network is added to determine the corresponding result of the feature map. The Res-Unet network structure is shown in Fig. 1.

With the accession of the residual network to the U-Net network, the number of layers increases, and the training parameters increase accordingly. At the same time, the degradation problem of the performance of deep convolutional neural networks with many layers is solved, thanks to the properties of residual networks.

3.3 Optimization Function

The optimization function helps the model to adjust the weights at training time so that the model weights are optimally adjusted to minimize the loss function. In this experiment, the Adam optimization function is used. Adam algorithm is an optimization algorithm, proposed by Diederik Kingma and Jimmy Ba in 2014 [9]. Compared with other optimization algorithms, ADAM has advantages of high computational efficiency, less memory consumption, and good handling of non-stationary models.

4 Experimental Methods

4.1 Dataset

Clear Voice: The clear voices are mainly gathered from LibriSpeech [8], an ASR corpus based on public domain audio books, and those audios have been carefully segmented and aligned.

Noise: The noise data set of this experiment mainly includes two kinds of noise, namely, engine noise and fan noise. The noise data are gathered from the actual sound recordings of the Automatic plastering robot and the ESC-50 datasets [12]. The ESC-50 dataset is a labeled collection of 2000 environmental audios, consisting of 5-second-long noise recordings.

To create the datasets for training/validation/testing, the audios are sampled at 8 kHz. The noise intensity is randomized between 20% and 80% and then mixed into the clear sound to give the model some generalization ability.

4.2 Feature Transformation

First, audio signals are sampled at 8 kHz and silent frames are removed from the signal. Then, spectral vectors are computed, using a 256-point Short Time Fourier Transform with a window shift of 64-point (128 Windows). Next, 256-point STFT magnitude vectors are reduced to 128-point by removing the symmetric half. Last, each frame is converted to a spectrogram matrix of size 128×128 and normalized to have zero mean and unit variance.

4.3 Training Schemes

The filter weights were initial as he_normal [14], It draws samples from a truncated normal distribution centered on 0 with where fan_in is the number of input units in the weight tensor. This method makes the variances of the input signals of different layers approximately equal, In order to ensure the effective flow of information during forwarding and backward propagation. The Res-Unet was trained from scratch. The Adam gradient descent optimization was used for back-propagation with a batch size of 20. The training rate started from 0.0001, and the Huber loss function was 0.2.

4.4   Evaluation Score

Perceptual Evaluation of Speech Quality (PESQ): The PESQ algorithm [15] requires a noisy attenuated signal and an original signal for reference. After level adjustment, input filter filtering, time alignment and compensation, and auditory transformation of the two speech signals to be compared, the parameters of the two signals are extracted respectively, and the PESQ score is obtained by integrating the time-frequency characteristics, which is finally mapped to the subjective average opinion score (MOS). PESQ scores range from −0.5 to 4.5. The higher the score is, the better the speech quality becomes.

Short Time Objective Intelligibility (STOI): Short-time Objective Intelligibility (STOI) [16] is one of the important indicators of speech intelligibility. For a word in a speech signal, it can only be understood or not. From this perspective, intelligibility can be considered binary. Therefore, the value range of STOI is also quantified from 0 to 1, which represents the percentage of the word being correctly understood. When the value is 1, the speech can be fully understood. Compared with PESQ, the calculation process of STOI is relatively simple.

5   Experimental Results

To make a comparison between performances, four methods are proposed as below:

OMLSA: A well-known statistical-based spectral amplitude estimator, which is used [17] for speech enhancement.

U-Net: A U-Net architecture trained with conventional magnitude spectrum images, represented by linear frequency and logarithm amplitude domains.

The Proposed Method(Res-Unet): A U-Net architecture integrated with ResNet, with the amplitude spectrum image as input for speech enhancement.

Clear voice with two kinds of noise at different SNR levels are mixed, ranging from a very low SNR (−7.5 dB) to a high one (12.5 dB). Then the mixed sound was put in a different model, and the processing results are calculated by PESQ and STOI respectively.

Regarding Tables 1 and 2, the following results can be achieved:

All methods using U-net neural networks have significantly outperformed the statistical-based enhancer. According to the speech quality metric PESQ, more improvements are obtained at higher signal-to-noise levels by deep learning methods compared with OMLSA. In the case of the speech intelligibility index (i.e. STOI), better performance is achieved at a lower signal-to-noise ratio.

Compared with U-Net, Res-Unet has achieved better results. In particular, at low SNR (−7.5 dB), the PESQ of Res-Unet has increased from 1.50 at U-Net

Table 1. Comparisons of methods based on **PESQ** measure

Method	SDR(dB)						
	−7.5	−2.5	2.5	7.5	12.5	17.5	Avg
Noisy	1.10	1.21	1.42	1.75	2.2	2.68	1.73
OMLSA	1.30	1.56	1.92	2.27	2.64	3.02	2.12
U-Net	1.50	1.94	2.53	2.87	3.14	3.37	2.56
Res-Unet	**1.82**	**2.19**	**2.55**	**2.88**	**3.17**	3.37	**2.66**

Table 2. Comparisons of methods based on **STOI** measure

Method	SDR(dB)						
	−7.5	−2.5	2.5	7.5	12.5	17.5	Avg
Noisy	0.35	0.52	0.67	0.79	0.88	0.93	0.69
OMLSA	0.49	0.61	0.73	0.82	0.89	0.92	0.74
U-Net	0.52	0.69	0.81	0.87	0.91	0.93	0.79
Res-Unet	**0.63**	**0.75**	**0.82**	**0.88**	**0.92**	**0.94**	**0.82**

to 1.82, with an increase of 21%, and the average of all SNR PESQ has increased from 2.56 to 2.66 as well. This experiment verifies that adding ResNet to the encoder and decoder can make great progress in speech enhancement.

6   Conclusion

The noise generated by industrial robots will affect the performance of speech recognition. In order to solve this problem, a Res-Unet model for industrial robot speech enhancement is proposed in this paper. In this model, residual learning is integrated into the U-Net architecture and the Huber function is used as the loss function. Finally, it is observed that the integration of ResNet has significantly improved the speech quality and the intelligible measures, yielding better results in speech enhancement. The model will have greater uses by adding an attention gate mechanism or using more complete industrial noise data set for training, and that is the future work.

References

1. Liu, W.J., Nie, S., Liang, S., et al.: Deep learning based speech separation technology and its developments. Acta Automatica Sinica **42**(6), 819–833 (2016)
2. Wang, Y., Wang, D.L.: Towards scaling up classification-based speech separation. IEEE Trans. Audio Speech Lang. Process. **21**(7), 1381–1390 (2013)
3. Narayanan, A., Wang, D.L.: Ideal ratio mask estimation using deep neural networks for robust speech recognition. In: 2013 IEEE International Conference on Acoustics, Speech and Signal Processing, pp. 7092–7096. IEEE (2013)

4. Park, S.R., Lee, J.: A fully convolutional neural network for speech enhancement. arXiv preprint arXiv:1609.07132 (2016)
5. Jansson, A., Humphrey, E., Montecchio, N., et al.: Singing voice separation with deep u-net convolutional networks (2017)
6. Oord, A., Dieleman, S., Zen, H., et al.: Wavenet: a generative model for raw audio. arXiv preprint arXiv:1609.03499 (2016)
7. Ronneberger, O., Fischer, P., Brox, T.: U-Net: convolutional networks for biomedical image segmentation. In: Navab, N., Hornegger, J., Wells, W.M., Frangi, A.F. (eds.) MICCAI 2015. LNCS, vol. 9351, pp. 234–241. Springer, Cham (2015). https://doi.org/10.1007/978-3-319-24574-4_28
8. Panayotov, V., Chen, G., Povey, D., et al.: Librispeech: an ASR corpus based on public domain audio books. In: 2015 IEEE International Conference on Acoustics, Speech and Signal Processing (ICASSP), pp. 5206–5210. IEEE (2015)
9. Kingma, D.P., Adam, B.J.: A method for stochastic optimization. arXiv preprint arXiv:1412.6980 (2014)
10. He, K., Zhang, X., Ren, S., et al.: Deep residual learning for image recognition. In: Proceedings of the IEEE Conference on Computer Vision and Pattern Recognition, pp. 770–778 (2016)
11. Huber, P.J.: Robust estimation of a location parameter. In: Kotz, S., Johnson, N.L. (eds.) Breakthroughs in Statistics, pp. 492–518. Springer, New York (1992). https://doi.org/10.1007/978-1-4612-4380-9_35
12. Piczak, K.J.: ESC: dataset for environmental sound classification. In: Proceedings of the 23rd ACM International Conference on Multimedia, pp. 1015–1018 (2015)
13. Wang, Y., Narayanan, A., Wang, D.L.: On training targets for supervised speech separation. IEEE/ACM Trans. Audio Speech Lang. Process. **22**(12), 1849–1858 (2014)
14. He, K., Zhang, X,. Ren, S., et al.: Delving deep into rectifiers: surpassing human-level performance on imagenet classification. In: Proceedings of the IEEE International Conference on Computer Vision, pp. 1026–1034 (2015)
15. Rix, A.W., Beerends, J.G., Hollier, M.P., et al.: Perceptual evaluation of speech quality (PESQ)-a new method for speech quality assessment of telephone networks and codecs. In: 2001 IEEE International Conference on Acoustics, Speech, and Signal Processing. Proceedings (Cat. No. 01CH37221), vol. 2, pp. 749–752. IEEE (2001)
16. Taal, C.H., Hendriks, R.C., Heusdens, R., et al.: A short-time objective intelligibility measure for time-frequency weighted noisy speech. In: 2010 IEEE International Conference on Acoustics, Speech and Signal Processing, pp. 4214–4217. IEEE (2010)
17. Cohen, I., Berdugo, B.: Speech enhancement for non-stationary noise environments. Signal Process. **81**(11), 2403–2418 (2001)

Surface Defect Detection and Classification Based on Fusing Multiple Computer Vision Techniques

Min Zhu[1], Bingqing Shen[1], Yan Sun[1], Chongyu Wang[1], Guoxin Hou[1], Zhijie Yan[2], and Hongming Cai[1(✉)]

[1] Software School, Shanghai Jiao Tong University, Shanghai, China
{ericzhumin,sunniel,sun_yan,chongyuwang,houguoxin,
hmcai}@sjtu.edu.cn
[2] Management Information Systems, L'Oreal APAC Operations, Shanghai, China

Abstract. Computer vision techniques are widely used for automated quality control in production line, which can identify defects in from collected images. Due to unclear features, diverse product shapes, and small defect sample size, a single computer vision method can hardly achieve the task of product surface defect detection with high accuracy and high efficiency. Thus, we proposed a novel approach based on fusing multiple computer vision methods, and combining online models with offline models. The proposed approach can achieve high detection accuracy in real-time over the whole production process. We also implemented the system and obtained excellent results in a case study of surface defect detection of lipstick. This research shows that new information and intelligence technologies have their importance in the quality control of manufacturing.

Keywords: Surface defect detection · Computer vision · Deep learning · Model fusion

1 Introduction

In manufacturing, automated surface inspection (ASI) is a very important component to achieve dynamic quality assurance, improve production management and promote production efficiency. ASI is an image-processing technique for automated quality control in production line. At present, the ASI systems based on computer vision have widely replaced manual quality inspection in various industrial product manufacturing, including steel plates [1], screen glasses [2], textile fabrics [3], weld parts [4], metal products [5], ceramic tiles [6] and other industries.

There are two main classes of computer vision, the conventional methods and the deep learning method. The conventional methods leverage traditional digital image processing algorithms or machine learning classifiers with artificial features. [7] presents a method with traditional digital image processing algorithms for reading multi-dial gauges. Nevertheless, the accuracy of conventional methods is limited. Deep learning

© Springer Nature Switzerland AG 2022
H. Fujita et al. (Eds.): IEA/AIE 2022, LNAI 13343, pp. 51–62, 2022.
https://doi.org/10.1007/978-3-031-08530-7_5

methods based on convolutional neural networks (CNN) are widely used for surface defect detection of industrial products [8]. It needs a lot of samples to train the model to increase the accuracy. However, in the real industrial production process, the number of the defect samples is always insufficient. To address this issue, the data augmentation techniques can be used to gain more data. [9] proposed a new GAN-based surface defect generative adversarial network (SDGAN). In product surface defect detection and classification, the effectiveness of data augmentation has been verified and applied in [10, 11].

In production, identifying product defects in real time is sometimes critical to production for finding technical or process faults in the early stage, which can largely reduce production cost and even time-to-market. Deep learning methods certainly cannot serve this purpose, due to the lack of sufficient samples during the early stage. Thus, it is impossible to use a single method to detect and classify product defects in real production.

This paper proposed a multi-stage integration of methods to meet the requirements of industrial production. In this approach, the conventional computer vison and deep learning techniques are combined to build an online detection model for real-time online defect detection. Meanwhile, the deep learning technique is used to build a classification model for offline defect classification. The first stage can achieve real-time defect detection on production line with a small number. In the second stage, the detected results will be classified into different concrete defect types with a deep learning method. In production, meanwhile, the collected positive samples and marked negative samples with data augmentation constantly feed both online and offline models to improve detection and classification accuracy, and a knowledge graph is employed for training cold boot.

The main contributions are as follows:

1. A framework of industrial product surface defect detection and classification is proposed based on a multi-stage approach.
2. An online defect detection method is proposed based on combining multiple computer vision method.
3. An offline product surface defect classification system is devised based on the deep learning method.
4. A real system for lipstick defect detection is devised with experiments, which shows how to implement the framework in practices.

This paper is organized as follows. Section 2 proposes the overall technical framework. Section 3 describes the online detection approach. Section 4 describes the offline classification approach. Section 5 introduces a case study with experimental results, and lastly, Section 6 concludes the work and provides some future directions.

2   Technical Framework

In industrial product surface defect detection and classification, both efficiency and accuracy and efficiency are critical to productivity and product quality. Thus, to reconcile the requirement of defect detection speed from quality control on production line and the requirement of defect classification accuracy from quality management off production line, a novel multi-stage surface defect detection and classification framework is proposed. It consists of three components: including online defect detection, offline defect classification, and model training cold boot. The detect results and classification results can be used for production applications. The entire framework is shown in Fig. 1.

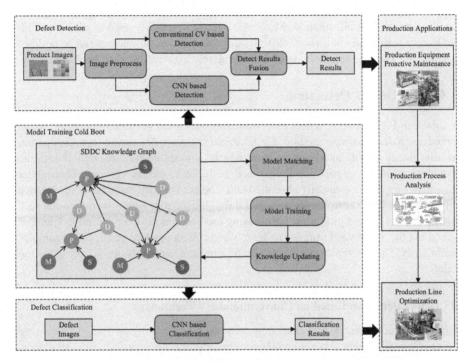

Fig. 1. Overall technical framework.

The online defect detection includes image preprocess, conventional computer vision based detection, CNN based detection, detect results fusion. Firstly, the product images will be preprocessed and detected by a conventional computer vision based detector and a CNN based detector in parallel. Finally, the detect results will be incorporated to generate the final fusion detection result, including defect data and defect images.

In offline defect classification, the defect images are classified by a CNN classifier, which generates the final defect classification results, including classification data and defect images. Meanwhile, the classifier trained by mask R-CNN is optimized with labeled negative samples collected from production line continuously.

Moreover, model training cold boot is introduced to increase the efficiency of modeling training for new products. This component is centered by a surface defect detection and classification (SDDC) knowledge graph for managing product defect related information. The nodes of the SDDC graph includes products (P), defects (D), shapes (S), and models (M) for surface defect detection & classification. The edges of the graph describe the "has-a" relation between a product type and its possible defects, shapes, and models.

In practice, when a new product comes, the model matching function searches for the most similar detection and classification model from the knowledge graph with the defect and shape information. Then, the model training function configures a new model with the matching result and train it with a few shorts of the new product sample. Lastly, the knowledge update function adds the new model into the knowledge graph, together with the new product information. Meanwhile, it can be quickly used in production for cold boot. Notably, any online or offline model optimized in production can be updated back into the knowledge graph, which can iteratively improve the cold boot capability.

3   Online Defect Detection

As shown in Fig. 2, online defect detection combines a conventional computer vision method and a deep learning method. First, sample images are retrieved from production line after image cutting and segmentation, which removes invalid area and divides the product image into four parts, including face, body, base, and background. Then, After the sample images are manually classified and labeled. Initially, the dataset of negative samples and positive samples are enlarged through data augmentation, due to the lack of samples. Next, the digital image processing method and the deep learning detection method use the augmented samples to defects and collect positive and negative samples continuously. These samples can be used for training and iteratively optimizing detection models.

3.1   Defect Detection Based on Conventional CV Technology

Conventional computer vision detection methods based on digital image processing needs to manually select features and chooses the appropriate algorithm to detect defects based the features. The normal part of a product body has fairly uniform intensity distribution. Therefore, we can extract the features of the homogeneous part to find the pixel clusters with abrupt changes, and finally locate the abnormal pixels. Simply, we can extract the edge features and corner features of a defect area.

To extract an edge, we firstly need to filter the image to remove the influence of noise firstly. Then, we need to finds the intensity gradient in the image to filter the false results and determine the suspected boundary. In addition, the boundaries of an edge needs to be tracked. As a reference, we can use the Canny algorithm [12] for edge feature extraction.

Moreover, a corner can be defined as follows: 1) it is the intersection point between the contours; 2) for the same target and scene, even if the perspective changes, the point still keeps the original features; 3) the pixels around the point have obvious changes in gradient direction and gradient amplitude. We can use the Harris algorithm [13] to detect corners.

3.2   Defect Detection Based on CNN

In the method of surface defect detection of industrial products based on deep learning, the problem can be transformed into object detection by framing the abnormal area. In this paper, a supervised deep learning network, R-CNN, is proposed to achieve feature extraction and frame selection of suspected area, as shown in Fig. 2. The R-CNN model combines FPN [14] and RPN [15]. FPN uses both deep and shallow feature maps to integrate the final feature maps, which improves the detection accuracy without significantly increasing the time cost. RPN is a fully convolutional network, which can generate high-quality region proposals.

In the R-CNN model, the shared convolutional layer of each scale outputs a feature map on which a small $n \times n$ window is used for sliding. Each sliding window is mapped to a low dimensional feature. By sliding window, we can get candidate areas according to the given prior anchor box.

Then, in the process of model training, anchors boxes are divided into two types through Softmax. A binary label is assigned to each anchor point to indicate whether it is a detection or not. Bounding Box regression is used to refine the proposals. In the process of defect detection, a set of feature images are obtained by a trained FPN network. Next, the suspected abnormal areas are detected and proposed by RPN.

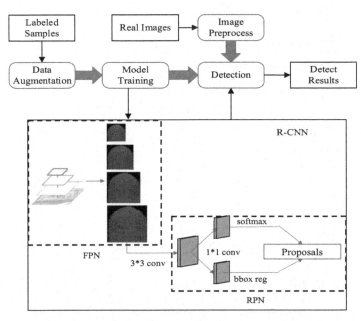

Fig. 2. Overall CNN-based defect detection procedure.

3.3 Detection Result Fusion

After the candidate regions has been generated from the two methods, they need to be fused to obtain a complete detection result. Algorithm 1 illustrates the process detection result fusion. First, the system calculates the overlap of all candidate regions. If there is an overlap (i.e., $t_1 > overlapThreshold$) and the centroid distance of the two regions are less than a threshold (i.e., $t_2 < distanceThreshold$), then the two candidate regions will be merged. Otherwise, the process will end.

Algorithm 1: Defect Detection Result Fusion

1. **Input**:
2.   Images to be detected in JPG format in path \tobedetected
3. **Output:**
4.   Defect detect fusion result in XML format in path \detectionresult
5.
6. Main Task
7.   Configure parameter *overlapThreshold*, *distanceThreshold*
8.   Images ← Load images from \tobedetected
9.   Initialize CCVSet, SDLSet, t1, t2, DetectionResult
10.   Async Run taskCCV
11.   Async Run taskSDL
12.   Timer ← 0
13.   While (CCVSet = ∅ ∨ SDLSet = ∅) ∧ Timer < 1000ms, **then**
14.   **Wait for 100ms**
15.   Timer ← Timer + 100ms
16.   **End While**
17.   t1 ← taskRegionOverlap (CCVSet, SDLSet)
18.   If t1 < *overlapThreshold*, then
19.   t2 ← taskRegionDistant (CCVSet, SDLSet)
20.   If t2 > *distanceThreshold*, then
21.   DetectionResult ← taskRegionMerge (CCVSet, SDLSet)
22.   End If
23.   End If
24.   Move Images to \detected
25.   Save DetectionResult to \detectionresult
26. End Main
27.
28. Task taskCCV
29.   EFs ← EdgeFeatureExtraction (Images)
30.   CFs ← CornerFeatureExtraction (Images)
31.   CCVSet ← DefectRegionIdentification (EFs, CFs)
32. End Task
33.
34. Task taskSDL
35.   FP ← FeaturePyramidExtraction (Images)
36.   RP ← RegionProposalExtraction (FP)
37.   SDLSet ← DefectRegionIdentification (RP)
38. End Task
39.

---

4   Offline Defect Classification

We propose an offline defect classification method by Mask R-CNN [16], as shown in Fig. 3. After training the model of Mask R-CNN with augmented defect samples, first, the features of the whole image are extracted by FPN to generate the feature map, which is shared for the subsequent RPN layer and the full connection layer. Secondly, the region to be detected is generated through a candidate region network (RPN). Then, FC layer is used as the network to classify the candidate regions. Meanwhile, the second adjustment of the target bounding box is carried out. At last, the output dimension of ROI align is expanded by a head to make the prediction mask more accurate, and then the mask of the detected target is obtained through a full convolution network.

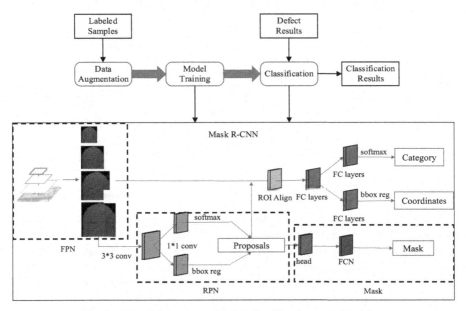

Fig. 3. Overall procedure of defect classification model training.

5   Case Study and Experiment

In this section, defective lipstick image detection and classification is studied as a typical case of industrial production. At present, lipstick quality inspection can only be carried out through artificial naked eye observation. There are some problems, including limited capacity improvement, high rate of missed inspection, long-term overuse of eyes causing eye damage and so on. Developing and deploying a lipstick surface defect detection system can improve the production efficiency, minimize the rate of missed inspection, and reduce the damage to human body. we can implement such system based on the proposed framework.

5.1 Overall System Architecture

The entire system is implemented with four subsystems: data acquisition, data storage, online defect detection, and offline defect classification, as shown in Fig. 4. Firstly, images are retrieved by the data acquisition module. They are stored in different folders of the data storage module for defect detection. Then, the online detection modules read the image from the respective folders and identify abnormal lipsticks from normal ones. The results will be sent back to the data storage module for offline defect classification.

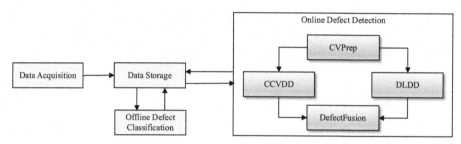

Fig. 4. Overall system architecture.

After data acquisition, the image group is uploaded for each production line, which is divided into three categories: the original lipstick image group, the abnormal lipstick image group, and the normal lipstick image group. They are separately stored in different folders for defect detection and future classification use. At present, there is no need to consider the influence of color. So the image is stored in the format of gray image to reduce the storage redundancy.

The offline defect classification subsystem employs a classifier trained with a large number of labeled defective lipstick images collected in the data storage subsystem. With the increase of sample data, the accuracy of classifier is continuously improved.

5.2 Data Acquisition

The overall design of the visual scheme is shown in Fig. 5(a), which includes four 1.3 mega-pixel cameras and four parallel lights. Cameras are placed per 90° relative to the lipstick. Meanwhile, each light is placed between two adjacent cameras. With this visual scheme, system can capture images of lipstick from production line as shown in Fig. 5(b).

5.3 Online Defect Detection

The online defect detection subsystem contains four modules: Computer Vision Prepro-cor (CVPrep), Conventional Computer Vision Defect Detector (CCVDD), Deep learning Detector (DLDD), and Defect Results Fuser (DefectFusion). The images collected by the visual scheme are preprocessed by module CVPrep for cutting and segmentation, and output the images as shown in Fig. 6(a).

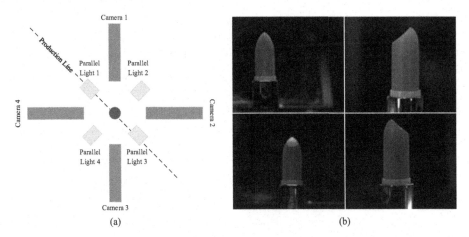

Fig. 5 (a) Visual scheme and (b) Images of lipstick.

Then, the subsystem calls CCVDD, DLDD, and DefectFusion to detect defect from preprocessed images by two detection method in parallel. CCVDD extracts edge features and corner features to locate defect regions with conventional computer vision method, as shown in Fig. 6(b) and Fig. 6(c). Meanwhile, DLDD locate defect regions with supervised deep learning methods. Once two detection results are generated, the DefectFusion module will fuse the detection results and provide the final detection result.

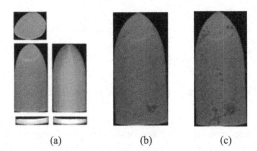

Fig. 6. (a) Segmentation results of lipstick, (b) Edge features of lipstick and (c) Corner features of lipstick.

To quantitatively evaluate the performance of the proposed online detection approach, we use F1 score to compare our approach to the CNN approach with different defect rates. Defect rate refers to the proportion of defective products in a batch of products. Different product types or production processes will lead to different defect rates. Different defect rates will also affect the recall and precision of the machine learning model, thus affecting the F1 score. Thus, defect rate brings important influence to defect detection.

We conduct experiments on data sets with different defect rates under the best parameter settings. All experiments use the same size of training data set (i.e., 800 samples).

The result is given in Table 1. From the results, the F1 scores of our fused approach are better than the CNN approach under the same defect rates. As the defect rates grows, the F1 score of both approaches increase, while our approach has higher increasing rate as shown in Fig. 7.

Table 1. Test results

Defect Rate	1%	4%	7%	10%	13%
F1 Score (CNN Approach)	0.14	0.31	0.33	0.36	0.36
F1 Score (Proposed Approach)	0.22	0.70	0.75	0.86	0.88

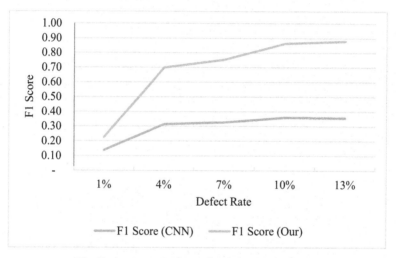

Fig. 7. Accuracy results under different defect rates

In online defect detection, the conventional computer vision method is more flexible for the analysis of image features and more sensitive to abnormal features. Only by analyzing the typical defect features can a targeted detection algorithm be designed, which is suitable for the detection of defect types and relatively stable detection environment. However, the problems of this method mainly include high false detection rate, incomplete feature extraction and poor generalization. Compared with the conventional computer vision method, CNN is more in-depth for defect extraction and cognition, but it needs a lot of training data support, so it can't carry out fast iteration in the early stage of detection model.

Our fused method can solve the problems in solely applying one computer vision method. Through conventional computer vision, the detection model can be trained quickly and applied to production. Meanwhile, with the increase of the quantity of samples during the production process, the CNN-based detection model is continuously trained and optimized, and gradually replaces the conventional computer vision detection model. The above experimental results have shown the advantage of our proposed method.

6 Conclusion

In order to solve the problems of production efficiency and high labor cost caused by artificial quality inspection in the quality control stage of lipstick production, this paper proposed a multi-stage industrial product surface defect detection approach by combining multiple computer vision techniques. It also presented a real system for lipstick surface defect detection in the case study. The experiment results show that the proposed approach can be used to satisfy the detection requirements of the lipstick industry. In the future, the proposed approach can be adapted to the surface defect detection and classification of other industrial products, which needs a further study.

Acknowledgments. This work was supported by the National Natural Science Foundation of China under Grant 61972243.

References

1. Soukup, D., Huber-Mörk, R. Convolutional neural networks for steel surface defect detection from photometric stereo images. In: Advances in Visual Computing, Bebis, G., et al. (eds.), 668–677, Springer International Publishing (2014). https://doi.org/10.1007/978-3-319-14249-4_64
2. Jian, C., Gao, J., Ao, Y.: Automatic surface defect detection for mobile phone screen glass based on machine vision. Appl. Soft Comput. **52**, 348–358 (2017)
3. Wei, B., Hao, K., Tang, X., Ding, Y.: A new method using the convolutional neural network with compressive sensing for fabric defect classification based on small sample sizes. Text. Res. J. **89**, 3539–3555 (2019)
4. Zhang, H., Chen, Z., Zhang, C., Xi, J., Le, X.: Weld defect detection based on deep learning method. In: 2019 IEEE 15th International Conference on Automation Science and Engineering (CASE), pp. 1574–1579 (2019). https://doi.org/10.1109/COASE.2019.8842998
5. Yun, J.P., et al.: Automated defect inspection system for metal surfaces based on deep learning and data augmentation. J. Manuf. Syst. **55**, 317–324 (2020)
6. Karimi, M.H., Asemani, D.: Surface defect detection in tiling Industries using digital image processing methods: analysis and evaluation. ISA Trans. **53**, 834–844 (2014)
7. Singh, H.V.P., Mahmoud, Q.H.: ViDAQ: A computer vision based remote data acquisition system for reading multi-dial gauges. J. Ind. Inf. Integr. **15**, 29–41 (2019)
8. Tao, X., Hou, W., Xu, D. A Survey of Surface Defect Detection Methods Based on Deep Learning. Acta Autom. Sin. 1–20 (2020)
9. Niu, S., Li, B., Wang, X., Lin, H.: Defect image sample generation with GAN for improving defect recognition. IEEE Trans. Autom. Sci. Eng. **17**, 1611–1622 (2020)
10. Hu, K., Wang, B., Shen, Y., Guan, J., Cai, Y.: Defect identification method for poplar veneer based on progressive growing generated adversarial network and MASK R-CNN model. BioResources **15**, 3041–3052 (2020)
11. Tao, X., et al.: Wire defect recognition of spring-wire socket using multitask convolutional neural networks. IEEE Trans. Components, Packag. Manuf. Technol. **8**, 689–698 (2018)

12. Canny, J.: A computational approach to edge detection. IEEE Trans. Pattern Anal. Mach. Intell. PAMI **8**, 679–698 (1986)
13. Derpanis, K.G.: The Harris Corner Detector. York Univ. 1–2 (2004)
14. Lin, T.-Y. et al.: Feature pyramid networks for object detection. In: Proceedings of the IEEE Conference on Computer Vision and Pattern Recognition (CVPR) (2017)
15. Ren, S., He, K., Girshick, R., Sun, J.: Faster R-CNN: Towards real-time object detection with region proposal networks. In: Advances in Neural Information Processing Systems, Cortes, C., Lawrence, N., Lee, D., Sugiyama, M., Garnett, R., (eds), vol. 28, pp. 91–99 (Curran Associates, Inc., 2015)
16. He, K., Gkioxari, G., Dollar, P., Girshick, R.: Mask R-CNN. In: Proceedings of the IEEE International Conference on Computer Vision (ICCV) (2017)

Development of a Multiagent Based Order Picking Simulator for Optimizing Operations in a Logistics Warehouse

Takuto Sakuma[1]([envelope]) [iD], Minami Watanabe[2], Koya Ihara[1,3] [iD],
and Shohei Kato[1,3] [iD]

[1] Computer Science Program, Department of Engineering, Graduate School
of Engineering, Nagoya Institute of Technology,
Gokiso-cho, Showa-ku, Nagoya-city, Aichi 466-8555, Japan
sakuma.takuto@nitech.ac.jp
[2] Creative Engineering Program, Department of Engineering, Graduate School
of Engineering, Nagoya Institute of Technology, Gokiso-cho, Showa-ku,
Nagoya-city, Aichi 466-8555, Japan
[3] Frontier Research Institute for Information Science, Nagoya Institute
of Technology, Gokiso-cho, Showa-ku, Nagoya-city, Aichi 466-8555, Japan

Abstract. In this study, we aim to develop a multiagent simulator for order picking in a logistics warehouse and to make it close to the actual data in the field. The product placement output by the optimization algorithm can be verified before the product placement is actually changed in the field, and more effective product placement can be proposed by using this simulator. The performance of the proposed simulator was evaluated based on the time from the creation of a product slip to the time when all the products in the slip are picked (survival time). It was confirmed that the simulation could be performed close to the actual survival time by using the actual data of 2020 in a real warehouse and selecting the appropriate parameters for each month.

Keywords: Multiagent · Simulation · Logistics warehouse

1 Introduction

The three main operations required to deliver products from producers to consumers in a logistics warehouse are storage, order picking, and packaging. Especially, order picking, in which workers travel around and collect products from shelves, accounts for about 55% [3] of the operations in a logistics warehouse. Various methods have been studied [3,6,8] to improve the efficiency of the picking process, which will reduce the overall cost of logistics operations.

Watanabe et al. [8] proposed an optimization system for product placement using BLPSO [5], a particle swarm optimization method. Watanabe et al. show that BLPSO can calculate better product placement than the current placement

© Springer Nature Switzerland AG 2022
H. Fujita et al. (Eds.): IEA/AIE 2022, LNAI 13343, pp. 63–73, 2022.
https://doi.org/10.1007/978-3-031-08530-7_6

in real warehouses or the product placement calculated by Class-based storage [3] or other particle swarm optimization methods. In addition, Watanabe et al. reported optimization results considering the cost of replacing products [9], which is state-of-the-art in research on product placement optimization. However, the goodness of the product placement is simply evaluated by the total distance of picking operation. Therefore, it is unclear whether the product placement is good when the picking operation is actually performed. For example, frequently ordered items were gathered near the picking station for shipping can reduce the travel distance. However, it is likely to cause traffic congestion as many carts gather at that location, making the picking process even more time-consuming. Since it is very costly to change the product placement actually, and the product placement proposed by the optimization algorithm cannot be easily tested in an actual warehouse, it is necessary to evaluate the optimized placement more accurately, taking into account the traffic jam problem.

The purpose of this study is to develop a multiagent order picking simulator that reproduces a cart's movement for picking products to evaluate the product placement more accurately. Figure 1 shows the relationship between the product placement optimization algorithm and our simulator. A simulation is performed based on the product placement output by the optimization algorithm, and the cost of picking is calculated. The degree of cost improvement from the current product placement can decide whether to re-optimize or apply to the field. For this reason, the simulator proposed in this paper is required to be able to perform evaluations similar to those when the product placement is applied to the actual warehouse, even if it takes time to calculate the evaluations, and the role of the simulator is different from the fitness function used in an optimization algorithm.

Studies related to constructing an order picking simulator include using Tecnomatix Plant Simulation 13 [1], focusing on replenishment of products [4], and targeting multiple forklifts [7]. In these studies, the ability of the simulator to reproduce actual warehouses was not verified, and some of the parameters used in the simulator, such as the picker's movement speed, were only determined in consultation with the data provider and the reproducibility was not validated [4]. In addition, the optimization of the parameters was not mentioned in the paper [1, 7], assuming that the user can arbitrarily determine them. In this paper, we report the results of an experiment in which we constructed an order picking simulator based on the warehouse data provided by YAHATA NEJI Corporation.

Notation

We use the following notation in the rest of the paper. For any natural number n, we define $[n] := \{1, \ldots, n\}$. For an n-dimensional vector $\boldsymbol{v}$ and a set $\mathcal{I} \subseteq [n]$, $\boldsymbol{v}_{\mathcal{I}}$ represents a subvector of $\boldsymbol{v}$ whose elements are indexed by $\mathcal{I}$. A sequence with length T is represented as $\langle g_1, g_2, \ldots, g_T \rangle$.

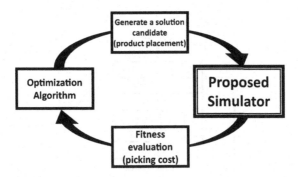

Fig. 1. The relationship between the product placement optimization algorithm and our simulator.

2   Order Picking Simulator

In order picking, a *picking group (PG)* is created as a list of products to be packed in the same box. We define *survival time* as when the PG is issued to when all the products in the PG are picked and assume that the closer the survival time is to the actual result, the simulation result is better. Therefore, we generated PGs at the same time as the actual results, measured the time until the cart finished picking all the items in the PG, and output the error from the actual results for each PG. Figure 2 shows the overall flow of the order picking simulation. By inputting the simulation term and data, the survival time of each PG is output for the simulation term. The simulation treats the period from 4:00 a.m. to 3:59 a.m. as a day, and PGs are extracted and processed one day at a time. Note that the simulator is reset every day, and any uncompleted PGs are discarded. Since the simulation is based on the PG data that could have been processed on that day, the larger the number of discarded PGs, the larger the discrepancy between the actual results and the simulator.

The PG is sent to the waiting cart, or if there is no waiting cart, it is held until the waiting cart appears. The cart assigned a PG collects the products in the PG according to the action decision algorithm described in the next section and reports the time taken to collect the products. In the simulator, several carts are running in parallel as multiagents. The simulation is performed by varying the number of carts per hour based on the hourly record of the actual number of workers.

2.1   Setting of Simulator

We construct an undirected graph $\mathcal{G}$ with all shelves set $s := \{s_i\}_{i \in [|s|]}$ and picking stations as nodes and Manhattan distance between two different nodes as edge weights, and transitions between nodes represent cart movement. A picking station is where products picked from shelves are packed and shipped, and carts must return to the picking station in certain conditions. In this paper,

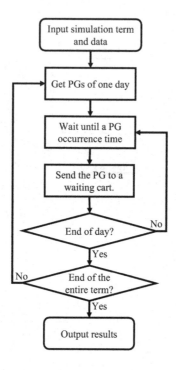

Fig. 2. The flow of order picking simulation

Table 1. Parameters used in the simulator

Name	Value	Note
Transfer coefficient	1.646	The number of seconds that it takes to move 1 m
Common operation time	35 s	Work on receiving PG
Box picking operation time	9.2 s	When the unit of picking is a box or case
Piece picking operation time	40.2 s	When the unit of picking is a piece
Base operation time	30 s	Unloading work after reaching the base

a cart returns to the picking station if it has finished collecting 4 PGs or if the weight of the products in the cart exceeds 100 kg when the currently collected PGs are finished. Since it is necessary to change the number of carts every hour based on the actual results, if the number of carts needs to be increased, the number of carts on standby is increased. If the number of carts needs to be decreased, the order of return to the picking station is dispatched and the carts are returned to the picking station as soon as the current PG is completed, and the carts are stopped after the base operation time has elapsed. Table 1 shows the various parameters used in the simulator. These parameters are all set based on the measurement results in the actual environment. The "piece picking operation time" is the time required for picking products in units of piece and is longer than the "box picking operation time" because it requires time to check whether the number of the piece is correct.

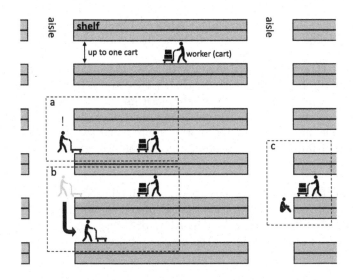

Fig. 3. Warehouse conceptual diagram and example of cart behavior decisions

2.2   Cart Behavior Decision Algorithm

Since the cart must traverse the entire set of shelves $s^g := \{s^g_i\}_{i \in [|s^g|]}$ containing the products in a given PG g, we consider it as a traveling salesman problem (TSP) and use an algorithm to find an approximate solution to the TSP to find the order to traverse in the shortest distance and create a target shelf sequence $\mathcal{S}^g := \langle s^g_j \rangle_{j \in [|s^g|]}$. We construct a complete undirected graph with the elements s^g_i of the shelf set s^g as nodes and the Manhattan distance between two different nodes as the edge weights. The algorithm for solving the TSP is the Christofides algorithm [2], and the cart moves according to the obtained target shelf sequence $\mathcal{S}^g$. The shortest path from the current node to the next target node is obtained using a graph $\mathcal{G}$ with the entire shelf set s as nodes and the A* algorithm. The heuristic function in the A* algorithm is the Euclidean distance between nodes. The cart transitions between nodes according to the shortest path calculated by the A* algorithm. The time required for node transitions is calculated by integrating the distance between nodes with the transfer coefficient.

Figure 3 shows a conceptual diagram of the logistics warehouse in this paper. The space between the shelves is so narrow that it is impossible to pass another cart because it is only as wide as one cart. However, some shelves face an aisle wide enough to be passed by other carts. In this paper, we call these aisle nodes, and they are mainly used for passing other carts and waiting areas. When a cart encounters another cart outside of an aisle node (Fig. 3a), it will either avoid the other route (Fig. 3b) or wait at the aisle node (Fig. 3c). If they are human, they will make the best choice according to the situation by talking to each other. For example, if the work on the blocked cart is likely to take a long time, prioritize picking another product to avoid it. Alternatively, if the blocking cart seems to be leaving the area quickly, wait in the aisle and watch for movement. Since the actual warehouse in this paper seldom waits and basically adopts an avoidance strategy, the following conditions determine the action of the cart.

1. If the next target node s_{j+1}^g exists within $\mathcal{S}^g$, move the current target node s_j^g to the tail of $\mathcal{S}^g$ and search for the shortest route from the current node to the target node s_{j+1}^g using the A* algorithm.
2. If it does not fall under condition 1 and there is a aisle node between the current node and s_j^g, search for another route using the A* algorithm.
3. If it does not fall under condition 2, wait for the nearest aisle node until clear of the obstructing cart.

After reaching the target node, the cart stops until all needed products have been picked. The time required for picking is calculated by adding the "box picking operation time" if the picking unit is a box or a case and the "piece picking operation time" for other units by the number of products to be picked.

3   Experiments for Simulator Performance Evaluation

3.1   Experimental Setting

The purpose of the order picking simulator in this paper is to reproduce the movements of an actual picker. The parameters shown in Table 1 are the target values in the field, and we assume that the work time of the actual picker is taken to be longer than that. In this experiment, we prepared conditions in which the coefficients were integrated for all the parameters listed in Table 1 in increments of 0.1 from 1.0 to 2.0. Note that the coefficient of 2.0 was set as the upper limit because it was found to be too slow in preliminary experiments. For example, in the experimental condition with a coefficient of 1.5, all the parameters shown in Table 1 were multiplied by 1.5 and then simulated. We investigated how much the mean survival time of PGs differed from the actual survival time in each experimental condition. It is assumed that the smaller the difference, the simulation has less deviation from actual results. Since only PGs that the simulator was able to process are used to calculate the mean survival time and compare it to the actual results, we assume that the simulation results with a large decrease in the number of PGs have a large eviation from the actual results. For some PGs, the actual survival time exceeds four hours. These are PGs that have been issued and left for some reason (e.g., there is a margin of delivery date) or PGs that have been interrupted by other PGs that should have been prioritized, and thus the process has been slowed down, which is clearly different from a normal picking operation. The PGs whose actual survival time was more than four hours were excluded from the evaluation of this simulator as outliers. For the experiments, we used the data provided by YAHATA NEJI Co. for four months (76,980 PG) from September to December 2020, and the simulations were performed for one month each. The warehouse to be simulated has an area of 2,058 m^2 and 635 shelves containing about 22,000 various products.

Algorithm 1. Cart behavior decision algorithm

Require: PG g

 1: When receiving PG, the cart stops on the spot during the common operation time.
 2: Find the shortest route that goes through all the shelf sets s^g that need to be visited in g using Christofides algorithm, and create a target shelf sequence $\mathcal{S}^g$ sorted by the shortest route.
 3: **for** $s_j^g \in \mathcal{S}^g$ **do**
 4:     Search for the shortest route from the current node to the next target node s_j^g using the A* algorithm.
 5:     **for** Node sequence of the root obtained by A* **do**
 6:         **if** The cart cannot move due to other carts. **then**
 7:             **if** s_{j+1}^g != NULL **then**
 8:                 Move s_j^g to the tail of $\mathcal{S}^g$ and use the A* algorithm to find the route from the current node to the target node s_{j+1}^g.
 9:             **else if** There is an aisle node between the current node and s_j^g. **then**
10:                 Search for another route with the A* algorithm.
11:             **else**
12:                 Wait until no more carts are obstructing the immediate aisle node.
13:             **end if**
14:         **end if**
15:         Move to the next node
16:     **end for**
17:     Stopped during picking operation time
18: **end for**
19: **if** Collected for 4 PG or total weight of products exceeds 100 kg or ordered to return. **then**
20:     Move to the picking station and stop during base operation time.
21: **else**
22:     Move to the nearest aisle node
23: **end if**

3.2 Results

Tables 2, 3, 4, 5 shows the average survival time of PGs that completed the process and the average survival time in the actual results of the simulation from September to December 2020, respectively. The column showing the difference in average survival time between the simulation and actual results, where the absolute value is the smallest, is bold.

Tables 2, 3, 4, 5 shows that the average survival time in the coefficient 1.0 condition is shorter than the actual survival time. In other words, the various parameters set as work targets (Table 1) were probably not achieved in the actual results. On the other hand, the average survival time in the coefficient 2.0 condition is longer than the actual results, and the number of PGs processed is much lower than in the 1.0 condition. It indicates that the time required to process each PG was too much, and a large number of uncompleted PGs remained at the end of the day, so the coefficient of 2.0 is not appropriate for the simulation. From Table 2, the coefficient with the smallest difference in average survival time

Table 2. Simulation results of September 2020

Coefficient	Survival time (s)		Actual survival time (s)		Difference in averages (s)	Number of PG
	average	SD	average	SD		
1.0	1539	2133	3405	3514	−1865	17577
1.1	1848	2579	3405	3514	−1557	17577
1.2	2270	3130	3405	3514	−1134	17577
1.3	2717	3607	3405	3514	−688	17576
1.4	3354	4087	3403	3513	**−49**	17550
1.5	3976	4654	3394	3500	581	17388
1.6	4640	5277	3381	3503	1259	17028
1.7	5460	5633	3376	3517	2085	16669
1.8	6306	6090	3390	3532	2916	16120
1.9	7118	6598	3364	3525	3755	15403
2.0	7611	6802	3301	3482	4310	14515

Table 3. Simulation results of October 2020

Coefficient	Survival time (s)		Actual survival time (s)		Difference in averages (s)	Number of PG
	Average	SD	Average	SD		
1.0	2505	2529	4264	3946	−1760	17025
1.1	3817	3854	4262	3939	−445	16963
1.2	4282	3705	4265	3946	**17**	17024
1.3	5382	4221	4264	3946	1117	17025
1.4	6616	4836	4265	3946	2351	17023
1.5	7957	5733	4261	3938	3696	16709
1.6	8928	6240	4265	3929	4664	16067
1.7	9937	6905	4225	3890	5712	15315
1.8	11031	7584	4212	3874	6819	14351
1.9	11834	8198	4220	3888	7614	13449
2.0	12433	8767	4222	3894	8212	12498

in September is 1.4. Tables 3, 4, 5 shows that the coefficient with the smallest difference in average survival time in October, November, and December is 1.2. On the other hand, the results of the coefficient 1.4 condition in September and the coefficient 1.2 condition in November show that the number of processed PGs is about 30 less than that of the coefficient 1.0 condition, which means that the simulator is not able to process PGs that were processed in the actual results.

4   Discussion

The experimental results found that, as we hypothesized, the target parameters in the field (Table 1) could not reproduce the actual results. The average survival time was found to be closest to the actual survival time by multiplying the parameter by 1.4 for September 2020 and by 1.2 for October, November, and December 2020. Since the purpose of this study is not to find a general-purpose coefficient, we would like to find the optimal coefficient for each simulation term. From the experimental results, we can simulate the actual performance of the actual warehouse by finding the best coefficients for each month. However, the simulations for September and November did not fully follow the actual results, and unprocessed PGs were generated, so the proposed simulator needs improvement. For example, as for the cart traffic jam, our simulator simply assumes that the carts cannot pass each other except at the aisle nodes. Therefore, when other carts are blocking (e.g., during a picking operation), the carts in the simulator search for another route (Fig. 3b) or wait at an aisle node until the blocking carts are gone (Fig. 3c). However, in actual warehouses, the workers deal with traffic jams flexibly, such as leaving the carts and having only human workers go to get the products or asking other workers to give up their places. Furthermore, the behavior of the cart and the time required for the operation are expected to vary greatly depending on the pecking order among the workers and their skill level in the field, but these factors are not incorporated in the simulator. In this experiment, coefficients were prepared to bring the average survival time for one month closer to the actual results. However, if the coefficients are adjusted on a daily basis, it is possible to absorb differences in workers and orders specific to that day with the coefficients and bring them closer to the actual results. Also, the coefficients can be adjusted individually to bring them closer to the actual results rather than multiplying a uniform coefficient for all parameters. Since these are beyond the limits of manual adjustment, we would like to consider using Bayesian optimization as a parameter optimization method to improve efficiency.

As another approach, we can consider a policy to make a processing time, measured from the start to the end of a PG processing, closer to the actual results instead of a survival time of PGs. As mentioned earlier, the survival time is not easily approximated by simulation because it is greatly affected by PGs that are left unattended after being issued and by changes in the processing order due to interruptions. We believe that the processing time can be adequately approximated by a simulation that purely emulates the behavior of a picking cart. Unfortunately, since there is no actual data on processing time, we could only use the survival time as a criterion at the moment, but we would like to continue to work with the data provider to gather data on processing time.

Table 4. Simulation results of November 2020

Coefficient	Survival time (s)		Actual survival time (s)		Difference in averages (s)	Number of PG
	Average	SD	average	SD		
1.0	2940	3835	4303	3916	−1363	14430
1.1	3551	4328	4304	3916	−753	14429
1.2	4315	5025	4307	3917	**8**	14401
1.3	5149	5516	4323	3921	826	14314
1.4	6189	6015	4326	3924	1863	14280
1.5	7354	6659	4340	3928	3013	14152
1.6	8178	6835	4337	3921	3841	13829
1.7	9187	7220	4293	3895	4894	13443
1.8	10112	7668	4294	3879	5818	12715
1.9	10751	8048	4284	3866	6467	11925
2.0	11195	8413	4238	3810	6957	11310

Table 5. Simulation results of December 2020

coefficient	Survival time (s)		Actual survival time (s)		Difference in averages (s)	Number of PG
	Average	SD	Average	SD		
1.0	2730	3317	4330	3969	−1599	13183
1.1	3312	3882	4330	3969	−1018	13183
1.2	4220	4622	4330	3969	**−110**	13183
1.3	5064	5231	4330	3969	734	13183
1.4	5968	5829	4330	3969	1638	13183
1.5	6917	6525	4332	3973	2586	13095
1.6	8565	7512	4294	3959	4271	12736
1.7	8901	8148	4302	3962	4599	12500
1.8	9741	8396	4283	3939	5459	12036
1.9	10364	8735	4343	3960	6021	11526
2.0	11032	8909	4356	3976	6676	11071

Although the proposed simulator is currently designed specifically for the data provider's warehouse, it can simulate any warehouse with the various parameters shown in Table 1 and shelf coordinate data. However, it is crucial to tune the simulation as shown in this paper to reflect the unique circumstances of the warehouse to make it more similar to actual performance.

5   Conclusion

This study aims to develop an order picking simulator close to the actual data in the field to evaluate the product placement output by the optimization algorithm more accurately. In this paper, we conducted experiments using the developed simulator under several conditions based on the actual data from September to December 2020 and tried to find parameters close to the actual data. As a result, we found the parameter that the average survival time of PGs is closest to the actual data in each month. Using this simulator, the user can perform detailed verification before actually changing the product placement in the field, which will help the optimization algorithm propose a more effective product placement.

Acknowledgments. This work was supported in part by the Ministry of Education, Culture, Sports, Science and Technology-Japan, Grant–in–Aid for Scientific Research under grant #JP19H01137, #JP19H04025, #JP20H04018, #JP20J14182, and #JP20K19905. We are grateful to YAHATA NEJI Corporation for providing us with the real logistic operation data.

References

1. Bučková, M., Krajčovič, M., Edl, M.: Computer simulation and optimization of transport distances of order picking processes. Procedia engineering **192**, 69–74 (2017)
2. Christofides, N.: Worst-case analysis of a new heuristic for the travelling salesman problem. Carnegie-Mellon Univ Pittsburgh Pa Management Sciences Research Group, Tech. rep. (1976)
3. De Koster, R., Le-Duc, T., Roodbergen, K.J.: Design and control of warehouse order picking: A literature review. European journal of operational research **182**(2), 481–501 (2007)
4. Gagliardi, J.P., Renaud, J., Ruiz, A.: A simulation model to improve warehouse operations. In: 2007 Winter Simulation Conference. pp. 2012–2018. IEEE (2007)
5. Ihara, K., Kato, S.: A novel sampling method with lévy flight for distribution-based discrete particle swarm optimization. In: IEEE Congress on Evolutionary Computation (CEC). pp. 2281–2288. IEEE (2021)
6. Lee, I.G., Chung, S.H., Yoon, S.W.: Two-stage storage assignment to minimize travel time and congestion for warehouse order picking operations. Computers & industrial engineering **139**, 106129 (2020)
7. Tarczynski, G.: Warehouse real-time simulator-how to optimize order picking time. Available at SSRN 2354827 (2013)
8. Watanabe, M., Ihara, K., Kato, S., Sakuma, T.: Initialization effects for pso based storage assignment optimization. In: 2021 IEEE 10th Global Conference on Consumer Electronics (GCCE). pp. 494–495. IEEE (2021)
9. Watanabe, M., Ihara, K., Sakuma, T., Kato, S.: Optimizing storage allocation for order picking considering product replacement operations using pso. In: The Twenty-Seventh International Symposium on Artificial Life and Robotics 2022 (AROB 2022). pp. 937–940 (2022)

Health Informatics

Predicting Infection Area of Dengue Fever for Next Week Through Multiple Factors

Cong-Han Zheng[1], Ping-Yu Hsu[1], Ming-Shien Cheng[2(✉)], Ni Xu[1], and Yu-Chun Chen[1]

[1] Department of Business Administration, National Central University, No. 300, Jhongda Road., Taoyuan County, Jhongli City 32001, Taoyuan, Taiwan (R.O.C.)
984401019@cc.ncu.edu.tw
[2] Department of Industrial Engineering and Management, Ming Chi University of Technology, No. 84, Gongzhuan Road., Taishan District, New Taipei City 24301, Taiwan (R.O.C.)
mscheng@mail.mcut.edu.tw

Abstract. Death rate of dengue fever is low, because dengue fever become severe illness only when second infection happened. However, global warming is getting severe recently, which make the infection distribution of dengue fever different. Common method of previous studies used climate factors combined with social or geographic factors to predict dengue fever. However, recent study did not use combination of these three factors into dengue fever prediction. We proposed a method that combines these three factors with data of Taiwanese dengue fever and uses the secondary area divided by the population as the granularity. Random Forest (RF) and XGBoost (XGB) are used for prediction model of weekly dengue fever infection area. Experimental results showed that the Receiver Operator Characteristic (ROC)/Area Under the Curve (AUC) of RF and XGB are both higher than 93%, and the Recall rate is higher than 80%. With the result, government can determine which area should do disinfection process to reduce the infection rate of dengue infection. Because of accurate prediction and disinfection process, the personnel cost can be reduced and it can prevent waste of medical recourse.

Keywords: Dengue fever · Random forest · XGBoost · Imbalanced data

1 Introduction

According to the World Health Organization (WHO, 2020), there have been a total of 4.2 million dengue infections in 2020, an eight-fold increase from 505,000 cases in 2000, and the number of deaths has increased from 960 to 4,032. There are 3.9 billion people in the world who are in the infected area of dengue fever, 70% of them are in Asia. Taiwan is located in a subtropical region with a climate just right for mosquitoes, which is also a high-risk area for dengue fever. According to Centers for Disease Control (CDC) of Taiwan (Department of Disease Control, Ministry of Health and Welfare, 2015), the number of dengue fever infections in Taiwan was 139 in 2000. However, in 2015, the number of people infected by the dengue fever outbreak reached 43,784, which was 315 times more than in 2000.

© Springer Nature Switzerland AG 2022
H. Fujita et al. (Eds.): IEA/AIE 2022, LNAI 13343, pp. 77–88, 2022.
https://doi.org/10.1007/978-3-031-08530-7_7

There is no appropriate vaccine for dengue fever in the world yet (Deng et al., 2020). As long as the environment is suitable for the growth of dengue virus, there is a possibility of another outbreak. There are many overseas studies that use machine learning to predict the spread of dengue fever, and the most popular ones use climate factors as characteristics, such as temperature and rainfall which are among the conditions for the growth of dengue fever (Xu et al., 2020). Some scholars have also used social factors as an entry point to predict the spread of dengue fever by using population size, income, education, etc. (Zhao et al., 2020). Many studies have shown that geographic factors are also associated with dengue fever epidemic, such as natural woodlands, residential areas, industrial areas, etc. (Francisco, et al., 2020). However, after searching many academic websites such as Google Scholar and Scopus, we did not find any scholar who combined the three factors for discussion and prediction.

Therefore, the motive of this study is as follows: to determine whether the use of three factors, namely geographic, climatic, and social factors, in combination with dengue fever case data will provide better prediction results than the use of single or two factors in the past. This study investigates the characteristics of dengue fever and verifies whether the above-mentioned three factors from both domestic and overseas studies are equally important in predicting dengue fever in Taiwan.

To investigate the above issues, this study focused on a total of 1234 areas in secondary areas of Kaohsiung City by population density, as well as the infection status for a total of 104 weeks in two years. We investigated whether using three factors (social factor, geographic factor, and climate factor) with machine learning techniques of Random Forest (RF) and XGBoost (XGB) training can yield a suitable model to effectively predict the dengue infection area in the next week, and whether it is better than using a single climate factor or dual factors such as combining social and climate factors or combining geographic and climate factors. Finally, we explored the important features in the prediction of dengue fever in Taiwan and validated the features proposed by other scholars.

This paper organized as follow: (1) Introduction: Research background, motivation and purpose. (2) Related work: Review of scholars' researches on dengue fever in Taiwan. (3) Research methodology: Content of research process in this study. (4) Result analysis: Experimental results and the discussion of the test results. (5) Conclusion and future research: Contribution of the study, and possible future research direction is discussed.

2   Related Work

2.1   Study on the Factor of Dengue Fever Model

Stolerman et al. (2019) used rainfall frequency and mean temperature as characteristics of climatic factors to predict the incidence of dengue fever and found that their study was most effective in winter. Salim, et al. (2021) used various climatic characteristics such as temperature, wind speed, humidity, rainfall, and various machine learning methods to predict the spread of dengue fever in Malaysia. Due to the imbalance in the dataset, the authors used data under-sampling to reduce multiple classes of data, forcing increased accuracy and resulting in less than 30% recall. Anno, et al. (2019) also used rainfall and temperature data as features of climate factors. In particular, he used the temperature of the neighboring ocean surface and designed the drop out and 8-Folder cross-validation to avoid model overfitting. Guo et al. (2017) used a multiple regression machine learning algorithm to predict dengue fever using weekly mean temperature, weekly mean humidity, and rainfall as characteristics.

Fewer studies have used geographical and social factors compared to climatic factors. Zhao Charland, et al. (2020) used random forest and Artificial Neural Networks (ANN) to predict dengue fever. In addition to climatic factors, they used population density, education level, and per capita income. The results showed that random forest was more accurate than ANN. In Taiwan, Huang (2020) also used data on social factors such as sex ratio, household volume, elderly support ratio, and dengue cases to predict the area of dengue infection in 2015 by the area of dengue infection in 2014, and the results showed good prediction of the epidemic. Francisco, Carvajal (2020) analyzed the relationship between dengue fever and climatic and geographical factors using MOdel Based recursive partitioning (MOB) regression method and geographical factors such as forest distribution, residential areas, and industrial areas in addition to climatic factors. The results of the study showed that dengue fever is more likely to occur in residential areas and industrial areas with high population density.

From the above studies, we can see that almost all dengue studies used climatic factors, while geographical and social factors were used less frequently. This study aims to use three factors for machine learning training to predict the areas that are likely to be infected every other week. The following is a detailed description of the research process, pre-training data processing, training and validation methods.

3   Research Methodology

The process of this study has six main components, namely data collection, data normalization, data splitting, data balancing, data modeling, and model validation (Fig. 1).

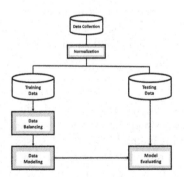

Fig. 1. The research process

3.1 Research Characteristics

The characteristics used in this study included dengue case-related characteristics, climate-factor-related characteristics, social-factor-related characteristics, and geography-related characteristics, for a total of 39 characteristics. In addition to the characteristics of these three factors, the characteristics of dengue fever cases are the most important characteristics, because dengue infection is diffuse and the situation in neighboring areas can easily influence the prediction of infection in that area. In this study, information about the neighborhood of the infected area is a very important feature, but it can only be easily obtained through geographic information system. In this study, we used Quantum Geographic Information System (QGIS) to obtain the location of the neighboring areas using public data from Taiwan CDC of the Ministry of Health and Welfare and secondary output areas in the Open Government Data Platform. QGIS is a geographic information system that can accept multiple file inputs and is commonly used to display, edit, and analyze geographic data.

In addition to the common features of climate factors, this study uses three additional climate-related features, namely aws, tempVariable, and wind. Scholars Jury (2008) pointed out in the research that stronger wind speeds will inhibit the occurrence of dengue fever. Lambrechts, et al. (2011) pointed out that a larger temperature difference will reduce the survival rate of dengue fever carriers, that is, mosquitoes. In addition, wind is because mosquitoes will fly farther than usual with the wind. Therefore, this study will combine wind direction and dengue fever cases to add this feature.

In previous studies, the main factors that used social factors as predictive features were population density, per capita income and other characteristics. Based on privacy issues, Taiwan did not release relevant information about per capita income and other secondary release areas. This study also uses the characteristics of low-and middle-income households mentioned by scholars (Yu, H.-L., et al., 2012) from Taiwan as one of the social factors of this time.

The geographic factors were also based on a study in Taiwan, in which Yu H.-L., et al. (2012) listed a number of characteristics related to dengue fever through a linear regression analysis of various land use data in Taiwan, as well as the residential and industrial areas mentioned in the literature, in addition to a few characteristics selected by this study. After all, Table 1 listed the all characteristics of this study.

3.2 Model Scoring

This section describes the scoring method used in this study. The training data used in this study is balanced data constructed by the Synthetic Minority Oversampling Technique (SMOTE) data balancing method. However, the test set was based on real data, so SMOTE was not used, and therefore the data was unbalanced, with an 8-fold difference between 0 and 1. Therefore, in this study, it is not possible to distinguish the model prediction results by confusion matrix, accuracy, and other metrics, otherwise an accuracy paradox would occur, Valverde-Albacete and Peláez-Moreno (2014). In this study, for example, the difference between 0 and 1 is nearly eight times, and the whole prediction model is designed to predict the absence of dengue fever every other week. The accuracy of the training result is nearly 90%, but it is not a good prediction model, which is the "accuracy paradox". Therefore, this study will use the confusion matrix as a reference, and use Recall, ROC/AUC, and F1-Measure as the main scoring methods.

Table 1. The all characteristics list

Case-related Factors			Geographic-related Factors		
Variable Name	**Variable Description**	**Source of Data**	**Variable Name**	**Variable Description**	**Source of Data**
quantity	Number of dengue fever infections	O.D.A	L0101	Farming area	O.D.A
neighbors_DC	The number of dengue fever infections in the vicinity	O.D.A with QGIS	L0102	Aquaculture area	O.D.A
Weather-related Factors			L0103	Livestock area	O.D.A
aws	The average wind speed of the neighboring weather stations	O.D.A with QGIS	L0201	Natural forest area	O.D.A
Temp	The average temperature of the neighboring weather stations	O.D.A with QGIS	L0304	Port area	O.D.A
RH	The average humidity of the neighboring weather stations	O.D.A with QGIS	L0401	River area	O.D.A
Precp	The average Rainfall of the neighboring weather stations	O.D.A with QGIS	L0402	Ditch area	O.D.A
tempVariable	The temperature difference between current week and the previous week	O.D.A with QGIS	L0403	Reservoir area	O.D.A
wind	The wind direction of the infection release area in the previous week	O.D.A with QGIS	L0404	The area of sandbar beaches	O.D.A
Social-related Factors			L0405	Area of water conservancy structures	O.D.A
M_F_RAT	Male to female ratio	O.D.A	L0501	Residential area	O.D.A
P_H_CNT	Average population per household	O.D.A	L0502	Commercial area	O.D.A
P_DEN	Population density	O.D.A	L0503	Industrial area	O.D.A
DEPENDENCY_RAT	Dependency Ratio	O.D.A	L0601	area of district government agencies	O.D.A
A0A14_A15A65_RAT	the ratio of the child	O.D.A	L0602	Area of schools	O.D.A
A65UP_A15A64_RAT	the ratio of the elderly	O.D.A	L0603	Medical care area	O.D.A
A65_A0A14_RAT	aging index	O.D.A	L0606	Area of environmental protection facilities	O.D.A
RLH2_CNT	Number of low-income households	O.D.A	L0902	Area of wetland	O.D.A
RLH2_H_RAC	Proportion of low-income households	O.D.A	L0903	Grassland area	O.D.A
RLH_CNT	Number of medium and low-income households	O.D.A	L0908	Vacant land area	O.D.A
RLH_H_RAC	Proportion of medium and low-income households	O.D.A	* O.D.A: Open Data Access		

To resolve the above accuracy contradiction, this study adopts Operator Characteristic (ROC)/Area Under the Curve (AUC) as the main measure. ROC mainly consists of two metrics in the confusion matrix, Recall as the X-axis and Specificity as the Y-axis.

4 Research Experiment

The data source for this study was publicly available government data from the Internet. This chapter describes the data collection process and sources, the machine learning hyperparameter tuning process, and the evaluation and analysis of the results of this experiment using the scoring method introduced in the previous chapter.

4.1 Data Collection

Since the dengue outbreak in Taiwan occurred mainly in 2014 and 2015 and was concentrated in Kaohsiung, this study collected 34,827 cases of dengue fever in Kaohsiung through public government data. The granularity of the prediction is based on the government's secondary output area by population size, with a total of 1,234 areas. A total of 104 weeks of data were collected over the two years, and 128,336 cases of dengue fever were created and placed in the data set on a weekly basis.

The climate factor data in this study were obtained from Cwb Observation Data and inquire System (CODiS), an observatory data query provided by the Taiwan government, using a web crawler. The data provided by the observatory are in hourly units and must be converted to weekly units.

The social and geographic factors were obtained from Social Economic Geographic Information System (SEGIS), a free online open-source information platform that provides easy access to data at a specific granularity. There is no larger unit of information on low-income households in the social factors than the secondary output area, which is one of the reasons why the secondary output area was chosen for this study.

4.2 Data Preprocessing

To improve the accuracy of the model, eliminate the effect of unitization, and speed up the convergence of the model, this study does data normalization by scaling the data so that each feature has the same weight (Singh and Singh, 2020). In this study, we used RobustScaler to perform the normalization process as shown in Eq. (1).

$$v_i' = \frac{v_i - median}{IQR} \qquad (1)$$

In Eq. (1), v_i' is the value after the RobustScaler standardization process. v_i is the value to be standardized for a particular feature, median is the median of the column, and IQR is the value of the 3rd quartile minus the 1st quartile of the feature.

After normalization, the values will fall between $[-1, 1]$. The reason for using Robust Scaler (RS) for standardization is that if we use Z-Score and other methods to perform standardization, it may cause interpretation errors when outliers are encountered, especially in the case of Min-Max and other methods. RS can handle outliers better. In this study, outliers are not dealt with except for the general outliers. The reason is that, for example, the natural forest in the geographic factor feature is several times more than the urban area in Taoyuan District of Kaohsiung City, but it is not considered as an outlier and cannot be excluded. That is why RS is used for standardization here.

In this study, the original dataset was split according to the 80/20 rule, where 80% was the training set and 20% was the test set. After splitting, the training set was sent to SMOTE for data balancing. The test set was unbalanced to maintain the truthfulness of the data. The volumes of data after splitting were shown in Table 2.

In the data collected in this study, 109,680 data were labeled as 0, while only 13,720 data were labeled as 1. The 8-fold difference between the two is too large. If training is performed with unbalanced data, the accuracy will decrease and the prediction will

Table 2. The distribution of original data set

Data Set	1-Positive	0-Negative
Train	87,697	11,023
Test	21,983	2,697

be biased towards 0. If the accuracy is forced to increase, it is likely to cause model overfitting (Padmaja, et al., 2007). Therefore, in this study, the SMOTE + Edited Nearest Neighbor (ENN) integrated sampling method is used for data balancing.

The reason for using SMOTE + ENN in this study is that after testing three data balancing combinations, SMOTE, SMOTE + ENN, and SMOTE + Tomek links, SMOTE + ENN has the highest score. Therefore, SMOTE + ENN is used for data balancing in this study.

4.3 Model Parameter Adjustment

In the parameter adjustment, since the classification of this study is binary, LogLoss is also used to adjust the parameters of RF and XGB. In this study, the minimum parameter of LogLoss is used as the hyperparameter of this study in conjunction with the Python-Scikit-learn suite GridSearchCV.

In this study, LogLoss is used to adjust the parameters of RF. The parameters used in the RF model of this study are shown in Table 3. In this study, GridSearchCV is used to find the minimum value of LogLoss to adjust the parameters. The parameters were found to be independent of each other during the adjustment process, so they were adjusted independently. In detail, we use a large range to find the approximate ranges of the parameters, and then gradually reduce the ranges to find the most suitable parameters. In this study, except for the three parameters introduced in Table 3, the other parameters did not change much during the experiment.

In the case of RF training, the performance does not decrease as the number of trees increases, not quite consistent with overfitting. After a certain number of trees increases, the performance tends to level off and finally stays at a specific value. This situation only affects the training time, not the score. Therefore, in this study, we use LogLoss for the basic parameter adjustment, and then use the ROC/AUC between the training set and the test set as a reference to adjust the parameter using the distance between the two scores.

The parameters used in the XGB model in this study are also shown in Table 3. XGB also uses GridSearchCV to find the minimum LogLoss value for parameter adjustment. The parameters are adjusted separately because they do not affect each other. By using GridSearchCV to adjust the parameters, the minimum LogLoss score in the data set can be found as the best parameter. However, since GridSearchCV can only process one data set at a time, this study compares the LogLoss of the training set with that of the test set. It is found that if the XGB model is trained based on the LogLoss parameters of the training set, there will be a serious overfitting, and the gap between the two sets will get farther and farther. Therefore, in this study, the overfitting of LogLoss and ROC/AUC

Table 3. The value of parameter of RF and XGB

Random Forest		XGBoost	
parameter	value	parameter	value
max_features	log2	learning_rate	0.05
n_estimators	90	n_estimators	150
max_depth	6	max_depth	5
		reg_alpha	25
		reg_lambda	55
		subsample	0.0123
		colsample_bytree	1
		scale_pos_weight	0.38

of the training and test sets are continuously adjusted, such as n_estimators and L1 and L2.

4.4 Experimental Results and Analysis

In this study, we use three characteristic factors and dengue-related case characteristics to train the model to predict the infection status of every other week using two machine learning methods, such as RF and XGB. It is important to note that the test set is not balanced to ensure the authenticity of the data, so the low Precision is normal. There were 21,983 negative data and 2,697 positive data in this study. The following are the results of the model built by climate factor alone, the model combining climate factor and social factor, the model combining climate factor and geographical factor, and finally the model combining the three factors.

In this study, four types of factors were used to model the data, including prediction by climate factors, prediction by climate factors and social factors, prediction by climate factors and geographical factors, and prediction by all three factors. Each model used RF and XGB as learning methods, so a total of eight models were built.

Let's first analyze this experiment from the perspective of two models. From the above experimental results, we can see that the scores of the 8 models are not very different in terms of the most important metric ROC/AUC, but the scores of RF are all higher than the scores of XGB, and they all obtain a good score of 92% or more. From the perspective of recall, RF scored at least 6 percentage points higher than XGB. This suggests that RF has a better grasp of positive infection than XGB in this study. However, if we look at the results of F1-Measure and Precision together, we can see that XGB is higher than RF, especially Precision, which is the opposite of Recall. In other words, for positive infection status, XGB is more accurate. Although it is generally lower than 50%, this is due to the imbalance of data in the test set. Compared with RF, XGB can predict the infected area more accurately and reduce unnecessary waste and labor cost.

From these four categories of factors, the prediction of dengue fever by climatic factors alone is not effective. In terms of precision, its accuracy is too low. If we take it as a real prediction model, the cost and the waste of manpower is the greatest compared to the other three types. However, if we look at the combination of climatic and social factors, and the combination of climatic and geographical factors, we can see that the scores of these two combinations are very close in the random forest trained model. The higher scores indicate that these two combinations are more effective than the climate factor alone. Both social and geographical factors have some correlation with dengue prediction. In the case of XGB, the differences are not significant, but there is only a slight improvement compared to climate factors alone, except for accuracy and F1-Measure, and recall even regresses in the combination with geographic factors. However, since the difference in scores was only in the second and third decimal places and the opposite was observed when the experiment was conducted multiple times, it is presumed that the difference was caused by random sampling during training.

Finally, the prediction model built by combining the three factors shows a small regression in the scores of both ROC/AUC and recall from the RF model, and the situation is similar to XGB in terms of the geographic factor combination. The results were showed as Table 4. Although it can be argued that this difference is due to the sampling in training, it also indirectly shows that the inclusion of social and geographic features, respectively, does not improve the classification effect. That is, the inclusion of these two features does not help much in classification. The situation is similar for XGB, where there is an improvement, but it is not significant.

However, for all the results, the ROC/AUC could reach 92% on average, which indicates the good classification ability of the model developed in this study. For the prediction of dengue fever, the higher the recall rate, the better. In this study, the recall rate was over 80%, which could detect the infected area more comprehensively. However, the predictions of the model built from the combinations of the three factors were not satisfactory, one of the reasons being the poor prediction of a small number of features, which led to ineffective tree partitioning when the model was built. The importance of all characteristics used in the model will be explained in the next section.

Compared with the original three models, the two models did show significant improvements after removing the features with lower contribution rates and readjusting the parameters and overfitting. The overall performance of the adjusted models was better than that of the pre-adjusted models.

Table 4. Prediction result of RF and XGB for three factors

Random Forest Result				XGBoost Result			
ROC/AUC	Recall	F1-Measuree	Precision	ROC/AUC	Recall	F1-Measure	Precision
94.2%	88.7%	57.7%	42.8%	93.0%	80.8%	62.0%	50.3%

4.5 Important Characteristics of the Model

There are 39 characteristics in this study, of which 0–10 are social factor characteristics, 11–30 are geographical factor characteristics, 31–35 (36) are climate characteristics, and (36) 37–38 are infection case characteristics. In this section, we present the important characteristics of the two machine learning models and which characteristics have low or even no relevance in the experiments.

The important characteristics of the model obtained in this study are based on the gini coefficient of Mean Decrease Impurity (MDI) calculated from the tree models of RF and XGB. Excluding important characteristics that can be inferred from common sense, such as wind direction in the infected area during the week (36), number of infections during the week (37), and number of infections in the vicinity (38), the first five important features have been discussed in the previous introduction and literature review, and the other features were not particularly prominent. The features that contributed less were natural forests (14), harbors (15), rivers (16), ditches (17), waterfront sandbar beaches (19), hydraulic structures (20), wetlands (28), and grasslands (29). After excluding the three characteristics related to dengue fever cases, XGB is different from RF in that it has a higher contribution from the geographic factor of residential land, and the average wind speed is more significant in XGB. The other characteristics with lower correlation are gender ratio (0), dependency ratio (3), child support ratio (4), elderly support ratio (5), aging index (6), and vacant land (30), besides the wetland, which is the same for RF (Figs. 2, 3).

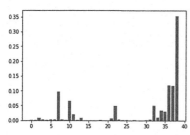

Fig. 2. Factor Importance of RF

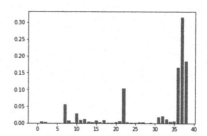

Fig. 3. Factor Importance of XGB

4.6 Adjusted Model Results and Analysis

From the results of the study, it can be found that geographic factors such as ports, rivers, and ditches do not contribute much to the prediction of dengue fever in random forest learning, while XGB, on the contrary, has a higher contribution of geographic factors. Considering that the characteristics with low contribution will reduce the prediction results of classification tree, this study will eliminate the characteristics with low contribution in each method and build a new model. The results were showed as Table 5.

Table 5. Prediction result of RF and XGB for adjusted three factors

Random Forest Result				XGBoost Result			
ROC/AUC	Recall	F1-Measure	Precision	ROC/AUC	Recall	F1-Measure	Precision
94.3%	89.5%	60.3%	45.5%	93.5%	81.2%	62.6%	52.0%

5 Conclusion and Future Research

In this study, two machine learning algorithms, namely RF and XGB, were used to predict dengue fever in an area every other week using various factors such as climatic factors, social factors, geographical factors, and dengue case association factors. Three research objectives were proposed in this study. The first objective was to develop a prediction model using these three types of factors. After data normalization, data balancing and data splitting, these features were imported into RF and XGB, respectively, and then the parameters were adjusted to obtain a suitable model after several tests. The final results showed that the ROC/AUC of RF was 94.3%, while the ROC/AUC of XGB was 93.5%, Recall RF was 89.5%, XGB was 81.2%, Precision RF was 45.5%, and XGB was 52.0%. The prediction results of both models were good.

The second objective was to determine whether projections using three factors were better than those using a single climate factor or two dual factors. In this study, both the single climate factor and the other two dual factors were used in the modeling. The final adjusted experimental results proved to be better than using a single factor or the other two dual factors.

The third objective was to investigate what are the important characteristics for dengue fever prediction. Several characteristics mentioned in the literature and methods, such as low- and middle-income households in the social factor, residential areas in the geographical factor, and wind speed in the climatic factor, are all important characteristics. In this study, we successfully combined all three types of factors and verified that the characteristics proposed by other scholars and even the wind speed proposed by foreign scholars are also suitable for the prediction of dengue fever in Taiwan. The combination of the three factors gives better prediction results than other methods that use only one or a combination of the two characteristics.

Since no vaccine for dengue has been developed, the only method of suppression is awareness and early spraying against the disease. The dengue prediction model developed in our study can narrow down the area of spraying to a secondary area, which is finer than the granularity of a village. In addition, it is easier and more cost effective to pinpoint potential areas of infection by analyzing characteristics.

The climate factors used in this study are collected from the data of meteorological stations in different areas. We cannot collect detailed climate information for secondary output areas or even villages, so we can only refer to the nearby meteorological stations. However, some of the weather stations are located in high altitude areas, and the characteristics of temperature and humidity obtained may be inaccurate. This is one of the possible reasons why the climate factor is not the most important characteristic in this study.

In fact, as mentioned in the literature, temperature differences can cause a decrease or increase in the growth cycle of dengue larvae, but this was not considered in this study. It is suggested that future studies may consider how to include this variable. In addition, the study did not obtain detailed information on the costs associated with dengue fever, such as the cost of resources needed for infected individuals and the cost of spraying an area to combat the disease. If such information is available in the future, a cost matrix can be created so that we can better infer which model is more effective. Finally, this study only used 2014 and 2015 data from Kaohsiung City for model validation and prediction, and if subsequent studies are to use the model in other areas, it is recommended to retrain the model and pay extra attention to the choice of granule size.

References

WHO, Dengue and Severe Dengue (2020)

Department of Disease Control, Ministry of Health and Welfare. "Statistics on Infectious Diseases" (2015)

Deng, S.-Q., et al.: A review on dengue vaccine development. Vaccines **8**(1), 63 (2020)

Xu, Z., et al.: Projecting the future of dengue under climate change scenarios: progress, uncertainties and research needs. PLoS Negl. Trop. Dis. **14**(3), e0008118 (2020)

Zhao, N., et al.: Machine learning and dengue forecasting: comparing random forests and artificial neural networks for predicting dengue burden at national and sub-national scales in Colombia. PLoS Negl. Trop. Dis. **14**(9), e0008056 (2020)

Francisco, M.E., et al.: Dengue disease dynamics are modulated by the combined influence of precipitation and landscapes: a machine learning-based approach. Cold Spring Harbor Laboratory (2020)

Stolerman, L.M., et al.: Forecasting dengue fever in Brazil: an assessment of climate conditions. PLoS ONE **14**(8), e0220106 (2019)

Salim, N.A.M., et al.: Prediction of dengue outbreak in Selangor Malaysia using machine learning techniques. Scientific Reports **11**(1), 1-9 (2021)

Anno, S., et al.: Spatiotemporal dengue fever hotspots associated with climatic factors in Taiwan including outbreak predictions based on machine-learning. Geospatial Health **14**(2) (2019)

Guo, P., et al.: Developing a dengue forecast model using machine learning: A case study in China. PLoS Negl. Trop. Dis. **11**(10), e0005973 (2017)

Huang, S.-H.: Application of geographical exploration technology to predict dengue fever spreading area. National Central University (2020)

Jury, M.R.: Climate influence on dengue epidemics in Puerto Rico. Int. J. Environ. Health Res. **18**(5), 323–334 (2008)

Lambrechts, L., et al.: Impact of daily temperature fluctuations on dengue virus transmission by Aedes aegypti. Proc. Natl. Acad. Sci. **108**(18), 7460–7465 (2011)

Yu, H.-L., et al.: Research on the establishment of a prediction model for the spatial and temporal distribution of dengue fever in Taiwan under climate change (2012)

Valverde-Albacete, F.J., Peláez-Moreno, C.: 100% classification accuracy considered harmful: the normalized information transfer factor explains the accuracy paradox. PLoS ONE **9**(1), e84217 (2014)

Singh, D., Singh, B.: Investigating the impact of data normalization on classification performance. Appl. Soft Comput. **97**, 105524 (2020)

Padmaja, T.M., et al.: An unbalanced data classification model using hybrid sampling technique for fraud detection. In: International Conference on Pattern Recognition and Machine Intelligence. Springer (2007) https://doi.org/10.1007/978-3-540-77046-6_43

Hospital Readmission Prediction via Personalized Feature Learning and Embedding: A Novel Deep Learning Framework

Yuxi Liu[(✉)] and Shaowen Qin

College of Science and Engineering, Flinders University, Tonsley, SA 5042, Australia
{liu1356,shaowen.qin}@flinders.edu.au

Abstract. Hospital readmissions are frequent and costly events. Early risk prediction can lead to more effective resource planning and utilization. This paper presents a deep learning framework for predicting the risk of 30-day all-cause readmission given a patient journey dataset. The problem is posed as a binary classification. A novel personalized self-adaptive feature learning and embedding strategy is applied to learn the representations of patient journeys. We first introduce a *Variable Attention* module to capture the interdependencies of clinical features and generate attention feature representations. We then place a convolutional neural network (CNN) on the generated feature representations to estimate outcome probabilities. Demographic features, including sex and age, are then incorporated into a personalized representation used for adaptively fixing the output of CNN by modifying the network loss function. We successfully predict 30-day all-cause risk-of-readmission with area-under-receiver-operating-curve (AUROC) ranging between 0.838 to 0.858 and overall maximum accuracy of 77.34%.

Keywords: Readmission · Attention mechanism · Deep learning

1 Introduction

Hospital readmission is a costly event, which imposes a tremendous burden on a nation's healthcare system. In the USA, there were 7.8 million (19.6%) of hospital-discharged patients readmitted from 2003 to 2004, which accounted for $17.4 billion of hospital payments [1]. A recent study focused on the readmission rate of atherothrombotic disease in Western Australia reported that the cost of readmissions (A$30 million) accounted for 42% of the original admissions cost (A$71 million) [2]. Moreover, high readmission rates cause a disruption to the normality of hospital management, particularly in critical resources allocation such as inpatient beds. Thus, predicting readmission is critically important for more effective healthcare resource planning and utilization.

© Springer Nature Switzerland AG 2022
H. Fujita et al. (Eds.): IEA/AIE 2022, LNAI 13343, pp. 89–100, 2022.
https://doi.org/10.1007/978-3-031-08530-7_8

Employing a predictive model is one of the useful strategies to reduce the hospital readmission rate [3]. Specifically, machine learning (ML) or deep learning (DL) algorithms can be adopted to identify high-risk patients from electronic health records (EHRs) data so that corresponding preventive approach may be developed to minimize their risk-of-readmission. To this end, this paper investigates the issues and challenges associated with the prediction task, and addresses these by developing a DL-based predictive framework to predict the 30-day all-cause risk-of-readmission given a patient journey dataset.

In practice, it is challenging to learn the representations of patient journeys. Specifically, each patient journey includes two aspects: patient visit and feature levels. Further, the feature level consists of demographic and clinical features. As shown in Fig. 1, two anonymous patients visited the emergency department (ED). They were admitted to different care units. The diagnoses and procedures performed at each visit were recorded as the documented content for a single patient. Each clinical feature records an independent observation, while a set of features can represent the medical conditions of a patient at a given time point.

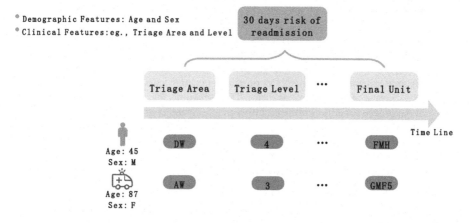

Fig. 1. Patient journey samples. A patient journey is often a consequence of clinical patterns that are associated with specific sequences of clinical events. Usually, a patient walks in or takes an ambulance to the ED for medical treatment. The ED has a dedicated triage area, where nurses and doctors follow a specialized triage process and triages a patient on the basis of how critical the illness or injury is at the time of presentation to the ED. As a result, a patient is either admitted to the corresponding unit or discharged after medical treatments.

When predicting the risk-of-readmission, we aim to automatically model the contexts of patient journeys and generate new patient journey representations. The benefit of modeling patient journey contexts is that such a consideration can help us capture patient medical conditions for effective risk-of-readmission prediction. However, the patient visit process is often patient-specific, which is a consequence of a range of health problems or the environment of clinical (the capacity of the hospital to admit patients). As a result, a part of clinical features within a patient journey is irrelevant to the target prediction, and should be treated as noise for the risk-of-readmission. This issue has been largely disregarded in the existing studies on patient journey learning, which leads to significantly reduced accuracy of the prediction results. Thus, in order to capture patient medical conditions correctly, a good risk prediction method should be able to learn patient journey contexts by distinguishing the importance of features in each patient journey.

Another noticeable challenge stems from the heterogeneity in the disease and demographic features [4]. In practice, it is difficult to predict risk-of-readmission because of the heterogeneity in the disease and demographic features. Specifically, demographic features are usually regarded as a health context before admission. To fully model a patient journey, researchers usually examine the demographic features and combine them with the clinical features. However, if two patients were admitted with heart failure, one is 20-year-old, another 80, then the corresponding health status may have significant differences, resulting in different risk-of readmission. Additionally, due to the complexity and diversity of all cause inpatient data, a certain amount of diseases are patient-specific.

To jointly tackle the above issues, in this paper, we propose a novel DL predictive framework, which can automatically learn the representations of the patient journey and effectively perform risk-of-readmission prediction. Firstly, it proposes a *Variable Attention* module, which is composed of a 1D-CNNs and the self-attention mechanism [5]. The 1D-CNN is developed to capture the interdependencies of clinical features to model each patient journey context and generate feature representations, which are then used as query and key vectors in the self-attention mechanism. The key-query pairs are used to compute the inner dependency weights, then used to update the values. A series of attention feature representations are generated for each patient journey. We place a customized CNN on attention feature representations to perform the predictive modeling. Secondly, it introduces a personalized characterization representation to fix the output of the neural network adaptively, which is achieved by adding two additional terms into the network loss function. The personalized characterization representation is formed by exploiting a standard logistic function to automatically and adaptively learn demographic feature distribution and importance and then embed the newly generated feature representations into one overall representation. This framework enables learning and embedding of demographic and clinical features self-adaptively take place so that their respective contributions to final outputs are captured.

Our major contributions are as follows:

1. We propose a novel deep learning framework for predicting 30-day all-cause risk-of-readmission by fully learning patient journey representations.
2. We designed a personalized feature learning and embedding strategy to incorporate demographic and clinical features. Meanwhile, we modify the network loss function to adjust their contributions in the framework.

3. We conduct extensive experiments on a real patient journey dataset to validate our proposed framework. The empirical results demonstrated significant
prediction performance improvement across the task.

2   Basic Notation and Problem Definition

Our patient journey dataset consists of patients' time-ordered visiting records,
which is denoted by V, i.e., $V = \{V_1, V_2, ..., V_T\}$. Each visiting record V_t consists
of demographic and clinical features. Demographic features contains age and
sex. For each V_t, we use $D_t \in R^{1 \times M}$ and $C_t \in R^{1 \times N}$ separately to denote
demographic and clinical features at time t. For each V_t, we provided the visit
event-level label for binary task, $Y_t = 1$ denotes a patient is readmitted within
30-day, otherwise $Y_t = 0$. The goal of this task is to predict Y_t by learning
$V = \{V_1, V_2, ..., V_T\}$ from the given dataset.

3   The Proposed Framework

To tackle the aforementioned challenges, in this work, we propose a novel deep
learning framework. Figure 2 provides an overview of the proposed framework.
In the rest of this section, we introduce the modules of the framework separately.

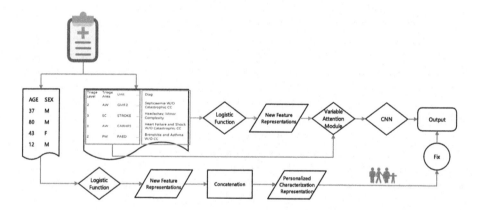

Fig. 2. An overview architecture of the proposed framework.

3.1   Personalized Feature Learning and Embedding

Clinical Features. Learning new feature representations for diagnosis is at
the heart of healthcare analytics [6]. In particular, our diagnostic information
contains a medical diagnosis, syndrome, or symptom. To model its impact, we
used a logistic function as follows:

$$f_{Diag}(C, Y; W_{Diag}) = \{1 + exp\{-W_{Diag} \cdot (C_{Diag} - \varphi)\}\}^{-1},$$

(1)

where C_{Diag} denotes diagnosis and $W_{Diag} \in R$ is the specific parameter to model the corresponding impact for the prediction task. φ is a predefined scalar and set $\varphi = 90$ (see Sect. 4.2, there are 181 types of diagnosis).

Capturing the Interdependencies of Clinical Features and Generating Attention Feature Representations. In this subsection, we propose *Variable Attention* module, which consists of a 1D-CNN and the self-attention mechanism [5]. Specifically, we first apply a 1D convolution operation on ***clinical features*** to learn the interdependencies of clinical features and generate the new feature representations. Then, the max-pooling operation [7] is used to extract the most important feature representations. Last, the pooling results are used as query and key vectors in the attention mechanism. Note that 1D convolution and max-pooling operations separately work in the horizontal and vertical directions.

With the help of convolution operation, the attention mechanism can learn the interdependencies of clinical features at a larger nonlinear space. Moreover, the obtained max-pooling results provide the most important feature representations used for the attention mechanism. Furthermore, one notable advantage of the module is that an interpretable attention map is given after the training, which gives valuable information about the target variables on how much they are correlated to each other.

Mathematically, in the 1D-CNN, the shape of output matrix corresponds to the shape of input matrix as follows: $L_{out} = (L_{in} - filter\_size + 2 \times padding)/stride + 1$. We define the stride is 1, and $filter\_size = 2 \times padding + 1$. For clinical features C, we have the following 1D convolution operation:

$$H = Conv1D(C), \tag{2}$$

where $Conv1D(\cdot)$ denotes the 1D convolution operation and $H \in R^{T \times N}$ denotes the new feature representations. Both have the same shape. The max-pooling operation extracts the most important feature representations from H. The pooling result is used as the query vector $q \in R^{1 \times N}$ and key vector $k \in R^{1 \times N}$ for the attention mechanism.

The query-key pairs are used to compute the inner dependency weights and then update the values. Mathematically, the formula is defined as below:

$$\alpha = softmax(q^\top k),$$
$$Attention(C) = C \cdot \alpha, \tag{3}$$

where $\alpha \in R^{N \times N}$ generated for a patient journey can explain the causative and associative relationships between diagnoses and procedures performed at each patient journey. For simplicity, we use *AttC* to denote *Attention(C)* in the following sections.

Demographics Features. In this work, we mainly consider patient age and gender when predicting risk-of-readmission [8,9]. Moreover, age was categorized by the Australian Institute of Health and Welfare (AIHW[1]) into below 1 year,

---

[1] https://www.aihw.gov.au/reports-data/myhospitals/sectors/admitted-patients.

1–4 years, 5–14 years, 15–24 years, 25–34 years, 35–44 years, 45–54 years, 55–64 years, 65–74 years, 75–84 years, over 85 years. In the same vein, we used logistic function again to model age and sex in order to learn their distributions and importance as follows:

$$f_{AGE}(D, Y; W_{AGE}) = \{1 + exp\{-W_{AGE} \cdot (D_{AGE} - \varphi)\}\}^{-1}, \qquad (4)$$

where D_{AGE} denotes age and $W_{AGE} \in R$ is the specific parameter to model the impact of age for prediction task. *The parameter is used to regularize variable outputs in order to achieve its optimization.* φ is a predefined scalar. We used age groups instead of patients' actual ages and set $\varphi = 5$ (based on above 11 subgroups of age).

$$f_{SEX}(D, Y; W_{SEX}) = \{1 + exp\{-W_{SEX} \cdot (D_{SEX})\}\}^{-1}, \qquad (5)$$

where D_{SEX} denotes sex and $W_{SEX} \in R$ is the specific parameter to model the impact of sex for prediction task. φ is a predefined scalar and set $\varphi = 1$.

To better characterize age and sex, we embed them into a personalized characterization representation.

$$f_{base} = W_{base}^{emb} \cdot base, \qquad (6)$$

where W_{base}^{emb} is an embedding matrix and base consists of f_{AGE} and f_{SEX}.

3.2   Personalized Prediction

We customize a CNN with a 1D convolutional layer and a max-pooling layer. The convolutional layer is placed on the **attention feature representations**. We use a combination of m filters with s different window sizes. We use l to denote the size of feature window and $AttC_{j:j+l-1}$ denote the concatenation of l clinical features from $AttC_j$ to $AttC_{j+l-1}$. A filter $W_f \in R^{1 \times l}$ is applied on the window of l clinical features to produce a new feature $f_j \in R$ with the ReLU activation function as follows:

$$f_j = ReLU(W_f \cdot AttC_{j:j+l-1} + b_f), \qquad (7)$$

where $b_f \in R$ is a bias term and ReLU(f) = max(f, 0).

This filter is applied to each possible window of clinical features in the whole $\{AttC_{1:l}, AttC_{2:l+1}, ..., AttC_{N-l+1:N}\}$ to generate a feature map $f \in R^{N-l+1}$ as follows:

$$f = [f_1, f_2, ..., f_{N-l+1}]. \qquad (8)$$

The max-pooling layer is placed on the feature map f. Each filter produces a feature. Since we have m filters with s different window sizes, the final vector representation of $AttC_t$ can be obtained by concatenating all the extracted features, e.g., $z_t \in R^{ms}$. A fully connected softmax layer is used to estimate outcome probabilities as follows:

$$\hat{Y}_t = softmax(W_Y \cdot z_t + b_Y), \qquad (9)$$

where $W_Y \in R^{n \times ms}$ and $b_Y \in R^n$ are the learnable parameters and n is the number of target labels (e.g., 0 or 1 in binary classification).

Let θ be the set of all the parameters in CNN. $\hat{Y}_t$ and Y_t are separately prediction probability vector and ground truth. The cross-entropy between ground truth Y_t and outcome probabilities $\hat{Y}_t$ is used to estimate the loss. The objective function can be defined as follows:

$$\mathcal{L}(\theta) = -\frac{1}{T} \sum_{t=1}^{T} (Y_t^\top \cdot log(\hat{Y}_t) + (1 - Y_t)^\top \cdot log(1 - \hat{Y}_t)). \tag{10}$$

Given the input data $_t$ to predict its true label vector Y_t, we can obtain outcome probability vector $\hat{Y}_t = p(Y_t|C_t; \theta))$. Now we use the proposed personalized characterization representation f_{base}^t (**demographic features**) to fix the output of CNN adaptively. The objective function can be rewritten as follows:

$$\mathcal{L}(\theta, \mathcal{W}) = \mathcal{L}(\theta) + \alpha \frac{1}{T} \sum_{t=1}^{T} KL(f_{base}^t | P(Y_t|C_t; \theta))$$
$$+ \beta \cdot \mathcal{L}'(\mathcal{W}), \tag{11}$$

where α, β are the hyper-parameters and $\mathcal{L}'(\mathcal{W})$ is the average cross entropy between f_{base}^t and ground truth Y_t. The $\mathcal{L}'(\mathcal{W})$ is defined as follows:

$$\mathcal{L}'(\mathcal{W}) = -\frac{1}{T} \sum_{t=1}^{T} (Y_t^\top \cdot log(f_{base}^t)$$
$$+ (1 - Y_t)^\top \cdot log(1 - f_{base}^t)). \tag{12}$$

In summary, Eq. (12) incorporates two additional loss terms, both of which are relevant with demographic features.

$\alpha \frac{1}{T} \sum_{t=1}^{T} KL(f_{base}^t | P(Y_t|C_t; \theta))$ is the KL loss between personalized characterization representation and prediction distributions, which is used to fix the prediction results achieved by CNN.

Another loss function of $\mathcal{L}'(\mathcal{W})$ represents the self-adaptive process for demographic features. It provides a bridge between distributions of demographic features and ground truth, where each demographic feature can achieve its optimization by updating its values with the learning process.

4   Experimental Setup

4.1   Dataset Description

We presented one case study on the risk-of-readmission prediction, using a patient journey dataset from a metropolitan hospital in Australia. The data set being used is the administrative part of the EHRs from the hospital for the whole of 2018 and 2019. It contains demographic, admission (emergency or elective, location of care and treating clinical team) and discharge data, and detailed diagnostic information. It also contains time and location stamp information that records every occasion a patient is moved between locations within the hospital. Ethical approval was obtained for access to the dataset. Table 1 presents an overview of all the selected features.

Table 1. Overview of the feature groups

Demographic features
AGE
SEX
Clinical features
Ambulance (yes or not)
Triage AREA
Triage Level (ESI: 1–5)
ED_HOURS
FIRST_BED
UNIT
FINAL_WARD
FINAL_UNIT
Diag (medical diagnosis, syndrome, or symptom)

4.2 Data Preprocessing

We selected the patient data that was readmitted within 30-day from the discharge based on the standardized HRRP readmission measure [10]. Each patient has a unique patient URN (ID) for distinguishing multiple admissions such as readmission. We examined the number of days between the current admission and him/her previous discharge for a given patient and checked if it is less than 30-day. Moreover, to accurately learn the common patient behaviors, we selected a diagnosis with more than 100 visits from the dataset. The final dataset includes 181 diseases and 61264 records (roughly 83% of the total data). Besides, we also used the undersampling approach [11] to address the label imbalance problem.

4.3 Baseline Approaches

To validate the performance of the proposed framework for our prediction task, we compare the proposed framework with logistic regression (LR), Gaussian Naive Bayes (GNB), Support Vector Machine (SVM), Random Forest (RF), AdaBoost, Explainable Boosting Machine (EBM) [12], and two CNN models. EBM is a strong baseline that has been applied in 30-day risk-of-readmission prediction and reported state-of-the-art accuracy. Besides, in ablation experiments, we present one variant of the proposed framework (CNN+). It only incorporates the attention mechanism [5].

4.4   Implementation Details and Evaluation Strategies

We implement the proposed framework with PyTorch 0.2.0. We implement ML approaches with scikit-learn[2] and EBM[3]. We use the same settings for the proposed framework and other CNN models. Specifically, we set the size of filter windows (l) from 2 to 5 with s = 100 filter maps. Moreover, we propose using 1 as the size of filter windows that can introduce more nonlinearities without modifying the size of the input and thus enhance the expression ability of the neural network [13]. We also use regularization (l_2 norm with the coefficient 0.001), and drop-out strategies (drop-out rate is 0.5) for all DL approaches. For training models, we employ a standard train/test/validation split. The validation set is used to select the best values of parameters. To evaluate binary outcomes, we calculate AUROC, Accuracy, F1 Score, and Brier Score Loss (BSL). We perform 100 repeats for all used approaches and report the average performance.

5   Results and Discussion

5.1   Performance Evaluation

Table 2 shows the performance of all approaches on the dataset. The results indicate that the proposed framework can outperform other baseline methods. Specifically, the risk-of-readmission was predicted with an AUROC of 0.8480 and one standard deviation of 0.010. Moreover, we find that the proposed framework can outperform CNN. Therefore, it indicates that the proposed personalized feature learning and embedding can improve the accuracy of risk-of-readmission prediction. Another important finding was that the proposed framework could outperform the CNN+. Therefore, it indicates that incorporating the self-attention mechanism and 1D-CNN is better for improving predictive accuracy than using the self-attention mechanism alone.

Table 2. Results of 30-day risk-of-readmission prediction

Model	AUROC	Accuracy	F1 Score	BSL
LR	0.7543(0.012)	0.7000(0.013)	0.6888(0.014)	0.2016(0.005)
GNB	0.7403(0.013)	0.6767(0.012)	0.6884(0.011)	0.2315(0.008)
SVM	0.7527(0.012)	0.6939(0.012)	0.6892(0.013)	0.2016(0.005)
RF	0.8150(0.011)	0.7488(0.012)	0.7369(0.013)	0.1751(0.006)
Adaboost	0.8123(0.011)	0.7435(0.012)	0.7311(0.013)	0.2464(0.000)
EBM	0.8292(0.010)	0.7588(0.011)	0.7468(0.012)	0.1677(0.005)
CNN	0.8180(0.011)	0.7463(0.012)	0.7341(0.013)	0.1734(0.005)
CNN+	0.8274(0.010)	0.7575(0.012)	0.7452(0.012)	0.1686(0.005)
The proposed framework	0.8480(0.010)	0.7734(0.011)	0.7573(0.013)	0.1579(0.005)

---

[2] http://scikitlearn.org/stable/.
[3] https://github.com/interpretml/interpret.

5.2   Clinical Feature Interdependencies

One aspect of the proposed method is that it explicitly captures the interdependencies of clinical features and generates attention feature representations. This is achieved by applying *Variable Attention* module to each patient journey. The module provides an interpretable attention map after the training, which gives valuable information about the target variables on how much they are correlated to each other. Therefore, this makes the proposed method explainable. To showcase this feature, we visualized four patient journeys, which correspond to patients A-D. These examples come from the test dataset. Patients A and B were readmitted with Heart Failure and Shock but with significant differences in other clinical features. Patients C and D were readmitted with Chronic Obstructive Airways Disease but with minor differences in other clinical features.

Figure 3 shows the clinical feature interdependencies of patients A-D. The attention scores calculated by the proposed *Variable Attention* module are shown. Note that for each attention score, we round up to 2 decimal places. The ordinates of the figure are the Query features, and the abscissas are the Key features. The boxes in the figures show how much each Key feature responds to the Query when a Query feature makes a query. We find that *Variable Attention* module can figure out clinical feature interdependencies for four patients. These feature interdependencies can be explained in part by attention scores. e.g., *Variable Attention* module figures out there are relatively high interdependencies between most clinical features and Diag. Additionally, *Variable Attention*

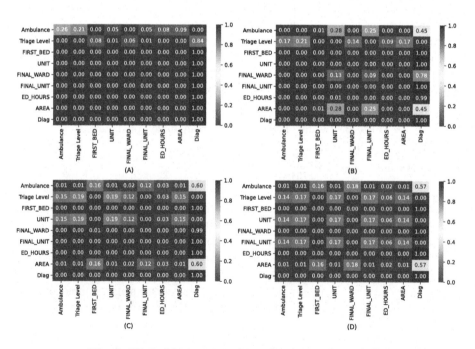

Fig. 3. Clinical feature interdependencies: patients A–D

module figures out that the part of the clinical features is also likely to respond strongly to themselves, which is denoted by the diagonal of four matrices.

6   Conclusion

In this work, we have presented a novel DL framework of personalized learning and embedding features with the aim of predicting risk-of-readmission. Experiments on a real patient journey dataset show that our framework demonstrated significant prediction performance improvement. The findings from this study make several contributions to the current literature. Firstly, we have proven that prediction of 30-day all-cause risk-of-readmission in hospitals is possible using patient journeys. Secondly, personalized feature learning and embedding contribute in several ways to our understanding of the importance of clinical and demographic features and provide a basis for further research. Lastly, it is a scalable framework, which can readily be of broad use to the scientific and health care communities. A number of possible future studies using the same experimental setup are apparent such as admission prediction at the time of triage and hospital admission location prediction.

References

1. Jencks, S.F., Williams, M.V., Coleman, E.A.: Rehospitalizations among patients in the medicare fee-for-service program. N. Engl. J. Med. **360**(14), 1418–1428 (2009)
2. Atkins, E.R., Geelhoed, E.A., Knuiman, M., Briffa, T.G.: One third of hospital costs for atherothrombotic disease are attributable to readmissions: a linked data analysis. BMC Health Serv. Res. **14**(1), 1–9 (2014)
3. Zhou, H., Della, P.R., Roberts, P., Goh, L., Dhaliwal, S.S.: Utility of models to predict 28-day or 30-day unplanned hospital readmissions: an updated systematic review. BMJ Open **6**(6), e011060 (2016)
4. Reddy, B.K., Delen, D.: Predicting hospital readmission for lupus patients: an RNN-LSTM-based deep-learning methodology. Comput. Biol. Med. **101**, 199–209 (2018)
5. Vaswani, A., et al.: Attention is all you need. In: Advances in Neural Information Processing Systems, pp. 5998–6008 (2017)
6. Cai, X., Gao, J., Ngiam, K.Y., Ooi, B.C., Zhang, Y., Yuan, X.: Medical concept embedding with time-aware attention. In: Proceedings of the 27th International Joint Conference on Artificial Intelligence, pp. 3984–3990 (2018)
7. Collobert, R., Weston, J., Bottou, L., Karlen, M., Kavukcuoglu, K., Kuksa, P.: Natural language processing (almost) from scratch. J. Mach. Learn. Res. 12(ARTI-CLE), 2493–2537 (2011)
8. Berry, J.G., et al.: Age trends in 30 day hospital readmissions: us national retrospective analysis. BMJ **360** (2018)
9. Maali, Y., Perez-Concha, O., Coiera, E., Roffe, D., Day, R.O., Gallego, B.: Predicting 7-day, 30-day and 60-day all-cause unplanned readmission: a case study of a Sydney hospital. BMC Med. Inform. Decis. Mak. **18**(1), 1–11 (2018)
10. McIlvennan, C.K., Eapen, Z.J., Allen, L.A.: Hospital readmissions reduction program. Circulation **131**(20), 1796–1803 (2015)

11. Lemaître, G., Nogueira, F., Aridas, C.K.: Imbalanced-learn: a Python toolbox to tackle the curse of imbalanced datasets in machine learning. J. Mach. Learn. Res. **18**(1), 559–563 (2017)
12. Nori, H., Jenkins, S., Koch, P., Caruana, R.: InterpretML: a unified framework for machine learning interpretability. arXiv preprint arXiv:1909.09223 (2019)
13. Lin, M., Chen, Q., Yan, S.: Network in network. arXiv preprint arXiv:1312.4400 (2013)

Intelligent Medical Interactive Educational System for Cardiovascular Disease

Sheng-Shan Chen[1], Hou-Tsan Lee[2], Tun-Wen Pai[1,3]([⊠]), and Chao-Hung Wang[4]

[1] Department of Information Engineering, National Taipei University
of Technology, Taipei, Taiwan
t110598037@ntut.org.tw, twp@ntut.edu.tw

[2] Department of Information Technology, Takming University of Science and Technology,
Taipei, Taiwan

[3] Department of Computer Science and Engineering, National Taiwan Ocean University,
Keelung, Taiwan

[4] Department of Cardiac and Cardiovascular, Keelung Chang Gung Memorial Hospital,
Keelung, Taiwan

Abstract. Cardiovascular disease is the number one killer of global deaths, accounting for 30% of total deaths every year. To arouse the public's attention to cardiovascular health, it is bound to provide sufficient education materials for cardiovascular patients to improve health and reduce medical costs. Currently, there is no systematic and patient orientated guidelines in Chinese regarding cardiovascular diseases. Therefore, this research proposed a smart medical interactive education system based on natural language processing and chatbot intelligence. We adopted Line Bot chat mode to provide pertinent medical knowledge for patients. All medical media are designed based on patient-oriented mechanisms. The Patient Educational Material Evaluation Tool (PEMAT) was applied to evaluate the quality of medical documents and video clips. In addition, relationship among various media elements were constructed based on collected keywords to ensure those messages not correctly processed through natural language processing could be successfully converted into a single response by text mining techniques. This research is the first constructive and intelligent interactive education chatbot system for patients with cardiovascular diseases.

Keywords: Line bot · NLP · Cardiovascular disease · Information retrieval · Medical interaction

1 Introduction

With accessible medical information from web resources and increased levels of citizenship education, the authoritative relationship between doctors and patients has gradually shifted to a patient-centered service model. Sufficient medical information can provide appropriate training and education for patients and healthcare staff [1], to guide patients to learn how to respond to emergencies independently, to clearly express their expectations from medical services, and even to assist patients' choices of medical treatment.

© Springer Nature Switzerland AG 2022
H. Fujita et al. (Eds.): IEA/AIE 2022, LNAI 13343, pp. 101–111, 2022.
https://doi.org/10.1007/978-3-031-08530-7_9

Such enhanced medical knowledge and good communication could promote mutual respect between doctors and patients and might reduce medical disputes [2].

However, most of healthcare education materials are designed based on medical professional. In addition, medical doctors or staffs often think they have already provided sufficient explanations to patients during their treatment in hospital, but patients may feel suspicious or even overwhelmed after leaving the hospital. Hence, patients may not be able to take care of themselves appropriately and make correct medical decision-making during self-homecare stages. It may result in medical disputes due to misunderstanding between medical staffs and patients [3].

According to a death statistics report by Ministry of Health and Welfare of Taiwan in 2019, the incidence and mortality of cardiovascular disease were increased significantly with age [4], and an aging society has obviously become an important medical policy issue. Over the past 20 years, superior medical systems have successfully extended average life expectancy for Taiwanese populations. Life-threatening diseases such as acute myocardial infarction (AMI), or chronic diseases such as hypertension, diabetes and stroke, can be well taken care through advanced medical technologies. However, patients with these coronary and vascular interventions possess high risk of transferring into heart failure conditions. This may be due to reasons of aging issues and refined foods among heart disease patients [5].

The popularity of the Internet has speeded up circulation of information, and a plenty of medical associated knowledge can be obtained only by searching on the Internet. However, these searched materials might be provided by various websites, and t medical related materials might be incorrect information provided by non-experts. Those query results are rich and improperly classified, and lengthy professional terminologies are difficult to be understood [6]. Therefore, cross-disciplinary cooperation between information developers and professional medical staffs is required to design an effective and convenient medical interaction system. All healthcare materials must be designed and evaluated to meet the minimum requirements according to patient's needs [7].

The Line Bot services provided by Line Corporation is widely used in daily life in Asian countries, with as many as 21 million users in Taiwan [8]. Due to willingness of downloading Apps has decreased, resulting in a relatively high cost for launching new App services. Therefore, this study tried to design a medical chatbot system based on Line Bot environments, and to combine natural language processing (NLP) technologies to analyze semantic information provided by patients during self-homecare stages, so as to refine the amount of medical information and improve the relevance of automatic responses. Lower the entry requirements for patients to learn medical education by themselves.

2   Materials and Methods

2.1   Medical Teaching Materials

To develop a medical interactive education system for patients with cardiovascular disease, a series of patient-oriented scene videos and medical illustration documents were used to avoid difficulty of learning for elderly patients and their families. We have created a total of 222 videos and 45 medical documents by medical experts. The medical documents for each heart related disease include disease introduction, examination, diagnosis, treatment strategy, postoperative care, prognosis information, and all related healthcare issues. The medical contents are designed at introductory level, suitable for patients with no common medical knowledge. Patients can clearly understand prognosis of the disease, and learn how to prevent the recurrence of hospitalization events. Healthcare documents were post-edited and reviewed by doctors, and all contents are evaluated by two experts according to the 26 evaluation items defined by the Patient Educational Material Assessment Tool (PEMAT) guidelines to ensure the quality of educational contents [9]. The video introduction covers coronary artery disease, aortic disease, peripheral vascular obstruction, congenital heart disease and arrhythmia, etc.; the examination methods include electrocardiogram (ECG), computed tomography (CT), angiography, etc.; treatment chapters include surgery, prescription drugs, etc.; self-care chapters contain practical instructions for symptom observation, postoperative care.

2.2 Patient-Orient Healthcare Documents

Language is a basic and most friendly communication tool. Humans use language to express their needs in life, and use words to describe reality. It can be represented by words, symbols, or images. As humans rely on computer systems more and more in daily life, natural language processing becomes an important technique to communicate between machines and human beings, and it is expected that machines may hold language understanding ability as normal people.

There are mainly two ways to enable machines to process language understanding. One is to establish rules of language communication (rule-based approach). This method is easy to be built and can quickly establish a response mechanism for interactive systems. However, to deal with various expressions of users, it is necessary to enumerate a large number of sentences. As the scope of knowledge expands, maintenance and scalability are relatively poor due to insufficient data collection. Another method is to solve the disadvantage of rule-based approach, and it allows the machine learns association modes of sentences and vocabularies by machine itself. Details will be introduced in the next section.

2.2.1 Natural Language Understanding

Natural Language Understanding is to explore mutual distances among words within a large amount of corpus, such as daily conversations, online messages, and web articles. Based on the metric analysis, it can parse and integrate the input sentence for identifying intent, entity and context of the message, so the computer could understand a sentence comprehensively and translates meaningful information.

Intent refers to determine the purpose of an input sentence, such as "I want to know coronary artery disease." and "What is coronary artery disease?". Though these two different sentences, they are exactly the same in meaning with the intent of understanding the disease of coronary artery disease.

An entity is a word or a set of combined words that clearly refers to the object within a sentence. For example, "briefing", "what is", and "introduction" all refer to the word "introduction"; "coronary artery disease", "CAD", "ischemic heart disease", "Coronary heart disease" refers to coronary artery disease as the corresponding entity.

When an entity is sufficient to cooperate with an intent, the corresponding answer can be executed. Once the corresponding response is replied, perhaps immediate responses such as "So what should I pay attention to?" and "How should I prevent it?", etc., would occurred for further responses. Hence, the system must understand the context of the dialogue and provide appropriate responses for the following questions. For example, assuming one user only said "how to prevent" to the chatbot system, and the system will trace back to previous dialogue with "coronary artery disease" as an entity. Hence, the system understands the user's intention and entity, and responds with an appropriate answer to the user. This research applied the Dialogflow API as the agent of natural language understanding for the proposed system, and a set of human-computer interaction (HCI) interface based on natural language understanding were developed by Google, which facilitates a user to retrieve specific keywords from dialogue sentences for further applications. Developers could set suitable reactions according to the retrieved specific keyword [10].

2.2.2 Natural Language Understanding

Considering inconveniences for patients or caregivers to type texts when using the chatbot software, an alternative approach using speech recognition technology is applied to convert speech contents into texts. This module will use Google Speech-to-Text API and machine learning technology to train a neural network model for speech recognition. The developed recognition model can recognize more than 100 different languages and convert them into corresponding texts.

2.3 System Design

In this study, the proposed education system mainly aimed at patient-oriented healthcare training for patients with cardiovascular disease. The education contents were designed and stored in a Directed Acyclic Graph (DAG) structure. Figure 1 illustrates the data flow and processes of the constructed system [11]. The main DAG structure integrated with medical professional advices through keyword analysis, and the titles and attributes for each medical videos/document were manually labelled and checked. The DAG structure was designed as an extensible and composable tree according to relations between the DAG structure and document files. Especially, the required medical content elements were designed through integrating medical images, real case scenarios, and entry-level video animations without using lengthy text messages to convey important medical knowledge to patients and caregivers.

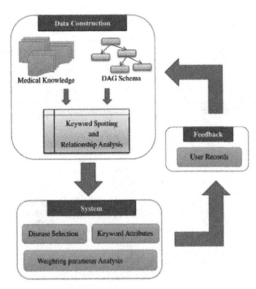

Fig. 1. System flow and procedures for constructing data in the proposed interactive education system.

Patients or medical staffs input information on user interface through Heroku, a platform as a service (PaaS - platform as a service), and the query messages are transmitted to Dialogflow for semantic analysis and determination of intent. Figure 2 illustrates the proposed system architecture, which aims to converge the results to a single answer responds to the user. There are three possible scenarios to satisfy this condition as follows:

1. If a user's query messages are not enough to be fully determined for the intention, the entity will be supplemented with a guiding question, and the corresponding answer will be returned if requirements are met.
2. If the intent cannot be completely determined twice in a series processes, the keyword will be used to calculate the relevance relationship, and the top three ranked results with the highest relevance measurements will be returned for the user to choose.
3. If the result of the second scenario yet to be empty, then end the topic or ask the user whether to change the query messages.

Content-oriented is the major advantage of this proposed approach. It can improve user's experiences, prolong dialogue life, increase users' stickiness, and guide users to achieve their goals. In addition, it is important to rely on word frequency technology identify the keywords embedded in each education document, and to converge a situational problem to avoid difficulty communicating situations between a user and a robot.

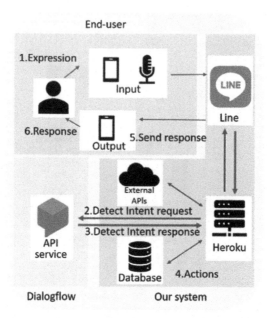

Fig. 2. Intelligent medical interactive educational system for cardiovascular disease

2.4 DAG Structure

Multimedia medical contents were previously constructed and manually annotated and verified by medical professionals. The principles of validation processes are mainly based on disease types and corresponding content characteristics, such as disease descriptions, diagnosis and treatment strategies, rehabilitation after cardiac surgery, and healthcare prevention programs. However, different disease types may have the same clinical symptoms and similar postoperative recovery guidelines. If a traditional tree structure is used to store the above characteristics, it will not only cause a lot of redundancy, but also make it difficult to maintain and reduce scalability. Therefore, a hierarchical DAG tree structure was designed to reduce a lot of redundant data and to replace the traditional tree structure in our proposed patient education interaction system.

In the DAG architecture, all edges possess directional characteristics and clearly express their hierarchical relationship between two nodes. Each node has multiple parent nodes that can be connected and traversal among nodes are allowed for multiple routes. If two different diseases have a same treatment strategy or post-operative rehabilitation, only one related multimedia element is required to be stored in this architecture and the powerful function of DAG can be utilized to realize an alternative path selection. In this system, the initial DAG structure is manually added according to the elements stored in the relational database (RDB), and then information of nodes and edges would be stored in separate tables to solve the DAG characteristics with variable number of nodes. This method avoids the confusion between parent nodes and child nodes in the RDB structure, and it can be well reflected in the scalability and integrity of adding new disease terms or multimedia elements. Figure 3 illustrates an example of a DAG substructure for coronary artery disease. The cardiovascular magnetic resonance imaging (MRI) and coronary interventional therapy (PCI) video contents can be shared by different types of cardiovascular disease [11].

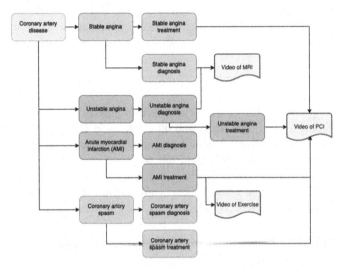

Fig. 3. Examples of DAG substructures in coronary artery disease

Each multimedia element has a defined ontology term in the DAG structure, and all multimedia contents possess disease, examination, treatment and storage attributes. Disease attribute describes the basic knowledge of different types, hierarchical relationships among different diseases, classifications; examination attribute refers to clinical diagnostic tests for cardiovascular disease, such as blood tests, ultrasound, computed tomography; treatment attribute represents all relevant medical treatment strategies, including cardiac intervention, postoperative treatment, prognostic selfcare; the last term of storage attribute contains information corresponding to multimedia elements. The relationship between two diseases is dynamically calculated for the shortest distance and associations between the two terms. These attributes play a key role in the proposed retrieval mechanism for identifying associated multimedia elements in the DAG structure.

2.5 Keyword Statistics Architecture

If a user's intention cannot be determined several times in the NLP mode, in order to guide the user to find useful education materials, keyword search will be used. Each media element is marked with corresponding keywords, and weighting techniques commonly used in information retrieval and text mining will be applied. TF-IDF, as shown in Eq. (1), the term frequency (tf) is used to identify important keywords in each multimedia [12], and it will be multiplied by the inverse document frequency (idf) to filter non-important keywords.

The numerator $n_{i,A}$ of the first item *tf* represents the number of times the word *ti* appears in the media element A. The denominator $\sum_k N_{k,A}$ represents the total number of words in A. The numerator $|D|$ of the second item *idf* represents the total number of media elements in the corpus, and the denominator represents the number of elements containing the word *ti* (i.e. $n_{i,A} \neq 0$). If the constant 1 was not added, the denominator might occur with zero which represents no such word in the media. According to the evaluation results from TF-IDF, for example, the word "disease" would appear in most of the document files, even if it has a high frequency in certain media, this word would not be considered as a representative keyword for the specific document.

$$\text{tfidf}_{i,A} = \frac{n_{i,A}}{\sum_k N_{k,A}} \times log \frac{|D|}{1 + |\{A : t_i \in d_A\}|} \tag{1}$$

In addition, part of speech (POS) is also used to remove meaningless words such as adverbs, prepositions, conjunctions, and interjections in Chinese. Then, the system compares the relevance relationship among various documents and search for those documents with higher keyword relevance. In this proposed interactive education system, users can not only learn healthcare knowledge in a relaxed chat environment, but also the medical information appeared within hierarchically organized results based on correlation measures can encourage users to learn interesting medical issue actively. Patients can get the most relevant medical knowledge through interaction experiences and avoid being misled by a huge amount of mixed and unauthorized medical information.

3   Result and Discussion

The smart medical interactive education learning system is a major breakthrough for patients with cardiovascular disease. All the medical contents used here are related to diagnosis and prevention for cardiovascular related diseases, and these healthcare documents are important for patients, patient's family, and caregivers, as well as the initial training for the professional medical staffs. On the user side, the use of speech recognition as input interface provides a good coping strategy for users who cannot type texts as query messages. Users may also be unfamiliar with professional disease items, and natural language recognition techniques are applied to effectively solve this situation. Even if users already have their own specific keywords, they can obtain accurate medical documents effectively through keyword searching mode. The DAG structure is used for the reliability of data retrieval, and the teaching materials are evaluated through PEMAT assessment tool for checking understandability of the designed patient education materials.

3.1   Develop a Patient-Centered Educational Interaction System

The system provides two different modes to medical document query, including the NLP mode and the keyword search mode, both of modes support voice input for speech recognition. In any case, users can be guided to find the related topics or medical information according to their interests. For example, a user can use a normal tone to express her/his curiosity or question about pre-auditory heart disease. Then, the robot will respond immediately to the only solution. Users can directly click on the weblink to watch the corresponding medical multimedia (a video clip or a document). In addition, in order to prevent users lost in navigation. For the unknown fields, a menu list was designed for users to select, but it does not mean that users must act according to certain criteria. Under the interaction between the system and a user, a situation like a real person conversation is created, creating an atmosphere of happy learning environment.

Not every user is good at or prefer to chat in the interactive system, or a user already has her/his own clear and precise keywords. After switching to the keyword searching mode through the system menu, a user can search results based on the data corpus and the previously constructed synonyms. The system will automatically search and display the searched results which were ranked according to the weight of related words. In any case, improving user experiences in the interactive medical education system is one of the major aims in the proposed system. For users with limited mobility or special needs, they can use the smartphone in the Line communication software to input personal voice, and convert it into text through the system for identification. Afterall, users may not have sufficient medical or computer related background knowledge to use the patient education interactive system, so we are committed to designing a highly flexible and patient-friendly system for them to use.

3.2 Evaluation of Cardiovascular Health Education Data

The Patient Education Materials Assessment Tool (PEMAT) is a tool for systematic evaluation of health education materials developed by the American Medical Quality Research Center AHRQ [12]. It can be divided into two issues of understandability and actionability. The former evaluated the understandability of the documents for patients with various literacy levels to understand the medical documents, and the latter refers to the patients with different backgrounds who are able to act learning from the healthcare guidelines. There are 19 assessments for understandability and 7 assessments items for actionability. A total of 26 items were applied to evaluate document materials (P-printable document) and audio-visual materials (A/V-audiovisual clips).

PEMAT is used by professionals who would like to provide healthcare contents for patients. The designed medical multimedia clips and documents in this interactive system were evaluated by two professional medical staffs individually. Each document can be annotated with "Agree", "Disagree", or "Not Applicable" regarding the 26 evaluation items to provide a final score for each medical document/clip through PEMAT evaluation. All obtained scores were summed up and divided by the highest possible score to obtain a comprehension percentage. If the PEMAT suggested score is higher than 70%, the document/clip can be considered as understandable and actionable. According to PEMAT evaluation, the average understandability score of our designed healthcare multimedia clips achieved 91.3%, and the actionability achieved 48%, while the understandability score for healthcare document achieved 87.5%, and actionability of documents achieved an average of 81.9%. It can be observed that multimedia clips are less actionable. That is due to three assessment criteria including whether the teaching materials clearly indicate at least one action that users can follow to do (PEMAT item 20), whether the documents directly facing users when describing the actions (item 21), and whether actions of the teaching materials are decomposed into possible manageable actions (item 22). However, some of the video clips were designed at introduction level and focused on explaining cardiovascular disease and corresponding surgical procedures. Therefore, patients do not need to take appropriate actions to simulate surgical procedures, so actionability cannot be satisfied at a certain level. If these items can be selected as "not applicable" instead of "Agree" or "Disagree", the actionability can be expected to reach more than 80%. This phenomenon can also be applied to healthcare document. Since meaningful illustrations and animations could improve the understandability and actionability, multimedia clips and documents with lower PEMAT scores will be revised and re-produced in the near future.

4 Future Work

In this study, a patient-oriented intelligent medical interactive education system for patients with cardiovascular disease has been developed, which has been actually tested by users for continuous evaluations and improvement. The keywords or questions defined by users sometimes varied greatly, and 80% of users asked the common questions of only account for 20% of all questions, as long as the robot could answer these 20% questions, it is enough to use the interactive modes of questions and answers to shorten the range of accurate answers. The proposed system has added TF- IDF weighting technology and coping mechanism to avoid infinite dead-loop dialogues. This system effectively solves 24-h medical professional consultation, and most healthcare documents meet the minimum requirements for most patients according to PEMAT evaluation., A continuous enhancement on weak documents or multimedia clips will be revised to improve the quality of healthcare education materials. However, there are still some unresolved issues and challenges. A big gap between patients and professional medical staffs may lead to misunderstandings and even create medical disputes. To overcome uncertainty of user dialogues, increment of the training corpus to improve accuracy of intent identification becomes crucial, and an improved DAG structure to store a larger amount of education document is another important issue. These problems can be solved gradually step by step. In the future, more medical staffs and computer engineers will cooperate in developing a global healthcare system, and it is expected to improve the satisfaction for both patients and medical staffs with comprehensive medical services. Improved quality of medical care and immediate responses to actual and urgent needs of medical care can be gradually achieved through realizing efficient and effective chat-bot applications.

References

1. Webber, G.C.: Patient education. A Rev. Issues, Med. Care **28**(11), 1089–1103 (1990)
2. Driscoll, A., Davidson, P., Clark, R., Huang, N., Aho, Z.: Tailoring consumer resources to enhance self-care in chronic heart failure. Aust. Crit. Care **22**(3), 133–140 (2009)
3. Kelly, P.A., Haidet, P.: Physician overestimation of patient literacy: a potential source of health care disparities. Patient Educ. Couns. **66**(1), 119–122 (2007)
4. Analysis of Statistical Results of Death Causes in 108 (108 Edition) [Data File]. Ministry of Health and Welfare: Statistics Division
5. Anand, S.S., Hawkes, C., de Souza, R.J., et al.: Food Consumption and its impact on cardiovascular disease: importance of solutions focused on the globalized food system: a report from the workshop convened by the world heart federation. J. Am. Coll. Cardiol. **66**(14), 1590–1614 (2015)

6. Durani, P., Croft, G.P., Kent, P.J.: Sources used by patients seeking information about peripheral vascular disease: is the Internet relevant? Surg. Pract. **9**(2), 46–49 (2005)
7. Safeer, R.S., Keenan, J.: Health literacy: the gap between physicians and patients. Am. Fam. Physic. **72**(3), 463–468 (2005)
8. Corporate Announcement (2019/03/29) LINECONVERGE 2019 春季記者會。2020/08/01 https://linecorp.com/zh-hant/pr/news/zh-hant/2019/2651
9. Shoemaker, S.J., Wolf, M.S., Brach, C.: Development of the patient education materials assessment tool (PEMAT): a new measure of understandability and actionability for print and audiovisual patient information. Patient Educ. Couns. **96**(3), 395–403 (2014)
10. Google Cloud (n. d.). Dialogflow documentation. August 1, 2020 https://cloud.google.com/dialoflow/docs
11. Huang, J.-L., Chen, C.-M., Pai, T.-W., Liu, M.-H., Wang, C.-H.: Directed acyclic graph-based patient education system for cardiovascular patients in Taiwan. J. Mechanics Med. Biol. **16**(1), 1640011 (2016)
12. Bramer, M.: Principles of Data Mining. Springer, London (2013) https://doi.org/10.1007/978-1-4471-7307-6
13. Sarah, J., Shoemaker, P.D., Wolf, M.S.: The Patient Education Materials Assessment Tool (PEMAT) and User's Guide.(2014) https://www.ahrq.gov/ncepcr/tools/self-mgmt/pemat.html

Evolutionary Optimization for CNN Compression Using Thoracic X-Ray Image Classification

Hassen Louati[1][(✉)] ⓘ, Slim Bechikh[1] ⓘ, Ali Louati[2] ⓘ, Abdulaziz Aldaej[2] ⓘ, and Lamjed Ben Said[1] ⓘ

[1] SMART Lab, University of Tunis, ISG, Tunis, Tunisia
hassen.louati@stud.acs.upb.ro, slim.bechikh@fsegn.rnu.tn,
lamjed.bensaid@isg.rnu.tn
[2] Department of Information Systems,
College of Computer Engineering and Sciences,
Prince Sattam bin Abdulaziz University, Al-Kharj 11942, Saudi Arabia
{a.louati,a.aldaej}@psau.edu.sa

Abstract. Computer Vision, as an area of Artificial Intelligence, has recently achieved success in tackling numerous difficult challenges in health care and has the potential to contribute to the fight against several lung diseases, including COVID-19. In fact, a chest X-ray is one of the most frequent radiological procedures used to diagnose a variety of lung illnesses. Therefore, deep learning researchers have recommended that deep learning techniques can be used to build computer-aided diagnostic systems. According to the literature, there are a variety of CNN structures. Unfortunately, there are no guidelines for compressing these architectural designs for any particular task. For these reasons, this design is still very subjective and hugely dependent on data scientists' expertise. Deep convolution neural networks have recently proven their capacity to perform well in classification and dimension reduction tasks. However, the problem of selecting hyper-parameters is essential for these networks. This is due to the fact that the size of the search space rises exponentially with the number of layers, and the large number of parameters requires extensive calculations and storage, which makes it unsuitable for application in low-capacity devices. In this paper, we present a system based on a genetic method for compressing CNNs to classify radiographic images and detect the possible thoracic anomalies and infections, including the case of COVID-19. This system uses pruning, quantization, and compression approaches to minimize the network complexity of various CNNs while maintaining good accuracy. The suggested technique combines the use of genetic algorithms (GAs) to execute convolutional layer pruning selection criteria. Our suggested system is validated by a series of comparison experiments and tests with regard to relevant state-of-the-art architectures used for thoracic X-ray image classification.

Keywords: Deep CNN compression · Genetic algorithms · Thorax disease · Chest X-ray

ⓒ Springer Nature Switzerland AG 2022
H. Fujita et al. (Eds.): IEA/AIE 2022, LNAI 13343, pp. 112–123, 2022.
https://doi.org/10.1007/978-3-031-08530-7_10

1   Introduction

Due to its visual appearance, Covid 19, an infectious disease caused by Severe Acute Respiratory Syndrome [1], is also called a corona virus [2]. The war against COVID 19 has prompted researchers around the world to explore, understand and innovate new diagnostic and treatment technologies to crown this threat to our generation. In fact, a chest X-ray is one of the most widely used radiological examinations for the identification of a variety of lung illnesses. Actually, numerous X-ray imaging studies are saved and gathered in various image archiving and communication systems in many modern hospitals. An unanswered challenge is how a database holding essential image information may be used to aid data-starved deep learning models in the development of computer-assisted diagnostic systems. There have been few study attempts published in the literature to determine the chest radiograph image view [27]. Deep learning has demonstrated quick and significant improvement in a variety of computer vision problems.

CNN has been the most important benefit of deep learning [3,4,15]. Researchers had to manually construct the features responsible for detection in order to achieve object identification accuracy [16,17,25]. As previously works, CNN is made up of stacked convolution layers followed by a fully connected layer. Each convolutional layer is made up of convolutional filters that are used to identify features. Each filter is made up of stacked 2D kernels in C channels that reflect the image's depth as showing in Fig. 1, the number of channels in an RGB image is 3, and the kernel is a 2D structure that is applied to each input channel pixel to generate the output.

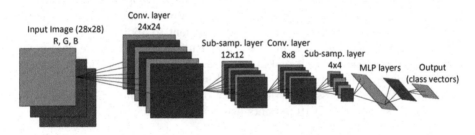

Fig. 1. RGB image; channels number = 3; kernel 2D structure [5]

The convolutional layer over an input image is made up of several planes per channel. A feature map, is the output of each of these planes. The output of the convolutional layer will be in the shape of a dimension (batch size (N), number of channels (Cout), height and width of each output (Hout, Wout)). Convolution weights have four dimensions 4D (F, C, K_H, K_W), with F denoting the number of filters in the layer, C denoting the number of channels, and K_H and K_W denoting the height and width of the kernel.

Different CNN designs have been created and have demonstrated excellent accuracy in image recognition and localization tasks [28], such as AlexNet,

VggNet, and ResNet, the designs are subjected to what is known as "the training process," which is a time-consuming procedure that considers suitable settings for the weights, activation functions, and kernels used.

Deep network compression is one of the most significant strategies for resizing a deep learning model by combining the removal of ineffective components. However, compressing deep models without considerable loss of precision is a key challenge. Recently, Many studies have been focused on discovering new techniques to minimize the computational complexity of CNNs based on EAs while retaining their performance. Evolutionary computation approaches have been effectively applied to deep convolution neural networks for over two decades, but they do not scale well to contemporary deep neural networks owing to complex designs and enormous numbers of connection weights. Although such images might be acquired at a reasonable cost, the scarcity of competent radiologists limits the use of X-Ray imaging technologies.

Motivated by this observation, the proposed approach called **Compression-CNN-Xray**, combines the following CNN Compressing features to establish a trade-off between pruning the convolution layers with acceptable accuracy for categorizing X-ray pictures and recognizing probable thoracic anomalies and infections, including the COVID-19 case.

This work's primary contributions can be summarized as follows:

- Genetic algorithms are used to perform channel pruning of convolutional layers, which affects the structure of CNN.
- Applying quantization for the obtained non-zero weights. Quantization will not only help reduce the number of bits required to store the weights; rather than 32 bits for floating-point weights, an integer representation of 5 bits would be used; however, this will assist to generate frequent duplicate weights, which will aid in making the compression efficient to apply later.
- Compression technique will be used, and Huffman is a suitable choice for lossless compression since it makes use of the statistical characteristics gained after compression to produce a CNN with minumum complexity.
- Examine the usefulness and adaptability of **Compression-CNN-Xray** in the diagnosis of COVID-19 patients using X-ray images.

2   Related Work

2.1   CNN for Xray Images Classification

Many computational approaches for detecting various thoracic illnesses utilizing chest X-ray. Wang et al. [6] developed a weakly supervised multi-label unified classification framework that takes into account diverse DCNN multi-label losses and pooling techniques. To improve classification accuracy, Islam et al. [7] designed a collection of different advanced network topologies. Rajpurkar et al. [8] proved that a common DenseNet architecture can diagnose pneumonia more accurately than radiologists. Yao et al. [9] devised a method that makes better use of statistic label dependencies and therefore improves performance.

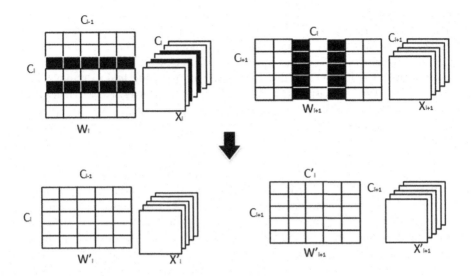

Fig. 2. An illustration of how channel pruning works [29]

Irvin et al. [10] created CheXNet, a deep learning network that uses dense connections and batch normalization to make optimization manageable. Prabira et al. [11] extracted a collection of deep features using nine pre-trained CNN models, which were then passed to the SVM (Support Vector Machines) classifier.

2.2   Channel Pruning

Channel pruning [12] is a different sort of weight pruning approach than neuron pruning. Channel pruning is a notion that refers to decreasing the number of channels in the input supplied to the CNN model's intermediate layers. The data fed into the CNN model is first channelized to produce an appropriate input. For example, an image has three channels (RGB). Each layer of the CNN model's output comprises a variety of channels that boost model efficiency while increasing storage and processing. So, it is desirable to remove unnecessary channels in order to reduce computation and storage needs. Figure 2 illustrates a scenario using channel-pruning.

Recently, various training-based channel pruning techniques [12] have been suggested, which include adding regularization terms to weights during the training stage. Many studies [12, 13] are provided to conduct channel pruning on pretrained models with varied pruning criteria, ignoring the pre-training procedure. Unfortunately, even existing techniques still have a lot of space for improvement when it comes to reducing model redundancy. Furthermore, most research [12, 13] exclusively accelerates networks during the inference stage, with few focusing on off-line pruning efficiency. Following that, various inference-based channel pruning techniques [14] were presented, with the core of these methods defining selection criteria. Li et al. [13] argued that filters with lesser weights usually yield weaker activations and so may be eliminated. However, the criteria may exclude certain useful filters, particularly in shallow layers. Hu et al. [14] assessed the significance of each channel based on its sparsity and eliminated channels with more zero values in their output activations, indicating poor performance at convolution layers. Thinet [12] proposed using a greedy method to remove filters based on statistical information collected from its next layer. However, a greedy method would not be the ideal technique to solve the combinatorial optimization issue, especially if the solution space is very large. Furthermore, its off-line pruning procedure takes a long time since it must traverse the full training sets at each iteration step.

Weight quantization decreases both the storage and computing requirements of the DCNNs model [18–20]. Han et al. [18] suggested a weight quantization approach for compressing deep neural networks by reducing the number of bits needed to encode weight matrices. The authors attempt to decrease the number of weights which should be stored in memory. The identical weights are removed as a result, and numerous connections are derived from a single remaining weight. During inference, the authors utilized integer arithmetic, while during training, they used floating-point operations. Jacob et al. [20] presented a quantization technique based on integer arithmetic for inference. Integer arithmetic is more efficient than floating-point arithmetic and requires fewer bits to represent. Additionally, the authors construct a training step that mitigates the accuracy penalty associated with the conversion of floating-point operations to integer operations. As a result, the suggested technique eliminates the trade-off between on-device latency and accuracy degradation caused by integer operations. The authors performed inference using integer arithmetic and training using floating-point operations. Quantization is a technique that creates an affine mapping between integers Q and real numbers R, i.e., of the type

$$R = W(Q - T) \tag{1}$$

where, Eq. 1 denotes the quantization method with the parameters W and T. For instance, Q is set to 8 for 8-bit quantization. W is an arbitrary positive real number, and T has the same type as variable Q. Existing work [18–20] explore quantization techniques for compressing DNN models. The approaches address model reduction by giving optimum weight matrix configurations.

A **Huffman code** is an efficient prefix code that is widely used for lossless data compression (Van Leeuwen, 1976) [21]. Schmidhuber et al. [22] compressed text files from a neural prediction network using Huffman coding. To encode the quantized weights, Han et al. [23] employed a three-stage compression approach consisting of pruning, quantization, and lastly, Huffman coding [21]. Ge et al. [24] presented a hybrid model compression method using Huffman coding to represent the trimmed weights' sparse nature. Among all variable length prefix codings, Huffman codes are the best. However, Elias and Golomb encoding [26] can benefit from several intriguing features, such as the recurrent occurrence of specific sequences, to obtain superior average code-lengths.

3   Proposed Method

Genetic algorithms imitate the evolution of biological genetic populations, it begins with an initial population that comprises certain chromosomes that are suggested as a solution to the optimization problem, then, using crossover and mutation operators, an offspring population with chromosomes from the new generation is generated, this goes through a fitness function to pick just those individuals who converge on the problem solution. According to the literature, genetic algorithms are coupled with convolutional neural networks to automatically build efficient deep neural networks [27,28]. However, in the current paper, the genetic algorithm is used for a different reason, assisting in the pruning of the convolutional layers in CNN.

The following question inspires our approach:

- How could we design a less complex effective CNN architecture with a minimum Number of convolution hyperparameters for X-Ray images?

Falling on the very extremely high dimensional space, we need to look for the best number of selected channels for CNN architecture to classify X-Ray images and detect the possible thoracic anomalies and infections. We propose an approach named **Compression-CNN-Xray** to discover an efficient evolutionary strategy for the reconstruction of the CNN hyperparameters by searching for the best CNN with the least complexity to identify various thoracic diseases. Figure 3 illustrates our **Compression-CNN-Xray** approach, which will be discussed in the next subsections.

3.1   Compression-CNN-XRAY

Each layer has n_i channels for the i_{th} layer to choose from, while certain layers in AlexNet have 96 channels, which expands to 384 at the output of the CONV-3 layer. So, choosing which channels to prune at each layer to achieve the best solution is a difficult task. This motivates the current approach to use genetic algorithms to perform heuristic search with the goal of discovering a near optimum solution after a sufficient number of iterations. The genetic algorithms for each channel work as follows:

- Only C chromosomes I $\{\theta_i, i = 1, 2, ..., c\}$) are used in the targeted layer of size $L*L*C$, therefore only C possibilities out of the 2^C possibilities are employed.
- Each chromosome M_i is encoded as a C length bit sequence of 0's and 1's, using Bernoulli probability density with mean = p, where p denotes the fraction of channels to be pruned.

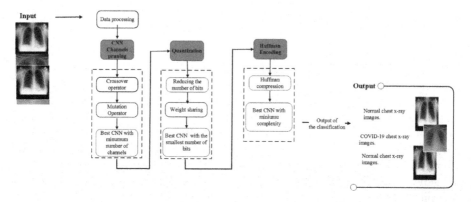

Fig. 3. The working principle of Compression-CNN-XRAY on X-Ray images based on GA.

- Therefore, if there are 3 channels, the possible initializations are $\{1,1,0\}$, $\{0,1,1\}$ and $\{1,0,1\}$, p = 0.666, the first chromosome $\{1,1,0\}$ results during the first kernel channel being pruned while the other two kernels are preserved.
- Crossover operator: We apply the two-point crossover operator to vary the population since it allows us to change all of the chromosomes at the same time. Each parent solution in this procedure is a set of binary strings. Two cutting points are used to make two of each parent. The bits between the cuts are then switched to provide two offspring solutions. Actually, we applied two-point crossover to maintain as much of the chromosome's local structure and solution feasibility as feasible.
- Mutation operator: Similarly, with the crossover operator, the solution is binary string encoding followed by one-point mutation.
- The Fitness Function:
  - The error is represented by the layer wise - error difference, the primary aim is to find a pruned chromosome (Wp) that minimizes the term E(Yp), where the error is determined by an equation.

$$E(Y) = \frac{1}{N} \sum_{i=1}^{N} \|Y - Y_i\| \tag{2}$$

  - However, after applying the Taylor expansion of the above equation and doing some approximation as detailed in Optimal Brain Damage (OBD), the goal function is reduced to a simpler version.

$$\delta E = \frac{1}{2} \delta W^T H \delta W \tag{3}$$

A quantization of 32-bit floating point values into 5-bit integer levels is used to further reduce the stored size of the weights file. The quantization part are spread linearly between Wmin and Wmax because it produces higher accuracy results than density-based quantization, thus even if a weight occurs with a

Table 1. Summary of parameter Settings.

Categories	Parameters	Value
-	Batch size	128
Gradient descent	Epochs	50/350
	SGD learning rate	0.1
	Momentum	0.9
	Weight decay	0.0001
-	# of generation	40
Search strategy	Population size	60
	Crossover propability	0.9
	Mutation propability	0.1

low probability, it may have a high value and therefore a high influence, and if quantized to be less than its real value. This stage produces a compressed sparse row of quantized weights.

Due to the statistical characteristics of the quantization output, Huffman compression might be used to further reduce the weights file. However, this adds the additional hardware needs of a Huffman decompressor and a compressed sparse row to weights matrix converter.

4 Experiments

4.1 Experiment Configuration and Setup

In order to assure the fairness of the comparisons, the parameters of the compared algorithms are set using the frequently used trial and error approach. Table 1 summarizes the parameter settings we utilized in our studies. The Python programming language (version 3.5) is used in the TensorFlow framework for implementation. The experiments done to evaluate the CNN architectures obtained from the test data were carried out on eight Nvidia 2080Ti GPU cards. The holdout validation technique is used to determine the Test error by randomly picking 80% of the data records for training and 20% for testing.

4.2 Results and Discussion

To evaluate the proposed **Compression-CNN-Xray**, different CNN architectures shown in Table 2 were pruned 20 times for Xray image classification. In fact, the results using the Xray image classification datasets are presented, where we provide the best test error and the percentage reduction in the number of FLOPs for each CNN used to classify Xray images. In terms of precision, **Compression-CNN-Xray** can provide better test error. Indeed, channel pruning reduces the flops number of the manual architecture used to classify the

Table 2. Obtained *Test error* Chest X-Ray images values.

Detection method	Reference	Test error	Test error (compression-CNN-Xray	FLOPs ↓
ResNet101	[18]	10.74%	10.19%	6.12%
Inceptionv3		8.92%	8.91%	7.28%
GoogleNet		8.56%	7.85%	11.7%
VGG16		7.24%	6.31%	8.87%
VGG19		7.09%	6.98%	5.7%
XceptionNet		6.08%	5.88%	2.41%
Inceptionresnetv2		6.68%	6.61%	2.28%
AlexNet		6.68%	5.81%	6.14%
DenseNet		6.12%	5.52%	4.17%

Xray images, which reduces the architectural complexity. The compression test error obtained for each of the severals pruned CNNs, ResNet101, Inceptionv3, Google Net, VGG16, VGG19, XceptionNet, AlexNet and DenseNet is presented in Table 2. Based on Table 2, the test error of manual architecture design is lying between 6.08% and 10.74%. Always on Table 2, with a 12% pruned limit for all architectures, ResNet101 has a test error of 10.74% and a 10.19% by using **Compression-CNN-Xray**, and achieving a 6.12% decrease in the number of Flops, Inceptionv3 achieves a 7.28% decrease in the number of Flops with a test error of 8.92% and an 8.91% by using **Compression-CNN-Xray**, GoogleNet reduces the number of Flops by 11.7% by using **Compression-CNN-Xray**, with a test error of 8.56% and a 7.85%. After employing **Compression-CNN-Xray**, VGG16 and VGG19 provide [6.31%–6.98%], respectively, with a decrease of 8.81% and 5.7% in the number of Flops. Always on Table 2, XceptionNet provides 6.08% with a 2.41% reduction in the number of flops via **Compression-CNN-Xray**, AlexNet provides 5.81% with a 6.14% reduction in the number of flops via **Compression-CNN-Xray**, and DenseNet provides 6.12% with a 4.17% reduction in the number of flops via **Compression-CNN-Xray**.

According to the results in Table 2, the proposed **Compression-CNN-Xray** can reduce the number of FLOPs in CNN architectures while maintaining good test errors. In terms of test error, **Compression-CNN-Xray** provides slightly better performance and accomplishes a reduction in design complexity by minimizing the number of flops. However, the DCNNs must be trained with strong and specific regularization in their method. Only the DCNN classification performance and computational complexity information are used to purne it. Thus, even without any regularization or other knowledge about the DCNN being pruned, **Compression-CNN-Xray** is capable of producing high-quality results. Furthermore, we discovered that ResNet is more difficult to compress than VGGNet. ResNet has less redundancy than VGGNet since its bottleneck structure prevents some layers from becoming pruned. In fact, there has been no attempt in the literature to prune DenseNet designs. Despite the fact that DenseNets already have lower computational complexity and greater classification performance than ordinary CNNs, Our results demonstrated that they may still be pruned while maintaining competitive classification performance.

To sum up, the proposed **Compression-CNN-Xray** algorithm can reduce the number of FLOPs in DCNN architectures significantly while preserving good test errors. To the best of our knowledge, **Compression-CNN-Xray** is the first EAs-based system in the literature capable of reducing manual architectural complexity for detecting COVID-19 infection.

5   Conclusion

Designing an adequate Deep CNN architecture is still a very interesting, challenging, and current topic. Many researchers have recently become interested in exploring novel approaches to lower the computational complexity of CNNs based on EAs. Following the channel pruning works the removal of the number of channels can also play a critical role in reducing the computational requirements of the DCNN model but it can fall into a very high dimensional space. In this work, We propose **Compression-CNN-Xray**, which incorporates the following steps into the compressing characteristics to achieve an acceptable trade-off between pruning the convolution layers and identifying potential thoracic abnormalities and infections, including the COVID-19 instance. This Pruning step employs genetic algorithms to minimize the absolute error difference fitness function that has been derived and approximated. The neurons are then pruned, which considerably reduces the size of the weights files. This is followed by quantization and various compression approaches to produce a final model that is faster in inference and less in size. Our approach achieves significant size reduction while taking a fair amount of time to create the pruning model, but even so, the test error decreases slightly. This decrease in test error could be considered acceptable in relation to the high compression obtained.

Future work, We conduct the pruning research process more effectively and efficiently, since pure channel pruning can lead to very large and very large space. By analyzing certain created structures and examining their common models, we are pursuing the pruning research process objective of providing an interaction model that allows the user to connect with filters and channels through an evolutionary process.

Acknowledgements. This project was supported by the Deanship of Scientific Research at Prince Sattam bin Abdulaziz University through the research project No. 2022/01/19780.

References

1. Paules, C.I., Marston, H.D., Fauci, A.S.: Coronavirus infections-more than just the common cold. JAMA **323**(8), 707–708 (2020)

2. Chen, Y., Liu, Q., Guo, D.: Emerging coronaviruses: genome structure, replication, and pathogenesis. J. Med. Virol. **92**(4), 418–423 (2020)

3. Louati, A., Louati, H., Li, Z.: Deep learning and case-based reasoning for predictive and adaptive traffic emergency management. J. Supercomput. **77**(5), 4389–4418 (2020). https://doi.org/10.1007/s11227-020-03435-3

4. Louati, A., Louati, H., Nusir, M., hardjono, B.: Multi-agent deep neural networks coupled with LQF-MWM algorithm for traffic control and emergency vehicles guidance. J. Ambient. Intell. Humaniz. Comput. **11**(11), 5611–5627 (2020). https://doi.org/10.1007/s12652-020-01921-3

5. Kiranyaz, S., Ince, T., Gabbouj, M.: Real-time patient-specific ECG classification by 1-D convolutional neural networks. IEEE Trans. Biomed. Eng. **63**, 664–675 (2016). https://doi.org/10.1109/TBME.2015.2468589

6. Wang, X., Peng, Y., Lu, L., Lu, Z., Bagheri, M., Summers, R.M.: ChestX-ray8: hospital-scale chest x-ray database and benchmarks on weakly-supervised classification and localization of common thorax diseases. In: IEEE Conference on Computer Vision and Pattern Recognition, pp. 3462–3471 (2017)

7. Islam, M.T., Aowal, M.A., Minhaz, A.T., Ashraf, K.: Abnormality detection and localization in chest X-rays using deep convolutional neural networks, CoRR, vol. abs/1705.09850 (2017)

8. Rajpurkar, P., et al.: Deep learning for chest radiograph diagnosis: a retrospective comparison of the CheXNeXt algorithm to practicing radiologists. PLoS Med. **15**(11), 1–17 (2018)

9. Yao, L., Poblenz, E., Dagunts, D., Covington, B., Bernard, D., Lyman, K.: Learning to diagnose from scratch by exploiting dependencies among labels. CoRR, vol. abs/1710.1050 (2017)

10. Irvin, J., et al.: Chexpert: a large chest radiograph dataset with uncertainty labels and expert comparison. In: Thirty-Third AAAI Conference on Artificial Intelligence, pp. 590–597 (2019)

11. Sethy, P.K., Behera, S.K.: Detection of coronavirus disease (Covid-19) based on deep features. Int. J. Math. Eng. Manag. Sci. **5**(4), 643–651 (2020)

12. Luo, J., Wu, J., Lin, W.: Thinet: a filter level pruning method for deep neural network compression, arXiv preprint arXiv: 1707.06342 (2017)

13. Liu, Z., Li, J., Shen, Z., Huang, G., Yan, S., Zhang, C.: Learning efficient convolutional networks through network slimming. In: International Conference on Computer Vision (ICCV), pp. 2755–2763 (2017)

14. Hu, H., Peng, R., Tai, Y., Tang, C.: Network trimming: a datadriven neuron pruning approach towards efficient deep architectures, arXiv preprint arXiv:1607.03250 (2016)

15. Louati, A.: A hybridization of deep learning techniques to predict and control traffic disturbances. Artif. Intell. Rev. **53**(8), 5675–5704 (2020). https://doi.org/10.1007/s10462-020-09831-8

16. Louati, A., Lahyani, R., Aldaej, A., Mellouli, R., Nusir, M.: Mixed integer linear programming models to solve a real-life vehicle routing problem with pickup and delivery. Appl. Sci. **11**(20), 9551 (2021). https://doi.org/10.3390/app11209551

17. Louati, A., Lahyani, R., Aldaej, A., Aldumaykhi, A., Otai, S.: Price forecasting for real estate using machine learning: a case study on Riyadh city. Concurr. Computa. Pract. Exp. **34**(6), e6748 (2022). https://doi.org/10.1002/cpe.6748

18. Han, S., Mao, H., Dally, W.J.: Deep compression: compressing deep neural networks with pruning, trained quantization and Huffman coding. arXiv preprint arXiv:1510.00149 (2015)

19. Chauhan, J., Rajasegaran, J., Seneviratne, S., Misra, A., Seneviratne, A., Lee, Y.: Performance characterization of deep learning models for breathing-based authentication on resource-constrained devices. In: Proceedings of IMWUT, vol. 2, no. 4, pp. 1–24 (2018)
20. Jacob, B., et al.: Quantization and training of neural networks for efficient integer-arithmetic-only inference. In: Proceedings of CVPR, pp. 2704–2713 (2018)
21. Han, S., Mao, H., Dally, W.J.: Deep compression: compressing deep neural networks with pruning, trained quantization and Huffman coding. In: ICLR (2016b)
22. Schmidhuber, J., Heil, S.: Predictive coding with neural nets: application to text compression. In: NeurIPS, pp. 1047–1054 (1995)
23. Han, S., Mao, H., Dally, W.J.: Deep compression: compressing deep neural network with pruning, trained quantization and Huffman coding. In: 4th International Conference on Learning Representations, ICLR 2016, San Juan, Puerto Rico, 2–4 May 2016, Conference Track Proceedings (2016)
24. Ge, S., Luo, Z., Zhao, S., Jin, X., Zhang, X.-Y.: Compressing deep neural networks for efficient visual inference. In: IEEE International Conference on Multimedia and Expo (ICME), pp. 667–672. IEEE (2017)
25. Bechikh, S., Said, L.B., Ghédira, K.: Negotiating decision makers' reference points for group preference-based Evolutionary multi-objective Optimization. In: 11th International Conference on Hybrid Intelligent Systems (HIS), pp. 377–382 (2011). https://doi.org/10.1109/HIS.2011.6122135
26. Gallager, R., van Voorhis, D.: Optimal source codes for geometrically distributed integer alphabets (Corresp.). IEEE Trans. Inf. Theory 21(2), 228–230 (1975). https://doi.org/10.1109/TIT.1975.1055357
27. Louati, H., Bechikh, S., Louati, A., Aldaej, A., Said, L.B.: Evolutionary optimization of convolutional neural network architecture design for thoracic xray image classification. In: IEA/AIE, vol. 32, no. 1 (2021)
28. Louati, H., Bechikh, S., Louati, A., Hung, C.-C., Said, L.B.: Deep convolutional neural network architecture design as a bi-level optimization problem. 655 Neurocomput. 439, 44–62 (2021)
29. Rahul, M., Gupta, H.P., Dutta, T.: A survey on deep neural network compression: challenges, overview, and solutions. arXiv, arXiv:2010.03954 (2020)

An Oriented Attention Model
for Infectious Disease Cases Prediction

Peisong Zhang[1]📧, Zhijin Wang[2](✉)📧, Guoqing Chao[3]📧, Yaohui Huang[4]📧,
and Jingwen Yan[5]📧

[1] School of Science, Jimei University, Yinjiang Road 185, Xiamen 361021, China
[2] Computer Engineering College, Jimei University,
Yinjiang Road 185, Xiamen 361021, China
zhijinecnu@gmail.com
[3] School of Computer Science and Technology, Harbin Institute of Technology,
2 West Culture Road, Weihai 264209, People's Republic of China
guoqingchao@hit.edu.cn
[4] College of Electronic Information, Guangxi University for Nationalities,
Daxue East Road 188, Nanning 530006, China
[5] College of Engineering, Shantou University,
University Road 243, Shantou 515063, China
jwyan@stu.edu.cn

Abstract. Effective infectious disease prediction supports the success
of infection prevention and control. Several attention-based predictive
models can be applied to undertake the prediction task. However, using
a single attention mechanism can only capture local information, i.e.
part of the temporal dynamics from time series. In this paper, we take
for the hypothesis that using multiple attention from different aspects
could improve prediction accuracy. An oriented attention model (OAM)
is proposed to draw temporal dynamics in several aspects, via oriented
attention units and their aggregation. Firstly, time series are represented
as oriented transformations. And then those representations are consolidated to connect with outputs. Intensive experiments on two real
infectious disease datasets show OAM's effectiveness.

Keywords: Infectious disease · Prediction · Oriented attention ·
Aggregation · Time series

1 Introduction

Infectious disease is always a health problem for human beings. For example,
about one million children are newly infected by hand, foot, and mouth disease
(HFMD), and more than one million people are newly infected by hepatitis beta

Supported in part by the Natural Science Foundation of Fujian Province (CN) (no.
2021J01859) and the Innovation School Project of Guangdong Province (CN) (no.
2017KCXTD015).

H. Fujita et al. (Eds.): IEA/AIE 2022, LNAI 13343, pp. 124–136, 2022.
https://doi.org/10.1007/978-3-031-08530-7_11

virus (HBV) every year in China [1]. The early warning system is conducive to managing the risk of infectious disease [18]. The predictive technique is the most critical part of this system [17], which supports the decision-making in healthcare and intervention strategies [10,13].

The infection cases prediction is commonly regarded as a time series prediction problem. Abundant methods have been used to predict different kinds of epidemics [6]. Owning to the success of attention mechanism [12] in time series prediction [9], it had been introduced to predicting infectious disease [22]. These attention mechanisms provide abilities in focusing on some important factors among different time intervals. In reality, temporal dynamics of infectious disease time series are usually complicated and changeable [15]. Hence, a single attention mechanism is commonly captured incomplete temporal dynamics. Meanwhile, most of these attention-based time series prediction methods are based on a single attention mechanism.

We take for the hypothesis that using multiple attention from different aspects could improve prediction accuracy. The goal is to investigate an attention representation and fusion model, which provides the ability to capture temporal dynamics of infectious disease time series from several aspects. There are two challenges as follows: (1) how to represent inputted time series in different aspects by using several attentions? (2) how to fuse those represented attentions?

An oriented attention model (OAM) is proposed to overcome the two challenges. Firstly, the inputted time series is normalized and split into a data cube of look-back windows. Secondly, an oriented attention unit (OAU) is exploited to represent the cube in four orientations. The OAU is designed to highlight information from past observations in several aspects. Finally, the oriented attentions are fused to connect with upcoming values. To evaluate the effectiveness of the proposed OAM, intensive experiments were conducted on HFMD and HBV datasets.

The main contributions of this paper are summarized as follows:

(1) The attention mechanism on time series can be represented in four aspects, other than one aspect before.
(2) Moreover, the feasibility of the fusion of different attentions has been validated.
(3) The proposed OAM significantly outperforms the state-of-the-art methods on two real datasets.

The rest of this paper is organized as follows. Section 2 reviews several relevant work. Section 3 defines the research problem. Section 4 illustrates the proposed OAM. Section 5 presents experiments and analyses. Finally, a conclusion is given in Sect. 6.

2   Related Work

The relevant researches are addressed at attention-based time series prediction methods [5,9] and temporal fusion methods [7,13,14,16].

Attention-Based Time Series Prediction Methods. These methods added attention components in the time series representation processes. These attentions provided the ability to strengthen or weaken observations within a time interval. These attentions have been reviewed in [3]. The major part of the attention mechanism is its score function, and they are divided into multiplicative function, additive function, and multiple layer perceptron (MLP). Besides, the multiplicative function consists of dot function, scaled dot function, and general form.

The attention mechanism was first proposed in [12]. The attention-based time series prediction was first applied to stock price prediction in [9], which captured the attentions in recurrent neural networks (RNN) from both NASDAQ index time series and the top 80 stock prices time series of NASDAQ. The stacked attentions were added after convolution neural networks (CNN) to represent temporal features in multivariate time series forecasting [5]. In summary, these attentions are used to highlight the observations of time intervals from inputted time series. However, this research believes the attentions are not only limited to the dimensionality of time steps. Technically, the dimensionality of observations should be inputted into the attention mechanism as well.

Temporal Fusion Methods. The temporal fusion layer is used to connect represented inputs with target values, a.k.a., model outputs. The temporal fusion problem can be regarded as the problem of mapping several input values to a target value. Methods in this category can be divided into additive operation [7], linear mapping and non-linear mapping.

To the best of our knowledge, the attention fusion from different aspects had not been considered in previous research. This research fuses attention into a temporal fusion layer to learn the temporal dynamics and make predictions.

3   Problem Definition

Time Series. A time series is used to denote an ordered observed sequence of outpatient cases. Let K be the length of a time series. The weekly infectious outpatient cases are denoted by $\boldsymbol{Z}$, where $\boldsymbol{Z} \in \mathbb{R}^{K \times 1}$.

Look-Back Window. A look-back window is used to describe observations in several consecutive time intervals. Let T be the window size. A look-back window is denoted by symbol $\boldsymbol{Z}_{t+1:t+T,1} \in \mathbb{R}^{T \times 1}$.

Time Series Prediction. Commonly, the time series prediction problem is formulated as:

$$\hat{Y}_{T+1,1} = F(\boldsymbol{Z}_{t+1:t+T,1}), \tag{1}$$

where $\hat{Y}_{T+1,1} \in \mathbb{R}$ is the model output, $\boldsymbol{Z}_{t+1:t+T,1}$ is the model input, and $F(\cdot)$ is a mapping.

Attention-Based Time Series Prediction. The problem of attention-based time series prediction problem is formulated as:

$$\hat{Y}_{T+1,1} = F(A(\boldsymbol{Z}_{t+1:t+T,1})), \tag{2}$$

where $A(\cdot)$ is an attention mechanism. It should be noted that the model inputs may be processed using neural networks before feeding into the attention component.

Oriented Attention-Based Time Series Prediction. The problem of oriented attention-based time series prediction problem is formulated as:

$$\hat{Y}_{T+1,1} = F(A^1(\boldsymbol{Z}_{t+1:t+T,1}), A^2(\boldsymbol{Z}_{t+1:t+T,1}), \cdots), \tag{3}$$

where $A^i(\cdot)$ is the oriented attention mechanism. It tells the oriented attention on the window size dimension and the time series dimension. The length of the time series dimension is 1.

The two dimensions for oriented attentions may be so limited. Hence utilizing attention on the dimension of predictive values is taken into consideration. Let B be the number of time consecutive batched inputs. Let $\boldsymbol{X} \in \mathbb{R}^{B \times T \times 1}$ be the time consecutive batched model inputs, and let $\boldsymbol{Y} \in \mathbb{R}^{B \times 1}$ be the outputs. Therefore, Eq. 3 is re-formulated as:

$$\hat{\boldsymbol{Y}} = F(A^1(\boldsymbol{X}), A^2(\boldsymbol{X}), \cdots). \tag{4}$$

According to the three dimensions, the oriented attentions are extended in several aspects. The main symbols are listed in Table 1.

Table 1. Symbols and meanings.

Symbol	Semantic
K	Time step number
T	Look-back window size
$\boldsymbol{Z}$	The time series of outpatient cases, $\boldsymbol{Z} \in \mathbb{R}^{K \times 1}$
B	The number of time consecutive batched inputs
$\boldsymbol{X}$	Batched input tensor $\boldsymbol{X} \in \mathbb{R}^{B \times T \times 1}$
$\boldsymbol{Y}$	Batched output tensor $\boldsymbol{Y} \in \mathbb{R}^{B \times T \times 1}$
$F(\cdot)$	A mapping
$A(\cdot), A^i(\cdot)$	Attention and oriented attention, respectively
S	The number of aspects for oriented attention
$\boldsymbol{M}^l$	The left mapped tensor of OAU
$\boldsymbol{M}^n$	The normalized tensor of OAU
$\boldsymbol{M}^r$	The right mapped tensor OAU
$\boldsymbol{P}$	Concatenated representation

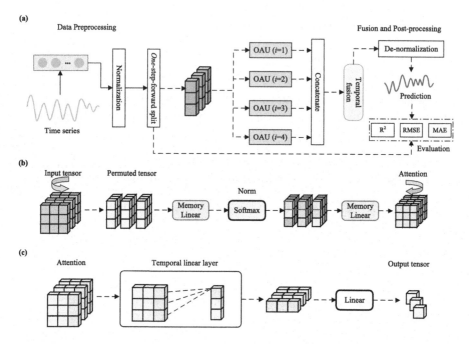

Fig. 1. The schematic illustration of the proposed oriented attention model (OAM). (a) The workflow. (b) Orientable attention unit (OAU). (c) Temporal fusion layer.

4   The Proposed OAM

The schematic illustration of the proposed OAM is displayed in Fig. 1. According to the workflow in Fig. 1(a), this model consists of three stages. The first stage is data processing. The inputted time series are normalized using Min-Max normalization [17]. There are some reasons for the normalization, such as computational speed and the computability of some neural networks. The normalized time series are transformed to supervised data using "*one*-step-forward split" [18]. The second stage is the generation of oriented attentions and their combinations. Those combinations are fed into a temporal attention layer, and model outputs are de-normalized [17] in the third stage.

4.1   Oriented Attention Unit (OAU)

The detailed process of the oriented attention unit (OAU) is plotted in Fig. 1(b). To capture the attention from several aspects and reduce the model complexity, the aspect selection component and the attention representation component are designed in OAU.

There are many ways to view the input tensor in different aspects. A simple way is to transpose the input tensor, which guarantees the attention mechanism can work well with two dimensions of it.

The *aspect selection component* views the input tensor by transposing its dimensions. Since the batch dimension may change, when the model optimizer changes. Hence, the batch dimension can not be put in the feature dimension. For the three dimensions of an input tensor, we have and only have four aspects.

Given an input tensor $\boldsymbol{X} \in \mathbb{R}^{B \times T \times 1}$, the four aspects of on it can be formulated as follows:

$$
\boldsymbol{X}^i = \begin{cases} \boldsymbol{X}, & i = 1 \\ \boldsymbol{X}.permute(0, 2, 1), & i = 2 \\ \boldsymbol{X}.permute(2, 0, 1), & i = 3 \\ \boldsymbol{X}.permute(1, 0, 2), & i = 4 \end{cases} \tag{5}
$$

where $\boldsymbol{X}^1 \in \mathbb{R}^{B \times T \times 1}$, $\boldsymbol{X}^2 \in \mathbb{R}^{B \times 1 \times T}$, $\boldsymbol{X}^3 \in \mathbb{R}^{1 \times B \times T}$, and $\boldsymbol{X}^4 \in \mathbb{R}^{T \times B \times 1}$ are the four kinds of aspects, respectively. The $permute(\cdot)$ function operates views on the input tensor by transposing dimensions.

The oriented attention mechanism works on the last two dimensions of an input tensor. For example, $\boldsymbol{X}^1$ is adopted to distinguish the impact of different time intervals by mapping values within a time interval. Technically, $\boldsymbol{X}^2$ is exploited to distinguish the impact of different time series by mapping the lock-back window of a time series. In this paper, only one target time series is considered. Especially, $\boldsymbol{X}^3$ is used to distinguish the impact of different instances of a batch by mapping values within a time interval. $\boldsymbol{X}^4$ is used to distinguish the impact of different instances of a batch by comparing different time series.

The *attention representation component* consists of the symmetric structure of two linear layers and normalized layers among them. To guarantee that the attention mechanism has a strong ability in learning different aspects and the fusion of multiple oriented attentions, the symmetric structure is employed. The detailed structure is drawn in Fig. 1(b).

Since OAU works with the last two dimensions of the input tensor, one aspect of the attention can be an example for other aspects. Taking the attention on X^2 for example, it is first mapped using the batch matrix multiplication. The mapping process is formulated below:

$$
\boldsymbol{M}^l = \boldsymbol{X}^2 \cdot \boldsymbol{W}^l, \tag{6}
$$

where $\boldsymbol{M}^l \in \mathbb{R}^{B \times 1 \times T}$ is the left mapped tensor, and $\boldsymbol{W}^l \in \mathbb{R}^{T \times T}$ is the linear weights.

And then a normalized layer is used to enlarge the differences of look-back windows. The normalized layer adopts the softmax operation and is formulated as below:

$$
M_{b,i,t}^n = \frac{\exp(M_{b,i,t}^l)}{\sum_{t=1}^T \exp(M_{b,i,t}^l)}, \tag{7}
$$

where $\boldsymbol{M}^n \in \mathbb{R}^{B \times 1 \times T}$ is the normalized tensor. The reason for using softmax to normalize is that it can enlarge the differences of a target dimension by considering its attributions.

The normalized tensor is mapped to generate attentions. It's formulated as below:

$$M^r = M^n \cdot W^r, \tag{8}$$

where $M^r \in \mathbb{R}^{B \times 1 \times T}$ is the right mapped tensor, and $W^r \in \mathbb{R}^{T \times T}$ is the linear weights. The attention $A^2 \in \mathbb{R}^{B \times 1 \times T}$ on aspect $s = 2$ is obtained by transposing the dimensions in M^r.

The oriented attention of four aspects are obtained using Eqs. 6–8. They can be denoted by symbols A^1, A^2, A^3, A^4.

4.2   Temporal Fusion Layer

The scheme of the temporal fusion layer is plotted in Fig. 1(c). The temporal fusion layer is used to aggregate several oriented attention, as well as inputs.

The four oriented attention and the input are concatenated as follows:

$$P = [A^1; A^2; A^3; A^4; X], \tag{9}$$

where $P \in \mathbb{R}^{B \times T \times 5}$ is the concatenated representations, and $[;]$ denotes the concatenation operation.

Several existing methods can map the representations to predictions $O \in \mathbb{R}^{B \times 1}$, such as global auto-regression (GAR), auto-regression (AR), and multiple linear regression (MLR). GAR is adopted to do the mapping since it's simple. The predictions O are de-normalized to compare with real values while training.

5   Experiments

5.1   Settings

Datasets. 49677 and 48359 real-world HFMD and HB outpatient records were collected and used to evaluate the proposed OAM and baseline methods. In the data collecting stage, the HB outpatient cases were reported when the transaminase exceeds the twice standard. The basic statistics of the two datasets are shown in Table 2. Consequently, there are certain errors in the outpatient case data of HB. The real-world HFMD and HB outpatient cases are shared by the Xiamen Center for Disease Control and Prevention (XCDC). The outpatient cases are divided into two parts, the first part from January 5, 2015 to December 23, 2019 is used to train the models; the remains were used to validate those trained models.

Baselines. To study the benefits of attention, several models have been developed and applied to the two real datasets. To study the benefits of oriented attentions, these models have been extended with self-attentions. Hence, those benefits can be observed by measuring the prediction performance of models.

Table 2. The basic description on HFMD and HB datasets. "STD" denotes standard deviation.

Dataset	Training size	Test size	Maximum	Average	Minimum	STD
Weekly HFMD cases	259	53	869	159.22	0	159.84
Weekly HB cases	259	53	282	154.99	14	38.98

The comparable models are listed as follows:

(1) Auto-Regression (AR) [8] generated predictions by mapping linear relationships between past observations and coming values.
(2) Long- and Short-Term Memory (LSTM) [4] exploited three gate units to capture the long-term dependency.
(3) Gated Recurrent Unit (GRU) [17] merged the hidden states and cell states of LSTM to reduce parameters.
(4) Encoder-Decoder (ED) [15] consisted of two RNN components in the encoder stage and decoder stage, respectively.
(5) Convolutional Neural Network (CNN) [2] used a CNN layer to extract temporal patterns and a linear layer to connect with outputs.
(6) CNNRNN [20] extracted the local sequential patterns to generate predictions, by leveraging a connected CNN and RNN network.
(7) Self-attention [11] is integrated into the above methods. And then we have LSTM-attn, GRU-attn, ED-attn, CNN-attn and CNNRNN-attn.

Model Configurations. For a fair competition, all the training-related constant parameters are set to the same values on a dataset. To reduce the complexity in presenting the experimental results, some training-related and data-related parameters are fixed according to cross-validations on parameters. For all experiments, the learning rate is set to 0.0015.

For HFMD and HB, according to the experiments on all the comparable methods, the optimal values of batch size B and look-back window size T are found at 4 and 2, respectively. Hence, $B = 4$ and $T = 2$ are both fixed in all the comparable experiments. For all the RNN-involved models, their hidden neuron sizes are both fixed at $\{32, 64\}$ to observe their prediction performance.

Performance Metrics. We follow the metrics in [17,19,21]. Those metrics are mean absolute error (MAE), root mean square error ($RMSE$), and correlation coefficient (R^2). For MAE and RMSE, the lower value has better performance. For R^2, the higher value has better performance.

5.2   Study on Attention Combinations

To study how oriented attention affect the prediction performance, the experiments on OAU combinations are conducted. The experimental results on their combinations are plotted in Fig. 2.

There are many situations of OAU combinations. To avoid searching for all situations, the greedy search is applied. Every aspect is done to get their corresponding performance, and the optimal aspect is selected to combine with every other aspect. Iteratively, the current optimal combination is selected to add every other aspect, until all the aspects are included.

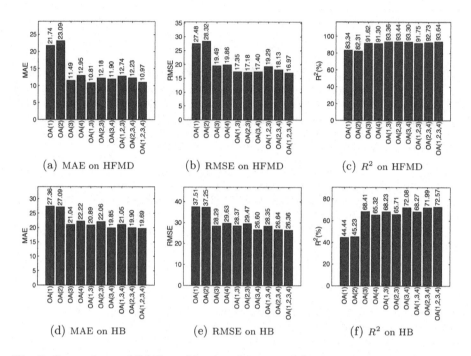

Fig. 2. Comparisons of OAU combinations in terms of three metrics on two datasets.

For easy presentation, let symbol OA(i) be the observation in aspect set $\{i\}$. For example, the optimal single aspect attention on the HFMD dataset is found at OA(3), and the experiments are done to measure what aspect works best with aspect $i = 3$. This process is recursive to include all four aspects, i.e., OA(1, 2, 3, 4).

The major observations from Fig. 2 are listed as below:

(1) For the HFMD dataset, see Figs. 2(a)–(c), the OA(1, 2, 3, 4) has the optimal RMSE and R^2 values, and OA(1, 3) achieves the optimal MAE value.
(2) For the HB dataset, see Figs. 2(d)–(f), the OA(3, 4) achieves the best performance in terms of the three metrics. The performance degrades when adding more OAUs.
(3) When observing at a single OAU, OA(3) achieves the best performance, and OA(4) gets the second-best performance.
(4) OA(3) and OA(4) play two important roles in reducing the prediction errors.
(5) OA(1) and OA(2) have slight effects on improving the performance.

For the two datasets, the OA$(1, 2, 3, 4)$ obtains optimal values in terms of RMSE and R^2. These results demonstrate that the fusion of attention mechanisms is effective. Whereas the OA$(1, 3)$ achieves the optimal MAE value on the HFMD dataset, the potential reason is that the HFMD and HB dataset both are uni-variate time series. Hence the effectiveness of the attention mechanism on observation in aspects $\{2, 4\}$ becomes weak.

As shown in Fig. 2, OA(3) or OA(4) significantly outperform OA(1) or OA(2). The possible reason is that OA(3) and OA(4) both can distinguish the impact of different instances of a batch and highlight the key instances among them, which assist the model capture the trends of time dynamics.

5.3   Performance Comparisons

Table 3. Comparable results of twenty-two methods on the two datasets in terms of three metrics.

Model	HFMD			HB		
	MAE	RMSE	R^2	MAE	RMSE	R^2
AR	22.6962	28.0246	0.8268	27.0550	37.1552	0.4550
LSTM-32	15.8524	25.6851	0.8545	26.1595	37.6786	0.4396
LSTM-attn-32	14.2515	24.5761	0.8668	23.3605	32.0211	0.5952
LSTM-64	15.9730	25.6855	0.8545	26.6344	37.7522	0.4374
LSTM-attn-64	13.5908	24.3641	0.8691	23.2384	32.0197	0.5953
GRU-32	19.0589	26.3326	0.8471	26.1140	36.8747	0.4632
GRU-attn-32	13.6181	24.3916	0.8688	22.7751	31.1019	0.6181
GRU-64	19.5851	26.5899	0.8441	26.1305	36.9358	0.4615
GRU-attn-64	13.2111	24.2959	0.8698	22.8632	31.1539	0.6169
ED-32	18.5562	26.2696	0.8478	26.2258	37.3032	0.4507
ED-attn-32	15.1897	25.4855	0.8568	23.1548	31.4827	0.6087
ED-64	17.4964	25.8206	0.8530	25.6824	37.0903	0.4569
ED-attn-64	14.8598	25.4567	0.8571	22.6006	30.6012	0.6300
CNN-32	20.9972	27.1071	0.8379	26.4590	37.2826	0.4513
CNN-attn-32	16.1284	25.9156	0.8519	26.0703	37.0805	0.4572
CNN-64	21.0921	27.4980	0.8332	26.0738	37.3023	0.4507
CNN-attn-64	16.6176	25.8872	0.8522	26.0183	37.0168	0.4591
CNNRNN-32	19.5186	26.6080	0.8439	27.1456	36.5898	0.4715
CNNRNN-attn-32	16.1384	25.8667	0.8524	25.8596	37.1470	0.4553
CNNRNN-64	18.7608	26.3040	0.8474	26.1287	36.4199	0.4764
CNNRNN-attn-64	15.4331	25.6848	0.8545	26.0657	37.2316	0.4528
OAM	**10.9700**	**16.9735**	**0.9364**	**21.7509**	**28.9388**	**0.6692**

The performance comparison is done to validate the effectiveness of OAM. The comparable results of twenty-two methods are shown in Table 3. For all the methods, parameter batch size B is fixed at 4. The parameter look-back window size T is set to 2 and 4 for the HFMD dataset and the HB dataset, respectively.

The main results are summarized as follows:

(1) OAM achieves the best performance in terms of three metrics on the two datasets. The improvement is significant.
(2) Self-attention improves the performance for all the benchmark methods.
(3) The number of hidden neurons in RNN-involved models does not affect the performance.
(4) For the HFMD dataset, GRU-attn-64 has the second-best performance.
(5) For the HB dataset, ED-attn-64 has the second-best performance.
(6) The AR has the worst performance on the two datasets.

According to the results on the HFMD dataset in Table 3, AR had the worst prediction performance, which indicated that the linear-based methods are unable to completely fit the transmission patterns of diseases. The self-attention significantly improves the predictive accuracy of all comparable methods. GRU-attn-64 has the second-best performance, but the hidden neurons slightly affect the results. A possible reason is that increasing the hidden size of the RNN-involved models can obtain a richer representation, but the accuracy is limited by the increment of the information.

According to results on the HB dataset in Table 3, the performance of CNN-involved methods was degrading when self-attention was added. A possible reason was that HB patterns are more complicated than HFMD in terms of trans-missions. CNN leads to information loss, which reduces the predictive accuracy. The performance of RNN-involved methods worked not well on the HB dataset.

The proposed OAM extracts the temporal dynamics from inputted time series by aggregating several oriented attention. Compare to other baseline methods, the proposed OAM achieved significant improvements. The MAE and RMSE values were decreased by 51.67% and 39.43% at most, respectively. The R^2 is increasing by 52.99% at most. This revealed that the fusion of oriented attention is feasible.

6   Conclusions

This paper proposed the oriented attention model (OAM) to predict the number of outpatient cases in the upcoming week, via fusing several oriented attentions. Intensive experiments on two real-world HFMD and HB data show the effectiveness of the proposed method. The attention mechanism had improved the prediction performance of several traditional methods. Moreover, the combination of oriented attention improved the performance of the single attention mechanism. This also reveals the feasibility of attention fusion and its essential in time series prediction.

In the future, the oriented attention mechanism will be further applied to multivariate time series prediction. The normalization methods, multi-instance learning will be further discussed with the attention mechanism as well.

References

1. Overview of the national epidemic situation of notifiable infectious diseases in 2020 (2022). http://www.nhc.gov.cn/jkj/s3578/202103/f1a448b7df7d4760976fea6d55834966.shtml. Accessed Jan 2022

2. Hoseinzade, E., Haratizadeh, S.: CNNPRED: CNN-based stock market prediction using a diverse set of variables. Expert Syst. Appl. **129**, 273–285 (2019)

3. Hu, D.: An introductory survey on attention mechanisms in NLP problems. In: INTELLISYS 2019, vol. 1038, pp. 432–448, September 2019

4. Hua, Y., Zhao, Z., Li, R., Chen, X., Liu, Z., Zhang, H.: Deep learning with long short-term memory for time series prediction. IEEE Commun. Mag. **57**(6), 114–119 (2019)

5. Huang, S., Wang, D., Wu, X., Tang, A.: DSANet: dual self-attention network for multivariate time series forecasting. In: CIKM 2019, pp. 2129–2132, November 2019

6. Keddy, K.H., et al.: Using big data and mobile health to manage diarrhoea disease in children in low-income and middle-income countries: societal barriers and ethical implications. Lancet Infect. Dis. (2021)

7. Lai, G., Chang, W., Yang, Y., Liu, H.: Modeling long- and short-term temporal patterns with deep neural networks. In: SIGIR 2018, pp. 95–104 (2018)

8. Mabrouk, A.B., Abdallah, N.B., Dhifaoui, Z.: Wavelet decomposition and autoregressive model for time series prediction. Appl. Math. Comput. **199**(1), 334–340 (2008)

9. Qin, Y., Song, D., Chen, H., Cheng, W., Jiang, G., Cottrell, G.W.: A dual-stage attention-based recurrent neural network for time series prediction. In: IJCAI 2017, pp. 2627–2633 (2017)

10. Shah, W., et al.: A machine-learning-based system for prediction of cardiovascular and chronic respiratory diseases. J. Healthc. Eng. (2021)

11. Shih, S., Sun, F., Lee, H.: Temporal pattern attention for multivariate time series forecasting. Mach. Learn. **108**(8–9), 1421–1441 (2019)

12. Vaswani, A., et al.: Attention is all you need. CoRR abs/1706.03762 (2017)

13. Wang, Y., Gu, J., Zhou, Z., Wang, Z.: Diarrhoea outpatient visits prediction based on time series decomposition and multi-local predictor fusion. Knowl.-Based Syst. **88**, 12–23 (2015)

14. Wang, Y., Li, J., Gu, J., Zhou, Z., Wang, Z.: Artificial neural networks for infectious diarrhea prediction using meteorological factors in Shanghai (China). Appl. Soft Comput. **35**, 280–290 (2015)

15. Wang, Z., Cai, B.: COVID-19 cases prediction in multiple areas via shapelet learning. Appl. Intell. **52**(1), 595–606 (2021). https://doi.org/10.1007/s10489-021-02391-6

16. Wang, Z., Huang, Y., Cai, B., Ma, R., Wang, Z.: Stock turnover prediction using search engine data. J. Circuits Syst. Comput. **30**(7), 2150122:1–2150122:18 (2021)

17. Wang, Z., Huang, Y., He, B.: Dual-grained representation for hand, foot, and mouth disease prediction within public health cyber-physical systems. Softw. Pract. Exp. **51**, 2290–2305 (2021)

18. Wang, Z., Huang, Y., He, B., Luo, T., Wang, Y., Fu, Y.: Short-term infectious diarrhea prediction using weather and search data in Xiamen, China. Sci. Program. **2020**, 8814222:1–8814222:12 (2020)

19. Wang, Z., Huang, Y., He, B., Luo, T., Wang, Y., Lin, Y.: TDDF: HFMD outpatients prediction based on time series decomposition and heterogenous data fusion in Xiamen, China. In: ADMA 2019, Dalian, China, pp. 658–667, November 2019

20. Wang, Z., Su, Q., Chao, G., Cai, B., Huang, Y., Fu, Y.: A multi-view time series model for share turnover prediction. Appl. Intell. **Early View** (2022)
21. Wang, Z., et al.: Prediction of HFMD cases by leveraging time series decomposition and local fusion. Wirel. Commun. Mob. Comput. **2021**, 5514743:1–5514743:10 (2021)
22. Zhu, X., et al.: Attention-based recurrent neural network for influenza epidemic prediction. BMC Bioinform. **20-S**(18), 575:1–575:10 (2019)

The Differential Gene Detecting Method for Identifying Leukemia Patients

Mingzhao Wang, Weiliang Jiang, and Juanying Xie[✉]

School of Computer Science, Shaanxi Normal University,
Xi'an 710119, People's Republic of China
xiejuany@snnu.edu.cn

Abstract. Leukemia is one of the cancers threatening human being for many years. There has not been any general method for identifying acute myeloid leukemia (AML) and acute lymphoblastic leukemia (ALL) accurately. This paper try to investigate a new way to detect the differential genes using which to identify ALL and AML easily and accurately as far as possible. We try to learn the differential genes automatically without any assistant knowledge by defining the local density and distance for each gene and representing all genes in the 2-dimensional space with density as x-axis and distance as y-axis, and detecting the density peaks of genes in the up-right corner far away from those in the down-left corner to comprise the differential genes. Four types of distance metrics are used to measure the distances between genes when defining the local densities and distances for genes, so as to detect the efficient differential genes using training data. DPC (Clustering by fast search and find of density peaks) algorithm is adopted to cluster samples from test subset to test the identifying capability of detected differential genes. Here the samples from test subset are only with the detected differential genes. The experimental results demonstrate that the detected differential genes are efficient, and the clustering accuracy is up to 91.18%, and the Rand index, *Jaccard* coefficient and Adjusted Rand Index (ARI) of clustering results are 0.8342, 0.7832 and 0.6185, respectively. This study provides clue to the studies of other cancers.

Keywords: Feature selection · Leukemia · Clustering · Cancers

1 Introduction

Leukemia is a malignant tumor disease caused by gene mutation [1]. The treatment of different types of leukemia varies greatly. There will be unimaginable very serious consequences causing by any inappropriate or error diagnoses. The differences between different types of leukemia are often subtle in cell morphology and biochemical indices [1], such that the traditional diagnostic methods are often failed. However, the rapid development of gene sequencing technology in recent years has brought about a large amount of data related to leukemia, making it possible to diagnose leukemia quickly. But these leukemia gene datasets

© Springer Nature Switzerland AG 2022
H. Fujita et al. (Eds.): IEA/AIE 2022, LNAI 13343, pp. 137–146, 2022.
https://doi.org/10.1007/978-3-031-08530-7_12

are always having small number of samples while having tens of thousands of genes per sample. This leads to the very high dimensional sparse space resulting in the fact that some traditional methods cannot be used, such as the tradition clustering algorithm based on Euclidean distance cannot work efficiently. Therefore, detecting the differential genes to reduce the dimensionality becomes the first thing to do when analyzing this kind of data. Once differential genes are detected, the identification accuracy will be guaranteed for the leukemia patients, and the misdiagnosis and missed diagnosis will be reduced as well, which will be greatly beneficial to the human being.

Feature selection [2–4] is the usual way to detect the differential genes (features), but the general feature selection algorithms often suck in time bottleneck problem when dealing with this kind of data having very-high-dimensions while with very small number of samples [5–9]. Therefore the unsupervised learning methods are introduced to feature selection procedure, so as to reduce the time overhead of feature selection to some extent [4,10–15]. The very common way using unsupervised learning method in feature selection is to cluster feature so that similar features are in the same cluster and dissimilar ones are in different clusters. Then the representative features are selected from various clusters to comprise the feature subset [13–15], ensuring that features in the feature subset are as irrelevant as possible and representative as well. The classification capability of the selected features are tested by building classifier using features included in the feature subset. Although this framework can solve the time bottleneck problem of feature selection to some extent, the efficiency of unsupervised learning algorithm must be considered, and the specific method need selecting to evaluate the representation of a feature to its cluster, especially the classification capability of the feature subset are always tested by the classifier built on them, which indicate the labels are necessary even feature selection process does not need them. If the labels are not available, what could we do to realize the feature selection? Fortunately, there is the type of feature selection algorithm which is completely unsupervised [13,14]. This kind of algorithm not only make the feature selection process unsupervised, but also make the following test to the classification capability of the feature subset unsupervised too. This paper will focus on this kind of feature selection idea and investigate an efficient differential gene detection method for identifying various leukemia patients.

For this destination, the density peaks idea is introduced to realize the feature selection. It is inspired by the DPC (clustering by fast search and find of density peaks) [16] algorithm. DPC is based on the assumptions that cluster centers are always surrounded by neighbors with lower local density and they are at a relatively large distance from any points with a higher local density. That is to say, the cluster centers are always with higher local density than the points around them, and apart from each other. DPC first detects density peaks of points in a dataset, then assigns the remaining points to the same cluster as the nearest neighbor with higher density. The density peaks are representative of clusters where they are, and they are different to each other [16–21]. Therefore, we are inspired to detect the density peaks of genes and use them as differential genes to identify the ALL and AML patients. We detect differential genes by finding the density peaks of genes directly using training dataset without clustering genes first and detecting the representative genes from clusters. The samples from testing dataset are grouped into clusters only using the differential genes for each sample. The clustering performance such as accuracy will test the identifying capability of detected differential genes. The experimental results demonstrate that our idea is valid. The differential genes we found can tell ALL patients from AML patients.

The rest of the paper is organized as followings. Section 2 will describe our ideas to detect the differential genes in detail. Section 3 will show the results of our algorithm and compare them to the available ones. Conclusions come in Sect. 4.

2   Proposed Method

DPC algorithm [16–21] is based on the assumption that the local density of the center point of a cluster is higher than that of points around it, and the distance from a center point to any other points with higher local densities is relatively far. The most contribution of DPC is that the local density ρ and its distance δ of a point are defined, and all the points are plotted in the 2-dimensional space with density ρ and distance δ as x-axis and y-axis, respectively, that is what so called decision graph. As a result, those points in the up-right corner far away from those in the down-left corner of the decision graph will comprise the density peaks because of their relatively higher local densities and distances.

Inspired by the density peaks idea, we define the local densities and distances for genes, and represent all genes in the 2-dimensional space using density as x-axis and distance as y-axis, so as to detect those density peaks of genes in the up-right corner of the 2-dimensional space as the differential genes. The local density ρ_i of gene i is define in (1). That essentially is the number of genes in the neighborhood of gene i with the radius of d_c, where d_{ij} is the distance between genes i, j and d_c is a parameter given manually. In our experiment we set d_c to be the 2% position distance by sorting all distances between genes in ascending order.

$$\rho_i = \|\{j|d_{ij} < d_c\}\| \tag{1}$$

The distance δ_i of gene i is defined in (2). That means the distance δ_i of gene i is given to the maximum value as far as possible when it has got the maximum local density, which will guarantee that the gene i will be detected as the differential gene when its local density is maximal; otherwise its distance is defined as the distance between it and the nearest gene with higher local density than it, which indicate its distance is given to the maximum as far as possible.

$$\delta_i = \begin{cases} max_j d_{ij}, & i = argmax\{\rho_k|k = 1, \cdots, n\} \\ min_{j:\rho_j > \rho_i} d_{ij}, & otherwise \end{cases} \tag{2}$$

From the above definitions we can say that the distance metric will greatly influence the local density and distance of a gene, which will further influence the detected differential genes, so we will investigate four distance metrics in detecting the differential genes. These four distance metrics are as follows. In order to describe the distance metrics we assume the two vectors $\mathbf{x} = (x_1, \cdots, x_n)$, $\mathbf{y} = (y_1, \cdots, y_n)$, then the Minkowski distance between them is defined in (3).

$$d_{\mathbf{xy}} = (\sum_{k=1}^{n} |x_k - y_k|^p)^{\frac{1}{p}} \tag{3}$$

This Minkowski distance will be the very common used Euclidean distance in (4) when $p = 2$, and become the Manhattan Distance in (5) when $p = 1$.

$$d_{\mathbf{xy}} = (\sum_{k=1}^{n} |x_k - y_k|^2)^{\frac{1}{2}} \tag{4}$$

$$d_{\mathbf{xy}} = \sum_{k=1}^{n} |x_k - y_k| \tag{5}$$

It will become the Chebychev distance in (6) when $p \to \infty$, and the Chebychev distance is usually expressed as that in (7).

$$d_{\mathbf{xy}} = lim_{p \to \infty}(\sum_{k=1}^{n} |x_k - y_k|^p)^{\frac{1}{p}} \tag{6}$$

$$d_{\mathbf{xy}} = max_{k=1,\cdots,n} |x_k - y_k| \tag{7}$$

The fourth distance metric we will investigate is correlation distance which is based on the Pearson coefficient between two vectors in (8), where $\bar{x}$, $\bar{y}$ are the means of elements of $\mathbf{x}$ and $\mathbf{y}$, respectively. Then the correlation distance is defined in (9). The detail procedure of detecting differential genes is described in Algorithm 1.

$$r_{\mathbf{xy}} = \frac{\sum_{i=1}^{n} (x_i - \bar{x})(y_i - \bar{y})}{\left(\sum_{i=1}^{n} (x_i - \bar{x})^2\right)^{\frac{1}{2}} \left(\sum_{i=1}^{n} (y_i - \bar{y})^2\right)^{\frac{1}{2}}} \tag{8}$$

$$d_{\mathbf{xy}} = 1 - |r_{\mathbf{xy}}| \tag{9}$$

3   Experiments and Results

We first do differential gene selection using Algorithm 1 on training subset. Because four distance metrics are used to measure the distance between genes, so there are four differential gene subsets will be obtained. Then we test the capability of the detected differential genes in each differential gene subset by grouping the samples in testing subset of leukemia data using original DPC algorithm proposed in [16]. It should be noted that the samples now only contains

---

Algorithm 1.   Density peaks based differential genes selection

---

Input: Training data $\mathbf{X} = \{\mathbf{x}_i | i = 1, \cdots, m\} \in R^{m \times n}$, percent $= 2\%$.

Output: The gene set $\mathbf{S}$ comprising the differential genes.

1: Initialize $\mathbf{S} = \phi$, $\mathbf{G} = \mathbf{X}^T$;

2: calculate distance matrix $\mathbf{D} = (d_{ij})_{n \times n}$, where d_{ij} is the distance between genes i,
     j and distance is one of the four distance metrics in (4), (5), (7), or (9);

3: calculate the density of gene $i(i = 1, \cdots, n)$ using (1);

4: calculate the distance for gene $i(i = 1, \cdots, n)$ using (2);

5: display genes in the 2-dimensional space with density as x-axis and distance y-axis;

6: select the genes in the up-right corner to comprise the differential gene set $\mathbf{S}$.

---

the differential genes from a differential gene subset. The distance between samples now is only Euclidean distance. The clustering results are evaluated in terms of clustering accuracy (ACC), Rand index (RI), $Jaccard$ coefficient (JC), and Adjusted Rand Index (ARI). To further verify the validity of doing differential gene selection, we merge training and testing subsets together and do clustering to the whole dataset using DPC algorithm without doing any differential gene selection. The clustering results on the whole dataset and on the testing subset are compared in Table 1. The samples for the former clustering contain all genes while the samples for the latter clustering are only with differential genes. The bold fonts indicate the best results. The differential gene selection process are displayed in Fig. 1 based on four distance metrics used to measure the distance between genes. The specific detected differential genes are shown in Table 2, that is, the colored genes in Fig. 1.

Furthermore in Table 3 we compared the capability of the differential genes detected by Algorithm 1 to that of the features selected by some other feature selection algorithms in identifying ALL and AML patients. At the same time the numbers of the detected features and whether using the labels to assistant are also shown in Table 3 as well.

To avoid the influences from far various scales in genes to the experimental results, we normalize the original data in (10). The $g_{i,j}$ is the expression value of sample i on gene j. The $\max(g_j)$ is the maximum value of gene j among all samples, while $\min(g_j)$ is the minimum value of gene j among all samples.

$$g_{i,j} = \frac{g_{i,j} - min\,(g_j)}{max\,(g_j) - min\,(g_j)} \qquad (10)$$

The leukemia dataset comprises 72 samples among which there are 47 samples from ALL and 25 from AML. There are 7129 genes for each sample. These 72 samples were partitioned into training and testing subset. There are 38 samples in training subset, among which 27 come from ALL type, and 11 from AML type. There are 34 samples in testing subset, among which there are 20 samples from ALL and 14 from AML.

The results in Table 1 show that the differential genes detected by using the Chebychev distance metric has obtained the strongest identification capability. The results in Table 2 show that there are four differential genes detected by

Table 1. Clustering results testing the detected differential genes

Gene similarity metrics	ACC	RI	JC	ARI
Manhattan	0.6471	0.5294	0.4762	−0.0533
Euclidean	0.2647	0.5989	0.5455	0.0771
Chebychev	**0.9118**	**0.8342**	**0.7832**	**0.6185**
Pearson coefficient	0.3529	0.5294	0.4762	−0.0533
All genes included	0.7222	0.5931	0.5336	0.0912

Table 2. Differential genes detected using different distance metrics to measure the distance between genes

Gdistance metrics	Gene description	Gene accession number
Manhattan	(p23) mRNA	L24804_at
	Atpase, Na+/K+ Transporting, Alpha 1 Polypeptide	HG1034-HT1034_f_at
Euclidean	KIAA0181 gene, partial cds	D80003_at
	Dihydropyrimidine dehydrogenase mRNA	U09178_s_at
	Testis specific RNA binding protein (SPGYLA) mRNA	U66726_at
Chebychev	KIAA0130 gene	D50920_at
	Proteasome inhibitor hPI31 subunit	D88378_at
	Kinase A anchor protein	X97335_at
	IREB1 Iron-responsive element binding protein 1	Z11559_at
Pearson coefficient	JUN V-jun avian sarcoma virus 17 oncogene homolog	U65928_at
	HMGI-C	X92518_s_at

Algorithm 1 when using Chebychev distance to measure the distance between genes. They are KIAA0130 gene, Proteasome inhibitor hPI31 subunit, Kinase A anchor protein, and IREB1 Iron-responsive element binding protein 1. These four genes together can obtain the best clustering results to tell ALL patients from AML patients with 91.18% clustering accuracy. The clustering results in terms of Rand index, *Jaccard* coefficient, and ARI are also superior to that based on differential genes detected by using Manhattan, Euclidean and Pearson coefficient distances. The clustering results without differential gene selection process rank in the second place in Table 1. The clustering results using the differential genes detected by Manhattan distance metric is the third rank in telling ALL patients from AML patients. The differential genes detected by Euclidean distance metrics has obtained the worst capability in telling ALL patients from AML patients.

Further analysis can find why the Chebychev distance metric can find the best differential genes. It is due to the Chebychev distance record the maximum difference between two genes of samples. The Manhattan distance can find the second good differential genes because it summarizes all of the differences between two genes on all training samples, which may reduce the main difference between two genes. It is the similar reasons that make the Euclidean and Pearson coefficient distance cannot strengthen the main difference between genes, resulting in the clustering results inferior to that without gene selection process. Although the clustering results based on the differential genes detected by Manhattan distance are inferior to that without gene selection, the former only uses two genes including ((p23) mRNA) and (Atpase, Na+/K+ Transporting, Alpha 1 Polypeptide) shown in Table 2, while the latter clustering results come from using 7129 genes.

Form the above analyses we can say that our Algorithm 1 is good enough in finding the differential genes to tell ALL patients from AML patients by using the Chebychev distance to measure the distance between genes when calculating the local density and distance for genes.

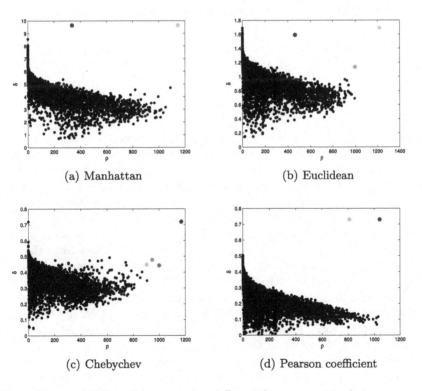

(a) Manhattan   (b) Euclidean

(c) Chebychev   (d) Pearson coefficient

Fig. 1. Detected differential genes using different distance metrics between genes

The results in Fig. 1 show that two differential genes, that is, the density peaks, can be found when using Manhattan or Pearson coefficient distance to measure the distance between genes, while the Euclidean distance can detect three differential genes, and Chebychev distance can find four differential genes to be used to tell ALL patients from AML patients.

Table 3. Comparison of the capability and the numbers of differential genes detected by different feature selection algorithms

Algorithms [sources]	Whether using labels and when	Accuracy	#genes
SVM-RFE [5]	Yes & feature selection process	0.82/1.00	4/8
MRMR+SVM [6]	Yes & feature selection process	0.9583	5
Weight+Pearson+SVM [10]	Yes & feature selection process	1.0000	11
FSDI+KNN [11]	Yes & feature selection process	0.9256	4
SCEFS+KNN [12]	Yes & training classifier	0.9709	13
AMID+SVM [13]	Yes & feature selection process	0.9887	43
FKNN-DPC [17]	No	0.8889	7
AVC+K-ELM [22]	Yes & feature selection process	0.9457	6
ARCO+K-ELM [23]	Yes & feature selection process	0.9143	36
LLE Score+K-ELM [24]	Yes & feature selection process	0.9305	30
Lasso+SVM [25]	Yes & feature selection process	0.9289	24
This paper	No	0.9118	4

Table 3 shows that only two algorithms are completely unsupervised feature selection algorithms. One is Algorithm 1 proposed in this paper, and the other is based on the FKNN-DPC algorithm. The results in Table 3 show that nearly all feature selection algorithms using labels to assistant feature selection process have obtained better accuracy than our algorithm in this paper except for the very famous SVM-RFE when it selecting same number of genes as our algorithm. Furthermore, although SCEFS is an unsupervised feature selection algorithm and has got good performance, the accuracy comes from the KNN classifier trained by using labels of training samples after finding the differential features, while the results of FKNN-DPC and our algorithm do not come from any classifier trained by using assistant information from labels of samples from training data. Even though FKNN-DPC is an advanced version of DPC, it cannot find the differential genes with better capability to identify AML and ALL patients than our algorithm in this paper can do.

Considering the classification capability and the number of differential genes and whether or not need labels to assistant, we can say that our algorithm inspired by the density peaks idea is good enough for it is completely unsupervised and can detect differential genes with strong classification capability.

4   Conclusions

The unsupervised differential gene selection algorithm for identifying leukemia patients is proposed using density peaks finding idea. The experimental results show that the Chebychev distance can detect the differential genes with the strongest capability in identifying ALL patients from AML patients. The clustering results are up to 0.9118, 0.8342, 0.7832 and 0.6185 in terms of clustering accuracy, Rand index, *Jaccard* coefficient and *ARI*. It is so far the best results obtained without any assistant knowledge. Although the results may not as good as that from some supervised methods, they are from a completely unsupervised method. This study can support other related studies especially when the labels of samples are difficult to access.

However, we can see from Table 2 that the differential genes detected using different distance metrics are different. How to get the stable differential genes using different distance metric need further studying. Furthermore, the results in Table 3 indicate that our algorithm can be further improved, so as to find the differential genes with far stronger identification capability.

Acknowledgements. This work is supported by the National Natural Science Foundation of China under Grant No. of 12031010, 62076159, and 61673251, and is also supported by the Fundamental Research Funds for the Central Universities under Grant No. of GK202105003 and 2018TS078. We also acknowledge those researchers who published the related studies for us to reference in this research.

References

1. Golub, T.R., et al.: Molecular classification of cancer: class discovery and class prediction by gene expression monitoring. Science **286**(5439), 531–537 (1999)
2. Venkatesh, B., Anuradha, J.: A review of feature selection and its methods. Cybern. Inf. Technol. **19**(1), 3–26 (2019)
3. Wan, J., Chen, H., Li, T., Huang, W., Li, M., Luo, C.: R2CI: information theoretic-guided feature selection with multiple correlations. Pattern Recogn. **127**, 108603 (2022)
4. Solorio-Fernández, S., Carrasco-Ochoa, J.A., Martínez-Trinidad, J.F.: A review of unsupervised feature selection methods. Artif. Intell. Rev. **53**(2), 907–948 (2019). https://doi.org/10.1007/s10462-019-09682-y
5. Guyon, I., Weston, J., Barnhill, S., Vapnik, V.: Gene selection for cancer classification using support vector machines. Mach. Learn. **46**(1), 389–422 (2002)
6. Ding, C., Peng, H.: Minimum redundancy feature selection from microarray gene expression data. J. Bioinform. Comput. Biol. **3**(02), 185–205 (2005)
7. Peng, H., Long, F., Ding, C.: Feature selection based on mutual information criteria of max-dependency, max-relevance, and min-redundancy. IEEE Trans. Pattern Anal. Mach. Intell. **27**(8), 1226–1238 (2005)
8. Xie, J., Xie, W.: Several feature selection algorithms based on the discernibility of a feature subset and support vector machines. Chin. J. Comput. **37**(8), 1704–1718 (2014)

9. Xie, J., Wang, M., Zhou, Y., Gao, H., Xu, S.: Differential expression gene selection algorithms for unbalanced gene datasets. Chin. J. Comput. **42**(6), 1232–1251 (2019)

10. Xie, J., Gao, H.: Statistical correlation and K-means based distinguishable gene subset selection algorithms. J. Softw. **25**(9), 2050–2075 (2014)

11. Xie, J., Wang, M., Zhou, Y., Li, J.: Coordinating discernibility and independence scores of variables in a 2D space for efficient and accurate feature selection. In: Huang, D.-S., Han, K., Hussain, A. (eds.) ICIC 2016. LNCS (LNAI), vol. 9773, pp. 116–127. Springer, Cham (2016). https://doi.org/10.1007/978-3-319-42297-8_12

12. Xie, J., Wang, M., Xu, S., Huang, Z., Grant, P.W.: The unsupervised feature selection algorithms based on standard deviation and cosine similarity for genomic data analysis. Front. Genet. **12**, 684100 (2021)

13. Xie, J., Wang, Y., Wu, Z.: Colon cancer data analysis by chameleon algorithm. Health Inf. Sci. Syst. **7**(1), 1–8 (2019). https://doi.org/10.1007/s13755-019-0085-1

14. Xie, J., Wu, Z., Xia, Q., Ding, L., Fujita, H.: The differential feature detection and the clustering analysis to breast cancers. In: Fujita, H., Fournier-Viger, P., Ali, M., Sasaki, J. (eds.) IEA/AIE 2020. LNCS (LNAI), vol. 12144, pp. 457–469. Springer, Cham (2020). https://doi.org/10.1007/978-3-030-55789-8_40

15. Xie, J., Ding, L., Wang, M.: Spectral clustering based unsupervised feature selection algorithms. J. Softw. **31**(4), 1009–1024 (2020)

16. Rodríguez, A., Laio, A.: Clustering by fast search and find of density peaks. Science **344**(6191), 1492–1496 (2014)

17. Xie, J., Gao, H., Xie, W., Liu, X., Grant, P.W.: Robust clustering by detecting density peaks and assigning points based on fuzzy weighted K-nearest neighbors. Inf. Sci. **354**, 19–40 (2016)

18. Xu, X., Ding, S., Wang, L., Wang, Y.: A robust density peaks clustering algorithm with density-sensitive similarity. Knowl.-Based Syst. **200**, 106028 (2020)

19. Yu, H., Chen, L., Yao, J.: A three-way density peak clustering method based on evidence theory. Knowl.-Based Syst. **211**, 106532 (2021)

20. Xu, X., Ding, S., Wang, Y., Wang, L., Jia, W.: A fast density peaks clustering algorithm with sparse search. Inf. Sci. **554**, 61–83 (2021)

21. Wang, Y., et al.: McDPC: multi-center density peak clustering. Neural Comput. Appl. **32**(17), 13465–13478 (2020). https://doi.org/10.1007/s00521-020-04754-5

22. Sun, L., Wang, J., Wei, J.: AVC: selecting discriminative features on basis of AUC by maximizing variable complementarity. BMC Bioinform. **18**(3), 73–89 (2017)

23. Wang, R., Tang, K.: Feature selection for maximizing the area under the ROC curve. In: Proceedings of the 2009 IEEE International Conference on Data Mining Workshops, pp. 400–405. IEEE Computer Society, Florida (2009)

24. Li, J., Pang, Z., Su, L., Chen, S.: Feature selection method LLE score used for tumor gene expressive data. J. Beijing Univ. Technol. **41**(8), 1145–1150 (2015)

25. Tibshirani, R.: Regression shrinkage and selection via the lasso. J. Roy. Stat. Soc. Ser. B (Methodol.) **58**(1), 267–288 (1996)

Epidemic Modeling of the Spatiotemporal Spread of COVID-19 over an Intercity Population Mobility Network

Yuxi Liu[1]([✉]), Shaowen Qin[1], and Zhenhao Zhang[2]

[1] College of Science and Engineering, Flinders University,
Tonsley, SA 5042, Australia
{liu1356,shaowen.qin}@flinders.edu.au
[2] College of Life Sciences, Northwest A&F University,
Yangling 712100, Shaanxi, China
zhangzhenhow@nwafu.edu.cn

Abstract. Intercity traveling has been recognized as a leading cause for the continuation of the COVID-19 global pandemic. However, there lacks credible prediction of the spatiotemporal spread of COVID-19 with humans traveling between metropolitan areas. This study attempts to establish a novel framework to simulate human traveling and the spread of virus across an intercity population mobility network. A Markov process was introduced to capture the stochastic nature of travelers' migration. A backward derivation algorithm was adopted and the Nelder-Mead simplex optimization method applied to overcome the limitation of existing deterministic epidemic models, including the difficulties in estimating the initial susceptible population and the optimal hyper-parameters required for simulation. We conducted two case studies with data from 24 cities in China and Italy. Our framework yielded state-of-the-art accuracy while being modular and scalable, indicating the addition of population mobility and stochasticity significantly improves prediction performance compared to using epidemic data alone. Moreover, our results revealed that transmission patterns of COVID-19 differ significantly with different population mobility, offering valuable information to the understanding of the correlation between traveling activities and COVID-19 transmission.

Keywords: Epidemic modeling · COVID-19 · Spatiotemporal analysis · Population mobility

1 Introduction

It is a common belief that the spread of COVID-19 across the world is through humans traveling over an open and interwoven network of population flow. Recent studies confirmed that human mobility is a major factor driving the spatial spread of COVID-19 [1–3]. Similarly, [4] revealed that mobility patterns

© Springer Nature Switzerland AG 2022
H. Fujita et al. (Eds.): IEA/AIE 2022, LNAI 13343, pp. 147–159, 2022.
https://doi.org/10.1007/978-3-031-08530-7_13

strongly correlated with COVID-19 growth rates, based on data from the most affected counties in the U.S. However, it is challenging to predict the extent and speed of virus transmission with credible accuracy. Further and deeper investigation on the spatiotemporal spreading patterns of COVID-19 utilizing a population mobility network beyond the traditional Susceptible-Exposed-Infectious-Removed (SEIR) model [5] needs to be undertaken. Such an investigation is not only essential for understanding the mechanism of the spread in a population mobility network involving intercity traveling, but also beneficial for devising measures for spread prevention and control on a global basis.

The art, also the science, of mathematical modeling, is to construct the simplest model that captures the salient features of the system. Virus transmission is a dynamic process that involves many stochastic components. Consequently, it is necessary to seamlessly incorporate population mobility and stochastic processes into the classical SEIR model to predict the spatiotemporal effect of virus transmission. Additionally, COVID-19 shows some variation from the typical progression of infectious disease transmission, in the sense that it is also infectious during the incubation period [6], which needs to be considered through modification of the classical SEIR model.

To our best knowledge, there does not exist any applicable mainstream network topology structure to simulate COVID-19 spread associated with population mobility. Also, some inadequacies in the deterministic epidemic models remain unaddressed. First, previous network models with clustering are context-specific hence not scalable [7,8]. Consequently, results generated from such models may not be generalizable [9]. Second, we propose an initial high-risk population to accurately represent the size of the initial susceptible population, which is pivotal for epidemic modeling [10]. Prior studies suggested using the city population base as the initial susceptible population [6,11]. However, such methods overestimate the infection cases at the early stage of epidemic transmission, which varies based on the demographic background of different cities. Thus, the infection rate is at risk of being underestimated as the size of infected populations being the product of the initial susceptible population and the infection rate. Third, the epidemic models usually are not parameter-free. Besides the essential initial susceptible population, initial parameters like the infection rate and recovery rate also need to be provided [12]. However, due to the lack of a comprehensive review of historical incidence data, it is hard to derive specific parameters for different survey sites.

In this paper, we use fine-grained spatiotemporal population mobility data, in conjunction with epidemic data, to construct an urban network epidemic framework, which is closer to real-world scenarios. The use of the framework offers a scalable and credible solution compared with the traditional SEIR model and strong baseline models such as Metapopulation SIR model [13] and Susceptible-Undiagnosed-Infected-Removed (SUIR) model [14]. Additionally, some challenges of the deterministic epidemic model are better addressed. Extensive experiments were conducted to assess the performance of our approach, using a real-world COVID-19 epidemic dataset of 12 cities in Hubei Province, China, and 12 cities in Italy.

2   The Proposed Approach

We propose an urban network epidemic framework (M-Urb-SEIR), a novel approach that incorporates population mobility and stochasticity for accurate COVID-19 confirmed case forecasting.

2.1   SEIR Model (Single-Network)

We adopt the SEIR model [5] to describe the dynamic process of epidemic propagation. Criteria for dividing the subjects are as follows: (P) represents the total population, Susceptible (S) is for the susceptible individuals, Exposed (E) for the exposed individuals, previously susceptible who have been exposed to the virus but may not be infected; Infected (I) for the infective individuals capable of transmitting the disease, this includes a non-symptomatic infectious period, and Recovered (R) for recovered individuals who were previously infected but have become immune. If the immune period is limited, R can be converted into S again. The relation between all variables is shown below:

$$\frac{dS}{dt} = -\alpha I(t)S(t)/N, \tag{1}$$

$$\frac{dE}{dt} = \alpha I(t)S(t)/N - \mu E(t), \tag{2}$$

$$\frac{dI}{dt} = \mu E(t) - \gamma I(t), \tag{3}$$

$$\frac{dR}{dt} = \gamma I(t), \tag{4}$$

$$S(t) + E(t) + I(t) + R(t) = N, \tag{5}$$

where α represents the rate of conversion for the exposed become infected; μ represents the rate of transformation for the incubation period to a patient; and γ represents the probability of recovery.

2.2   M-Urb-SEIR (Urban Network Epidemic Framework)

The traditional SEIR model assumes a single infection network among individuals and only model epidemic propagation in a single dimension. We extend the traditional SEIR model to the scenario of urban networks. We assume that there are N cities in a city network of interests. Eligibility criteria required individuals to be divided into four states: S, E, I, and R. To assess the city n, the city population base was used P_n, and the number of S, E, I, and R in the city at time t is $S_n(t)$, $E_n(t)$, $I_n(t)$, and $R_n(t)$. This study assumes a mobility intensity (W_{nm}) between city n and m, representing the average number of visitors from the city n to m. Recent evidence suggests that cases with the latent period are infectious [6]. We therefore set out to assess the effect of the infection rate of the infected individual and the effect of infection rate of the latent individual (infectious and lag onset). α_1 represents the infection rate of the infected individual; α_2 represents the infection rate of the latent individual; μ represents the rate of transformation of the incubation period to patients; γ represents recovery rate of patients.

The transmission of and recovery from infection are intrinsically stochastic processes, and the deterministic epidemic model does not account for fluctuations [15]. To tackle this issue, we assume the process is Markovian on the relevant time scales, the dynamic variations between states of the four populations at t are summarized as follows:

$$
\begin{aligned}
\frac{dS_n(t)}{dt} = & -\alpha_1 S_n(t) \sum_{m=1}^{N} \left(\frac{W_{mn}}{P_m} + \frac{W_{nm}}{P_n}\right) I_m(t)/P_n \\
& -\alpha_2 S_n(t) \sum_{m=1}^{N} \left(\frac{W_{mn}}{P_m} + \frac{W_{nm}}{P_n}\right) E_m(t)/P_n \\
& +\sqrt{\alpha_1 S_n(t) I_n(t)/P_n} \cdot P_t(S_n \xrightarrow{\alpha_1} E_n) \\
& +\sqrt{\alpha_2 S_n(t) E_n(t)/P_n} \cdot P_t(S_n \xrightarrow{\alpha_2} E_n) \\
& +\sqrt{\alpha_1 S_n(t) \sum_{m \neq n}^{N} \left(\frac{W_{nm}}{P_n}\right) I_m(t)/P_n} \\
& \quad \cdot P_t(S_n \xrightarrow{\alpha_1} E_n) \\
& +(\sqrt{\alpha_2 S_n(t) \sum_{m \neq n}^{N} \left(\frac{W_{nm}}{P_n}\right) E_m(t)/P_n} \\
& \quad \cdot P_t(S_n \xrightarrow{\alpha_2} E_n) \\
& +\sqrt{\sum_{m \neq n}^{N} \alpha_1 S_n(t) \left(\frac{W_{nm}}{P_n}\right) I_m(t)/P_n} \\
& \quad \cdot P_t(S_n \xrightarrow{\alpha_1} E_m).
\end{aligned}
\tag{6}
$$

$$
\begin{aligned}
\frac{dE_n(t)}{dt} = & \; \alpha_1 S_n(t) \sum_{m=1}^{N} \left(\frac{W_{mn}}{P_m} + \frac{W_{nm}}{P_n}\right) I_m(t)/P_n \\
& +\alpha_2 S_n(t) \sum_{m=1}^{N} \left(\frac{W_{mn}}{P_m} + \frac{W_{nm}}{P_n}\right) E_m(t)/P_n \\
& -\mu \cdot E_n(t) - (\sqrt{\alpha_1 S_n(t) I_n(t)/P_n} \\
& \quad \cdot P_t(S_n \xrightarrow{\alpha_1} E_n) \\
& -(\sqrt{\alpha_2 S_n(t) E_n(t)/P_n} \cdot P_t(S_n \xrightarrow{\alpha_2} E_n) \\
& -(\sqrt{\alpha_1 S_n(t) \sum_{m \neq n}^{N} \left(\frac{W_{nm}}{P_n}\right) I_m(t)/P_n} \\
& \quad \cdot P_t(S_n \xrightarrow{\alpha_1} E_n) \\
& -(\sqrt{\alpha_2 S_n(t) \sum_{m \neq n}^{N} \left(\frac{W_{nm}}{P_n}\right) E_m(t)/P_n} \\
& \quad \cdot P_t(S_n \xrightarrow{\alpha_2} E_n) \\
& -(\sqrt{\sum_{m \neq n}^{N} \alpha_1 S_n(t) \left(\frac{W_{nm}}{P_n}\right) I_m(t)/P_n} \\
& \quad \cdot P_t(S_n \xrightarrow{\alpha_1} E_m) \\
& -(\sqrt{\sum_{m \neq n}^{N} \alpha_2 S_n(t) \left(\frac{W_{nm}}{P_n}\right) E_m(t)/P_n} \\
& \quad \cdot P_t(S_n \xrightarrow{\alpha_2} E_m) \\
& +(\sqrt{\mu E_n(t)} \cdot P_t(E_n \xrightarrow{\mu} I_n).
\end{aligned}
\tag{7}
$$

$$
\begin{aligned}
\frac{dI_n(t)}{dt} = & \; \mu \cdot E_n(t) - \gamma \cdot I_n(t) \\
& (\sqrt{\mu E_n(t)} \cdot P_t(E_n \xrightarrow{\mu} I_n) \\
& +\sqrt{\gamma I_n(t)} \cdot P_t(I_n \xrightarrow{\gamma} R_n).
\end{aligned}
\tag{8}
$$

$$\frac{dR_n(t)}{dt} = \gamma \cdot I_n(t) - \sqrt{\gamma I_n(t)} \cdot P_t(I_n \xrightarrow{\gamma} R_n). \tag{9}$$

$P_t (S_n \xrightarrow{\alpha_1} E_n)$, the probability that individuals in S state (city n) will be transformed into E state (city n) at t time, which is caused by the contact between S (city n) and I (city n). $P_t (S_n \xrightarrow{\alpha_2} E_n)$, the probability that individuals in S state (city n) will be transformed into E state (city n) at t time, which is caused by the contact between S (city n) and E (city n). $P_t (S_n \xrightarrow{\alpha_1} E_m)$, the probability that the individuals in S state (city n) will be transformed into E state (city m) at t time due to the contact between individuals in S (city n) and I (city m). $P_t (S_n \xrightarrow{\alpha_2} E_m)$, the probability that the individuals in S state (city n) will be transformed into E state (city m) at t time due to the contact between individuals in S (city n) and E (city m). $P_t (E_n \xrightarrow{\mu} I_n)$, the probability that individuals in E state (city n) transforms into I state (city n) at t time. $P_t (I_n \xrightarrow{\gamma} R_n)$, the probability of individuals in I state (city n) transforms into R state (city n) at time t.

In an urban network, there are three behaviors for susceptible populations in urban n to access the incubation period.

(1) Internal transmission of city n:

$$\begin{aligned}
&\alpha_1 \cdot S_n(t) \cdot I_n(t)/P_n + \alpha_2 \cdot S_n(t) \cdot E_n(t)/P_n \\
&- \sqrt{\alpha_1 S_n(t) I_n(t)/P_n} \cdot P_t(S_n \xrightarrow{\alpha_1} E_n) \\
&- \sqrt{\alpha_2 S_n(t) E_n(t)/P_n} \cdot P_t(S_n \xrightarrow{\alpha_2} E_n)
\end{aligned} \tag{10}$$

(2) Transmission caused by the flow of infected and exposed individuals from city m to n:

$$\begin{aligned}
&\alpha_1 \cdot S_n(t) \cdot \sum_{m \neq n}^{N} \left(\frac{W_{mn}}{P_m}\right) \cdot I_m(t)/P_n \\
&+ \alpha_2 \cdot S_n(t) \cdot \sum_{m \neq n}^{N} \left(\frac{W_{mn}}{P_m}\right) \cdot E_m(t)/P_n \\
&- \sqrt{\alpha_1 S_n(t) \sum_{m \neq n}^{N} \left(\frac{W_{nm}}{P_n}\right) I_m(t)/P_n} \cdot P_t(S_n \xrightarrow{\alpha_1} E_n) \\
&- \sqrt{\alpha_2 S_n(t) \sum_{m \neq n}^{N} \left(\frac{W_{nm}}{P_m}\right)/P_n} \cdot P_t(S_n \xrightarrow{\alpha_2} E_n)
\end{aligned} \tag{11}$$

(3) The susceptible population from city n flows into m and infected:

$$\begin{aligned}
&\sum_{m \neq n}^{N} \alpha_1 \cdot S_n(t) \left(\frac{W_{nm}}{P_n}\right) \cdot I_m(t)/P_n \\
&+ \sum_{m \neq n}^{N} \alpha_2 \cdot S_n(t) \left(\frac{W_{nm}}{P_n}\right) \cdot E_m(t)/P_n \\
&- \sqrt{\sum_{m \neq n}^{N} \alpha_1 S_n(t) \left(\frac{W_{nm}}{P_n}\right) I_m(t)/P_n} \cdot P_t^2(S_n \xrightarrow{\alpha_1} E_m) \\
&- \sqrt{\sum_{m \neq n}^{N} \alpha_2 S_n(t) \left(\frac{W_{nm}}{P_n}\right) E_m(t)/P_n} \cdot P_t^2(S_n \xrightarrow{\alpha_2} E_m)
\end{aligned} \tag{12}$$

The proposed framework is implemented by an overall algorithm as follows:

Algorithm 1: M-Urb-SEIR

Require: $\mathcal{R}_0$, (α', γ'), $(\alpha_1, \alpha_2, \mu, \gamma)$, W, P, cumulative confirmed $\hat{C}(t)$, T
Ensure: $S(t), E(t), I(t), R(t)$
1: **Initialize:** $\alpha', \gamma', \alpha_1, \alpha_2, \mu, \gamma$
2: **Backward derivation $\mathcal{E}_0$:**
3:    Apply $\mathcal{R}_0$, α', γ', P on Eq. (3.13), then use the Nelder-Mead simplex optimization method to obtain the optimal $\mathcal{E}_0$ from $I(T) + R(T) = \mathcal{E}_0$
4:    Apply $(\alpha_1, \alpha_2, \mu, \gamma)$, W, P on Markov stochastic process, and infer the P_t ($S_n \xrightarrow{\alpha_1} E_n$), P_t ($S_n \xrightarrow{\alpha_2} E_n$), P_t ($S_n \xrightarrow{\alpha_1} E_m$), P_t ($S_n \xrightarrow{\alpha_2} E_m$), P_t ($E_n \xrightarrow{\mu} I_n$), P_t ($I_n \xrightarrow{\gamma} R_n$) (see Table 3)
5: **Estimation:**
6:    Apply $(\alpha_1, \alpha_2, \mu, \gamma)$, W, P, $\mathcal{E}_0(S(0) = \mathcal{E}_0)$ and $P_t(\cdot)$ on Eq. (3.6-3.9), obtain $I(t)$, $R(t)$, and $C(t) = I(t) + R(t)$
7:    Optimize $(\alpha_1, \alpha_2, \mu, \gamma)$ by using the Nelder-Mead simplex optimization method
8: **Simulation:**
9: Obtain the relative error from $C(t)$ and $\hat{C}(t)$
10: **for** $t = 1$ to T **do**
11:    Apply $(\alpha_1, \alpha_2, \mu, \gamma)$ on Eq. (3.6-3.9)
12:    update $S(t)$, $E(t)$, $I(t)$, $R(t)$
13: **end for**

2.3   Addressing the Challenges of a Deterministic Epidemic Model

(1) Our proposed framework is scalable. Once the original and target domain are located and marked, the actual cross-domain propagation of COVID-19 can be simulated. (2) We propose a backward derivation algorithm to derive the initial susceptible population $\mathcal{E}_0$. Specifically, we first used the way of [16] to obtain $\mathcal{R}_0$ sequences from time-series data of confirmed cases. We then established a basic Susceptible-Infected-Recovered (SIR) model [11] as shown in Eq. (3.13).

$$\mathcal{R}_0 = \frac{\alpha \cdot P}{\gamma}, \tag{13}$$

where α represents the infection rate, P represents the total population in the area, and γ represents the recovery rate. Based on Eq. (3.13), we first initialized the infection rate (α) and recovery rate (γ). We then adopted the Nelder-Mead simplex optimization method to optimize α and γ to make the total number of infected individuals (I) and recovered individuals (R) as close as possible to the real number of confirmed cases. Last, the total number of infected individuals (I) and recovered individuals (R) were used as $\mathcal{E}_0$ of the urban network epidemic framework. Moreover, the optimal α and γ were used as the initial hyper-parameters of the urban network epidemic framework. Note that we used $\mathcal{E}_0$ instead of the city population base in the urban network epidemic framework, and the Markov process used the difference between the city population base and $\mathcal{E}_0$ to incorporate stochasticity. (3) We used the Nelder-Mead simplex optimization method again to optimize the α, γ, and μ (the rate of transformation of the incubation period to patients) of the urban network epidemic framework.

3   Experimental Settings

We present datasets, competitors, and evaluation metrics for our experiments.

3.1   Datasets

We adopted the epidemic data from the National Health Commission of the People's Republic of China[1] (daily confirmed new cases for each city between January 23, 2020 and February 29, 2020) and Italian epidemic data[2] (daily confirmed new cases for each city between February 24, 2020 and April 15, 2020). The statistical data includes the cumulative number of infected, recovered, and death cases. Chinese migration data were obtained with consent from Baidu migration, and the most recent data are available on the website (https://qianxi. baidu.com). The dataset includes the immigration scale index, the emigration scale index, and intracity travel intensity. The migration scale index is converted according to the absolute value of the number of individuals who move in/out, reflecting the population scale of the cities. The intensity of intracity travel is the exponential result of the number of individuals who have traveled in the city and the city's resident population, reflecting the intracity mobility scale. Similarly, Italian migration data were obtained from [17].

3.2   Competitors

To fairly compare different approaches, we compare our approach with the following deterministic epidemic models.

1. SEIR is an epidemiological model used to simulate the spread of infectious disease.
2. Metapopulation SIR model [13] (SIGKDD 2018) extends the SIR model to a metapopulation SIR model that allows visitors transmission between any two sub-populations.
3. SUIR model [14] (SIGKDD 2020) incorporates a unique 'undiagnosed' state of the COVID-19 on the basis of the SIR model.
4. Urb-SEIR (without Markov process) is one variant of the proposed framework.

3.3   Evaluation Metrics

This study assumes that the prediction date between t and T, and the relative error is defined as:

$$e(t, T) = \frac{|C(t + T) - \hat{C}(t + T)|}{\hat{C}(t + T)}, \tag{14}$$

where $C(t)$ represents the cumulative confirmed cases obtained from the baselines and our proposed methods, and $\hat{C}(t)$ represents the cumulative truth cases based on the database.

---

[1] https://github.com/CSSEGISandData/COVID-19.
[2] https://github.com/pcm-dpc/COVID-19.

4   Experimental Results

Most research works use the number of city population base as the initial susceptible population, however, Table 1 shows the initial high-risk population base deduced by the backward derivation algorithm is much less than city population but more reasonable. Furthermore, the optimal hyper-parameter identified in these responses are summarized in Table 2. Table 3 shows the inference formulation by Markov stochastic process.

Figures 1, 2, 3 and 4 illustrate the prediction error bars of models in Chinese and Italian cities. The graphs show that our M-Urb-SEIR performs well in the 28 days forecast period of 12 cities in China compared with baseline models. Moreover, we find that the M-Urb-SEIR outperforms the Urb-SEIR. This benefits from incorporating random effects into epidemic models, which is critical for improving prediction accuracy. Besides, we had three critical findings from the experimental results in Italian cities. First, the Metapopulation SIR model prediction result performs the best on most days of the first week, and the suboptimal model is the traditional SEIR model. Second, by contrast, Urb-SEIR and M-Urb-SEIR perform better with a longer prediction horizon. Usually, they will perform better than other approaches when the prediction horizon is longer than 1 or 2 weeks. Third, the performance of the Urb-SEIR is better than M-Urb-SEIR. Compared to China, where 'Chunyun' leads to dramatic population migration, Italy's strength of population movement is much milder. Therefore, M-Urb-SEIR, which considers more about the stochastic effect, performs worse than Urb-SEIR. These figures suggest that the prediction of COVID-19 should be customized, and contextual information should be considered. In different application scenarios, the model should be able to be extended and modulized. Specifically, for the high randomness effect such as the 'Chunyun' event (the largest periodic human migration in China), models should take the Markov

Table 1. Number of the initial high-risk population obtained from backward derivation algorithm

Site	Ezhou	Enshi	Huanggang	Huangshi	Jingzhou	Shiyan
City Population	1059.7k	3390k	6333k	2471.7k	5570.1k	3398k
High-Risk Population	663	180	460250	682	7474	538
Site	Suizhou	Wuhan	Xiantao	Xiangyang	Xiaogan	Yichang
City Population	2221k	11212k	1140.1k	5680k	4921k	4137.9k
High-Risk Population	919	304800	687	582	4355	1770
Site	Roma	Milano	Brescia	Torino	Monza	Bologna
City Population	4342.2k	3293.7k	1266k	2259.7k	830.5k	1014.6k
High-Risk Population	17513	55678	129944	10214	62505	35390
Site	Firenze	Catania	Verona	Bergamo	Trieste	Napoli
City Population	1011.3k	1107.7k	926.5k	1114.6k	234.5k	3084.9k
High-Risk Population	4117	29981	19017	21485	1443	2865

Table 2. Optimal hyper-parameters of outbreak cities in China and Italy

Site	Ezhou	Enshi	Huanggang	Huangshi	Jingzhou	Shiyan
α_1	0.0971	0.0801	0.0965	0.0958	0.0967	0.0897
α_2	0.0764	0.0615	0.0751	0.0756	0.0637	0.0699
μ	0.0833	0.2882	0.1667	0.1053	0.4082	0.1942
γ	0.4261	0.3012	0.3661	0.4052	0.4753	0.3521
Site	Suizhou	Wuhan	Xiantao	Xiangyang	Xiaogan	Yichang
α_1	0.0828	0.1534	0.0459	0.0963	0.0732	0.0735
α_2	0.0512	0.0785	0.0403	0.0855	0.0614	0.0632
μ	0.1333	0.1429	0.2857	0.101	0.295	0.3846
γ	0.3253	0.3656	0.3081	0.4233	0.3031	0.4146
Site	Roma	Milano	Brescia	Torino	Monza	Bologna
α_1	0.1525	0.1585	0.1371	0.1734	0.1513	0.1539
α_2	0.1317	0.1457	0.1164	0.1611	0.1261	0.1372
μ	0.1316	0.1887	0.1811	0.0813	0.1942	0.1361
γ	0.4131	0.4373	0.4513	0.4461	0.4333	0.4717
Site	Firenze	Catania	Verona	Bergamo	Trieste	Napoli
α_1	0.1683	0.1746	0.1768	0.1476	0.1853	0.1736
α_2	0.1505	0.1681	0.1367	0.1346	0.1641	0.1623
μ	0.1357	0.3142	0.1392	0.2665	0.1047	0.0899
γ	0.4338	0.4162	0.4743	0.4526	0.4297	0.4529

Table 3. Inference formulation by Markov stochastic process

Site	S_n	$E_{n\alpha_1}$	$E_{n\alpha_2}$	I_n	R_n
S_m	$\frac{W_{mn}}{P_m}$	$\frac{W_{mn}}{P_m}\alpha_1$	$\frac{W_{mn}}{P_m}\alpha_2$	$\frac{W_{mn}}{P_m}(\alpha_1 + \alpha_2\mu)$	$\frac{W_{mn}}{P_m}(\alpha_1 + \alpha_2)\mu\gamma$
$E_{m\alpha_1}$	0	$\frac{W_{mn}}{P_m}$	0	$\frac{W_{mn}}{P_m}\mu$	$\frac{W_{mn}}{P_m}\mu\gamma$
$E_{m\alpha_2}$	0	0	$\frac{W_{mn}}{P_m}$	$\frac{W_{mn}}{P_m}\mu$	$\frac{W_{mn}}{P_m}\mu\gamma$
I_m	0	0	0	$\frac{W_{mn}}{P_m}$	$\frac{W_{mn}}{P_m}\gamma$
R_m	0	0	0	0	$\frac{W_{mn}}{P_m}$

stochastic process into account; however, in the context of regular population movements (Italian cities), the results highlighted the need to use the Urb-SEIR as a predictive tool.

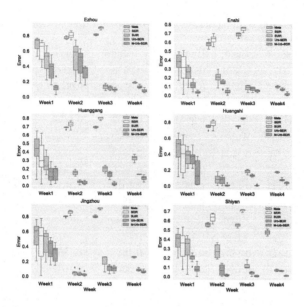

Fig. 1. Prediction error bars of models after T days.

With regard to the research methods, some limitations need to be acknowl-edged. The principal limitation of this analysis was the variance to estimate R_0 [18]. Another major source of uncertainty is in the backward derivation algorithm used to calculate the initial susceptible population. The latent infectivity popula-tion and other external factors were not accounted for in the derivation process. Although there are limitations in the backward derivation, it contributed to the infinite approach to the real world's transmission state. An additional uncon-trolled factor is the effect of 'Chunyun' [19], which is hard to be measured in the prediction process. Furthermore, the summary of error results is subject to inevitable fluctuation. The fluctuation phenomenon is intrinsically one of the challenges of the deterministic epidemic model, which reflects the likelihood that the precision of a deterministic epidemic model will vary across the 'life cycle' of an epidemic outbreak when analyzed using a set of fixed parameters. This will clearly influence the results across a long forecast period; Therefore, we pro-vided a 28 days prediction horizon, which is much longer than most prior stud-ies. Future studies that adopt a 'stage' forecast in the life cycle of an epidemic are clearly indicated. Despite its limitations, this study indicates the effective-ness of incorporating population mobility and random effects into the epidemic simulation.

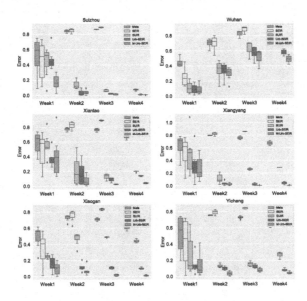

Fig. 2. Prediction error bars of models after T days.

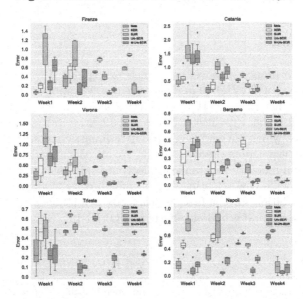

Fig. 3. Prediction error bars of models after T days.

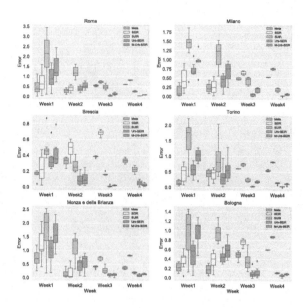

Fig. 4. Prediction error bars of models after T days.

5   Conclusion

In this paper, we propose a novel urban network epidemic framework to study the spread pattern of COVID-19 in different cities. We applied the framework to simulate and predict the COVID-19 confirmed cases in 'epicenter' Wuhan and other 11 cities in Hubei Province of China and 12 cities in Italy with severe epidemic situations, which outperforms other deterministic state-of-the-art models in the COVID-19 spreading prediction task. We also demonstrated that incorporating population mobility and random effect into epidemic models is necessary. Our findings provide new scientific evidence for further epidemic model design and offer a foundation for conducting other research studies, such as assessing the long-term social and economic effects of COVID-19.

References

1. Zhong, P., Guo, S., Chen, T.: Correlation between travellers departing from Wuhan before the spring festival and subsequent spread of Covid-19 to all provinces in China. J. Travel Med. **27**(3), taaa036 (2020)
2. Tian, H., et al.: An investigation of transmission control measures during the first 50 days of the Covid-19 epidemic in china. Science **368**(6491), 638–642 (2020)
3. Du, Z., et al.: Risk for transportation of coronavirus disease from Wuhan to other cities in China. Emerg. Infect. Diseas. **26**(5), 1049 (2020)
4. Badr, H.S., Du, H., Marshall, M., Dong, E., Squire, M.M., Gardner, L.M.: Association between mobility patterns and Covid-19 transmission in the USA: a mathematical modelling study. Lancet Infect. Disea. **20**(11), 1247–1254 (2020)

5. Cooke, K.L., Van Den Driessche, P.: Analysis of an SEIRS epidemic model with two delays. J. Math. Biol. **35**(2), 240–260 (1996)
6. Li, R., et al.: Substantial undocumented infection facilitates the rapid dissemination of novel coronavirus (SARS-COV-2). Science **368**(6490), 489–493 (2020)
7. Ball, F., Britton, T., Sirl, D.: A network with tunable clustering, degree correlation and degree distribution, and an epidemic thereon. J. Math. Biol. **66**(4), 979–1019 (2013)
8. Maki, Y., Hirose, H.: Infectious disease spread analysis using stochastic differential equations for sir model. In: 2013 4th International Conference on Intelligent Systems, Modelling and Simulation, pp. 152–156. IEEE (2013)
9. Pellis, L., et al.: Eight challenges for network epidemic models. Epidemics **10**, 58–62 (2015)
10. O'Dea, E.B., Pepin, K.M., Lopman, B.A., Wilke, C.O.: Fitting outbreak models to data from many small norovirus outbreaks. Epidemics **6**, 18–29 (2014)
11. Kermack, W.O., McKendrick, A.G.: A contribution to the mathematical theory of epidemics. Proc. Roy. Soc. Lond. Ser. A (Containing Papers of a Mathematical and Physical Character) **115**(772), 700–721 (1927)
12. Wang, N., Fu, Y., Zhang, H., Shi, H.: An evaluation of mathematical models for the outbreak of Covid-19. Precis. Clin. Med. **3**(2), 85–93 (2020)
13. Wang, J., Wang, X., Wu, J.: Inferring metapopulation propagation network for intra-city epidemic control and prevention. In: Proceedings of the 24th ACM SIGKDD International Conference on Knowledge Discovery and Data Mining, pp. 830–838 (2018)
14. Wang, J., Lin, X., Liu, Y., Feng, K., Lin, H., et al.: A knowledge transfer model for Covid-19 predicting and non-pharmaceutical intervention simulation. In: Proceedings of the 26th ACM SIGKDD International Conference on Knowledge Discovery and Data Mining (2020)
15. Hufnagel, L., Brockmann, D., Geisel, T.: Forecast and control of epidemics in a globalized world. Proc. Natl. Acad. Sci. **101**(42), 15124–15129 (2004)
16. Cintrón-Arias, A., Castillo-Chávez, C., Betencourt, L., Lloyd, A.L., Banks, H.T.: The estimation of the effective reproductive number from disease outbreak data. Technical report, North Carolina State University, Center for Research in Scientific Computation (2008)
17. Pepe, E., et al.: Covid-19 outbreak response, a dataset to assess mobility changes in Italy following national lockdown. Sci. Data **7**(1), 1–7 (2020)
18. Adam, D.: A guide to R - the pandemic's misunderstood metric. Nature **583**(7816), 346–349 (2020)
19. Jia, J.S., Lu, X., Yuan, Y., Xu, G., Jia, J., Christakis, N.A.: Population flow drives spatio-temporal distribution of Covid-19 in china. Nature **582**(7812), 389–394 (2020)

Skin Cancer Classification Using Different Backbones of Convolutional Neural Networks

Anh T. Huynh[1,2], Van-Dung Hoang[3], Sang Vu[2,4], Trong T. Le[1,2],
and Hien D. Nguyen[2,5(✉)]

[1] Faculty of Software Engineering, University of Information Technology, Ho Chi Minh City, Vietnam
{anhht,tronglt}@uit.edu.vn
[2] Vietnam National University, Ho Chi Minh City, Vietnam
[3] Faculty of Information Technology, HCMC University of Technology and Education, Ho Chi Minh City, Vietnam
dunghv@hcmute.edu.vn
[4] Faculty of Information System, University of Information Technology, Ho Chi Minh City, Vietnam
sangvm@uit.edu.vn
[5] Faculty of Computer Science, University of Information Technology, Ho Chi Minh City, Vietnam
hiennd@uit.edu.vn

Abstract. Melanoma is the deadliest of many different types of skin cancer. Clinical screening is followed by dermoscopic analysis and histopathological examination in the diagnosis of melanoma. Melanoma is a type of skin cancer that is highly curable if caught early. A visual examination of the affected area of the skin is the first step in melanoma skin cancer diagnosis. Dermatologists use a high-speed camera to take dermatoscopic images of skin lesions, which have an accuracy of 65–80% in melanoma diagnosis without any additional technical support. This research shows how to classify skin cancer using skin lesion photos using an automated classification approach based on image processing techniques. By studying images of skin lesions, the classification system will be able to determine whether or not a patient has melanoma. The contribution of this paper includes testing many different backbones and input sizes on the CNN models to evaluate the accuracy of the model on the siim-isic dataset. The overall prediction rate of melanoma diagnosis was raised to 82–86% on Sensitivity.

Keywords: Skin cancer · Benign · Malignant · Melanoma · CNN · Deep learning

1 Introduction

Melanoma is the least prevalent type of skin cancer, but it is responsible for 75% of all skin cancer deaths [1]. Early detection is critical. Skin imaging is being facilitated by the International Skin Imaging Collaboration (ISIC) to reduce melanoma mortality. Melanoma can be curable if caught early and treated properly. The utilization of digital

H. Fujita et al. (Eds.): IEA/AIE 2022, LNAI 13343, pp. 160–172, 2022.
https://doi.org/10.1007/978-3-031-08530-7_14

skin lesion imaging to construct a remote chemo autonomous diagnostic system that can enhance clinical decision-making is possible [2].

Artificial Intelligence technologies brings many benefits in healthcare. It has been changed medical practice and biomedical applications [3]. Deep learning has transformed the future because of its ability to solve complex problems [4]. This approach has been applied in many fields, such as autonomous vehicle control [5], sentiment analysis [6], estimating Air Quality Index (AQI) [7], and extracting information for resume filter [8]. In the processing of medical image, the feature learning is necessary to determine specified features for prediction [9, 10]. The goal was to create a system that would help dermatologists improve their diagnostic accuracy by merging contextual photos and patient-level data and eliminating bias in the model's predictions [11]. The purpose is to evaluate the capacity of deep learning with the performance of highly educated dermatologists. Overall, the mean findings suggest that all deep learning models excelled dermatologists.

The overarching purpose is to aid in the fight against skin cancer fatalities. The utilization of advanced image classification technologies for human well-being is the project's main driving factor. Machine learning and deep learning in computer vision have achieved significant progress that is scalable across domains. We hope to bridge the gap between diagnosis and therapy with the help of this effort. The dermatological clinic's job can be aided by completing the research with more precision on the dataset. The model's improved accuracy and efficiency could help detect melanoma earlier and decrease wasteful biopsies.

The next section presents some related works about the methods for skin cancer classification. Section 3 describes information about the datasets we are using in this paper, data pre-processing and data augmentation in these datasets. Section 4 presents the model configuration of a skin cancer classification system, we used different backbones of Convolution Neural Networks (CNN) to find out what's model has met expectations with the best evaluation indicators. Experimental results of our proposed methods are shown in Sect. 5. The last section concludes the paper and gives some future works.

2   Related Work

There are several studies for solving the problem about skin cancer using deep learning. They proposed many automated systems with different approaches. In [12], authors used a deep-CNN architecture that has been tuned, as well as proprietary mini-batch logic and a loss function to achieve promising performance on the problem of Skin Cancer Classification. They proposed a hybrid method for dealing with skin-disease classification class imbalances. It combines the data level method of balanced mini-batch logic followed by real-time image augmentation with the algorithm level method of the newly designed loss function to address the problem of slow learning of minority classes in networks. Besides using for detecting diabetic retinopathy [13], Deep CNN has been also used to classify skin lesions in several recent studies, but there are still many challenges due to data limitations and imbalance issues [14].

In [15], they present their SIIM-ISIC Melanoma Classification Challenge winning solution. It's a collection of convolutional neural network (CNN) models with various backbones and input sizes, the majority of which are image-only models, with a few incorporating image- and patient-level metadata. The following factors contributed to our victory: (1) a stable validation scheme; (2) a well-chosen model target; (3) a finely tuned pipeline; and (4) assembly with a wide range of models. The winning entry received a cross validation AUC of 0.9600 and a private leaderboard AUC of 0.9490.

In [16], the ISIC2020 dataset is used to train and evaluate their proposed model. Furthermore, experimental results demonstrate that our method outperforms other deep-learning approaches. With the AUC metric, our DenseNet model achieves 0.925, which is higher than approaches using VGG and ResNet backbones. In [17], They also proposed an InSiNet architecture to detect benign and malignant lesions. They used algorithm and other machine learning techniques (GoogleNet, DenseNet-201, ResNet152V2, EfficientNetB0, RBF-support vector machine, logistic regression, and random forest) and achieved high accuracy in ISIC 2018, 2019, and 2020 datasets. In [18], they used CNN as an approach to help build intelligent systems as a foundation, combined with a proposed new method for aggregating results called adaptive thresholds for data specifically, along with the optimization model to make reliable conclusions, their method is also applied by us in our method.

This study proposes a method of using multiple backbones in combination with the use of both ISIC 2019, 2020 datasets and data augmentation to reduce the imbalance between the two layers of benign and malignant and use TFRecords together with TPU. combine logical data arrangement, network structure with loss function we use Binary Focal Loss mentioned in [19].

3 Dataset

The dataset is open to the public on Kaggle (SIIM-ISIC Melanoma Classification, 2020) [20]. JPEG and TFRecord image formats are also available (in the jpeg and tfrecords folders, respectively). The TFRecord image has been scaled to a uniform size of 1024×1024. This dataset is split into two parts: a training set (33,126 images) and a testing set (10,982 images). However, a considerable class imbalance in the training dataset was discovered by our fast investigation. 31,956 images were classified as benign (non-cancerous), whereas just 575 images were classified as cancerous (Cancer). That is, just 1.76 percent of the malignant images are included in the training set (Fig. 1).

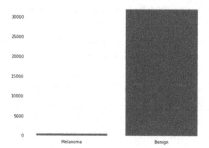

Fig. 1. ISIC 2020 target distribution

Classifiers often face challenges when trying to maximize both precision and recall, which is especially true when working with imbalanced datasets. To address the issue of data imbalance, we combined data from the 2019 competition (ISIC 2019) [21–23] with data from the 2020 competition. There are a total of 25,331 images in the 2019 dataset, with the Training set (25,331 images) and Testing set (8,238 images), in the Training set, 4,522 images are positive (17.85%) (Fig. 2).

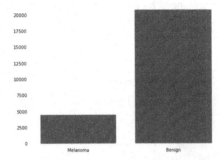

Fig. 2. ISIC 2019 target distribution

Figure 3 and Fig. 4 show some images in the dataset about malignant and benign melanoma.

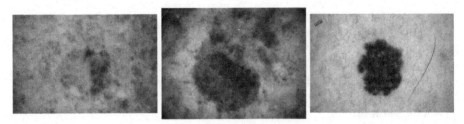

Fig. 3. Malignant melanoma

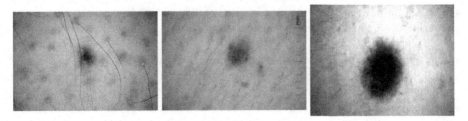

Fig. 4. Benign melanoma

Data Pre-Processing

The training set of the ISIC 2020 and 2019 datasets (58,457 images) from the ISIC dataset is used for training. This set will be divided into two training and validation sets, with the training set accounting for 80% and the validation set accounting for 20%. The train set will contain 42,212 benign images and 4,077 malignant images, while the validate set will contain 10,553 benign images and 1,020 malignant images. These sets will be center cropped and resized to 256×256 pixels, and the 4,077 malignant images in the training set will be enhanced by ten times to 40,770 pixels to reduce the loss. To train TensorFlow models, there is a total of 82,982 images for the training set and 11,573 images for the validation set. Figure 5 shows the final dataset:

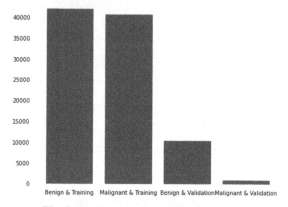

Fig. 5. Final training and validation dataset

The benign and malignant images are saved ordinarily in the training set's TFRecords to ensure that the model sees both positive and negative cases in the same batch.

Data Augmentation

The augmentation will help the model predict more accurately is chosen. The chosen augmentations are as follows:

- Random Crop: Crop images to the same size making the model easier to use.
- Transpose: A spatial level transformation that transposes an image by swapping rows and columns.
- Flip: When an image is flipped horizontally or vertically, it is referred to as a spatial level transition. Images are randomly flipped horizontally or vertically to make the model more robust.
- Rotate: This is a spatial level transformation that rotates images at random to ensure that they are distributed evenly. Due to random rotation, the model becomes invariant to the object orientation.
- Random Brightness: This is a pixel-level adjustment that changes the brightness of an image at random. Because we don't always have ideal lighting conditions in real life, this enhancement aids in simulating those conditions.

- Gaussian Blur: a pixel-level transformation that uses a gaussian filter to blur the input image.
- Motion Blur: This is a pixel-level transformation that applies motion blur using a random-sized kernel.
- Random Contrast: This is a pixel-level transformation that randomly changes the contrast of the input image. Because we don't always have ideal lighting conditions in real life, this enhancement aids in simulating those conditions.
- Median Blur: This is a pixel-level transformation that blurs the input image using a median filter.
- Gauss Noise: Adds Gaussian noise to an image at the pixel level. This augmentation will imitate measurement noise while the photos are being taken.
- Optical Distortion (lens error): It imitates the lens distortion effect.

Figure 6 shows the augmented image before and after. The training dataset is the only one that is augmented, while the validation datasets are only normalized.

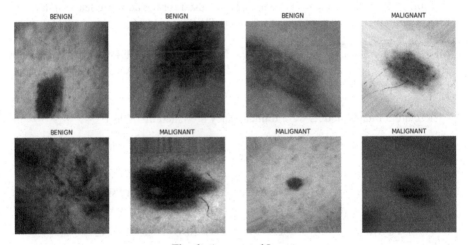

Fig. 6. Augmented Images

4   Model Configuration

Until now, typical machine learning and deep learning algorithms have been designed to work in isolation. These algorithms have been programmed to solve certain problems. When the feature-space distribution changes, the models must be rebuilt from the ground up. Transfer learning is the concept of breaking free from the isolated learning paradigm and applying what you've learned to solve related problems [24].

Traditional learning is isolated and entirely based on specific tasks and datasets, with isolated models trained on them separately [24, 25]. There is no knowledge that can be transferred from one model to another. Transfer learning allows you to use previously trained models' knowledge (features, weights, and so on) to train newer models and even solve problems like having less data for more recent tasks.

In this study, we used the backbones of the models EffecientNetB6, VGG16, ResNet152V2, InceptionResNetV2, InceptionV3. The input will be the dataset above, and the output will include two layers: the first is GlobalAveragePooling2D, and the second is Dense with units = 1, activation = 'sigmoid', and a bias initializer initialized with np.log([pos/neg]).

To get the final prediction, the ensemble terminology is used to train various models and then took the average probability ranks of the models. The model is setup as follows:

- Backbone Pre-trained CNN Model: Use the EffecientNetB6, VGG16, ResNet152V2, InceptionResNetV2, InceptionV3 as they have achieved higher accuracy on ImageNet competition.
- Targets: All in the model are trained on binary categories.
- The original images have been cropped to 256×256 pixels in the center. This cropping gets rid of the random noise and the black border around the edges of the images.
- Optimizer: Because of the sparse data, adam is used for the adaptive learning rate.
- Loss function: Binary Focal Loss function in [19] is used to train the model because of its higher accuracy when using Binary Cross Entropy.
- Metrics: some measures are used, such as True Positives, False Positives, True Negatives, False Negatives, Binary Accuracy, Precision, Recall, Area Under the Curve, Precision-Recall Curve.

There are 5 image models were trained. They are listed in Table 1:

Table 1. Image Models

Model	Backbone	Target	Data	Image size	Epochs
1	EffecientNetB6	2c	2019–2020	256×256	40
2	VGG16	2c	2019–2020	256×256	40
3	ResNet152V2	2c	2019–2020	256×256	40
4	InceptionResNetV2	2c	2019–2020	256×256	40
5	InceptionV3	2c	2019–2020	256×256	40

5 Experimental Results

There are 40,962,441 features were extracted after using the EfficientNetB6 model to solve the problem of skin cancer identification through image datasets. The result is shown below after 40 epochs (Fig. 7):

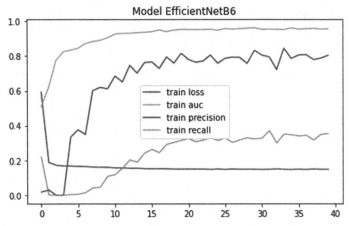

Fig. 7. Training results of EfficientNet B6 model.

Overall, the experimental results with the EfficientNetB6 model produced positive results for the following indicators: loss: 0.1480 - auc: 0.9559 - specificity: 0.74 - sensitivity: 0.84. At epoch 20, the model clearly converges, and the model's results do not improve any further. The model, on the other hand, is not yet overfit. Since epoch 2, the loss index has remained relatively constant.

The total number of features extracted is 14,715,201 after using the VGG16 model. As shown below, the result is obtained after 40 epochs (Fig. 8):

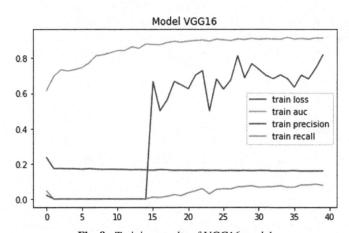

Fig. 8. Training results of VGG16 model.

In general, the experimental results with the VGG16 model achieved good results with the following indicators: loss: 0.1593 - auc: 0.9139 - specificity: 0.75 - sensitivity: 0.84. The pattern peaked at epoch 27 and barely improved after this epoch. On the other hand, the model has not been overfit yet. The loss index fell sharply from epoch 2 and has remained virtually unchanged since. Precision increases sharply at epoch 15.

After using the ResNet152V2 model, the total number of features extracted is 58,333,697. The result is obtained after 40 epochs, as shown below (Fig. 9):

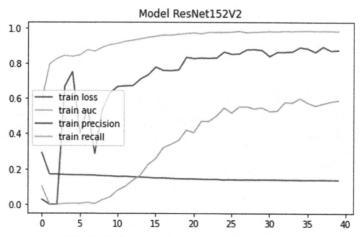

Fig. 9. Training results of ResNet152V2 model.

In general, the experimental results with the ResNet152V2 model were good, as evidenced by the following indicators: loss: 0.1365 - auc: 0.9806 - specificity: 0.67 - sensitivity: 0.86. At epoch 24, the converging pattern is clearly visible, and it doesn't get much better after that. The model, on the other hand, has not yet been overfit. Since epoch 2, the loss index has dropped sharply and has remained virtually unchanged. Starting at epoch 8, precision and recall metrics are stable and incremental.

After using InceptionResNetV2 model, the total number of features extracted is 54.338.273. The result is obtained after 40 epochs, as shown below (Fig. 10):

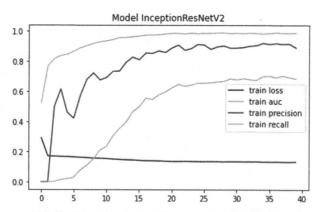

Fig. 10. Training results of InceptionResNetV2 model.

In general, the InceptionResNetV2 model experimental results were good in terms of the following indicators: loss: 0.1326 - auc: 0.9878 - specificity: 0.81 - sensitivity:

0.84. At epoch 21, the converging pattern is clearly visible, and it doesn't get much better after that. The model, on the other hand, has not yet been overfit. Since epoch 2, the loss index has dropped sharply and has remained virtually unchanged. Starting with epoch 5, precision and recall metrics are stable and incremental.

Total of 21,804,833 features extracted from the image dataset after using the InceptionV3 model. As shown below, the result is obtained after 40 epochs (Fig. 11).

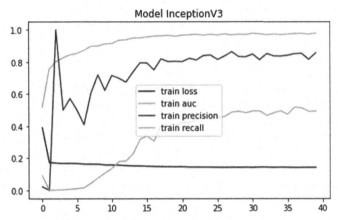

Fig. 11. Training results of InceptionV3 model.

In general, the experimental results with the InceptionV3 model achieved good results with the following indicators: loss: 0.1410 - auc: 0.9788 - specificity: 0.79 - sensitivity: 0.82. The converging pattern is clearly visible at epoch 17 and barely improves after this epoch. On the other hand, the model has not been overfit yet. The loss index fell sharply from epoch 2 and since then there has been almost no significant change. Precision and recall metrics are stable and incremental starting from epoch 6.

After applying 5 models, after 40 epochs, the following table summarizes the results (Table 2):

Table 2. The table summarizes the results of the models after 40 epochs.

Model	Loss	AUC	Specificity	Sensitivity
EfficientNetB6	0.1480	0.9559	0.74	0.84
VGG16	0.1593	0.9139	0.75	0.84
ResNet152V2	0.1365	0.9806	0.67	0.86
InceptionV3	0.1410	0.9788	0.79	0.82
InceptionResNetV2	0.1326	0.9878	0.81	0.84

The InceptionResNetV2 and VGG16 models have the highest and lowest results, respectively, among the selected models, based on the obtained results. This is not surprising given that the VGG16 model was released in 2014, while the InceptionRes-NetV2 model was released later that year (2016) and was also created by combining two intensive CNN models, ResNet and InceptionV3.

6 Conclusion and Future Work

In this paper, the method to classify skin cancer images through five different backbones is proposed on the ISIC 2020 and 2019 datasets and obtained the published results. The image preprocessing by combining more malignant images from the 2019 dataset with enhancing the dataset by performing augmentation and saving the dataset in order of interlacing classes into a TFRecords file which helps the dataset becomes more balanced. Using TFRecords also helps the training process take advantage of the power of TPU Kaggle, making the training process faster than using the GPU (about less than 3 h for every model). Using the Binary Focal Loss function also makes the results more accurate than using other loss functions.

Finally, experimenting on different backbones finds out that the InceptionResNetV2 model has met expectations with the best evaluation indicators. In addition, there are internal factors such as the methods still have some shortcomings and the data sets have different characteristics, so it requires the research team to be flexible, continuously testing many models in many cases to achieve the desired result. In the next studies, the ways to overcome the shortcomings would be studied and optimized to be more accurate than the current method. Moreover, the integrating between CNN and a knowledge-based system [26], as Graph Neural Network [27], will be also an emerging approach to design the explanation system for diagnosing in skin cancer.

Acknowledgment. This research is supported by the VNUHCM-University of Information Technology's Scientific Research Support Fund.

References

1. Carvajal, R.D., Marghoob, A., Kaushal, A., et al.: Melanoma and Other Skin Cancers. Cancer Network (2015) https://www.cancernetwork.com/view/melanoma-and-other-skin-cancers
2. Habuza, T., Navaz, A., Hashim, F., et al.: AI applications in robotics, diagnostic image analysis and precision medicine: current limitations, future trends, guidelines on CAD systems for medicine. Informatics in Medicine Unlocked **24**, 100596 (2021). https://doi.org/10.1016/j.imu.2021.100596
3. Yu, K.H., Beam, A., Kohane, I.: Artificial intelligence in healthcare. Nature Biomedical Eng. **2**, 719–731 (2018)
4. Goodfellow, I., Bengio, Y., Courville, A.: Deep Learning. The MIT Press (2016)
5. Huynh, A., Nguyen, B.T., Nguyen, H.T, et al.: A method of Deep Reinforcement Learning for Simulated Autonomous Vehicle Control. In: Proceedings of 16th International Conference on Evaluation of Novel Approaches to Software Engineering (ENASE 2021), Online streaming (2021).
6. Nguyen, H.D., Huynh, T., Hoang, S., Pham, V., Zelinka, I.: Language-oriented Sentiment Analysis based on the grammar structure and improved Self-attention network. In: Proceedings of 15th International Conference on Evaluation of Novel Approaches to Software Engineering (ENASE 2020), Prague, Czech Public (2020)

7. Duong, D., Le, Q., Nguyen-Tai, T.L., et al.: An effective AQI estimation using sensor data and stacking mechanism. In: Proceedings of 20th International Conference on Intelligent Software Methodologies, Tools, and Techniques (SOMET 2021), Cancun, Mexico. FAIA 337, pp. 405418 (2021) IOS Press
8. Phan, T., Pham, V., Nguyen, H., et al.: Ontology-based resume searching system for job applicants in information technology. In: Proceedings of 34th International Conference on Industrial, Engineering & Other Applications of Applied Intelligent Systems (IEA/AIE 2021), Kuala Lumpur, Malaysia. LNAI 12798, pp. 261 – 273 (2021). Springer
9. Nguyen, D., Nguyen, T., Vu, H., et al.: TATL: task agnostic transfer learning for skin attributes detection. Med. Image Anal. **78**, 102359 (2022)
10. Nguyen, H., Sakama, C.: Feature learning by least generalization. In: Proceedings of the 30th International Conference on Inductive Logic Programming (ILP 2021), Online streaming. LNCS, vol. 13191, pp. 193202 (2021). Springer, Cham
11. Pham, V., Nguyen, H., Pham, B., et al.: Robust engineering-based unified biomedical imaging framework for liver tumor segmentation. Current Medical Imaging (2022). https://doi.org/10.2174/1573405617666210804151024
12. Pham, T.C., Doucet, A., Luong, C.M., et al.: Improving skin-disease classification based on customized loss function combined with balanced mini-batch logic and real-time image augmentation. IEEE Access **8**, 150725–150737 (2020)
13. Nguyen, H., Tran, V., Pham, V., et al.: Design a learning model of mobile vision to detect diabetic retinopathy based on the improvement of MobileNetV2. Int. J. Digit. Enterp. Technol. (IJDET) **2**(1), 38–53 (2022)
14. Pham, T.C., Luong, C.M., Hoang, V.D., Doucet, A.: AI outperformed every dermatologist: Improved dermoscopic melanoma diagnosis through customizing batch logic and loss function in an optimized deep CNN architecture. Scientific Reports **11**, 17485 (2021)
15. Ha, Q., Liu, B., Liu, F.: Identifying Melanoma Images using EfficientNet Ensemble: Winning Solution to the SIIM-ISIC Melanoma Classification Challenge (2020). https://arxiv.org/pdf/2010.05351v1.pdf
16. Zhang, Y., Wang, C.: SIIM-ISIC melanoma classification with DenseNet. In: Proceedings of the IEEE 2nd International Conference on Big Data, Artificial Intelligence and Internet of Things Engineering (ICBAIE 2021), pp. 1417, Nanchang, China (2021). https://doi.org/10.1109/ICBAIE52039.2021.9389983
17. Reis, H.C., Turk, V., Khoshelham, K., Kaya, S.: InSiNet: a deep convolutional approach to skin cancer detection and segmentation. Med Biol Eng Comput **60**(3), 643–662 (2022)
18. Le, T.H.V., Van, H.T., Tran, H.S., Nguyen, P.K., Nguyen, T.T., Le, T.H.: Applying convolutional neural network for detecting highlight football events. In: Cong Vinh, P., Rakib, A. (eds.) ICCASA 2021. LNICSSITE, vol. 409, pp. 300–313. Springer, Cham (2021). https://doi.org/10.1007/978-3-030-93179-7_23
19. Lin, T., Goyal, P., Girshick, R., et al.: Focal loss for dense object detection. IEEE Trans. Pattern Anal. Mach. Intell. **42**(2), 318–327 (2020)
20. Rotemberg, V., Kurtansky, N., Betz-Stablein, B., et al.: A patient-centric dataset of images and metadata for identifying melanomas using clinical context. Sci Data **8**, 34 (2021). https://doi.org/10.1038/s41597-021-00815-z
21. Codella, N., Gutman, D., Celebi, M., et al.: Skin lesion analysis toward melanoma detection: a challenge at the 2017 International Symposium on Biomedical Imaging (ISBI), Hosted by the International Skin Imaging Collaboration (ISIC). In: Proceedings of IEEE 14th International Symposium on Biomedical Imaging (ISBI 2017), Washington DC, USA (2018)
22. Combalia, M., Codella, N., Rotemberg, V., et al.: BCN20000: Dermoscopic Lesions in the Wild. (2019) https://core.ac.uk/download/pdf/286456448.pdf

23. Tschandl, P., Rosendahl, C., Kittler, H.: The HAM10000 dataset, a large collection of multi-source dermatoscopic images of common pigmented skin lesions. Sci. Data **5**, 180161 (2018). https://doi.org/10.1038/sdata.2018.161
24. Olivas, E., Guerrero, J., Sober, M., et al.: Handbook Of Research On Machine Learning Applications and Trends. Information Science Reference (2009)
25. Gollapudi, S.: Deep learning for computer vision. In: Learn Computer Vision Using OpenCV, pp. 51–69. Apress, Berkeley, CA (2019). https://doi.org/10.1007/978-1-4842-4261-2_3
26. Nguyen, H.D., Do, N.V., Pham, V.T.: A method for designing knowledge-based systems and application. In: Elgnar, A., et al. (eds.) Applications Computational Intelligence Multi Disciplinary Research, Academic Press, Elsevier (2022)
27. Scarselli, F., Gori, M., Ah Chung, T., et al.: The graph neural network model. IEEE Trans. Neural Networks **20**(1), 61–80 (2009)

Cardiovascular Disease Detection on X-Ray Images with Transfer Learning

Nguyen Van-Binh[1] and Nguyen Thai-Nghe[2(✉)]

[1] Tam Binh High School, Vinh Long, Vietnam
[2] Can Tho University, 3-2 Street, Can Tho, Vietnam
ntnghe@cit.ctu.edu.vn

Abstract. Cardiovascular disease is one of the most dangerous and common diseases in Vietnam and the World today. More worrying is that this disease commonly happened in young people in recent years. Especially in the context of the complicated developments of the COVID-19 pandemic, people with cardiovascular disease are at high risk of being infected by the Corona virus. Therefore, the identification and early diagnosis of cardiovascular disease are important and necessary research to help the patients. In this work, we propose using a transfer learning approach to detect and identify two common types of cardiovascular diseases, which are *cardiomegaly disease* and *aortic aneurysm disease*, through X-ray chest images. Specifically, this study used the transfer learning method with the pre-trained VGG16 deep learning model, combined with data pre-processing to identify cardiovascular diseases. Experiments are performed on a dataset that has been labeled by experts in the field of cardiology using three scenarios. Experimental results from three scenarios show that this approach is satisfactory with the accuracy of 0.95, 0.96, and 0.70, respectively.

Keywords: Cardiovascular disease · Aortic aneurysm · Cardiomegaly · Deep learning · VGG16

1 Introduction

According to statistics of the Ministry of Health in Vietnam, every year, there are about 200,000 people who died from diseases related to cardiovascular and more than 20,000 people died from cardiovascular diseases. This number is higher than the number of deaths caused by cancer. Cardiovascular diseases are the leading causes of death and the number of people suffering from these diseases at a young age is increasing. These are also the main causes of stroke and sudden death syndrome, both of which have increased in number and are being happened in young people recently. People with these syndromes are often due to cardiovascular or pulmonary diseases that are not detected and treated in time, which can lead to death. Especially in the context of complicated developments of the COVID-19 pandemic, people with cardiovascular disease are at high risk of being infected with Corona virus. However, the diagnosis of these diseases is made through X-ray images and the accuracy is mainly based on the professional ability and the knowledge of the doctors who read these images.

© Springer Nature Switzerland AG 2022
H. Fujita et al. (Eds.): IEA/AIE 2022, LNAI 13343, pp. 173–183, 2022.
https://doi.org/10.1007/978-3-031-08530-7_15

To reduce the load and better support the doctors when reading X-ray images, especially when working with high intensity and pressure when reading a lot of images in a long time, as well as to increase the accuracy in diagnosing cardiovascular disease, building an application that can detect and identify the cardiovascular disease is very necessary for practice.

This work proposes an approach in diagnosing cardiovascular disease based on chest X-ray images by using transfer learning. For instance, we perform the identification of two common types of cardiovascular diseases which are *cardiomegaly disease* and *aortic aneurysm disease*. This study first pre-processes the input images, then, uses the transfer learning model from the pre-trained VGG16 deep learning model to identify cardiovascular diseases. Three scenario experiments on a dataset, which have been labeled by experts in the field of cardiology, show very positive results.

This work consists of 5 parts. Sect. 1 is an introduction and definition of the problem. Related studies will be presented in Sect. 2. Section 3 is the implementation methodology. The next part is the experiments. Finally, the conclusion is presented.

2   Related Work

There are many related studies in the field of cardiovascular disease identification, in this section, we will summarize some recent studies.

In the study [1], the authors proposed a deep learning-based technique to detect and quantify abdominal aortic aneurysms (AAA). This disease, which leads to more than 10,000 deaths each year in the United States, is asymptomatic, often discovered incidentally, and often missed by radiologists. The proposed model architecture is a modified 3D U-Net incorporating an elliptical joint that performs aortic segmentation and AAA detection. The study used 321 abdominal-pelvic CT exams performed by the radiology department of Massachusetts General Hospital for training and testing. The model was then further tested for generalizability on a set of 57 separate tests incorporating different patient demographics compared to the original data set. The model has achieved high performance on both test datasets.

In the study [2], the authors used a deep convolutional neural network to detect enlarged hearts in digital X-rays. First, the author used and connected several deep CNN architectures to detect the presence of cardiac pathology in chest X-rays. Next, they introduced a pre-trained model. The authors used two publicly available datasets, the NLM-Indiana, and NIH-CXR datasets. With this method, they obtained certain results with an accuracy of 0.88, an F1 measure of 0.89 on the refined VGG16 model.

In the study [3], the authors proposed an automated method based on deep learning to calculate the thoracic ratio and detect the presence of cardiomegaly from chest radiographs. They developed two separate models to delineate the cardiac and thoracic regions in X-ray images using bounding boxes and used their outputs to calculate thoracic ratios. They obtained an accuracy of about 96%.

In the study [4], the authors used 2000 chest X-ray images with 1000 normal images and 1000 images of cardiomegaly. An optimized logistic regression algorithm was used to classify the disease. With the VGG19 model, the accuracy obtained is about 84%. In the study [5], the author's team performed a hyper-parameter search on the training models, the parameters including architecture, learning, … were systematically optimized. The authors have tested two methods, segmentation and classification. The obtained results show that segmentation provides higher accuracy than classification. The AUC values of segmentation and classifier are 0.977 and 0.941, respectively.

In the study [6], the authors proposed an automated method to detect, segment, and classify abdominal aortic aneurysms (AAA) in computed tomography (CT) imaging. The DBN network was applied for AAA detection and severity classification in this study. The optimal parameters for training the DBN are determined for the training data from the selected dataset. The AAA region can be successfully segmented from CT images and the results are comparable with currently available methods. In the study [7], three neural network models (ResNet, VGG-16, and AlexNet) were adapted for 3D classification and applied to a dataset consisting of 187 heterogeneous CT scans. The ResNet 3D model performed better than the other two. The result achieved an accuracy of 0.856 and an AUC of 0.926.

In the study [8], a proposed algorithm named CardioXNet uses deep learning with U-NET method and thoracic ratio to diagnose cardiomegaly by chest X-ray. U-NET learns the segmentation task, OpenCV is used to eliminate and maintain the accuracy of the region of interest when minor errors occur. Thoracic ratio (CTR) was calculated as a criterion to determine cardiomegaly from the U-net segments. This study has shown that the feasibility of combining deep learning segmentation and medical criteria to automatically recognize heart disease on medical images with high accuracy with ACC and F1 is 93.75% and F1 respectively. 94.34% and consistent with the laboratory results. In the study [9], the authors applied a deep learning method to identify enlarged heart disease from X-ray images. They tested the algorithms on a dataset containing 600 images and obtained the best performance with an AUC of 0.87 and an accuracy of 0.65 to 0.79 on different CNN network models, using the transfer learning method [14]. This result demonstrates the feasibility of developing computer-aided diagnostic systems for various pathologies from X-rays using deep learning techniques.

In this study, we propose using the transfer learning method from the pre-trained VGG16 model and retrain this model on pre-processing data to diagnose two common types of cardiovascular diseases based on X-ray chest images.

3 Proposed Method

To identify and classify cardiovascular diseases, this study will compare two implementation methods which are recognition from original data and recognition from the preprocessed data.

3.1 Data Pre-processing

The problem in this work is that we will process on X-ray images. This kind of image is transparent and difficult to extract the features or signs of the diseases and usually requires experts or specialized doctors to be able to recognize. Thus, this type of image will also have a significant impact on machine learning algorithms. In this study, we perform contrast improvement of the image to clarify the features and signs inside the image as shown in Fig. 1. However, performing this pre-processing also has certain difficulties, because increasing the contrast may lead to distortion of the features of the original signal leading to the decreasing of training accuracy. Therefore, we need to perform many times with many different methods to evaluate the obtained results to find the method that gives the highest efficiency.

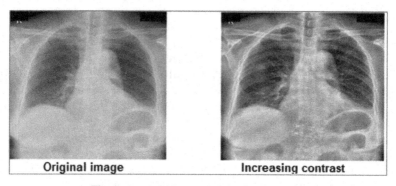

Fig. 1. Image before and after pre-processing

Besides, this study also used the data augmentation method to increase the number of images from the input data set, such as the image can be rotate 10 degree in both horizontal and vertical. The augmentation parameter values are set as follows: <rotation_range = 10, zoom_range = 0.1, shear_range = 0.1, width_shift_range = 0.1, height_shift_range = 0.1, horizontal_flip = True, vertical_flip = True>.

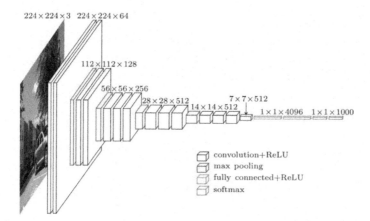

Fig. 2. VGG16 architecture [13] (source: neurohive.io/en/popular-networks/vgg16)

3.2 Proposed Model for Cardiovascular Disease Detection

To identify two common cardiovascular diseases which are aortic aneurysm and cardiomegaly, this study proposes using the transfer learning method by reusing the pre-
trained the VGG16 model [13]. The overall architecture as well as the parameters of the
VGG16 model are presented in Fig. 2. Moreover, for the re-training process to have the
best effect, this study also used fine-tuning of the model parameters. These two steps
are described in the following.

- **Transfer learning stage**
  The architecture of the proposed model is illustrated in Fig. 3. In this transfer learning
  model, we reuse the parameters of the pre-trained VGG16 model. This model has
  been pre-trained on the ImageNet dataset consisting of 1.2 million images and 1000
  classes, so the transfer learning will take advantage of the available parameters of this
  model in case the input data is small. To use in the problem of cardiovascular disease
  identification, the pre-trained VGG16 model will be removed from the last layer and
  replaced with new fully-connected layers and an output layer to classify 2 types of
  diseases which are *aortic aneurysm, cardiomegaly,* and the normal image.
- **Fine-tuning stage**
  We reused the hyper-parameters of the pre-trained VGG16 model, however, during
  the model calibration phase, we continue to fine-turn the hyper-parameters to help the
  model achieve the highest accuracy. The hyper-parameters used for the calibration

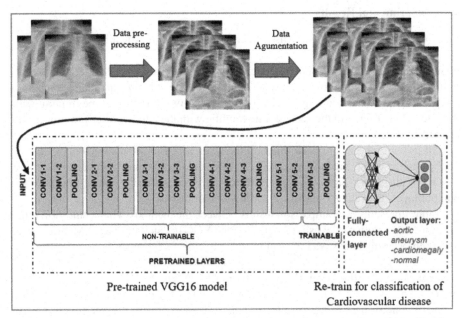

Fig. 3. Transfer learning model with VGG16 network for cardiovascular disease identification.
The last layer of the VGG16 model will be removed and replaced with new fully-connected layers
and an output layer for 2 types of cardiovascular diseases.

process are as follows: (a) Epochs are from 10 to 120 for finding a suitable epoch threshold; (b) Batch sizes are from 16 to 512; (c) Hidden layers are from 64 to 1024; (d) Learning rates are 0.001, 0.0001, and 0.00001.

4   Experiments

4.1   Data Set

The dataset used is the VinBigData dataset about abnormalities on chest X-ray images provided by VinGroup Research Institute [15]. The dataset has been pre-labeled by specialists/radiologists. This dataset includes approximately 12,000 chest X-ray images (7,000 images associated with thoracic aortic aneurysm and 5,000 associated with cardiomegaly in DICOM format). All of the images are resized to 256×256 before feeding to the model.

4.2   Evaluation Methods and Baselines

To evaluate the model, this study divides the data set into three subsets for training, testing, and evaluation with a ratio of 60-20-20. The Accuracy (acc) and F1 measures were used to compare with other models such as MobileNet [12] and Xception [10, 11].

4.3   Experimental Results

Three experimental scenarios are used to compare the accuracy results of the models.

a. Scenario 1: Identification of Thoracic Aortic Aneurysm Disease
In this scenario, we train the model with two classes which are *aortic aneurysm* and no disease (*normal*). The training process is carried out twice, the first time training with the original data set, and the second time training with the data set after pre-processing by increasing the contrast.

All training phases are performed on the same set of hyper-parameters: epochs = 20, batch-size = 32, hidden units = [512, 128], hidden layers = number of layers of VGG16 with its last layers are removed and replaced with new fully-connected and out_put layers to predict 2 classes of cardiovascular diseases.

Table 1 presents the comparison results between the methods. This result shows that the transfer learning model using the pre-trained VGG16 can improve the accuracy after data preprocessing and model fine-tuning.

b. Scenario 2: Identification of Cardiomyopathy Disease
In this scenario, we train the model with two classes of data, which are cardiomegaly disease and no disease (normal). The training process is done twice, the first time with the original data set, and the second time with the pre-processed data set by contrast enhancement.

Table 1. Result comparison on scenario 1

Phase	Model	Original data			Pre-processing data		
		Train acc	Test acc	F1	Train acc	Test acc	F1
Transfer learning	VGG16	0.97	0.94	0.94	0.97	**0.95**	**0.95**
	MobileNet	0.98	0.93	0.93	0.96	0.93	0.93
	Xception	0.96	0.92	0.92	0.97	0.88	0.89
Fine turning	VGG16	0.98	0.95	0.95	0.99	**0.95**	**0.95**
	MobileNet	0.98	0.92	0.92	0.97	0.92	0.92
	Xception	0.98	0.88	0.88	0.99	0.88	0.88

Table 2. Result comparison on scenario 2

Phase	Model	Original data			Pre-processing data		
		Train acc	Test acc	F1	Train acc	Test acc	F1
Transfer learning	VGG16	0.97	0.93	0.93	0.93	**0.95**	**0.94**
	MobileNet	0.98	0.94	0.94	0.98	0.94	0.94
	Xception	0.96	0.91	0.91	0.94	0.91	0.91
Fine turning	VGG16	0.99	0.94	0.94	0.98	**0.96**	**0.96**
	MobileNet	0.98	0.94	0.93	0.98	0.95	0.95
	Xception	0.97	0.88	0.88	0.98	0.89	0.89

Table 2 presents the comparison results between the methods in scenario 2. Similar to scenario 1, the results also show that the transfer learning approach with the pre-trained VGG16 model can improve accuracy after data preprocessing and fine-tuning.

Scenario 3: Identification of Both Aortic Aneurysms and Cardiomegaly
In this scenario, we train the model with three layers of data, which are cardiomegaly disease, aortic aneurysm disease, and normal images without the disease. The training process is done twice, the first time with the original data set, the second time with the data set after preprocessing by increasing the contrast. Table 3 presents the comparison results between the methods in scenario 3.

Similar to other problems in machine learning, as the number of classes to be predicted increases, the accuracy will also decrease. However, the transfer learning model still can improve the accuracy after data preprocessing and model refinement compared with other models.

4.4 Discussion on Experimental Results

Through experimental scenarios, we can see that pre-processing of the data combined with the transfer learning methods have helped to significantly improve the accuracy

Table 3. Result comparison on scenario 3

Phase	Model	Original data			Pre-processing data		
		Train acc	Test acc	F1	Train acc	Test acc	F1
Transfer learning	VGG16	0.59	0.63	0.59	0.60	0.61	0.52
	MobileNet	0.80	0.64	0.65	0.65	0.63	0.64
	Xception	0.76	0.61	0.61	0.69	0.62	0.62
Fine turning	VGG16	0.75	0.64	0.63	0.67	**0.70**	**0.70**
	MobileNet	0.84	0.62	0.63	0.86	0.65	0.65
	Xception	0.90	0.59	0.59	0.86	0.59	0.58

results compared to the other approaches. The summary of the comparison results of the proposed transfer learning using the VGG16 model is presented as shown in Fig. 4.

Moreover, to check whether the model is overfitting, we also looked at the accuracy and the loss during training and testing phases as presented in Fig. 5 and Fig. 6 (for scenario 3). These results show that the model works well.

In addition, the obtained results in scenario 2 can relatively be compared to other studies such as the study presented in [3] with an accuracy of 0.88 and the F1 of 0.89. The study in [4] has an accuracy of 0.85.

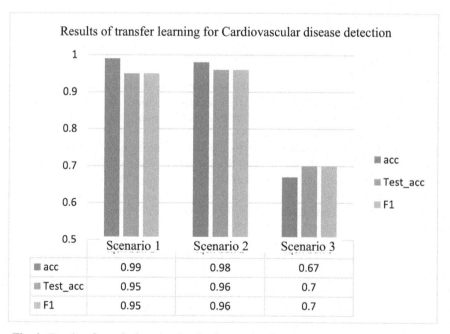

Fig. 4. Results of transfer learning for Cardiovascular disease detection on three scenarios

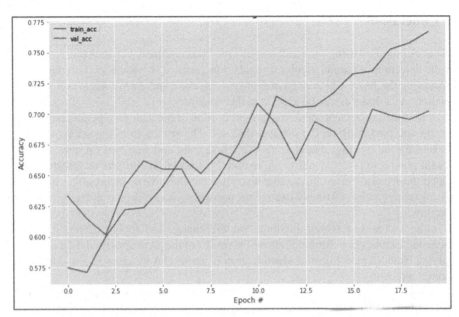

Fig. 5. Accuracy (acc) during training and testing phase – scenario 3 (two other scenarios are similar)

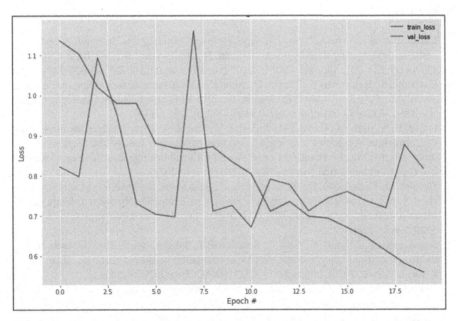

Fig. 6. Loss during training and testing phase – scenario 3 (two other scenarios are similar)

5   Conclusion

This work proposed using a transfer learning method combined with data pre-processing for the detection of cardiovascular diseases. For instance, we focus on two dangerous diseases of the heart which are *cardiomegaly disease* and *aortic aneurysm disease*, however, other diseases can be applied in the same way.

In all three experiment scenarios, pre-processing of the data gave better results than training and predicting on the original data set. The results from the scenarios show that reusing the existing model with VGG16 to train and fine-turn on new data has brought good results compared to the remaining models. The results of this study can support the detection of cardiovascular diseases from chest X-ray images. This may help patients receive early and timely treatments, prevent stroke syndrome and reduce patient mortality.

In the upcoming research direction, we will change and apply different data pre-processing techniques to further improve the results of the predictive model. In addition, applying Yolo version 5 in the identification of these two diseases is also a positive research direction.

Besides, it is necessary to collect more data on existing disease classes and expand other cardiovascular disease classes. Data augmentation also has the potential to improve model accuracy.

References

1. Lu, J.-T., et al.: DeepAAA: clinically applicable and generalizable detection of abdominal aortic aneurysm using deep learning. In: Shen, D., et al. (eds.) MICCAI 2019. LNCS, vol. 11765, pp. 723–731. Springer, Cham (2019). https://doi.org/10.1007/978-3-030-32245-8_80
2. Candemir, S., Rajaraman, S., Thoma, G., Antani, S.: Deep learning for grading cardiomegaly severity in chest X-rays: an investigation. In: 2018 EEE Life Sciences Conference (LSC), pp. 109–113 (2018). https://doi.org/10.1109/LSC.2018.8572113
3. Gupte, T., Niljikar, M., Gawali, M., Kulkarni, V., Kharat, A., Pant, A.: Deep learning models for calculation of cardiothoracic ratio from chest radiographs for assisted diagnosis of cardiomegaly. In: 2021 International Conference on Artificial Intelligence, Big Data, Computing and Data Communication Systems (icABCD), pp. 1–6 (2021)
4. Bougias, H., Georgiadou, E., Malamateniou, C., Stogiannos, N.: Identifying cardiomegaly in chest X-rays: a cross-sectional study of evaluation and comparison between different transfer learning methods. Acta Radiol. **62**(12), 1601–1609 (2021). https://doi.org/10.1177/028418 5120973630. PMID: 33203215
5. Sogancioglu, E., Murphy, K., Calli, E., Scholten, E.T., Schalekamp, S., Van Ginneken, B.: Cardiomegaly detection on chest radiographs: segmentation versus classification. IEEE Access **8**, 94631–94642 (2020). https://doi.org/10.1109/ACCESS.2020.2995567
6. Hong, H.A., Sheikh, U.U.: Automatic detection, segmentation and classification of abdominal aortic aneurysm using deep learning. In: 2016 IEEE 12th International Colloquium on Signal Processing and Its Applications (CSPA), pp. 242–246 (2016). https://doi.org/10.1109/CSPA. 2016.7515839
7. Golla, A.K., et al.: Automated screening for abdominal aortic aneurysm in CT scans under clinical conditions using deep learning. Diagn. (Basel). **11**(11), 2131 (2021). https://doi.org/ 10.3390/diagnostics11112131. PMID: 34829478; PMCID: PMC8621263

 8. Que, Q., et al.: CardioXNet: automated detection for cardiomegaly based on deep learning. In: Annual International Conference of the IEEE Engineering in Medicine and Biology Society, pp. 612–615, July 2018. https://doi.org/10.1109/EMBC.2018.8512374. PMID: 30440471
 9. Zhou, S., Zhang, X., Zhang, R.: Identifying cardiomegaly in ChestX-ray8 using transfer learning. Stud. Health Technol. Inf. **21**(264), 482–486 (2019). https://doi.org/10.3233/SHT I190268. PMID: 31437970
10. Chollet, F.: Xception: deep learning with depthwise separable convolutions. In: Proceedings of the IEEE Conference on Computer Vision and Pattern Recognition (2017)
11. Szegedy, C., et al.: Rethinking the inception architecture for computer vision. In: Proceedings of the IEEE Conference on Computer Vision and Pattern Recognition (2016)
12. Howard, A.G., et al.: MobileNets: efficient convolutional neural networks for mobile vision applications. arXiv preprint arXiv:1704.04861 (2017)
13. Simonyan, K., Zisserman, A.: Very deep convolutional networks for large-scale image recognition. arXiv:1409.1556 (2014)
14. Chang, H., Han, J., Zhong, C., Snijders, A.M., Mao, J.-H.: Unsupervised transfer learning via multi-scale convolutional sparse coding for biomedical applications. IEEE Trans. Pattern Anal. Mach. Intell. **40**(5), 1182–1194 (2018). https://doi.org/10.1109/TPAMI.2017.2656884
15. VinBigData. https://www.kaggle.com/c/vinbigdata-chest-xray-abnormalities-detection/data

Causal Reasoning Methods in Medical Domain: A Review

Xing Wu[1,2]([⊠]), Jingwen Li[1], Quan Qian[1,3], Yue Liu[1,2], and Yike Guo[4]

[1] School of Computer Engineering and Science,
Shanghai University, Shanghai 20444, China
xingwu@shu.edu.cn
[2] Shanghai Institute for Advanced Communication and Data Science,
Shanghai University, Shanghai 20444, China
[3] Materials Genome Institute, Shanghai University, Shanghai 20444, China
[4] Hong Kong Baptist University, Hong Kong, China

Abstract. Causal reasoning has been a key topic in medical domain with many applications, in which the core problem is to infer the causal effects of medical treatments with data mining. However, there are obstacles such as unstable identification and false associations when applying traditional machine learning methods dealing with the effect estimation about medical treatments due to the large-scale and high-dimensionality of medical data. Furthermore, there is no thorough survey of causal reasoning methods for medical domain problems, which is an emerging research direction. To meet the challenge, the causal reasoning in medical domain is surveyed to systematically classify and summarize causal reasoning methods in two dimensions: four categories of core ideas and three levels of causal structure. The thorough review demonstrates that causal reasoning methods have theoretical and practical significance in medical domain, which is a research field full of potential.

Keywords: Causality · Causal reasoning · Model-based reasoning · Causal effect estimation · Automated reasoning

1 Introduction

Machine learning methods have achieved great success in medical domain, but most of them require that the training data and test data should be independently and identically distributed, so predictions based on such models are particularly unstable. The main reason is that current machine learning methods are association-driven and do not distinguish between causal and spurious associations in data. Coupled with the high-dimensional characteristics of medical data, it is not enough to make a decision diagnosis in medical domain only by correlation analysis.

Causal reasoning has been regarded as the missing part of machine learning, providing robust identification of causal estimates of interest and effective methods to reveal causalities for real-world problems in a variety of fields, especially

© Springer Nature Switzerland AG 2022
H. Fujita et al. (Eds.): IEA/AIE 2022, LNAI 13343, pp. 184–196, 2022.
https://doi.org/10.1007/978-3-031-08530-7_16

in medical domain. Many scholars have done some work related to it. Rubin [1] pointed out that a central problem in medical domain is how to draw inference about the causal effects of treatment from random and non-random data [2], and put forward the Potential Outcomes Model (RCM) [3]; Pearl pioneered the graphical causal inference model, proposed the Structural Causal Model (SCM) and a three-level causal hierarchy across all causal problems [4]; Robins [5] focused on causal reasoning without model, with model and with complex longitudinal data; Cui et al. [6] combined causal inference with machine learning and introduced the causal effect estimation of traditional methods like matching, and advanced representation learning methods, such as deep neural network.

Nowadays, numerous studies have focus on the causal reasoning methodologies, such as traditional machine learning [7], deep learning [8], reinforcement learning [9,10], and so on. However, there is no thorough survey of causal reasoning methods in the medical context. Therefore, an in-depth discussion of causal reasoning methods for medical domain problems in two dimensions is provided in this paper. In the horizontal aspect, according to the core ideas of different methods, causal reasoning methods can be divided into four categories: probability-based method, model-based method, regression-based method and balancing-based method. In the vertical aspect, depending on the causal information used for the learning task and according to the three-level causal hierarchy proposed by Pearl, the capabilities possessed by causal reasoning methods can be classified from low to high into three categories: Association, Intervention, and Counterfactual.

Our main contributions are as follows:

- A high degree of consistency between causal reasoning and medical problems is explored.
- A road map of causal reasoning methods in medical domain is built with three-level hierarchy and four types.
- A "router" bridging between causal reasoning methods and medical cases is constructed.

The remaining of the paper is organized as follows. Some probability-based methods are given in Sect. 2. Then in Sect. 3, we describe typical model-based methods in detail. And regression-based methods in medical domain are given in Sect. 4. In Sect. 5, the balancing-based methods are introduced. Finally, we do the conclusion and discussion in Sect. 6.

2   Probability-Based Reasoning Methods

2.1   Causal Bayesian Networks

In medical domain, clinicians often ask semantic questions about data related to uncertainty, which leads to the emergence and analysis of Bayesian Networks (BN) [11].

As an example, take a BN that models the relationship between disease, symptoms, and risk factors, as shown in Fig. 1. If we consider the arrows in the figure as pointing from cause to effect, that is, $A \rightarrow B$ as A leads to B, BN seems to express causality. However, BN itself cannot distinguish the direction of causality. Causality and interaction can be captured in the topology of the BN to create the Causal Bayesian Networks (CBN) [12], where causality is expressed as a conditional probability between nodes. By using the causal links learned in BN and CBN, it is possible to evolve the medical oncology and its hierarchy [13], enabling us to infer under uncertainty.

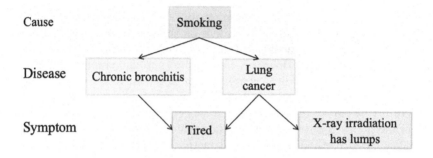

Fig. 1. Simple Bayesian network in medical example.

2.2   Causal Graph

When multiple variables are involved, the causality between them forms a causal graph [14]. Causal graphs are similar to BNs, but all arrows in a causal graph have causal implications rather than simple conditional probabilities. Causal graphs use existing contextual, empirical, and theoretical knowledge to objectively specify probabilistic causal relationships between covariates, and associated exposure and outcome variables, relying on temporality. By plotting causal graphs, it is possible to reduce the bias of confounders by mitigating adjustments to covariates to support causal reasoning.

Common causal relationships in medical domain can be divided into three categories: indirect cause, common cause, and joint action. The first category is shown in Fig. 2(a), where it is assumed that A indirectly affects C by affecting B. For example, prolonged exposure to sunlight (A) directly causes damage to skin cells (B) thus leading to skin cancer (C). In this case, B is the mediator, which completely shields the effect of A on C. A is an indirect cause of C, that is, A is independent of C under the condition of a given B. The second category is shown in Fig. 2(b), assuming that A is a common cause of B and D. In this case, A is called a confounding factor, creating an association between B and D. That is, given A, B and D are independent of each other. The third category is shown in Fig. 2(c), where B is assumed to be a common cause of A and E. In this case, B is called a collider, and A and E are independent a priori. That is, the link between A and E is introduced under the condition of a given B.

In general, the causal graph of a typical medical example is shown in Fig. 2(d), where R represents confounders, D represents diseases, and S represents symptoms. There are direct and common causes between diseases and symptoms, which cannot be simply solved. For example, elderly smokers are among those with a high prevalence of emphysema, and when patients present with symptoms such as chest pain, nausea, and fatigue, physicians diagnose angina pectoris. At this point, although emphysema cannot cause the above symptoms, they are all associated with an underlying common cause (in this case, elderly smokers), and without this condition, an association between emphysema and the above symptoms would not occur.

Although causal graphs have become a powerful form of representation, the data collection process that generates them is fraught with sample selection bias, which leads to spurious correlations between entities.

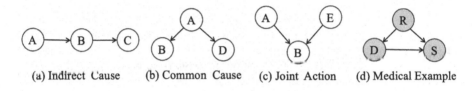

(a) Indirect Cause   (b) Common Cause   (c) Joint Action   (d) Medical Example

Fig. 2. Common causal relationships in medical domain.

2.3 Probability Tree

The probability tree [15] is used to represent the probability space. Each node in the tree corresponds to a potential state of the process and represents an event and its probability. Compared with CBN, the probability tree can model context-specific causal dependencies and solve the problems of association, intervention, and counterfactual in the three-level causal hierarchy.

3 Model-Based Reasoning Methods

To solve these problems of causal reasoning from observed data, we cannot obtain meaningful estimates only through the data itself, but usually need to supplement the data with models. The graphical models [16] developed for probabilistic reasoning [17] and causal analysis discussed above perform well only if the predictions are accurate, while the SCM and the RCM [18] can consistently represent prior causal knowledge, assumptions, and estimates.

3.1 SCM

SCM relies on the use of causal graph and Structural Equation Model (SEM) [19]. Each SCM M has a corresponding causal graph G. Each node in G represents each variable in M, and the edges in G represent functions f in M. When a causal graph and a set of structural equations are given, we can specify the causal effects represented by directed edges.

$$x = f_x(\varepsilon_x), t = f_t(x, \varepsilon_t), y = f_y(x, t, \varepsilon_y) \tag{1}$$

Equation (1) represents each variable as a deterministic function of its direct causes and unobserved exogenous noises, where ε_x, ε_t and ε_y represent the noise of the observed variables. If we impose an intervention on the SCM, this operation is represented in the structural equation as setting t to t' and on the causal graph as removing all arrows pointing to t'. Thus, the causal graph under the intervention is shown in Fig. 3(b), the structural equation can be expressed as:

$$x = f_x(\varepsilon_x), t = t', y = f_y(x, t, \varepsilon_y) \tag{2}$$

This model can answer both intervention and counterfactual questions [20], so SCM has been widely used in medical domain. For example, medical problems generate large amounts of data, such as brain imaging, the use of SCM can address the complexity of neuroimaging data that cannot be handled by traditional statistical analysis methods because SCM contains both observable variables and potential variables that cannot be directly observed.

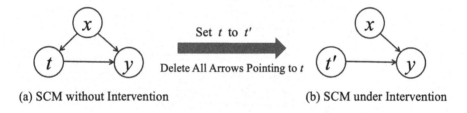

(a) SCM without Intervention   (b) SCM under Intervention

Fig. 3. SCM under intervention and without intervention.

3.2   RCM

RCM and SCM are logically equivalent. In contrast to SCM, RCM allows the modeling of causal effects without knowing the full causal graph, and it aims to estimate potential outcomes and then calculate causal effects [21]. The causal effect of an intervention on an outcome is the difference between the symptoms of the same group of people when they are in the treated group and the control group. In other words, the causal effect can be defined as the difference between the "Observed Outcome" and the "Counterfactual Outcome" shown in Fig. 4, hence the name of Counterfactual Framework.

The counterfactual reasoning is located at the top of the causal hierarchy and is the opposite of "Factual Statements". In medical domain, take the effect of a drug treatment as an example, if a group of patients are assigned to the treated group to receive new drug treatment. The reduction or worsening of symptoms in patients is a "Fact" that we can observe, and the "Counterfactual" is the hypothesis of what the symptoms would have been like if the same group of patients had been assigned to the control group.

A basic challenge for healthcare systems is to provide accurate diagnoses. However, existing diagnostic algorithms rely on associative inference [22,23], which cannot separate correlation from causation [24]. Causal reasoning methods based on the estimation of potential outcome concepts can draw inferences about the causal effects of interventions and effectively control for confounders to eliminate confounding bias. The introduction of a counterfactual diagnostic algorithm for causal reasoning into medical diagnosis has greatly improved the accuracy of diagnosis compared to a correlation-based Bayesian diagnostic algorithm [25].

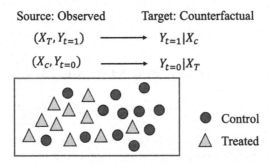

Source: Observed   Target: Counterfactual

$(X_T, Y_{t=1})$ $\longrightarrow$ $Y_{t=1}|X_c$

$(X_c, Y_{t=0})$ $\longrightarrow$ $Y_{t=0}|X_T$

● Control

△ Treated

Fig. 4. The definition of causal effect.

3.3   MSM

The Marginal Structural Model (MSM) [26] provides a effective solution to the widespread problem of time-dependent confounding in medical studies, allowing us to unbiased estimation of causal effects. The basic idea of MSM is to construct a virtual population that equalizes the distribution of confounders among the groups and the effect of observable confounders on the treatment is eliminated. MSM provides a practical and effective solution to the widespread problem of time-dependent confounders in medical field research.

4   Regression-Based Reasoning Methods

4.1   Granger Causality Test

It is a typical regression-based causal reasoning method, the basic idea of which is the combination of the Vector Auto-Regressive (VAR) model and Granger Causality [27]. Among them, the former is mainly used to construct time series models to test the time relationship between variables. The latter is able to establish the dependence between two time series and assess whether there is a causal relationship between two different series of data in time, i.e., if the joint prediction error of two time series X and Y is smaller than the prediction error of X itself, then Y has Granger causality on X.

In medical domain, it is possible to use Granger causality analysis to assess whether there is a causal relationship between patients' participation in discussions about prenatal checkups and their overall emotions expressed during the subsequent course of their lives. The first time series represents the average sentiment of patients, the second time series measures the number of posts about the topic of prenatal screening shared by patients per day, and the third time series represents the overall sentiment expressed by patients over the course of their lives beyond that. Then, according to these time series, it can be concluded that there is a Granger causality between patients on the given medical topic and their general sentiment.

It can be seen that the key to the analysis of Granger causality is to have a stable time series, so one of its limitations is that it is sensitive to the selection of time and has poor robustness.

5   Balancing-Based Reasoning Methods

As shown in Fig. 5, the central problem of causal reasoning is to eliminate the confounding bias caused by the unbalanced distribution of confounders. A classic question in medical domain is whether smoking leads to lung cancer? At this point, age is associated with both the study factor and the study disease, and if unevenly distributed in the comparison population, it can lead to a wrong estimate of the relationship between smoking and lung cancer. The balancing-based method ensures that the distribution of confounding variables is consistent across the treated group and control group of the assessed data.

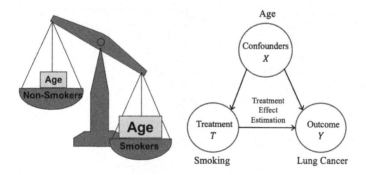

Fig. 5. Unbalanced Distribution of Confounders.

5.1   Propensity Score Matching

The basic logic of PSM [28] is to ensure that the propensity values of the matched individuals are equal or approximate, so that causal effect estimation is based on different outcomes between comparable individuals. Propensity score refers to the conditional probability assigned to a specific treatment given the observation covariate vector, as defined by Eq. (3).

$$e(x) = P_r(T = 1|X = x) \tag{3}$$

where T represents a specific treatment, X represents a given covariate, and $e(x)$ represents the propensity score. This method implements matching by calculating the probability of samples in the treated group. Samples with the same probability in the treated group are matched, and the distribution of confounding variables of the matched samples is also theoretically guaranteed to be balanced.

In medical domain, taking the study of the efficacy of a drug in the treatment of asthma as an example, PSM is to pair asthmatic patients who do and do not receive drug treatment, and ensure that their propensity value (here refers to the probability of receiving drug treatment) is the same or similar. Since the existing confounding variables have been controlled in the pairing process based on propensity value, the differences between the two groups can only be attributed to whether they received drug treatment, rather than other confounding variables, thus curbing the confounding bias.

5.2   Re-weighting

The basic principle of re-weighting is to create a pseudo population by assigning appropriate weights to each sample so that the distribution of the treatment group and the control group is consistent [29]. The Inverse Propensity Weighting (IPW) [30] method first assigns to each sample the probability of receiving the treatment and then the opposite probability by which weighting each observation. So, the weight assigned to each unit is:

$$r = T/e(x) + (1 - T)/(1 - e(x)) \tag{4}$$

where T is the treatment and $e(x)$ is the propensity score.

In medical domain, it is assumed that the treatment variable is the mother's smoking status during pregnancy and the outcome is the birth weight of the baby. Because of the greater proportion of younger smokers and older nonsmokers, the IPW method was used, weighting the observations for older smokers by $1/e(x)$ and for younger nonsmokers by $1/(1 - e(x))$ so that the weights would be larger when the probabilities were small. As shown in Fig. 6.

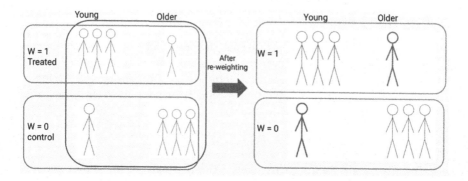

Fig. 6. Medical example of IPW.

5.3 Confounder Balancing

In the real world, the structure of our model between variables is unknown, we cannot eliminate confounding bias by preformulating models when conducting causal effect estimation [31]. Therefore, there is a trend to balance the confounding distributions by directly optimizing the potential confounding weights without modeling or estimating propensity scores [32].

The Causally Regularized Logistic Regression (CRLR) [33] algorithm uses the idea of global confounder balancing to identify causal features. In addition, not all variables are confounding variables in the big data scenario, and thus, the Data-Driven Variable Decomposition (D2VD) [34] method decomposes all the observed variables into confounding, adjusting, and irrelevant variables. And because the confounding bias caused by different confounding variables is unequal. Therefore, compared with the previous methods of balancing all variables, the Differentiated Confounder Balancing (DCB) [35,36] algorithm focuses on differentiating the weights of confounding factors in balancing.

From the above study, a comprehensive investigation of causal reasoning methods in the medical context and a specific description of the pros and cons of those methods are provided. In the vertical aspect, depending on the causal information used for the learning task and according to the three-level causal hierarchy, the capabilities possessed by causal reasoning methods can be classified from low to high into three categories: Association, Intervention, and Counterfactual. The higher-level model possess stronger capability, so models that can answer counterfactual questions can also answer questions about interventions and predictions. Based on this, we have summarized the classification of causal reasoning methods in medical domain as shown in Table 1.

Table 1. Summary of causal reasoning methods in medical domain.

	Association	Intervention	Counterfactual
Probability-based	Casual graph [37],	Casual graph [37],	
	BN [11],	CBN [39],	Probability tree [15]
	Probability tree [38]	Probability tree [15]	
Model-based	SCM [4],	SCM [41],	SCM [43],
	RCM [2],	RCM [3],	RCM [44],
	MSM [40]	MSM [42]	MSM [5]
Regression-based	Granger causality test [27]		
Balancing-based	PSM [28],	Confounder balancing [46],	
	Confounder balancing [45],	IPW [47]	PSM [49]
	IPW [30],	PSM [48]	

6   Conclusion and Discussion

To our knowledge, this is the first paper to provide a thorough review of causal reasoning methods in medical domain. In general, our main contributions are summarized below. First, a high degree of consistency between causal reasoning and medical problems is explored. The association, intervention, and counterfactual problems of causal reasoning are discussed in medical cases to overcome the shortcomings of traditional methods. And, in the vertical aspect, a road map of causal reasoning methods in medical domain is built with three-level hierarchy, in which different causal reasoning methods are analyzed from different perspective. Furthermore, in the horizontal aspect, the causal reasoning methods are grouped into four types: probability-based, model-based, regression-based, and balancing-based. To sum up, in the medical context, a "router" bridging between causal reasoning methods and medical cases is constructed.

Causal problems can be defined into two categories: causal discovery and causal reasoning. The latter is the area of research focused on in this paper, while the former, determining the existence and direction of causal relationships from data, belongs to the field of causal discovery, which is an active but challenging area of research.

Acknowledgements.. This work is supported by the National Natural Science Foundation of China (Grant No. 62172267), the Natural Science Foundation of Shanghai, China (Grant No. 20ZR1420400), the State Key Program of National Natural Science Foundation of China (Grant No. 61936001).

References

1. Rubin, D.B.: 2 statistical inference for causal effects, with emphasis on applications in epidemiology and medical statistics. Handb. Statist. **27**, 28–63 (2007)
2. Rubin, D.B.: Estimating causal effects of treatments in randomized and nonrandomized studies. J. Educ. Psychol. **66**(5), 688 (1974)

3. Rubin, D.B.: Causal inference using potential outcomes: design, modeling, decisions. J. Am. Stat. Assoc. **100**(469), 322–331 (2005)
4. Judea, P.: An introduction to causal inference. Int. J. Biostat. **6**(2), 1–62 (2010)
5. Hernán, M.A., Robins, J.M.: Causal inference (2010)
6. Cui, P., et al.: Causal inference meets machine learning. In: Proceedings of the 26th ACM SIGKDD International Conference on Knowledge Discovery and Data Mining, pp. 3527–3528 (2020)
7. Schölkopf, B.: Causality for machine learning. arXiv preprint arXiv:1911.10500 (2019)
8. Xia, K., Lee, K.Z., Bengio, Y., Bareinboim, E.: The causal-neural connection: expressiveness, learnability, and inference. Adv. Neural Inf. Process. Syst. **34** (2021)
9. Dasgupta, I., et al.: Causal reasoning from meta-reinforcement learning. arXiv preprint arXiv:1901.08162 (2019)
10. Rezende, D.J., et al.: Causally correct partial models for reinforcement learning. arXiv preprint arXiv:2002.02836 (2020)
11. Pearl, J.: Bayesian networks (2011)
12. Messaoud, M.B., Leray, P., Amor, N.B.: Semcado: a serendipitous strategy for causal discovery and ontology evolution. Knowl.-Based Syst. **76**, 79–95 (2015)
13. Hu, H., Kerschberg, L.: Evolving medical ontologies based on causal inference. In: 2018 IEEE/ACM International Conference on Advances in Social Networks Analysis and Mining (ASONAM), pp. 954–957. IEEE (2018)
14. Pearl, J.: Causal diagrams for empirical research. Biometrika **82**(4), 669–688 (1995)
15. Genewein, T., et al.: Algorithms for causal reasoning in probability trees. arXiv preprint arXiv:2010.12237 (2020)
16. Lauritzen, S.L.: Graphical Models, vol. 17. Clarendon Press, Oxford (1996)
17. Pearl, J.: Probabilistic Reasoning in Intelligent Systems: Networks of Plausible Inference. Elsevier, Amsterdam (2014)
18. Imbens, G.W., Rubin, D.B.: Causal Inference in Statistics, Social, and Biomedical Sciences. Cambridge University Press, Cambridge (2015)
19. Pearl, J.: Graphs, causality, and structural equation models. Soci. Methods Res. **27**(2), 226–284 (1998)
20. Shpitser, I., Pearl, J.: What counterfactuals can be tested. arXiv preprint arXiv:1206.5294 (2012)
21. Aliprantis, D.: A distinction between causal effects in structural and Rubin causal models (2015)
22. Pearl, J.: Theoretical impediments to machine learning with seven sparks from the causal revolution. arXiv preprint arXiv:1801.04016 (2018)
23. Greenland, S.: For and against methodologies: some perspectives on recent causal and statistical inference debates. Eur. J. Epidemiol. **32**(1), 3–20 (2017)
24. Gigerenzer, G., Marewski, J.N.: Surrogate science: the idol of a universal method for scientific inference. J. Manag. **41**(2), 421–440 (2015)
25. Richens, J.G., Lee, C.M., Johri, S.: Improving the accuracy of medical diagnosis with causal machine learning. Nat. Commun. **11**(1), 1–9 (2020)
26. Cafri, G., Wang, W., Chan, P.H., Austin, P.C.: A review and empirical comparison of causal inference methods for clustered observational data with application to the evaluation of the effectiveness of medical devices. Stat. Methods Med. Res. **28**(10–11), 3142–3162 (2019)
27. Granger, C.W.: Causality, cointegration, and control. J. Econ. Dyn. Control **12**(2–3), 551–559 (1988)

28. Rosenbaum, P.R., Rubin, D.B.: The central role of the propensity score in observational studies for causal effects. Biometrika **70**(1), 41–55 (1983)
29. Kuang, K., Xiong, R., Cui, P., Athey, S., Li, B.: Stable prediction with model misspecification and agnostic distribution shift. In: Proceedings of the AAAI Conference on Artificial Intelligence, vol. 34, pp. 4485–4492 (2020)
30. Mansournia, M.A., Altman, D.G.: Inverse probability weighting. BMJ **352** (2016)
31. Hernán, M.A.: Beyond exchangeability: the other conditions for causal inference in medical research (2012)
32. Zubizarreta, J.R.: Stable weights that balance covariates for estimation with incomplete outcome data. J. Am. Stat. Assoc. **110**(511), 910–922 (2015)
33. Shen, Z., Cui, P., Kuang, K., Li, B., Chen, P.: Causally regularized learning with agnostic data selection bias. In: Proceedings of the 26th ACM International Conference on Multimedia, pp. 411–419 (2018)
34. Kuang, K., Cui, P., Li, B., Jiang, M., Yang, S., Wang, F.: Treatment effect estimation with data-driven variable decomposition. In: Proceedings of the AAAI Conference on Artificial Intelligence, vol. 31 (2017)
35. Kuang, K., et al.: Treatment effect estimation via differentiated confounder balancing and regression. ACM Trans. Knowl. Discov. Data (TKDD) **14**(1), 1–25 (2019)
36. Kuang, K., Cui, P., Li, B., Jiang, M., Yang, S.: Estimating treatment effect in the wild via differentiated confounder balancing. In: Proceedings of the 23rd ACM SIGKDD International Conference on Knowledge Discovery and Data Mining, pp. 265–274 (2017)
37. Greenland, S., Pearl, J., Robins, J.M.: Causal diagrams for epidemiologic research. Epidemiology, 37–48 (1999)
38. Yao, L., Chu, Z., Li, S., Li, Y., Gao, J., Zhang, A.: A survey on causal inference. arXiv preprint arXiv:2002.02770 (2020)
39. Ellis, B., Wong, W.H.: Learning causal Bayesian network structures from experimental data. J. Am. Stat. Assoc. **103**(482), 778–789 (2008)
40. Robins, J.M., Hernan, M.A., Brumback, B.: Marginal structural models and causal inference in epidemiology (2000)
41. Pearl, J.: Causal inference in statistics: an overview. Stat. Surv. **3**, 96–146 (2009)
42. Hernán, M.Á., Brumback, B., Robins, J.M.: Marginal structural models to estimate the causal effect of zidovudine on the survival of HIV-positive men. Epidemiology, 561–570 (2000)
43. Pearl, J., et al.: Models, Reasoning and Inference, vol. 19. Cambridge University Press, Cambridge (2000)
44. Pearl, J.: Probabilities of causation: three counterfactual interpretations and their identification. Synthese **121**(1), 93–149 (1999)
45. Hainmueller, J.: Entropy balancing for causal effects: a multivariate reweighting method to produce balanced samples in observational studies. Polit. Anal. **20**(1), 25–46 (2012)
46. Athey, S., Imbens, G.W., Wager, S.: Approximate residual balancing: debiased inference of average treatment effects in high dimensions. arXiv preprint arXiv:1604.07125 (2016)
47. Austin, P.C.: An introduction to propensity score methods for reducing the effects of confounding in observational studies. Multivar. Behav. Res. **46**(3), 399–424 (2011)

48. Guo, S., Fraser, M.W.: Propensity Score Analysis: Statistical Methods and Applications, vol. 11. SAGE Publications, Thousand Oaks (2014)
49. Brookhart, M.A., Wyss, R., Layton, J.B., Stürmer, T.: Propensity score methods for confounding control in nonexperimental research. Circul. Cardiovasc. Qual. Outcomes **6**(5), 604–611 (2013)

Optimization

Enhancing a Multi-population Optimisation Approach with a Dynamic Transformation Scheme

Shengqi Dai, Vincent W. L. Tam[✉], Zhenglong Li, and L. K. Yeung

Department of Electrical and Electronic Engineering, The University of Hong Kong, Pokfulam Road, Hong Kong SAR, China
vtam@eee.hku.hk

Abstract. The adaptive multi-population optimisation (AMPO) algorithm is an intelligent meta-heuristic search method utilising multiple search groups to conduct a diversity of search strategies in evolutionary algorithms or swarm intelligence. With the careful design of different search operators, the AMPO algorithm has achieved outstanding performance in many optimisation problems including two sets of benchmark functions when compared to some latest approaches including the hybrid firefly and particle swarm optimisation for continuous optimisation. Yet there are still opportunities to enhance the adaptability of its search mechanism in various aspects. Therefore, a more adaptive AMPO (AMPO$^+$) algorithm is considered in this work in which the probability of the transformation between specific search groups can be more flexibly adjusted during the different stages of the search process. In this way, the AMPO$^+$ can better adapt its search efforts to specific search groups through revising its search strategies so as to effectively solve many challenging optimisation problems. To carefully examine the search effectiveness of the enhanced framework, the proposed AMPO$^+$ algorithm is evaluated against the original AMPO and other sophisticated meta-heuristic algorithms on a set of well-known benchmark functions of different dimensions in which impressive results are attained by the AMPO$^+$. More importantly, the proposed adaptive search framework sheds light on many possible directions for further investigation.

Keywords: Meta-heuristic algorithms · Adaptive search strategies · Continuous optimisation

1 Introduction

Optimisation problems occur widely in many different disciplines of studies including Engineering [15], Humanities and Science [7]. Yet most problems are computationally intensive or sometimes infeasible for derivative-based techniques to find optimal solutions in a limited period of time, especially when the concerned problem has many local optima. Therefore, to efficiently solve nonlinear, non-differentiable and non-convex optimisation problems, there are various

© Springer Nature Switzerland AG 2022
H. Fujita et al. (Eds.): IEA/AIE 2022, LNAI 13343, pp. 199–210, 2022.
https://doi.org/10.1007/978-3-031-08530-7_17

meta-heuristic search approaches that are often easy-to-implement, and do not require a lot of computational resources for computing the gradient or Hessian matrix of the underlying objective function as in other numerical methods.

With the potential advantages of flexibility and fast convergence, nature-inspired meta-heuristics have attracted much attention from many researchers in Artificial Intelligence (AI) [6] or relevant fields. Genetic algorithms are originated from the intent to imitate the natural evolution mechanism that has inspired many intelligent optimisation techniques to be developed later. Examples include the immune algorithms (IA) [18] which are based on the learning and cognitive function of the biological immune system. In addition, ant colony optimisation (ACO) algorithms [6] are based on the collective path-finding behaviour of ants whereas the particle swarm optimisation (PSO) approach [11] tries to imitate the intelligent swarm behaviour of birds or fishes. Furthermore, simulated annealing (SA) algorithms [16] are based on the annealing process of solid matter while tabu search (TS) approach [5] is originated from simulating the memory process of human brains. Among the latest swarm intelligence optimisation algorithms, the whale optimisation algorithm (WOA) [15] is inspired by the unique predation behaviour of humpback whales to perform intelligent search strategies through social groups exhibiting swarm intelligence. All these meta-heuristic approaches and other hybrid or multi-population search methods such as the hybrid firefly and particle swarm optimisation (HF-PSO) [2] and multi-population ensemble differential evolution (MPEDE) [19] have been extensively applied in various fields of studies with considerable successes attained in many real-world applications.

The adaptive multi-population optimisation (AMPO) [12] is a newly proposed meta-heuristic algorithm inspired by the transformation among multiple search groups to flexibly adapt the search efforts for solving continuous optimisation problems. In the AMPO framework, there are different search groups or sub-populations to perform local search, global search, random search, resets in local minima, and the differential evolutionary (DE) operator. In the AMPO approach, each individual group member as an array of bits, integers or floating-point numbers denotes a potential solution to the current search or optimisation problem. The AMPO algorithm is targeted to simulate the competition among the multiple search groups to find the globally best solution, and also the spreading of such globally best individual solutions to different search groups to continue the search through applying the corresponding operator for each specific search groups. For instance, local search is applied to individual members of the local search group whereas the DE operator is utilised in individual members of the migrating group. On the other hand, when the whole population is converged or being stuck in certain local minima for several iterations, the reset mechanism will be activated among a pre-determined portion like 60% of the whole population to copy the globally best candidate(s) from the DE group while the remaining part of the whole population is randomised. Clearly, when the globally best candidate is the globally optimal solution, the AMPO algorithm will be halted since the concerned optimisation problem is solved. Alternatively, when the preset number of iterations is exceeded, the AMPO algorithm will also be stopped. Through the various operators, the search strategy of the AMPO algorithm is diversified so that the quality of the best solution may be improved over successive iterations. In a previous work, the AMPO algorithm attains very impressive results in tackling a set of benchmark functions when compared to other existing meta-heuristic approaches such as the ACO, PSO and WOA, and other hybrid or multi-population search methods such as the HF-PSO and MPEDE. Yet the AMPO approach may still suffer from several shortcomings as revealed by some previous investigations. Similar to other meta-heuristic or hybrid search approaches, one of its main shortcomings is the requirement to determine or tune a relatively larger number of parameters when compared to those of other AI or numerical methods.

Specifically, this paper considers the transformation probability of the original AMPO approach, and carefully examines how the transformation process can be better adapted to enhance the search ability of the algorithm with "more dynamic search strategies". Accordingly, an enhanced AMPO (AMPO$^+$) algorithm is proposed. To demonstrated the search effectiveness of the enhanced framework, the proposed AMPO$^+$ algorithm is evaluated against the original AMPO and other sophisticated meta-heuristic algorithms on a set of well-known benchmark functions of different dimensions in which impressive results are attained by the AMPO$^+$. More importantly, the enhanced (AMPO$^+$) framework sheds light on various directions for future investigation.

The rest of this paper is organised as follows. Section 2 describes the improved AMPO$^+$ in detail. Then, the AMPO$^+$ is applied to deal with numerical benchmark functions, with the obtained results of the AMPO$^+$ carefully compared with those of other sophisticated meta-heuristic or hybrid search approaches in Sect. 3. Lastly, Sect. 4 gives some concluding remarks of the work.

2 Related Work

2.1 The Original AMPO Algorithm

The original AMPO algorithm [12] is proposed for solving unconstrained, static and single-objective continuous optimisation problems. Essentially, the AMPO approach is a multi-population-based meta-heuristic that tries to diversify the search strategies to enhance its search and optimisation capability. The algorithm consists of individual candidate solutions that can be further categorised into five search groups or sub-populations including the **random search group, global search group, local search group, leader group** and **migrating group**. Besides the **leader** and **migrating groups**, the sizes of all other search groups can be dynamically changed during the search process. Besides, various search strategies are assigned to the individual(s) in each group.

Figure 1 clearly depicts the logical flow of the AMPO approach. First, the user-controlled parameters of the algorithm and the objective function of the optimisation problem are inputed to initialise the AMPO algorithm. Then, the following operations are executed consecutively during the optimisation process:

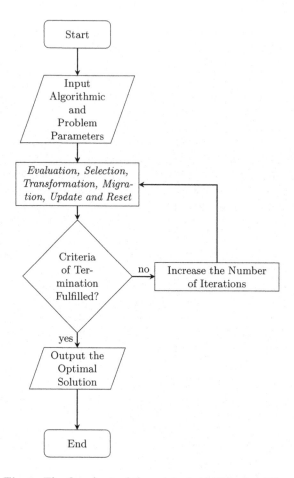

Fig. 1. The flowchart of the original AMPO algorithm

1. the *evaluation* is used to calculate the results of the objective function as the fitness values for all individuals;
2. the *selection* operation is conducted to select the individuals with the highest fitness values;
3. the *transformation* operation is executed to opportunistically transform an individual from one group to another;
4. the *migration* is to possibly migrate a better solution from the *migrating* group to the *leader group*;
5. the *update* operation is utilised to search solutions using different search strategies;
6. the *reset* operation is triggered when the size of the *random search group* is shrunk to zero.

The above optimisation process is repeated until the predefined criteria of termination is satisfied, i.e. with the maximum number of iterations exceeded. Lastly, the AMPO algorithm will return the best or optimal solution obtained so far. It is worth noting that the transformation probabilities among the five search groups have to be predetermined in the original AMPO approach, and more importantly being treated as fixed probabilities during the whole search process, thus adversely affecting both the applicability and flexibility of the AMPO approach to tackle any real-life application. In the subsequent discussion, we will consider some possible enhancements to remedy these shortcomings of the original AMPO approach.

2.2 Other Metaheuristic Algorithms

As inspired by the nature, the particle swarm optimisation (PSO) approach [11] tries to imitate the intelligent swarm behaviour of birds or fishes. Intrinsically, it is a global optimisation technique to tackle any optimisation problem in which the solution can be represented as a point or surface in a multi-dimensional space. All the search particles of the PSO are typically randomly generated, or produced through hypotheses about the potential solutions in the underlying search space. Each particle will then be assigned with an initial velocity to move through the search space, and evaluated according to some fitness function after each iteration. Over successive iterations, particles will be accelerated towards those particles within their communication group with better fitness values. When compared with other global optimisation approaches such as the simulated annealing [16], the PSO approach tends to be more resilient to the problem of local minima due to its relatively large number of search particles to build up the whole particle swarm, thus achieving impressive performance in many challenging and real-world optimisation problems.

Another example of metaheuristic search approaches is the whale optimisation algorithm (WOA) [15] as inspired by the unique predation behaviour of humpback whales to perform intelligent search strategies through social groups exhibiting swarm intelligence (SI). The WOA is based on simulating the bubble-net hunting technique of humpback whales to solve continuous optimisation problems. Due to its simple structure and fast convergence speed, the WOA has been widely adopted to tackle many real-world engineering applications. For instance, Aljarah et al. [1] consider the uses of the WOA to optimise the connection weights in artificial neural networks as one of the most difficult challenges in machine learning. In addition, Medani et al. [14] apply the WOA to tackle the challenging optimal reactive power dispatch as the minimisation of an objective function to represent the total active power losses in the electrical networks in a case study of the Algerian power system.

Furthermore, there are some recent research investigations in the past few years trying to hybridise these metaheuristic or other evolutionary approaches for which their successes much depends on the balance of their exploration and exploitation capabilities, and the potential gains out of the hybridisation when solving the challenging optimisation problems at hand. In the hybrid firefly and particle swarm optimisation (HFPSO) approach [2], the proposed hybrid algorithm tries to exploit the strengths of both particle swarm and firefly algorithms in which the HFPSO works to determine the start of the local search process by checking the previously global best fitness values. The proposed HFPSO algorithm has gained remarkable successes in the evaluation results against those of the standard particle swarm, firefly and other recent hybrid algorithms on various computationally intensive functions of both the CEC 2015 and CEC 2017 benchmark problems. The evaluation results reveal that the proposed HFPSO algorithm can quickly give outstanding solutions that excels those of other SI or hybrid approaches in the unimodal, simple multimodal, hybrid, and composition categories of the involved benchmark problems.

3   The Enhanced Search Framework

As carefully examined in Sect. 2.1, the original AMPO approach suffers from two major pitfalls. First, the transformation probabilities among the five search groups have to be predetermined before the start of the search process in the original AMPO approach. In some cases, such transformation probabilities can be arbitrarily set by random guesses. More importantly, these transformation probabilities are considered as fixed values throughout the search process. Clearly, this will adversely affect the flexibility of transforming each individual between the different search groups, and ultimately the dynamic search ability of the whole population. In particular, it is revealed through some empirical observations that certain transformation probability like the one for changing from the random search group to the migrating group can be very significant since it may affect whether each individual may have the "opportunity" to be transformed so as to make possible improvements to the currently best solution of the whole population in solving various challenging optimisation problems. When such critical transformation probability is inappropriately set and being fixed throughout the search process, the ultimate search efficiency and solution quality of the original AMPO approach will surely be adversely affected.

Accordingly, the following three major mechanisms are proposed to improve the original AMPO approach as the enhanced AMPO (AMPO$^+$) algorithm so as to remedy the existing shortcomings.

1. For the initialisation of the transformation probabilities, the AMPO$^+$ algorithm will make use of some default values as obtained from previous tests on other benchmark problems such as a set of 16 classical benchmark functions commonly used for evaluating the performance of metaheuristic algorithms, instead of randomly presetting all the transformation probability values;
2. Besides, to flexibly change specific transformation probability such as the critical one to switch from the random search group to the migrating group according to the available resource, a new concept called the "turning point" (tp) is introduced. Intrinsically, tp is defined as a ratio ranging from 0.1 to 0.9 with respect to the maximum number of iterations allowed for the AMPO$^+$ algorithm such that the enhanced AMPO will flexibly adapt the specific transformation probability to a new value according to the available resources. For instance, when tp is set to 0.5 with the maximum number of iterations as

10, 000, the enhanced AMPO algorithm will dynamically revise the concerned transformation probability to a new probability after $0.5 \times 10,000 = 5,000$ iterations. Essentially, after a specified portion of the whole search process, a new change to some critical transformation probability will be triggered;

3. Lastly, an adaptive search strategy within the enhanced AMPO framework is readily defined by the value of tp, and the pair of the original and new transformation probabilities for a specific transformation process between two search groups.

Obviously, introducing the values of tp and new transformation probability in the enhanced AMPO search framework will definitely increase the number of parameters to be determined as compared to that of the original AMPO algorithm. Nevertheless, such increase in the number of parameters is unavoidably necessary for promoting the flexibility of the overall search framework. Besides, the number of increased parameters is kept at the minimum as 2 for the simplicity to define each adaptive search strategy in the enhanced framework. More importantly, the enhanced AMPO framework provides a very "systematic" revision scheme to dynamically "focus" the search for more appropriate values of the critical transformation probability value(s) during the different stages of the search process. As clearly revealed by the empirical evaluation of the enhanced AMPO framework in the upcoming section, this systematic scheme will definitely produce a more efficient and also effective search strategy when compared to *the brute force approach trying to guess the appropriate values for all the fixed transformation probabilities* in the original AMPO algorithm.

Figure 2 shows the logical flow of the enhanced AMPO ($AMPO^+$) algorithm in which the added key component as "Revise the Transformation Probabilities According to Available Resources and Configurations" is highlighted. This key component clearly summarises the essence of the 3 mechanisms through revising the transformation probability as adaptive search strategies to remedy the shortcomings of the original AMPO framework. Clearly, there can be many other possible mechanisms like increasing the number of turning points $tp's$ with more adaptive search strategies to adjust the transformation probabilities during different stages of the search process, or directly adjusting the population sizes of the different search groups in the original AMPO framework. Yet for both the simplicity and ease of analysis of the resulting $AMPO^+$ framework, it is our intent to keep the number of turning points as one with a pair of the transformation probabilities to define the simplest adaptive search strategy for the subsequent performance evaluation of our proposed framework.

4   The Empirical Evaluation

To evaluate the performance of our enhanced AMPO framework, the proposed AMPO+ algorithm is applied to solve the CEC 2014 benchmark set [13] involving 30 uni-modal, multi-modal, hybrid or composition functions. The results are compared against those of the original AMPO [12] and other meta-heuristic

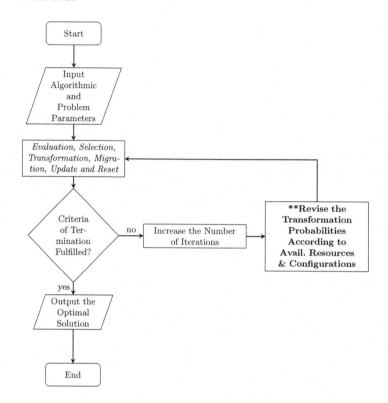

Fig. 2. The flowchart of the enhanced AMPO (AMPO$^+$) algorithm

algorithms including the particle swarm optimisation (PSO) [11], genetic algorithm (GA) [7], differential evolution (DE) [17], artificial bee colony (ABC) [10] and whale optimisation algorithm (WOA) [15]. For a fair comparison, all algorithms use the recommended parameter settings as reported in the literatures [3,4,8,9,11] for which all the population sizes are consistently set as 50. All algorithms are run 30 times independently on each function with the maximum number of generations set to 10,000 on the 10-dimensional function, and 15,000 on the 30-dimensional function respectively.

For the 10-dimensional functions, the numerical solutions reveal that the AMPO+ approach attains the best results on 23 out of the 30 benchmark functions for which the detailed evaluation results are summarised as follows.

– For the uni-modal functions of CEC1-CEC3, the AMPO+ algorithm outperforms all the other algorithms under comparison. The rotation complexity of the involved benchmark functions leads to a significant difference between the obtained numerical results and the globally optimal solution for all the evaluated algorithms on the CEC1 function. The performance of the AMPO+ and that of the original AMPO algorithm are on the same order of magnitude, i.e. 10^4, that is basically better than those of the other optimisers. The performance of most optimisers on the CEC2 function is similar to that of the CEC1 function. It is worth noting that both the AMPO+ and original AMPO successfully find the globally optimal value on the CEC3 function. Yet the standard deviation of the AMPO+ algorithm is much more closer to 0;

- For the multi-modal functions of CEC4-CEC16, it is evident that the AMPO+ algorithm achieves better results than those of other algorithms. The AMPO+ approach is ranked first in terms of the mean of fitness values in more than half of the functions including the CEC4, CEC6, CEC7, CEC8, CEC9, CEC11, CEC13, CEC14 and CEC15 whereas the original AMPO gives the best fitness values in 7 functions;
- For the hybrid functions of CEC17-CEC22, the AMPO+ algorithm has achieved the best performance in terms of the fitness values except the CEC18 function on which the AMPO+ still gives the second best;
- For the composition functions of CEC23-CEC30, the AMPO+ algorithm excels all the other algorithms in the first six functions of CEC23-CEC28. This clearly demonstrates that the proposed AMPO+ approach has a superior capability when compared with the other approaches in dealing with complex optimisation problems.

Table 1 shows a summary of all the evaluation results obtained on the 10-dimensional CEC 2014 benchmark of 30 different functions in which the average ranking of each algorithm is computed. The AMPO+ algorithm is ranked first on the fitness values yet as the last on the computational time probably due to the relatively large computational overheads in adapting the search strategies for the different benchmark functions at hand.

Table 1. A summary of results on the 10-dimensional CEC-2014 benchmark functions

Algorithm	Avg. fitness rank	Avg. time rank	Overall fitness rank	Overall time rank
ABC	3.87	1.93	3	2
DE	5.23	4.83	6	5
GA	4.13	4.50	4	4
PSO	5.60	1.20	7	1
WOA	4.73	2.90	5	3
AMPO	2.53	6.20	2	6
AMPO+	1.60	6.43	1	7

Table 2 shows a summary of all the evaluation results obtained on the 30-dimensional CEC 2014 benchmark of 30 different functions, in which the average ranking on all functions of each algorithm is calculated. The AMPO+ approach is ranked first on the fitness values, and the fifth on the computational time. It is worth noting that the computational overheads of the AMPO+ approach is relatively lower than those of the GA or DE approach on these benchmark functions involving a higher dimension of 30, especially when compared to the previous ranking results as presented in Table 1 on the 10-dimensional CEC 2014 benchmark functions. This may possibly demonstrate the potential benefit in the computational overheads of the proposed AMPO+ approach after being extended to solve complex optimisation problems of higher dimensions, for which more vigorous evaluation tests should be conducted in the future investigation to validate this potential strength of the AMPO+ search framework.

Table 2. A summary of results on the 30-dimensional CEC-2014 benchmark functions

Algorithm	Avg. fitness rank	Avg. time rank	Overall fitness rank	Overall time rank
ABC	4.73	2.23	5	2
DE	5.07	6.83	6	7
GA	3.13	5.83	3	6
PSO	6.00	1.27	7	1
WOA	4.67	2.93	4	3
AMPO	2.37	4.00	2	4
AMPO+	1.67	4.90	1	5

5 Concluding Remarks

As an intelligent meta-heuristic search method, the adaptive multi-population optimisation (AMPO) algorithm [12] employs multiple search groups to perform a variety of search strategies as originated from evolutionary algorithms [7] or swarm intelligence (SI) [11]. With a careful design of both the evolutionary and SI search operators, the AMPO algorithm has attained very impressive performance in many optimisation problems including two sets of benchmark functions when compared to some recent approaches including the whale optimisation algorithm (WOA) [15] and the hybrid firefly and particle swarm optimisation [2] for continuous optimisation. Yet there are still many opportunities to enhance the adaptability of its search mechanism in various aspects. Therefore, a more adaptive AMPO (AMPO$^+$) algorithm is proposed in this work in which the probability of the transformation between specific search groups can be more flexibly adjusted during the different stages of the search process. As a result, the proposed AMPO$^+$ framework can better adapt its search efforts to specific search groups through revising its search strategies in order to effectively solve many challenging optimisation problems. To carefully evaluate the search effectiveness of the enhanced search framework, the proposed AMPO$^+$ algorithm is compared against the original AMPO approach and other sophisticated meta-heuristic algorithms on two sets of the well-known 10-dimensional and 30-dimensional CEC 2014 benchmark functions involving a diversity of function types in which impressive results are attained by the AMPO$^+$ approach.

Essentially, the AMPO+ algorithm is ranked first in terms of the fitness values for the 10-dimensional CEC 2014 benchmark functions yet being the last on the computational time probably due to the relatively large computational overheads in adapting the search strategies for the different benchmark functions at hand. For the 30-dimensional CEC 2014 benchmark functions, the proposed AMPO+ approach is also ranked first on the fitness values, and the fifth on the computational time. This may demonstrate the potential strength of the AMPO+ framework in the overall computational overheads in tackling complex optimisation problems of higher dimensions, thus demanding more vigorous tests to validate this potential advantage of the AMPO+ framework.

More importantly, the adaptive AMPO+ search framework sheds light on many possible directions for further investigation. First, it is interesting to thoroughly evaluate the search effectiveness of the AMPO+ framework through applying to other challenging benchmark functions and real-world applications such as the portfolio optimisation problems in finance, or the flight scheduling problems in logistics, etc. Besides, it is worth exploring the potential benefit(s) or pitfalls of the proposed AMPO+ framework in the overall computational, memory or other overheads in tackling complex optimisation problems of higher dimensions when compared to those of the existing meta-heuristic search algorithms. Last but not least, it is worthwhile to consider other possible schemes such as increasing the number of turning points or constructing more adaptive search strategies so as to further enhance the adaptability of the overall AMPO+ search framework in tackling different optimisation problems.

References

1. Aljarah, I., Faris, H., Mirjalili, S.: Optimizing connection weights in neural networks using the whale optimization algorithm. Soft Comput. **22**(1), 1–15 (2018)
2. Aydilek, İ.B.: A hybrid firefly and particle swarm optimization algorithm for computationally expensive numerical problems. Appl. Soft Comput. **66**, 232–249 (2018). https://doi.org/10.1016/j.asoc.2018.02.025, https://www.sciencedirect.com/science/article/pii/S156849461830084X
3. Borisenko, A., Gorlatch, S.: Comparing GPU-parallelized metaheuristics to branch-and-bound for batch plants optimization. J. Supercomput. **75**(12), 7921–7933 (2018). https://doi.org/10.1007/s11227-018-2472-9
4. Borisenko, A., Gorlatch, S.: Efficient GPU-parallelization of batch plants design using metaheuristics with parameter tuning. J. Parallel Distrib. Comput. **154**, 74–81 (2021). https://doi.org/10.1016/j.jpdc.2021.03.012
5. Bortfeldt, A., Gehring, H., Mack, D.: A parallel Tabu search algorithm for solving the container loading problem. Parallel Comput. **29**(5), 641–662 (2003)
6. Dorigo, M., Di Caro, G.: Ant colony optimization: a new meta-heuristic. In: Proceedings of the 1999 Congress on Evolutionary Computation-CEC99 (Cat. No. 99TH8406), vol. 2, pp. 1470–1477. IEEE (1999)
7. Holland, J.H., et al.: Adaptation in Natural and Artificial Systems: An Introductory Analysis with Applications to Biology, Control, and Artificial Intelligence. MIT Press, Cambridge (1992)

8. Huang, C., Li, Y., Yao, X.: A survey of automatic parameter tuning methods for metaheuristics. IEEE Trans. Evol. Comput. **24**(2), 201–216 (2019). https://doi.org/10.1109/TEVC.2019.2921598

9. Joshi, S., Bansai, J.: Parameter tuning for meta-heuristics. Knowl.-Based Syst. **189**, 105094 (2020). https://doi.org/10.1016/j.knosys.2019.105094

10. Karaboga, D., Gorkemli, B., Ozturk, C., Karaboga, N.: A comprehensive survey: artificial bee colony (ABC) algorithm and applications. Artif. Intell. Rev. **42**(1), 21–57 (2014)

11. Kennedy, J., Eberhart, R.: Particle swarm optimization. In: Proceedings of ICNN 1995-International Conference on Neural Networks, vol. 4, pp. 1942–1948. IEEE (1995)

12. Li, Z., Tam, V., Yeung, L.K.: An adaptive multi-population optimization algorithm for global continuous optimization. IEEE Access **9**, 19960–19989 (2021). https://doi.org/10.1109/ACCESS.2021.3054636

13. Liang, J.J., Qu, B.Y., Suganthan, P.N.: Problem definitions and evaluation criteria for the CEC 2014 special session and competition on single objective real-parameter numerical optimization. Technical report 201311, Zhengzhou University, Henan Province, China (2014)

14. ben oualid Medani, K., Sayah, S., Bekrar, A.: Whale optimization algorithm based optimal reactive power dispatch: a case study of the Algerian power system. Electr. Power Syst. Res. **163**, 696–705 (2018)

15. Mirjalili, S., Lewis, A.: The whale optimization algorithm. Adv. Eng. Softw. **95**, 51–67 (2016)

16. Selim, S.Z., Alsultan, K.: A simulated annealing algorithm for the clustering problem. Pattern Recogn. **24**(10), 1003–1008 (1991)

17. Storn, R., Price, K.: Differential evolution - a simple and efficient heuristic for global optimization over continuous spaces. J. Global Optim. **11**(4), 341–359 (1997)

18. Wang, L., Pan, J., Jiao, L.c.: The immune algorithm. Acta Electronica Sinica **28**(7), 74–78 (2000)

19. Wu, G., Mallipeddi, R., Suganthan, P., Wang, R., Chen, H.: Differential evolution with multi-population based ensemble of mutation strategies. Inf. Sci. **329**, 329–345 (2016). https://doi.org/10.1016/j.ins.2015.09.009, https://www.sciencedirect.com/science/article/pii/S0020025515006635, special issue on Discovery Science

A Model Driven Approach to Transform Business Vision-Oriented Decision-Making Requirement into Solution-Oriented Optimization Model

Liwen Zhang[1,2(✉)], Hervé Pingaud[3], Elyes Lamine[2], Franck Fontanili[2], Christophe Bortolaso[1], and Mustapha Derras[1]

[1] Berger-Levrault, Jean Rostand. 64, 31670 Labège, France
{liwen.zhang,christophe.bortolaso,
mustapha.derras}@berger-levrault.com
[2] CGI, University of Toulouse-IMT Mines Albi, Campus Jarlard, 81000 Albi, France
{liwen.zhang,elyes.lamine,franck.fontanili}@mines-albi.fr
[3] CNRS LGC, University of Toulouse-INU Champollion, Place de Verdun, 81012 Albi, France
herve.pingaud@univ-jfc.fr

Abstract. Currently in our highly connected society, there is a strong requirement for decision-makers in organizations to coordinate and schedule their activities. Frequently, there are various uncertain factors, multiple objectives, many business knowledge and requirements, which heavily increase the difficulty of decision-making process regarding these issues. Therefore, a decision-maker will appreciate having control over the formulation of decision-making models and being able to adapt to highly dynamic situation. In this paper, we study a Model Driven Engineering (MDE) approach to link the business requirement defined by a model with solution-oriented logical models, which are codes that could be submitted to a combinatorial optimization solver. The design of our proposal follows the principles of three-levels Model Driven Architecture (MDA) and is based on a cognitive process for decision-making systems. Then, several transformation rules between models are explained to realize automatic Model to Model Transformation (M2M) with a special emphasis on the Platform Independent Model (PIM) to Platform Specific Model (PSM) part. To make a proof of our model transformation chain efficiency, a classical Travelling Salesman Problem (TSP) is chosen as a use case.

Keywords: Model driven architecture · Model to model transformation · Decision-making support · Scheduling problem

1 Introduction

Currently in a connected society with worldwide relationships, there is a strong incentive on productive organizations to regularly coordinate their activities within their environment when managing their products and service flows. For them, if it is necessary to achieve the right objectives (e.g., performance control and risk mitigation), such

H. Fujita et al. (Eds.): IEA/AIE 2022, LNAI 13343, pp. 211–225, 2022.
https://doi.org/10.1007/978-3-031-08530-7_18

objectives are often evolving and varying facing dynamics of the industrial and market contexts. Such modifications can occasionally obstruct autonomous decision-making abilities.

Due to the strategic nature of such problems, a regular debate about management culture within the company has emerged in recent years. Especially about the ability to fulfill the demands of multiple networks into which the organization is involved focusing on supply chain management approaches. Thus, when continuous improvement permeates the quality process, it is not only limited to operational and support processes, but also impacts those of management. It raises concerns about methods of thinking, decision-making activities, and decision-making aids.

The industrial engineering community has worked intensively and for some while to promote progress on such decision-making support systems. Planning and scheduling, for example, are not only strategic functions, but must be upgraded to strategic means. The challenge became to write "personalized" formulations of the relevant problems, considering local specificities in a given organization as real opportunities. Doing so, the trend is to include more and more business information in the decision-making process. Saving time being of prime interest, the short-term demands that company must manage are rising. Not only should the supply chain aim for economic and social performance goals, but it should also be able to deal with these unexpected changes in demand and know exactly how to meet them. As the importance of these uncertainties grows in the decision-making process, the utilization of decision support technologies becomes more critical. For all these reasons, a decision-maker will appreciate having control over the formulation of decision-making models and being able to adapt them when the situation is highly dynamic.

Mathematical solvers are tools (e.g., CPLEX, OptaPlanner) that have been widely disseminated. These digital technologies have historically been commonly used by experts in the decision-making support domain, including the Operations Research (OR) community. For tackling large scale constrained optimization problems, most solvers create with libraries of mathematical algorithms (e.g., branch and bound for CPLEX or simulated annealing for OptaPlanner, for example). After carefully implementing them using a descriptive language (e.g., respectively Optimization Programming Language-OPL for CPLEX, and Drools for OptaPlanner), users should be able to get a personalized solution quite effortlessly. Our research tries to facilitate the knowledge management for solving decision support problems using models. The objective is to investigate how to give the decision-maker an environment where the ability to fluently specify his/her problems online will be effective and where any change management would require minimal mathematical expertise.

This method of thinking prompts a reconsideration of modelling decision-making in business models. A transition must change from a high-level business knowledge to the knowledge required by a solver to perform calculations in order to optimize decisions. We discuss and develop our decision support environment with this goal in mind. The approach for making the transition concrete is based on Model Driven Engineering (MDE). We apply an MDE process by mapping (1) business knowledge and expertise available at the Computer Independent Model (CIM) level of the Model

Driven Architecture recommended by OMG [1] (2) formal addressing of a decision-making problem in mathematical transcription at the Platform Independent Model (PIM) level (3) ability to make decisions by a solver at the Platform Specific Model (PSM) level. Hereafter, we will further study these conceptual links between mathematical decision support models and logical solution-oriented models.

The following content is divided into five sections. In the section two, we briefly remind the foundation of MDE, its basic concepts and what may be a lifecycle of an MDE process. Then, we proceed to a literature review about past references on the same subject. In section three, an overview of our MDE approach is explained. Then, we put the emphasis on the PIM to PSM part, which extends some of our previous results [2]: the transformation from a Graph-based Operations Research Model (GORM) to a Solution-oriented mathematical Model (SM). The two meta-models and the transformation rules defined between them to achieve the Model-to-Model transformation (M2M) are discussed in the next section. We have chosen the Travelling Salesman Problem (TSP) as a use case to illustrate our study throughout the paper. Finally, we deliver conclusions about the study and draw some perspectives for future research on this field.

2 Past Related Studies

2.1 Theorical Foundation of MDE

MDE is motivated by the idea of valuing models for a certain purpose [3]. OMG (Object Management Group) is a well-known nonprofit international organization who promoted this usage more than twenty years ago. One of their recommendation for MDE is a standard named Model Driven Architecture (MDA) which was published at the end of 2001 [1]. This standard has been a reference framework for a long time. This initiative delivers two concrete learning outcomes. Firstly, it clarifies the relationships between a model and a meta-model located at two different levels into the MDA pyramid. A meta-model is a model that defines the language used to express a model [4]. Second, the MDA requires that a software production design system be divided into three components, resulting in a two-stage engineering life cycle. As a direct result, three different models are identified, one for each component of the life cycle: Computer Independent Model (CIM), Platform Independent Model (PIM), Platform Specific Model (PSM) [5].

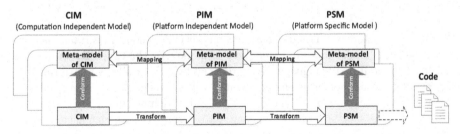

Fig. 1. Model-to-model transformation based on Model Driven Architecture

Model-to-Model transformation (M2M) is a key feature. As shown in Fig. 1, M2M begins with a business model which is independent of the technique (CIM), to generate another model which is independent of a development platform (PIM) during the first stage of the life-cycle. During the second stage, the PIM model is a source and its transformation ends to a model which is specific to an executing platform (PSM). According to [6], M2M performs globally an automatic generation of a target PSM model from a CIM source model, following two successive transformation by specifying mapping rules at the meta-model level. Mapping rules could be summarized as translation directives of models expressed by meaningful links between meta-models. In our application, let us remind that the ultimate target is a source code for a solver, written in its implementation language.

2.2   Previous Experiences in M2M

In recent years, many M2M transformation methods have been investigated by the scientific community working on MDE for engineering purposes. In researching applications in this engineering field, we have found many traces of such experiments. In healthcare, a prototype is designed to generate a human-machine interface on mobile technologies (target model, PSM), starting from the modeling of a care plan (source model, CIM) [7]. In [8], the authors explain how to transform a CIM model entitled "International Electrotechnical Commission Common Information Model" to a target PSM mathematical formal "Modelica" model. This target model is used to perform a dynamic simulation of a complex physical system. In [9], the authors use the BPMN notation to model business processes from a CIM (source model) to a PIM (target model). The target model includes static, dynamic and functional views, which are expressed using three appropriate UML diagram types. In terms of industrial applications, enterprise systems interoperability between partners involved in different configurations of collaborative networks have been extensively addressed by MDE. We refer to research works on Mediation Information System Engineering (MISE) [10]. The approach collects the knowledge about a company active in a collaboration network, at the CIM level through an organizational model. The PIM level represents the technical knowledge that is a collaborative process model in BPMN language. Finally, the PSM performs an extraction of knowledge from the collaborative process model to make a digital service orchestration in a bus of enterprise services [11].

The key element in model transformation is a Model Transformation Language (MTL) [12]. Initially, OMG proposed the Query/View/Transformation (QVT) as a model transformation language candidate [13]. Visual Model Transformation Language (VMTL) [14] is another MTL initiative that supports endogenous transformations (the source and target model conform to the same meta-model). The Graph Rewriting And Transformation (GReAT) language was introduced by [15], it allows transformations from one domain to another using heterogeneous meta-models. ATL (Atlas Transformation Language) is an MTL of the QVT type [16], which is extensively cited in the literature. ADOxx specific Scripting language (AdoScript) is a script language with the characteristics of a MTL, which is integrated in the ADOxx meta-modeling platform developed by OMILAB [17]. This language allows to access and process the information stored in the models, and then to perform multiple functions such as M2M, model visualization and simulation.

As far as we know, no reference exists about an application of MDE for the transformation of business models into a mathematical model that feeds a combinatorial optimization-oriented solver to solve scheduling problems. We would like to emphasize that our approach allows business experts with no background in operations research to easily build optimization models, aiming at providing them with decision-making supports. For MTL, we select AdoScript which is embedded in the ADOxx meta-modeling platform [18], because it integrates both design of modeling languages and use of M2M features. ADOxx is extensively used to build Proof of Concept (PoC) in research projects. The OMILAB community [17] shares experiences and provides efficient technical supports developed with ADOxx for a purpose of reusability.

3 MDE for Decision-Making Process Design

3.1 Cognitive Process for Decision-Making System

We will describe the decision-making system through a knowledge chain that evolves from a preliminary stage of business requirements analysis, to a decision-making support model that meets these requirements. Three types of models are sequentially used in this chain. As shown in Fig. 2, this breakdown in many models is similar to a cognitive process that a decision-making process designer will apply. It is resumed in three steps: (1) decision-making requirements explanation for business expert's perspective, (2) decision-making support-based formulation and (3) preparation of solution calculation for OR expert's perspective. At the end of each step, the MDE tool produces a new model which is an input for the next step.

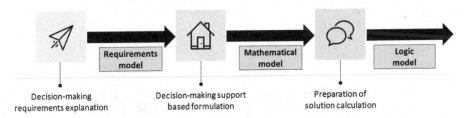

Fig. 2. Cognitive process of a decision-making system

3.2 Cognitive Process-Based Model Driven Architecture

Figure 3 extends the conceptual view of Fig. 2 following the MDA principles depicted in Fig. 1. Three types of decision-making models are considered in this new representation:

- The CIM level is the specification of a decision-making need. **A requirement model** entitled "Business-oriented Conceptual Model (**BCM**)" captures the relevant knowledge before addressing the decision-making problem (e.g., the TSP).

- The PIM level is typically devoted to what the operations research will perform to translate this BCM into a mathematical model of the problem. The resulting **mathematical model** is formalized by a graph and entitled "Graph-based Operations Research Model (**GORM**)".
- The PSM level follows and has to prepare the mathematical problem-solving computations to be performed within a specific execution environment. It consists in a coding of the GORM using a solver's specific language. Therefore, this last **logic model** is entitled "Solution-oriented mathematical Model (**SM**)".

To complete the generation of the PSM from the CIM, a collection of transformation rules is an input of the MTL that has to be defined, and we made it with AdoScript. As shown in Fig. 3, three types of transformation rules are distinguished:

- **Mapping rules:** these rules give birth to one or several element(s) of the target model from one or several elements of the source model.
- **Browsing rules:** these rules are defined in target meta-model, by processing locally some elements. Their aim is to check consistency of the new elements with the relevant meta-model.
- **Parsing rules:** these rules also operate local treatments, but only in model level of PIM and PSM. The production of code lines must be compliant with the syntax of the programming language (e.g., OPL for computation of the TSP by CPLEX).

In [2], the transformation between CIM (BCM) and PIM (GORM) has been explained. Therefore, the content of this paper is complementary to this previous publication. We focus hereby on the transformation from PIM (GORM) to PSM (SM), providing technical information about it. Theory and practice are still illustrated in the same use case (TSP).

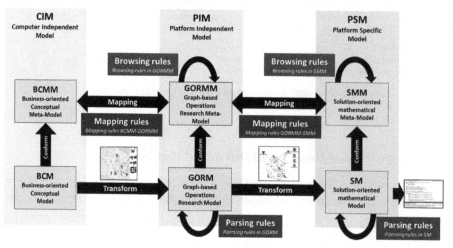

Fig. 3. Overview of the cognitive process developed following MDA principles

4   PIM to PSM Transformation Applied to TSP

The transformation from PIM to PSM requires (1) the declaration of relevant data (set of parameters and set of decision variables), and (2) the definition of the constraints and the objective function.

Figure 4 shows respectively both ends of this transformation work: the GORM as an input model and the SM as an output for a given TSP use case. This example concerns four cities over a metropolitan area. The objective function is the travel distance we want to minimize. The left side of Fig. 4 is a graph of the cities to be visited by the traveler, with many annotations related to the possible pathways. The right side of Fig. 4 is the OPL code with the OPL keywords written in blue color.

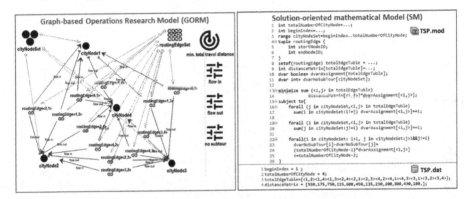

Fig. 4. GORM and SM on a case study

Because the GORM will conform to the meta-model called Graph-based Operation Research Meta-Model (**GORMM**), the reader is invited to refer to the previous paper [2] for further details on the related concepts. Similarly, as our SM must conform to a meta-model called Solution-oriented mathematical Meta-Model (**SMM**, see Fig. 3), we will just give some details about the concepts defined in this SMM in the next section.

4.1   Specification of Solution-Oriented Mathematical Meta-model (SMM)

The design of the SMM is inspired by [19] with the two expected parts: (1) data declaration view and (2) formulation building view. Describing each part independently in the following, we use figures (from 1 to 6, in black circles) to map the literal description with the graphical representation of the SMM.

For data declaration, we list all the concepts in Table 1 including the data types in the OPL syntax.

For formulation building, the transformation is supported by a library of mathematical relation patterns from which the OPL formulas of the constraint and the objective functions can be adapted consistently. Thus, our transformation is guided by the need to identify the better appropriate formulation pattern within the library for each constraint

Table 1. Overview of concepts in SMM: data declaration view

❶	**Data Declaration**	It represents the relationship established between data and their types in OPL. The attributes include: • name: the name retained to reference a data to be declared. • type: the data type in OPL. • isDecisionVar: a Boolean indicating if a data is a decision variable, which means that this type of data can be modified during optimization process in the relevant decision space. • indexRep: this attribute is enumeration type including three forms of index declaration in scalar, couple or triplet: "i or j or k", "\<i,j\>" and "\<i,j,k \>".
❷	**AbstractType**	A generic descriptor for data type (scalar, vector or hybrid structure) in OPL whose subclasses can be derived with an implementation by a keyword. Inheriting this generic concept, the other child concepts are extracted from the work of [19].

Table 2. Overview of concepts in SMM: formulation building view

❸	**Formulation Building**	An abstract concept which is a container of formulation patterns written in OPL. It is characterized by: • name: same function as the identifier (name) of the "Constraint" or "Objective Function" concept in GORM. • pattern content: this concept aims to identify the characteristics of OPL patterns before the allocation of the data required by the calculation.
❹	**Constraint Building**	This concept inherits the "Formulation Building" concept. It has an attribute "equation in CPLEX" aiming at displaying the complete formulation of a constraint once the components of the concerned pattern have been assembled.
❺	**OF Building**	This concept inherits the "Formulation Building" concept. It performs the similar function to "Construction Building", but for the Objective Function.
❻	**Formulation Component**	This concept contains the attributes to indicate the formulation components for the relevant pattern, based on the library that stores these patterns. It is characterized by the name of concerned pattern (name), parameters and decision variables with an adapted number: parameter 1 [P1], parameter 2 [P2], parameter 3 [P3], decision variable 1 [D1], decision variable 2 [D2].

and the objective function. Then, it is to perform an appropriate instantiation with the right problem data. Table 2 summarized the formulation building view of the SMM.

Figure 5 is a representation of this SMM as an UML class diagram. Note that the descriptors for the data structures (the "AbstractType" concept along with all its sub-concepts) draw directly from the work we have been inspired [19].

4.2 Transformation Process

The transformation process operates in three times:

1. Firstly, the M2M transformation is carried out through inferring the mapping rules between GORMM and SMM.
2. Then, a browsing takes place inside the SM at the PSM level. This is an automatic processing to set up the adapted call means for the two formulations "constructors".

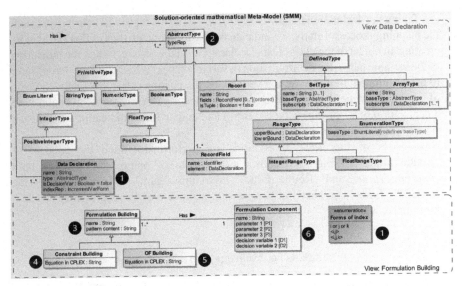

Fig. 5. Solution-oriented mathematical Meta-Model (SMM)

3. The third time is a data compiling in respect to the OPL syntactic rules. A parser developed in Java is making the coded lines assembly, which is partly directed by an interpretation of the OPL keywords.

At the end of the inference of this set of rules, we get the final model that can be submitted to CPLEX for running the search of an optimum result. We mean that this SM OPL-based model can technically be submitted to the CPLEX software to find the solution of a TSP. In the following, we will provide further details about the three categories of rules introduced above, regarding these three times of the transformation concerning the TSP use case.

Mapping Rules Between GORMM and SMM
Data Declaration
The translation of GORM data into the SM correct type is focused on two concepts called "sets": the "CityNodeSet" and the "RoutingEdgeSet". This translation is performed towards the declarative part of the OPL code with a differentiation between parameters and decision variables belonging to the model. Each of them is then assigned its correct keyword regarding the data type required by OPL. This translation is made in two parts:

1. The declaration of the "CityNodeSet" starts by choosing the primary type: "RangeType", then primitive integer types are chosen for two components: its upper bound (*"totalNumberOfCityNode"*) and lower bound (*"beginNodeIndex"*). Finally, the declaration of the decision variable *"dvarNoSubtour"* is made. Notice that the expression of variable *"dvarNoSubtour"* in OPL is as follows:

```
dvar int+ dvarNoSubtour [CityNodeSet]
```

Taking the declaration of *"dvarNoSubTour"* as an example will help us to explain the mapping process. Firstly, we indicate the basic characteristics of this variable by creating an object of the class "Data Declaration" in the SM, including the indication of the variable name, the determination of variable type (*"isDecisionVar* = true" refer to the assignment of `dvar` for the variable by the parsing rules). Secondly, the type of *"dvarNoSubtour"* is a one-dimensional vector array ("Array" type). It begins with a declaration of its primitive type (`int+`) and ends with an indication of the array size into brackets ("CityNodeSet").

2. Next, a set of successive translations are performed around the concept of "RoutingEdgeSet". Figure 6 shows an example of mapping to declare both *"totalEdgeTuple"* and *"RoutingEdge"*. From this figure, the declaration of the *"totalEdgeTuple"* is realized in two steps to achieve its OPL expression, which is formulated as follows:

```
setof(routingEdge) totalEdgeTuple
```

Then, the declaration of the "RoutingEdge" is done in three steps to obtain the following expression in OPL:

```
tuple routingEdge{
int startNodeID;
int endNodeID;}.
```

Formulation Building

This type of transformation uses pointers to build the code lines related to the mathematical patterns in the executable statements of the SM. Therefore, the transformation should start by building firstly a body of an equation in a library, including keywords and operators, the order must be respected one each other of an equation. Afterwards, the pointer is used to indicate the elected mathematical formula in the library. The library is indexed to map to the "Constraint" and "Objective Function" concepts defined in the GORM, and written as function types in AdoScript. After this selection performed, the pattern of each formulation is recorded in the attribute "pattern content" of the concept "Formulation building".

After, a rule is used to generate components by completing the patterns with the relevant parameters or decision variables of the mathematical relations. This phase is based on a catalog of specification for the patterns belonging to the library. Briefly speaking, the catalog supports the necessary distribution of components (see "Formulation Component" in Fig. 6) over a given pattern in the library.

Browsing Rules in SM

The local browsing aims to make an assembly of all the results inherited from the previous work "Formulation Building", together with those from "Formulation Component". The assembly process consists of (1) components-pattern assembly to build formulations, (2) final assembly assigned to the *"equation in CPLEX"* attribute of "Constraint Building" or "OF Building" concept, respectively.

Parsing Rules for GORM and SM

This operation performs the role of a parser for the GORM and SM. An OPL model in CPLEX consists of two files: a data file (.dat) and a model file (.mod). On one hand, the parser extracts the target ".mod" file from the SM. On the other hand, the parser performs the export of the GORM to the target ".dat" file, which contains the values of the relevant parameters defined in the ".mod" file (e.g., *"beginNodeIndex"*, *"totalNumberOfCityNode"*, *"totalEdgeTuple"*, *"distanceMatrix"*).

As models (BCM, GORM and SM) are created in ADOxx and, as so, can be exported in an XML format, the parser was developed in Java using Java Architecture for XML Binding (JAXB). To complete the generation of the OPL model, a series of rules developed in Java fulfill the function of keyword assignment, for example:

- The decision variables *"dvarNoSubTour"* and *"dvarAssignment"* must have a keyword " dvar" as a prefix.
- The set of constraints must be surrounded by " subject to {...}".

In addition, this phase allows the formatting of two output files, for example:

- The " =...;" that follows the declared parameters.

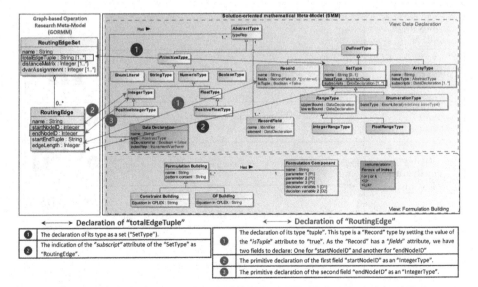

Fig. 6. Mapping rules for declaring *"totaEdgeTuple"* and "RoutingEdge"

- The " { } " for data of type "tuple".
- The " ... " connector for data of type "range".
- The separator " ; " or the space (signified by " \n" in regular expression in Java) to isolate the data and the specified formulations.

5 Case Study

Let us go back to the BCM using the graphical operational semantics of our BCMM, which is recalled in Table 3. The same notation is used to report the results of the computation made by the solver.

In this case, we have four cities to be visited by the "salesman 1" which are: Toulouse, Paris, Lyon and Nice. Three constraints are considered and symbolized by an operational semantics (i.e., ⬛, ⬛ and ⬛). The objective is to minimize the salesman's total routing distance. All the mentioned contextual information is modeled by business experts using BCM at CIM level.

As described in the upper part of Fig. 7, the three MDA models are shown from left to right, i.e., CIM/PIM/PSM. BCM is first transformed into GORM at the PIM level, and then into SM at the PSM level through two transformation process: CIM to PIM, PIM to PSM. The PSM has been compiled by CPLEX correctly. The result given by CPLEX is for a TSP with four cities to visit and "Salesman 1" will travel is the following optimal order: Toulouse → Paris → Lyon → Nice → Toulouse. This

Table 3. Graphical representation of the concepts included in GORMM

Name of concept	Enumeration element for "type" attribute	Graphical rep.
Salesman		
City		
Business constraints (BuConst)	Each city reached from exactly one other city by the salesman.	
	From each city, there is only a departure to another city.	
	There is only one tour covering all the cities to be visited.	
Goal	Minimize the salesman's total routing distance	

result is visualized graphically by simply changing the color of the optimal path. The achieved goal is 1405 kms qualifying a minimal total distance along with a very low computational time on this small-size problem. The specified business constraints are all well-satisfied, as demonstrated by the annotation to the question "satisfied?" for each business constraint.

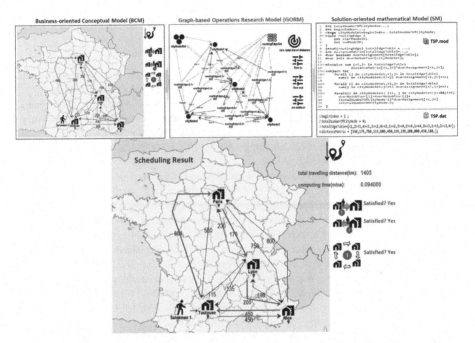

Fig. 7. Use case specification of TSP: the three models and the final scheduling result

6 Conclusion and Research Perspectives

This research work addresses a model driven approach to transform a business vision-oriented decision-making model into a solution-oriented optimization model, following the three levels (CIM-PIM-PSM) of the Model-Driven Architecture (MDA). Considering the result of our previous work [2], we have focused on the Model to Model transformation (M2M) between a Graph-based Operations Research Model (GORM) at PIM level and a Solution-oriented mathematical Model (SM) at PSM level. The AdoScript Model Transformation Language (MTL) has been used to perform the PSM generation using Model to Model translation. In the core of this M2M machinery, three categories of transformation rules have been described. To make a proof of concept of the MDE approach, we decided to apply it on the well-known Travelling Salesman Problem (TSP).

Considering new perspectives, future work will address a problem that is more difficult to deal with than a TSP. We have chosen to extend our approach towards this new case study of the Home Health Care Routing and Scheduling Problem (HHCRSP)

[20], considered as a clustered multi-TSP. This problem is the subject of many variations related to different organizations in charge of home health care coordination of caregivers, including the diversity of practices, caregiver skills, means of transportation, work schedules and so on. It appears to be particularly prone to make extension of our approach. Upcoming studies of an agile model-based environment to support efficiently decision-making for HHCRSP are planned.

References

1. Poole, J.D.: Model-driven architecture: vision, standards and emerging technologies. In: Workshop on Metamodeling and Adaptive Object Models, ECOOP. Citeseer (2001)
2. Zhang, L., Fontanili, F., Lamine, E., Bortolaso, C., Derras, M., Pingaud, H.: A systematic model to model transformation for knowledge-based planning generation problems. In: Fujita, H., Fournier-Viger, P., Ali, M., Sasaki, J. (eds.) IEA/AIE 2020. LNCS (LNAI), vol. 12144, pp. 140–152. Springer, Cham (2020). https://doi.org/10.1007/978-3-030-55789-8_13
3. OMG: MDA Guide revision 2.0 (2014)
4. Benaben, F.: Conception de Système d'Information de Médiation pour la prise en charge de l'Interopérabilité dans les Collaborations d'Organisations (2012). https://hal-mines-albi.arc hives-ouvertes.fr/tel-01206234/document
5. Truyen, F.: The fast guide to model driven architecture: the basics of model driven architecture. Whitepaper, Architecture Oriented Services, Cephas Consulting Corp. (2006)
6. Kleppe, A.G., Warmer, J., Warmer, J.B., Bast, W.: MDA Explained: The Model Driven Architecture: Practice and Promise. Addison-Wesley Professional, Reading (2003)
7. Khambati, A., Grundy, J., Warren, J., Hosking, J.: Model-driven development of mobile personal health care applications. In: Proceedings of the 2008 23rd IEEE/ACM International Conference on Automated Software Engineering, pp. 467–470. IEEE Computer Society, Washington, DC (2008). https://doi.org/10.1109/ASE.2008.75
8. Gómez, F.J., Vanfretti, L., Olsen, S.H.: CIM-compliant power system dynamic model-to-model transformation and Modelica simulation. IEEE Trans. Industr. Inf. **14**, 3989–3996 (2018). https://doi.org/10.1109/TII.2017.2785439
9. Rhazali, Y., Hadi, Y., Mouloudi, A.: Model transformation with ATL into MDA from CIM to PIM structured through MVC. Procedia Comput. Sci. **83**, 1096–1101 (2016). https://doi.org/10.1016/j.procs.2016.04.229
10. Mu, W., Benaben, F., Boissel-Dallier, N., Pingaud, H.: Collaborative knowledge framework for mediation information system engineering. Sci. Program. **2017**, 1–18 (2017). https://doi.org/10.1155/2017/9026387
11. Wang, T., Truptil, S., Benaben, F.: An automatic model-to-model mapping and transformation methodology to serve model-based systems engineering. Inf. Syst. E-Bus Manage. **15**(2), 323–376 (2016). https://doi.org/10.1007/s10257-016-0321-z
12. Bézivin, J., Büttner, F., Gogolla, M., Jouault, F., Kurtev, I., Lindow, A.: Model transformations? Transformation models! In: Nierstrasz, O., Whittle, J., Harel, D., Reggio, G. (eds.) MODELS 2006. LNCS, vol. 4199, pp. 440–453. Springer, Heidelberg (2006). https://doi.org/10.1007/11880240_31
13. Kurtev, I.: State of the art of QVT: a model transformation language standard. In: Schürr, A., Nagl, M., Zündorf, A. (eds.) AGTIVE 2007. LNCS, vol. 5088, pp. 377–393. Springer, Heidelberg (2008). https://doi.org/10.1007/978-3-540-89020-1_26
14. Acreţoaie, V., Störrle, H., Strüber, D.: VMTL: a language for end-user model transformation. Softw. Syst. Model. **17**(4), 1139–1167 (2016). https://doi.org/10.1007/s10270-016-0546-9

15. Agrawal, A., Karsai, G., Neema, S., Shi, F., Vizhanyo, A.: The design of a language for model transformations. Softw. Syst. Model. **5**, 261–288 (2006). https://doi.org/10.1007/s10270-006-0027-7

16. Jouault, F., Allilaire, F., Bézivin, J., Kurtev, I., Valduriez, P.: ATL: a QVT-like transformation language. In: Companion to the 21st ACM SIGPLAN Symposium on Object-Oriented Programming Systems, Languages, and Applications, pp. 719–720 (2006)

17. Fill, H.-G., Karagiannis, D.: On the conceptualisation of modelling methods using the ADOxx meta modelling platform. Enterp. Model. Inf. Syst. Archit. (EMISAJ) **8**, 4–25 (2013). https://doi.org/10.18417/emisa.8.1.1

18. Karagiannis, D., Kühn, H.: Metamodelling platforms. In: Bauknecht, K., Tjoa, A.M., Quirchmayr, G. (eds.) EC-Web 2002. LNCS, vol. 2455, p. 182. Springer, Heidelberg (2002). https://doi.org/10.1007/3-540-45705-4_19

19. Assouroko, I., Denno, P.O.: A metamodel for optimization problems. National Institute of Standards and Technology (2016). https://doi.org/10.6028/NIST.IR.8096

20. Zhang, L.: De la vision métier à la génération assistée de plannings pour la coordination centralisée de services de soins à domicile. Ph.D. thesis manuscript (2021). https://tel.archives-ouvertes.fr/tel-03405854

A Hybrid Approach Based on Genetic Algorithm with Ranking Aggregation for Feature Selection

Bui Quoc Trung, Le Minh Duc, and Bui Thi Mai Anh[(✉)]

School of Information and Communication Technology,
Hanoi University of Science and Technology, Hanoi, Vietnam
{trungbq,anhbtm}@soict.hust.edu.vn, duc.lm173035@sis.hust.edu.vn

Abstract. Recently, feature selection has become challenging for many machine learning disciplines. The success of most existing approaches depends on the effectiveness of searching strategies to select the most salient features from the original feature space. Unfortunately, these approaches may become impractical when dealing with high-dimensional datasets. In order to overcome this problem, recent studies rely on a boundary scheme (i.e., fixing a number of selected features) to reduce the searching space or a ranking scheme (e.g., features with less correlated scores) to guide the selection phase. However, choosing the best-fitted size for the feature subset is also a hard problem, and relying on one feature comparison criteria may ignore important features. In this paper, we propose a genetic algorithm that aims to optimize the feature subset and the appropriate number of selected features to maximize the performance of an Artificial Neural Network (ANN) classifier. To improve the efficiency of the selection phase, we combine the proposed GA with a local search algorithm based on a ranking aggregation approach. Our objective is to speed up the searching algorithm by taking advantage of different feature scoring criteria. We have assessed the performance of our approach over three categories of datasets: *small, medium* and *high* in terms of feature dimensionality (e.g., the smallest and the largest datasets include 8 and 7129 features, respectively). The empirical results have shown that our proposed approach outperforms the other state-of-the-art works when dealing with medium- and high-dimensional datasets and is comparable to them in the case of small-dimensional datasets.

Keywords: Feature selection · Genetic algorithm · Ranking aggregation · Local search

1 Introduction

Feature Selection (FS) is an important stage of data preprocessing, which is applied in many machine learning applications. As real-world data contains irrelevant, redundant, and noisy features, the main objective of feature selection is data-dimensionality reduction, in which the smallest-size subset of features is

© Springer Nature Switzerland AG 2022
H. Fujita et al. (Eds.): IEA/AIE 2022, LNAI 13343, pp. 226–239, 2022.
https://doi.org/10.1007/978-3-031-08530-7_19

selected to maximize the model performance. Additional benefits of feature selection include the ability to increase the computational efficiency while decreasing memory storage as well as better generalizing models [4].

From the selection strategy perspective, feature selection methods are mainly categorized as a wrapper, filter, and embedded approaches [10]. Wrapper techniques rely on a predefined learning algorithm to assess the performance of different subsets of features [20]. In contrast, filter methods are independent of any learning algorithms. Based on a statistical analysis of features, the selection process is carried out through two steps: (i) assessing the importance score of each feature and producing a ranking list (ii) filtering out features with a low ranking score. In comparison with the former, the latter is typically more efficient in computing performance and can provide a generic selection of features that are not tuned for a given learning algorithm. However, due to the lack of a specific algorithm of filter approaches, the selected features may not be optimal for target learning algorithms. As a trade-off between filter and wrapper approaches, embedded methods integrate the feature selection into the learning process and take advantage of both previous techniques. However, embedding the feature selection phase into the learning procedure is not a simple task, which typically requires altering the learning algorithm [22].

The success of different feature selection approach depends particularly on searching strategies over the feature space. Sequential strategies, including forwarding and backward approaches, have been early applied to the feature selection process [8]. However, the generated feature subset cannot be altered once a feature is added (forward approach) or removed (backward approach). Random search strategies have been considered to overcome this issue [19]. These strategies, otherwise, may suffer from a local optimum due to a partial search over the solution space. As a result, global search algorithms have been applied to select salient features, in which the Genetic Algorithm (GA) has shown many promising results [4,16,21].

We argue that most searching strategies become impractical when dealing with high-dimensional datasets. In order to tackle this problem, certain studies aim to narrow the searching space by fixing a number of selected features, improving the computational performance of the searching process [16] considerably. Nevertheless, choosing the best-fitted size for the feature subset is also a hard problem. Other studies focus on the selection phase to build a candidate subset of salient features based on their statistical relationship [15]. As such, subsets of features are formed from only less correlated ones. Although this strategy avoids redundant information coming from positively correlated features, the most important features may be ignored. This paper proposes a hybrid approach that combines both filter and wrapper methods for feature selection. Precisely, we apply a filter approach that relies on correlation information and other ranking criteria to score features. A ranking list is obtained according to each ranking criterion, which provides an ordering of features. We then aggregate these ranking lists to obtain an optimal ranking order which allows taking advantage of all the criteria. The proposed wrapper method is based on a genetic algorithm in which the aggregation ranking list is used as a local operator during the selection phase to highlight features with high ranking scores. We employ an Artificial Neural Network (ANN) model as the target model in our wrapper approach, whose performance assists in fulfilling the evolutionary process of GA. Finally, to deal with the problem of high-dimensional searching space, in our study, the size of the selected feature subset is also an optimization target which is solved by our proposed GA.

The rest of the paper is organized as follows: Sect. 2 describes the related works to the proposed techniques and algorithms. The details of our filter-wrapper combined approach for feature selection will be presented in Sect. 3. Sections 4 and 5 introduce the benchmark settings and the empirical results of our approach compared to other state-of-the-art methods. Finally, Sect. 6 concludes with some future directions.

2   Related Work

We briefly review some related works on the filter approach based on ranking aggregation and hybrid approaches for feature selection. The ranking technique is one of the most practical methods to filter out irrelevant features. Different ranking criteria have been proposed, such as Pearson correlation coefficient, Information Gain, Relief, Chi-square, Fisher score, etc. [20]. According to each measurement, the relevancy of features is computed and ordered, producing a different ranking list. Ranking aggregation methods combine different ranking results into a better one. Early work in this field was carried out by Borda, who computed the final rank of a data property as the sum of all positional scores from all the rankers [3].

The Borda count and its variants (e.g., median rank aggregation), acknowledged as *positional* methods, have been widely applied in feature selection due to its simplicity [14]. An alternative approach, called *majority* ranking, determines an outcome in terms of the number of votes for each feature by individual rankers, such as Condorcet criterion [6], Kemeny rule [17]. However, these approaches are seldom used in practice due to their low computational effectiveness. Bouaguel et al. [4] employed a genetic algorithm to optimize the aggregation ranking list by minimizing the distance to all individual ranking lists. This method has shown promising results in terms of filter technique for feature selection.

While filter methods rely on feature relevance criteria, wrapper methods employ searching strategies to build feature subsets to optimize the performance of a specific learning model. Many wrapper models have been proposed for feature selection using meta-heuristic searching algorithms such as Genetic Algorithm (GA) [16], Particle Swarm Optimization (PSO) [26], etc. The effectiveness of optimization algorithms concern two aspects: *exploration* and *exploitation*. Exploration mechanism explores promising sources in the search space, while exploitation technique aims to locate more potential sources using these explored sources.

Compared to other meta-heuristic algorithms, GA has been widely employed in feature selection thanks to the proper trade-off between exploration (i.e., through crossover) and exploitation (i.e., through mutation). As random mutation may lead to a poor exploitation capability, recent studies focused on hybridizing GA with a local search operator. This operator plays the role of a filter technique to guide the GA process. Huang et al. proposed to integrate Mutual Information (MI) into GA to rank candidate solutions (i.e., feature subsets) [12]. Kabir et al. combined GA with a ranking criterion based on correlation information [16]. This paper enhances the traditional GA with a filter operator that aggregates different ranking criteria. To the best of our knowledge, most studies in the field of feature selection employ a particular ranking technique.

3   Proposed Approach

This section introduces our proposed algorithm, namely RA-GA, which combines a wrapper-based genetic algorithm with a ranking aggregation filter technique to address the feature selection issue.

3.1   The Filter Based Ranking Aggregation

As mentioned in the previous sections, ranking aggregation allows merging all individual ranking lists to create a more consensus one. In this paper, we explore four filter criteria as follows:

(a) **Chi-Square** (CS) aims to evaluate the independence degree of each feature from the categorical attribute using the *chi-square* statistical measure [20].

(b) **Relief** (REF) assigns to each feature a score representing the ability to distinguish its neighboring instances [20]. Given a feature, its REF score is calculated by selecting a random number of samples from the dataset then taking into account the distance between instances of the same and different classes.

(c) **Fisher Score** (FS) is a supervised ranking approach that orders features based on discriminant ability [9]. The Fisher score maximizes the between-class scatter and minimizes the within-class scatter.

(d) **Maximum-relevance and Minimum-redundancy** (mRmR) is an extension of mutual information feature selection [25]. A greedy heuristic approach is applied to select a subset of features S based on the following criterion:

$$mRmR = \max_{i \in Q \setminus S} \left(I(x_i, c) - \frac{1}{|S|} \sum_{s \in S} I(x_i, x_s) \right) \tag{1}$$

where Q is the feature set and $|S|$ is the size of the feature subset S. Given that $I(x, y)$ denotes the mutual information between two feature x and y. mRmR aims to maximize the relevancy $I(x_i, c)$ of the selected feature x_i to the target class c and to minimize the redundancy amongst the selected features $I(x_i, x_s)$.

While the three first filter criteria (CS, REF and FS) explore the salience of each feature individually, the last one (mRmR) considers the overlapping information amongst the features. Each ranking method orders features differently. We then apply our proposed ranking aggregation algorithm to create a final ranking that is as coherent as possible with all the individual rankings.

Algorithm 1: The proposed Ranking Aggregation Algorithm

   Input   : N normalized position ranking vector
$$r^{<k>} = \{r_{CS}, r_{REF}, r_{FS}, r_{mRmR}\}$$
   Output: The final ranking vector $\mathcal{R} = \{\hat{r}_1, \hat{r}_2, ..., \hat{r}_N\}$

1 **foreach** $r^{<k>}$, $k \in [1..N]$ **do**
2 | $n = 4$ is the number of preference ranking list
3 | re-ordering $r^{<k>} = \{r_{(1)}, ..., r_{(n)}\}$ where $r_{(i)} \leq r_{(j)}$ with $i < j$
4 | **foreach** $j \in [1..n]$ **do**
5 | | compute the jth order statistic $\hat{r}_{(j)} = \sum\limits_{\ell=j}^{n} \binom{n}{\ell} r_{(j)}^{\ell} (1 - r_{(j)})^{n-\ell}$
6 | **end**
7 | $\hat{r}_k = min\{\hat{r}_{(j)}\}$ with $j \in [1..n]$
8 **end**
9 **return** *The vector* $\mathcal{R} = \{\hat{r}_1, \hat{r}_2, ..., \hat{r}_N\}$

Inspired by the simplicity and efficiency of positional approaches which have been employed in many works [11,18], the proposed ranking aggregation algorithm, as depicted in Algorithm 1, assigns to each feature a statistical score based on its position in preference ranking lists. Given a set of N features and four individual filters, $r^{<k>} = \{r_{CS}, r_{REF}, r_{FS}, r_{mRmR}\}$ denotes the normalized position ranking vector of the kth feature ($k \in [1..N]$) where $r_{\mathcal{F}}$ are the position of the kth feature in the corresponding filter $\mathcal{F}$, divided by N. We then compute the order statistic [5] of $n = 4$ independent random variables uniformly distributed over the range $[0..1]$ (lines 2–6). The final ranking score of the kth feature is the minimum of all the order statistic values (line 7). The output consensus ranking list is then used as a local operator in our proposed wrapper approach for feature selection.

3.2 The RA-GA Algorithm

A Genetic Algorithm is a meta-heuristic search algorithm that mimics the theory of natural evolution. Given a problem, GA encodes candidate solutions as individuals of a population and evolves this population to reach the best solution. An internal structure characterizes an individual in GA, so-called *chromosomes*. For the feature selection problem, a chromosome is represented as a binary vector in which each element 1 indicates that the corresponding feature is selected and vice versa: $\mathcal{I} = \{x_1, x_2, ..., x_N\}$ where $x_i \in \{0, 1\}$, $i \in [1..N]$ and N is the number of features. Our proposed GA approach is depicted in Algorithm 2.

Algorithm 2: The RA-GA Algorithm

    Input   : The set of N features $\mathcal{S} = \{f_1, f_2, ..., f_N\}$ and the ranking list
        $\mathcal{R} = \{\hat{r}_1, \hat{r}_2, ..., \hat{r}_N\}$

    Output: The subset $\hat{\mathcal{S}}$ includes the most salient features

1   $currentPop \leftarrow initializePop(popSize)$

2   $nbGen \leftarrow 1$

3   **while** $nbGen < maxGen$ **do**

4   compute $fitness[\mathcal{I}] \; \forall \mathcal{I} \in currentPop$

5       $newPop \leftarrow RWSelection(currentPop)$

6   **while** $size(newPop) < popSize$ **do**

7           $performCrossover(newPop)$

8           $performMutation(newPop)$

9           $performLocalSearch(newPop, \mathcal{R})$

10   **end**

11      $currentPop \leftarrow newPop$

12      $nbGens \leftarrow nbGens + 1$

13 **end**

14 compute $fitness[\mathcal{I}] \; \forall \mathcal{I} \in currentPop$

15 $\mathcal{I}^* \leftarrow bestFitnessIndividual(currentPop)$

16 $\mathcal{S} \leftarrow getSelectedFeatures(\mathcal{I}^*)$

17 **return** *The subset* $\hat{\mathcal{S}}$

The process of GA first initializes a population randomly with a pre-defined number of individuals (line 1). The evolution is performed over the population until the allocated resource has been used up (i.e., after a number of generations) (lines 3–13). At each iteration of the evolution process, candidate solutions will be evaluated using the fitness function (Eq. 2) (line 4). Genetic operators, including selection, crossover, and mutation (lines 5–8). We apply our proposed local search operator to the new offspring sequentially to adjust 1-bits according to the aggregation ranking list $\mathcal{R}$ (line 9). The selected feature subset is built from the best individual from the final generation (lines 14–16).

Fitness Function. As a wrapper-based approach, we adopt a neural network (NN) classification as a predictor in this study. The fitness function is formalized using two components: (i) the NN model's percentage classification accuracy (CA) and the ratio between the size of the feature subset and the total number of features.

$$fitness(\mathcal{I}) = \alpha(1 - CA(\mathcal{I})) + (1 - \alpha)k/N \tag{2}$$

where k is the number of 1-bits in the given candidate solution $\mathcal{I}$, N is the total number of features, $CA(\mathcal{I})$ represents the accuracy of the NN model[1] regarding the solution $\mathcal{I}$. The parameter $\alpha \in (0..1]$ indicates the significance of the model accuracy. By incorporating the size of the feature subset into the solution evaluation mechanism, the proposed GA algorithm aims to optimize two

---

[1] The accuracy of the classification model is calculated using the testing dataset.

Algorithm 3: The proposed Local Search Operator

Input   : An offspring $\mathcal{I} = \{x_1, x_2, ..., x_N\}$ and the ranking list $\mathcal{R} = \{\hat{r}_1, \hat{r}_2, ..., \hat{r}_N\}$

Output   : New chromosome of $\mathcal{I} = \{\hat{x}_1, \hat{x}_2, ..., \hat{x}_N\}$

Parameter: The maximum number of selected features ϵ and the number of local search process k^*

1 while $(\sum\limits_{i=1}^{N} x_i > \epsilon)$ **do**

2 $\quad$ find the index of the lowest ranking selected feature $j \leftarrow minRankedFeature(\{x_i = 1 | i \in [1..N]\}, \mathcal{R})$

3 $\quad$ $x_j \leftarrow 0$

4 end

5 $count \leftarrow 1$

6 while $(count < k^*)$ **do**

7 $\quad$ $j \leftarrow minRankedFeature(\{x_i = 1 | i \in [1..N]\}, \mathcal{R})$

8 $\quad$ $k \leftarrow maxRankedFeature(\{x_i = 0 | i \in [1..N]\}, \mathcal{R})$

9 $\quad$ replace j and k: $x_j \leftarrow 0, x_k \leftarrow 1$

10 end

11 return *New chromosome* $\mathcal{I} = \{\hat{x}_1, \hat{x}_2, ..., \hat{x}_N\}$

---

goals: (1) maximize the accuracy of the classification model and (2) find the most appropriate number of selected features.

Genetic Operators. The current population is evolved through three operators: (1) selection, (2) crossover, and (3) mutation. We apply the Roulette Wheel selection to select the best parents from the population [30]. As such, the ith individual is assigned a probability according to its fitness value f_i, $pr_i = \frac{f_i}{\sum_{j=1}^{popSize} f_j}$.

To be selected, the cumulative probability of the ith individual $p_{cum_i} = \sum\limits_{j=1}^{i} pr_j$ should exceed a random value in the range $[0..1]$. To generate new offspring from two-parent individuals, we adopt the one-point crossover strategy [30]. This strategy randomly picks a point on both parent chromosomes to exchange their part. Finally, for mutation, we flip the value of the randomly chosen element from the parent chromosome (i.e., from 0 to 1 and vice versa).

Local Search Operator. The objective of the local search operator (LSO) is to alter the structure of offspring according to the aggregation ranking information. The predictor model can benefit the general information from the dataset and exhibit better generalization ability. Although the size of the feature subset is optimized through GA, to improve the efficiency of the feature selection process, we employ an upper bound ϵ to limit the dimensional reduction. The local search operator is introduced in Algorithm 3.

Given an offspring $\mathcal{I}$, the proposed LSO first verifies whether the number of selected features exceeds ϵ. If it is the case, all the lowest ranking features will

Table 1. Characteristics of benchmark datasets

Datasets	Categories	#features	#classes	#observations	Partitions		
					Training	Validation	Testing
Diabetes	Small	8	2	768	384	192	192
Cancer	Small	9	2	699	349	175	175
Glass	Small	9	6	214	108	53	53
Vehicle	Medium	18	4	846	424	211	211
Hepatitis	Medium	19	2	155	77	39	39
Sonar	Medium	60	2	208	104	52	52
Splice	Medium	60	3	3170	1584	793	793
Usps	High	256	10	9298	4649	2324	2324
Madelon	High	500	2	2600	1300	650	650
Isolet	High	617	26	1560	780	390	390
Coil20	High	1024	20	1440	720	360	360
Colon	High	2000	2	62	31	15	15
Leukemia	High	7129	2	72	36	18	18

be removed from $\mathcal{I}$ (lines 1–4). LSO replaces the remaining selected feature with the lowest ranking by the highest-ranking feature that has not been selected (lines 7–9). This process is repeated with some k^* times.

4 Empirical Settings

In order to evaluate the performance of our proposed approach, we experiment on widely used benchmark datasets including Diabetes, Cancer, Glass, Vehicle, Sonar, Splice, Colon cancer, Leukemia [16]. We also consider four large-dimensional datasets, including Usps, Madelon, Isolet, and Coil20 with thousands of observations [7]. Table 1 summaries the characteristics of these selected benchmark datasets. We also categorize these datasets into three groups according to their number of features: (i) small, (ii) medium, and (iii) high.

Our proposed approach is a filter wrapper using a classification neural network model as a predictor. Parameters used in the training of the neural network and other parameters (e.g., parameters of the local search, parameters used in the genetic algorithm) were empirically turned in and figured out in Table 2.

The algorithms were implemented using the Python programming language. All experiments were carried out independently by a Personal Computer with a CPU of 8 cores/1.7 GHz, RAM of 12 GB under the Ubuntu Operating System of version 20.04.

5 Experimental Results

We present our experimental results through the following research questions.

Table 2. Parameter settings for RA-GA according to the category of datasets

Categories of datasets	Parameters	Values
Small	Population size	20
Medium	Population size	40
High	Population size	50
All categories	Probability of crossover operation	0.6
All categories	Probability of mutation operation	0.02
All categories	The number of generation of the GA process	30
All categories	k^* - The repetition times for local search operator	3
All categories	Initial weights of the NN model	[−1.0, 1.0]
All categories	Learning rate of the NN model training	0.1–0.2
All categories	The number of epochs for partial training	40

Table 3. Comparable results of HGAFS and RA-GA for all datasets

Datasets	Without FS			HGAFS				RA-GA			
	$\#f$	$\%acc$	$std(\%acc)$	$\#f$	$std(f)$	$\%acc$	$std(\%acc)$	$\#f$	$std(f)$	$\%acc$	$std(\%acc)$
Diabetes	8	75.67	0.88	4.05	0.97	**76.56**	1.04	2.35	0.67	76.12	0.83
Cancer	9	98.40	0.22	4.30	1.10	**98.77**	0.51	2.05	0.39	97.31	0.50
Glass	9	73.30	6.17	3.65	1.15	**80.09**	2.69	2.55	0.82	78.27	1.14
Vehicle	18	68.83	10.2	3.65	1.19	75.75	1.06	7.15	0.98	**82.6**	1.3
Hepatitis	19	64.74	4.99	3.95	1.16	82.94	2.84	7.55	1.23	**93.50**	2.34
Sonar	60	77.88	5.72	4.65	2.28	82.98	2.59	11.9	1.12	**95.36**	2.53
Splice	60	75.33	2.01	4.35	1.72	83.32	1.60	13.8	1.36	**92.45**	0.56
Usps	256	53.81	12.91	21.2	12.25	70.98	12.5	50	0	**83.57**	1.94
Madelon	500	57.07	4.46	31	6.04	58	2.65	50	0	**64.72**	0.9
Isolet	616	10.42	6.15	46.45	9.64	30.12	5.8	60	0	**40.98**	4.83
Coil20	1024	28.56	12.12	57.6	32.7	56.64	2.33	120	0	**68.23**	5.62
Colon	2000	57.5	14.7	5.55	3.12	82.18	2.23	100	0	**94.76**	1.32
Leukemia	7129	63.33	8.72	7.25	3.73	99.44	1.67	100	0	**99.72**	1.34

5.1 RQ1: How Does the Proposed Approach Perform Comparing with Some State-of-the-Art Methods?

We first conduct a comparison between our proposed approach RA-GA and HGAFS proposed by Kabir et al. [16] since this work is close to our approach. We have re-implemented HGAFS to experiment with our benchmark datasets.

The experiment is carried out on three approaches: (i) Without feature selection (i.e., using the whole datasets with the same neural network model); (ii) HGAFS approach; and (iii) our RA-GA approach. We compare the performance over the average number of selected features ($\#f$), the average accuracy of the classification model ($\%acc$), the standard deviation of both the number of selected features and the model accuracy over 20 executions ($std(f)$, $std(\%acc)$). The result is shown in Table 3.

Table 4. Compare result with different threshold subset size

Datasets	Best of RA-GA		Benchmark		
----------	#features	%acc	#features	%acc	reference
	#features	%acc	#features	%acc	reference
Diabetes	4	78.13	4	91.65	[31]
Cancer	2	97.72		98.24	[13]
Glass	5	81.49		85.98	[24]
Vehicle	11	85.23	17	85.26	[23]
Hepatitis	7	97.44		100	[24]
Sonar	11	98.08	60	97.11	[24]
Splice	14	93.24	..	96.48	[1]
Usps	50	86.88	100	92.69	[27]
Madelon	50	66.78	214	80.3	[29]
Isolet	60	48.97	300	69	[20]
Coil20	120	78.89	..	100	[20]
Colon	100	94.86	..	98.4	[28]
Leukemia	100	99.83	..	100	[2]

It is observable that both RA-GA and HGAFS can effectively reduce the number of features while improving the accuracy of the classification model. For example, considering the dataset Coil20 with 1024 features, reducing the feature space (e.g., from 1024 to 57.6 and 100 with HGAFS and RA-GA, respectively), the model's accuracy improves about 28.08% with HGAFS and 39.67% with RA-GA. Our method is 11.6% higher than HGAFS in terms of model performance.

As can be seen in Table 3, RA-GA outperformed HGAFS for all medium and high dimensional datasets in terms of the classification accuracy (CA). Indeed, the average accuracy of Vehicle, Hepatitis, Sonar, Splice increases sharply from 6.85% up to 12.58% when compared between HGAFS and RA-GA. Regarding Usps, Madalon, Isolet, Coil20, Colon and Leukemia, RA-GA significantly improves the average accuracy by about 10.68% in average compared to HGAFS.

It is also arguable that compared to HGAFS, RA-GA has not worked effectively for the small-dimensional datasets in terms of CA (i.e., regarding the datasets Diabetes, Cancer, and Glass). However, the difference in CA between RA-GA and HGAFS is considerably tiny, while the numbers of selected features produced by HGAFS are relatively higher than those produced by RA-GA for all these three datasets. As one of the goals of RA-GA is to optimize the number of selected features, regarding small-dimensional datasets, the small number of selected features may ignore important information therefore reducing the model accuracy.

Finally, the results also show that the numbers of selected features given by RA-GA are much higher than those given by HGAFS for all medium- and high-dimensional datasets. The correlation between the total number of features and

Table 5. Experimental results for different size of feature subset produced by RA-GA

Dataset	Percentage	#f	std(f)	%acc	std(acc)	Dataset	percentage	#f	std(f)	%acc	std(acc)
Coil20	5%	60	0.00	63.85	4.60	Isolet	5%	30	0.00	33.53	4.30
	10%	120	0.00	68.23	5.62		10%	60	0.00	40.98	4.83
	15%	180	0.00	68.30	8.31		15%	90	0.00	43.77	3.80
	20%	240	0.00	**69.69**	5.59		20%	120	0.00	44.39	4.53
	25%	300	0.00	69.47	6.14		25%	149.45	1.19	**50.55**	5.67
Madelon	5%	25	0.00	63.92	0.58	Usps	5%	25	0.00	79.11	1.40
	10%	50	0.00	64.68	0.93		10%	50	0.00	83.57	1.94
	15%	75	0.00	65.04	0.40		15%	71.2	4.49	84.51	1.36
	20%	99.55	0.76	65.11	0.60		20%	95.86	5.23	86.68	1.82
	25%	122.8	3.56	**65.39**	0.71		25%	116.27	6.74	**86.94**	3.61
Colon	5%	50	0.00	**96.24**	3.19	Leukemia	5%	50	0.00	99.4	0.50
	10%	100	0.00	94.74	2.29		10%	100	0.00	99.62	0.34
	15%	150	0.00	94.76	0.60		15%	150	0.00	99.72	2.3
	20%	200	0.00	93.81	0.04		20%	200	0.00	99.82	1.25
	25%	250	0.00	93.84	0.01		25%	250	0.00	**99.91**	0.02

the number of features selected by RA-GA is high, while it is not valid in the case of HGAFS. Furthermore, the standard deviation values for RA-GA considering the number of selected features and the model accuracy are much smaller than those for HGAFS, indicating that RA-GA is more deterministic and stable than HGAFS.

These above observations lead to the following conclusion: RA-GA is an improved version of HGAFS for large datasets. The improvement comes from the number of selected features and the selected features themselves.

We then conducted a short survey for the highest CA considering each dataset separately to observe the gap between RA-GA and other state-of-the-art feature selection algorithms. As can be seen in Table 4, the difference in terms of CA between RA-GA and the state-the-art algorithms is not significant for Cancer, Vehicle, Sonar, Leukemia. However, this difference is huge for other datasets, especially with many observations. The observation shows that RA-GA can get an effective CA for medium-dimensional datasets. However, the performance of RA-GA is still far from the benchmark result for high-dimensional datasets.

5.2 RQ2: What is the Impact of the Subset's Size Produced by RA-GA?

In order to evaluate the impact of the feature subset's size on the accuracy of the classification task, we set up five scenarios in which the expected numbers of selected features are equal to 5%, 10%, 15%, 20%, and 25% of the total number of available features considering all the benchmark datasets. The experimental result is summarized in Table 5. We can observe a strong positive correlation between the size of the subset of selected features and CA for Isolet, Madelon, Usps, and Leukemia. It shows that the larger the subset of selected features given by RA-GA, the higher classification accuracy we received. Additionally, the difference in CA between different scenarios for almost datasets is not significant, excepting Isolet and Colon. Finally, the variant of the number of features selected by RA-GA is very low, particularly for high-dimensional datasets.

In summary, RA-GA prefers to select a subset of 10% of features. This ratio is a reasonable value for the trade-off between CA and the number of selected features. However, the variant of the size of the selected feature subset produced by RA-GA is still weak; this leads to difficulty searching for the optimal size.

6 Conclusion

Feature Selection has a vital role in many machine learning applications. However, this task faces many challenges for high-dimensional datasets. This paper proposed a genetic algorithm to optimize the feature subset and the appropriate number of selected features to maximize the performance of an artificial neural network classifier. The selection phase was significantly improved by combining the proposed GA and a local search based on a ranking aggregation approach. The experimental results showed that the approach of this manuscript improved some existing works for medium and high dimensional datasets.

As future work, we will accelerate the approach proposed in this paper by combining it with some unsupervised techniques to give some prior knowledge for genetic algorithms and local search.

References

1. Abellan, J., Mantas, C.J., Castellano, J.G., Moral-Garcia, S.: Increasing diversity in random forest learning algorithm via imprecise probabilities. Expert Syst. Appl. **97**, 228–243 (2018)
2. Bhola A, Tiwari, A.K.: Machine learning based approaches for cancer classification using gene expression data. Mach. Learn. Appl. Int. J. (MLAIJ) (2015)
3. Borda, J.d.: Mémoire sur les élections au scrutin. Histoire de l'Academie Royale des Sciences pour 1781 (Paris, 1784) (1784)
4. Bouaguel, W., Brahim, A.B., Limam, M.: Feature selection by rank aggregation and genetic algorithms. In: Proceedings of the International Conference on Knowledge Discovery and Information Retrieval and the International Conference on Knowledge Management and Information Sharing, pp. 74–81 (2013)
5. David, H.A., Nagaraja, H.N.: Order Statistics. Wiley, Hoboken (2004)
6. De Condorcet, N.: Essai sur l'application de l'analyse à la probabilité des décisions rendues à la pluralité des voix. Cambridge University Press, Cambridge (2014)
7. Dua, D., Graff, C.: UCI machine learning repository (2017). http://archive.ics.uci.edu/ml
8. Gasca, E., Sánchez, J.S., Alonso, R.: Eliminating redundancy and irrelevance using a new MLP-based feature selection method. Pattern Recogn. **39**(2), 313–315 (2006)
9. Gu, Q., Li, Z., Han, J.: Generalized fisher score for feature selection. arXiv preprint arXiv:1202.3725 (2012)
10. Guyon, I., Elisseeff, A.: An introduction to variable and feature selection. J. Mach. Learn. Res. **3**(Mar), 1157–1182 (2003)

11. Hancer, E., Xue, B., Zhang, M.: Differential evolution for filter feature selection based on information theory and feature ranking. Knowl.-Based Syst. **140**, 103–119 (2018)

12. Huang, J., Cai, Y., Xu, X.: A hybrid genetic algorithm for feature selection wrapper based on mutual information. Pattern Recogn. Lett. **28**(13), 1825–1844 (2007)

13. Ibrahim, S., Nazir, S., Velastin, S.A.: Feature selection using correlation analysis and principal component analysis for accurate breast cancer diagnosis. J. Imaging **7**(11), 225 (2021)

14. Jia, L.: A hybrid feature selection method for software defect prediction. In: IOP Conference Series: Materials Science and Engineering, vol. 394, pp. 32–35. IOP Publishing (2018)

15. Kabir, M.M., Islam, M.M., Murase, K.: A new wrapper feature selection approach using neural network. Neurocomputing **73**(16–18), 3273–3283 (2010)

16. Kabir, M.M., Shahjahan, M., Murase, K.: A new local search based hybrid genetic algorithm for feature selection. Neurocomputing **74**(17), 2914–2928 (2011)

17. Kemeny, J.G.: Mathematics without numbers. Daedalus **88**(4), 577–591 (1959)

18. Kolde, R., Laur, S., Adler, P., Vilo, J.: Robust rank aggregation for gene list integration and meta-analysis. Bioinformatics **28**(4), 573–580 (2012)

19. Lai, C., Reinders, M.J., Wessels, L.: Random subspace method for multivariate feature selection. Pattern Recogn. Lett. **27**(10), 1067–1076 (2006)

20. Li, J., Tang, J., Liu, H.: Reconstruction-based unsupervised feature selection: an embedded approach. In: IJCAI, pp. 2159–2165 (2017)

21. Liu, X.Y., Liang, Y., Wang, S., Yang, Z.Y., Ye, H.S.: A hybrid genetic algorithm with wrapper-embedded approaches for feature selection. IEEE Access **6**, 22863–22874 (2018)

22. Maldonado, S., López, J.: Dealing with high-dimensional class-imbalanced datasets: embedded feature selection for SVM classification. Appl. Soft Comput. **67**, 94–105 (2018)

23. Mokdad, F., Bouchaffra, D., Zerrouki, N., Touazi, A.: Determination of an optimal feature selection method based on maximum shapley value. In: 2015 15th International Conference on Intelligent Systems Design and Applications (ISDA), pp. 116–121. IEEE (2015)

24. Nekkaa, M., Boughaci, D.: A memetic algorithm with support vector machine for feature selection and classification. Memetic Comput. **7**(1), 59–73 (2015). https://doi.org/10.1007/s12293-015-0153-2

25. Peng, H., Long, F., Ding, C.: Feature selection based on mutual information criteria of max-dependency, max-relevance, and min-redundancy. IEEE Trans. Pattern Anal. Mach. Intell. **27**(8), 1226–1238 (2005)

26. Qiu, C.: A novel multi-swarm particle swarm optimization for feature selection. Genet. Program Evolvable Mach. **20**(4), 503–529 (2019). https://doi.org/10.1007/s10710-019-09358-0

27. Quanquan Gu, Zhenhui Li, J.H.: Generalized fisher score for feature selection. In: UAI 2011: Proceedings of the Twenty-Seventh Conference on Uncertainty in Artificial Intelligence (2011)

28. Rahman, M.A., Muniyandi, R.C.: Feature selection from colon cancer dataset for cancer classification using artificial neural network. Int. J. Adv. Sci. Eng. Inf. Technol. (2018)

29. Taherkhani, A., Cosma, G., McGinnity, T.M.: Deep-FS: a feature selection algorithm for deep Boltzmann machines. Neurocomputing **322**, 22–37 (2018)

30. Thede, S.M.: An introduction to genetic algorithms. J. Comput. Sci. Coll. **20**(1), 115–123 (2004)
31. Li, X., Zhang, J., Safara, F.: Improving the accuracy of diabetes diagnosis applications through a hybrid feature selection algorithm. Neural Process. Lett. (2021)

A Novel Type-Based Genetic Algorithm for Extractive Summarization

Bui Thi Mai Anh, Nguyen Thi Thu Trang[(✉)], and Tran Thi Dinh

School of Information and Communication Technology,
Hanoi University of Science and Technology, Hanoi, Vietnam
{anhbtm,trangntt}@soict.hust.edu.vn, dinh.tt173015@sis.hust.edu.vn

Abstract. Automatic text summarization has been an growingly important task since a huge amount of textual information needs to be processed on the Internet. Genetic Algorithm (GA) is an efficient approach for extractive text summarization, which aims to find out the best summary with an optimized fitness function through the evolution of generations. This paper proposes a novel extractive summarization method using GA with two types of individuals based on their internal chromosomal structure. Each individual may have one or two full chromosomes, where a chromosome represents a candidate summary. In this type-based GA, good summaries are better kept through generations, the mutation more likely happens with more flexible strategies and prominent summaries are more likely found in the solution space. The mutation can occur in two levels: off-springs can be obtained by changing their parents' type or flipping some genes, i.e. multi-point, in their parents' chromosomes. Our proposed approach has been experimented on DUC2001, DUC2002 and CNN/DailyMail datasets, outperforming all other extractive state-of-the-art methods by all three Rouge points. Indeed, the Rouge-1 and Rouge-L scores considerably improve from 10% to 20%, while the Rouge-2 has the highest performance.

Keywords: Genetic algorithm · Extractive summarization

1 Introduction

Automatic text summarization has been increasingly important to address the ever-growing amount of text data available online. Summaries can be constructed either by generating new (and human-like written) sentences (i.e. abstractive summarization) or selecting the most important sentences from the original document (i.e. extractive one) [5].

The first approach requires advanced language generation techniques such as paraphrasing, generalizing, or incorporating human knowledge but receives unsatisfactory results. Recently, abstractive summarization has a considerable improvement with deep learning models, e.g. Pointer-generator network [15], Transformer [23]. However, these approaches require a huge training dataset

© Springer Nature Switzerland AG 2022
H. Fujita et al. (Eds.): IEA/AIE 2022, LNAI 13343, pp. 240–252, 2022.
https://doi.org/10.1007/978-3-031-08530-7_20

including human-written summaries from the original documents, very costly and time-consuming.

The second approach, i.e. extractive summarization, is typically investigated as a binary classifying task in which the sentences from the original text are categorized into two groups: *in-summary* and *not-in-summary*. Supervised methods for this task, as the modern abstractive approach, require a large training set of human-generated summaries for machine learning or deep learning models [12]. In contrast, an unsupervised learning model does not require training corpus or specific thesaurus but discovers the relation between sentences inside the original document in order to assign a score to each sentence. Many unsupervised techniques were adopted for text summarization such as clustering, Hidden Markov Model [22], graph-based model like Text Rank [10], optimization algorithms such as Genetic Algorithm (GA) [2,9], Particle Swarm Optimization [1], Multi-Objective Artificial Bee Colony Algorithm (MOABC) [14] etc.

GAs have been proposed for text summarization and have shown promising results [2,9,20]. Genetic Algorithms (GAs) were used in many research disciplines as a solution for global search optimization [19]. In the context of extractive text summarization, GAs have been typically adopted as a clustering method which searches in the space of document sentences, those who are the most relevant for the summary. More concisely, given an input document $\mathcal{D}$ consisting of N sentences, the objective of GA is to find a subset of ℓ sentences (with $\ell < N$) which contains the main information of $\mathcal{D}$. GA performs a heuristic search by imitating the process of natural evolution. Its routine is as follows:

1. Initialize a population whose individuals represent different solutions of the problem (i.e., candidate summaries).
2. Assess all individuals of the current population by using a *fitness function* and produce a ranking list of individuals according to their fitness value.
3. Evolve the current population by: (i) selecting the top individuals from the ranking order as the parent set; (ii) applying genetic operators including *crossover* and *mutation* on this set to generate new offspring. The next generation includes the parents and their offspring.
4. Repeating the step 2. on the next population.

The process of GA is repeated until the satisfactory solution is reached or some stop-criteria are triggered. Most studies adopted this traditional process of GA to solve the extractive summarization. However, the nature shows that gender-based selection through mate choice is more likely to guide the evolution than natural selection [11]. Recent studies in the field of optimization have focused on improving standard genetic algorithms by splitting the population into groups based on gender or type of individuals. Some proposed gender-based [3] and type-based GAs [17] for other problems outperformed the classic GAs both in terms of solution quality and performance. Indeed, the enhancement of standard GAs with gender-based selection helps speed up the evolutionary and increase the diversification of the population, consequently reaching better candidate. In this work, we propose a novel extractive single-document summarization method with type-based GA. An individual in the population can contain one or two full chromosomes, where a chromosome represents a summary. The use of different types allows to maintain better summaries as well as to retain the ability to explore more effectively the solution space.

The rest of the paper is organized as follows. Section 2 presents in detail our proposal with the type-based GA for the extractive summarization. Section 3 discusses the related works to the proposed algorithm and techniques. In Sect. 4, we introduce the empirical settings, benchmark datasets as well as evaluation metrics to assess the performance of our model. Section 5 discusses the experimental results conducted on three datasets DUC2001, DUC2002 and CNN/DailyMail comparing to other state-of-the-art methods. Finally, Sect. 6 draws some conclusions and perspectives of the future work.

2 Our Proposed Type-Based GA for Extractive Summarization

The first part of this section introduces different components of the proposed type-based GA including the chromosome encoder and the fitness function. The details of the proposed algorithm together with the genetic operators are depicted in the second part.

2.1 Chromosome Encoder

An individual in the population (i.e., a candidate summary) needs to be encoded using a chromosome encoder. Given a document with N sentences:

$$\mathcal{D} = \{S_1, S_2, S_3, ..., S_N\}$$

where S_i corresponds to the i-th sentence of the document. A solution to the extractive text summarization is a subset $\mathcal{S} \subseteq \mathcal{D}$. We encode a candidate summary of $\mathcal{D}$ as a binary string in which 0/1-elements denote that the corresponding sentences are not-included/included in the summary. For example, a document of five sentences, a candidate solution represented by $\{0, 1, 1, 0, 0\}$ means that the summary contains the second and the third sentences. In order to improve the diversification of the population, individuals are distinguished from each other by their internal structure, so called *type*. In this work, we consider two types, namely T_A and T_B, where:

1. An individual of T_A, denoted by I_A, has two chromosomes $(\mathcal{S}_1, \mathcal{S}_2)$ where $\mathcal{S}_1$ and $\mathcal{S}_2$ are encoded as mentioned above.
2. An individual of T_B, denoted by I_B, has the chromosomal structure as $(\mathcal{S}, null)$.

2.2 Fitness Function

The objective of a fitness function is to assess the quality of candidate summaries. In the context of text summarization, the fitness function is built based on salient features extracted from the original document. In this study, we adopt features which have been widely used in recent studies [2, 21] including sentence position, similarity to the topic sentence, sentence length, and the number of proper nouns. As sentence-level features, they are computed for candidate summaries by considering only within-summary sentences.

Given a candidate summary represented by a binary vector $\mathcal{S} = \{x_i\}_{i=1}^N$ where N is the number of sentences of the original document. The quality of $\mathcal{S}$ is assessed by maximizing the fitness function $f(\mathcal{S})$ according to Eq. 1.

$$f(\mathcal{S}) = \alpha \mathcal{R}_\mathcal{S} + \beta \mathcal{L}_\mathcal{S} + \gamma \mathcal{P}_\mathcal{S} + \sigma \mathcal{N}_\mathcal{S} \tag{1}$$

where $\mathcal{R}_\mathcal{S}, \mathcal{L}_\mathcal{S}, \mathcal{P}_\mathcal{S}$ and $\mathcal{N}_\mathcal{S}$ represent four aforementioned features, respectively[1]. $\alpha, \beta, \gamma, \sigma$ show the contribution of each feature to assess the quality of the summary $\mathcal{S}$, provided that Eq. 2 is satisfied.

$$\alpha + \beta + \gamma + \sigma = 1 \tag{2}$$

As an individual of the type-based GA population can be either the type T_A or T_B, denoted as $I_A = (\mathcal{S}_1, \mathcal{S}_2)$ or $I_B = (\mathcal{S}, null)$, respectively, its fitness value is defined as follows:

$$f(I) = \begin{cases} max(f(\mathcal{S}_1), f(\mathcal{S}_2)) \text{ if}(I_A) \\ f(\mathcal{S}) \qquad\qquad \text{if}(I_B) \end{cases} \tag{3}$$

2.3 The Proposed Type-Based GA

The proposed algorithm is depicted in Algorithm 1. The first stage of the process is to initialize randomly the population (line 1). Comparing to traditional GA, the initialization of type-based GA population is quite different. Given a pre-defined population size $\mathcal{N}$ and a pre-fixed length of summary ℓ, individuals of types T_A and T_B are created so that the population contains at least one individual of each type.

$$\mathcal{P} = \{I_i : type(I_i) = T_A || type(I_i) = T_B\}_{i=1}^{\mathcal{N}}$$

where $I_A = (\mathcal{S}_1, \mathcal{S}_2)$ and $I_B = (\mathcal{S}, null)$. For each chromosome, the summary length constraint needs to be checked based on Eq. 4.

$$\sum_{i=1}^N x_i = \ell \tag{4}$$

The evolution process is performed until the allocated resources have been used up (i.e., number of generations) (line 3). At each iteration of the evolution process, the fitness value of all individuals are computed (line 4) according to Eq. 1. The type-based selection operator is then performed to select the strongest parents from the current population for the reproduction process using crossover

---

[1] The formula of these features are detailed in the work of Anh et al. [2].

Algorithm 1: The proposed type-based GA for extractive summarization

---

    Input   : Document D with N sentences
    Output: Summary of ℓ sentences
1 $currentPop \leftarrow initializePop(popSize)$
2 $nbGen \leftarrow 1$
3 **while** $nbGen < maxGen$ **do**
4   compute $fitness[I] \; \forall I \in currentPop$
5      $newPop \leftarrow RWSelect(currentPop, \alpha)$
6   **while** $size(newPop) < 2popSize$ **do**
7         $I_A \leftarrow rndChoose(newPop, T_A)$
8         $I_B \leftarrow rndChoose(newPop, T_B)$
9         $offspring \leftarrow performCrossover(I_A, I_B)$
10        $newPop \leftarrow newPop \cup offspring$
11        $performMutation(newPop)$
12   **end**
13      $currentPop \leftarrow elitismSelect(newPop, popSize)$
14      $nbGen \leftarrow nbGen + 1$
15 **end**
16 **return** *The best summary of D*

---

and mutation operators (lines 5–12). Finally, the algorithm carries out an elitism selection to create the next generation including the best individuals (in terms of highest fitness values) (line 13).

The following sections introduce our proposed genetic operators for type-based GA.

(a) Type-Based Selection Operator. In order to create the next generation, we choose top ν percent ($\nu < 100$) from the current population with the constraint that both best individuals of T_A and T_B should be selected (line 5, Algorithm 1). The selection strategy adopted in this work is the Roulette Wheel Selection [19]. Each individual i ($i \in [1..\mathcal{N}]$), either I_A or I_B, is assigned by a probability $p_i = \frac{f_i}{\sum_{i=1}^{N} f_i}$ where f_i is the fitness value of the i-th individual from the population $\mathcal{P}$ (see Eq. 3). To be selected, the cumulative probability of the i-th individual $p_{cum_i} = \sum_{j=1}^{i} p_j$ should exceed a random value in the range $[0..1]$. By this selection strategy, $\nu\mathcal{N}$ individuals will be selected for the next generation.

(b) Cross-over Operator. Type-based mating is performed between an individual of T_A and an individual of T_B (lines 7–9, Algorithm 1). The original type-based algorithm of Sizov and Simovici [17] focused only on one-point cross-over at the chromosome level. As such, the type A individual provides the first chromosome while the type B individual contributes the second chromosome to their offspring. In order to improve the diversification of the next generation, our work enables the crossover at both chromosome and gene levels. Given two individuals, $I_A = (S_1^A, S_2^A)$, $I_B = (S_1^B, null)$, the chromosomal structure of their offspring $I_C = (S_1^C, S_2^C)$ is defined as follows:

$$S_1^C = \text{uniform}(S_1^A, S_1^B)$$

$$S_2^C = \begin{cases} \text{uniform}(S_2^A, S_1^B) & \text{if}(pt \geq 0.5) \\ null & \text{otherwise} \end{cases}$$

where pt is a random value in the range of $[0..1]$. To perform the gene-level crossover, we adopt the uniform strategy. This strategy considers that each gene from either parent has an equal probability of being chosen. Given a random mixing ratio a ($a \in [0..1]$) and two parents chromosomes S^A, S^B, the uniform crossover is built as :

$$S_c[i] = \begin{cases} S^A[i] & \text{if } (p_i \geq a) \\ S^B[i] & \text{otherwise} \end{cases}$$

where p_i is randomly picked between 0 and 1 for each gene. For each offspring, the summary length constraint needs to be checked following Eq. 4.

(c) **Mutation Operator.** The mutation process aims to alter one or more genes of a chromosome, replaces their values with other information. The mutation process is carried out at both chromosome and gene levels. At chromosome level, an individual of T_A may drop its second chromosome to become T_B with some small probability (m_c). At the gene level, each gene of a chromosome has an equal probability to be mutated. Its value will be flipped (i.e., from 0 to 1 or vice versa):

$$S_{new}[i] = \begin{cases} \text{flip}(S[i]) & \text{if } (p_i \geq a_{mut}) \\ S[i] & \text{otherwise} \end{cases}$$

where a_{mut} is the gene-level mutation probability, p_i is a real value randomly picked in the range $[0..1]$ for the i-th gene. The mutation process is performed only on the generated offspring from the crossover operator. The mutation probability $a_{mut} = \frac{1}{2N}$ or $a_{mut} = \frac{1}{N}$ for offspring of type T_A and T_B, respectively. Before mutating, the summary length constraint (see Eq. 4) is checked. If the restriction is not met, the gene is not mutated.

3 Related Works

Genetic Algorithm (GA). Mendoza et al. proposed a GA approach for extractive summarization [9] in which solutions are encoded as a binary string. Meanwhile, Simon et al. studied another problem representation method which is based on sentence indices [16]. Population initialization in GA is mostly randomly generated [9,16] with genetic operators: 1-point crossover [9,16], 2-point crossover [18], multi-bit mutation [18], insertion mutation [16]; selection strategy: Ranking Selection [9], Tournament selection [16] etc. Another work of Bui et al. aimed to evaluate the role of some sentence features to improve the fitness function for GA extractive summarization [2].

Type-Based Genetic Algorithm. Type-based GA is an improved algorithm from the traditional GA proposed by Sizov and Simovici [17]. They introduced a genetic algorithm based on the classification of individuals and their applications to two well-known problems: the N-queen problem and finding the global minimum of the Rosenbrock function. This paper aims to address the automatic extractive summarization of a single document as a binary optimization problem, similar to other GA approaches. However, in this work, we propose a novel method that uses the type-based GA for this problem. We build two types of individuals based on their internal chromosomal structure. With this novel method, good summaries are better kept through generations, the mutation more likely happens with more flexible strategies and prominent summaries are more likely found in the solution space.

Fitness Function. The fitness function plays an important role in the context of meta-heuristic search algorithms. Garcia and Ledeneva proposed a combination of word frequency and sentence position features to build the fitness function [4]. Meena et al. proposed a feature set that measures the importance of each sentence and optimizes this set with the acceptable weights [8]. However, the main shortcoming of this approach is time consuming. Recent studies indicated that relevance and surface-level sentence features have been shown competitive results when extracting salient sentences to build summaries [9,20]. Various features are introduced such as similarity to the topic sentence, similarity between sentences, sentence position, sentence length, term frequency, term weight, coverage similarity between summaries and original source document, etc.

In this work, we refine the fitness function with four features including sentence position, the relation of sentences with title, sentence length, and the number of proper nouns. These features are proved to be efficient in scoring the importance of sentences in the document.

4   Empirical Settings

4.1   Dataset

There are several datasets for the automatic text summarization task. Recent studies typically focus on CNN/DailyMail [6] - a well-known corpus which provides abstractive summaries from news articles. Another famous corpus for this field is Document Understanding Conference (DUC) [13], which was collected from newspapers and news agencies. There are several years that DUC provides datasets for text summarization task. We do not experiment on the DUC2004 and DUC2007 since they only include (or supplement) summaries for the multi-document summarization task.

We therefore evaluate the proposed model on three datasets DUC2001, and DUC2002 and CNN/DailyMail. The DUC2002 dataset includes 567 documents categorized into 59 sets while the DUC2001 consists of 30 sets with 309 documents. Each set from both datasets involves reference 100-word summaries for single and multiple documents. The joint CNN/DailyMail contains 286,722 training documents, 13,362 validation documents and 11,480 test documents. Each document with 28 sentences on average associates with a reference summaries of 3–4 sentences (about 56 words).

4.2 Evaluation Metrics

The performance of our proposed model has been evaluated based on different variants of the Rouge metric [7] with respect to ground-truth summaries. We compute three scores Rouge-1, Rouge-2 and Rouge-L by counting the number of overlapping words between the generated summaries and the reference ones. The higher the Rouge metrics are, the better performance and higher effectiveness the extractive summarization approach will offers.

4.3 Tuning Parameters

The parameters for our proposal algorithm are set as follows: population size $\mathcal{N} = 100$, maximum number of generations $maxGen = 300$. At the selection phase, we keep top $\nu = 30\%$ of the population for creating the next generation. We choose the summary length parameter (in sentences) $\ell = 5$ for the datasets DUC2001, DUC2002 and $\ell = 4$ for the dataset CNN/DailyMail.

The value of feature weights (see Eq. 1) is determined empirically by a grid search on training data. In order to create disjoint training and test data, for each dataset (DUC2001 and DUC2002), documents are split into equally sized folds: three folds for DUC2001, five folds for DUC2002. The feature weight parameters are evaluated on $fold_i$ to maximize the Rouge scores and tested on $fold_{i+1}$. The parameters evaluated on the last fold will be used to test the first fold. In the end, we collect the results from all test folds and compute the overall performance of the model. About the CNN/DailyMail corpus, we use the training dataset to evaluate the weighted parameters and evaluate the performance of the proposed model on the test dataset.

5 Results

To measure the performance of the proposed method, we did some experiments with other state-of-the-art works on extractive single-document summarization.

- TextRank [10]: This work proposes a graph-based ranking model for text processing which can be used in order to find the most relevant sentences in text and also to find keywords.
- SummaRuNNer [12]: The supervised algorithm uses a Recurrent Neural Network (RNN) based sequence model for extractive summarization of documents.
- GA-Features [2]: This work aims to evaluate the role of some sentence features for the fitness function and genetic operators in GA extractive summarization.

Table 1. Performance of various models on the dataset DUC2001

Method	Rouge scores (%)		
	Rouge-1	Rouge-2	Rouge-L
TextRank [10]	44.50	18.665	
GA-Features [2] (re-run)	45.37	18.94	38.04
MA-SingleDocSum [9] (published)	44.86	20.14	
MA-SingleDocSum [9] (re-implement)	44.17	20.68	38.67
TBGA-BKSum (our proposal method)*	**57.60**	**26.52**	**49.25**

Table 2. Performance of various models on the dataset DUC2002

Method	Rouge scores (%)		
	Rouge-1	Rouge-2	Rouge-L
TextRank [10]	48.79	21.52	44.81
SummaRuNNer [12]	46.60	**23.10**	43.03
GA-Features [2] (re-run)	44.98	18.49	38.47
MA-SingleDocSum [9] (published)	48.28	22.84	
MA-SingleDocSum [9] (re-implement)	48.33	20.60	41.93
TBGA-BKSum (our proposal method)*	**53.40**	23.07	**46.17**

- MA-SingleDocSum [9]: This method bases on genetic operators and guided local search. A memetic algorithm is used to integrate the own-population-based search of evolutionary algorithms with a guided local search strategy.
- TBGA-BKSum: Our proposed method with type-based GA for single document extractive summarization. GA operations on two chromosome types are more diverse and flexible, hence it is promising to obtain prominent summaries.

The experimental results of our proposal and other methods are summarized in Tables 1, 2 and 3 with respect to the DUC2001, DUC2002 and CNN/DailyMail datasets, respectively.

We can see that our proposed method TBGA-BKSum outperforms all other state-of-the-art methods by all three Rouge points on DUC2001 and DUC2002 datasets (see Tables 1 and 2. For instance, regarding the DUC2002 dataset, the Rouge-1 score is higher about 0.51 to 0.85 points, hence improves about 10,5% to 18,7% compared to other methods[2]. It is also observable that our model slightly improves the Rouge-2 score comparing to MA-SingleDocSum (i.e., 0,23 points), TextRank (i.e., 1,55 points) and is comparable with SummaRuNNer. In terms of Rouge-L score, our model consistently achieves better performance comparing

---

[2] The improvement score is calculated as the ratio of the difference on the Rouge Score between our proposed method and the comparing method and the Rouge score of this method.

Table 3. Performance of various models on the entire joint CNN/DailyMail dataset

Method	Rouge scores (%)		
	Rouge-1	Rouge-2	Rouge-L
TextRank [10]	34.32	15.68	33.54
SummaRuNNer [12]	**39.60**	16.20	**35.30**
GA-Features [2] (re-run)	35.30	14.01	32.83
MA-SingleDocSum [9] (re-implement)	33.63	13.91	31.98
TBGA-BKSum (our proposal method)*	38.70	**16.72**	34.61

Gold Summary:
Bryan Redpath has ended his eight-year association with Sale Sharks.
Redpath spent five years as a player and three as a coach at Sale.
He has thanked the owners, coaches and players for their support.

SummaRuNNer	TBGA-BKSum
The 43-year-old Scot ends an eight-year association with the Aviva Premiership side, having spent five years with them as a player and three as a coach.	The 43-year-old Scot ends an eight-year association with the Aviva Premiership side, having spent five years with them as a player and three as a coach.
Redpath returned to Sale in June 2012 as director of rugby after starting a coaching career at Gloucester and progressing to the top job at Kingsholm.	Redpath returned to Sale in June 2012 as director of rugby after starting a coaching career at Gloucester and progressing to the top job at Kingsholm.
Redpath spent five years with Sale Sharks as a player and a further three as a coach.	Bryan Redpath has thanked Sale Sharks' owners, coaches and players after leaving Sale Sharks.
	Redpath spent five years with Sale Sharks as a player and a further three as a coach.

Fig. 1. Example on generated summaries from SummaRuNNer and our approach

to all other approaches, 20% higher than GA-Features [2] and 10,1% higher than MA-SingleDocSum [9]. Regarding the DUC2001 dataset, TBGA-BKSum even improves the Rouge-L score by 37%, comparing to MA-SingleDocSum. As SummaRuNNer does not public their experiments on DUC2001, we only carry out the comparison with their approach on DUC2002.

We then compare the performance of TBGA-BKSum with the other approaches on the CNN/DailyMail dataset. While two datasets DUC2001 and DUC2002 provide extractive gold summaries, the CNN/DailyMail dataset comes up with only abstractive reference summaries (i.e., human-written summaries). As most previous studies, those are based on GA approach, typically focused on DUC2001 and DUC2002 datasets. We therefore re-implement TextRank [10], GA-Features [2] and MA-SingleDocSum [9] to explore their performance on the CNN/DailyMail dataset. The results are showed in Table 3. Our proposed model significantly outperforms TextRank [10], GA-Features [2], MA-SingleDocSum [9] and is comparable with SummaRuNNer [12] in terms of Rouge-1, Rouge-2 and Rouge-L. It can be observable that our proposed approach performed better than SummaRuNNer on DUC2001 and DUC2002 but not on CNN/DailyMail. The main reason is that our model needs to fix the number of sentences for generated summaries to encode GA-chromosome (i.e., for DUC2001 and DUC2002, $\ell = 5$, for CNN/DailyMail, $\ell = 4$) while SummaRuNNer selects sentences depending on their probabilities and evaluates Rouge metrics on the same number of gold summaries' sentences. As Rouge-metrics use *word-overlap* for comparison between the reference and generated summaries, dynamically adjusting the length of generated summaries may further improve the performance of the model. Figure 1 shows an example of generated summaries from our proposed approach and SummaRuN-Ner comparing to the reference summary. As can be seen in this example, all the selected sentences of SummaRuNNer appear in our generated summary.

6   Conclusion

In this paper, we propose a novel extractive summarization model with type-based genetic algorithm. We focus on building all the components of GA for two types of individuals containing one or two full chromosomes, where a chromosome represents a summary. The mutation can occur in two levels: off-springs can be obtained by changing their parents' type (type level) or flipping some genes, i.e. multi-point, in their parents' chromosomes (chromosome level). In addition, we refine the fitness function with four features including sentence position, the relation of sentences with title, sentence length, and the number of proper nouns. Our proposal is experimented to compare with state-of-the-art methods using Rouge measures on the datasets DUC2001, DUC2002 and CNN/DailyMail. The empirical results have shown that the proposed type-based GA for extractive text summarization gives much better accuracy than all other studied methods. The Rouge-1 and Rouge-L scores considerably improve from 10% to 20%, while the Rouge-2 has the highest performance. We intend to examine type-based genetic algorithms in multi-document extractive summarization with more advanced genetic operators.

References

1. Al-Abdallah, R.Z., Al-Taani, A.T.: Arabic single-document text summarization using particle swarm optimization algorithm. Procedia Comput. Sci. **117**, 30–37 (2017)
2. Anh, B.T.M., My, N.T., Trang, N.T.T.: Enhanced genetic algorithm for single document extractive summarization. In: Proceedings of the Tenth International Symposium on Information and Communication Technology, pp. 370–376 (2019)
3. Ansótegui, C., Sellmann, M., Tierney, K.: A gender-based genetic algorithm for the automatic configuration of algorithms. In: Gent, I.P. (ed.) CP 2009. LNCS, vol. 5732, pp. 142–157. Springer, Heidelberg (2009). https://doi.org/10.1007/978-3-642-04244-7_14
4. García-Hernández, R.A., Ledeneva, Y.: Single extractive text summarization based on a genetic algorithm. In: Carrasco-Ochoa, J.A., Martínez-Trinidad, J.F., Rodríguez, J.S., di Baja, G.S. (eds.) MCPR 2013. LNCS, vol. 7914, pp. 374–383. Springer, Heidelberg (2013). https://doi.org/10.1007/978-3-642-38989-4_38

5. Hahn, U., Mani, I.: The challenges of automatic summarization. Computer **33**(11), 29–36 (2000)
6. Hermann, K.M., et al.: Teaching machines to read and comprehend. In: Advances in Neural Information Processing Systems, pp. 1693–1701 (2015)
7. Lin, C.Y.: Rouge: A package for automatic evaluation of summaries. In: Text Summarization Branches Out, pp. 74–81 (2004)
8. Meena, Y.K., Gopalani, D.: Evolutionary algorithms for extractive automatic text summarization. Procedia Comput. Sci. **48**, 244–249 (2015)
9. Mendoza, M., Bonilla, S., Noguera, C., Cobos, C., León, E.: Extractive single-document summarization based on genetic operators and guided local search. Expert Syst. Appl. **41**(9), 4158–4169 (2014)
10. Mihalcea, R., Tarau, P.: Textrank: bringing order into text. In: Proceedings of the 2004 Conference on Empirical Methods in Natural Language Processing, pp. 404–411 (2004)
11. Miller, G.F., Todd, P.M.: The role of mate choice in biocomputation: sexual selection as a process of search, optimization, and diversification. In: Banzhaf, W., Eeckman, F.H. (eds.) Evolution and Biocomputation. LNCS, vol. 899, pp. 169–204. Springer, Heidelberg (1995). https://doi.org/10.1007/3-540-59046-3_10
12. Nallapati, R., Zhai, F., Zhou, B.: Summarunner: A recurrent neural network based sequence model for extractive summarization of documents. In: Thirty-First AAAI Conference on Artificial Intelligence (2017)
13. Over, P., Liggett, W.: Introduction to DUC: an intrinsic evaluation of generic news text summarization systems. In: Proceedings of DUC (2002). http://wwwnlpir.nist.gov/projects/duc/guidelines/2002.html
14. Sanchez-Gomez, J.M., Vega-Rodríguez, M.A., Perez, C.J.: A decomposition-based multi-objective optimization approach for extractive multi-document text summarization. Appl. Soft Comput. **91**, 106231 (2020)
15. See, A., Liu, P.J., Manning, C.D.: Get to the point: summarization with pointer-generator networks. arXiv preprint arXiv:1704.04368 (2017)
16. Simón, J.R., Ledeneva, Y., García-Hernández, R.A.: Calculating the significance of automatic extractive text summarization using a genetic algorithm. J. Intell. Fuzzy Syst. **35**(1), 293–304 (2018)
17. Sizov, R., Simovici, D.A.: Type-based genetic algorithms. In: Kotenko, I., Badica, C., Desnitsky, V., El Baz, D., Ivanovic, M. (eds.) IDC 2019. SCI, vol. 868, pp. 170–176. Springer, Cham (2020). https://doi.org/10.1007/978-3-030-32258-8_19
18. Suanmali, L., Salim, N., Binwahlan, M.S.: Genetic algorithm based sentence extraction for text summarization. Int. J. Innov. Comput. **1**(1) (2011)
19. Thede, S.M.: An introduction to genetic algorithms. J. Comput. Sci. Coll. **20**(1), 115–123 (2004)
20. Vázquez, E., Arnulfo Garcia-Hernandez, R., Ledeneva, Y.: Sentence features relevance for extractive text summarization using genetic algorithms. J. Intell. Fuzzy Syst. **35**(1), 353–365 (2018)
21. Wong, K.F., Wu, M., Li, W.: Extractive summarization using supervised and semi-supervised learning. In: Proceedings of the 22nd International Conference on Computational Linguistics, vol. 1, pp. 985–992. Association for Computational Linguistics (2008)

22. Yang, L., Cai, X., Zhang, Y., Shi, P.: Enhancing sentence-level clustering with ranking-based clustering framework for theme-based summarization. Inf. Sci. **260**, 37–50 (2014)
23. Zhang, J., Zhao, Y., Saleh, M., Liu, P.: Pegasus: pre-training with extracted gap-sentences for abstractive summarization. In: International Conference on Machine Learning, pp. 11328–11339. PMLR (2020)

Dragonfly Algorithm for Multi-target Search Problem in Swarm Robotic with Dynamic Environment Size

Mohd Ghazali Mohd Hamami[1](✉) ⓘ and Zool H. Ismail[2] ⓘ

[1] School of Mechanical Engineering, College of Engineering, Universiti Teknologi MARA Cawangan Johor, Kampus Pasir Gudang, 81750 Masai, Johor, Malaysia
ghazali.hamami@uitm.edu.my
[2] Malaysia Japan International Institute of Technology, Universiti Teknologi Malaysia, Jalan Sultan Yahya Petra, 54100 Kuala Lumpur, Malaysia

Abstract. Target search elements are very important in real-world applications such as post-disaster search and rescue missions, and pollution detection. In such situations, there will be time limitations, especially under a dynamic environment size which makes multi-target search problems are more demanding and need a special approach and intention. To answer this need, a proposed multi-target search strategy, based on Dragonfly Algorithm (DA) has been presented in this paper for a Swarm Robotic application. The proposed strategy utilized the DA static swarm (food hunting process) and dynamic swarm (migration process) to achieve the optimized balance between the exploration and exploitation phases during the multi-target search process. For performance evaluation, numerical simulations have been done and the initial results of the proposed strategy show more stability and efficiency than the previous works.

Keywords: Multi-target search problem · Swarm robotics · Dragonfly Algorithm

1 Introduction

Swarm robotics (SR) strategy inspired by natural swarm animals behavior such as birds or bees [1] emerged and nurture as Swarm Intelligence (SI) new sub-domain in the last two decades. SR strategy with its robust, flexible, and scalable characteristic, has significant advantages and outperforms the individual robots, especially in time restriction challenges such as target search problems. The target search problem is one of the real-world demanding problems which continuously increased researcher attention over time. Solving the target search problem using the SR strategy can be traced back to the work of [2] which proposed a PSO algorithm for collective robotic search in 2004. More real-world applications have emerged since then, particularly in 3D (Dangerous, Dirty, and Dull) mission-related such as search and rescue, natural disaster monitoring, and pollution detection.

Target search problems can be divided into single and multiple target searches. Previous research leaned towards single-target search problems due to the increase of

© Springer Nature Switzerland AG 2022
H. Fujita et al. (Eds.): IEA/AIE 2022, LNAI 13343, pp. 253–261, 2022.
https://doi.org/10.1007/978-3-031-08530-7_21

complexity and computationally burden (NP-hard problem) of multi-target search tasks. In the domain of target search problems, Particle Swarm Optimization (PSO) method received significant attention in the research community [3]. Although the PSO method is efficient, it has a high tendency for trapping in local minimum which turns out the PSO method is not recommended towards the multiple target search problem. The PSO method also has no obstacle avoidance characteristic embedded in the algorithm and needs to combine with other methods such as Artificial Potential Fields (APF) [4] which increase the algorithm complexity and computational intensity.

To counter the above limitations, this paper proposes the multi-target search strategy based on Dragonfly Algorithm (DA) [5] strategy. It aims to perform more efficiently in terms of targets detection period and less algorithm complexity and computational intensity. In the next section, some related works are briefly presented. Then simulation studies and environment setup are explained in Sect. 3, followed by simulation results and comparison in Sect. 4. Finally, a conclusion is made in Sect. 5.

2   Related Works

One of the important and high-impact applications of SR is target search problems. This application domain research has continued for several decades and more civilian applications such as post-disaster rescue operations [6], and most recent covid 19 outbreak monitoring [7]. As mentioned in the previous Sect., PSO received the most interest from the research community for solving the target search problems using the SR strategy. Inspired by bird flocking behavior, the PSO method was initiated by Kennedy and Eberhart in 1995 [8]. The mechanism of the PSO method is being explained in Table 1.

Traceback into 2004, Doctor et al. adapted the PSO principle into SR for performing the collective robotic search task [2]. The method implemented the dynamic inertia weight to optimize the PSO parameters for better searching output but lack of the obstacle's consideration in its algorithm. The consideration of the obstacles constrain has been introduced in Extended PSO (EPSO) method [9] with the integration of the Braitenberg obstacle avoidance algorithm into the PSO algorithm. Research in Robotic Darwinian PSO (RDPSO) [10] considered the communication between the swarm population approach as the required element in the target search problems. Even though this approach improves the search efficiency by being able to improve the scalability of the swarm, it required more computational energy and reduces the robot's diversity element. Extension towards the RDPSO method, Dadgar et al. with the Repulsion based RDPSO (RbRDPSO), [11] utilized the ion-based repulsion mechanism as its obstacle avoidance mechanism. This improved the swarm convergence rate, but the method is only suitable for a small swarm population ($n < 20$). All the mentioned methods also only focus on a single target search problem. Only recently, in the research of Exploration Enhance RPSO (E2RPSO) [12] and Mechanical PSO [13] they shifted their focus towards multi-target search problems but still required to combine with other methods such as Artificial Potential Field (APF) for the obstacle avoidance requirement. Based on the above explanation, although the PSO method is able to tackle the target search problems it still needs to improve in terms of local minimum issues and real-world constraints (obstacle avoidance requirement). This is where the new approach is required to tackle the multi-target search problems. This research proposed the Dragonfly Algorithm (DA) for this purpose. Table 1 portrays the suitability (advantages) of the DA strategy based on the limitations of the PSO method.

Table 1. Mechanism, advantages, and limitations of PSO and DA strategy

Strategy	Mechanism	Advantages	Limitations
Particle Swarm Optimization (PSO)	1. Mimicking the behavior of birds flocking 2. Agents moved within the search space searching for the possible best result position 3. Particle's velocity, best position, and global best position are updated for each iteration during the search process	1. Algorithm implementation is simple 2. Computational burden is on a minimum level 3. Consideration of each particle's best position from the overall swarm population's position	1. High tendency for trap in a local minimum 2. There is no obstacle avoidance characteristic embedded in the algorithm 3. There is no convergence guaranteed especially in global search problems
Dragonfly Algorithm (DA)	1. Inspired by PSO algorithm iteration basis 2. Have five swarm behaviors of Dragonfly which separation, alignment, cohesion, food attraction, and enemy distraction 3. Step vector and positioning of each agent will be updated for each iteration	1. Algorithm simplicity which leads to outcome efficiency 2. Obstacle avoidance behavior has been embedded in the algorithm 3. Have an excellent neighborhood search characteristic to prevent premature convergence 4. Suitable for wide-area search	1. The need of searching the balance between the exploration and exploitation

The history of the Dragonfly Algorithm (DA) can be traced back to 2016, which has been developed by [5] as a new meta-heuristic strategy for optimization problems. The mechanism of the DA strategy has been listed in Table 1. The DA strategy consists of the main PSO mechanism which is separation, alignment, and cohesion with the inclusion of food extraction and enemy distraction swarm behavior. The inclusion of enemy distraction as obstacle avoidance behavior is where the main difference of algorithm implementation between the DA strategy and the PSO method. Besides the advantages of DA strategy that can complement the PSO method limitations, the Dragonflies swarming inherit the core requirements of optimization phases which exploration (migration process) and exploitation (food hunting process). Since the DA naturally inherit the two most needed optimization requirements, is it plausible to implement the DA into optimization problems which in this study is a multi-target search problem. In 2018, a study of the Robotic Binary Dragonfly Algorithm (RBDA) which the application of SR based on DA strategy for cooperative search and rescue has been introduced [14]. The study implemented a swarm of robots between 5 to 15 robots to search for a single target with the availability of obstacles in the search environment. The obtained results were encouraged but still, it is only focusing on a single target search problem.

3 Methodology

In this section, the approached methodology is explained in detail. First, the simulation setup with the selected suitable parameter for both the PSO method and the proposed DA strategy for the multi-target search problem will be clarified. Then, the environment setup for the performance measure of DA strategy in comparison to the PSO method will be described. All the simulation experiments have been simulated in a MATLAB R2021b running on a Lenovo Legion Y520 laptop with Intel Core i7 7th Gen, 2.8 GHz powered processor with 16 GB of RAM.

3.1 Simulation Parameters Setup

DA strategy is inheriting the advantages of the PSO method with the inclusion of obstacle avoidance in its basic algorithm. The implemented DA strategy algorithm is as per below:

DA Algorithm
1: Dragonflies population initialization
2: Step vectors initialization
3: **while** the termination condition is not satisfied
4:      Calculate the objective (fitness) function for all dragonflies
5:      Update separation weight, alignment weight, cohesion weight, food factor weight, and enemy factor weight.
6:      Calculate separation element, alignment element, cohesion element, food factor, and enemy factor.
7:      Update neighboring radius
8:      **if** available at least one
9:      Update velocity vector
10.   Update position vector
11.   **else**
12.   Update position vector
13.   **end if**
14.   Verified and align the new positions based on the variable's boundaries
15.**end while**

Table 2 shows the parameters setup for both PSO and the proposed DA algorithm strategy. Selected PSO parameters setup are based on the empirical study of [15] which work very well in general including this multi-target search problem in terms of agents convergence towards multi-target positions. On the other hand, the selected DA parameters are based on a heuristic process based on research outcomes of [16] which highlighted the speed vs accuracy tradeoff in exploration-exploitation collective decision making.

Table 2. Parameter setting for the simulation experiments

PSO parameters setup		DA parameters setup	
Inertia weight (ω)	0.7298	Inertia weight (ω)	0.5
		Separation weight (s)	0.1
Cognitive scaling factor (c_1)	1.49618	Alignment weight (a)	0.1
Social scaling factor (c_2)	1.49618	Cohesion weight (c)	0.7
r_1 and r_2	Random value between [0, 1]	Food factor (f)	1
		Enemy factor (e)	1

3.2 Environment Setup

The environment setup is selected to replicate the multi-target search problems. The purpose of the experiment is to measure the performance of the proposed DA strategy in comparison to the PSO method in terms of optimizing the searching process with the obstacle avoidance capability intact. The research focus and observe the convergence time with the different sets of swarm population (robots) and search environment size with the following scopes:

- The target is more than a single target (multiple targets).
- The location of the targets and available obstacles are unknown to the robots.

Figure 1 portrays the multi-target environment setup. The red plus symbol represents the search agents (robots), the two green squares are the target's position, and the obstacles are represented with the red square, blue triangle, and orange pentagon.

The chosen environment parameters are as per Table 3. The swarm population quantities are been selected based on bucket brigade swarm performance experiments [17] which highlighted the swarm density from n = 15 until n = 30 is the optimal swarm density for the optimal swarm outcome. The objective function (fitness function) is $f_1(x, y) = \sin\left(\frac{\pi}{5}x\right) + \sin\left(\frac{\pi}{10}y\right)$, adapted from Tang et al. [13] research which also shows interest in the multi-target search problem. The iterations number is set to 500 iterations for the statically steady conditions [18].

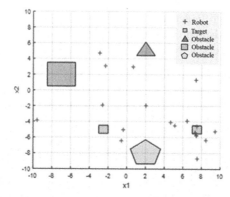

Fig. 1. The multitarget environment setup in MATLAB (Color figure online)

Table 3. The environment parameters

1. Swarm population (robots):			
n = 15	n = 20	n = 25	n = 30
2. Search environment size:			
A = 10 x 10	A = 20 x 10		A = 20 x 20

4   Results and Discussion

The numerical simulation result will be presented in this section. Both PSO and DA algorithms have been running with the parameters and environment setup as described in Sect. 3. Figure 2 shows the comparison of the PSO algorithm with the proposed DA algorithm during the multitarget search task in terms of convergence speed (iterations) as averaging over 30 independent simulations.

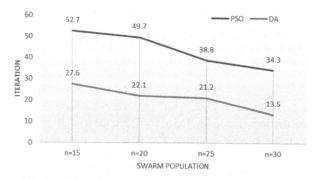

Fig. 2. PSO and DA convergence performance comparison

The simulations have been running with the swarm population of 15, 20, 25, and 30 robots. Based on the figure data, the increase of the swarm population contributed towards the decreasing of the time needed to find the multi-target for both strategies. However, the DA strategy produces more targets search efficiency for all swarm populations. For example (n = 20), the DA only required 22.1 iterations compared to the PSO which needed 49.7 iterations for the targets search task. This proved that the DA strategy able to find the targets faster and consume less energy while doing it so.

The effect of the search environment size on the target searching efficiency also has been studied. The simulation took place with a swarm population of 20 robots. The obtained data shows in Fig. 3. From the data, it can conclude that the DA strategy remains efficient in targets searching compared to the PSO strategy. This is because the DA strategy has embedded with obstacle avoidance behavior and has neighboring radius consideration in its basic algorithm while the PSO strategy needs to combine with other algorithms for the same purpose making it require more computational demand. From the data trend, we can also understand that the larger the search area the more time required to find the targets with the exception of the PSO data in 20 × 20 environment size. It shows the decreasing trend from 117.8 iterations (PSO, 20 × 10) to 49.7 iterations (PSO, 20 × 20). These conditions happen due to for both PSO, 10 × 10 and PSO, 20 × 10 environment size, the position of the targets is located exactly at the search boundaries making the search process slow compared to larger search environment (PSO, 20 × 20). The same conditions do not affect the DA strategy during the multi-target search task.

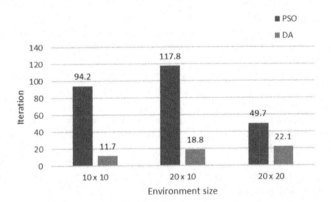

Fig. 3. Search environment size effect

5 Conclusion

In this study, a Swarm Robotic (SR) strategy based on Dragonfly Algorithm (DA) has been proposed for multi-target search problems. To validate the DA strategy performance, we had run several numerical simulations. Compared to the more established PSO method, the DA strategy exhibit more efficient result in terms of convergence speed towards targets and computational requirement during the multi-target search task. The results data also show the same trends with the different search environment sizes. For future consideration, the proposed DA strategy can be optimized and improved by implementing a new approach such as dynamic inertial weight function, sub-swarm algorithm, or dynamic tuning of its parameter.

References

1. Beni, G.: From swarm intelligence to swarm robotics. In: Şahin, E., Spears, W.M. (eds.) SR 2004. LNCS, vol. 3342, pp. 1–9. Springer, Heidelberg (2005). https://doi.org/10.1007/978-3-540-30552-1_1
2. Doctor, S., Venayagamoorthy, G.K., Gudise, V.G.: Optimal PSO for collective robotic search applications. In: Proceedings of the 2004 Congress on Evolutionary Computation (IEEE Cat. No. 04TH8753), pp. 1390–1395. IEEE (2004). https://doi.org/10.1109/CEC.2004.1331059
3. Ismail, Z.H., Hamami, M.G.M.: Systematic literature review of swarm robotics strategies applied to target search problem with environment constraints. Appl. Sci. **11**, 2383 (2021). https://doi.org/10.3390/app11052383
4. Khatib, O.: Real-time obstacle avoidance for manipulators and mobile robots. In: Proceedings of the 1985 IEEE International Conference on Robotics and Automation, pp. 500–505. Institute of Electrical and Electronics Engineers (1986). https://doi.org/10.1109/ROBOT.1985.1087247
5. Mirjalili, S.: Dragonfly algorithm: a new meta-heuristic optimization technique for solving single-objective, discrete, and multi-objective problems. Neural Comput. Appl. **27**(4), 1053–1073 (2015). https://doi.org/10.1007/s00521-015-1920-1
6. Quenzel, J., et al.: Autonomous fire fighting with a UAV-UGV team at MBZIRC 2020. In: 2021 International Conference on Unmanned Aircraft Systems (ICUAS), pp. 934–941. IEEE (2021). https://doi.org/10.1109/ICUAS51884.2021.9476846
7. Houacine, N.A., Drias, H.: When robots contribute to eradicate the COVID-19 spread in a context of containment. Prog. Artif. Intell. **10**(4), 391–416 (2021). https://doi.org/10.1007/s13748-021-00245-3
8. Kennedy, J., Eberhart, R.: Particle swarm optimization. In: Proceedings of ICNN 1995 - International Conference on Neural Networks, pp. 1942–1948. IEEE (1995). https://doi.org/10.1109/ICNN.1995.488968
9. Pugh, J., Martinoli, A.: Inspiring and modeling multi-robot search with particle swarm optimization. In: 2007 IEEE Swarm Intelligence Symposium, pp. 332–339. IEEE (2007). https://doi.org/10.1109/SIS.2007.367956
10. Couceiro, M.S., Figueiredo, C.M., Rocha, R.P., Ferreira, N.M.F.: Darwinian swarm exploration under communication constraints: Initial deployment and fault-tolerance assessment. Rob. Auton. Syst. **62**, 528–544 (2014). https://doi.org/10.1016/j.robot.2013.12.009
11. Dadgar, M., Couceiro, M.S., Hamzeh, A.: RbRDPSO: repulsion-based RDPSO for robotic target searching. Iran. J. Sci. Technol. Trans. Electr. Eng. **44**(1), 551–563 (2019). https://doi.org/10.1007/s40998-019-00245-z
12. Yang, J., Xiong, R., Xiang, X., Shi, Y.: Exploration enhanced RPSO for collaborative multi-target searching of robotic swarms. Complexity **2020**, 1–12 (2020). https://doi.org/10.1155/2020/8863526
13. Tang, Q., Yu, F., Xu, Z., Eberhard, P.: Swarm robots search for multiple targets. IEEE Access **8**, 1 (2020). https://doi.org/10.1109/ACCESS.2020.2994151
14. Abuomar, L., Al-Aubidy, K.: Cooperative search and rescue with swarm of robots using binary dragonfly algorithm. In: 2018 15th International Multi-Conference on Systems, Signals & Devices (SSD), pp. 653–659. IEEE (2018). https://doi.org/10.1109/SSD.2018.8570410

15. Eberhart, R.C., Shi, Y.: Comparing inertia weights and constriction factors in particle swarm optimization. In: Proceedings of the 2000 Congress on Evolutionary Computation. CEC00 (Cat. No. 00TH8512), pp. 84–88. IEEE (2000). https://doi.org/10.1109/CEC.2000.870279

16. Raoufi, M., Hamann, H., Romanczuk, P.: Speed-vs-accuracy tradeoff in collective estimation: an adaptive exploration-exploitation case. In: 2021 International Symposium on Multi-Robot and Multi-Agent Systems (MRS), pp. 47–55. IEEE (2021). https://doi.org/10.1109/MRS50823.2021.9620695

17. Hamann, H.: Introduction to swarm robotics. In: Hamann, H. (ed.) Swarm Robotics: A Formal Approach, pp. 1–32. Springer, Cham (2018). https://doi.org/10.1007/978-3-319-74528-2_1

18. Kwa, H.L., Kit, J.L., Bouffanais, R.: Optimal swarm strategy for dynamic target search and tracking. In: Proceedings of the International Joint Conference on Autonomous Agents and Multiagent Systems, AAMAS, pp. 672–680 (2020). https://dl.acm.org/doi/abs/10.5555/3398761.3398842

Video and Image Processing

Improved Processing of Ultrasound Tongue Videos by Combining ConvLSTM and 3D Convolutional Networks

Amin Honarmandi Shandiz[✉] and László Tóth

Institute of Informatics, University of Szeged, Szeged, Hungary
{shandiz,tothl}@inf.u-szeged.hu

Abstract. Silent Speech Interfaces aim to reconstruct the acoustic signal from a sequence of ultrasound tongue images that records the articulatory movement. The extraction of information about the tongue movement requires us to efficiently process the whole sequence of images, not just as a single image. Several approaches have been suggested to process such a sequential image data. The classic neural network structure combines two-dimensional convolutional (2D-CNN) layers that process the images separately with recurrent layers (e.g. an LSTM) on top of them to fuse the information along time. More recently, it was shown that one may also apply a 3D-CNN network that can extract information along both the spatial and the temporal axes in parallel, achieving a similar accuracy while being less time consuming. A third option is to apply the less well-known ConvLSTM layer type, which combines the advantages of LSTM and CNN layers by replacing matrix multiplication with the convolution operation. In this paper, we experimentally compared various combinations of the above mentions layer types for a silent speech interface task, and we obtained the best result with a hybrid model that consists of a combination of 3D-CNN and ConvLSTM layers. This hybrid network is slightly faster, smaller and more accurate than our previous 3D-CNN model.

Keywords: Silent speech interface · Convolutional neural network · 3D convolution · ConvLSTM · Ultrasound tongue video

1 Introduction

The area of Silent Speech Interfaces (SSI) deals with the problem of converting articulatory recordings to speech signals [1]. The studies in this field have considered various types of articulatory signals as input, such as Electroencephalography (EEG), Electromagnetic Articulography (EMA), Ultrasound Tongue Video Imaging (UTI) and so on. SSIs could provide a great amount of help for those disabled people who can not talk loud, but are able to silently articulate the speech. Converting the signals recorded during articulation to speech would allow these people to interact with others. SSI solutions could also be used in some other

© Springer Nature Switzerland AG 2022
H. Fujita et al. (Eds.): IEA/AIE 2022, LNAI 13343, pp. 265–274, 2022.
https://doi.org/10.1007/978-3-031-08530-7_22

conditions where normal communication is not possible, for example in certain military situations or in very noisy industrial environments where people could barely understand each other. In this paper, we work with Ultrasound Tongue Video Imaging (UTI) Data [2–4] recorded from Hungarian speakers.

In SSI systems, the classic approach of estimating the speech signal from the articulatory data consisted of two steps: the first one is to convert the input to text using speech recognition, and the second is to synthesize speech based on the text. Nowadays, however, directly converting the articulatory signals to speech is more popular, as it is less time consuming and seems to be more suitable for real-time applications. This direct approach has been made viable by the Deep Neural Network (DNN) technology, which also revolutionized many other speech-related fields, for example speech recognition [5], and speech synthesis [6]. Here, we also follow the direct approach and apply DNNs [3,7,8].

The given articulatory-to-acoustic mapping task can be addressed by applying simple fully connected DNNs [4,7]. However, as we are working with images, using convolutional neural networks (CNN) [9] is more reasonable, and they have been successfully applied to the SSI task by several authors [3,10]. A very important further aspect is that our input consists of a sequence of images (an ultrasound video), so it is not effective to simply process these images separately. Thus, we can apply a Recurrent Neural Network (RNN) such as an LSTM [11] to extract information from a sequence. And when the sequence consists of images, like in our case, an obvious solution is to combine an LSTM with a 2D CNN that extracts the information from individual video frames [12]. Alternatively, it would be possible to extend the 2D convolution to 3D by adding the time axis as an extra dimension [13]. This approach was followed in [3], and it was found that the two approaches can achieve very similar performances. However, there is a third option: for the processing of image sequences, Shi et al. proposed the ConvLSTM layer type [14], which combines the advantages of convolutional and recurrent processing in one layer. For some reason, however, the ConvLSTM construct is not widely known. To knowledge, it was applied to UTI data only in one case [15], but even in that paper the task was different from ours. The goal of this paper is to experiment with ConvLSTM models, and compare their performance with the previous 2D-CNN+LSTM and 3D-CNN approaches. We also try to combine the three types of layers, resulting in hybrid models, and our results show that the hybrid approach yields the best performance for our task.

The paper is organized as follows. In Sect. 2, we briefly introduce the concept of the Convolutional LSTM that we are going to use. In Sect. 3, we explain the data acquisition and processing steps for our input and output data. In Sect. 4, we present our experimental setup, while in and following Sect. 5, the experimental results are discussed and explained. Finally, in Sect. 6 our main conclusions are given.

2   Convolutional LSTM for SSI

SSI systems synthesize speech from articulatory videos by learning the mapping between the input ultrasound image sequence and the output audio signal. SSI is

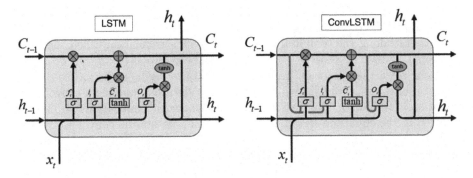

Fig. 1. Internal structure of a standard LSTM cell and its extended version (with extra peephole connections) used in Convolutional LSTMs [17,18].

LSTM

$$i_t = \sigma(W_{xi}x_t + W_{hi}h_{t-1} + W_{ci} \circ c_{t-1} + b_i)$$
$$f_t = \sigma(W_{xf}x_t + W_{hf}h_{t-1} + W_{cf} \circ c_{t-1} + b_f)$$
$$c_t = f_t \circ c_{t-1} + i_t \circ \tanh(W_{xc}x_t + W_{hc}h_{t-1} + b_c)$$
$$o_t = \sigma(W_{xo}x_t + W_{ho}h_{t-1} + W_{co} \circ c_t + b_o)$$
$$h_t = o_t \circ \tanh(c_t)$$

ConvLSTM

$$i_t = \sigma(W_{xi} * \mathcal{X}_t + W_{hi} * \mathcal{H}_{t-1} + W_{ci} \circ \mathcal{C}_{t-1} + b_i)$$
$$f_t = \sigma(W_{xf} * \mathcal{X}_t + W_{hf} * \mathcal{H}_{t-1} + W_{cf} \circ \mathcal{C}_{t-1} + b_f)$$
$$\mathcal{C}_t = f_t \circ \mathcal{C}_{t-1} + i_t \circ \tanh(W_{xc} * \mathcal{X}_t + W_{hc} * \mathcal{H}_{t-1} + b_c)$$
$$o_t = \sigma(W_{xo} * \mathcal{X}_t + W_{ho} * \mathcal{H}_{t-1} + W_{co} \circ \mathcal{C}_t + b_o)$$
$$\mathcal{H}_t = o_t \circ \tanh(\mathcal{C}_t)$$

Fig. 2. The equations behind the operation of Long Short Term Memory (LSTM) versus Convolutional LSTM neurons [14].

a sequential task, as both the input and the output are sequences, with a strong correlation between consecutive elements of the sequence. As in our case the input data consist of ultrasound images, convolutional networks (CNN) seems to be a proper tool for processing the input, as they are known to perform well when working with images [9], and also in particular with SSI ultrasound tongue images [12,16]. As our input is a sequence, the information content along the time axis of the data can be extracted by applying Recurrent Neural Networks (RNNs). In particular, a variant of recurrent networks called the Long-Short Term Memory (LSTM) is known to be more effective in extracting long-term dependencies in the input sequence [11]. These networks have special gates in their internal implementations which improve their abilities to handle large-distance relations between time-related features.

The data flow in the standard implementation of an LSTM is shown in Fig. 1. Some implementations also contain extra connections, such a so-called "peephole" variant in shown on the right side of Fig. 1, In both cases, the input of the LSTM consist of a sequence of vectors. As in our case we want to process a sequence of images, a straightforward solution is to combine a CNN with an LSTM. In the trivial arrangement the images of the input are first processed by a (2-dimensional) CNN, and the sequence of CNN outputs are integrated over time by using an LSTM. This approach was shown to work fine in SSI implementations [12]. And we will shortly refer to this scheme as the CNN+LSTM approach. However, this type of processing requires the combination of two layers – a 2D-CNN to process the data along the two spacial axes, and an LSTM to process it along the temporal axis. A more efficient solution called the Convolutional LSTM, or shortly ConvLSTM has been proposed by Shi et al. As the name says, their solution performs the two processing steps in one. It can extract spatio-temporal features from the input data by applying a convolution operation in the inner steps LSTM instead of matrix multiplication [14]. This is reflected in the equations of Fig. 2. Where "$*$" represents the convolution operation, and "$\circ$" stands for gating. Note that, apart from the convolution, the equations are exactly the same as those of the (peephole) LSTM. The convolution allows the more efficient processing of image sequences, resulting in better performance with much fewer parameters. For example, Kwon et al. successfully applied a hierarchical ConvLSTM for speech recognition [19]. Recently, Zhao et al. used ConvLSTMs for predicting subsequent ultrasound images in an SSI task [15].

Processing a sequence of images is also viable by extending the convolution operation to the time axis, resulting in a three-dimensional convolution (3D-CNN) model. The main advantage of this approach is that it is faster, as it applies only convolution operations. Convolution also allows the skipping of input images, which is not possible in an LSTM framework. In a previous paper, the 3D-CNN model gave results that were comparable or slightly better than those with the more conventional CNN+LSTM approach [3]. Here, we extend this earlier comparison to the ConvLSTM model, and we are also going to experiment with hybrid models that combine 3D-CNN and ConvLSTM layers.

3   Data Acquisition and Preprocessing

The ultrasound data was collected from a Hungarian female subject (42 years old) while she was reading sentences aloud. Her tongue movement was recorded in a midsaggital orientation – placing the ultrasonic imaging probe under the jaw – using a "Micro" ultrasound system by Articulate Instruments Ltd. The transducer was fixed using a stabilizer headset. The 2–4 Mhz/64 element 20 mm radius convex ultrasound transducer produced 82 images per second. The speech signals were recorded in parallel with an Audio-Technica ATR 3350 omnidirectional condenser microphone placed at a distance of 20 cm from the lips. The ultrasound and the audio signals were synchronized using the software tool provided with the equipment. Altogether 438 sentences (approximately half an hour) were recorded from the subject, which was divided into train, development and test sets in a 310-41-87 ratio. We should add that the same data set was used in several earlier studies [2,3,7], and the data set is publicly available[1]. The ultrasound probe records 946 samples along each of its 64 scan lines. The recorded data can be converted to conventional ultrasound images using the software tools provided. However, due to its irregular shape, this image is harder to process by computers, while it contains no extra information compared to the original

---

[1] The dataset is available upon request from csapot@tmit.bme.hu.

scan data. Hence, we worked with the original 964×64 data items, which were downsampled to 128×64 pixels. The intensity range of the data was min-max normalized to the $[-1, 1]$ interval before feeding it to the network.

The speech signal was recorded with a sampling rate $11025\,\mathrm{Hz}$, and then converted to a 80-bin mel-spectrogram using the SPTK toolkit (http://sp-tk. sourceforge.net). The goal of the machine learning step was to learn the mapping between the sequence of ultrasound images and the sequence of mel-spectrogram vectors. As the two sequences are perfectly synchronized, it was not necessary to apply a sequence-to-sequence learning strategy. We simply defined the goal of learning as an image-to-vector mapping task, using the mean squared error (MSE) as the loss function in the network training step. The 80 mel-frequency coefficients served as training targets, from which the speech signal was reconstructed using WaveGlow [20]. To facilitate training, each of the 80 targets were standardized to zero mean and unit variance. The input of training consisted of a block of 25 consecutive. This allowed all DNN variants to involve the time axis in the information extraction process. The whole SSI framework followed our earlier study [3].

4   Experimental Setup

We implemented our networks using Keras with a tensorflow back-end [21]. We applied three different network architectures that can process 3-dimensional blocks of data. In the tables, "3D-CNN" refers to the fully convolutional model proposed in [3]. This model does not have any LSTM component. "3D-CNN + BiLSTM" refers to a combination which applies a BiLSTM layer as the topmost hidden layer to integrate the temporal features extracted by the previous 3D-CNN layers. The final model referred to as "3D-CNN + ConvLSTM" replaces the tompost 3D-CNN and BiLSTM layers by a ConvLSTM layer. Notice that the ConvLSTM technique fuses the convolution and the LSTM operations into one layer, so here we can also spare one hidden layer by this substitution. In the following, we give a more detailed description of the three configurations.

3D Convolutional Neural Network(3D-CNN): This model was described in detail in [3], and its network layers are shown in Table 1. The networks processes the input sequence of 25 video frames in 5-frame blocks using 3D convolution. The overlap between these blocks is minimized by setting the stride parameter s of the time axis to 5. These blocks are processed further by 3 additional Conv3D layers, with pooling layers after every second convolution layer. Finally, the output is flattened and integrated over the time axis by a dense layer as the topmost hidden layer. The output hidden layer is a linear layer with 80 neurons, corresponding to the 80 spectral parameters given as training targets. This special network structure was motivated by Tran et al., who found that for the best result the processing should focus on the two spacial axes first, performing the integration over the temporal axis only afterwards [13]. Toth at al. also obtained the best result with performing the 3D convolution in this decomposed,

"(2D+1)" form [3]. Compared to that study, we achieved slightly better results with the same architecture by switching to the Adam optimizer instead of SGD, and by adjusting some meta-parameters, for example the dropout rate.

3D CNN + BiLSTM: As the output of the four layers of 3D convolution, the 3D-CNN network produces a sequence of 5 matrices, which are combined by a simple dense layer (cf. Table 1). Our first modification was to replace this fully connected layer with a (bidirectional) LSTM, which required us to reshape the matrices into vectors. The LSTM is a more sophisticated solution to extract the information from a temporal sequence, so we hoped to get slightly better results from this approach. As Table 1 shows, we set the *return_sequences* parameter of the LSTM to False, so the output is a simple vector, which serves as the input of the subsequent dense linear output layer.

3D CNN + ConvLSTM: Our main goal in this paper was to examine the efficiency of the ConvLSTM layer for this task. In the first experiment we applied it only at the topmost hidden layer of our 3D-CNN model (see Table 1). As the ConvLSTM layer implements the operation of a convolutional and an LSTM layer in one, we replaced the uppermost Conv3D and LSTM layers by a ConvLSTM layer, reducing the number of neural hidden layers from 5 to 4. Also, as the ConvLSTM layer works with matrices and also outputs matrices, the reshaping was required after the layer and not before it.

5   Results and Discussion

In the first experiment we compared the performance of the baseline 3D-CNN model we the two hybrid solutions proposed in the previous chapter and in Table 1. In Table 2 we report two simple objective metrics of the quality of training, the mean squared error (MSE) and the R^2 score, which is popular in regression tasks implemented with neural networks (for R^2 a higher value means better performance). As can be seen, replacing the dense layer by the LSTM layer already brings a slight but consistent improvement in the results, both on the development and on the test set. Fusing the uppermost Conv3D and the LSTM layer into a ConvLSTM layer resulted in further error reduction of about the same rate, even though the network depth is decreased. This clearly proves the efficiency of the ConvLSTM layer. However, we also observed a drawback, namely that the ConvSLTM layer has much more trainable parameters than the Conv3D layer. Hence, we had to reduce the filter size in the ConvLSTM layer, in order to keep the number of parameters in the original range. Theoretically, similar to the LSTM layer, the ConvLSTM layer can also be made bidirectional. However, we ran into the same problem that it tremendously increased the number of parameters while yielding only a marginal improvement. Thus, we stuck with using the unidirectional variant. Finally, to fuse the Conv3D and the LSTM layers, we had to remove the second MaxPooling layer. We also tried to insert it back after the ConvLSTM layer, but the results did not change considerably.

Table 1. The layers of the 3D-CNN, the 3D-CNN + BiLSTM and the 3D-CNN + ConvLSTM networks in Keras implementation, along with their most important parameters. The differences are highlighted in bold.

3D-CNN	3D-CNN + BiLSTM
Conv3D(30, (5,13,13), strides=(s, 2,2))	Conv3D(30,(5,13,13), strides=(s,2,2))
Dropout(0.3)	Dropout(0.3)
Conv3D(60, (1,13,13), strides=(1,2,2))	Conv3D(60,(1,13,13),strides=(1,2,2))
Dropout(0.3)	Dropout(0.3)
MaxPooling3D(pool_size=(1,2,2))	MaxPooling3D(pool_size=(1,2,2))
Conv3D(90, (1,13,13), strides=(1,2,1))	Conv3D(90,(1,13,13),strides=(1,2,1))
Dropout(0.3)	Dropout(0.3)
Conv3D(85, (1,13,13), strides=(1,2,2))	Conv3D(85, (1,13,13), strides=(1,2,2))
Dropout(0.3)	Dropout(0.3)
MaxPooling3D(pool_size=(1,2,2))	MaxPooling3D(pool_size=(1,2,2))
Flatten()	**Reshape((5, 340))**
Dense(500)	**Bidirectional(LSTM(320,**
Dropout(0.3)	**ret_seq=False))**
Dense(80, activation='linear')	Dense(80, activation='linear')

3D-CNN + ConvLSTM
Conv3D(30,(5,13,13),strides=(s,2,2))
Dropout(0.35)
Conv3D(60,(1,13,13),strides=(1,2,2))
Dropout(0.35)
MaxPooling3D(poolsize =(1,2,1))
Conv3D(90,(1,13,13),strides=(1,2,2))
Dropout(0.35)
ConvLSTM2D(64, (3,3), Strides=(2,2), ret_seq=False)
Flatten()
Dense(80,activation=linear)

Obviously, many other possible configurations exist that combine Conv3D and ConvLSTM layers. In the second experiment we tried out further combinations of these two layers. We experimented with 4 hidden layer constructs, and we fixed the uppermost layer to be a ConvLSTM, as it convincingly proved to be the better setup in the previous experiment. Table 3 summarizes the architectures we experimented with. As regards ConvLSTM layers, the *return_sequences* parameter was set to True for intermediate layers, and set to False only when the ConvLSTM layer was the topmost hidden layer. The meta-parameters were always chosen so that the global count of the free parameters stayed similar to that of the baseline model.

Seeing the good performance of the ConvLSTM layer in the previous experiment, we first tried to build a fully ConvLSTM model. However, as the first row of Table 3 shows, we obtained no improvement. As the ConvLSTM layer proved to be more efficient than the Conv3D layer earlier, next we tried to create a

Table 2. The MSE and mean R^2 scores obtained with the various network configurations for the development and test sets, respectively. The best results are highlighted in bold.

Network type	Dev		Test	
	MSE	Mean R^2	MSE	Mean R^2
3D-CNN	0.292	0.714	0.293	0.710
3D-CNN + BiLSTM	0.285	0.721	0.282	0.721
3D-CNN + ConvLSTM	**0.276**	**0.727**	**0.276**	**0.73**

Table 3. The MSE for different combinations of Conv3D and ConvLSTM layers in the four hidden layers of the network. The best results are highlighted in bold.

Layer 1	Layer 2	Layer 3	Layer 4	Dev	Test
ConvLSTM	ConvLSTM	ConvLSTM	ConvLSTM	0.31	0.31
ConvLSTM	ConvLSTM	ConvLSTM	—	0.29	0.3
Conv3D	ConvLSTM	ConvLSTM	ConvLSTM	0.31	0.31
Conv3D	Conv3D	ConvLSTM	ConvLSTM	0.36	0.35
Conv3D	**Conv3D**	**Conv3D**	**ConvLSTM**	**0.27**	**0.27**
Conv3D	ConvLSTM	Conv3D	ConvLSTM	0.3	0.3
ConvLSTM	Conv3D	Conv3D	ConvLSTM	0.34	0.34

network of just 3 ConvLSTM layers instead of 4. As shown in the second row of the table, the results became slightly better, but still worse than the baseline. This result reinforces our previous observation that ConvLSTM networks do not require the same depth as a convolutional network.

Conv3D layers have the advantage that they can easily downsample the time axis using a stride parameter larger than 1. On the contrary, ConvLSTM units cannot easily skip elements of their input sequence, due to their recurrent nature, which results in a large parameter count and a slow training. Hence, it seemed to be more efficient to put a Conv3D layer into the first hidden layer. We tried to place 1-2-3 Conv3D layers in the lower layers, and 3-2-1 ConvLSTM layers in the remaining layers. The middle block of Table 3 shows that the optimal solution is to have just one ConvLSTM layer, as in our original experiment. Lastly, we tried two further configurations with alternating Conv3D and ConvLSTM layers, motivated by papers like [15, 19], but we did not receive any better results.

6   Conclusion

Here, we were seeking the optimal neural network architecture for the articulatory-to-acoustic mapping task of SSI systems. The task involves the processing of 3D data blocks – sequences of images – for which one can apply 3D-CNN models, such as in [3]. Alternatively, one may apply a ConvLSTM model proposed by [14]. Besides comparing the purely convolutional and ConvLSTM models, we also experimented with hybrid architectures where the two layers types are mixed. The 3D-CNN + ConvLSTM hybrid model obtained the best results, better than the baseline 3D-CNN model, and it also outperformed other models with a different order of layers, as applied in [19] for emotion recognition, and in [15] for the prediction of the subsequent ultrasound image. Applying the ConvLSTM layer in the uppermost hidden layer even made the model smaller (with one hidden layer) and slightly faster to train. The optimal model arrangement consists of three Conv3D layers and a ConvLSTM on top of them, which illustrates that it is worth combining the ConvLSTM layer with other layer types such as the Conv3D to extract spatio-temporal features from videos – in our case, to better capture the tongue movement. The winning architecture also shows that the Conv3D blocks are more efficient in extracting local spectro-temporal information, while ConvLSTM is more efficient in fusing these pieces of information along the time axis. Interestingly, this coincides with the observation of Tran et al. about the optimal order of feature extraction for 3D video blocks [13]. In the future we plan to extend our research to transformer models that apply two separate networks for encoding and decoding, which would allow us to experiment with different network types for the decoder and encoder components, similar to the UNET [22] architecture.

Acknowledgments. Project no. TKP2021-NVA-09 has been implemented with the support provided by the Ministry of Innovation and Technology of Hungary from the National Research, Development and Innovation Fund, financed under the TKP2021-NVA funding scheme, and also within the framework of the Artificial Intelligence National Laboratory Programme. The RTX A5000 GPU used in the experiments was donated by NVIDIA.

References

1. Schultz, T., Wand, M., Hueber, T., Krusienski, D.J., Herff, C., Brumberg, J.S.: Biosignal-based spoken communication: a survey. IEEE/ACM Trans. Audio Speech Lang. Process. **25**(12), 2257–2271 (2017)

2. Csapó, T.G., Grósz, T., Gosztolya, G., Tóth, L., Markó, A.: DNN-based ultrasound-to-speech conversion for a silent speech interface. In: Proceedings of InterSpeech, pp. 3672–3676 (2017)

3. Tóth, L., Shandiz, A.H.: 3D convolutional neural networks for ultrasound-based silent speech interfaces. In: Rutkowski, L., Scherer, R., Korytkowski, M., Pedrycz, W., Tadeusiewicz, R., Zurada, J.M. (eds.) ICAISC 2020. LNCS (LNAI), vol. 12415, pp. 159–169. Springer, Cham (2020). https://doi.org/10.1007/978-3-030-61401-0_16

4. Jaumard-Hakoun, A., Xu, K., Leboullenger, C., Roussel-Ragot, P., Denby, B.: An articulatory-based singing voice synthesis using tongue and lips imaging. In: ISCA Interspeech 2016. vol. 2016, pp. 1467–1471 (2016)

5. Hinton, G., et al.: Deep neural networks for acoustic modeling in speech recognition: the shared views of four research groups. IEEE Sig. Process. Mag. **29**(6), 82–97 (2012)

6. Ling, Z.H., et al.: Deep learning for acoustic modeling in parametric speech generation: a systematic review of existing techniques and future trends. IEEE Sig. Process. Mag. **32**(3), 35–52 (2015)
7. Grósz, T., Gosztolya, G., Tóth, L., Csapó, T.G., Markó, A.: F0 estimation for DNN-based ultrasound silent speech interfaces. In: 2018 IEEE International Conference on Acoustics, Speech and Signal Processing (ICASSP), pp. 291–295. IEEE (2018)
8. Young, T., Hazarika, D., Poria, S., Cambria, E.: Recent trends in deep learning based natural language processing. IEEE Comput. Intell. Mag. **13**(3), 55–75 (2018)
9. Krizhevsky, A., Sutskever, I., Hinton, G.E.: ImageNet classification with deep convolutional neural networks. Adv. Neural. Inf. Process. Syst. **25**, 1097–1105 (2012)
10. Saha, P., Liu, Y., Gick, B., Fels, S.: Ultra2Speech - a deep learning framework for formant frequency estimation and tracking from ultrasound tongue images. In: Martel, A.L., et al. (eds.) MICCAI 2020. LNCS, vol. 12263, pp. 473–482. Springer, Cham (2020). https://doi.org/10.1007/978-3-030-59716-0_45
11. Hochreiter, S., Schmidhuber, J.: Long short-term memory. Neural Comput. **9**(8), 1735–1780 (1997)
12. Juanpere, E.M., Csapó, T.G.: Ultrasound-based silent speech interface using convolutional and recurrent neural networks. Acta Acust. Acust. **105**(4), 587–590 (2019)
13. Tran, D., Wang, H., Torresani, L., Ray, J., LeCun, Y., Paluri, M.: A closer look at spatiotemporal convolutions for action recognition. In: Proceedings of CVPR (2018)
14. Shi, X., Chen, Z., Wang, H., Yeung, D.Y., Wong, W.K., Woo, W.C.: Convolutional LSTM network: a machine learning approach for precipitation nowcasting. arXiv preprint arXiv:1506.04214 (2015)
15. Zhao, C., Zhang, P., Zhu, J., Wu, C., Wang, H., Xu, K.: Predicting tongue motion in unlabeled ultrasound videos using convolutional LSTM neural networks. In: ICASSP 2019–2019 IEEE International Conference on Acoustics, Speech and Signal Processing (ICASSP), pp. 5926–5930. IEEE (2019)
16. Kimura, N., Kono, M., Rekimoto, J.: SottoVoce: an ultrasound imaging-based silent speech interaction using deep neural networks. In: Proceedings of the 2019 CHI Conference on Human Factors in Computing Systems, pp. 1–11 (2019)
17. Convolutional LSTM (2019). https://medium.com/neuronio/an-introduction-to-convlstm-55c9025563a7
18. Recurrent neural networks and LSTMs with keras (2020). https://blog.eduonix.com/artificial-intelligence/recurrent-neural-networks-lstms-keras
19. Kwon, S., et al.: CLSTM: deep feature-based speech emotion recognition using the hierarchical ConvLSTM network. Mathematics **8**(12), 2133 (2020)
20. Prenger, R., Valle, R., Catanzaro, B.: WaveGlow: a flow-based generative network for speech synthesis. In: Proceedings of ICASSP, pp. 3617–3621 (2019)
21. Chollet, F., et al.: Keras (2015). https://github.com/fchollet/keras
22. Behboodi, B., Rivaz, H.: Ultrasound segmentation using U-net: learning from simulated data and testing on real data. In: 2019 41st Annual International Conference of the IEEE Engineering in Medicine and Biology Society (EMBC), pp. 6628–6631. IEEE (2019)

Improvement of Text Image Super-Resolution Benefiting Multi-task Learning

Kosuke Honda[1]([✉])(iD), Hamido Fujita[2](iD), and Masaki Kurematsu[1]

[1] Faculty of Software and Information Science, Iwate Prefectural University,
Iwate, Japan
`g231t031@s.iwate-pu.ac.jp`
[2] Regional Research Center, Iwate Prefectural University, Iwate, Japan

Abstract. Text image super-resolution is a pre-processing of scene text recognition, which aims to improve the visual quality of text from low-resolution images. However, existing super-resolution (SR) models designed for general images have difficulty in recovering text from low-resolution images in real scenes. There are several reasons for this, including the fact that the models do not consider text-specific properties and that the background is not important for text images SR. In this paper, we propose a multi-task learning model for reconstruction and SR termed TRSRT using a transformer for text images. Compared to the super-resolution model, the reconstruction model is better at denoising and tends to have structural information about the text. Focusing on this point, the proposed method utilizes these properties of the reconstructed model to the SR model through the transformer. In addition, we attempt to acquire a text-specific model by training with three loss functions including feature-driven loss using a text recognizer. Experimental results on TextZoom show that the proposed method achieves performance comparable to state-of-the-art methods and prove the advantages of multi-task learning.

Keywords: Text image super-resolution · Multi-task learning · Scene text recognition · Transformer

1 Introduction

Super-resolution (SR) for scene text image, which is a preprocessing of text recognition, aims to reconstruct text shapes from low-resolution (LR) text images. In recent years, this task has received a lot of attention due to its various applications such as environmental recognition of license plates and signs [1], ID card recognition [2], and text retrieval [3]. However, the recognition is still challenging for LR text images that exist in real scenes due to some issues such as blurring, poor illumination.

© Springer Nature Switzerland AG 2022
H. Fujita et al. (Eds.): IEA/AIE 2022, LNAI 13343, pp. 275–286, 2022.
https://doi.org/10.1007/978-3-031-08530-7_23

With the recent development of deep learning, SR methods for generic images have achieved significant progress. In several studies [4,5], these SR methods are adopted for text image SR. Tran *et al.* [5] employ a Laplacian-pyramid network as the backbone to combine features from several middle layers and capture the text font details. However, these methods were trained on synthetic SR datasets that LR images were generated from high-resolution (HR) images using degradation methods such as bicubic down-sampling and Gaussian blur. In addition, these previous works don't consider text image-specific characteristics such as text fonts and layouts. Therefore, it is difficult to reconstruct more complex degraded images of the real world with these previous works. Wang *et al.* [7] proposed TextZoom, which is the first dataset that contains the real-world LR text images, and the Text super-resolution network (TSRN). TSRN is based SRResNet [8] and uses BLSTMs [9] to capture sequential information of text images. This network significantly boosts the recognition accuracy for LR text images in the real scene compared to previous SR methods. However, the recognition accuracy for harder-level LR images is low. Therefore, scene text image SR is still a difficult open task.

Recently, multi-task learning is attracting attention due to its successful application in various tasks such as natural language processing [10], computer vision [11], and speech recognition [12]. It aims to improve learning efficiency, model flexibility, and generalization performance by sharing feature representations among related tasks. In several works [13–15], multi-task learning is applied for single image SR. In [13], multi-task learning is used for simultaneously SR and semantic segmentation to extract and use semantic information for SR. T2Net [15] is the multi-task learning model of MRI reconstruction and SR. It applies the complementary capacity of the reconstruction model such as stronger noise removal capacity and information of correct MRI structure for SR.

Inspired by these works, we propose an end-to-end Text Reconstruction and Super-Resolution Transformer (TRSRT). This network aims to get complimentary feature representations of text image reconstruction and SR by multi-task learning of these. The attention mechanism of the transformer is used to share the features of the reconstructed branch with the SR branch. In addition, we design the feature-driven loss function to construct a more text recognition-oriented SR. The loss function in previous works focused on every pixel of the text image. However, the quality of the background is not important in this task because the goal is to achieve high quality in the text region. Therefore, it is not necessary to consider every pixel in the training. For feature-driven loss, the output image from TRSRT and the high-resolution image, which is the label, are input to a pre-trained text recognizer, and the loss is obtained from two feature maps. We then attempt to train an efficient model that focuses on the text domain. The main contributions of this work are as follows:

1. We propose a text-focused image restoration framework based on multi-task learning of reconstruction and SR using transformers as shown in Fig. 1.
2. We attempted to improve the efficiency of learning and the generalization performance of the model by sharing the complementary feature

representations of each branch through multi-task learning of reconstruction and SR of text images.

3. The loss function, which uses the feature map output from the pre-trained scene text recognizer, is designed for training to capture text-specific features into account.

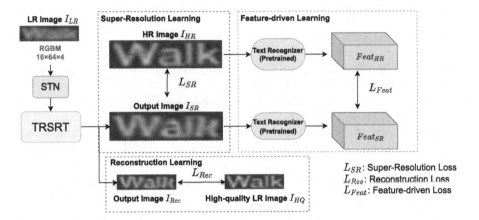

Fig. 1. The overall architecture of the proposed method during learning. For text-specific Super-Resolution, TRSRT is trained using three loss functions: L_{SR}, L_{Rec}, and L_{Feat}.

2   Related Work

2.1   Scene Text Recognition (STR)

In recent years, with the development of deep learning, Scene text recognition (STR) techniques have significant progress. Traditional methods [16–18] attempt to STR by extracting text from characters into words in a bottom-up manner. The disadvantage of these methods is that they do not provide sequential information such as context. CRNN [19] combines CNN and RNN to extract sequential information from the whole text image without splitting the characters. In addition, it improves the accuracy of prediction using the CTC [20] loss function for training. ASTER [21] employs a Spatial Transformer Network [22] for image rectification and the attention mechanism to obtain robustness for various shapes of text images. Recently, many advanced methods have been proposed, such as the method using Attention in 2D direction [23] and using Transformer for STR [24], and they have achieved high performance in the STR benchmark. However, LR text images in real scenes, such as those included in TextZoom [7], are difficult to recognize even with these advanced methods. In our method, CRNN is used to compute the feature-driven loss. We use ASTER to evaluate the performance of the SR model.

2.2   Scene Text Image Super Resolution

Scene text image SR is a task to improve the resolution and quality of LR text images that exist in real scenes. There are not yet many studies on SR models specific to text images using deep learning. Several studies [4,5] adapt the general single image SR method to text images. In [4], SRCNN [6] is extended to text images SR as a backbone network. In [5], the deep laplacian pyramid network is employed as the backbone to upsample LR text images. Most of these methods use LR images as training, which are downsampled from HR images using bicubic or bilinear. This means that they are not effective for LR text images in real scenes. In addition, these directly use a generic SR model and do not take into account text-specific features such as font and character level. Wang et al. [7] proposed TextZoom, which is the first dataset that contains the real-world LR text images, and the Text super-resolution network (TSRN). TSRN [7] uses BLSTMs [9] to capture sequential information of text images. Chen et al. [25] proposed a transformer-based text image SR model and a text-specific learning method. TSRGAN [26] is a GAN-based model that introduces adversarial loss and wavelet loss and attempts to reconstruct sharp text edges.

2.3   Multi-task Learning for Super Resolution

Multi-task learning has been applied in machine learning models in various fields such as computer vision [11], speech recognition [12], and natural language processing [10]. Several studies [13–15] apply multi-task learning in SR as well. Saeed et al. [13] employs multi-task learning for simultaneously SR and semantic segmentation. It aims to extract and apply semantic information in the image for SR. The decoder only uses one shared deep network for two task-specific output layers to search for categorical information during training. In [14], the SR model and the image category (e.g., natural image, cartoon image, text image, etc.) classification model are trained simultaneously to perform category-aware SR. T2Net [15] employs multi-task learning for MRI reconstruction and SR. It uses the complementary capacity of the reconstruction model such as stronger noise removal capacity and the information of correct MRI structure for SR. The reconstruction branch and SR branch simultaneously learn and send the complementary information from the reconstruction branch to the SR branch through a transformer using the attention mechanism.

Inspired by these studies, multi-task learning is employed for text image reconstruction and SR. We attempt to apply the text structure information and denoising capacity of the reconstruction model to SR.

3   Methodology

In this paper, we propose the end-to-end Text Reconstruction and Super-Resolution Transformer (TRSRT). The overall architecture of the proposed method during learning is shown in Fig. 1. First, the input image is rectified using a Spatial Transformer Network (STN) [22]. The rectified image is input to the T2Net [15] based multi-task model, TRSRT, which outputs two images, I_{SR} and I_{Rec}. I_{SR} is a SR image and I_{Rec} is a reconstructed image of the same size as the input. In feature-driven learning, I_{SR} and the SR label image I_{HR} are input to the pre-trained STN model CRNN [19] respectively, and the loss between the output feature maps is calculated. Our input image is RGBM 4-channel, which is concatenated binary mask with an RGB channel following [7]. In this section, Details of the TRSRT mechanism and training method are introduced.

3.1 Text Reconstruction and Super Resolution Transformer (TRSRT)

The overall architecture of the multi-task model is shown in Fig. 2. The proposed framework consists of a Super-Resolution branch, a Reconstruction branch, and a Transformer module. Reconstruction (Rec) branch and Transformer module. The SR branch and Rec branch extract features are specific to each task. The transformer module learns the shared features of the two tasks and brings generality to the network. The final output of the SR branch is a high-quality SR image with reconstructed text, while the final output of the Rec branch is a denoised image of the same size as the input image.

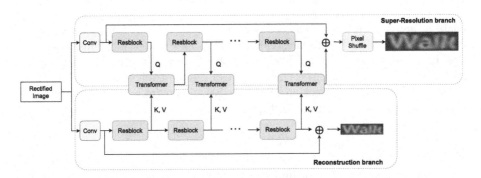

Fig. 2. Architecture of the proposed multi-task model. It consists Super-Resolution branch, Reconstruction branch, and Transformer module.

Super-Resolution Branch. The SR branch aims to generate the SR image I_{SR} from the LR image rectified by the STN [22]. First, shallow features are extracted from the input image using the convolutional layer as shown in Fig. 2. Then, they are sent to Resblock, which is the backbone to extract the features for SR. The features extracted from the Resblock are sent as the query to the transformer module and it outputs enhanced features using the features obtained from the Reconstruction task. Finally, the SR image I_{SR} is generated by upsampling using pixel shuffling.

Reconstruction Branch. The Reconstruction branch aims to generate a denoised and high-quality text image I_{Rec} with the same size as the input. Image reconstruction methods contain stronger noise removal capacity and information of correct text structure than SR methods. Only the capability of the SR branch is not sufficient to generate a high-resolution image from a blurry, LR text image that exists in the real scene. Therefore, we attempt to generate a high-quality SR image by fusing the features of the Rec branch and the SR branch through the Transformer module. The backbones of this Rec branch is employed the same design as the backbones of the SR branch. Features extracted from these backbones are sent transformer module as the key and the value.

Transformer Module. The overall diagram of the Transformer module is shown in Fig. 3. As mentioned above, the query (Q) is features extracted from the SR branch, and the key (K) and value (V) are features extracted from the Rec branch. These are basic elements of the attention mechanism and are used to calculate the relevance of each branch.

First, the feature map is flattened. Then, the similarity between Q and K is estimated and the relevant information from the Rec branch is embedded. The relevance $R_{q,k}$ between the SR branch and the Rec branch is defined in Eq.(1).

$$R_{q,k} = \{\frac{q_i}{||q_i||}, \frac{k_j}{||k_j||}\}, \tag{1}$$

where q_i and k_j are the patches of Q and K, respectively. Next, the Transfer attention is calculated from the relevance $R_{q,k}$ and value (V). This attention aims to transfer the structural features of the text from the Rec branch to the SR branch. In order to transfer the features of the most relevant positions of the Rec branch, a transfer attention map T is calculated from the relevance $R_{q,k}$. Each position t_i of T is defined in Eq.(2).

$$t_i = \underset{j}{\text{argmax}}(R_{i,j}) \qquad i \in [1, \frac{h}{2} \times \frac{w}{2}], \tag{2}$$

where h and w are the height and width of the input image, respectively. The T is used to represent the position of the feature map extracted from the Rec branch that is highly relevant to the i-th position of the feature map extracted from the SR branch. Then, the denoised feature C is obtained from V using t_i as an index.

Finally, the features extracted from the two branches are fused. First, for each position of C as well as T, a soft attention map S is represented, which is a highly relevant position. Each position s_i of S is $max_j(r_{i,j})$. The final output F_t of the Transformer module is defined in Eq. (3).

$$F_t = Q \oplus Conv_{out}(Z) \otimes S, \tag{3}$$

where $\oplus$ is the element-wise summation, $\otimes$ is the element-wise multiplication, and Z is the concatenation of C and Q. The output F_t is sent to the SR branch to enable high-quality SR.

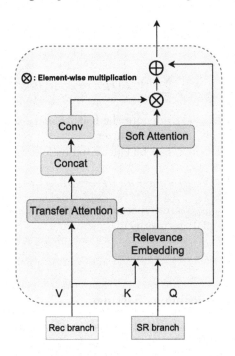

Fig. 3. The architecture of the Transformer module. By fusing the features of the Rec branch with those of the SR branch, it utilizes the properties of the Rec branch, such as noise reduction capability and text structure information.

3.2 Learning with Three Loss Function

For text-specific learning, the proposed method is used three loss functions: SR loss L_{SR}, reconstruction loss L_{Rec}, and feature-driven loss L_{Feat}. Following [7], we adopt a loss that combines Gradient Profile Prior (GPP) [28] loss and L2 loss for L_{SR} and L_{Rec}. Using GPP makes output images sharpen the boundaries of the characters. L_{SR} and L_{Rec} are defined in Eq.(4) and Ep.(5).

$$L_{SR} = \alpha \cdot L2(I_{HR}, I_{SR}) + \beta \cdot \mathbb{E}_x ||\nabla I_{HR}(x) - \nabla I_{SR}(x)||_1 \quad (x \in [x_0, x_1]), \quad (4)$$

$$L_{Rec} = \alpha \cdot L2(I_{HQ}, I_{Rec}) + \beta \cdot \mathbb{E}_x ||\nabla I_{HQ}(x) - \nabla I_{Rec}(x)||_1 \quad (x \in [x_0, x_1]), \quad (5)$$

where $L2$ denotes L2 loss, ∇ denotes the gradient fields, I_{HR} denotes the high-resolution images, and I_{HQ} denotes the LR high-quality images. The α and β denote the hyperparameters. In accordance with the experimental results of [7], α and β are 1 and 0.0001, respectively, in this work.

In [29], the pre-trained object detector is used to compute the loss function for object detection-specific SR. Inspired by this work, we design a feature-driven loss L_{Feat} using a text recognizer. The SR image I_{SR} output from the network and the high-resolution image I_{HR} of the label are input to the pre-trained text recognizer, respectively. Then, the MSE of each feature map output from the text recognizer is calculated. In this work, CRNN [19] is used as a text recognizer. L_{Feat} is defined as given in Eq. (6).

$$L_{Feat} = \frac{1}{N} \sum_{i=1}^{N} ||F_{ext}(I_{HR}) - F_{ext}(I_{SR})||^2 \tag{6}$$

where $F_{ext}(I_{HR})$ and $F_{ext}(I_{SR})$ denote the feature maps obtained by inputting I_{HR} and I_{SR}, respectively, to the recognizer.

The overall loss function L can be calculated as follows:

$$L = \lambda_{SR} L_{SR} + \lambda_{Rec} L_{Rec} + \lambda_{Feat} L_{Feat} \tag{7}$$

where λ_{SR}, λ_{Rec}, and λ_{Feat} are hyperparameters to balance three loss. These have a maximum value of 1, respectively.

4 Experiment

4.1 Datasets and Experiment Settings

TextZoom [7] is used for training and evaluation of the proposed method. This dataset contains 17,367 LR-HR paired images as training data and 4,373 LR-HR paired images as test data. These are collected by lens zooming of the camera in real scenes. The test data is divided into three difficulty levels: easy, medium, and hard. The proposed framework is implemented using the Pytorch library. The PC used is equipped with an Intel Core i7-9700k CPU and two NVIDIA GeForce RTX2070 super GPUs in parallel. In the experiments, the size of all input images is 16×64 and the size of output images is 32×128. The optimizer for training is Adam, the learning rate is 2e-4, and the batch size is 64. We set λ_{SR} as 0.5, λ_{Rec} as 0.5, and λ_{Feat} as 0.0001. Ten groups of Resblocks are used in the network. For evaluation, the accuracy of text recognizers such as ASTER [21], MORAN [30], and CRNN [19] on the generated images are used. In addition, Peak Signal-to-Noise Ratio (PSNR) and Structural Similarity Index Measure (SSIM) are used as supplementary evaluation metrics.

4.2 Experimental Results

The performance of the proposed method is compared with other representative SR models to prove its superiority. Table 1 shows the comparison of the recognition accuracy on the generated images. Table 2 shows the comparison of image quality. The compared methods include SRCNN [6], SRResNet [8], EDSR [27], LapSRN [31], TSRN [7], TSRGAN [26], and TBSRN [25]. Here, methods other than TSRN, TSRGAN, and TBSRN are not SR models designed for text images. In Table 1, TRSRT-EDSR and TRSRT-BLSTM indicate that the backbones are Resblock of EDSR [27] and Resblock with BLSTM [9] added, respectively. The bold fonts mean the highest value in the column.

The experimental results show that multi-task learning of SR and reconstruction can improve the performance of SR. For example, TRSRT-EDSR improves the accuracy of ASTER by 4% on average compared to the original EDSR [27]. In particular, there is a significant improvement in the performance of levels medium and hard, indicating that the model is successful in learning to focus on the text. Furthermore, the result shows that TRSRT-BLSTM has the first or second highest accuracy in each condition. This means the same or better performance than the state-of-the-art text-specific SR models TSRN [7], TBSRN [25], and TSRGAN [26]. Using BLSTM [9] can train sequential information in the vertical and horizontal directions like TSRN [7]. Therefore, TRSRT-BLSTM achieves high performance due to sequential information of BLSTM and the improved noise removal capability of multitask learning.

Table 1. Comparison of recognition accuracy between the proposed method and existing methods on the TextZoom dataset [7].

Method	ASTER [21]			MORAN [30]			CRNN [19]		
	Easy	Medium	Hard	Easy	Medium	Hard	Easy	Medium	Hard
BICUBIC	64.7%	42.4%	31.2%	60.0%	37.9%	30.8%	36.4%	21.1%	21.1%
SRCNN [6]	69.4%	43.4%	32.2%	63.2%	39.0%	30.2%	38.7%	21.6%	20.9%
SRResNet [8]	69.4%	50.5%	35.7%	60.9%	42.9%	32.6%	39.7%	27.6%	22.7%
EDSR [27]	72.3%	48.6%	34.3%	63.6%	45.4%	32.2%	42.7%	29.3%	24.1%
LapSRN [31]	71.5%	48.6%	35.2%	64.6%	44.0%	32.2%	46.1%	27.9%	23.6%
TSRN [7]	75.1%	56.3%	40.1%	70.1%	53.3%	37.9%	52.5%	38.2%	31.4%
TSRGAN [26]	75.7%	57.3%	40.9%	72.0%	54.6%	39.3%	56.2%	42.5%	32.8%
TBSRN [25]	**75.6%**	**59.5%**	41.7%	**74.1%**	57.0%	**40.8%**	**59.6%**	**47.1%**	35.2%
TRSRT-EDSR	72.1%	55.6%	39.5%	69.8%	54.3%	37.9%	51.7%	39.6%	31.2%
TRSRT-BLSTM	74.8%	**59.5%**	**42.0%**	72.5%	**57.2%**	40.2%	57.3%	45.6%	**35.4%**

Table 2. Comparison of image quality between the proposed method and existing methods on the TextZoom dataset.

Method	PSNR			SSIM		
	Easy	Medium	Hard	Easy	Medium	Hard
BICUBIC	22.35	18.98	19.39	0.7884	0.6254	0.6592
SRCNN [6]	23.13	**19.57**	19.56	0.8152	0.6425	0.6833
SRResNet [8]	20.65	18.90	19.53	0.8176	0.6324	0.7060
EDSR [27]	24.26	18.63	19.14	0.8633	0.6440	0.7108
LapSRN [31]	24.26	18.63	19.14	0.8633	0.6440	0.7108
TSRN [7]	**25.07**	18.86	19.71	**0.8897**	0.6676	0.7302
TSRGAN [26]	24.22	19.17	19.99	0.8791	0.6770	0.7420
TBSRN [25]	23.82	19.17	19.68	0.8660	0.6533	**0.7490**
TRSRT-EDSR	23.62	18.90	**20.07**	0.8545	0.6644	0.7349
TRSRT-BLSTM	23.47	19.10	19.94	0.8664	**0.6796**	0.7456

The results in Table 2 show that the PSNR and SSIM values are not proportional to the recognition accuracy in Table 1. This is because these metrics consider all pixels in the image. Therefore PSNR and SSIM are affected by the quality of the background. Since the proposed method is not trained to improve the quality of the background, PSNR and SSIN are inevitably not high. The results show that they are not important metrics for scene text image SR. We visualize several examples of SR images using the TRSRT-BLSTM and the prediction result by the ASTER [21] in Fig. 4.

Fig. 4. Examples of SR images generated using TRSRT and results of prediction by ASTER [21] in TextZoom [7]. The red text is the result of misrecognition.

5   Conclusions

In this paper, we propose TRSRT, which is a text-specific multi-task learning model for SR and reconstruction. Through the transformer, the TRSRT utilizes the properties of the reconstruction branch, such as denoising capability and structural information of the text, to the SR branch. The proposed method attempts to obtain a text-specific model by learning with three loss functions including feature-driven loss using a text recognizer. Experimental results on the TextZoom [7] dataset show that the proposed method significantly improves the performance of the base method and achieves state-of-the-art performance to the best of our knowledge. This result proves the superiority of multi-task learning of reconstruction and SR in the scene text image SR task.

Acknowledgements. This study is supported by JSPS/JAPAN KAKENHI (Grants-in-Aid for Scientific Research) #JP20K11955.

References

1. Qadri, M.T., Asif, M.: Automatic number plate recognition system for vehicle identification using optical character recognition. In: 2009 International Conference on Education Technology and Computer, pp. 335–338 (2009). https://doi.org/10.1109/ICETC.2009.54
2. Tian, Z., Huang, W., He, T., He, P., Qiao, Yu.: Detecting text in natural image with connectionist text proposal network. In: Leibe, B., Matas, J., Sebe, N., Welling, M. (eds.) ECCV 2016. LNCS, vol. 9912, pp. 56–72. Springer, Cham (2016). https://doi.org/10.1007/978-3-319-46484-8_4
3. Wang, Z., et al.: CAMP: cross-modal adaptive message passing for text-image retrieval. In: ICCV, pp. 5763–5772 (2019). https://doi.org/10.1109/ICCV.2019.005
4. Dong, S., Zhu, X., Deng, Y., Loy, C.C., Qiao, Y.: Boosting optical character recognition: a super-resolution approach, arXiv preprint arXiv:1506.02211 (2015)
5. Tran, H.T.M., Ho-Phuoc, T.: Deep laplacian pyramid network for text images super-resolution. In: RIVF, pp. 1–6 (2019)
6. Dong, C., Loy, C.C., He, K., Tang, X.: Learning a deep convolutional network for image super-resolution. In: Fleet, D., Pajdla, T., Schiele, B., Tuytelaars, T. (eds.) ECCV 2014. LNCS, vol. 8692, pp. 184–199. Springer, Cham (2014). https://doi.org/10.1007/978-3-319-10593-2_13
7. Wang, W., et al.: Scene text image super-resolution in the wild. In: ECCV (2020)
8. Ledig, C., et al.: Photorealistic single image super-resolution using a generative adversarial network. In: CVPR, pp. 4681–4690 (2017)
9. Graves, A., Schmidhuber, J.: Framewise phoneme classification with bidirectional LSTM and other neural network architectures. Neural Netw. **18**(5–6), 602–610 (2005)
10. Collobert, R., Weston, J.: A unified architecture for natural language processing: deep neural networks with multitask learning. In: Proceedings of the 25th International Conference on Machine Learning, pp. 160–167 (2008)
11. Liu, S., Johns, E., Davison, A.J.: End-to-end multi-task learning with attention. In: Proceedings of the IEEE/CVF Conference on Computer Vision and Pattern Recognition, pp. 1871–1880 (2019)
12. Kim, S., Hori, T., Watanabe, S.: Joint CTC-attention based end-to-end speech recognition using multi-task learning. In: 2017 IEEE International Conference on Acoustics, Speech and Signal Processing (ICASSP), pp. 4835–4839. IEEE (2017)
13. Rad, M.S., et al.: Benefiting from multitask learning to improve single image super-resolution. Neurocomputing **398**, 304–313 (2020). https://doi.org/10.1016/j.neucom.2019.07.107
14. Urazoe, K., Kuroki, N., Kato, Y., Ohtani, S., Hirose, T., Numa, M.: Multi-category image super-resolution with convolutional neural network and multi-task learning. IEICE Trans. Inf. Syst. **E104**.D(1), 183–193: Released January 01, 2021, Online ISSN 1745–1361. Print ISSN 0916–8532 (2021). https://doi.org/10.1587/transinf.2020EDP7054
15. Feng, C.-M., Yan, Y., Fu, H., Chen, L., Xu, Y.: Task transformer network for joint MRI reconstruction and super-resolution. In: de Bruijne, M., Cattin, P.C., Cotin,

S., Padoy, N., Speidel, S., Zheng, Y., Essert, C. (eds.) MICCAI 2021. LNCS, vol. 12906, pp. 307–317. Springer, Cham (2021). https://doi.org/10.1007/978-3-030-87231-1_30

16. Wang, K., Babenko, B., Belongie, S.: End-to-end scene text recognition. In: 2011 International Conference on Computer Vision, pp. 1457–1464. IEEE (2011)

17. He, P., Huang, W., Qiao, Y., Loy, C.C., Tang, X.: Reading scene text in deep convolutional sequences. arXiv preprint arXiv:1506.04395 (2015)

18. Jaderberg, M., Vedaldi, A., Zisserman, A.: Deep features for text spotting. In: Fleet, D., Pajdla, T., Schiele, B., Tuytelaars, T. (eds.) ECCV 2014. LNCS, vol. 8692, pp. 512–528. Springer, Cham (2014). https://doi.org/10.1007/978-3-319-10593-2_34

19. Shi, B., Bai, X., Yao, C.: An end-to-end trainable neural network for image-based sequence recognition and its application to scene text recognition. IEEE Trans. Pattern Anal. Mach. Intell. **39**(11), 2298–2304 (2016)

20. Graves, A., Fernandez, S., Gomez, F., Schmidhuber, J.: Connectionist temporal classification: labelling unsegmented sequence data with recurrent neural networks. In: ICML, pp. 369–376 (2006)

21. Shi, B., Yang, M., Wang, X., Lyu, P., Yao, C., Bai, X.: Aster: an attentional scene text recognizer with flexible rectification. IEEE Trans. Pattern Anal. Mach. Intell. **41**(9), 2035–2048 (2018)

22. Jaderberg, M., Simonyan, K., Zisserman, A., et al.: Spatial transformer networks. In: NeurIPS, pp. 2017–2025 (2015)

23. Li, H., Wang, P., Shen, C., Zhang, G.: Show, attend and read: a simple and strong baseline for irregular text recognition. In: AAAI, vol. 33, pp. 8610–8617 (2019)

24. Yang, L., Wang, P., Li, H., Li, Z., Zhang, Y.: A holistic representation guided attention network for scene text recognition. Neurocomputing **414**, 67–75 (2020)

25. Chen, J., Li, B., Xue, X.: Scene text telescope: text-focused scene image super-resolution. In: IEEE/CVF Conference on Computer Vision and Pattern Recognition (CVPR) 2021, pp. 12021–12030 (2021). https://doi.org/10.1109/CVPR46437.2021.01185

26. Fang, C., Zhu, Y., Liao, L., Ling, X.: TSRGAN: real-world text image super-resolution based on adversarial learning and triplet attention. Neurocomputing **455**, 88–96 (2021). https://doi.org/10.1016/j.neucom.2021.05.060. ISSN 0925-2312

27. Lim, B., Son, S., Kim, H., Nah, S., Lee, K.M.: Enhanced deep residual networks for single image super-resolution. In: CVPR (2017)

28. Sun, J., Sun, J., Xu, Z., Shum, H.: Gradient profile prior and its applications in image super-resolution and enhancement. In: TIP (2011)

29. Wang, B., Lu, T., Zhang, Y.: Feature-driven super-resolution for object detection. In: 2020 5th International Conference on Control, Robotics and Cybernetics (CRC), pp. 211–215 (2020). https://doi.org/10.1109/CRC51253.2020.9253468

30. Luo, C., Jin, L., Sun, Z., Moran: a multi-object rectified attention network for scene text recognition. Pattern Recogn., 109–118 (2019)

31. Lai, W., Huang, J., Ahuja, N., Yang, M.: Deep laplacian pyramid networks for fast and accurate super-resolution. In: CVPR (2017)

Question Difficulty Estimation with Directional Modality Association in Video Question Answering

Bong-Min Kim and Seong-Bae Park[✉]

Department of Computer Science and Engineering, Kyung Hee University,
Yongin 17104, South Korea
{klbm126,sbpark71}@khu.ac.kr

Abstract. The questions in question-answering (QA) tasks have a different level of difficulty. Thus, a number of methods have been proposed to estimate a difficulty level of the questions. However, the existing methods estimate the difficulty based only on text information, and thus loose other modalities if a QA task is intrinsically multi-modal. To solve this problem, this paper proposes a novel question difficulty estimator for multi-modal QAs. The proposed estimator is designed to solve the question difficulty estimation for video QA but is not limited to. That is, it is capable of managing both a text and a video as a sequence of images. In addition, it models the directional influence of one modality to the other modality with Directional Modality Association Transformer (DiMAT). Inspired by the transformer, the directional influence in DiMAT is expressed through a directional attention layer and a feed-forward network layer. Then, the representations for the directional influence are used together with the representations of each modality to determine the difficulty of the questions. The experiments on two benchmark video QA data sets show that the proposed question estimator outperforms the SOTA modality interaction models, which proves the effectiveness of the proposed model.

Keywords: Video question answering · Question difficulty estimation · Directional modality association · Transformer · Multi-modality

1 Introduction

The questions in most question-answering (QA) tasks have different levels of difficulty. Some questions are easy to answer, but some are difficult. The easy questions can be answered through a simple inference, but the difficult questions require a complex process of inference. Therefore, identifying the difficulty level of a question is, in general, the first step to answer the question [21].

The previous studies have proposed a number of methods to determine question difficulty in many QA tasks. Desai et al. and Huang et al. respectively tried to determine

---

This paper received the best Student Award, entitled with registration fee paid by ACM SIGAI as co-sponsored of the IEA/AIE2022.

H. Fujita et al. (Eds.): IEA/AIE 2022, LNAI 13343, pp. 287–299, 2022.
https://doi.org/10.1007/978-3-031-08530-7_24

the difficulty of questions in reading comprehension [6,9], and Qiu et al. predicted the question difficulty in medical examinations [17]. All these studies manage only text information, since questions, answers, and clues for the questions are all given in a text form. However, some QAs are multi-modal intrinsically. For instance, in visual QA, an image is given as a reference material while a question and its answer are texts. Similarly, in video QA, a sequence of images are given as a reference for a text question. Thus, it is of importance to process the multi-modality in such QAs. Since video QA can be understood as an extension of visual QA, this paper focuses only on predicting question difficulty in video QA.

The previous studies on video QA expressed each modality as a vector independently, and then composed the modality vectors into a single vector to represent the association among the modalities. For instance, Lei et al. composed text information and image information with the hadamard product [15], while Khan et al. first concatenated all modality vectors and then composed them with the transformer encoder [10]. On the other hand, Kim et al. composed the representations of modalities with a LSTM network after generating the modality representations through the heterogeneous reasoning network (HAM) [11]. However, the associations among modalities are not represented strongly enough by HAM, since they are expressed implicitly through a concatenation of modality vectors. Therefore, for more accurate estimation of question difficulty, each modality should not only be represented for its own idiosyncratic characteristic, but also should express its association to other modalities explicitly.

This paper proposes a novel question-difficulty estimator which considers both text and video information for video QA. Since video QA provides a reference video for every question, the proposed estimator encodes individual modality as a vector representation according to its own characteristic. That is, the proposed estimator has two types of modality encoders. One is a video encoder which expresses a reference video to a vector, and the other is text encoder for representing a text information such as a question and the subtitle of the reference video. In addition to them, a directional association from one modality to the other modality is expressed through the Directional Modality Association Transformer (DiMAT). That is, DiMAT generates a directional association vector of each modality to convey the directional influence from a video to a text and vice versa. The difficulty of a question is then estimated with the concatenation of the modalities and the directional associations.

The proposed estimator is validated with two benchmark video QA data sets of DramaQA [3] and TVQA [15]. While DramaQA provides a manually-tagged difficulty level for every question, TVQA does not have difficulty information. Thus, we tagged a difficulty level to the questions of TVQA using the base model of TVQA. According to the experimental results, the performance of the question-difficulty estimation is improved by adopting a video modality as well as a text modality. In addition, it is also shown that DiMAT helps the performance of question-difficulty estimation improve. As a result, the proposed estimator with DiMAT outperforms the current the state-of-the-art baseline models.

The contributions of this paper are two folds. One is that this is the first attempt, at least to our knowledge, to adopt multiple modalities for question-difficulty estimation. Some QA tasks are intrinsically multi-modal. Thus, we have shown that it is of importance to consider the multi-modality even in question difficulty estimation for the QA tasks. The other is that we have proposed an effective way to express the directional influence between a text modality and a video modality. As a result, the proposed difficulty estimator gets able to represent both the information from the modalities and their interactions, which leads to a performance improvement.

The rest of this paper is organized as follows. Section 2 reviews the related work of question difficulty estimation and Sect. 3 explains how the proposed difficulty estimator works. Then, Sect. 4 proves the effectiveness of the proposed estimator by performing the experiments on two video QA data sets. Finally, Sect. 5 draws the conclusions.

2   Related Work

Question difficulty estimation has been studied for various QA-related tasks [6, 8, 9, 14, 17]. The previous studies have developed their own difficulty estimator by considering the characteristics of a target QA task. For instance, Huang et al. studied question difficulty prediction in reading comprehension [9] and Liu et al. estimated it for community QA services [14]. They all determined the difficulty of a question with the question itself and a set of candidate answers. Ha et al. estimated a question difficulty in medical examinations [8]. Since the problems in the medical examinations are multiple choice ones, they did not distinguish medical examinations from reading comprehension. On the other hand, Qiu et al. considered a special type in which every question is accompanied by a set of question-related documents as well as candidate answers [17]. Thus, they predicted two kinds of difficulties of (i) searching relevant documents for a question and (ii) confusion degree among candidate answers. After that, they presented the weighted sum of the difficulties as a final difficulty.

The proposed question difficulty estimator manages two modality types: a text and a video which is a sequence of images. Since a texts and an image are basic materials to deliver human information, there have been a great number of studies for image processing and text processing. VGG [22] and ResNet [7] achieve good performances in image processing by allowing a deep neural structure. Thus, they are often used as a base model for image processing. Especially, they work well as a feature extractor in object detection [19, 25, 26]. In text processing, transformer-based networks such as BERT [4] and RoBERTa [16] overwhelm other methods due to their high performance in many natural language processing (NLP) tasks. When multiple modalities are considered, the composition of multi-modal information is one of the key tasks. The hadamard product was used for the composition at the earlier studies [15], but it looses the relatedness information among individual modalities. To solve this problem, Yu et al. adopted co-attention [24] and Khan et al. used the transformer with a special token [10]. On the other hand, Kim et al. proposed the heterogeneous reasoning network for expressing the relatedness among the modalities [11]. However, the information for individual modalities is not kept well even in this network.

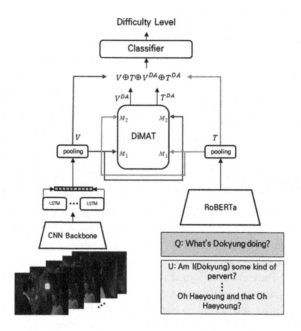

Fig. 1. The overall structure of the proposed question-difficulty estimator.

3 Question Difficulty Estimator for Multi-modality

3.1 Overall Structure

The question-difficulty estimation for video QA is a task to determine a difficulty level of a question. The task assumes that a reference video, $V = \langle v_1, \dots, v_n \rangle$ and its subtitle $U = \{u_1, \dots, u_m\}$ are given for every question $Q = \{q_1, \dots, q_l\}$. That is, V consists of n sampled images, and U and Q have m and l words, respectively. It is common in video QAs that a separate subtitle U is given together with V, a set of images. This is because the subtitle can provide some information about the entities in the video such as characters or objects. The difficulty estimation of a question is then to classify Q into an optimal difficulty level $d^* \in D$ using all Q, U, and V, where $D = \{d_1, \dots, d_k\}$ is a set of all possible difficulty levels. The answer for a question is usually given in video QAs, but it is discarded because question difficulty should be determined before inferring an answer.

Figure 1 shows the overall structure of the proposed question difficulty estimator. This estimator consists of a video modality network, a text modality network, a directional modality association transformer (DiMAT), and a classifier to determine the question difficulty. The video modality network encodes V into a vector representation, $\mathbf{V} \in R^h$, and the text modality network represents text information of Q and U as another vector, $\mathbf{T} \in R^h$. DiMAT models a directional influence of one modality to the other modality. Thus, it generates two new representations of $\mathbf{V}^{DA}$ and $\mathbf{T}^{DA}$, where $\mathbf{V}^{DA} \in R^h$ is a video representation attended by the text modality and $\mathbf{T}^{DA} \in R^h$ is a text representation attended by the video modality. Then, $\mathbf{V}$, $\mathbf{T}$, $\mathbf{V}^{DA}$, and $\mathbf{T}^{DA}$ are all concatenated into $\mathbf{I}$. That is,

$$\mathbf{I} = \langle \mathbf{V} \oplus \mathbf{T} \oplus \mathbf{V}^{DA} \oplus \mathbf{T}^{DA} \rangle,$$

where $\oplus$ implies vector concatenation. After that, the classifier, a simple MLP, takes $\mathbf{I}$ as its input and determines the difficulty level of Q from $\mathbf{I}$. That is, the difficulty level, d^* is determined by

$$d^* = softmax(\mathbf{W}_d \mathbf{I}^\mathsf{T} + b_d),$$

where $\mathbf{W}_d \in R^{k \times 4h}$ and b_d are trainable parameters of the classifier and k is the number of difficulty levels.

3.2 Video Modality Network

The aim of the video modality network is to represent the reference video V as a h-dimensional vector $\mathbf{V} \in R^h$. Many previous studies showed that a hybrid of the convolutional neural network (CNN) and the long short-term memory (LSTM) is effective in extracting temporal and spatial features of sequential data [1,20]. Especially, Cai et al. showed that such combination is still effective in video processing [2]. Inspired by these studies, the video modality network combines a CNN backbone network with a bi-directional LSTM. In this paper, ResNet [7] trained with ImageNet [5] is used as a CNN backbone.

The CNN backbone extracts spatial features, V_{emb} from V.

$$V_{emb} = ResNet(V).$$

Then, the bi-directional LSTM adds temporal information to V_{emb}. That is, it computes two directional representations of $\hbar_{v1}$ and $\hbar_{v2}$ by

$$\hbar_{v1} = \overrightarrow{LSTM}(V_{emb}) \in R^{\frac{h}{2}},$$
$$\hbar_{v2} = \overleftarrow{LSTM}(V_{emb}) \in R^{\frac{h}{2}}.$$

The final video representation, $\mathbf{V}$, is then obtained through a pooling layer of which input is the concatenation of $\hbar_{v1}$ and $\hbar_{v2}$. That is,

$$\mathbf{V} = \tanh(\mathbf{W}_v(\hbar_{v1} \oplus \hbar_{v2})^\mathsf{T} + b_v), \tag{1}$$

where $\mathbf{W}_v \in R^{h \times h}$ and b_v are trainable parameters.

3.3 Text Modality Network

RoBERTa [16] is adopted for the text modality network since it shows higher performance than BERT in many QA tasks such as SQuAD [18], GLUE [23], and RACE [12]. The concatenation of Q and U separated by a special token, [SEP], is used as

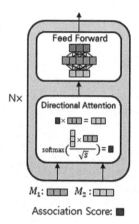

Fig. 2. The structure of the directional modality association transformer.

an input for the text modality network. RoBERTa summarizes the input text into $\hbar_t$, a vector for the zeroth token, [CLS], by

$$\hbar_t = RoBERTa(Q, U)[0].$$

Then, as in the video modality network, the final text representation, $\mathbf{T}$, is obtained by pooling $\hbar_t$. That is,

$$\mathbf{T} = \tanh(\mathbf{W}_l \hbar_t^{\mathsf{T}} + b_l)), \tag{2}$$

where $\mathbf{W}_l \in R^{h \times h}$ and b_l are also trainable parameter matrices.

3.4   Directional Modality Association Transformer

The directional modality association transformer (DiMAT) computes a directional influence of one modality to the other modality. It has a similar architecture with the transformer as shown in Fig. 2. That is, it consists of N blocks of a directional attention layer and a feed-forward network layer. Assume that the representations of two modalities, M_1 and M_2, are given. The directional attention layer represents M_1^a, a new presentation of M_1 influenced by M_2, where M_1^a is

$$M_1^a = softmax\left(\frac{M_2^{\mathsf{T}} M_1}{\sqrt{s}}\right) M_1,$$

where s is a scaling factor. The role of the feed-forward network layer is equivalent to that of the transformer. The final $DiMAT(M_1, M_2)$ then produces a new representation of M_1 attended by M_2 by passing M_1^a through the feed-forward network. That is,

$$DiMAT(M_1, M_2) = FFN(M_1^a)$$
$$= FFN\left(softmax\left(\frac{M_2^{\mathsf{T}} M_1}{\sqrt{s}}\right) M_1\right).$$

Table 1. The statistics on the data sets used for the experiments.

Data sets	TVQA	DramaQA
No. of training instances	40,000	21,113
No. of validation instances	5,000	4,385
No. of test instances	5,000	4,463
No. of difficulty levels	two	two (memory complexity) four (logic complexity)

Note that the residual connection is excluded from DiMAT since it allows one modality to affect the other modality excessively.

Similarly, $DiMAT(M_2, M_1)$ is

$$DiMAT(M_2, M_1) = FFN(M_2^a),$$

where M_2^a is

$$M_2^a = softmax\left(\frac{M_1^{\mathsf{T}} M_2}{\sqrt{s}}\right) M_2.$$

A video QA has a text modality and a video modality. Thus, DiMAT for video QA computes the video representation influenced by a text and the text representation influenced by a video which are computed respectively as

$$\mathbf{V}^{DA} = DiMAT(\mathbf{V}, \mathbf{T}),$$
$$\mathbf{T}^{DA} = DiMAT(\mathbf{T}, \mathbf{V}),$$

where $\mathbf{V}$ and $\mathbf{T}$ are computed by Eq. (1) and (2), respectively.

Many previous studies have shown that the use of modality interactions is effective in answering a question at multi-modal QAs. For this, Kim et al. have proposed the heterogeneous attention mechanism (HAM) [11]. HAM has only an attention layer. As a result, it has less potential to obtain a rich representation of modalities and to resist against noises than DiMAT. On the other hand, Yu et al. performed the co-attention for visual QA by making the transformer encoder process a query and the decoder manage an image [24]. Thus, the attention in their method is made only from the query to the image, while the influences from a text to a video and that from a video to a text are all considered in DiMAT.

4   Experiments

4.1   Data Sets

Two benchmark data sets are used for the validation of the proposed question-difficulty estimator. The first one is TVQA data set [15]. It is the most widely used video QA data set based on six TV shows. It consists of QA pairs from captioned video clips, but it does not contain any difficulty information. Thus, the difficulty level for every question is tagged as follows. The base model of TVQA [15] is adopted as a labeler of the question difficulty. This model is not perfect in that its accuracy of predicting an answer is just 68.48%. Therefore, the questions to which the model gives a wrong answer are regarded as difficult ones, while the questions for which the model makes a correct answer are easy ones. As a result, the difficulty level for this data set is binary. Original TVQA has 137,292 questions, but 87,292 among them are used to train the base model. The remaining 50,000 questions are used for evaluation of the proposed model, where 40,000 questions are used for training, 5,000 are for validation, and the remaining 5,000 are for test.

The other data set is DramaQA data set [3]. It is a video QA data set based on a Korean TV show, "Another Miss Oh". Every QA pair in this data set is followed by a video clip, some visual metadata, and the coreference-resolved script, but only the video clip is used in this paper for the consistency with TVQA data set. The difficulty of each question is labeled by two criteria of memory complexity and logic complexity. The memory complexity has two levels according to the memory length required to answer a question. On the other hand, the logic complexity measures the amount of inference level and auxiliary information needed to answer a question. It has four levels. A simple statistics on both data sets is given in Table 1.

The dimension of representation vectors, h, is set to be 768 as in RoBERTa, and the number of blocks in DiMAT, i.e. N in Fig. 2, is four. All models are optimized with AdamW [13] with a learning rate $2e - 5$. Training batch size and the maximum input length are 64 and 256 respectively.

4.2   Baselines

The proposed question difficulty estimator is compared empirically with two baselines. These two baselines are the methods for combining multiple modalities. Thus, they can replace DiMAT in the proposed estimator. The first baseline is the multi-modal fusion transformer (MMFT) [10] that was proposed to solve video QAs, not to estimate question difficulty estimation in video QA. It uses a simple method to represent the combination of multiple modalities. It simply concatenates all modalities and then computes a relative importance of each input feature with a self-attention. For this purpose, it adopts the transformer network since the relative importances of all modalities can be obtained in a space to which the transformer network maps the modalties.

The other baseline is the heterogeneous attention mechanism (HAM) [11] that was proposed also to solve video QA. HAM aims at representing both intra- and inter-modality interactions among modalities assuming that there are three modalities of a question, a subtitle, and a video. The intra-modality interaction is first represented by a self-attention at each modality, and then the inter-modality interaction is obtained by an attention between different intra-modality interactions. It shares a similar goal to DiMAT in that it tries to represent inter-interactions among the modalities. Thus, it can be used in the proposed model instead of DiMAT and become a baseline. The main difference of DiMAT from HAM is that the inter-modality interaction is expressed with multiple layers of the transfomer network in DiMAT while all interactions are represented with the dot-product attention in HAM.

Table 2. The F1-measures of the proposed model and its baselines.

Used Model	DramaQA-Memory			DramaQA-Logic			TVQA		
	R	P	F1	R	P	F1	R	P	F1
T (RoBERTa)	85.32	83.93	84.61	77.27	81.32	78.82	52.91	59.32	55.92
V ⊕ **T**	95.86	95.55	95.67	84.91	86.33	85.44	52.65	60.74	56.41
V ⊕ **T** ⊕ **V**^{DA} ⊕ **T**^{DA}	**96.46**	**96.82**	**96.64***†	**85.93**	**87.28**	**86.55***†	**54.01**	61.75	**57.62***
HAM	96.45	95.41	95.93	48.69	32.28	38.75	50.00	34.52	40.84
MMFT	96.18	95.31	95.74	84.92	86.62	85.61	51.39	**62.93**	56.51

Table 3. The accuracies of the proposed model and its baselines.

Used Model	DramaQA-Memory	DramaQA-Logic	TVQA
	Acc (%)	Acc (%)	Acc (%)
T (RoBERTa)	90.69	79.43	68.63
V ⊕ **T**	97.37	85.58	68.80
V ⊕ **T** ⊕ **V**^{DA} ⊕ **T**^{DA}	**98.01***†	**86.44***	**69.75***†
HAM	97.55	67.17	69.04
MMFT	97.43	85.80	69.14

4.3 Experimental Results

Table 2 and 3 summarize the performances of the proposed question difficulty estimator according to the information used. In these tables, $\mathbf{V} \oplus \mathbf{T} \oplus \mathbf{V}^{DA} \oplus \mathbf{T}^{DA}$ is the proposed model and it achieves the best performances. Its accuracy and F1-score in DramaQA are 98.01% and 96.64 for memory complexity and 86.44% and 86.55 for logic complexity, and those in TVQA are 69.75% and 57.62, respectively. The performance of the proposed model for memory complexity is much higher than that for logic complexity. This is because memory complexity has binary difficulty classes while logic complexity has four classes. The performance in TVQA is lower than those in DramaQA even if TVQA has only two classes. The difficulty labels in TVQA are not tagged by human beings, but by the base model of TVQA. As a result, the difficulty labels are somewhat noisy and inconsistent, which leads to low performance of the proposed model in this data set.

RoBERTa trained only with **T** achieves the worst performance, which implies that text information is insufficient for accurate question difficulty prediction. On the other hand, when both text and video information are used ($\mathbf{V} \oplus \mathbf{T}$), the improvement in F1-score for DramaQA is, on average, 9.88 and that for TVQA is 1.70. The accuracy improvement is 7.17% for DramaQA and 1.12% for TVQA. These facts support the rationale for the use of both text and video information in the proposed model. The reason why '**V** (CNN-LSTM)' is excluded from the tables is that it is impossible to estimate the difficulty of a question only with **V** since a query Q is given in a text form.

The performances of $\mathbf{V} \oplus \mathbf{T} \oplus \mathbf{V}^{DA} \oplus \mathbf{T}^{DA}$ are slightly higher than those of $\mathbf{V} \oplus \mathbf{T}$, the proposed model without DiMAT, for both complexities of DramaQA and

Table 4. F1-score comparison between separation Q from U and their concatenation.

	DramaQA-Memory	DramaQA-Logic	TVQA
Concatenation	**96.64**	**86.55**	**57.62**
Separation	95.51	85.32	55.77

TVQA. The performance improvement by DiMAT is statistically significant at the 5% level ($p \leq 0.05$). Therefore, the directional associations by DiMAT help improve the performance of the proposed model. One thing to note is that the improvement for memory complexity in DramaQA is greater than that for logic complexity. This is also due to the different number of difficulty class labels.

HAM and MMFT can be used instead of DiMAT. When they replace DiMAT, the performances get worse. The average F1-scores of HAM are 67.34 for DramaQA, and 40.84 for TVQA, while those of MMFT are 90.68 and 56.51 respectively. They are about 0.92 $\sim$ 24.26 lower in DramaQA than that of the proposed model and 1.11 $\sim$ 16.78 lower in TVQA. A similar phenomenon is found for accuracy. These results prove the effectiveness of DiMAT in representing the modality associations. The symbol '*' in Tables 2 and 3 indicates that the performance difference between the proposed model and HAM is statistically significant at the 5% level, and '†' imples that the difference between the proposed model and MMFT is statistically significant at the same level. Thus, the improvement over HAM is significant, but the significance of the improvement over MMFT is confirmed only for the memory complexity of DramaQA.

Especially, HAM shows just 38.75 of F1-score for logic complexity. This is because HAM adds the information of Q to U and V, but does not use Q directly. That is, HAM uses Q to generate a composed vector representation through an attention mechanism. There is no way to provide the information within Q directly. When estimating logic complexity that requires auxiliary facts and inference, the performance is degraded because the information that can be obtained only from Q disappears. On the other hand, its F1-score for the memory complexity is comparable to DiMAT since this complexity depends mainly on the video length and the effect of Q is small.

In Fig. 1, RoBERTa processes the concatenation of a query Q and a subtitle U since both are texts. However, Q has a different characteristic from U. Thus, it seems natural to separate Q from U as an independent modality. That is, the question difficulty could be estimated with three modalities of Q, U, and V. Table 4 shows how the performance changes after separating Q from U. The F1-score of separating Q from U becomes 95.51 for memory complexity of DramaQA, 85.32 for logic complexity, and 55.77 for TVQA, while it is 96.64, 86. 55, and 57. 62 respectively when Q and U are concatenated. That is, if the modalities are separated, the performance gets worse. This performance drop can be explained as follows. Q is usually much shorter than other modalities. Therefore, if Q is separated from U, the information within Q is apt to be insufficient so that its information is often neglected at the final vector representation.

5 Conclusions

This paper has proposed a novel estimator of question difficulty in video QA. The key characteristic of video QA is that it has a multi-modal input of a text and a video. Thus, the proposed estimator encodes the video information with a CNN-backbone and the text information with RoBERTa. In addition, it models the directional association from one modality to the other modality with DiMAT. Since DiMAT generates a representation of one modality attended by the other modality, the proposed estimator gets able to predict the difficulty of a question by considering the directional influences between modalities as well as each individual modality. As a result, the proposed question-difficulty estimator determines the difficulty level of a question by considering not only a reference video and a text information such as the question itself and a subtitle, but also their directional associations. The empirical findings through the experiments on DramaQA and TVQA data sets are two folds. One is that the text information itself is insufficient for determining the difficulty of questions. The other is that it improves the performance of question difficulty estimation additionally to consider the directional modality associations as well as text and video information. These findings are proven by an experiment in which it improves the F1-score and accuracy much to add video information and modality associations to the proposed model. It is also shown that DiMAT outperforms existing modality interaction models, which implies that DiMAT is an outstanding method to represent the text and video information including their associations.

Acknowledgement. This work was supported by the National Research Foundation of Korea (NRF) grant funded by the Korea government(MSIT) (No. 2020R1A4A1018607) and by the Institute of Information and Communications Technology Planning and Evaluation (IITP) Grant funded by the Korea Government (MSIT) (Artificial Intelligence Innovation Hub) under Grant 2021-0-02068 and Institute of Information and Communications Technology Planning and Evaluation (IITP) grant funded by the Korea government (MSIT) (No. 2013-0-00109, WiseKB: Big data based self-evolving knowledge base and reasoning platform).

References

1. Akilan, T., Wu, Q.-J., Safaei, A., Huo, J., Yang, Y.: A 3D CNN-LSTM-based image-to-image foreground segmentation. IEEE Trans. Intell. Transp. Syst. **21**, 959–971 (2019)
2. Cai, J., Hu, J., Tang, X., Hung, T.-Y., Tan, Y.-P.: Deep historical long short-term memory network for action recognition. Neurocomputing **407**, 428–438 (2020)
3. Choi, S.-H., et al.: DramaQA: character-centered video story understanding with hierarchical QA. In: Proceedings of the 35th AAAI Conference on Artificial Intelligence, pp. 1166–1174 (2021)
4. Devlin, J., Chang, M.-W., Lee, K., Toutanova, K.: BERT: pre-training of deep bidirectional transformers for language understanding. In: Proceedings of the 2019 Conference of the North American Chapter of the Association for Computational Linguistics: Human Language Technologies, pp. 4171–4186 (2019)
5. Deng, J., Dong, W., Socher, R., Li, L.-J., Li, K., Fei-Fei, L.: ImageNet: a large-scale hierarchical image database. In: Proceedings of the 2009 IEEE Conference on Computer Vision and Pattern Recognition, pp. 248–255 (2009)

6. Desai, T., Moldovan, D.: Towards predicting difficulty of reading comprehension questions. In: Proceedings of the 32nd International Florida Artificial Intelligence Research Society Conference, pp. 8–13 (2019)

7. He, K., Zhang, X., Ren, S., Sun, J.: Deep residual learning for image recognition. In: Proceedings of the 2016 IEEE Conference on Computer Vision and Pattern Recognition, pp. 770–778 (2016)

8. Ha, L.-A., Yaneva, V., Baldwin, P., Mee, J.: Predicting the difficulty of multiple choice questions in a high-stakes medical exam. In: Proceedings of the 14th Workshop on Innovative Use of NLP for Building Educational Applications, pp. 11–20 (2019)

9. Huang, Z., et al.: Question difficulty prediction for reading problems in standard tests. In: Proceedings of the 31st AAAI Conference on Artificial Intelligence, pp. 1352–1359 (2017)

10. Khan, A.-U., and Mazaheri, A., Lobo, N., Shah, M.: MMFT-BERT: multimodal fusion transformer with BERT encodings for visual question answering. In: Proceedings of the 2020 Conference on Empirical Methods in Natural Language Processing, pp. 4648–4660 (2020)

11. Kim, J., Ma, M., Pham, T., Kim, K., Yoo, C.: Modality shifting attention network for multimodal video question answering. In: Proceedings of the 2020 IEEE Conference on Computer Vision and Pattern Recognition, pp. 10103–10112 (2020)

12. Lai, G., Xie, Q., Liu, H., Yang, Y., Hovy, E.: RACE: large-scale ReAding comprehension dataset from examinations. In: Proceedings of the 2017 Conference on Empirical Methods in Natural Language Processing, pp. 785–794 (2017)

13. Loshchilov, I., Hutter, F.: Decoupled weight decay regularization. In: Proceedings of the 7th International Conference on Learning Representations (2019)

14. Liu, J., Wang, Q., Lin, C.-Y., Hon, H.-W.: Question difficulty estimation in community question answering services. In: Proceedings of the 2013 Conference on Empirical Methods in Natural Language Processing, pp. 85–90 (2013)

15. Lei, J., Yu, L., Bansal, M., Berg, T.: TVQA: localized, compositional video question answering. In: Proceedings of the 2018 Conference on Empirical Methods in Natural Language Processing, pp. 1369–1379 (2018)

16. Liu, Y., et al.: RoBERTa: a robustly optimized BERT pretraining approach. arXiv preprint arXiv:1907.11692 (2019)

17. Qiu, Z., Wu, X., Fan, W.: Question difficulty prediction for multiple choice problems in medical exams. In: Proceedings of the 28th ACM International Conference on Information and Knowledge Management, pp. 139–148 (2019)

18. Rajpurkar, P., Zhang, J., Lopyrev, K., Liang, P.: SQuAD: 100,000+ questions for machine comprehension of text. In: Proceedings of the 2016 Conference on Empirical Methods in Natural Language Processing, pp. 2383–2392 (2016)

19. Ren, S., He, K., Girshick, R., Sun, J.: Faster R-CNN: towards real-time object detection with region proposal networks. In: Proceedings of the 28th International Conference on Neural Information Processing Systems, pp: 91–99 (2015)

20. Shi, B., Bai, X., Yao, C.: An End-to-end trainable neural network for image-based sequence recognition and its application to scene text recognition. IEEE Trans. Pattern Anal. Mach. Intell. 39, 2298–2304 (2017)

21. Song, H.-J., Yoon, S.-H., Park, S.-B.: Question Difficulty Estimation based on Attention Model for Question Answering. Technical report CSE-2021-1. Kyung-Hee University (2021)

22. Simonyan, K., Zisserman, A.: Very deep convolutional networks for large-scale image recognition. In: Proceedings of the 3rd International Conference on Learning Representations (2015)

23. Wang, A., et al.: SuperGLUE: a stickier benchmark for general-purpose language understanding systems. In: Proceedings of the 32nd Conference on Neural Information Processing Systems, pp. 3261–3275 (2019)

24. Yu, Z., Yu, J., Cui, Y., Tao, D., Tian, Q.: Deep Modular Co-Attention Networks for Visual Question Answering. In: Proceedings of the 2019 IEEE Conference on Computer Vision and Pattern Recognition, pp. 6281–6290 (2019)
25. Zhao, Q, et al.: M2Det: a single-shot object detector based on multi-level feature pyramid network. In: Proceedings of the 33rd AAAI Conference on Artificial Intelligence, pp. 9259–9266 (2019)
26. Zhu, X., Su, W., Lu, L., Li, B., Wang, X., Dai, J.: Deformable DETR: deformable transformers for end-to-end object detection. arXiv preprint arXiv:2010.04159 (2020)

Natural Language Processing

Improving Neural Machine Translation by Efficiently Incorporating Syntactic Templates

Phuong Nguyen[1], Tung Le[2,3], Thanh-Le Ha[4], Thai Dang[5], Khanh Tran[5], Kim Anh Nguyen[5], and Nguyen Le Minh[1(✉)]

[1] Japan Advanced Institute of Science and Technology, Nomi, Japan
{phuongnm,nguyenml}@jaist.ac.jp
[2] Faculty of Information Technology, University of Science, Ho Chi Minh, Vietnam
lttung@fit.hcmus.edu.vn
[3] Vietnam National University, Ho Chi Minh City, Vietnam
[4] Karlsruhe Institute of Technology, Karlsruhe, Germany
thanh-le.ha@partner.kit.edu
[5] Vingroup Big Data Institute, Hanoi, Vietnam
{v.thaidt4,v.khanhtv13,v.anhnk9}@vinbigdata.org

Abstract. In the success of Transformer architecture in Neural Machine Translation, integrating linguistic features into the traditional systems gains a huge interest in both research and practice. With less increase in computational cost as well as improving the quality of translation, we propose an abstract template integration model to intensify the structural information in source language from syntactic tree. Besides, the previous works have not considered the effect of the template generating mechanism, while this is an essential component of template-based translation. In this work, we investigate various template generating methods and propose two prominent abstract template generation techniques based on the POS information. Together with the strength of Transformer, our proposed approach allows to effectively incorporate and extract the linguistics features to enrich the information in encoding phase. Experiments on several benchmarks prove that our approaches achieve competitive results against the competitive baselines with less effort in training time. Furthermore, our results reflect that syntactic information is the rich fertile ground to have benefited greatly in neural machine translation. Our code is available at https://github.com/phuongnm-bkhn/multisources-trans-nmt.

Keywords: Machine translation · Syntactic template · Transformer

1 Introduction

Neural machine translation models (NMT) have been taken much attention in machine translation domain. The key idea of NMT is based on encoder-decoder

© Springer Nature Switzerland AG 2022
H. Fujita et al. (Eds.): IEA/AIE 2022, LNAI 13343, pp. 303–314, 2022.
https://doi.org/10.1007/978-3-031-08530-7_25

models which have been upgraded with Transformer models [11]. One of the promising research directions is to incorporate linguistic information into the encoder representation and guide the decoder to enhance the generation. The recent work on using target syntactic templates with NMT [15] has shown that utilizing soft template prediction could lead to large translation gains (Fig. 1). However, the authors only considered the syntactic template of target side that is generated by a pruning technique based on length of target sentence. Obviously, the performance of these approaches is based on the quality of target template prediction from the source sentence. However, extracting the general instruction of target sentences in low-resource language is a great challenge in both research and practice.

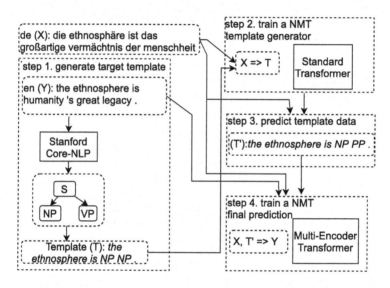

Fig. 1. Flow of NMT system using soft target template.

On the other hand, many previous works have claimed that using source tree (phrase) or tree structure in the encoder component by modifying the self-attention layers of Transformer architecture would help to improve the translation quality [2,4,6,16]. Indeed, it is more straightforward to extract and integrate the structure of source sentences into the encoder phase. In this paper, inspired from the work of [15], our work is proposed to answer some natural questions: *(1) how the different kind of templates affect to the performance - template on abstract or detailed levels? (2) what is the better between two ways of integrating syntactic information - on the source side or on the target side? (3) how to inject syntactic information into the NMT model effectively?*

In the most related works against our approach, [15] showed that soft target template is potential to guide the decoding process for improving the performance of NMT systems. While they encoded the target template as a second source representation, there are, however, not any constraints related to the POS tags in the templates (e.g. NP, VP) in decoding phase. With our observation in their approach, the generated soft templates may adversely affect the translation quality. The enhancement of this model comes from the generalization of template prediction phase. Obviously, it depends on both the performance of parser in target side and the strength of predictor. In addition, [15] need to perform an external component in order to produce the template. This becomes a 2-fold process and might suffer from error propagation.

In less modification and promising performance, we propose a direct approach to consolidate the context of source sentence via structural information in the source template instead of the target template. With our proposed model, the process of learning and integrating the syntactic structure into machine translation model is done continuously. On the other hand, we also inherit advanced technologies in the language understanding. Without any external components, our model avoids the error transmission against the previous approach in parsing and predicting the target template. In our model, the structure of source sentence comes from its natural characteristics from the syntactic parser. With our integration, our model is powerful to intensify the structure of sentence that is highly useful for translation. Obviously, it exists the corresponding structure between the source and target sentence. Therefore, our proposed approach with the intensification of source template is the promising guidance for translating phase. Especially, in the case that the target side is in low-resource languages, our proposed method is more effective and applicable than previous works utilizing the target template.

Besides the side of template, the structure of template is critical to maintain the meaning and syntactic of sentence. Therefore, to reflect the effect of template extraction into the NMT system, we conducted experiments using different kinds of templates from an abstract level containing constituent POS tags of a sentence to a detailed level containing a mixture of both POS tags and words. To prove the strength of our proposed model, this investigation is simultaneously done in both target-based and source-based approaches. Experimental results in several popular MT benchmarks showed that our approach achieves the promising results against both competitive baselines and target-based method. Especially, through our detailed comparison, it also emphasize the strength of our proposed model in low-resource language.

2   Related Works

Many works have been considered to utilize the linguistic structure representation for improving NMT, both in the encoder and decoder components. [14] indicates that a source phrase representation can be applied for boosting the performance of NMT. [13] introduces tree encoder architecture for Transformer. The works presented in [10,15] demonstrate that the use of soft template prediction can improve NMT. Besides, Template-based machine translation also typically are applied in the Semantic parsing field to deal with the complicated logic syntax [1], the various entity names problem [5], or support to generate response in a Dialog system [3]. Based on the success of previous works in this area, our work is inspired by the approach of [15].

3 Proposed Template Integration

In this section, we would like to sketch the main ideas of how to extract syntactic templates to be used in NMT architectures. In the work of [15], the templates are generated from the syntactic tree of the target sentences based on some length-based heuristic. Based on observation and assumption from the linguistic features, we also propose two other approaches to generate templates from the syntactic tree. Besides, our proposed techniques are deployed on both the source and target sides to evaluate the effect of template extraction and the strength of proposed frameworks.

3.1 Template Generating Methods

Given the syntactic tree, each POS tag is a non-terminal node, and each leaf node is a terminal node containing one tokenized word. Based on one of the below methods, some non-terminal nodes may be pruned by removing their child nodes and become terminal nodes. Finally, the template is the list of all leaf nodes of the pruned syntactic tree.

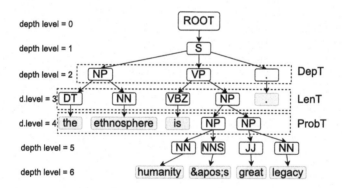

Fig. 2. Different depth levels in syntactic tree of three types of template: LENT, PROBT, DEPT.

*Length-based Template (*LENT*)* is proposed by [15], the template depends on length of input sentence: $d = \min(\max(L \times \lambda, \gamma_1), \gamma_2)$ where d is the depth level for pruning; γ_1, γ_2 is lower and upper bound depth that are extracted from parsed syntactic tree in each sentence, respectively; λ is hyper-parameter reflects the dependency between the pruning depth level with the length of sentence.

For example in Fig. 2, the sentence length (L) is 8, with $\lambda = 0.15$, $\gamma_1 = 3$ and $\gamma_2 = 6$, therefore, the pruning depth level is 3.

Probability-based Template (PROBT) is based on the average of probabilities of POS tags at each tree level to choose the best depth level for representing the template. Coming from the lack of POS consideration in previous work, we propose to utilize these information to extract the abstract template via the distribution of POS tags in languages. In particular, we analyze to obtain the probabilities of all POS tags in the training data. In our assumption, the higher probabilities the level obtains, the less noises we avoid against the rare POS tags (e.g. CVZ) in the original template.

$$d^* = \underset{\gamma_1 \le d \le \gamma_2}{\mathrm{argmax}}(\mathrm{mean}(p_{d,1}, p_{d,2} \ldots, p_{d,i}))$$

where $p_{d,i}$ is the probability that of POS tag i^{th} in the depth level d. After this step, we find the best depth (d^*) for pruning and get the soft template. For example in Fig. 2, d^* is chosen in the range from 3 to 6, and the depth level 4 is the level containing the most frequent POS tags.

Depth-based Template (DEPT) is extracted from the first depth level of *simple declarative clause* tag (i.e. "*S*"). We aim to get the highest abstract level of template for the sentence POS tags representation. The depth level for pruning (d) is fixed by formula: $d = d_S + 1$ where d_S is the first depth level of sentence POS tag. With this method, we expect that it is easier than others to model generalize structure of a natural sentence. For example in Fig. 2, $d_S = 1$, and this template is generated from the syntactic tree depth level 2.

3.2   Template Sides

Besides the template extraction techniques, we also emphasize the importance of template sides in the machine translation. Specifically, given a pair of sentences (X, Y), there are two different types of templates: templates of the source sentence (X) or template of the target sentence (Y).

Source Template. In this setting, the source template is generated from a syntactic tree based on one of the template generating methods. After that, both source sentence and the template are used to decode the target sentence.

Target Template. In this setting, similar to [15], the translation process is split into 2 phases: (1) decoding the target template from the source sentence; (2) incorporating decoded target template with the source sentence to decode the target sentence. For example, the translation of this German→English sentence pair (X, Y): "*die ethnosphäre ist das großartige vermächtnis der menschheit.*"→ "*the ethnosphere is humanity's great legacy.*" will be split into two steps: decoding $X \to T$ and then decoding $(X, T) \to Y$ where T is "*the ethnosphere is NP NP.*"

4 Model Architecture

Transformer (Baseline). Together with the recent works, we consider Transformer model [12] as a competitive baseline for the machine translation task.

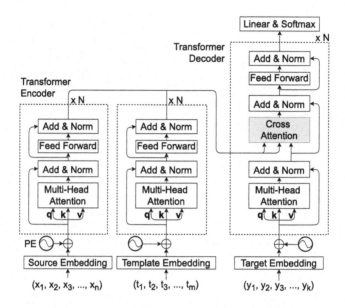

Fig. 3. Transformer model with multi encoders.

Transformer Multi Encoders (TME). For incorporating target template information, we re-implement a architecture similar to [15]). The model consists of the source encoder, template encoder, and target decoder components which are based on the Transformer architecture. The Cross-Attention learns the attention scores of both source and template, separately (Fig. 3).

$$H^{xy} = \text{Attention}(H^x; H^x; H^y) \tag{1}$$
$$H^{zy} = \text{Attention}(H^z; H^z; H^y) \tag{2}$$
$$r = \text{Sigmoid}(W_1 H^{xy} + W_2 H^{yz}) \tag{3}$$
$$H^y = r \cdot H^{xy} + (1 - r) \cdot H^{zy} \tag{4}$$

where H^{xy}, H^{zy} are the incorporating hidden states of the *source-target* and *template-target*, respectively; r is the impacting coefficient of source and template; H^y is the hidden state of the target language that contains both source and template information; Attention is the function similar to [12] that flows information from encoder to decoder.

Drop Template Mechanism. We follow the observation of [15] that the model achieves better performance when dropping the soft target template by a dropping probability (e.g. 0.5). We also randomly replaced the Eq. 4 by $H^y = H^{xy}$ as in the baseline model. It means ignoring template information and keeping original source sentence.

Source Template Concatenation (STC). Intuitively, the implicit relations between the source and the template may be useful for the translation process. Therefore, we use a simple method to concatenate the source and the template via a *[SEP]* token and then proceed with the concatenation as a normal input in our Transformer model (Fig. 4). In this way, the relationship between source and template can be learned in the self-attention mechanism of the Transformer Encoder.

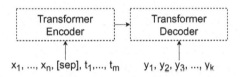

Fig. 4. Transformer model with source sentence and template concatenation via *[SEP]* token where x_i is source words, t_i is template tokens (words or POS tags), y_i is target words.

5   Experimental Results

Dataset. In order to prove the strength of our work, we conducted experiments on the four datasets: IWSLT 2014 German - English[1], IWSLT 2015 English - Vietnam, IWSLT 2017 English - French[2] and WMT 2014 English - German. The statistics of these data are shown in Table 1.

Table 1. Statistic information of NMT datasets.

Information	IWSLT			WMT
	de-en	en-vi	en-fr	en-de
# training examples	160K	133K	230K	4M
# development examples	7.3K	1.5K	1K	40K
# testing examples	6.8K	1.2K	1K	3K
# BPE operators	10K	10K	10K	40K

---

[1] Test set is merged from dev2010, dev2012, tst2010, tst2011, tst2012.

[2] Test set identify is tst2015.

Settings. To verify the performance of the proposed methods, we trained the following systems with the same settings: (1) the baseline NMT translation methods (TRANSFORMER) with 6 Self-Attention layers for the encoder and decoders; 8 heads for WMT14 dataset, and 4 heads for others; model size is 512 and hidden size is 2048; dropout is 0.3; (2) the NMT translation using both source template and target template where the template is generated from length-based (LENT) [15], our probability-based (PROBT), or our depth-based (DEPT) methods; the drop template threshold is selected in {0.5; 0.6; 0.7; 0.8; 0.9} similar to [15]. All datasets are pre-processed with the standard Moses toolkit[3]. We evaluated performance by averaging 5 latest checkpoints and compute BLEU score via SacreBleu[4] [7] on IWSLT 2017, WMT 2014 datasets and use multi-bleu script[5] on IWSLT 2014, 2015 datasets for comparable with previous published results.

Template Encoding Methods. To compare the effectiveness of TME and STC methods, we conducted experiments 5–10 using three types of generating target templates LENT, PROBT and DEPT on IWSLT 2014 dataset (Table 2). The TME method beats the STC method on all experiments of IWSLT 2014 de-en and en-de. With our observation, the reason is that TME method seems to have a *gating component* (Eq. 4) that automatically select the useful information from template via cross-attention. The STC model used a simple *"[SEP]"* token to separate source sentence and template, and this model always utilizes these features for translating process while the TME model use learn-able parameters to adjust what information should be used.

Besides, to prove the effectiveness of *gating component* and *drop template mechanism*, we also conducted an ablation experiment on IWSLT 2014 de-en dataset (Table 3). Comparing to run 2 (Table 2) with runs 11, 12 (Table 3) and run 8 (Table 2) with runs 13, 14 (Table 3), we found that the performance of the NMT system is hurt a little bit, particularly in removing the *gating component*. These results are homologous to [15] conducted experiments about the *drop template mechanism*.

Template Types. Firstly, we consider the effectiveness of three types of templates: LENT, PROBT, DEPT. Our proposed DEPT template is generated as the highest abstract level representation of a sentence. Therefore, we argue that it contains useful structure information for the encoding process, especially on the source side. The evidence for this observation is shown in setting 10 (Table 2) with a stable improvement when compared to the competitive baseline Transformer model on all datasets. With setting 4 using DEPT on the target side, the result is just slightly improved in en-de datasets and decrease in others. We found that the quality of the prediction target template in the first phase of previous approach does not actually work well because it is tremendously challenging to predict the

---

[3] https://github.com/moses-smt/mosesdecoder.

[4] https://github.com/mjpost/sacrebleu.

[5] https://github.com/moses-smt/mosesdecoder/blob/master/scripts/generic/multi-bleu.perl.

Table 2. Translation results for test sets of IWSLT 2014 German↔English, IWSLT 2015 English→Vietnam, IWSLT 2017 English→French and WMT 2014 English→German. The numbers in the pair of brackets are different values when compared to baseline model Transformer. The marked (*) result in WMT 2014 dataset refers to the run using Transformer-big setting [12] while the others use Transformer-base setting. The method ST-NMT by [15] is equal to setting 2 in our implementation.

Methods	IWSLT14		IWSLT15	IWSLT17	WMT14
	de-en	en-de	en-vi	en-fr	en-de
Previous works					
Transformer [12]	34.42	28.35	–	–	27.30
TreeTransformer [6]	35.96	29.47	–	–	28.40
BPE-dropout [8]	–	–	33.27	40.02	28.01
ST-NMT [15]	35.24	–	–	–	29.68*
Our implementation					
1. Transformer	35.93	29.63	32.20	39.37	27.27
Target side template					
2. TME +LenT	$36.07_{(+0.14)}$	$29.77_{(+0.14)}$	$31.81_{(-0.39)}$	$39.40_{(+0.03)}$	$27.07_{(-0.20)}$
3. TME +ProbT	$36.00_{(+0.07)}$	$29.72_{(+0.09)}$	$31.50_{(-0.70)}$	$38.97_{(-0.40)}$	$26.99_{(-0.28)}$
4. TME +DepT	$36.04_{(+0.11)}$	$29.70_{(+0.07)}$	$31.73_{(-0.47)}$	$38.92_{(-0.45)}$	$27.09_{(-0.18)}$
Source side template					
5. STC +LenT	$35.79_{(-0.14)}$	$29.33_{(-0.30)}$	–	–	–
6. STC +ProbT	$35.85_{(-0.08)}$	$29.43_{(-0.20)}$	–	–	–
7. STC +DepT	$35.84_{(-0.09)}$	$29.55_{(-0.08)}$	–	–	–
8. TME +LenT	$35.99_{(+0.06)}$	$29.69_{(+0.06)}$	$32.48_{(+0.28)}$	$39.11_{(-0.26)}$	$27.10_{(-0.17)}$
9. TME +ProbT	$36.04_{(+0.11)}$	$29.70_{(+0.07)}$	$\mathbf{32.50}_{(+0.30)}$	$\mathbf{39.56}_{(+0.19)}$	$\mathbf{27.34}_{(+0.07)}$
10. TME +DepT	$\mathbf{36.19}_{(+0.26)}$	$\mathbf{29.80}_{(+0.17)}$	$32.36_{(+0.16)}$	$39.45_{(+0.08)}$	$27.16_{(-0.11)}$

Table 3. Translation results for ablation experiments removing *gating component* or *drop template mechanism* on test sets of IWSLT 2014 German→English.

Methods	IWSLT14 de-en
Target template	
11. TME +LenT -GATING	35.17
12. TME +LenT -DROP	35.77
Source template	
13. TME +LenT -GATING	35.67
14. TME +LenT -DROP	35.82

abstract representation of the target sentence based on the source sentence, and the output is usually repeated with some popular DepT templates. Differently, the LenT template proposed by [15] is more suitable in target side (settings 2, 8). The LenT and ProbT templates are the mixture of target words and POS

Table 4. Computation resource comparison between NMT system using a template on source side with target side. The values in the table present the number of learnable parameters (M = million) and training time in hours.

Method	Templ. generator	Target generator	Total
Source side	0M (0 h)	59.5M (9.7 h)	59.5M (9.7 h)
Target side	37.1M (7.8 h)	58.8M (9.3 h)	95.9M (17.1 h)

tags, that is punched in a more detailed level than DEPT. Although it is hard to predict the correct template in the first phase, the NMT systems have the advantage of the predicted words in the template for the final target sentence prediction.

Secondly, we consider the effect of template sides (source or target sides) on settings 2, 3, 4, and 8, 9, 10 described in Table 2. These results show that our proposed methods using templates on the source side are more effective than ones on the target side. Since the target templates need to be learned by an NMT model, the quality of the target templates prediction is lower than the source templates extraction, and the overall translation performance is decreased. In the IWSLT 2015 dataset, the Stanford Core-NLP tool does not support the Vietnamese language for syntactic parsing tasks, so we utilize a *spaCy* to parse constituent tree from a natural sentence. Therefore, the performance on the target side of this pair of languages drops sharply compared to the source side as well as other pairs of languages. Obviously, in this case, our proposed methods are more suitable and adaptive than the target-based templates for low-resource language (e.g. Vietnamese). The reason of this phenomena comes from the performance of syntactic parser in these kinds of languages and the error transmission in the original approach of [15]. The detailed comparison in Table 2 proves the strength of our model to deal with the low-resource language in machine translation. In the large-scaled dataset (WMT 2014), the template integration did not show clear improvement. We argue that the structure information of DEPT template in large scale dataset is less meaningful due to the repetition in abstract template extracting from the syntactic shallow level.

Computation Resource. Besides the performance, the other important aspects of the NMT system that affect the practicality are the training time and model size. With the approaches using target template, the NMT system has to contain two internal sub-modules which consists of one module to predict the target template, and another to predict the final sentence from source sentence and target template. Therefore, the computational time and model size is almost two times larger than our proposed approaches using a direct source template. Particularly, Table 4 shows the model size and training time of setting TME +DEPT on IWSLT 2014 de-en dataset on both source and target sides for comparison.

Previous Works Comparison. Our method (TME +DEPT) achieves the state-of-the-art result on IWSLT 2014 German→English with an 0.95-BLEU-score improvement when compare with [15]. This method also shows the improvement compared to the strong baselines in small datasets IWSLT 2014 English↔German, IWSLT 2015 English→Vietnam, and IWSLT 2017 English→French within the same settings. Comparing to the work of [6] on the WMT 2014 dataset, the TreeTransformer model can extract more structure information than methods using a template because it encodes all the constituent trees instead of a particular depth level. Comparing to the SOTA result [9] on IWSLT 2015 and IWSLT 2017, our method can be incorporated with BPE-dropout technical to improve, however, we leave it for our future work.

6   Conclusion

This paper presents our proposed framework to integrate the syntactic template from source sentences into NMT models. Besides, we also propose two different kinds of template extraction methods to determine the abstract template of sentence. To prove the strength and robustness of our models, we also conduct the empirical experiments using either source or target side in the various generating methods for conventional Transformer models. With our detailed comparison and evaluation, our proposed architecture obtains the potential results against the original approach and competitive baselines in many benchmarks. Besides, we also analyze in detail the effect of the template on the translation process to accentuate the appropriate method for incorporating syntactic information into the encoding process.

References

1. Dong, L., Lapata, M.: Coarse-to-fine decoding for neural semantic parsing. In: Proceedings of the 56th Annual Meeting of the Association for Computational Linguistics (Volume 1: Long Papers), Melbourne, Australia, pp. 731–742. Association for Computational Linguistics, July 2018
2. Eriguchi, A., Hashimoto, K., Tsuruoka, Y.: Incorporating source-side phrase structures into neural machine translation. Comput. Linguist. 45(2), 267–292 (2019)
3. Gupta, P., Bigham, J., Tsvetkov, Y., Pavel, A.: Controlling dialogue generation with semantic exemplars. In: Proceedings of the 2021 Conference of the North American Chapter of the Association for Computational Linguistics: Human Language Technologies, pp. 3018–3029. Association for Computational Linguistics, June 2021
4. Hao, J., Wang, X., Shi, S., Zhang, J., Tu, Z.: Multi-granularity self-attention for neural machine translation. In: Proceedings of the 2019 Conference on Empirical Methods in Natural Language Processing and the 9th International Joint Conference on Natural Language Processing (EMNLP-IJCNLP), Hong Kong, China, pp. 887–897. Association for Computational Linguistics, November 2019
5. Nguyen, P.M., Than, K., Nguyen, M.L.: Marking mechanism in sequence-to-sequence model for mapping language to logical form. In: 2019 11th International Conference on Knowledge and Systems Engineering (KSE), pp. 1–7 (2019)

6. Nguyen, X.-P., Joty, S., Hoi, S., Socher, R.: Tree-structured attention with hierarchical accumulation. In: International Conference on Learning Representations (2020)
7. Post, M.: A call for clarity in reporting BLEU scores. In: Proceedings of the Third Conference on Machine Translation: Research Papers, Belgium, Brussels, pp. 186–191. Association for Computational Linguistics, October 2018
8. Provilkov, I., Emelianenko, D., Voita, E.: BPE-dropout: simple and effective subword regularization. In: Proceedings of the 58th Annual Meeting of the Association for Computational Linguistics, pp. 1882–1892. Association for Computational Linguistics, July 2020
9. Sennrich, R., Haddow, B., Birch, A.: Neural machine translation of rare words with subword units. In: Proceedings of the 54th Annual Meeting of the Association for Computational Linguistics (Volume 1: Long Papers), Berlin, Germany, pp. 1715–1725. Association for Computational Linguistics, August 2016
10. Shang, W., Feng, C., Zhang, T., Xu, D.: Guiding neural machine translation with retrieved translation template. In: 2021 International Joint Conference on Neural Networks (IJCNN), pp. 1–7 (2021)
11. Vaswani, A., et al.: Attention is all you need. In: Guyon, I., et al. (eds.) Advances in Neural Information Processing Systems, vol. 30. Curran Associates Inc. (2017)
12. Vaswani, A., et al.: Attention is all you need. In: Guyon, I., et al. (eds.) Advances in Neural Information Processing Systems, vol. 30, pp. 5998–6008. Curran Associates Inc. (2017)
13. Wang, Y., Lee, H.-Y., Chen, Y.-N.: Tree transformer: integrating tree structures into self-attention. In: Proceedings of the 2019 Conference on Empirical Methods in Natural Language Processing and the 9th International Joint Conference on Natural Language Processing (EMNLP-IJCNLP), Hong Kong, China, pp. 1061–1070. Association for Computational Linguistics, November 2019
14. Xu, H., van Genabith, J., Xiong, D., Liu, Q., Zhang, J.: Learning source phrase representations for neural machine translation. In: Proceedings of the 58th Annual Meeting of the Association for Computational Linguistics, pp. 386–396. Association for Computational Linguistics, July 2020
15. Yang, J., Ma, S., Zhang, D., Li, Z., Zhou, M.: Improving neural machine translation with soft template prediction. In: Proceedings of the 58th Annual Meeting of the Association for Computational Linguistics, pp. 5979–5989. Association for Computational Linguistics, July 2020
16. Zhang, Z., Wu, Y., Zhou, J., Duan, S., Zhao, H., Wang, R.: SG-Net: syntax guided transformer for language representation. IEEE Trans. Pattern Anal. Mach. Intell. **44**(6), 3285–3299 (2022)

Forensic Analysis of Text and Messages in Smartphones by a Unification Rosetta Stone Procedure

Claudio Tomazzoli[1(✉)], Simone Scannapieco[2], and Matteo Cristani[3]

[1] CITERA Interdepartmental Centre, Sapienza University of Rome, Rome, Italy
claudio.tomazzoli@uniroma1.it
[2] R&TD Department, Real T S.r.l., Verona, Italy
[3] Dipartimento di Informatica, Università di Verona, Verona, Italy

Abstract. In this paper we introduce an innovative application of translation techniques applied to the problem of forensics analysis of smartphones. This analysis has the specific objective of determining which messages (either text or vocal), transmitted from and received by a specific device, seized for forensic analysis, may contain data that are relevant in a criminal investigation. The problems that make this analysis difficult are three: (1) the content could be written in a language that is not spoken by the analyst, (2) the number of messages actually containing pertinent and relevant traits is a small percentage on a potentially quite large space and (3) texts could be rather noisy in terms of content, for they could contain emoticons, language loans, and slang terms (beyond the fact that they could also be written in obscure languages such as specific dialects or languages spoken by small communities). We adopt a machine translation approach by providing an algorithm that takes messages of a smartphone as input, and processes them to a target language in an innovative way. We then show that the application is effective when applied to a set of real world cases, demonstrating a performance increase in terms of accuracy that could exceed 30% when compared to traditional approaches.

1 Introduction

The seizure of smartphones is one of the preferred means, employed by investigators, to identify traits of data and behaviours that could be fruitfully used in court. The existence of these data is indeed the result of a bold behaviour that is typical of low-level criminals who essentially use smartphones as they were not full of sensitive and evidential data, treating those technological means as their behaviour was actually honest. This does not happen with criminal organisations, where technologies are used in a sophisticated and prudential manner.

There exist many serious difficulties in the practice of forensics with data coming from smartphones. First of all, texts are often written in a language that is not actually spoken by the forensic analyst, and, rather frequently, it

© Springer Nature Switzerland AG 2022
H. Fujita et al. (Eds.): IEA/AIE 2022, LNAI 13343, pp. 315–326, 2022.
https://doi.org/10.1007/978-3-031-08530-7_26

can be also the case that the language is not even immediately detected. Many languages, for instance Dravidic ones, are quite different in terms of syntax and lexicon but are written with the same writing system of other languages in the same group. Typically, an European analyst is not even aware of the writing system of these languages. In a number of cases, the naive behaviours described above may result in a large percentage of text and voice messages whose pertinence with respect to a specific forensic activity results null, like greetings, short messages about family situation, or random small talk conversations with friendly or love content, and, in a large number of cases, empty messages.

Processing the above messages by hand is a complex and unfruitful activity in most cases. It would be therefore critical to develop a technical solution that satisfies the following conditions:

- Text and voice messages are treated in a unified fashion;
- The language (or languages) spoken by the authors of such messages should be considered unknown *a priori*; in particular, text messages may be potentially written in a mixture of languages;
- Single messages should be considered written in source language with language loans, symbols (emoticons, gif, acronyms) and slang terms.

We can therefore devise a methodology for forensic analysis of text messages that is a sequence of the following steps and is essentially an extension of what discussed by Anglano et al. in a multiple years study path [2–4], discussed a bit wider in Sect. 2.

Step 1: Forming the corpus. The messages (text or audio converted into text by a speech-to-text procedure) are collected into a document repository and archived in a database with all their metadata, coming from the analytical procedure of gathering.

Step 2: Translating the messages. All the messages are translated by using a modified translation procedure that converts texts into a Rosetta language and then to the target language while using a specific step, discussed in Sect. 4, and introduced in the analytical procedure reported in Sect. 3.

Step 3: Cleaning the message corpus. The messages are then filtered, either by hand or by one automation method, possibly based on text analytics, that we discuss briefly in Sect. 5 as a possible further work.

Consequence of the above assumptions is that we should consider as *run-out* those messages that we cannot process into a target language, for the distinction of messages resulting pertinent is rather general, but choosing those that are relevant is another issue. In this paper we only focus upon Step 2 of the above devised analytical methodology, whilst the implementation of Step 3 is left to further work, as discussed in Sect. 5.

To go into a detail, that is quite relevant for the forensic purpose, the data flow is made in a way that, starting from a language not necessarily known to the analyst, a text is translated into a language the analyst commands. The main issue with the existing translation technologies is that they work rather well when the source or the target language is English, but work rather inappropriately when neither of them is English. Clearly, we can pass through an English translation, and this is exactly the approach we adopted, although an issue (discussed below) still remains open.

To make a translation to a language that is not known to the analyst, we need to provide the following translation path:

1. Detect the language L in which the text T is written;
2. Translate the text T into a Rosetta stone language (English - E) obtaining the text T';
3. Translate the text T' into a target language (for the purposes of the application in the case in discussion, Italian I).

This approach contains some risks, namely we may introduce errors in the translation that results in approximate texts. For the purpose of this investigation, this is not an issue, for the final goal is to identify those messages that constitute *potential evidence of some crime*. As the precision is relevant, even a rather imprecise analysis, with high recall, based on a pure probabilistic model, could be very fruitful. There is however another risk, that is the consequence of the imprecision in the translation process, and in turn could result in the impossibility of translating the message, as discussed in Sect. 3. In this case, we adopt a solution that is innovative, and it is based on a cycle of translations, back and forth onto the source language.

In other terms, we can classify the run-out messages into three different categories: (a) messages that are empty, that should be dismissed, (b) messages with erroneous translation, in which the actual meaning is not obtained in the Rosetta language, and therefore in the target language as well, that could be recovered, and finally (c) messages that are fine in terms of translation after all the translation steps but cannot be considered for they are irrelevant in forensic terms, that are kept in the corpus after Step 2 of the above devised methodology, but should be eliminated by Step 3.

With the concepts expressed in the schema above we developed a prototype that has shown to perform in a rather good way. In particular, with respect to the run-out elements of two corpora, that we shortly discuss in Sect. 4.1, the obtained results are improved, on Step 2, to a 30% extent.

The rest of the paper is organised as follows: Sect. 2 discusses related work, and Sect. 3 provides a schema of the application from an architectural and applicative viewpoint. The core of the architecture is the algorithm presented in Sect. 4 where the technique is illustrated, analysed from a computational viewpoint and discussed in terms of its application to real-world cases. Section 4.1 discusses the application of the algorithm to a specific case study context, where the measures of the performance in the mentioned context are also presented and discussed. Finally Sect. 5 takes some conclusions and sketches further work.

2   Related Work

There is a variety of other studies involving smartphones and natural language processing, for applications other than forensics that could be taken as examples for some specific issues that are relevant to natural language processing techniques in those contexts, but the references are too many to be completely reviewed here. Let us just refer the basic ideas.

First of all, we have considered methods based on Google Machine Translation technologies [22]. Secondly, we have framed our study in the applicability of concepts of multi-language analysis [8]. Recent advances in these fields could be referred as few. An advanced sophisticated investigation has been provided, that could be generalised with some effort (for the specific type of context) by Tsai [20] in the context of text analysis for English as a Foreign Language test. Chen et al. have focused on the issues of multi-language *search* [8], while Farhan et al. [15] discuss the issues related to dialectal texts, that is very relevant to our investigation for what we discussed already in Sect. 1.

Forensics on smartphones assisted with technological means is a rather recent idea in artificial intelligence applications. Mylonas et al. [16] provided a schema to treat the materials discovered during a forensic analysis. Their methodology is the inspiration of some further investigations; on the other hand, in an apparently independent context, Pieters and Olivier [18] and Anglano [2] introduced methods that make use of specific analytical technologies. Specific aspects related to the analysis in the dark web have also been further studied by Anglano et al. [3,4]. Wu et al. studied the chat platform that is more common in the chinese domain WeChat in their investigation [21]. General issues on Android devices have been dealt with by Zhang et al. [23]. Similar issues have been also discussed in [10].

Further studies have been taken on the topic of social relationships that could be identified thanks to the forensic analysis of smartphones, as in Peng et al.'s investigation [17], analogously to the study of Choi et al. [9] and by Awan [6], although with a different focus on what to be actually measured. The techniques employed in these investigations dealt with similar issues of those dealt with in [11,14].

The problem of *forensic artefacts*, namely those modifications to the source content of a digital object that could be relevant in an analysis, is the focus of Alyahya and Kausar study [1]. Among all the above studies, only recently scholars have put their attention on the problems of audio, and in particular audio descriptions and audio messages that could be relevant from a forensics viewpoint (as in Rumsey [19]).

On top of the above investigations, we have developed in a more practical and directly usable way the ideas of pertinence and relevance of text and image analysis discussed in [12,13].

3   Framework

In this section we discuss the adopted approach and show how it is relevant to the problem of forensics. First of all, let us discuss the main issue on which the approach is founded: the usage of a Rosetta Language (English) and its relevance and sufficiency to the problem we deal with.

Accuracy of translation methods that could be available in practical applications is not yet analysed in a systematic way, but there is a significant consensus, and some preliminary results (see, for instance [7]) confirm it, that *short text* translations, though may suffer in a heavier way of the context bias, tend to be better than those of long texts. Clearly, text messages, audio messages and chat messages in smartphones are short in general, thus the risk of adopting an automated translator as a means for forensic analysis is low. Moreover, it is also acknowledged that common pairs of languages generate better translations than less common ones. Consequently, the adoption of a Rosetta Language is worth when source and target languages in a pair are not common (especially when we can rely to a good extent on one of the two translation steps).

There is a long history of using automated methods for discovering deleted material, to identify traits in simple subdomains, such as pictures, contacts, and other access points in the material contained in a smartphone. On the other hand, there is a lower focus on the analysis of texts; this is due to a number of factors, but the most crucial one is that texts have to be read, activity that is extremely time-consuming. However, texts typically contain a *very limited number* of tokens that result decisive. Once we have a good target language we can query the languages not containing those traits. We shall give below some query samples in this direction. Moreover, we can run a garbage analysis on the resulting texts, then eliminate the large majority, which contains greeting, love messages, single emoticons, and so on.

A schema of the application framework is presented in Fig. 1. The novelty stands entirely in the step where the technology filters the result by means of a *parametric* evaluation. The evaluation steps rely on the probability of the translation to work appropriately in the specific context, for which we may provide different parameter values.

The most common event determining the largest part of errors in the translation process, consists in the *usage of wrong keyboard set*. For instance, when someone uses a smartphone for text messaging in Italy, the Italian keyboard is typically adopted. Most of the times, when writing a message in the source language, a user changes the keyboard from Italian to the desired one. However, users messaging in a mixed fashion (as often happens when communicating not with nationals, but with other strangers or locals), may stay on Italian keyboard and send a message transliterated into the pronunciation of the source in a phonetic form. This causes a mistranslation, that occurs when the text is submitted to the Google Machine Translation service. For instance, the text **Tathtata tawn akata giya awahama gannam**, whose translation (that contains an error, evidently) is **Daddy went to town and I'll take it**, is not detected in terms of language, and therefore it is passed as it is to the translator from

English to Italian, where it finally lands in its source form. We can therefore devise the following aspects:

- When the text is transliterated into Latin alphabet, Google translation technologies does not detect the language and therefore leaves the text completely untouched in both steps;
- When the text, instead, is transliterated into Latin alphabet and Google detects some English terms, then it goes to the Rosetta translation with a few words that remained unchanged, but these are correctly translated into the target language;
- Therefore, if we get the message and evaluate the translation as *null* for both steps had been without effect, we can consider the language to be undetected.

On the other hand, in a corpus generated by analytically act on a single smartphone, we very likely detect the *preferred language*, that is the language the owner of the phone usually speaks. Given that language we can provide a translation of the *source message* into the target language, by using machine translation techniques. This provides a new source text, that is detected to belong to the preferred language and passes through the two translation steps towards the target language effectively.

Schema of the application data flow is presented in Fig. 1, whilst the architecture of the technology is presented in Fig. 2.

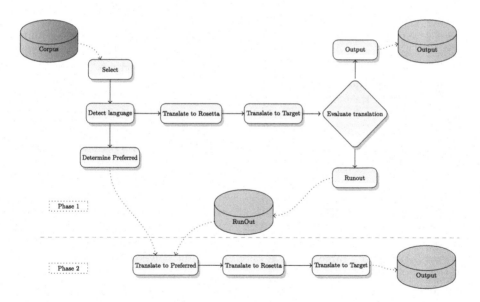

Fig. 1. The schema of the application of translation for text analytics in forensic of smartphones

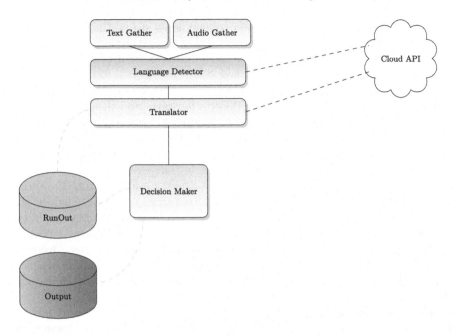

Fig. 2. System architecture

4   An Algorithm for Machine Translation with Rosetta Language Reference, and Double Step-Translation

Algorithm 1 implements the analytical framework discussed in Sect. 3. It can be splitted into four basic functions, two of which (**read** and **translate**) employ the machine learning remote web services of Google Cloud API, that act as devised below.

textSelect: selects all files which may have a text message as file content. The selection is based on the recognition of a sufficiently large number of sufficiently long patterns that are unlikely artefact texts resulting from the streaming of a binary file.

read: reads the file content and extract the candidate text message (a scan procedure employed to remove at least a part of the garbage around a text).

isMessage: verifies whether the message has a non null or trivial content, in particular by comparing the text to a list of emoticons.

detectLanguage: identifies the language of a text and translates into the Rosetta Language (English).

translate: translates the text in the Rosetta Language to the target language.

Theoretically speaking it is easy to see that the complexity of the method is linear in the number of messages, and is in general quadratic in the number of words for each message (for this is the typical cost of translation methods adopted in Google APIs). Performance analysis is provided in Sect. 4.1 for some case studies and it is very encouraging in terms of scalability of the method.

In practice, processing a single smartphone is not a real-time task, and can require hours of execution, due to the length of the extraction and translation processes. This is not a problem for the issue has to be addressed only once for each seized device.

The algorithm takes into consideration the case discussed in Sect. 3, when a single text has been translated to the Rosetta language and then into the target language without any effects of the translation; this means that the text has been transliterated (this is true for roughly 50–60% of the cases) and therefore, we pass it to the translator as a text to be translated into the preferred language of the smartphone. The evaluation measure is that we *do not* find even one word changed from the source text onto the Rosetta and then into the target. It could indeed happen that the source text contains loans from English, that therefore are translated from Rosetta (English) onto the target language. In this case, even though it would be possible to distinguish some cases that result ambiguously translated and get back to the actual words that come from the source language, we discard the possible correction, because it could be beneficial in a very limited number of cases (preliminary investigation can drop a maximum of 5% of potential matches missed). We leave this step for further work.

The quality of the second step we introduced is essentially limited into the domain of machine translation. Generally speaking, there is a basic issue in the development of these techniques as applied to the solution of the presented issues: the translation accuracy is very limited, being, for the languages not belonging to the family of Indo-European ones, fundamentally a syntactic linear translation method, that takes every single word as a single token. There is a process of *context* analysis that is performed in order to limit the side effects of polysemy of languages, by means of the reduction of it when a context is focused. This context analysis is not particularly effective, when text messages in smartphones are the source for a corpus formation. These texts are small, and typically the context formed in this way is very limited, therefore with a little consequence on the reduction of polysemy.

4.1   A Case Study: Running the Algorithm

In Table 1 we present a typical situation that gives rise to the need of a novel method to be used to deliver message recovery from runout situations, occurring when no actual translation process has been performed. The example show that among these six cases, only a few present the potential of being evidential in a forensic process. In particular, it is clear that example 2., 3., and 7. are out of any interest of a criminal investigation. Examples 1., 5. and 6. can attain at a potential *secret language* or at the usage of a conventional word substitution in a codified communication, while example 4. shows the potential of being an element of evidence.

We have taken the message corpus of a seized iPhone SE. The device has been customarily analysed in a way that gives the results below:

– The address book contains 954 contacts;

Algorithm 1. BABEL: Given a list of names of files *candidates* which might contain voice or text messages, returns a list of translated *messages*

Input: a list of file names *candidates* containing either voice or text messages Θ ;
Output: a list of *Messages* $\mathcal{M}$ where $m = \langle t_o, L, t_e, t_i \rangle$ (original text, language, english text, translated text);
$\Psi^T \leftarrow textSelect(\Theta); \mathcal{M}^T \leftarrow \emptyset; \mathcal{M} \leftarrow \emptyset;$
repeat
  $filename \leftarrow pop(\Psi^T);$
  $lines \leftarrow leggi(filename);$
  repeat
    $line \leftarrow pop(lines);$
    if $isMessage(line)$ **then**
      $m \leftarrow new();$
      $m.t_o \leftarrow line;$
      push($\mathcal{M}^T$,m);
    end if
  until $lines = \emptyset$;
until $\Theta = \emptyset$;
repeat
  $m^T \leftarrow pop(\mathcal{M}^T);$
  $m^T.L \leftarrow detectLanguage(m^T.t_o);$
  if $m^T.L \notin \{targetLanguage, rosettaLanguage\}$ **then**
    $m^T.t_e \leftarrow translate(m^T.t_o, rosettaLanguage);$
    $m^T.t_i \leftarrow translate(m^T.t_e, targetLanguage);$
  end if
  push($\mathcal{M}'$,m);
until $\mathcal{M}^T = \emptyset$;
$\gamma \leftarrow preferredLanguage(\mathcal{M}');$
repeat
  $m' \leftarrow pop(M');$
  if $m'.L \in \{targetLanguage, rosettaLanguage\}$ **then**
    push($\mathcal{M}$,m');
  else
    if $m'.t_o \neq m'.t_e$ **then**
      push($\mathcal{M}$,m');
    else
      $s \leftarrow translate(m'.t_o, \gamma);$
      $m'.t_e \leftarrow translate(s, rosettaLanguage);$
      $m'..t_i \leftarrow translate(m'..t_e, targetLanguage);$
      push($\mathcal{M}$,m');
    end if
  end if
until $M' = \emptyset$;
return $\mathcal{M};$

- All the classic chat technologies have been considered (sms, WhatsApp and other systems, including the deployment by speech-to-text technology based on Google translate API);

– Out of 50 messages, 19 have been recovered thanks to the above described technology, namely roughly 38%.

Table 1. Sample messages from a seized iPhone.

1	Tathtata tawn akata giya awahama gannam	→	Daddy went to town and I'll take it
2	Niyamai nee	→	That's right
3	Subama suba aluth awruddak wewa oyala sematama.	→	Happy New Year to you all
4	5000 eken 2 k ada yanawaù	→	5000 2 of which go today
5	Suji aiye cud eka danna puluwanda den	→	Can you tell me about Suzy's brother?
6	Aiye dala thiyenne	→	My brother has left
7	Mava mathakada danne na	→	I do not remember

Preliminary analysis of other devices have shown that the percentage above is a typical recovery performance in those ones presenting the mentioned phenomenon of transliteration of messages. We estimate that for those devices presenting such a phenomenon typical percentage of recovery would roughly be on 30%. Again on preliminary data, when the spoken language is coming from Asia, in particular from the Indian subcontinent, the phenomenon of text transliteration is ubiquitous.

5   Conclusions and Further Work

We have dealt with the problem of making the process of text analytics for forensics purposes on smartphones as effective as possible. We have been dealing with the following issues:

– Messages could be either text or audio messages;
– Due to the effects of globalisation and migrations, the source language of a text/audio message could be mashups, slang terms, emoticons, loan terms;
– Again for the same effects of the above, we may have more languages to deal with, different alphabets, and also different writing systems, such as logographic languages, abjads, abugidas, syllabaries;
– Reliability of translations from one language to another varies in a significant way, and tends to be higher for English as a target language, for evident reasons related to the diffusion of the language technologies in the western world.

Due to the above challenges we developed a technology that is based on English as a Rosetta language, where we employ a methodology for text relevance analysis that is able to improve greatly the accuracy of the translation.

Further work regards mainly the possibility of implementing the desiderata expressed in the description of Step 3 of the methodology provided in Sect. 1. We now give an example of how such an implemented solution could work. Let us assume that we have a corpus in the target language. Since the language standard messages are a few, and very frequent, we can just list them:

- Hi;
- Kiss;
- Happy birthday;
- See you;
- See you later;
- Good morning;
- Love;
- ...

Lists of messages such as the aforementioned could be easily extracted from a corpus formed by messages taken from a collection of individual streams of messages, by a combined analysis of *length of messages* and *frequency*. It could be the case that we can anticipate this schema to every language, but this is clearly more expensive in terms of activities to be performed and probably useless given the length and number of messages typically contained in a single phone. Therefore, acting on the target language, or possibly on the Rosetta one is worth.

The resulting analytical schema is therefore implementable by means of a text analytic method, based upon measures of relevance and pertinence. One further step to be employed regards the possibility of analytically take into consideration diversity of channel communications, for texts in a specific chat could be different than other ones.

Furthermore, we will investigate the performance of the algorithm according to a bias vs. accuracy perspective with methods (such as $\langle \phi, \delta \rangle$ diagrams [5]) that take into account sets of data, like smartphones under forensics investigation, with imbalance in data between negative and positive samples.

References

1. Alyahya, T., Kausar, F.: Snapchat analysis to discover digital forensic artifacts on Android smartphone. Procedia Comput. Sci. **109**, 1035–1040 (2017)
2. Anglano, C.: Forensic analysis of WhatsApp Messenger on Android smartphones. Digit. Investig. **11**(3), 201–213 (2014)
3. Anglano, C., Canonico, M., Guazzone, M.: Forensic analysis of the ChatSecure instant messaging application on Android smartphones. Digit. Investig. **19**, 44–59 (2016)
4. Anglano, C., Canonico, M., Guazzone, M.: Forensic analysis of Telegram Messenger on Android smartphones. Digit. Investig. **23**, 31–49 (2017)

5. Armano, G., Giuliani, A.: A two-tiered 2D visual tool for assessing classifier performance. Inf. Sci. **463–464**, 323–343 (2018)
6. Awan, F.: Forensic examination of social networking applications on smartphones, pp. 36–43 (2016)
7. Birkenbeuel, J., et al.: Google translate in healthcare: preliminary evaluation of transcription, translation and speech synthesis accuracy. BMJ Innov. **7**(2), 422–429 (2021)
8. Chen, J., Bao, Y.: Cross-language search: the case of Google language tools. First Monday **14**(3) (2009)
9. Choi, J., Lee, S.: A study of user relationships in smartphone forensics. Multimed. Tools Appl. **75**(22), 14971–14983 (2016). https://doi.org/10.1007/s11042-016-3651-4
10. Cristani, M., Burato, E., Santacá, K., Tomazzoli, C.: The spider-man behavior protocol: exploring both public and dark social networks for fake identity detection in terrorism informatics, vol. 1489, pp. 77–88 (2015)
11. Cristani, M., Fogoroasi, D., Tomazzoli, C.: Measuring homophily, vol. 1748 (2016)
12. Cristani, M., Tomazzoli, C.: A multimodal approach to exploit similarity in documents. In: Ali, M., Pan, J.-S., Chen, S.-M., Horng, M.-F. (eds.) IEA/AIE 2014. LNCS (LNAI), vol. 8481, pp. 490–499. Springer, Cham (2014). https://doi.org/10.1007/978-3-319-07455-9_51
13. Cristani, M., Tomazzoli, C.: A multimodal approach to relevance and pertinence of documents. In: Fujita, H., Ali, M., Selamat, A., Sasaki, J., Kurematsu, M. (eds.) IEA/AIE 2016. LNCS (LNAI), vol. 9799, pp. 157–168. Springer, Cham (2016). https://doi.org/10.1007/978-3-319-42007-3_14
14. Cristani, M., Tomazzoli, C., Olivieri, F.: Semantic social network analysis foresees message flows, vol. 1, pp. 296–303 (2016)
15. Farhan, W., et al.: Unsupervised dialectal neural machine translation. Inf. Process. Manag. **57**(3), 102181 (2020)
16. Mylonas, A., Meletiadis, V., Tsoumas, B., Mitrou, L., Gritzalis, D.: Smartphone forensics: a proactive investigation scheme for evidence acquisition. In: Gritzalis, D., Furnell, S., Theoharidou, M. (eds.) SEC 2012. IAICT, vol. 376, pp. 249–260. Springer, Heidelberg (2012). https://doi.org/10.1007/978-3-642-30436-1_21
17. Peng, L., Zhu, X., Zhang, P.: An efficient model for smartphone forensics using SMS spam filtering, pp. 166–169 (2020)
18. Pieterse, H., Olivier, M.: Smartphones as distributed witnesses for digital forensics. In: Peterson, G., Shenoi, S. (eds.) DigitalForensics 2014. IAICT, vol. 433, pp. 237–251. Springer, Heidelberg (2014). https://doi.org/10.1007/978-3-662-44952-3_16
19. Rumsey, F.: Audio forensics keeping up in the age of smartphones and fakery. AES: J. Audio Eng. Soc. **67**(7–8), 617–622 (2019)
20. Tsai, S.-C.: Using Google translate in EFL drafts: a preliminary investigation. Comput. Assist. Lang. Learn. **32**(5–6), 510–526 (2019)
21. Wu, S., Zhang, Y., Wang, X., Xiong, X., Du, L.: Forensic analysis of WeChat on Android smartphones. Digit. Investig. **21**, 3–10 (2017)
22. You, Y., Zhang, Z., Hsieh, C.-J., Demmel, J., Keutzer, K.: Fast deep neural network training on distributed systems and cloud TPUs. IEEE Trans. Parallel Distrib. Syst. **30**(11), 2449–2462 (2019)
23. Zhang, H., Chen, L., Liu, Q.: Digital forensic analysis of instant messaging applications on Android smartphones, pp. 647–651 (2018)

Relation-Level Vector Representation for Relation Extraction and Classification on Specialized Data

Camille Gosset[1,2(✉)], Mokhtar Boumedyen Billami[1], Mathieu Lafourcade[2], Christophe Bortolaso[1], and Mustapha Derras[1]

[1] Berger-Levrault, 64 Rue Jean Rostand, 31670 Labège, France
{camille.gosset,mb.billami,christophe.bortolaso,
mustapha.derras}@berger-levrault.com
[2] LIRMM, 161 Rue Ada, 34095 Montpellier, France
mathieu.lafourcade@lirmm.fr

Abstract. During this last decade, word embeddings models learning continuous vector representations of words have been established and integrated in several applications of Natural Language Processing (NLP). These models have been subsequently extended to learn representations of other textual objects such as word senses/definitions, fragments of textual documents or even whole texts. In this paper, we focus on the creation of continuous vector representations for relations. We propose a model where vectors embeddings of relations are deduced from (multi-)words embeddings. The training of these representations is carried out from a business corpus referring to several specialized domains such as health, justice, urbanism, or elections. The quality of these representations is evaluated on the task of identifying/classifying lexical-semantic relations from texts with binary classifiers (Machine Learning techniques). This task consists in classifying from a relation embedding in input, the relation type of this representation. The obtained results are good and surpass the performances for a recent one state-of-the-art system dedicated to the creation of relations representations.

Keywords: Word2Vec · Relation embeddings · Relation type identification and classification · Natural language processing

1 Introduction

Nowadays, vector representation methods of lexical items have been proven to be a significant task-solving manner regarding automatic natural language processing. The lexical items representation is a main concept to support machine learning oriented systems, and these representations can preserve more semantic and syntactic information about words. This improves performance in almost all NLP-related tasks such as information extraction, word sense disambiguation, and automatic text summarization.

Over time, creation of word representations as continuous vectors has been studied extensively [1, 2]. In the last ten years, the popularity of these works has exploded

© Springer Nature Switzerland AG 2022
H. Fujita et al. (Eds.): IEA/AIE 2022, LNAI 13343, pp. 327–338, 2022.
https://doi.org/10.1007/978-3-031-08530-7_27

with the use of neural networks for training word embeddings. In 2013, a significant work was proposed regarding the utilization of two layers neural networks to reconstruct word's linguistic context [3]. In this work, an algorithm called Word2Vec is presented to create word embeddings. Based on the work presented by Mikolov and his co-workers, representation techniques of other lexical item types have been proposed, which aim at a higher level, such as sentence representations [4, 5], paragraph representations [6] and full text document representations [7, 8], more than just word representations. Moreover, lexical items vector representations are widely used due to multiple use cases. In [6], Le and Mikolov illustrated the capabilities of their method and the quality of the generated paragraph vectors on text document classification and sentiment analysis tasks. Furthermore, Galke and al. have shown in [9] that lexical embeddings play an essential role to improve document search and information retrieval. In this paper, we focus on another level of representation: the lexical-semantic relations representation that can take place between words (or terms). To clarify, relation representations creation is part of a problem concerning relations extraction, which is extensively utilized in ontology development. These relations can be identified and extracted from unstructured textual documents. We aim to learn relation representations for the purpose of automatically identifying these lexical relations from unstructured documents.

Next, a comparative study between two systems is presented to see the quality of the relations representation vector: (1) refer to our proposed system where vectors embeddings of relations are deduced from vectors of terms and (2) refer to the state-of-the art system RELATIVE (*RELations as LATent dIscourse VEctors*) [10] where relation embedding vectors are learned. This system is a latent variable model that aims to explicitly determine what words from the given sentences best characterize the relations to be extracted. Our approach allows us to learn word embeddings for multi-words and terms/phrases contained in our specialized French data corpus. We work with key terms and not just single words. The underscore ("_") is used to separate tokens in a multiword expression (e.g., *united_states*) in the corpus. Based on the existing expressions in the corpus, we generate relations between them and validate them with a lexical network in French, called JeuxDeMots [11][1].

The following content is divided into 4 sections. In Sect. 2, we present a state of the art on the relation vector representation methods and on the relation extraction methods. Then, in Sect. 3, for the purpose of the creation of relation vector representations, we describe the exploited data as well as our methodological process used on these data. Finally, we discuss in Sect. 4 our obtained experimental results and compare them to a state-of-the-art system before concluding and presenting the perspectives of our work in Sect. 5.

2 Previous Works

2.1 Vector Representations of Relations

Several works have been proposed that allow the learning of relation vectors for a set of word pairs (a, b), based on sentences in which these words coexist. For example, a

---

[1] http://www.jeuxdemots.org/diko.php.

method was introduced and called "Latent Relational Analysis (LRA)" [12], which relies first on the identification of a set of sufficiently frequent lexical patterns. This method constructs a matrix for which the frequency of lexical patterns is measured for each pair of words. The relation vectors are then obtained using the singular value decomposition (SVD). Furthermore, an approach inspired by the GloVe model (*Global Vectors for Word Representation*) [13] was proposed in [14], which is a word embeddings model. Their idea is to learn relation vectors by considering the co-occurrence statistics between a target pair of words (a, b) and other words.

Washio and Kato have proposed in [15] a completely different approach to those mentioned above. They trained a neural network to predict dependency paths from a pair of words. Their approach uses standard word vectors as input, so relational information is encoded implicitly in the weights of the neural network, rather than as relation vectors (although the output of this neural network for a pair of words can still be considered as a relation vector). Pair2Vec [16] is a model where the focus is on learning relation vectors that can be used for cross-sentence attention mechanisms in tasks such as question-answer systems or textual implication. One of the problems with implementing these approaches is the time-consuming consumption of resources. For example, by using standard physical hardware, Pair2Vec will need a training time of between 7 and 10 days on a representative corpus size [17].

Other approaches have been proposed to improve the representation of relation vectors using knowledges from external databases. These approaches are based on the fact that two words are similar (i.e., their vectors are close in the representation space) only if they are related to each other in a given knowledge graph [18]. The advantage of these approaches is that they work on typed relations, for example: hypernymy (*generalization*), hyponymy (*specification*) or synonymy (*equivalence*).

2.2 Relation Extraction

Several research works on relation extraction have been proposed, two main categories of approaches exist [19]: (1) lexical-syntactic pattern-based approaches [20, 21]; and (2) machine learning/deep learning approaches [22].

The first research work on relation extraction was done by Hearst in [21]. This method is based on lexical-syntactic patterns and seeks to extract taxonomic relations only on English texts. Later, several researches were made on French corpus: patterns have been extended to limit a certain amount of noise in the French corpus [20]. Subsequently, this type of approach produces interesting results, particularly in terms of accuracy [23, 24]. However, this type of approach does not provide full coverage of the information. Indeed, the use of a predefined number of lexical-syntactic patterns is problematic because of the precision of certain relations and the management of the ambiguity of certain patterns.

To overcome the problem of information covering and thus to limit the use of lexical-syntactic patterns, approaches based on machine learning have been proposed, either by supervision [22] or by no-supervision [25]. One of the first approaches of this category was proposed by [26]. It is based on a technique of "*bootstrapping*" which consists in selecting patterns with semi-supervised learning to build a learning base. Along the same lines, distributional word selection and semantic feature selection techniques have been used to identify relations between named entities [27]. Semi-supervised methods were also used and were trained on annotated corpus to automatically learn linguistic properties signaling the targeted relations. Several learning algorithms were used such as Support Vector Machines (SVM) [28], Logistic Regression [29], Statistical Parsing [30] and Conditional Random Fields (CRF) [31].

It's possible to combine the two types of approaches to extract relations from a corpus of one million definitions from two resources [32], namely: "TLFi, *Trésor de la Langue Française informatisé*" and Wikipedia. This new approach combines the precision of lexical-syntactic patterns and the recall of statistical methods by distributional analysis and is applied to several relations. The aim is to automatically obtain a semantic resource for contemporary French from a corpus of texts.

Through the review of the related works, we observe a lack of consideration of specialized corpus, especially in the legal field. We ensure that we have key terms from a specific domain and therefore relationships between key terms in the domain exist. The aim is to obtain vectors of relations according to the specific domain. We work with key terms that may be phrases such as "Project of construction subject to public enquiry (*Projet de construction soumis à enquête publique*)" or singular words such as "provider (*Prestataire*)". Using models trained on standard corpus does not necessarily reflect the provision of embeddings trained on key phrases/terms identified in advance by experts. In other words, we are interested in learning embeddings for multi-words, terms, and phrases. To our knowledge, if one does not have thesauri or ontological resources what is often done is to determine the average vector from the different components of an expression.

3   Methodology

In this section, the presentation begins with the corpus and the way in which experts annotate the corpus (i.e., Subsect. 3.1). We then detail the data preprocessing, i.e., the experts' annotation representation unification (i.e., Subsect. 3.2). Thereafter, we will explain the core of the relation vector representation in 2 subparts: (1) word embeddings are trained on a specialized corpus (i.e., Subsect. 3.3), and (2) relation vector representations are deducted (i.e., Subsect. 3.4).

After the implementation of this method, we can classify the type of possible relations between terms (in Subsect. 4.2) to, then, validate the approach by comparing with a state-of-the art system (in Subsects. 4.3 and 4.4).

3.1   Specialized Corpus and Expert Annotations

We use a French corpus containing 172 books and 12,838 online articles on legal and practical expertise. This corpus deals with 8 specialized areas of the public domain, namely: (1) 'Civil status and cemeteries', (2) 'Elections', (3) 'Public procurement', (4) 'Urban planning', (5) 'Local accounting and finance', (6) 'Territorial human resources', (7) 'Justice' and (8) 'Health'. In the following, we refer to the corpus by the name "BL.Corpus" to give meaning to the editorial database published by the Berger-Levrault company.

We have a set of annotations proposed by several experts from different fields of expertise. We did not need to identify target terms automatically because the business corpus is already annotated with these terms. The set of books contains 52,476 annotations, and the set of articles contains 8,014 annotations. These annotations are in two forms: (1) multi-words and expressions such as "Notice to the public" and (2) Singular words such as "Parking".

3.2 Annotation Representation Unification

The experts' annotations are manual, and experts did not hold a reference lexical resource at the time of annotation. They, therefore, do not annotate the corpus in the same way. The key terms were described with several inflected forms and sometimes with additional irrelevant information such as determiners (e.g., "some expenses"). A key term groups all the annotations by arrowed forms under a single representative form.

We retrieved all the annotations from the experts. From the different inflected forms, we want to move them to a representative form called "key term". To do this, we will preprocess the data to unify experts' annotations. We use Stanza [33] to parse the expression and eliminate stop words in first position (e.g., articles/determiners, conjunction, preposition, etc.). In the example "a light", the annotation is changed to "light". However, removing stop words can make it syntactically ambiguous (e.g., both noun and adjective). To remove this ambiguity, we use Stanza to provide the part-of-speech in context before removing the stop words. Stanza also allows us to provide the lemmatized form of a word. We take advantage of this to perform this operation at the same time so as not to repeat the same calculation. Thus, each annotation is associated with (1) its grammatical class and (2) its preprocessed lemmatized form. This combination represents the identifier of a key term. For multi-words, the governor class is the one assigned to the key term.

Based on these elements, we wish to unify the annotations of the same identifier under a single representative key term called "referent": a single form for terms with the same identifiers but different inflected forms. For example, "participation in concerted development areas", "participating in concerted development areas" and "participating in concerted development area" are all referring to the same key term.

To choose the right representative manner, we have divided our problem into two cases: simple terms and complex terms. In the context of complex terms, we have chosen to take the inflected form with the highest number of occurrences in BL.Corpus. To do this, a statistical analysis is carried out of the whole corpus to calculate the number of occurrences of each annotation on the whole corpus (annotations with their representation in the paragraphs). Thus, the most frequent form represents the substitute that can refer to a given key term. For simple terms, we favor a so-called standard canonical form because we wish to favor the singular and generic form over the plural form. To satisfy this need, we reuse the lemmatized forms previously retrieved using Stanza.

3.3 Word Embeddings Training

We work with key terms that may be expressions such as "Annex to the urban planning document" or singular words such as "fair". Using models trained on standard corpus does not necessarily reflect the provision of embeddings trained on expressions/key terms, which are identified in advance by experts. In other words, we are interested in learning embeddings for terms and multi-words/phrases.

From the constructed key terms previously, we train embeddings at term-level on the BL.Corpus to obtain continuous vector representations of key terms and other words. To do this, we first perform a lexical substitution in the whole corpus of each expert's annotation by its *referent key term*. With this prepared corpus, we can now train a Word2Vec model with a CBOW architecture (*Continuous Bag-of-Words*) [6]. We used Gensim[2] to satisfy this training (with a default setting). The lexical embeddings of our key terms are constructed by considering them as entities and not as an average of the words that compose a given key term. We seek to force the understanding of word sequences as an element (i.e., "public_spending (*dépense publique*)" for the key term "public spending").

3.4   Creation by Deduction of Typed Relation Vector Representations

After training our word/term embeddings with Word2Vec, we can derive vector representations of so-called typed relations. To do this, we retrieve a set of key term pairs for a given type of relation from a knowledge graph, according to the key terms of BL.Corpus. Our choice of knowledge graph is the French lexical-semantic network JeuxDeMots [11]. Nowadays, this knowledge graph represents one of the largest opensource French networks with 14 million of nodes and approximately 320 million of relations. It allows us to have the common and business meanings of French polysemous terms. The use of JeuxDeMots allowed us to retrieve a set of relations of different types such as hypernymy, hyponymy, synonymy and antonymy.

From the trained embedding vectors, we can derive vector representations of socalled typed relations. To do this, we produce a set of key term pairs linked by a given relation extracted from the JeuxDeMots network. Then, using an arithmetic operation, we derive a vector representation of a (typed) term pair. A relation is symmetrical if simultaneously (X, relation, Y) and (Y, relation, X) exist for each of the pairs, like synonymy for example. For asymmetric relations, we subtract the vectors of the two key terms of a relation to maintain the meaning of the relation, whereas for symmetric relations we take the absolute value of this difference because the meaning is bidirectional. Hypernymy "r_isa" and hyponymy "r_hypo" represent cases of non-symmetry. Taking a concrete example of a non-symmetric relation (here, hypernymy), if $V_{public\_procurement}$ is the vector of the key term "public procurement" and V_{market} is the vector of the key term "market" then $(V_{public\_procurement} - V_{market})$ is the vector of the relation (*public procurement, r_isa, market*). In the case of a symmetrical relation (here, synonymy), if $V_{statutes}$ is the vector of the key term "statutes" and $V_{regulations}$ is the vector of the key term "regulations" then $| V_{statutes} - V_{regulations}|$ is the vector of the relation (*statutes, r_syn, regulation*).

---

[2] https://radimrehurek.com/gensim/models/word2vec.html.

4   Evaluation

We compared our system, a Word2Vec-based model, with another system RELATIVE. Taking into consideration a raw textual corpus and a model of word embeddings, RELATIVE learns embeddings pairs of terms. We use binary classifiers to validate the quality of vector representations of typed relations. This classification phase uses several algorithms like k-NN (k-Nearest Neighbors), SVC (Support Vector Classification), RFC (Random Forest Classifier), and DT (Decision Trees). Each of which is used on two types of relations, namely: (*Hypernymy, Hyponymy*) and (*Synonymy, Antonymy*).

4.1   Dataset Balancing

We own a set of key term pairs linked by a given relation extracted from JeuxDeMots. For example, we have identified 4,669 pairs linked by antonymy, 25,889 pairs linked by synonymy and 8,720 pairs for each set of hyponyms and hyperonyms. However, we must take into consideration the problem of balance of our dataset. Indeed, a classification phase will be performed to validate or invalidate our vector representation of our key term pairs. It is therefore preferable to obtain a balanced corpus, i.e., the same number of term pairs for each type of relation (for each class). To do this, we decided to downgrade the number of pairs associated with a relation type. The Fig. 1 illustrates applied downgrade process in our dataset. Consequently, the number of pairs (denoted by $n_i = 4,669$) for antonymy relation regarding our dataset is smallest. Obviously, the other sets of relation pairs are more numerous. Finally, we randomly pick n_i pairs in each dataset. We repeat the operation five times to obtain five representative datasets of a given class (type of relation). In the end, each dataset is reduced to 4,669 pairs.

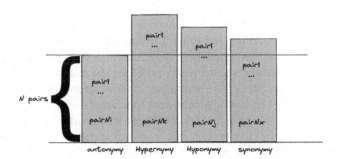

Fig. 1. Number of pairs in each dataset according to relation type

When we compare our system to RELATIVE, we retrieve representations of the same extracted term pairs as we can see in the Fig. 2. If one system has a pair that does not exist in the other system's dataset, we delete it. The list of pairs for evaluation must be the same for both systems by deleting the pairs that appear in one system and not in the other (i.e., Fig. 2).

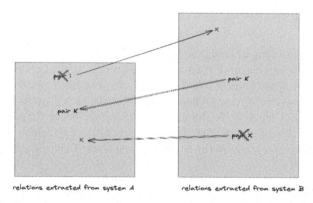

Fig. 2. Principle of pair deletion for levelling when comparing systems

4.2 Typed Relation Identification/Classification

We use binary classifiers to validate the quality of vector representations of a relation type. This is performed by identifying relations and predicting the type of relation between two terms. We partition the balanced set of semantic relations drawn from JeuxDeMots between learning and validation. This classification phase uses several binary classifiers and each of the classifiers is based on 2 pairwise relations and associates a relation type.

To validate our representation system, several binary classification models were created: from the use of decision trees and ensemble learning method to the use of Support Vector Machines. This allows us to check the most relevant model for classifying relations from relation vectors. When we compare our system with RELATIVE, we will evaluate on the set of classifiers and not on one to check the relevance in these representation methods.

We used Scikit-learn[3] [34] to instantiate different classifiers, one for each pair of relations. Here, we are interested in two pairs of relations. On the one hand, we group together for a binary classifier the relations of hypernymy and hyponymy. On the other hand, we group together for a second binary classifier, the relations of synonymy and antonymy. Each classifier predicts whether a pair of terms refers to an association either between a pair of the first type of relation in a package, or of the other type of relation in the same package. The embedding vectors of relations are provided to the classifiers. The aim is to use these binary classifiers to predict potential semantic relations not currently recognized by JeuxDeMots. Indeed, the coverage of JeuxDeMots on semantic relations, containing business domain terms remains to be improved. Thus, the development of classifiers for the prediction of new relations can be seen as a way of enriching such resources.

4.3 Comparison to a State-of-the-Art System

We wanted to compare our proposal to a state-of-the-art system for representing relation vectors. The chosen state-of-the-art system is RELATIVE, a recent 2019 method of

---

[3] https://scikit-learn.org/stable/.

learning relation vectors based on principles that have been proven to be an effective one. This method, using a textual corpus and a model of word embeddings, learns embeddings on pairs of terms that are semantically related. The approach used is different than ours since we learn relations embeddings, which are deduced by a simple arithmetic operation in a term-pairs-oriented vector representation extracted from JeuxDeMots. On the other side, RELATIVE learns directly vector representations of relations. We chose RELATIVE because it obtained very good results during its experimentation on representative data sets in the field of document classification such as 20-newsgroups [35] or *Reuters* [36]. In addition, this system provides a very complete experimentation space. We are intrigued to be able to reuse it to represent the relations in our proposed model.

4.4 Experimental Results

We applied a series of binary classifiers on the balanced dataset as explained in Sect. 4.2. Because of the balancing of dataset in Sect. 4.1, we run our classifiers on each representative dataset of our class (type of relation). For each type of relation, we have 5 datasets. To get the most accurate result, we average these five runs. For this study, we did not go into depth in selecting the best classifier parameters. We simply identified the best relation representations given the same classifier parameters.

For each of the five runs, we use the cross-validation to obtain reliable results. A function of cross-validation is available in *Scikit-learn*. To evaluate the quality of our relation vector representations as well as those of RELATIVE, we use two evaluation methods: (1) the accuracy rate, (2) the F-measure. These two methods were specified directly as an input parameter to the cross-validation function.

The Table 1 presents the results obtained for the pair of relations synonymies and antonymy on the one hand and hypernymy and hyponymy on the other. We are interested in these two cases for each of the systems (indicated in the table). The accuracy rate and the F-measure are abbreviated as "Acc" and "F". In the Table 1, we refer to our system as REM (Relation EMbeddings).

We find that our system performs better in accuracy and F-measure than RELATIVE. We observe good results for the random forest classifiers with an accuracy rate of 78% for each of the pairs of relations for the REM system. We therefore have relevant and balanced results. Both cases can be explained by our corpus, which is a specialized one. Thus, our system is modelled for this purpose. This is not the case for RELATIVE. Nevertheless, the overall results do not reach a very high percentage because RELATIVE found fewer relations between the key terms than our system. We therefore had to level down to obtain 4669 pairs for training and test dataset of each type. Consequently, test dataset of each relation type was reduced to 1,206 pairs of key terms. Since antonymy represents a term and its opposite, it was normal to find few matching key terms, and this contributed to the drop in the number of key term pairs.

Table 1. Classification results with RELATIVE and REM.

Classifier	Synonymy vs antonymy				Hypernymy vs hyponymy			
	REM		RELATIVE		REM		RELATIVE	
	Acc	F	Acc	F	Acc	F	Acc	F
SVC	0.734	0.734	0.548	0.508	0.746	0.746	0.566	0.55
DT	0.72	0.714	0.518	0.52	0.708	0.708	0.534	0.534
RFC	**0.78**	**0.776**	**0.554**	**0.554**	0.782	0.78	**0.572**	**0.572**
k-NN ($k = 1$)	0.744	0.744	0.536	0.53	**0.8**	**0.8**	0.532	0.53
k-NN ($k = 3$)	0.614	0.614	0.544	0.534	0.794	0.79	0.548	0.548
k-NN ($k = 5$)	0.628	0.634	0.538	0.528	0.786	0.78	0.558	0.558
k-NN ($k = 7$)	0.61	0.608	0.538	0.524	**0.8**	0.738	0.558	0.558

5   Conclusion and Perspectives

In this paper, we have proposed an approach to create vector representation of lexical-semantic relations. We are interested to realize a comparative study on the quality of the vector representation of lexical-semantic relations for our system and a state-of-the-art system, called RELATIVE. We have proposed embedding vectors for key terms (multi-words and expressions) and not only for singular words. The corpus of data that we use comes from the specialized public sectors. The vector representations that we create have been validated by the lexical network JeuxDeMots for which French is the language studied in our corpus. The results obtained were satisfactory when compared to RELATIVE.

In the future works, two perspectives are available to us: (1) we will use these representations to manipulate different types of relations in addition to those already exploited, for example, the relation type "has for characteristic" or "is part of". It will thus be necessary to dwell on the rules to be added to create representations for each type of relation. Indeed, to model the relation "is part of", the question is which operations between the two terms should be performed; (2) the second is about how to detach from the binary classification. For the work presented in this article, we have chosen to put two classes simultaneously. It would therefore be interesting to add a class to manage the non-membership of the two main classes. Moreover, a validation of threshold can also be introduced to validate the relation types.

References

1. Rumelhart, D.E., McClelland, J.L.: Distributed representations (1986)
2. Elman, J.L.: Finding structure in time. Cogn. Sci. **14**, 179–211 (1990)
3. Mikolov, T., Chen, K., Corrado, G.S., Dean, J.: Efficient estimation of word representations in vector space. In: Proceedings of the International Conference on Learning Representations (ICLR), pp. 1–12 (2013)

4. Hill, F., Cho, K., Korhonen, A.: Learning distributed representations of sentences from unlabelled data. In: Proceedings of NAACL, San Diego, California, pp. 1367–1377, June 2016. https://doi.org/10.18653/v1/N16-1162

5. Reimers, N., Gurevych, I.: Sentence-BERT: sentence embeddings using siamese BERT-networks (2019)

6. Le, Q.V., Mikolov, T.: Distributed representations of sentences and documents. In: International Conference on Machine Learning, pp. 1188–1196 (2014). https://arxiv.org/abs/1405.4053

7. Chen, M.: Efficient vector representation for documents through corruption (2017)

8. Wu, L., et al.: Word mover's embedding: from Word2Vec to document embedding (2018)

9. Galke, L., Saleh, A., Scherp, A.: Word embeddings for practical information retrieval. In: INFORMATIK 2017, pp. 2155–2167 (2017). https://doi.org/10.18420/in2017_215

10. Camacho-Collados, J., Espinosa-Anke, L., Shoaib, J., Schockaert, S.: A latent variable model for learning distributional relation vectors, Macau, China (2019)

11. Lafourcade, M.: Making people play for lexical acquisition with the JeuxDeMots prototype. In: SNLP 2007: 7th International Symposium on NLP, p. 7 (2007). https://hal-lirmm.ccsd.cnrs.fr/lirmm-00200883

12. Turney, P.D.: Measuring semantic similarity by latent relational analysis. In: Proceedings of IJCAI, pp. 1136–1141 (2005)

13. Pennington, J., Socher, R., Manning, C.D.: GloVe: global vectors for word representation. In: Proceedings of EMNLP, pp. 1532–1543 (2014)

14. Jameel, S., Bouraoui, Z., Schockaert, S.: Unsupervised learning of distributional relation vectors. In: Proceedings of ACL (Volume 1: Long Paper), Australia, pp. 23–33 (2018)

15. Washio, K., Kato, T.: Filling missing paths: modeling co-occurrences of word pairs and dependency paths for recognizing lexical semantic relations. In: NAACL, pp. 1123–1133 (2018)

16. Joshi, M., Choi, E., Levy, O., Weld, D.S., Zettlemoyer, L.: Pair2Vec: compositional word-pair embeddings for cross-sentence inference (2019)

17. Espinosa-Anke, L., Schockaert, S., Camacho-Collados, J.: Relational word embeddings (2019)

18. Chin, J., Havasi, C., Speer, R.: ConceptNet 5.5: an open multilingual graph of general knowledge. In: Proceedings of AAAI, pp. 4444–4451 (2017)

19. Granada, R., Vieira, R., Trojahn, C., Aussenac-Gilles, N.: Evaluating the Complementarity of Taxonomic Relation Extraction Methods Across Different Languages (2018). http://export.arxiv.org/pdf/1811.03245

20. Panchenko, A., Naets, H., Brouwers, L., Fairon, C.: Recherche et visualisation de mots sémantiquement liés. In: TALN-RÉCITAL, pp. 747–754 (2013)

21. Hearst, M.A.: Automatic acquisition of hyponyms from large text corpora. In: Proceedings of the 14th Conference on Computational Linguistics (COLING 1992), Stroudsburg, PA, USA, vol. 2, pp. 539–545 (1992)

22. Bunescu, R.C., Mooney, R.J.: A shortest path dependency kernel for relation extraction. In: Proceedings of the Conference on Human Language Technology and Empirical Methods in Natural Language Processing, pp. 724–731 (2005)

23. Panchenko, A., et al.: TAXI at SemEval-2016 task 13: a taxonomy induction method based on lexico-syntactic patterns, substrings and focused crawling (2016)

24. Bordea, G., Buitelaar, P., Faralli, S., Navigli, R.: SemEval-2015 task 17: taxonomy extraction evaluation (TExEval) (2015)

25. Mintz, M.D., Bills, S., Snow, R., Jurafsky, D.: Distant supervision for relation extraction without labeled data. In: Proceedings of the 47th Annual Meeting of the ACL and the 4th IJCNLP of the AFNLP, pp. 1003–1011 (2009)

26. Brin, S.: Extracting patterns and relations from the world wide web. In: Atzeni, P., Mendelzon, A., Mecca, G. (eds.) WebDB 1998. LNCS, vol. 1590, pp. 172–183. Springer, Heidelberg (1999). https://doi.org/10.1007/10704656_11

27. Etzioni, O., et al.: Methods for domain-independent information extraction from the web: an experimental comparison. In: American Association for Artificial Intelligence (AAAI), pp. 391–398 (2004)

28. Bunescu, R.C., Mooney, R.J.: Subsequence kernels for relation extraction (2005)

29. Kambhatla, N.: Combining lexical, syntactic, and semantic features with maximum entropy models for extracting relations, Morristown, NJ, USA (2004)

30. Miller, S., Fox, H., Ramshaw, L., Weischedel, R.: A Novel Use of Statistical Parsing to Extract Information from Text (2000). https://aclanthology.org/A00-2030

31. Culotta, A., McCallum, A., Betz, J.: Integrating probabilistic extraction models and data mining to discover relations and patterns in text. In: Proceedings of the Human Language Technology Conference of the North American Chapter of the ACL, pp. 296–303 (2006)

32. Cartier, E.: Extraction automatique de relations sémantiques dans les définitions: approche hybride, construction d'un corpus de relations sémantiques pour le français. In: Actes de la 22e conférence sur le TALN, pp. 131–145, June 2015. https://aclanthology.org/2015.jeptal nrecital-long.12

33. Qi, P., Zhang, Y., Zhang, Y., Bolton, J., Manning, C.D.: Stanza: a Python natural language processing toolkit for many human languages (2020)

34. Pedregosa, F., et al.: Scikit-learn: machine learning in Python. Machine Learning in Python, p. 6 (2011)

35. Lang, K.: NewsWeeder: learning to filter netnews. In: Prieditis, A., Russell, S. (eds.) Machine Learning Proceedings 1995, San Francisco, CA, pp. 331–339 (1995). https://doi.org/10.1016/B978-1-55860-377-6.50048-7

36. Lewis, D.D., Yang, Y., Rose, T.G., Li, F.: RCV1: a new benchmark collection for text categorization research. J. Mach. Learn. Res. **5**, 361–397 (2004)

SAKE: A Graph-Based Keyphrase Extraction Method Using Self-attention

Ping Zhu, Chuanyang Gong, and Zhihua Wei[✉]

College of Electronic and Information Engineering, Tongji University,
Shanghai, China
{pingzhu,gongchuanyang,zhihua_wei}@tongji.edu.cn

Abstract. Keyphrase extraction is a text analysis technique that automatically extracts the most used and most important words and expressions from a text. It helps summarize the content of texts and recognize the main topics discussed. The majority of the existing techniques are mainly domain-specific, which require application domain knowledge and employ higher-order statistical methods. Supervised keyphrase extraction requires a large amount of labeled training data and has poor generalization ability outside the training data domain. Unsupervised systems have poor accuracy, and often do not generalize well. This paper proposes an unsupervised graph-based keyphrase extraction model that incorporates the words' self-attention score. Specifically, the proposed approach identifies the importance of each source word based on a word graph built by the self-attention layer in the Transformer and further introduces a new mechanism to capture the relationships between words in different sentences. The experimental results show that the proposed approach achieves remarkable improvements over the state-of-the-art models.

Keywords: Keywords extraction · Self attention · Pre-trained model

1 Introduction

With the explosive growth of the amount of information, the complexity of text processing has further increased. Keyphrases can effectively summarize the main idea of the article and are the key part of the article. Keyphrases are useful in many tasks, such as information retrieval, text summarization, or document clustering. Although scientific articles usually provide them, most types of documents do not have corresponding keywords. Therefore, the automatic extraction of key phrases is a problem worth studying.

Automatic keyphrase extraction methods can be divided into two categories: supervised and unsupervised methods.

In the field of supervised research, keyphrase extraction is formulated as a binary classification problem, where candidate phrases are classified as positive or negative [1, 2].

In the field of unsupervised research, many different kinds of techniques are applied such as language modeling [3], clustering [4], or graph-based ranking

© Springer Nature Switzerland AG 2022
H. Fujita et al. (Eds.): IEA/AIE 2022, LNAI 13343, pp. 339–350, 2022.
https://doi.org/10.1007/978-3-031-08530-7_28

[5]. Graph-based ranking techniques being considered state-of-the-art [6]. These graph-based techniques build a word graph from each target document such that nodes correspond to words and edges correspond to word association patterns. Nodes are then ranked using graph centrality measures such as PageRank [5], and the top ranked phrases are returned as keyphrases. Since their introduction, many graph-based extensions have been proposed that aim to model various types of information.

In this paper, we present a new unsupervised method called SAKE. This new method is an improvement of the TextRank method applied to keyphrase extraction [5]. In the TextRank method, a document is represented by a graph where words are vertices and edges represent co-occurrence relations. A graph-based ranking model derived from PageRank [7] is then used to assign a significance score to each word. Here, we select the important words according to the self-attention score. The self-attention layer in the Transformer [8] builds a directed graph whose vertices represent the source words and edges are defined in terms of the relevance score between each pair of source words by dot-product attention [8] between the query Q and the key K. Further, given the text lengths that Transformer can process are limited and it cannot capture relationships between words that are far apart, we proposes an mechanism to deal with this situation.

The contributions of this paper are summarized as follow.

- We propose a keyphrase extraction method based on self-attention, which can select the important words according to the self-attention score.
- We further introduce a new mechanism to capture the relationships between words in different sentences.
- We achieve state-of-the-art on the most public keyphrase extraction dataset.

2   Related Works

Extracting relevant keywords from documents is a long-standing problem [9]. The solution of this problem is of great value to common NLP tasks such as text summarization, information retrieval, text classification, and recommendation systems. The research on keyword extraction in the past few years has mainly focused on two directions: (1) supervised methods and (2) unsupervised methods.

Supervised methods treat the keywords extraction as a binary classification problem, such as GenEx [10] and KEA [1]. They treat the frequency and the position of a phrase in a target document as the most important features and both of them become the baseline system for the subsequent methods.

Unsupervised methods view keywords extraction as an importance ranking problem, such as statistic-based methods TF-IDF [11], KP-MINER [12], RAKE [13], etc., which quantifying the weight, position, mutual information, and word span in the text. Graph-based ranking methods and centrality measures are considered state-of-the-art for unsupervised keywords extraction. The graph-based approaches include TextRank [5], SingleRank [14], TopicalRank [15] and PositionalRank [16], which take preprocessed words in the text as nodes and the relationship between words as edges, then link them into network diagrams and evaluate the importance of each node, finally select the words represented by the top-ranked nodes as keywords.

Recently, the pre-trained models using the self-supervised learning method achieve the best in many NLP tasks. Wang et al. [17] uses pre-trained word embeddings and the frequency of each word to generate weighted edges between words in the document. The weighted PageRank algorithm is used to calculate the final score of the word. Mahata et al. [18] uses a similar method, using phrase embedding to represent candidate words and calculating the semantic similarity and co-occurrence of phrases to rank the importance of phrases. Sun et al. [19] proposed SIFRank, integration of statistical models and pre-trained language models to calculate the correlation between candidates and document topics. Ding et al. [20] use the attention value in BERT to rank the importance of words, and used cross-attention to identify the correlation between candidates and documents to select keywords.

In contrast to the above approaches, we present a new unsupervised method called SAKE, aimed at using attraction scores of words to build the word graph, and then the weighted PageRank algorithm is used to calculate the final score of the word.

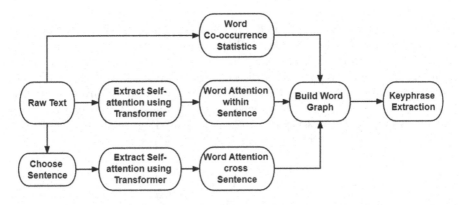

Fig. 1. The overview architecture of SAKE model

3   Algorithms

The main purpose of this research is to use the self-attention mechanism to extract key phrases from documents and identify the connections between phrases to construct a weighted word graph. Attention scores are used to rank noun phrases and then select key phrases. The following sections provide detailed descriptions of each component. The overview architecture of our model is shown in Fig. 1. Specifically, we use three different methods to capture the relationship between words. The first one is similar to textrank and counts the co-occurrence frequency between words as a score. The second is to calculate the attention score between words as a score based on the Transformer. The third method is to first select different sentence combinations and then calculate the attention score between words as the score according to the Transformer. Then, a word-graph is established according to the obtained scores, and finally the keyword extraction is performed.

3.1 Attention Score of Words Within a Sentence

The self-attention mechanism has three important components, Query (Q), Key (K) and Value (V) [8]. In the definition of self-attention, Q is K itself. dk is the dimension of Q and K. Scaled dot-product attention is widely used in self-attention networks. Scaling the dot product prevents the dot product from growing too fast, which leads to the gradient of softmax function (shown as Eq. (1)) being too small. Multi-head attention parallel computing self-attention. Each self-attention mechanism is called a "head". In our mechanism, we obtain the attention distributions with the sum of multiple heads.

$$\text{Attention}\ (Q, K, V) = \text{softmax} \left(\frac{QK^T}{\sqrt{d_k}} \right) V \tag{1}$$

We use the Transformer-based pre-trained model BERT to obtain self-attention weights. Because Bert uses BPE (Byte-Pair Encoding), a word may be encoded into multiple tokens, for example, "transformers" is encoded into "transform" and "##ers". What we get directly is a token-level attention matrix. In order to get the attention score between two words, we need to convert the token-level attention matrix into a word-level attention matrix. Given two words w_i and w_j, where w_i is consisted of n tokens $t_p, p \in [1, n]$, w_j is consisted of m tokens $t_q, q \in [1, m]$, we simply sum the attention scores.

$$Attn_{w_i w_j} = \frac{1}{n} \sum_{p=1}^{n} \sum_{q=1}^{m} Attn_{t_p t_q} \tag{2}$$

For a document d for extracting keyphrases, d is consisted of n sentences $sent_i, i \in [1, n]$, $sent_i$ represents the ith sentence. The attention score between two words w_p and w_q in the same sentence $sent_i$ is calculated as follows:

$$W_{<w_p, w_q>_{within}} = Attn_{w_p w_q}(sent_i), w_p \in sent_i, w_q \in sent_i \tag{3}$$

3.2 Attention Score of Words Cross Sentences

The self-attention mechanism based on Transformer can well capture the correlation of each word in one sentence, which is of great help in constructing the word graph. However, there are certain connections between two words in different sentences, and Transformer cannot deal with this situation because the input text that Transformer can process has a length limit. Therefore, we propose a mechanism that can capture the relations of two words in different sentences.

For the two words w_p and w_q located in different sentences $sent_i$ and $sent_j$, this method is no longer applicable. Therefore, we concatenate the two sentences $sent_i$ and $sent_j$ into one sentence and then use Transformer to calculate the weight. How to choose $sent_i$ and $sent_j$ is a problem worth considering. Here we design a discriminant function F to select sentences. If the interval between two sentences is less than a certain threshold (we choose 5), the discriminant is true.

$$W_{<w_p,w_q>_{cross}} = Attn_{w_p w_q}(Concat(sent_i, sent_j)), w_p \in sent_i, w_q \in sent_j \quad (4)$$

when calculating the attention score of concated sentence, score of words in one sentence will be calculated repeatedly in each sentences, so we simply ignored it and only calculate attention scores of words in different sentences.

3.3   Graph-Based Ranking

Graph Construction. Our approach to constructing the graph is similar to TextRank. Let d be a target document for extracting keyphrases. We first apply the part-of-speech filter and then select as candidate words only nouns and adjectives.

We build a word graph $G = (V, E)$. Two nodes v_i and v_j are connected by an edge $(v_i, v_j) \in E$ if the weights is not zero.

The weight of edge (v_i, v_j) is calculated as follows:

$$W_{<v_i,v_j>} = \alpha \times W_{<v_i,v_j>_{within}} + \beta \times W_{<v_i,v_j>_{cross}} + (1 - \alpha - \beta) \times W_{<v_i,v_j>_{textrank}} \quad (5)$$

$W_{<v_i,v_j>_{textrank}}$ is calculated according to the co-occurrence count of two words in consecutive tokens with window size w in document D, $W_{<v_i,v_j>_{within}}$ is the attention score between words in one sentence described in Eq. (3), and $W_{<v_i,v_j>_{cross}}$ is the attention score between words across sentences described in Eq. (4). α and β are the corresponding weights and we set them all to 0.45.

Node Ranking. Once the graph is created, the graph-based ranking algorithm PageRank is used to calculate the importance score of each node. The PageRank score of a vertex v_i, i.e., $S(v_i)$, can be obtained in an algebraic way by recursively computing the following equation:

$$S(v_i) = (1 - \lambda) + \lambda \times \sum_{v_j \in Adj(v_i)} \frac{W_{<v_j,v_i>} \times S(v_j)}{\sum_{v_k \in Adj(v_j)} W_{<v_j,v_k>}} \quad (6)$$

where λ is the dumping factor generally defined to 0.85, $w_{j,i}$ is the edge weight between v_i and any of the neighbor node v_j, $Adj(v_i)$ is the set of neighbor nodes connect to node v_i.

3.4   Forming Candidate Phrases

Candidate words that have contiguous positions in a document are concatenated into phrases. We consider noun phrases that match the regular expression (adjective)*(noun)+, of length up to three, (i.e., unigrams, bigrams, and trigrams). Finally, phrases are scored by using the sum of scores of individual words that comprise the phrase [14]. The top-scoring phrases are output as predictions (i.e., the predicted keyphrases for the document).

4   Experimental Settings

4.1   Datasets

To evaluate the effectiveness of our system, we tested it on three different English document collections: Inspec [2], WWW [21] and SemEval2017 [22]. Inspec consists of 2,000 abstracts of scientific journal papers from Computer Science collected between the years 1998 and 2002. The WWW collection is based on the abstracts of papers collected from the World Wide Web Conference (WWW) published during the period 2004–2014, with 1330 documents. The gold-keywords of these papers are the author-labeled terms. SemEval2017 consists of 500 paragraphs selected from 500 ScienceDirect journal articles. Each text has a number of keywords selected by one undergraduate student and an expert annotator.

A Summary of Datasets is provided in Table 1.

Table 1. A Summary of datasets

Dataset	Inspec	WWW	SemEval2017
Language	EN	EN	EN
Type of Doc	Abstract	Paper	Paragraph
Domain	Comp. Science	Comp. Science	Misc
Docs	2000	1330	493
Gold Keys	14.62	5.80	18.19
AveWords	128.2	84.08	178.22
AveSentences	5.37	7.36	3.71

4.2   Baselines

To evaluate the efficiency of our method, we compare SAKE with six most commonly used algorithms, e.g., TF-IDF [11], TextRank [5], SingleRank [14], PositionalRank [16], TopicRank [15], Rake [13], and Yake [9]. TF-IDF calculates the tf of each candidate keyword in the target document, and idf is estimated with our datasets. TextRank ranks the importance of words by utilizing the co-occurrence of words on the semantic graph. In SingleRank, similar to TextRank, the weight of edge is the co-occurrence number of words in the slide window. PositionalRank uses the information from all positions of a word occurrence to adjust the weight of edges. TopicRank represents documents as a graph, where nodes instead of terms are topic clusters of single and composed expressions. RAKE considers sequences of terms as candidate keywords, and based on this builds a matrix of term co-occurrences. YAKE is a light-weight unsupervised automatic keyword extraction method which rests on text statistical features extracted from single documents to select the most important keywords of a text.

4.3   Evaluation Methods

The precision Pkw, recall Rkw, and $F1$ score are used to evaluate the performance of methods, as defined:

$$P_{kw} = \frac{|A \cap B|}{|B|} \tag{7}$$

$$R_{kw} = \frac{|A \cap B|}{|A|} \tag{8}$$

$$F_1 = 2 * \frac{P_{kw} * R_{kw}}{P_{kw} + R_{kw}} \tag{9}$$

where A represents a collection of labeled keywords, B represents a collection of extracted keywords.

5   Results

As already mentioned, prior work on keyphrase extraction report results also in terms of precision (P), recall (R), and F1-score. Consistent with these works, in Table 2, we show the results of the comparison of SAKE with all baselines, in terms of P, R and F1 for top k = 5, 10, 15, 20 predicted keyphrases, on all three datasets. As can be seen from the table, SAKE achieve best results on most datasets.

For example, on Inspec SAKE performs better than other models when k is taken to all values. When K is taken to 15, SAKE achieves the best score of 27.02 compared to 25.27 achieved by SingleRank and 24.21 achieved by PositionRank. The score of other models are around 20 or lower. It behaves similarly when K takes other values.

Table 2. Baseline comparison with Precision (P), Recall (R), and F-score (F1)

Dataset	Algorithm	Top5			Top10			Top15			Top20		
		P (%)	R (%)	F1 (%)	P (%)	R (%)	F1 (%)	P (%)	R (%)	F1 (%)	P (%)	R (%)	F1 (%)
Inspec	PositionRank	31.42	13.47	18.85	25.58	20.87	22.99	22.60	26.06	24.21	20.65	29.40	24.26
	SingleRank	30.53	13.07	18.31	26.77	21.79	24.03	23.54	27.27	25.27	21.08	30.62	24.97
	TextRank	18.20	7.31	10.43	16.33	10.49	12.78	15.09	11.46	13.03	14.74	11.72	13.06
	TFIDF	17.75	8.20	11.21	15.95	14.32	15.09	14.84	19.57	16.88	14.02	23.90	17.67
	TopicRank	27.21	11.87	16.53	21.68	17.69	19.48	19.11	21.38	20.18	18.13	24.13	20.70
	YAKE	26.06	11.60	16.05	20.78	18.17	19.39	18.15	23.40	20.44	16.33	27.52	20.50
	SAKE(-cross)	32.45	13.83	19.39	27.54	22.39	24.70	24.26	28.00	26.00	21.61	31.31	25.57
	SAKE(-within)	32.41	13.87	19.42	28.17	22.87	25.25	24.80	28.52	26.53	22.01	31.76	26.00
	SAKE	**33.56**	**14.37**	**20.12**	**28.70**	**23.41**	**25.79**	**25.22**	**29.09**	**27.02**	**22.31**	**32.28**	**26.38**
WWW	PositionRank	13.11	12.73	12.92	**11.31**	15.95	13.23	**10.31**	17.44	**12.96**	**9.72**	18.41	**12.72**
	SingleRank	10.25	9.98	10.11	9.92	13.85	11.56	9.45	16.26	11.95	8.99	17.48	11.87
	TextRank	10.17	6.67	8.05	9.37	8.78	9.07	9.15	9.91	9.51	9.07	10.26	9.63
	TFIDF	**13.76**	**15.39**	**14.53**	10.45	**22.63**	**14.30**	8.22	**25.22**	12.40	7.16	**26.38**	11.27
	TopicRank	11.92	11.27	11.59	9.93	12.84	11.19	9.15	13.64	10.95	8.75	13.96	10.76
	YAKE	7.85	9.26	8.50	7.72	17.38	10.69	7.15	23.40	10.96	6.51	26.12	10.43
	SAKE(-cross)	11.21	11.07	11.14	10.37	14.86	12.22	9.66	16.89	12.29	9.23	18.37	12.28
	SAKE(-within)	11.03	10.80	10.92	10.26	14.61	12.05	9.70	17.03	12.36	9.22	18.33	12.27
	SAKE	11.39	11.26	11.32	10.55	15.06	12.41	9.86	17.30	12.56	9.39	18.69	12.50
SemEval2017	PositionRank	37.70	11.93	18.12	32.84	20.56	25.29	29.97	27.61	28.74	27.56	32.97	30.02
	SingleRank	35.01	11.38	17.17	32.92	20.78	25.47	30.67	28.45	29.52	28.78	34.97	31.57
	TextRank	19.12	5.99	9.12	17.54	10.42	13.07	16.18	12.70	14.23	15.90	13.85	14.80
	TFIDF	30.05	9.34	14.25	24.06	14.61	18.18	20.50	18.61	19.51	18.60	22.76	20.47
	TopicRank	35.62	11.35	17.22	28.41	17.77	21.86	24.82	22.75	23.74	22.82	26.91	24.69
	YAKE	24.83	8.19	12.32	23.61	15.37	18.62	21.94	20.93	21.42	20.28	25.39	22.55
	SAKE(-cross)	38.13	12.36	18.67	36.10	22.85	27.98	33.13	30.80	31.93	30.32	36.79	33.24
	SAKE(-within)	38.01	12.40	18.70	36.57	23.10	28.32	33.76	31.30	32.49	31.33	37.97	34.33
	SAKE	**41.51**	**13.35**	**20.20**	**38.53**	**24.29**	**29.79**	**35.10**	**32.71**	**33.86**	**32.22**	**39.00**	**35.29**

On the WWW dataset, the performance of SAKE is not the best, but it has similar performance to other excellent models. For example, when k is 15, The F1 score of SAKE is 12.56 and close to 12.96 achieved by PositionRank. When k is 20, The F1 score of SAKE is 12.50 and close to 12.72 achieved by PositionRank.

On the SemEval2017 dataset, we can see that the performance of SAKE obviously outperforms other methods. For example, when k is 15, the F1 score of SAKE is 33.86, which is nearly 4 points higher than 29.52 achieved by SingleRank. When k is 20, the F1 score of SAKE is 35.29, which is nearly 3 points higher than 31.57 achieved by SingleRank.

In addition, it can be seen from the Table 2 that the number of extracted keywords k has an important impact on the performance. When k is 5, the F1 score of SAKE is obviously much lower than when k takes other values, which is at least 5 points lower on the Inspec, at least 1 point lower on the WWW data set, and at least 5 points lower on the SemEval2017. Other models have similar performance. In order to achieve better performance, the number of keywords extracted should be more than 10.

It is worth noting that our model also introduces the method of word attention in one sentence and word attention between sentences to construct word graphs. In order to evaluate the contribution of these two components introduced in the model, we also conducted a component ablation experiment. As shown in Table 2, SAKE, SAKE(-cross) and SAKE(-within) are the results of the three experiments, SAKE(-within) is to remove the word attention score in the sentence, and SAKE(-cross) is to remove word attention score between sentences. It can be seen from the experimental results in the table that the complete model has a certain improvement compared with the control model, and removing any one of them will bring about a performance loss of 2%-6%. There is no significant difference between the results of the two control groups. In Inspec and SemEval2017 we can see that SAKE(-cross) has a slightly lower result than SAKE(-within) and the opposite in the WWW. Overall, the introduction of these two methods at the same time has a positive effect on the effect of the model.

The PR curve (shown in Fig. 2) was drawn using the top 20 candidate keywords generated by each model for overall comparison. The PR curve shows consistent results, i.e. SAKE's results are relatively poor on the WWW dataset and better than all benchmark models on the Inspec dataset and SemEval2017 dataset.

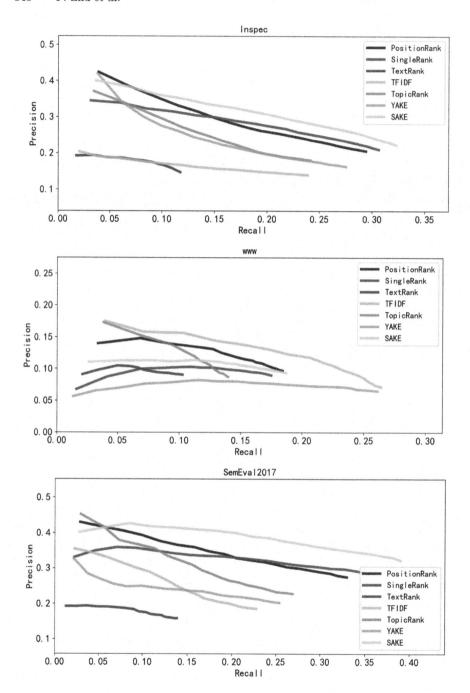

Fig. 2. The PR-curve evaluation of models performance

6   Conclusion

In this paper we presented SAKE, a self-attention guided unsupervised method for keyphrase extraction. We evaluate our method on six keywords extraction datasets and compare it with six most commonly used keywords extraction methods in order to validate the effectiveness of the proposed method. Results show that our method outperforms compared methods.

In the future, we will further investigate the keywords extraction method of text in other domains, and explore the supervised method in the presence of unlabeled data from the perspective of transfer learning. We also prepare to construct a single-text knowledge graph based on the keywords extracted and acquire the knowledge contained in the text more intuitively. As for the keyphrase selection, our experiments show that the current method does not provide the best solution. We plan to improve it in a more efficient way.

Acknowledgments. The work is partially supported by the National Nature Science Foundation of China (No. 61976160, 61906137, 61976158, 62076184, 62076182) and Shanghai Science and Technology Plan Project (No. 21DZ1204800) and Technology research plan project of Ministry of Public and Security (Grant No. 2020JSYJD01).

References

1. Witten, I.H., Paynter, G.W., Frank, E., Gutwin, C., Nevill-Manning, C.G.: KEA: practical automated keyphrase extraction. In: Design and Usability of Digital Libraries: Case Studies in the Asia Pacific, pp. 129–152. IGI Global (2005)
2. Hulth, A.: Improved automatic keyword extraction given more linguistic knowledge. In: Proceedings of the 2003 Conference on Empirical Methods in Natural Language Processing, pp. 216–223 (2003)
3. Tomokiyo, T., Hurst, M.: A language model approach to keyphrase extraction. In: Proceedings of the ACL 2003 Workshop on Multiword Expressions: Analysis, Acquisition and Treatment, pp. 33–40 (2003)
4. Liu, Z., Li, P., Zheng, Y., Sun, M.: Clustering to find exemplar terms for keyphrase extraction. In: Proceedings of the 2009 Conference on Empirical Methods in Natural Language Processing, pp. 257–266 (2009)
5. Mihalcea, R., Tarau, P.: TextRank: bringing order into text. In: Proceedings of the 2004 Conference on Empirical Methods in Natural Language Processing, pp. 404–411 (2004)
6. Hasan, K.S., Ng, V.: Conundrums in unsupervised keyphrase extraction: making sense of the state-of-the-art. In: Coling 2010: Posters, pp. 365–373 (2010)
7. Brin, S., Page, L.: The anatomy of a large-scale hypertextual web search engine. Comput. Netw. ISDN Syst. **30**(1–7), 107–117 (1998)
8. Vaswani, A., et al.: Attention is all you need. In: Advances in Neural Information Processing Systems, pp. 5998–6008 (2017)
9. Campos, R., Mangaravite, V., Pasquali, A., Jorge, A., Nunes, C., Jatowt, A.: YAKE! Keyword extraction from single documents using multiple local features. Inf. Sci. **509**, 257–289 (2020)
10. Turney, P.D.: Learning algorithms for keyphrase extraction. Inf. Retrieval **2**(4), 303–336 (2000). https://doi.org/10.1023/A:1009976227802

11. Sparck Jones, K.: A statistical interpretation of term specificity and its application in retrieval. J. Doc. **28**(5), 111–121 (1972)
12. El-Beltagy, S.R., Rafea, A.: KP-Miner: a keyphrase extraction system for English and Arabic documents. Inf. Syst. **34**(1), 132–144 (2009)
13. Rose, S., Engel, D., Cramer, N., Cowley, W.: Automatic keyword extraction from individual documents. In: Text Mining: Applications and Theory, vol. 1, pp. 1–20 (2010)
14. Wan, X., Xiao, J.: Single document keyphrase extraction using neighborhood knowledge. In: AAAI, vol. 8, pp. 855–860 (2008)
15. Bougouin, A., Boudin, F., Daille, B.: TopicRank: graph-based topic ranking for keyphrase extraction (2013)
16. Florescu, C., Caragea, C.: PositionRank: an unsupervised approach to keyphrase extraction from scholarly documents. In: Proceedings of the 55th Annual Meeting of the Association for Computational Linguistics (Volume 1: Long Papers), pp. 1105–1115 (2017)
17. Wang, R., Liu, W., McDonald, C.: Corpus-independent generic keyphrase extraction using word embedding vectors. In: Software Engineering Research Conference, vol. 39, pp. 1–8 (2014)
18. Mahata, D., Kuriakose, J., Shah, R., Zimmermann, R.: Key2Vec: automatic ranked keyphrase extraction from scientific articles using phrase embeddings. In: Proceedings of the 2018 Conference of the North American Chapter of the Association for Computational Linguistics: Human Language Technologies, Volume 2 (Short Papers), pp. 634–639 (2018)
19. Sun, Y., Qiu, H., Zheng, Y., Wang, Z., Zhang, C.: SIFRank: a new baseline for unsupervised keyphrase extraction based on pre-trained language model. IEEE Access **8**, 10896–10906 (2020)
20. Ding, H., Luo, X.: AttentionRank: unsupervised keyphrase extraction using self and cross attentions. In: Proceedings of the 2021 Conference on Empirical Methods in Natural Language Processing, Punta Cana, Dominican Republic, pp. 1919–1928. Association for Computational Linguistics, November 2021
21. Gollapalli, S.D., Caragea, C.: Extracting keyphrases from research papers using citation networks. In: Twenty-Eighth AAAI Conference on Artificial Intelligence (2014)
22. Augenstein, I., Das, M., Riedel, S., Vikraman, L., McCallum, A.: SemEval 2017 task 10: scienceie-extracting keyphrases and relations from scientific publications. arXiv preprint arXiv:1704.02853 (2017)

Synonym Prediction for Vietnamese Occupational Skills

Hai-Nam Cao[1], Duc-Thai Do[1], Viet-Trung Tran[1(✉)], Tuan-Dung Cao[1], and Young-In Song[2]

[1] Hanoi University of Science and Technology, Hanoi, Vietnam
{Nam.CH212239M,Thai.DD211260M}@sis.hust.edu.vn,
{trungtv,dungct}@soict.hust.edu.vn
[2] NAVER, Seongnam-si, Republic of Korea
song.youngin@navercorp.com

Abstract. To date, online job portals are the main media to connect job seekers and employers. These platforms offer job catalogues and search functionality where one can search for jobs that alphabetically match specific keywords. This setup has limitations in terms of retrieval and accuracy since keyword matching suffers from inconsistent representations of meaning such as typo, slang, abbreviations, and synonyms. Thus, to enable advanced search, it is needed to have a machine learning model that can automatically detect occupational skill synonyms. In this work, we propose a rational process to construct a practical labeled Vietnamese Skill Synonym (ViSki) dataset. We experiment with 2 approaches for the synonym prediction task: cosine similarity-based and classification-based. Our best model, XGBoost with LaBSE embedding, achieved 96.15% accuracy and an F1-score of 0.9285.

Keywords: Skill synonym · Word embedding · XGBoost · Siamese network

1 Introduction

Nowadays, The labor market in Vietnam has been developing in an imbalanced manner [12]: the recruitment demand is increasing, while the unemployment rate is also at the highest level. One of the root causes is the competency gap, in which occupational skills that workers are trained with do not match in quality and quantity with the actual need.

To address the problem, many platforms have recently been developed to bring job applicants and recruitment managers closer together by using simple search engines to complex recommendation systems. Unsurprisingly, occupational skills are used as one of the key matching factors since they technically can tell if the candidates are qualified for a specific job. However, current implementation approaches still have limitations in terms of retrieval and precision as the

Supported by NAVER corporation.

occupational skills can be represented differently with inconsistent terms in job postings and resumes, e.g., "OOP" and "Lập trình hướng đối tượng " (object-oriented programing in English). Therefore, it is essential to have a model that can automatically determine if two skills are synonyms.

In order to train these models in a supervised setting, a large amount of labeled Vietnamese synonym skill pairs is required. In this paper, we present a rational process for creating a labeled Vietnamese skill synonym dataset, named ViSkim and experimental results on training several synonyms prediction models. To construct the dataset, we leveraged Wikipedia as a synonym suggestion tool to minimize human effort. To build the prediction models, we leverage distributed representations of phrases and XGBoost models with various features engineering settings.

The rest of this paper is organized as follows. We review related literature on Sect. 2. Section 3 describes how we build our ViSki dataset in three main steps: skill collection, Wikipedia-based synonym suggestion, and synonym set verification. In Sect. 4, we detail the construction of the prediction models and various feature engineering configurations. Section 5 discusses the experimental results. We conclude and discuss the future direction in Sect. 6.

2   Related Work

Synonym Prediction. There are various approaches have been proposed for synonym prediction. [2,30] exploits a structured knowledge base such as query logs for synonym discovery. [9,28,34] considers each pair of synonymous words as a sentence to exploit lexical and textual patterns. [15,21,26] determines synonyms by training classification models on term embedding vectors. [24] integrates two kinds of mutually complementing signals: distributional features based on corpus-level statistics and textual patterns based on local contexts for synonym discovery. SurfCon [33] combined surface form information and distributional features for synonym discovery on privacy-aware clinical data.

Text Embedding. Our work is also related to text embedding techniques, which convert textual data into vector representations. Word2vec [17] and Glove [23] have been applied to detect relations in medical phrases [20,31]. Recently, BERT [6] and its variants are taken into account as the best methods for text embedding tasks, by achieving SOTA results in various sub-tasks in natural language processing. However, these methods only learn word representation from textual features and do not exploit knowledge bases for embedding learning. Pei et al. [22] employed Graph Attention Network to learn the distributional features of terms in a graph. They constructed a graph of terms and various edge types: term to term, synonym set to synonym set, and term to synonym set. A drawback of this approach is that it cannot determine the embeddings for the terms that do not appear in the training corpus (the out of vocabulary problem).

3   Vietnamese Skill Synonym Dataset

Figure 1 describes the construction of our dataset in three steps: Skill collection, Wikipedia-based synonym suggestion, and synonym set verification.

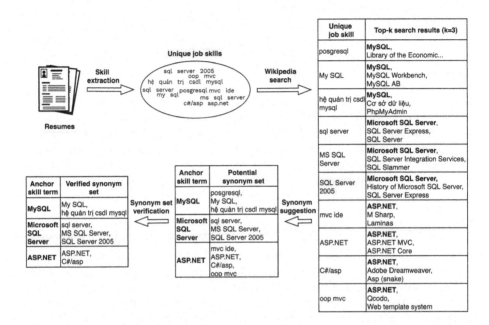

Fig. 1. ViSki dataset construction overview

3.1   Phase 1: Skill Collection

To collect Vietnamese occupational skills, we periodically crawled 12.629 Vietnamese resumes by our crawler system since 2020 Oct. As there are no Named Entity Recognition models for extracting Vietnamese occupational skills due to the lack of labeled data, we use crowd-sourcing to extract skills from the resumes. In detail, we divide the resumes into 20 batches and distribute them to 20 crowd-sourced workers. The workers are asked to extract only occupational skills from the resumes.

We obtain 10,245 unique job skills from the resumes. At the first glance, the quality of the skills is not as good as expected since we can observe many typos and residual words or symbols in the data. For example, some of our job skills are: "C #)", "với OpenCV", "java oop unit test".

3.2   Phase 2: Wikipedia-Based Synonym Suggestion

Wiki, NYT [24] are two public English benchmark datasets that are often used in synonym discovery tasks [22,24,27]. The authors apply Freebase[1] as the

---
[1] https://developers.google.com/freebase/.

knowledge base and DBpedia Spotlight[2] as the entity linker for Wiki and NYT datasets. For biomedical entities in [29], the authors mapped disease entities in the NCBI Disease Corpus [7] and BioCreative V CDR [14] into the MEDIC dictionary [5] and chemical entities in the BioCreative V CDR datasets into the Comparative Toxicogenomics Database [4].

These methods are effective in automatically constructing the labeled dataset. However, these techniques cannot be applied in our work as there are no such knowledge bases and entity linkers for Vietnamese occupational skills.

Differentiate from knowledge bases which are authoritative resources with a dedicated team of content producers and managers, Wikipedia is one of the largest and most popular sources for information on the Internet with many worldwide content editors and contributors. Wikipedia, although not being as accurate as knowledge bases, Wikipedia has a huge amount of information. It is continuously updated and supports many languages, including Vietnamese. Therefore, we propose using Wikipedia as a synonym suggestion tool to obtain Vietnamese job skill synonym sets. Our process consists of the two following steps:

Step 1 We search each skill in our collected skills on Wikipedia to get top-k results using the Wikipedia-provided API. We assume skills with no Vietnamese diacritics as English skills and search them on the English-language edition of Wikipedia because it is more complete. Otherwise, if the skills are with Vietnamese diacritics, we search them on the Vietnamese version for more accurate results. The results are often standard or canonical names of the skills. For example, searching "sql server" on Wikipedia returns "Microsoft SQL Server" as the top-1 result.

Step 2 We assume that if there are m matches between the terms' search results ($1 \leq m \leq k$), they are most likely synonyms. If m is big, we can only capture very similar lexical terms to be synonyms, which reduces the diversity of data. With m being fixed, choosing k is a trade-off between synonym set quality and a higher probability of missing some synonyms. We set $m = 1$ and experiment with $k = 1, 2, 3$. Finally, we select $k = 1$ as it gets us the best quality of synonyms sets without losing many synonyms. After this step, we can create the potential synonym sets, each set contains all terms that have the same top-1 search result. For instance, "sql server", "MS SQL Server", and "SQL Server 2005" have the same top-1 search result, so we consider them as potential synonyms and group them into the same set.

From our 10,245 unique skills, we obtain 1228 synonym sets corresponding to 5337 unique skills. The remaining 4908 unique skills don't belong to any sets.

3.3   Phase 3: Synonym Set Verification

With Wikipedia-based suggestions, the potential synonym sets are of good quality. We further verify these sets to obtain accurate data. A skill that should not

---

[2] https://github.com/dbpedia-spotlight/dbpedia-spotlight.

belong to its current synonym set is moved to a suitable existing set or deleted if there is no such set. For example, one of our potential synonym sets which is the output of phase 2 contains "machine learning", "machine learning big", "learning n3". We will remove incorrect terms "machine learning big", "learning n3" and move suitable terms to the set such as "ML", "học máy", " thuật"," toán học máy", "máy học". Sets with all skills within that are not synonymous are removed. By using Wikipedia as a synonym suggestion tool, we don't need to verify every skill pair but just potential synonym sets, which reduce human effort.

After the verification process, we remain 740 correct synonym sets corresponding to 3,787 unique skills. The sets have various sizes, of which the smallest size is 2 and the biggest size is 84 skills. These synonym sets will be used to generate data to train and test synonymous skill prediction models.

Table 1. Data statistics of synonym datasets

Dataset	Documents	Sentences	Entities	Synonym sets
Wiki [24]	100.000	6.839.331	98.664	4.920
NYT [24]	118.664	3.002.123	31.702	1.494
NCBI disease [7]	792	–	1.855	335
BC5CDR disease [14]	1.500	–	2.727	519
BC5CDR chemical [14]	1.500	–	1.734	393
ViSki	**12.629**	**121.958**	**10.245**	**740**

Table 1 gives some comparisons between our dataset and the other public datasets, which were used in previous researches in synonym discovery [10,18, 22,29]. The statistics of Wiki, NYT dataset can be found in [22]. About the NCBI Disease, BC5CDR Disease, and BC5CDR Chemical, we use the data at the BioSyn repository[3] for statistics.

4   Vietnamese Skill Synonym Prediction

4.1   Problem Statement

For an occupational skill, its synonym refers to a string that can be used as an alternative name to describe that skill. For instance, both "Microsoft SQL Server" and "ms sql" consider as relational database management systems developed by Microsoft. "Microsoft SQL Server" and "ms sql" are called a pair of synonym skills. Given a pair of skills, our problem is to identify if they are synonymous or not.

---

[3] https://github.com/dmis-lab/BioSyn.

4.2   Methodology

Cosine Similarity-Based: Distributed Representation of Phrases. This approach identifies the relation of two given skills based on the cosine similarity between embedding vectors of the inputs. If the cosine score is greater than the defined threshold (e.g., 0.7), the input skills are synonymous. To encode skills, we employ various embedding models, including:

- PhoBERT [19]: a Vietnamese pre-trained language model which was developed by VinAI. It is built based on Roberta architecture [16]. Model is trained in a 20GB pre-training dataset of Vietnamese uncompressed texts.
- LaBSE [8]: a massive language model supporting 109 languages including Vietnamese. LaBSE combines masked language model and translation language model [11] pretraining with a translation ranking task using bidirectional dual encoders [35].
- S-PhoBERT: a fine-tuning PhoBERT model based on the Sentence Transformer architecture [25].

Figure 2 depicts our network architecture to fine-tune the S-PhoBERT model. It is a Siamese network [1] where each branch consists of a pre-trained PhoBERT and a pooling layer. The pooling layer aims to produce distributed representation of input sentences by computing the mean of all token embedding vectors. We use the mean-square-error loss as the objective function.

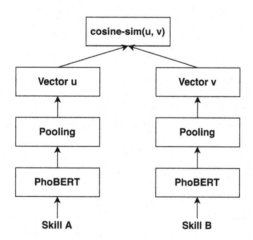

Fig. 2. S-PhoBERT architecture with classification objective function. The two S-PhoBERT networks have tied weights (Siamese network structure).

Classification-Based: Synonym Classification Model. The model adopted is XGBoost [3], a powerful model that can efficiently handle large-scale machine learning tasks. It can achieve superior performance on both regression and classification problems.

The input of synonym classifier-based XGBoost is a feature vector constructed from the two occupational skill inputs. Our feature vector consists of 9 features, as follows:

Is Initial: returns 1 if a skill is the same as the concatenation of the first character of each word in the other skill. Otherwise, return 0.

Levenshtein Distance: returns the Levenshtein distance score [13] of the two skills.

Jaro-Winkler Distance: returns the Jaro-Winkler distance score [32] of the two skills.

Character in common: returns the number of common characters between the two skills.

Token in common: returns the number of common tokens between the two skills.

Difference in numbers of tokens: returns the number of different tokens of the two skills.

Cosine similarity of embeddings: returns the cosine similarity of the two skill vectors. We experiment with three variants, Each leverages an embedding model mentioned in Sect. 4.2.

Co-occurrence-based: returns the co-occurrence frequency of the two skills, which are the number of times two terms co-occur in the same resume. The larger the co-occurrence value of the two skills is, the lower probability they are synonymous.

Common neighbor-based: returns the numbers of common neighbors of the two skills. A neighbor of a skill **s** is any skill that appears together with **s** in documents. The larger length of the common neighbor list of the two input skills is, the larger probability they are synonymous.

Table 2. Example of feature engineering for XGboost

Feature description	Skill pair → Output
is Initial	(Object Oriented Programming, OOP) → 1
Levenshtein Distance	(php, java) → 4
Jaro-Winkler Distance	(ms exchange 2013, microsoft exchange) → 0.74
Character in common	(java, angular) → 1
Tokens in common	(nodejs, angularJS) → 0
Difference in numbers of tokens	(nodejs, lập trình php) → 2
Cosine similarity of embeddings	(skillA, skillB) → $cosine(x_{skillA}, x_{skillB})$
Co-occurrence-based	co_occur[c++, php] = 166 → 166
Common in neighbor-based	neighbor[php] = [html, c #, java], neighbor[c++] = [oracle, java], common_neighbor(php, c++) → 1

One example of these features is depicted in Table 2. To train the XGBoost model, we feed the model with a training set as a list of $[x, y]$, where x is the feature vector of the input skill pair $(skill_1, skill_2)$ and y is the label that expresses if the two skills $skill_1$ and $skill_2$ are synonymous or not.

5   Experiments and Results

In this section, we describe the methodology for generating the synonyms skill from the synonym sets to be used in evaluation for the synonym prediction task. In addition, we report the models and the measures used for experimental evaluation.

5.1   Dataset Construction

Data used to train and evaluate the synonym prediction model was constructed based on synonym sets acquired in Sect. 3.3. In detail, we use a 50/30/20 split on these sets to get the training, validation, and test set respectively.

The combination of all skill terms within each set are collections of positive synonym pairs. For generating negative pairs, we randomly pick two skills from two different synonym sets. The number of negative pairs is selected to be 5 times the number of positive pairs.

Finally, our ViSki dataset contains:

A training set contains 76350 pairs (12725 positive pairs and 63625 negative pairs).
A validation set contains 53910 pairs (8985 positive pairs and 44925 negative pairs).
A test set contains 15294 pairs (2549 positive pairs and 12745 negative pairs).

Our data is available at: https://github.com/CaoHaiNam/ViSki.

5.2   Experiments

Performance Evaluation. We consider our task as a binary classification problem. Thus, the evaluation metric that we adopt includes accuracy (Acc), precision (P), recall (R), and f1-score (F1). The classification accuracy, precision, recall, f1-score are defined as:

$$Acc = \frac{TP + TN}{N} \tag{1}$$

$$P = \frac{TP}{TP + FP} \tag{2}$$

$$R = \frac{TP}{TP + FN} \tag{3}$$

$$F1 = 2 \times \frac{P \times R}{P + R} \tag{4}$$

where N is the number of the test samples. TP, TN, FP, FN are the number of true positive, true negative, false positive, false negative samples, respectively.

In these above metrics, accuracy is the primary metric to evaluate the quality of the classification model, while the precision, recall, f1-score are usually employed for imbalanced test sets.

Implementation Details

Cosine similarity-based. We use the Pytorch library to develop the S-PhoBERT model. The Siamese based S-PhoBERT network is optimized using Adam optimizer with initial learning rate = 2e-5, $\beta_1 = 0.9$, $\beta_2 = 0.999$, weight_decay = 0.01. We train a model for 25 epochs in the validation set. After accomplishing the training process, we use the S-PhoBERT model for encoding skills.

Classification-based. We use open-source xgboost[4] to implement the XGBoost model. Tree-based models are used as learners and the objective function is binary cross-entropy. We apply a grid search strategy to tune important parameters of the model to maximize the f1-score on the validation set. These tuning parameters include:

max_depth: The maximum depth per tree. A deeper tree might increase the performance, but also the complexity and chances to overfit.

learning_rate: The learning rate determines the step size at each iteration while the model optimizes toward its objective. A low learning rate makes computation slower and requires more rounds to achieve the same reduction in residual error. But it optimizes the chances to reach the best optimum.

n_estimators: The number of trees in the ensemble. It is equivalent to the number of boosting rounds.

colsample_bytree: Represents the fraction of columns to be randomly sampled for each tree. It might improve overfitting.

subsample: Represents the fraction of observations to be sampled for each tree. Lower values prevent overfitting but might lead to under-fitting.

The detailed hyperparameter configuration for the proposed model is given in Table 3. We apply the default values for other parameters of the model.

Table 3. Hyper parameter configuration for XGBoost

Parameter	Value
max_depth	3, 5, 10
learning_rate	0.01, 0.1, 0.5
n_estimators	100, 200
colsample_bytree	0.5, 1
subsample	0.5, 1

---

[4] https://github.com/dmlc/xgboost.

5.3   Results

In this section, we report the experimental results and for our proposed models. All experiments are conducted in the test set. The result is detailed in Table 4.

Table 4. Experimental results.

Method	Embedding model	Accuracy	Precision	Recall	F1-score
Cosine similarity-based	PhoBERT	0.8956	0.9192	0.6959	0.7497
	S-PhoBERT	0.8993	0.9397	0.7001	0.7568
	LaBSE	0.8926	0.9343	0.6804	0.7342
XGBoost	PhoBERT	0.9548	0.9456	0.8869	0.9129
	S-PhoBERT	0.9610	**0.9496**	0.9064	0.9262
	LaBSE	**0.9615**	0.9430	**0.9154**	**0.9285**

In general, the XGBoost models are considerably better than Cosine similarity-based approaches. In particular, the results between XGBoost models themselves are insignificantly different with the LaBSE embedding variant being the best model.

6   Conclusions and Future Work

In this paper, we present a practical process for creating a Vietnamese skill synonym dataset (ViSki) and a straightforward approach for synonym prediction models. We use Wikipedia as a synonym suggestion tool to obtain potential synonym sets with less human effort and utilize distributed representation of phrases and XGBoost for model building. Our ViSki dataset has 740 synonym sets corresponding to 3787 unique skills. XGBoost with LaBSE embedding is our best model, which achieved 96.15% accuracy and an f1-score of 0.9285.

In the future, we plan to apply different approaches for synonym discovery problems. We will leverage the work for skill normalization and skill taxonomy construction to support further analytic works such as job recommendation and labor market analysis.

Acknowledgments. This research is supported by NAVER corporation.

References

1. Bromley, J., Bentz, J.W., Bottou, L., Guyon, I., Lecun, Y., Moore, C., et al.: Signature verification using a "Siamese" time delay neural network. Int. J. Pattern Recognit. **07**(04), 669–688 (1993). https://doi.org/10.1142/s0218001493000339
2. Chaudhuri, S., Ganti, V., Xin, D.: Exploiting web search to generate synonyms for entities. In: Proceedings of the 18th International Conference on World Wide Web - WWW 2009, pp. 151–160. ACM Press (2009). https://doi.org/10.1145/1526709.1526731

3. Chen, T., He, T., Benesty, M., Khotilovich, V., Tang, Y., Cho, H., et al.: Xgboost: Extreme gradient boosting. R Package Version 0.4-2 **1**(4), 1–4 (2015)
4. Davis, A.P., et al.: The comparative toxicogenomics database: update 2019. Nucleic Acids Res. **47**(D1), D948–D954 (2019). https://doi.org/10.1093/nar/gky868
5. Davis, A.P., Wiegers, T.C., Rosenstein, M.C., Mattingly, C.J.: MEDIC: a practical disease vocabulary used at the comparative toxicogenomics database. Database **2012** (2012). https://doi.org/10.1093/database/bar065
6. Devlin, J., Chang, M.W., Lee, K., Toutanova, K.: Bert: Pre-training of deep bidirectional transformers for language understanding. arXiv preprint arXiv:1810.04805 (2018)
7. Doğan, R.I., Leaman, R., Lu, Z.: NCBI disease corpus: a resource for disease name recognition and concept normalization. J. Biomed. Inform. **47**, 1–10 (2014). https://doi.org/10.1016/j.jbi.2013.12.006
8. Feng, F., Yang, Y., Cer, D., Arivazhagan, N., Wang, W.: Language-agnostic BERT sentence embedding. arXiv preprint arXiv:2007.01852 (2020)
9. Hearst, M.A.: Automatic acquisition of hyponyms from large text corpora. In: Proceedings of the 14th conference on Computational linguistics. Association for Computational Linguistics (1992). https://doi.org/10.3115/992133.992154
10. Ji, Z., Wei, Q., Xu, H.: Bert-based ranking for biomedical entity normalization. AMIA Summits Transl. Sci. Proc. **2020**, 269 (2020)
11. Lample, G., Conneau, A.: Cross-lingual language model pretraining. arXiv preprint arXiv:1901.07291 (2019)
12. Le, Q.T.T., Doan, T.H.D., Nguyen, Q.L.H.T.T., Nguyen, D.T.P.: Competency gap in the labor market: evidence from Vietnam. J. Asian Finance, Econ. Bus. **7**(9), 697–706 (2020). https://doi.org/10.13106/jafeb.2020.vol7.no9.697
13. Levenshtein, V.: Levenshtein distance (1965)
14. Li, J., et al.: Biocreative V CDR task corpus: a resource for chemical disease relation extraction. Database **2016** (2016). https://doi.org/10.1093/database/baw068
15. Lin, D., Zhao, S., Qin, L., Zhou, M.: Identifying synonyms among distributionally similar words. In: IJCAI, vol. 3, pp. 1492–1493. CiteSeer (2003)
16. Liu, Y., et al.: RoBERTa: A robustly optimized BERT pretraining approach. arXiv preprint arXiv:1907.11692 (2019)
17. Mikolov, T., Sutskever, I., Chen, K., Corrado, G.S., Dean, J.: Distributed representations of words and phrases and their compositionality. In: Advances in Neural Information Processing Systems, pp. 3111–3119 (2013)
18. Mondal, I., et al.: Medical entity linking using triplet network. arXiv preprint arXiv:2012.11164 (2020)
19. Nguyen, D.Q., Nguyen, A.T.: PhoBERT: Pre-trained language models for Vietnamese. arXiv preprint arXiv:2003.00744 (2020)
20. Pakhomov, S.V., Finley, G., McEwan, R., Wang, Y., Melton, G.B.: Corpus domain effects on distributional semantic modeling of medical terms. Method. Biochem. Anal. **32**(23), btw529 (2016). https://doi.org/10.1093/bioinformatics/btw529
21. Pantel, P., Crestan, E., Borkovsky, A., Popescu, A.M., Vyas, V.: Web-scale distributional similarity and entity set expansion. In: Proceedings of the 2009 Conference on Empirical Methods in Natural Language Processing Volume 2 - EMNLP 2009, pp. 938–947. Association for Computational Linguistics (2009). https://doi.org/10.3115/1699571.1699635
22. Pei, S., Yu, L., Zhang, X.: Set-aware entity synonym discovery with flexible receptive fields. IEEE Trans. Knowl. Data Eng. (2021). https://doi.org/10.1109/tkde.2021.3087532

23. Pennington, J., Socher, R., Manning, C.: GloVe: global vectors for word representation. In: Proceedings of the 2014 Conference on Empirical Methods in Natural Language Processing (EMNLP), pp. 1532–1543. Association for Computational Linguistics (2014). https://doi.org/10.3115/v1/d14-1162

24. Qu, M., Ren, X., Han, J.: Automatic synonym discovery with knowledge bases. In: Proceedings of the 23rd ACM SIGKDD International Conference on Knowledge Discovery and Data Mining, pp. 997–1005. ACM, August 2017. https://doi.org/10.1145/3097983.3098185

25. Reimers, N., Gurevych, I.: Sentence-BERT: Sentence embeddings using Siamese BERT-networks. arXiv preprint arXiv:1908.10084 (2019)

26. Roller, S., Erk, K., Boleda, G.: Inclusive yet selective: supervised distributional hypernymy detection. In: Proceedings of COLING 2014, the 25th International Conference on Computational Linguistics: Technical Papers, pp. 1025–1036 (2014)

27. Shen, J., Lyu, R., Ren, X., Vanni, M., Sadler, B., Han, J.: Mining entity synonyms with efficient neural set generation. In: Proceedings of the AAAI Conference on Artificial Intelligence, vol. 33, pp. 249–256 (2019). https://doi.org/10.1609/aaai.v33i01.3301249

28. Sun, A., Grishman, R.: Semi-supervised semantic pattern discovery with guidance from unsupervised pattern clusters. In: ACL (2010)

29. Sung, M., Jeon, H., Lee, J., Kang, J.: Biomedical entity representations with synonym marginalization. arXiv preprint arXiv:2005.00239 (2020)

30. Tseng, X.W.F.P.H., Dumoulin, Y.L.B.: Context sensitive synonym discovery for web search queries. In: Proceeding of the 18th ACM Conference on Information and Knowledge Management - CIKM 2009 (2009). https://doi.org/10.1145/1645953.1646178

31. Wang, C., Cao, L., Zhou, B.: Medical synonym extraction with concept space models. In: Twenty-Fourth International Joint Conference on Artificial Intelligence (2015)

32. Wang, Y., Qin, J., Wang, W.: Efficient approximate entity matching using Jaro-Winkler distance. In: Bouguettaya, A., et al. (eds.) WISE 2017, Part I. LNCS, vol. 10569, pp. 231–239. Springer, Cham (2017). https://doi.org/10.1007/978-3-319-68783-4_16

33. Wang, Z., Yue, X., Moosavinasab, S., Huang, Y., Lin, S., Sun, H.: SurfCon. In: Proceedings of the 25th ACM SIGKDD International Conference on Knowledge Discovery and Data Mining, pp. 1578–1586. ACM, July 2019. https://doi.org/10.1145/3292500.3330894

34. Yahya, M., Whang, S., Gupta, R., Halevy, A.: ReNoun: fact extraction for nominal attributes. In: Proceedings of the 2014 Conference on Empirical Methods in Natural Language Processing (EMNLP), pp. 325–335. Association for Computational Linguistics (2014). https://doi.org/10.3115/v1/d14-1038

35. Yang, Y., et al.: Improving multilingual sentence embedding using bi-directional dual encoder with additive margin softmax. arXiv preprint arXiv:1902.08564 (2019)

A Survey of Pretrained Embeddings for Japanese Legal Representation

Ha-Thanh Nguyen[1]([✉]), Le-Minh Nguyen[1], and Ken Satoh[2]

[1] Japan Advanced Institute of Science and Technology, Nomi, Japan
nguyenhathanh@jaist.ac.jp
[2] National Institute of Informatics, Chiyoda City, Japan

Abstract. Pretrained embeddings have proven effective in legal problems in English. Even so, working well in one language does not guarantee that these models have an advantage in other languages. Understanding the characteristics of these models in a particular language helps us to make more accurate decisions when choosing technology for problems in that language. This paper provides an analytical perspective on pretrained embeddings in the legal field in Japanese. These models are measured on quantitative numbers as well as visualized in terms of their ability to represent Japanese legal terms. With such contributions, this paper may be useful to researchers and engineers who are building Japanese legal embeddings.

Keywords: Japanese · Pretrained embeddings · Legal domain · Evaluation

1 Introduction

Embedding is a technique that allows AI systems to process natural languages by converting them into vectors. Traditional word embedding such as GloVe [1] or Word2Vec [2] brought a revolution in natural language processing at that time. With these embeddings, we can understand the relationship between words without using rigid rules. Operations like *King-Man+Women=Queen* are feasible with this technique. Even so, these techniques do not yet allow models to understand a polysemy in the context in which it is used. Solving that problem, contextual embeddings were born. With the multiheaded attention mechanism, these models can look at a word through many different perspectives and find its best meaning through special pretraining tasks. The contextual embeddings are then finetune to optimize for downstream tasks. This approach has resulted in many breakthroughs in both the general domain and the legal domain [3–12].

Evaluating the embeddings in legal domain, Nguyen *et al.* [13] introduce two quantitative metrics *i.e.*, LVC and LECA. These two metrics measure legal vocabulary coverage and legal embedding centroid-based assessment. Assumption of the authors when giving these two metrics is that embeddings will effectively represent legal text if the legal terms coverage is high and the distance

© Springer Nature Switzerland AG 2022
H. Fujita et al. (Eds.): IEA/AIE 2022, LNAI 13343, pp. 363–369, 2022.
https://doi.org/10.1007/978-3-031-08530-7_30

between the vectors representing these term is small. The authors also propose a method of visualize embeddings in the legal aspect. As an extension of the previous work, in this paper, we use these tools to examine the legal pretrained embeddings for Japanese.

2 Experimental Settings

The resource we use to evaluate these models on LVC and LECA metrics are the legal terms and legal documents provided by the Japanese Law Translation website[1]. The number of different legal terms $\mathcal{L}$ provided by this website is 3,810. For the convenience of experiments, we build $\mathcal{D}$ with the legal sentences with the length from 100 to 500 characters with the number of 27,614. Our sizes of $\mathcal{L}$ and $\mathcal{D}$ are larger than in the original paper (*i.e.*, paper of Nguyen *et al.* [13]). We also use *cosine distance* as the distance metric function d.

First, we measure the same models as in the original paper's experiment including: FastText [14], GloVe [1], Law2Vec [15], BERT Base Uncased [3], LEGAL-BERT [16] and BERTLaw [9]. Next, we measure models trained on multilingual data and Japanese data. Multilingual models include: BERT Multilingual[2] (*i.e.*, a variant of BERT on multiple languages), XLM-RoBERTa [17] (the model was trained in over 100 different languages), ParaLaw Nets [11] (the model was further pretrained on Japanese and English bilingual legal data). Models on Japanese include BERT Kyoto[3] (pretrained with 18M Japanese sentences) and BERT Tohoku[4] (pretrained with 30M Japanese sentences).

3 Experimental Results

Table 1 shows the *LVC* and *LECA* scores of the models on Japanese legal text. BERT Kyoto achieves state-of-the-art results on LVC while NMSP gets the best LECA score. Trained on Japanese data, BERT Tohoku and BERT Kyoto have LVCs far ahead of other models. This means that the vocabulary of these two models contains more Japanese legal terms than the others. Although BERT Tohoku has a slightly smaller LVC than BERT Kyoto, its LECA is significantly superior. This shows that BERT Tohoku models legal terms better than BERT Kyoto when trained on more Japanese sentences (30M vs. 18M). Among the multilingual models, XLM-RoBERTa has the highest LVC. Despite being trained in 100 languages, the LVC of this model is still around 50% of the LVC of the two models trained in Japanese (*i.e.*, BERT Tohoku and BERT Kyoto).

Paralaw Nets and BERT Multilingual models have similar vocabulary sets, so their Japanese LVC scores have roughly the same results and are not significant. Interestingly, in their paper, the authors of ParaLaw Nets encountered a convergence problem when training with Japanese augmented data in the pretraining

---

[1] japaneselawtranslation.go.jp.

[2] https://github.com/google-research/bert/blob/master/multilingual.md.

[3] https://nlp.ist.i.kyoto-u.ac.jp/?ku_bert_japanese.

[4] https://github.com/cl-tohoku/bert-japanese.

Table 1. *LVC* (higher is better) and *LECA* (lower is better) scores of the models on Japanese legal text.

Embedding	LVC	LECA
GloVe [1]	0	–
FastText [14]	0	–
Law2Vec [15]	0	–
LEGAL-BERT [16]	0	–
BERTLaw [9]	0	–
BERT Base Uncased [3]	0.0016	0.8000
BERT Multilingual Base Uncased	0.0058	0.6327
BERT Multilingual Base Cased	0.0060	0.4806
ParaLaw Nets - NFSP [11]	0.0060	0.3040
ParaLaw Nets - NMSP [11]	0.0060	**0.1395**
XLM-RoBERTa [17]	0.0971	0.2284
BERT Tohoku	0.1861	0.4192
BERT Kyoto	**0.1913**	0.9166

phrase. Our results contribute to explaining this phenomenon. For LECA, NMSP has the best score, showing that this model is better able to model points in the space corresponding to the Japanese legal terms than the rest. This suggests that we can create a better pretrained model with NMSP's method on an original model with better LVC.

Interestingly powerful models in English like GloVe, FastText, Law2Vec, LEGAL-BERT and BERTLaw achieve zero LVC (*i.e.*, undefined LECA) on Japanese. BERT's Vocabulary contains only 6 out of 3,810 Japanese legal terms with an LVC of 0.0016 and a LECA of 0.8000. This is an important finding for us to avoid using pretrained models in English for Japanese legal text processing problems.

4   Further Analysis

To better visualize the power of models on Japanese legal text, we use their tokenizer to analyze a Japanese legal sentence: 年齢二十歳をもって、成年とする。 (The age of majority is reached when a person has reached the age of 20). In this sentence the word 成年 (majority) is a legal term. The results of tokenization are represented in Table 2.

From the results in Table 2, it is not surprising that models with zero LVC cannot detect any components in the sentence. This also shows that they have difficulty not only with Japanese legal terms but also with Japanese vocabulary in general. BERT Base Uncased even though not pretrained with the intention of multilingual embeddings still tokenizes the sentence quite well with only two

Table 2. Tokenization results for the sentence 年齢二十歳をもって、成年とする。 Unknown tokens are represented by ×.

Embedding	Tokenized text
GloVe [1]	[×][×][×][×][×][×][×][×][×][×][×][×][×][×][×][×]
FastText [14]	[×][×][×][×][×][×][×][×][×][×][×][×][×][×][×][×]
Law2Vec [15]	[×][×][×][×][×][×][×][×][×][×][×][×][×][×][×][×]
LEGAL-BERT [16]	[×][×][×][×][×][×][×][×][×][×][×][×][×][×][×][×]
BERTLaw [9]	[×][×][×][×][×][×][×][×][×][×][×][×][×][×][×][×]
BERT Base Uncased [3]	[年][×][二][十][×][歳][##も][##っ][##て][、][成][年][と][##す][##る][。]
BERT Multilingual Base Uncased	[年][齢][二][十][歳][を][も][っ][て][、][成][年][と][する][。]
BERT Multilingual Base Cased	[年][齢][二][十][歳][を][も][っ][て][、][成][年][と][する][。]
ParaLaw Nets - NFSP [11]	[年][齢][二][十][歳][を][も][っ][て][、][成][年][と][する][。]
ParaLaw Nets - NMSP [11]	[年][齢][二][十][歳][を][も][っ][て][、][成][年][と][する][。]
XLM-RoBERTa [17]	[年][齢][二][十][歳][を][も][っ][て][、][成][年][とする][。]
BERT Tohoku	[年][齢][二][十][歳][を][も][っ][て][、][成][年][と][##する][。]
BERT Kyoto	[年][齢][##二][##十][##歳][##を][##もっ][##て][、][##に][##つ][##いて][、][成][年][##と][##する][。]

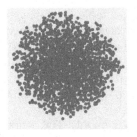

(a) BERT Tohoku   (b) BERT Kyoto

Fig. 1. Visualization of the embeddings pretrained in Japanese text (*i.e.*, BERT Tohoku and BERT Kyoto). Red points represent legal terms, blue points represent non-legal terms. (Color figure online)

unknown positions. The multilingual embeddings and Japanese embeddings have no unknown position. Even so, we can understand the ability of these embeddings by their tokenized text. XLM-RoBERTa groups 年齢 into a phrase, while this result does not appear in BERT Tohoku, which is a Japanese embedding. With legal term 成年 , only BERT Kyoto keeps it intact while other embeddings split it into two sub-kanjis. This analysis gives us some insights to improve Japanese legal embeddings in the future.

Next, we use the visualization strategy described in the original paper. For each embedding, the 2,000 most common words are extracted and represented in 3D with TSNE. We do not visualize models with an LVC of 0 because Japanese legal terms do not appear in their vocabulary. For multilingual models, to be able to visualize correctly, we need a mechanism to filter out the Japanese terms in the vocabulary. However, this is not possible because kanji in Japanese are used in many different languages. Therefore, we only visualize the pretrained embeddings on Japanese as BERT Kyoto and BERT Tohoku.

Figure 1 shows the visualization of BERT Kyoto and BERT Tohoku's embeddings. In both visualization, we can see a sphere in the space. The red points representing legal terms in BERT Kyoto's visualization are denser but more scattered than in BERT Tohoku's visualization. This visualization shows results that are compatible with the results in Table 1.

5 Conclusions

This paper provides detailed measurements for pretrained embeddings for Japanese legal text. Through the experimental results, we can confirm the issue of language adaptation for pretrained embeddings. A powerful embedding on English can become less effective if not pretrained on the adapted language. Based on the LVCs of the models, we can see that multilingual embedding may not be effective in Japanese legal text because their vocabulary is scattered across many languages. The LECA scores of the models show that further pretraining methods suitable for Japanese legal text such as method of ParaLaw Nets can enhance the embeddings' modeling ability in Japanese legal terms. We also provide further on Japanese pretrained embeddings to give readers an intuitive view of the problem. Different embeddings offer different ways of tokenizing input sentences and representing them in the vector space. These results are insights for improving Japanese legal embeddings in future works.

Ackowledgement. This work was supported by JSPS Kakenhi Grant Number 20H04295, 20K20406, and 20K20625.

References

1. Pennington, J., Socher, R., Manning, C.: Glove: global vectors for word representation. In: Proceedings of the 2014 Conference on Empirical Methods in Natural Language Processing (EMNLP), Doha, Qatar, pp. 1532–1543. Association for Computational Linguistics (2014). https://doi.org/10.3115/v1/D14-1162
2. Mikolov, T., Chen, K., Corrado, G., Dean, J.: Efficient estimation of word representations in vector space. arXiv preprint arXiv:1301.3781 (2013)
3. Devlin, J., Chang, M.-W., Lee, K., Toutanova, K.: Bert: pre-training of deep bidirectional transformers for language understanding. arXiv preprint arXiv:1810.04805 (2018)
4. Radford, A., Wu, J., Child, R., Luan, D., Amodei, D., Sutskever, I.: Language models are unsupervised multitask learners (2019)
5. Brown, T.B., et al.: Language models are few-shot learners. arXiv preprint arXiv:2005.14165 (2020)
6. Lewis, M., et al.: Bart: Denoising sequence-to-sequence pre-training for natural language generation, translation, and comprehension. arXiv preprint arXiv:1910.13461 (2019)
7. Rabelo, J., Kim, M.-Y., Goebel, R., Yoshioka, M., Kano, Y., Satoh, K.: COLIEE 2020: Methods for Legal Document Retrieval and Entailment (2020)
8. Hannes, W., Jaromir, S., Karim, B.: Paragraph similarity scoring and fine-tuned bert for legal information retrieval and entailment. In: COLIEE 2020 (2020)
9. Nguyen, H.-T., et al.: Jnlp team: deep learning for legal processing in COLIEE 2020. arXiv preprint arXiv:2011.08071 (2020)
10. Hsuan-Lei, S., Yi-Chia, C., Sieh-Chuen, H.: Bert-based ensemble model for the statute law retrieval and legal information entailment. In: COLIEE 2020 (2020)
11. Nguyen, H.-T., et al.: Paralaw nets-cross-lingual sentence-level pretraining for legal text processing. arXiv preprint arXiv:2106.13403 (2021)
12. Nguyen, H.-T., et al.: Transformer-based approaches for legal text processing. Rev. Socionetwork Strat. **16**, 1–21 (2022). https://doi.org/10.1007/s12626-022-00102-2
13. Thanh, N.H., Binh, D.T., Quan, B.M., Le Minh, N.: Evaluate and visualize legal embeddings for explanation purpose. In: 2021 13th International Conference on Knowledge and Systems Engineering (KSE), pp. 1–6. IEEE (2021)
14. Mikolov, T., Grave, E., Bojanowski, P., Puhrsch, C., Joulin, A.: Advances in pre-training distributed word representations. In: Proceedings of the International Conference on Language Resources and Evaluation (LREC 2018) (2018)
15. Chalkidis, I., Kampas, D.: Deep learning in law: early adaptation and legal word embeddings trained on large corpora. Artif. Intell. Law **27**(2), 171–198 (2018). https://doi.org/10.1007/s10506-018-9238-9

16. Chalkidis, I., Fergadiotis, M., Malakasiotis, P., Aletras, N., Androutsopoulos, I.: Legal-bert: the muppets straight out of law school. arXiv preprint arXiv:2010.02559 (2020)
17. Conneau, A., et al.: Unsupervised cross-lingual representation learning at scale. arXiv preprint arXiv:1911.02116 (2019)

Machine Reading Comprehension Model for Low-Resource Languages and Experimenting on Vietnamese

Bach Hoang Tien Nguyen[ID], Dung Manh Nguyen[ID], and Trang Thi Thu Nguyen[(✉)]

School of Information and Communication Technology,
Hanoi University of Science and Technology, Hanoi, Vietnam
`trangntt@soict.hust.edu.vn`

Abstract. Machine Reading Comprehension (MRC) is a challenging task in natural language processing. In recent times, many large datasets and good models are public for this task, but most of them are for English only. Building a good MRC dataset always takes much effort, this paper proposes a method, called UtlTran, to improve the MRC quality for low-resource languages. In this method, all available MRC English datasets are collected and translated into the target language with some context-reducing strategies for better results. Tokens of question and context are initialized word representations using a word embedding model. They are then pre-trained with the MRC model with the translated dataset for the specific low-resource language. Finally, a small manual MRC dataset is used to continue fine-tuning the model to get the best results. The experimental results on the Vietnamese language show that the best word embedding model for this task is a multilingual one - XLM-R. Whereas, the best translation strategy is to reduce context by answer positions. The proposed model gives the best quality, i.e. F1 = 88.2% and Exact Match (EM) = 71.8%, on the UIT-ViQuAD dataset, compared to the state-of-the-art models.

Keywords: Low-resource languages · Translated datasets · Pre-train layer

1 Introduction

Machine reading comprehension (MRC) is one of the most essential problems in Natural Language Processing (NLP), which requires computers to read, understand the text and be able to answer the questions related to it. Over the past few years, MRC has played an important role in many NLP systems such as search engines, artificial intelligence-powered virtual assistants, and dialogue systems. To improve MRC task results in many aspects, large-scale datasets, higher computing power, and modern deep learning techniques have been used. In our study, MRC is referred to as question answering (QA), however, the main purpose of

© Springer Nature Switzerland AG 2022
H. Fujita et al. (Eds.): IEA/AIE 2022, LNAI 13343, pp. 370–381, 2022.
https://doi.org/10.1007/978-3-031-08530-7_31

a QA system is to answer the given questions, while the main goal of our MRC system is to understand the meaning of the text.

MRC systems can be divided into two types based on the output: extractive (selective) answers and generative (abstractive) answers. In generative-answer systems, the output is auto-generated according to the context and its related question. On the other hand, the output of an extractive-answer system is a specific span of the given context. Our MRC system is an example of the extractive-answer system.

In recent years, there is an explosion in the number of MRC benchmark datasets, especially in English, such as MCTest [15], SQuAD1.1 [14] and TriviaQA [10]. This has given an impulse to the creation of a large number of new MRC models. For instance, after Richardson et al. [15] created MCTest [15] dataset in 2013, many researchers started to propose different machine learning models on this dataset. Another example is SQuAD1.1 [14], many new neural network MRC models were created to accomplish the task, such as BERT [5], XLM-RoBERTa [3] and mT5 [26]. The performance of the state-of-the-art models in related English datasets is approximate with human performance.

In contrast with the variety of MRC datasets in English, resources in low-resource languages for MRC tasks, however, are limited and meager. For instance, the public datasets for Vietnamese MRC task only include UIT-ViQuAD [13], UIT-ViNewsQA [24] for extractive MRC datasets, and ViMMRC [23] for multiple-choice datasets. These corpora include only about 2K–25K question-answer pairs in each dataset, making it difficult to build a complicated model. As a result, the models for Vietnamese MRC tasks are simple and were often created by using zero-shot learning in a multilingual model (mBERT, XLM-R) or training a BERT base model on the training set directly then evaluating it on the test set, which performs low results.

To solve this problem, we proposed an approach that utilizes the translated version of English MRC datasets for pre-training the model before fine-tuning it on a dataset of a specific low-resource language. We also apply our training strategy for different BERT-based multilingual models (e.g.: XLM-R [3], mBERT [6]) then evaluate them on the UIT-ViQuAD dataset.

The rest of the paper is organized as follows. Section 2 presents baseline model the along with the proposed joint one. Section 3 goes in-depth into the related works done in this field of research. In Sects. 4 and 5, the experiment setup and evaluation results will be reported and discussed with baseline models. Finally, the paper draws some conclusions from the work in Sect. 6.

2 Proposed Model

2.1 Baseline Model

We design our baseline model based on XLM-R model [3] and follow the same steps of Kiet Nguyen et al. [13]. The baseline model is depicted in Fig. 1. Tokens of question and the context are fed into an encoder (a BERT base word embedding model) with the top layer being a full connection layer that determines start and

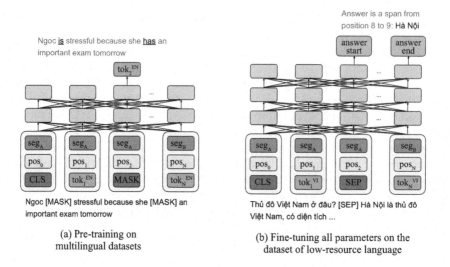

(a) Pre-training on
multilingual datasets

(b) Fine-tuning all parameters on the
dataset of low-resource language

Fig. 1. Baseline model for machine reading comprehension task

end positions of answer. On each step t, the encoder converts input tokens to word representations based on multi-head self-attention mechanism [25].

Word representation with dim_h are forward to the top layer with $dim_o = 2$ (i.e. fully connected layer) and the output is the logits of start and end positions. The loss function is total both loss Cross-Entropy of start and end positions:

$$l_{start} = \sum_{i=1}^{i=N_{samples}} y_{start_i} \cdot \log \hat{y}_{start_i} \tag{1}$$

$$l_{end} = \sum_{i=1}^{i=N_{samples}} y_{end_i} \cdot \log \hat{y}_{end_i} \tag{2}$$

$$Loss_{total} = l_{start} + l_{end} \tag{3}$$

We also compare our method with the zero-shot learning which proposal by Artetxe et al. [1] for evaluating performances.

2.2   Proposal of UtlTran Model

In this section, we propose a method for low-resource language MRC, called Utl-Tran, that can utilize the existing English datasets. The pipeline of the method is illustrated in Fig. 2. Our pipeline includes three steps: (i) Pre-training BERT base word embedding model on a large multilingual corpus with masked language modeling (MLM) and next sentence prediction (NSP) objectives. (ii) We apply a translation strategy to English MRC datasets then train all parameters on the translated datasets. (iii) final step, we fine-tune the model on a small low-language dataset to evaluate the effectiveness of the model.

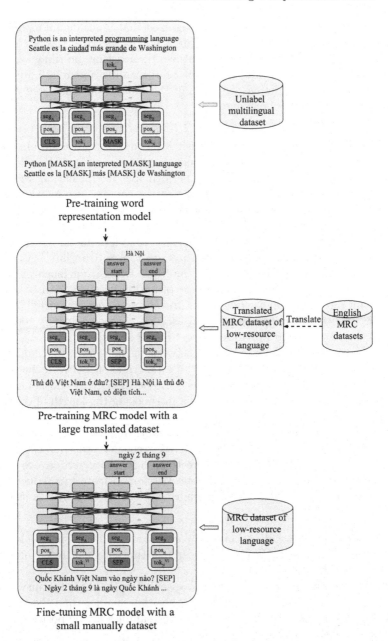

Fig. 2. Our proposal model include three steps which utilize English MRC datasets for Vietnamese MRC

Instead of using a general word embedding model like the baseline model, we use the multilingual model on translated datasets at step (ii) then fine-tune all of the parameters on a dataset of low resource language.

Word Representation Model. We recognized that multilingual models have better performance in different low-resource languages MRC tasks compared to English MRC models. Therefore, we utilize the existing multilingual models for initializing word representations. In place of training a general multilingual model, we focus on continuing pre-training the models on translated datasets for a specific low-resource language.

Translation Strategies. We use English MRC datasets to translate to low-resource languages. First, we transform English datasets to SQuAD version 1 format for evaluating MRC tasks. For good performance in translation, we remove samples that have more than 5000 characters in context. About the span of answers in context after translation, we are using two special tokens to mark them is [ans_s] and [ans_e].

To translate these datasets from English to Vietnamese, we use *deep-translator* - a flexible free and unlimited python tool to translate between different languages in a simple way using multiple translators (such as Google Translate, Microsoft Translator). There are other libraries that can be used for this task, but most of them are buggy, not free, limited, not supported anymore or complex to use. Hence, we decided to use this simple tool. It is 100% free, unlimited, easy to use and provides support for all languages.

After the translation phase, we align the value of the answer start and answer text based on the special tokens. We also have a trade-off between the number of samples and length of context to get a suitable dataset, we do not use long context samples. We show our experiments in choosing the length of context in Table 1.

Although BERT-based models are good in capturing the context of sentences they still do not work well in long contexts. We recognize that almost the context in the English MRC dataset is more than 2000 characters, so we have to reduce the context length in both datasets. We demonstrate two reducing context strategies to evaluate how context length can be captured by a multilingual model:

1. *Reducing context by length of context:* We set a threshold (λ) for the length of context in two datasets. We only keep samples that have the length of context smaller than the threshold.
2. *Reducing context by answer position:* We set a threshold (δ) for answer position, we keep samples that have answer position smaller than the threshold then we random truncate the context of the samples at the position smaller than another threshold (α).

Fine-Tuning on a Low-Resource MRC Dataset. In this phase, we need a standard low-resource dataset to fine-tune the model again to get the best results. This dataset is normally labeled by humans.

3   Related Works

Word Representation Model. Word representation models are used to present words to vectors based on the context, it is essential in NLP tasks. Some early methods used different algorithms to measure the similarity between sentences [9,17]. Another popular method is attention-based interaction between contexts like Bi-attention [18] and Attention over Attention [4].

Over the last few years, the deep contextual language models based on Transformers have shown effectiveness in various MRC tasks. Some remarkable models are BERT [5], XLM-RoBERTa [3] and mT5 [26]. There are two approaches for building contextual language models: Building a generative model like mT5 [26], GPT [2], BART [11], all models combine encoder and decoder of Transformers or applying many encoders from Transformers like BERT and variants of BERT. All of that models when applying to NLP tasks: PoS Tagging task, NER task, text summarization, text classification, and machine reading comprehension, it has good results in English. But the problem with that models is they are trained on a large English corpus and also fine-tuning on large English datasets for each task makes models learn a better representation than low-resource languages. We try to solve this problem by utilizing English datasets to contextual models that can represent words better in other languages.

Word embedding models for Vietnamese are PhoBERT [12], ViBERT [20], vELECTRA [20], and BARTPho [21] recently. Those models pre-train on a general dataset then fine-tune in many tasks: PoS Tagging task, NER task, and text summarization, it has good results but the models when downstream on MRC task it has low results. About the MRC task, the approach is focused on building a gold dataset [13], then public a baseline model for datasets. Our method provides a new approach to utilize BERT-based models and English datasets to get a higher result than the basic approaches.

Translated Datasets. Translation from a language resource to others is an accepted solution that aims to solve the lack of datasets problem in lots of languages. There are many datasets are created following this method, such as Business Scene Dialogue (BSD) Dataset [16] and MuST-C Dataset [7]. These translation corpora are widely used in various tasks of NLP. For example, Text Summarization [27] uses different translation datasets to enhance performance. They are translated to Chinese to augment Chinese datasets that make better results. Meanwhile, we use translated datasets for the machine reading comprehension task to enrich datasets for low-resource languages like Vietnamese.

Low-Resource MRC Dataset. For evaluating the effectiveness of MRC models, the high-quality datasets in many languages are created (even non-English languages) (UIT-ViQuAD [13], UIT-ViNewsQA [24]). These datasets are limited in the number of samples as it is labeled manually by professional workers only, however it is still suitable for evaluating purpose. We experiment with our model in Vietnamese by using UIT-ViQuAD dataset [13]. This dataset is created based on Wikipedia Vietnamese documents. Many researchers use this dataset for pre-training on other languages [19] or for evaluation [13].

4   Experiment Preparation

4.1   Datasets

In this part, we describe the datasets which are used for the pre-training and fine-tuning phase. There are two kinds of datasets: one is created by translating different English MRC datasets to Vietnamese and is suitable to use as pre-train datasets, another is a high-quality MRC dataset used for fine-tuning and evaluation.

Pre-trained Datasets. In this research, we use three popular English datasets for MRC task (TriviaQA [10], NewsQA [22], and XQuAD datasets [1]) to pre-train our model.

TriviaQA. Trivia dataset is a reading comprehension dataset containing over 650K question-answer-evidence triples. TriviaQA includes 95K question-answer pairs authored by trivia enthusiasts and independently gathered evidence documents. Documents are collected by Wikipedia and websites. We only use samples collected by Wikipedia.

NewsQA. NewsQA dataset is a crowd-sourced machine reading comprehension dataset that was collected through four different steps: (i) article curation, (ii) question sourcing, (iii) answer sourcing, (iv) validation. Documents are articles collected from CNN using the same script created by Hermann et al. [8]. These cover a variety of topics, including economics, science, sports, politics, and more. This dataset has total of 120K question-answer pairs.

XQuAD. XQuAD dataset is from SQuAD v1.1 translated into ten languages by professional translators for the test set and using an automatic machine translation system for the training set. We only use the training set for the pre-train phase.

With the XQuAD dataset, we use translated dataset of Artetxe et al. [1]. We also transform the original format of TriviaQA and NewsQA datasets into SQuAD v1.1 format to have an easier setup for evaluation. Detailed information of three datasets after applying two translation strategies from Sect. 2.2 is shown in Tables 1 and 2.

Fine-Tuned Dataset. The Vietnamese Question Answering Dataset (UIT-ViQuAD) is used to fine-tune our model. The dataset contains 23,074 question-answer pairs based on 174 articles, which are randomly chosen from the top 5,000 Vietnamese Wikipedia articles. The statistics of the dataset are enumerated in Table 3. The table contains detailed information about the number of articles, passages and questions, the average length of the passage, question and answers, vocabulary sizes of the train/dev/test set.

Table 1. Number of samples of three pre-trained datasets

	TriviaQA	NewsQA	XQuAD
Original dataset	110,647	92,549	87,187
Translation strategy 1 (context length) (TriviaQA: $\lambda = 4350$, NewsQA: $\lambda = 3500$)	42,284	40,805	–
Translation strategy 2 (answer position) ($\theta = 500, \alpha = 650$)	40,285	36,310	–

Table 2. Translated datasets VIT-MRC for Vietnamese with 2 different translation strategies (TS)

	Dataset	Article	Passage	Questions	Passage length average	Question length average	Answer length average
Strategy 1	TriviaQA	42,284	42,284	42,284	3782.40	83.59	37.11
	NewQA	40,805	40,805	40,805	2292.38	37.53	28.15
Strategy 2	TriviaQA	40,285	40,285	40,285	403.32	84.18	9.75
	NewQA	36,310	36,310	36,310	441.35	37.36	24.56
	XQuAD	87187	87187	87187	764.51	59.26	19.87

Table 3. UIT-ViQuAD dataset.

Dataset	Article	Passage	Questions	Passage length average	Question length average	Answer length average	Vocabulary size
Train	138	4101	18,579	153.9	12.2	8.1	36,174
Dev	18	515	2,285	147.9	11.9	8.4	9,184
Test	18	493	2,210	155.0	12.2	8.9	9,792
All	174	5,109	**23,074**	153.4	12.2	8.2	41,773

4.2   Evaluation Metrics

In this research, we use two popular metrics in MRC tasks (EM and F1-score) as evaluation metrics.

Exact Match (EM). For each question-answer pair, if the characters of the MRC system's predicted answer exactly match the characters of (one of) the gold standard answer(s), EM = 1, otherwise EM = 0. EM is a stringent all-or-nothing metric, with a score of 0 for being off by a single character. When evaluating against a negative question, if the system predicts any textual span as an answer, it automatically obtains a zero score for that question.

F1-Score. F1-score is a popular metric for natural language processing and is also used in machine reading comprehension. F1-score estimated over the individual tokens in the predicted answer against those in the gold standard answers. The F1-score is based on the number of matched tokens between the predicted and gold standard answers.

$$Precision = \frac{\#\ matched\ tokens}{\#\ tokens\ in\ the\ predicted\ answer} \tag{4}$$

$$Recall = \frac{\#\ matched\ tokens}{\#\ tokens\ in\ the\ gold\ standard\ answer} \tag{5}$$

$$F1\text{-}score = \frac{2 \times Precision \times Recall}{Precision + Recall} \tag{6}$$

4.3 Experimental Setup

We use a single NVIDIA Tesla V100 GPU via Google Colaboratory to train all MRC models. We utilize the pre-trained word embeddings models, including mBERT, XLM-R, PhoBERT to evaluate our strategies. We set $epochs = 1$ for pre-training and $epochs = 2$ for fine-tuning, $batch\ size = 4$. We also use the HuggingFace library for implementing models in our experiments with configurations that are the same with Nguyen et al. [13]: The maximum answer length to 300, the question length to 64, and the input sequence length to 384 for all the experiments on mBERT and XLM-R.

5   Experimental Results

5.1 Word Embedding Models

Evaluation effects of pre-training phase on models. We utilize two popular word embedding multilingual models, including mBERT, XLMR, and one Vietnamese model is PhoBERT to show the effect when adding one pretrain step with translated dataset still produce a better result than direct training on the UIT-ViQuAD dataset. The XLM-Roberta has the best results with the F1 score equal 88.17% and EM score being 71.83%. Detailed results of models are in Table 4.

Table 4. Experimental results of our UltTrans model (XQuAD pre-trained) on different word embedding models (UIT-ViQuAD dev set)

MRC model	EM (%)	F1 (%)
Baseline model with mBERT	62.20	80.77
Baseline model with XLM-R$_{large}$	69.18	87.14
Our UltTran model with mBERT	66.94	84.40
Our UltTran model with XLM-R$_{large}$	**73.88**	**89.59**
Our UltTran model with PhoBERT	68.43	86.74

5.2 Translation Strategies

Evaluation two translation strategies Based on the analysis in Sect. 2.2, we show that reducing by answer start position (translation strategy 2) is better than reducing based on context length (translation strategy 1) for both datasets. Although TriviaQA and fine-tuning dataset UIT-ViQuAD are the same resources from Wikipedia, in both strategies pre-training on NewsQA is better than pre-training on the TriviaQA dataset, which shows that the style of documents does not have a huge impact on results. Detailed results of the two strategies are shown in Table 5.

Table 5. Experimental results on translation strategies (UIT-ViQuAD dev set)

MRC model	EM (%)	F1 (%)
Baseline model (XLM-R$_{large}$) [13]	69.18	87.14
Our UltTran (TriviaQA pre-training) + translation strategy 1	68.96	86.24
Our UltTran (NewsQA pre-training) + translation strategy 1	72.25	88.89
Our UltTran (TriviaQA pre-training) + translation strategy 2	73.04	88.61
Our UltTran (TriviaQA+NewQA pre-training) + translation strategy 2	72.21	88.67
Our UltTran (NewsQA pre-training) + translation strategy 2	**74.23**	**89.83**

5.3 MRC Models

In Table 6, we demonstrate that our method using additional training makes better results than combining all training sets into a training set (VIT-MRC dataset), then fine-tuning on that. This is caused by the different domains between datasets, when we combine datasets that make noised information that makes results are lower than the baseline model. We also evaluate zero-shot learning [1], another method utilize the MRC English dataset to apply to other language but this method has low results than baseline models because the model does not train on a low-resource language dataset so the word representation ability of the model is limited than the models trained on low-resource language datasets. On the dev set, when using XquAD and NewQA datasets for pre-training, we get better results than baseline models so we use it to evaluate on the test set.

Table 6. Comparing our model UltTran with state-of-the-art models (UIT-ViQuAD test set)

MRC model	EM (%)	F1 (%)
Zero-shot learning: XLM-R [1]*	54.98	79.28
Baseline: XLM-R + UIT-ViQuAD fine-tuning [13]	68.98	87.02
Baseline: XLM-R + (VIT-MRC + UIT-ViQuAD) fine-tuning	66.58	85.94
Our UtlTran Model: XLM-R + NewQA pre-training + UIT-ViQuAD finetuning	70.47	87.92
Our UtlTran Model: XLM-R + XQuAD pre-training + UIT-ViQuAD finetuning	**71.83**	**88.17**

* We implement zero-shot learning for UIT-ViQuAD dataset ourselves.

6   Conclusions

In this work, we introduce a new approach to utilize the resources from the English MRC task for low-resource languages. In this approach, all available MRC English datasets are collected and translated into the target language. Tokens of question and context are initialized word representations using a word embedding model, then pre-trained with the MRC model with the translated dataset. Finally, they continue being finetuned with a small manual MRC dataset. The experimental results on the Vietnamese language show that the best word embedding model for this task is a multilingual one - XLM-R. The context before translating should be reduced by answer positions to get better quality. The proposed UltTran model gives a considerable improvement, from 1% to 3% for F1 and EM, on the UIT-ViQuAD dataset, compared to the state-of-the-art models.

References

1. Artetxe, M., Ruder, S., Yogatama, D.: On the cross-lingual transferability of mono-lingual representations. arXiv preprint arXiv:1910.11856 (2019)
2. Brown, T.B., et al.: Language models are few-shot learners. arXiv preprint arXiv:2005.14165 (2020)
3. Conneau, A., et al.: Unsupervised cross-lingual representation learning at scale. arXiv preprint arXiv:1911.02116 (2019)
4. Cui, Y., Chen, Z., Wei, S., Wang, S., Liu, T., Hu, G.: Attention-over-attention neural networks for reading comprehension. arXiv preprint arXiv:1607.04423 (2016)
5. Devlin, J., Chang, M.W., Lee, K., Toutanova, K.: Bert: pre-training of deep bidirectional transformers for language understanding. arXiv preprint arXiv:1810.04805 (2018)
6. Devlin, J., Chang, M.W., Lee, K., Toutanova, K.: BERT: pre-training of deep bidirectional transformers for language understanding. In: Proceedings of the 2019 Conference of the North American Chapter of the Association for Computational Linguistics: Human Language Technologies, Volume 1 (Long and Short Papers), pp. 4171–4186. Association for Computational Linguistics, Minneapolis, Minnesota, June 2019. https://doi.org/10.18653/v1/N19-1423, https://aclanthology.org/N19-1423
7. Di Gangi, M.A., Cattoni, R., Bentivogli, L., Negri, M., Turchi, M.: Must-C: a multilingual speech translation corpus. In: 2019 Conference of the North American Chapter of the Association for Computational Linguistics: Human Language Technologies, pp. 2012–2017. Association for Computational Linguistics (2019)
8. Hermann, K.M., et al.: Teaching machines to read and comprehend. Adv. Neural. Inf. Process. Syst. **28**, 1693–1701 (2015)
9. Hirschman, L., Light, M., Breck, E., Burger, J.D.: Deep read: a reading comprehension system. In: Proceedings of the 37th Annual Meeting of the Association for Computational Linguistics, pp. 325–332 (1999)
10. Joshi, M., Choi, E., Weld, D.S., Zettlemoyer, L.: Triviaqa: a large scale distantly supervised challenge dataset for reading comprehension. arXiv preprint arXiv:1705.03551 (2017)

11. Lewis, M., et al.: Bart: denoising sequence-to-sequence pre-training for natural language generation, translation, and comprehension. arXiv preprint arXiv:1910.13461 (2019)
12. Nguyen, D.Q., Nguyen, A.T.: PhoBert: pre-trained language models for vietnamese. arXiv preprint arXiv:2003.00744 (2020)
13. Nguyen, K., Nguyen, V., Nguyen, A., Nguyen, N.: A Vietnamese dataset for evaluating machine reading comprehension. In: Proceedings of the 28th International Conference on Computational Linguistics, pp. 2595–2605 (2020)
14. Rajpurkar, P., Zhang, J., Lopyrev, K., Liang, P.: Squad: 100,000+ questions for machine comprehension of text. arXiv preprint arXiv:1606.05250 (2016)
15. Richardson, M., Burges, C.J., Renshaw, E.: MCTest: a challenge dataset for the open-domain machine comprehension of text. In: Proceedings of the 2013 Conference on Empirical Methods in Natural Language Processing, pp. 193–203 (2013)
16. Rikters, M., Ri, R., Li, T., Nakazawa, T.: Designing the business conversation corpus. arXiv preprint arXiv:2008.01940 (2020)
17. Riloff, E., Thelen, M.: A rule-based question answering system for reading comprehension tests. In: ANLP-NAACL 2000 Workshop: Reading Comprehension Tests as Evaluation for Computer-Based Language Understanding Systems (2000)
18. Seo, M., Kembhavi, A., Farhadi, A., Hajishirzi, H.: Bidirectional attention flow for machine comprehension. arXiv preprint arXiv:1611.01603 (2016)
19. de Souza, L.R., Nogueira, R., Lotufo, R.: On the ability of monolingual models to learn language-agnostic representations. arXiv preprint arXiv:2109.01942 (2021)
20. The, V.B., Thi, O.T., Le-Hong, P.: Improving sequence tagging for Vietnamese text using transformer-based neural models. arXiv preprint arXiv:2006.15994 (2020)
21. Tran, N.L., Le, D.M., Nguyen, D.Q.: BartPho: pre-trained sequence-to-sequence models for Vietnamese. arXiv preprint arXiv:2109.09701 (2021)
22. Trischler, A., et al.: NewsQA: a machine comprehension dataset. arXiv preprint arXiv:1611.09830 (2016)
23. Van Nguyen, K., Tran, K.V., Luu, S.T., Nguyen, A.G.T., Nguyen, N.L.T.: Enhancing lexical-based approach with external knowledge for Vietnamese multiple-choice machine reading comprehension. IEEE Access 8, 201404–201417 (2020)
24. Van Nguyen, K., Van Huynh, T., Nguyen, D.V., Nguyen, A.G.T., Nguyen, N.L.T.: New Vietnamese corpus for machine reading comprehension of health news articles. arXiv preprint arXiv:2006.11138 (2020)
25. Vaswani, A., et al.: Attention is all you need. In: Advances in Neural Information Processing Systems, pp. 5998–6008 (2017)
26. Xue, L., et al.: mt5: a massively multilingual pre-trained text-to-text transformer. arXiv preprint arXiv:2010.11934 (2020)
27. Zhu, J., Zhou, Y., Zhang, J., Zong, C.: Attend, translate and summarize: an efficient method for neural cross-lingual summarization. In: Proceedings of the 58th Annual Meeting of the Association for Computational Linguistics, pp. 1309–1321 (2020)

Inducing a Malay Lexicon from an Unlabelled Dataset Using Word Embeddings

Ian H. J. Ho$^{(\boxtimes)}$ ⓘ, Hui-Ngo Goh, and Yi-Fei Tan

Multimedia University, 63100 Cyberjaya, Malaysia
ianhohengjin@gmail.com

Abstract. The Malay language in Malaysia is commonly mixed with English and slang words especially when observing social media text. In the past, researchers would develop Malay-English lexicons by translating widely accepted English lexicons and carrying over the labels. However, translation may produce words that have no sentiment or is rarely used in its translated form. In this paper, an end-to-end framework involving a two-phase sentiment classification method is proposed. The first phase presents a method where a domain-specific lexicon is induced from word embeddings trained on an unlabelled dataset. It was found that the induced lexicon-based classifier outperformed general and translated lexicons. However, lexicon-based methods still face low recall issues when documents have sentiment words not found in the lexicon. This can be mitigated by the second phase of the framework where a supervised classifier is trained on a filtered output of the induced lexicon-based method to produce a model that can classify new and unseen documents. This framework was shown to be applicable to any Malay or English unlabelled dataset through results with a f1-score of 0.81 on the MALAYA Twitter dataset and 0.82 on the IMDB movie review dataset.

Keywords: Lexicon induction · Domain-specific · Sentiment analysis

1 Introduction

1.1 Background

Sentiment analysis (SA) on text is usually performed with two main methods. The first method is machine learning. It involves training a supervised classifier using a labelled dataset. Generally, a word embedding technique is used to vectorize the dataset. This method's performance depends on the quality and size of labelled datasets available. Hence, it is challenging for low-resourced languages to apply this method.

The other common method uses a lexicon which is a collection of words that express sentiment. Usually, a static numerical sentiment score, positive or negative, is associated with each word in the lexicon. This method does not require any labelled datasets to develop. However, it is very time-consuming and difficult to curate relevant sentiment words. The performance of a lexicon-based method is dependent on the initial curation and labelling of words in the lexicon. If the lexicon-based classifier does not find any

© Springer Nature Switzerland AG 2022
H. Fujita et al. (Eds.): IEA/AIE 2022, LNAI 13343, pp. 382–394, 2022.
https://doi.org/10.1007/978-3-031-08530-7_32

sentiment words in a document, there is an ambiguity in the result: Does the document carry a neutral sentiment or does the document contain sentiment words that were not included in the lexicon?

The Malay language in the field of SA is considered a low-resource language [1, 2]. As a result, many studies on Malay SA in recent years have used lexicon-based methods [3–5]. These studies curate their own dataset and produce their own lexicons but do not release them publicly. Many researchers in this field resort to translating an existing well-known English lexicon into Malay and carrying their labels over to the translated word. Lexicons produced this way will have a lower quality to one that is manually curated, thus, producing less than optimal classification results [6].

Besides that, lexicons contain words that have differing polarity across domains. For example, the word 'insane' has a contrasting sentiment polarity between the context of medicine and sports, with the latter context being positive. The task of SA becomes more complicated when is coupled with the multilingual aspect of informal text data from Malaysia. For example, the word 'terror' is commonly associated with a negative sentiment in English. However, in informal Malay, 'terror' (commonly pronounced as 'terrer' /tɛrɛr/) is considered a slang and is synonymous with 'amazing'.

1.2 Motivation and Contribution

In this paper, an end-to-end framework is proposed to overcome these challenges summarized below:

1. SA on Malay text in Malaysia is difficult because of the lack of resources and common appearance of mixed Malay and English slang.
2. Lexicon-based classifiers commonly used for Malay SA have a high variance in their performance across different domains and contexts. Developing a lexicon for each scenario is time-consuming and difficult.
3. Lexicon-based classification of documents, that do not have sentiment words found in the lexicon, is ambiguous. This is the low recall issue in lexicon-based methods.

To overcome the challenges above, the contributions of this framework are:

1. A semi-supervised induction of a domain-specific Malay/English lexicon from an unlabelled dataset.
2. A supervised document sentiment classifier trained on 'pseudo-labelled' data. Pseudo-labelled data is obtained and filtered from the output of a lexicon-based classifier using the induced lexicon in #1.

To briefly explain at a high-level, the induction of a domain-specific lexicon will be performed through word embeddings and propagating sentiment polarities of a small set of seed words onto candidate words in the vector space generated from the embeddings. Inducing a domain-specific lexicon from unlabelled datasets would alleviate the time-consuming nature of manually curating one and maintenance will only require running the proposed framework again on new text data from the same domain.

The process of training a supervised document sentiment classifier on the output of a lexicon-based classifier will produce a classifier able to process new and unseen data from that domain. Also, it resolves the issue of lexicons not being comprehensive enough to account for all types of sentiment words and spelling differences. The paper is structured as follows. In Sect. 2, work related to this topic is reviewed. In Sects. 3 and 4, the proposed framework and results are presented. Finally, Sect. 5 presents a summary of this paper and its contributions.

2   Related Work

2.1   Lexicon-Based Classification Methodology

Many early studies done on the Malay language used lexicon-based classifiers to perform SA. The two commonly used methods for developing sentiment lexicons are manual curation and translation of existing well-known English lexicons into Malay. RojakLex [4] is a comprehensive, multilingual (English-Malay) lexicon that was manually curated. This lexicon included regular sentiment words, common slang words, emoticons, and a collection of short form words formally reported by Dewan Bahasa dan Pustaka in their report titled 'Panduan Singkatan Khidmat Pesanan Ringkas Bahasa Melayu' [7]. However, this lexicon was not released publicly.

Table 1. Example of sentiment words from AFINN-111 translated into Malay and addition of potential affixes.

English word	Translated word	Potential affix variation
intimidate	intimidasi	intimidasikan, mengintimidasi, mengintimidasikan, diintimidasi, diintimidasikan, berintimidasi, berintimidasikan, terintimidasi, terintimidasikan

As for studies that translated well-known English lexicons into Malay, Asyraf et al. [3] created a multilingual lexicon by translating the AFINN-111 lexicon into Malay and carrying over the labels. Furthermore, Asyraf et al. attached every possible affix to each Malay word produced from translation. An example of this can be seen in Table 1. The similar weaknesses shared by these approaches are that the lexicons formed are not easily maintainable and lack robustness.

More recent studies in Malay lexicon-based SA field have not shown much progress [5, 8, 9]. These studies create lexicons through translation of other well-known English lexicons and perform SA on very small-scale datasets that were self-curated.

2.2   Automatic Lexicon Generation

In terms of automatically generating sentiment lexicons, Darwich et al. proposed a framework for the Malay language [10]. They mapped WordNetBahasa onto the English WordNet to produce a more accurate translation between the two languages based on

WordNet's synsets. A set of seed words were then used to propagate sentiment across the lexical graph based on word synonyms and antonyms. This method produced a very general sentiment lexicon that should perform acceptably across many domains but not optimally compared to the framework proposed in this paper.

In 2016, Hamilton et al. [11] developed an SVD-based (Singular Value Decomposition) lexicon induction method for the English language named SentProp. The induction method used a random-walk algorithm on the word vector space to associate sentiment propagated from a set of seed words. Generally, SVD is resource heavy, and a random-walk method does not perform well at close decision borders between positive and negative sentiment words as reported by Mudinas et al. [12]. They instead proposed a computationally lighter framework, using word embeddings generated by word2vec [13] and a linear logistic regression model to classify the sentiment of words. These words were candidate words obtained from a collection of well-known English lexicons to ensure only words that had a high probability of carrying sentiment were included into the induced lexicon.

2.3   Word Embeddings

Word embedding is a technique that transforms words into numerical vectors that retains contextual information about each word and its surrounding. In this study, the skip-gram word2vec model [13] is one of the word embedding techniques used. This word2vec model generates a vector representation for a word from the model's hidden layers' weights learned by predicting the surrounding context of the word within a defined range as seen in Fig. 1.

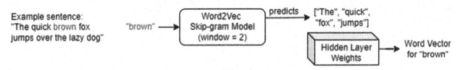

Fig. 1. Skip-gram word2vec. Weights learned from predicting the surrounding words represent the word vector.

Mudinas et al. showed that the cluster hypothesis in information retrieval [14], which states nodes that are clustered together are likely to be of the same class, is applicable to English sentiment words [12]. These sentiment clusters were clearly separated when the word embeddings and sentiment words originated from the same domain. For example, sentiment clusters of sentiment words from a finance lexicon were separated when the word embeddings were generated from text data from the finance domain.

To ensure that this relationship was present in the Malay language, we replicated the investigation in this study. A list of Malay and English sentiment words from the translated AFINN-111 lexicon by Asyraf et al. [3] was plotted on word2vec word embeddings of text data from Malaysia's Twitter domain[1]. Principle Component Analysis (PCA) was

---

[1] https://twitter.com/home.

used to visualize the word embeddings on a 2D plane. The clustered nature of each sentiment class was clearly seen in Fig. 2a. This was expected as AFINN-111 was developed
specifically for use in the social media domain.

As a comparison, the same sentiment words were plotted onto word embeddings
from formal Malaysian news articles[2]. Since the sentiment words that were curated for a
social media domain did not match the formal news article domain, the clusters formed
overlapped as seen in Fig. 2b. This forms the basis of proposed framework in this study.
A supervised classifier can be trained on these embeddings to learn the separation of the
sentiment clusters in a matching domain. The trained classifier can then be used to label
the sentiment of any given word, essentially inducing a sentiment lexicon that is specific
to the domain it was trained on. This is discussed with more depth in the next section.

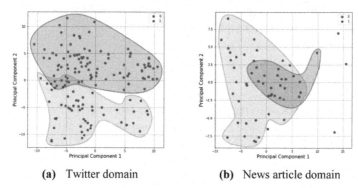

(a) Twitter domain   (b) News article domain

Fig. 2. We used the translated AFINN-111 lexicon to visualize the sentiment word vectors on
different domains' word embeddings. The green nodes (legend: 1) indicate a positive sentiment
word and red nodes (legend: 0) indicate a negative one. (Color figure online)

3 Proposed Framework and Methodology

In this section, the proposed framework is presented. The key aspect of this framework is
that it is applicable on unlabelled datasets, which the Malay language lacks. As shown in
Fig. 3, the framework is composed of three parts: (a) Preprocessing, (b) Lexicon induction and lexicon-based classification, and (c) Supervised document sentiment classifier
training and classification.

3.1 Preprocessing

Preprocessing was performed to remove noise in text. A text sample in Malay and
English, and their normalized form can be seen in Table 2.

---

[2] https://www.astroawani.com/.

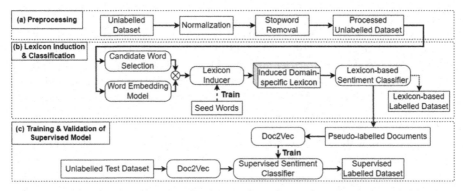

Fig. 3. Framework showing preprocessing, lexicon induction, and training a supervised document sentiment classifier using a subset of the output of the induced lexicon-based classifier.

Table 2. Text samples in Malay and English showing an unprocessed and normalized form.

Language	Unprocessed	Normalized
Malay	@mention Nak xnak berjual ps4, jarang2 main. Tapi nanti bosan pulak lahhh https://t.co/lbcLuUl0wH	nak tidak nak jual ps4 jarang main tapi nanti bosan pulak lah
English	There wasn't a lot of censorship when it showed in the cinema	there was not a lot of censorship when it show in the cinema

3.2 Lexicon Induction and Lexicon-Based Classification.

Manual development of a high-quality sentiment lexicon is time consuming because it requires careful curation of words that are relevant to the use case of the lexicon and accurate labelling of sentiment polarities expressed by these words. Inducing a lexicon is basically automating this process. There are three aspects to this process:

- **Seed words:** A seed word in this study is a word that is generally associated with only one sentiment polarity. The word 'good' is an example of a positive seed word.

Table 3 shows the seed words defined for the Malay Twitter domain and English IMDB domain. The English seed words were obtained from a study done by Hamilton et al. [11] and Mudinas et al. [12]. The Malay seed words were curated manually based on domain knowledge.

- **Candidate words:** A list of words that have a high probability of carrying sentiment. These words have no prior associated sentiment.

Mudinas et al. [12] extracted words from well-known English lexicons such as the General Inquirer and Bing-Liu Opinion Lexicon to populate their candidate word list.

However, this approach could not be emulated for this study due to the lack of high-quality and widely accepted Malay lexicons. Hence, a two-fold method was executed to produce a high-quality Malay-English candidate word list. First, translation of the Bing-Liu lexicon [15] and SentiWordNet [16] into Malay. Secondly, the cluster hypothesis that holds true for words in the same sentiment class was found to extend to words that do and do not express sentiment. Therefore, a word propagation algorithm was developed to use word vector similarity in the word embedding space to populate the candidate word list.

The propagation algorithm would iterate through the list of seed words and find the n-most-similar words in the word vector space. Through experimentation and manual observation of the results, it was determined that the optimal value for n was 3 for both datasets used in this study. If the most-similar search was extended past n = 3, the candidate word list would be populated with many words that have a vague sentiment.

- **Lexicon Inducer:** A supervised classifier trained on word embeddings of seed words. Used to classify candidate words with a positive or negative sentiment class.

To induce a sentiment lexicon, a supervised sentiment word classifier is required to perform a binary classification task (positive vs. negative) on the list of candidate words. The choice of a simple linear logistic regression model for this task was due to two factors: (1) A small number of seed words used for training, (2) Sentiment words in the same class are strongly clustered within the same domain in the word vector space as shown in Fig. 2a, thus, easily linearly separable.

Table 3. Seed words used for sentiment word polarity classifier training.

Domain	Positive	Negative
Malaysian Twitter	cinta, disayangi, suka, hebat, bagus, luar biasa, terbaik, betul, gembira, sihat, syukur	benci, dahsyat, jahat, mengerikan, teruk, sedih, sial, buruk, bodo, mampos, fitnah
English movie reviews	good, excellent, perfect, happy, interesting, amazing, unforgettable, genius, gifted, incredible	bad, bland, horrible, disgusting, poor, banal, shallow, disappointed, disappointing, lifeless, simplistic, bore

These three aspects are used to induce a domain-specific lexicon with sentiment words associated with either a positive (+1) or negative (−1) sentiment. The lexicon is then used with a simple lexicon-based classifier to perform sentiment classification discussed below.

Lexicon-Based Classification. The lexicon-based classifier implemented in this study considers sentiment polarity shifts through negation words and uses simple term-counting of sentiment words found in each document.

$$s = \sum \frac{w_s}{N_w} \tag{1}$$

$$f(s) = \begin{cases} negative, & s \le -0.5 \\ positive, & s \ge 0.5 \end{cases} \tag{2}$$

The sentiment score, s, is the sum of polarities of each sentiment word found, w_s, averaged by the total number of sentiment words, N_w as seen in Eq. 1. A score threshold of $0.5/-0.5$ was defined in Eq. 2 for a document to be successfully classified as positive or negative respectively. Any documents that have sentiment scores between 0.5 and -0.5 are not classified using the lexicon-based classifier and will be addressed by the supervised document sentiment classifier discussed in the next section.

3.3   Supervised Document Sentiment Classifier Training and Classification

Lexicon-based methods face a low recall issue when used in a sentiment classification task because a lexicon is a collection of words that have a static sentiment polarity attached to them. When a document contains a sentiment word that is not found in the sentiment lexicon, the lexicon-based classifier will not be able to capture and label the document appropriately. Additionally, the design of the lexicon-based classifier described in Sect. 3.2 results in dismissing documents that do not meet the thresholds set in Eq. 2, worsening the low recall issue.

Training a Supervised Document Sentiment Classifier. The framework in Fig. 3(c) proposes that the output of the induced lexicon-based classifier is used to train a supervised document sentiment classifier. The output of the induced lexicon-based classifier can be considered as a 'pseudo-labelled' dataset. To reduce the error propagated from the induced lexicon-based classifier, the median of the total number of sentiment words found in each document across the whole dataset was chosen as a threshold. The hypothesis here is:

– Documents with higher number of sentiment words than the median have a higher probability of belonging to its predicted sentiment label.

The f1-score of the pseudo-labelled dataset provides a measure of confidence on the quality of pseudo-labelled documents entering the training dataset.

The doc2vec model [17] was used to generate a single vector that represented a single document for the whole dataset. The doc2vec model was trained on the whole dataset of which the pseudo-labelled dataset is a subset of with a vector size of 300-dimensions. The trained doc2vec model was then used to infer the document vectors of the pseudo-labelled dataset. Subsequently, these vectors were used to train a linear logistic regression model as the supervised document sentiment classifier.

4   Experimental Results and Discussion

4.1   Dataset

A Malaysian Twitter dataset (primarily Malay) from MALAYA[3], labelled with sentiment (positive vs. negative) was used. To show that this framework is applicable across domains and writing styles, a commonly used labelled IMDB movie review dataset[4] was included in this study as well. Both datasets had no neutral sentiment class. Table 4 presents the characteristics of these two datasets.

Table 4. Characteristics of the MALAYA Twitter dataset and the IMDB movie review dataset.

Dataset	No. of documents	Language	Average no. of words per document
MALAYA Twitter	1.6M	Malay (primary) and English	15
IMDB movie reviews	50k	English	231

Baseline Tests. To form a baseline result, the two datasets used in this study were classified with a lexicon-based classifier using the following publicly available lexicons:

- The translated AFINN-111 lexicon [3] was used on the MALAYA Twitter dataset.
- The Bing-Liu lexicon [15] was used on the IMDB movie review dataset

The baseline results can be found in Table 5.

4.2   Lexicon Induction

Candidate Word List. In this study, a 4,614-candidate word list and 5000-candidate word list were produced for the MALAYA Twitter dataset and IMDB movie review dataset respectively.

Defining the Induction Threshold. When inducing a lexicon using the probabilistic output ($0 \leq p \leq 1$) from a logistic regression model, there is a typical trade-off between precision and recall when determining how many candidate words are allowed in the lexicon. Assuming a naïve threshold of 0.5 was set, all candidate words will be included into the induced lexicon. In this extreme case, candidate words at the decision border introduced into the induced lexicon will degrade its quality.

Increasing the probability threshold will mitigate this but also reduce the number of candidate words that enter the induced lexicon. The smaller the lexicon, the greater the

---

[3] https://malaya.readthedocs.io/en/latest/.

[4] https://www.kaggle.com/lakshmi25npathi/imdb-dataset-of-50k-movie-reviews.

recall issue when classifying documents due to less sentiment words being identified. However, since the aim is to induce a high-quality domain-specific lexicon and train a supervised document sentiment classifier on top of the lexicon-based classifier output, a high precision is emphasized. Thus, an induction probability threshold is defined to be:

$$w_s = \begin{cases} negative, & p \leq 0.3 \\ positive, & p \geq 0.7 \end{cases} \tag{3}$$

Induced Lexicon-Based Classification. The domain-specific lexicons induced were used in a lexicon-based classifier to classify their respective datasets. The results were compared against the baseline tests as seen in Table 5.

Results in Table 5 show that both domains, differing in language and writing style, produce similar results. The induced lexicon for the Malay Twitter domain is much smaller compared to the translated AFINN-111 lexicon but has a higher document recall as well as a higher f1-score. The translated AFINN-111 lexicon still performed reasonably well because it was originally developed to be a general social media domain lexicon. However, the induced domain-specific lexicon at roughly a tenth of the original translated AFINN-111's size was more optimal. Upon inspection of this induced lexicon, many slang words were captured through the word propagation algorithm discussed in Sect. 3.2. Therefore, maintaining and updating the induced lexicon with newly coined terms, an event that frequently occurs on social media, can be done with ease.

Table 5. Baseline vs. induced lexicon results

Dataset	Size	Lexicon	Lexicon size	F1-score	No. of classified documents
MALAYA Twitter	1.6M	AFINN-111 (translated)	26,0004	0.80	948,209
		MALAYA Twitter Induced Lexicon	3,035	0.88	1,039,341
IMDB movie reviews	50k	Bing-Liu Lexicon	6,788	0.88	13,493
		IMDB Induced Lexicon	1,984	0.88	21,115

A similar result is seen with the English IMDB dataset. The induced lexicon had twice as much document recall compared to the Bing-Liu lexicon while only having a third of the lexicon size in terms of number of sentiment words. Another point to note is that this framework was applicable on a dataset with 1.6M tweets as well as a dataset with only 50k movie reviews. However, in both datasets, roughly half of the documents were successfully classified, leaving the rest of the dataset unclassified. The next section will present a method on how to mitigate this low document recall issue.

Filtering for Pseudo-Labelled Documents. After performing sentiment classification on each dataset using their respective induced lexicons, the classified documents were filtered using the median of the number of sentiment words of each document across the whole dataset to produce the pseudo-labelled documents. As shown in Table 6, the f1-score of the pseudo-labelled documents is significantly higher than the total classified documents in Table 5, proving the hypothesis in Sect. 3.3.

Table 6. Number of pseudo-labelled documents resulting from the median threshold filter.

Dataset	Median no. of sentiment words	F1-score	No. of pseudo-labelled documents
MALAYA Twitter	2	0.90	16,604
IMDB movie reviews	10	0.95	1,519

4.3 Supervised Document Sentiment Classifier

The results of this supervised document sentiment classifier show a slightly worse performance in Table 7 compared to the induced lexicon-based classifier. However, the main difference is that the supervised model can classify all documents present in the dataset without any ambiguity with unclassified documents.

Table 7. Results of the supervised document sentiment classifier

Dataset	Model	F1-score
MALAYA Twitter	Pseudo-labelled trained classifier (this work)	0.81
	CNN classifier by [18]	0.78
IMDB movie reviews	Pseudo-labelled trained classifier (this work)	0.82
	LSTM classifier by [19]	0.89

As a comparison, two other studies that performed SA on the same datasets were recorded in Table 7. The methods used were more computationally expensive compared to the lightweight and linear classifier used in this work. Overall, the performance of this work's supervised document sentiment classifier is excellent when considering it was trained on pseudo-labelled documents that were classified with a lexicon induced from a small set of seed words.

5 Conclusion

SA in Malay text has been challenging due to the lack of resources. In this paper, the proposed framework was shown to be robust and applicable to two datasets differing in language, document length, dataset size and writing style. Two domain-specific lexicons with sizes of 3,035 and 1,984 words were successfully induced from a small set of seed words. These induced lexicons exceeded the performance of general lexicons in terms of f1-score and recall when considering the documents that were successfully classified.

As for the documents that were not classified due to the low recall issue faced by lexicon-based classifiers, a supervised document sentiment classifier was built on top of the induced lexicon-based classifier's output. The supervised classifier mitigated the low recall issue and was able to classify all documents at a cost of a slightly lower f1-score. In conclusion, this opens the possibility of producing labelled datasets autonomously across many domains to further advance the field of SA in Malay text.

References

1. Samsudin, N., Puteh, M., Hamdan, A.R.: Bess or xbest: mining the Malaysian online reviews. In: Conference on Data Mining and Optimization, pp. 38–43 (2011). https://doi.org/10.1109/DMO.2011.5976502
2. Lan, T.S., Logeswaran, R.: Challenges and development in Malay natural language processing. J. Crit. Rev. 7(3), 61–65 (2020). https://doi.org/10.31838/jcr.07.03.10
3. Azlan, A., Tan, Y.F., Lam, H.S., Soo, W.K.: Sentiment analysis for Telco popularity on Twitter big data using a novel Malaysian dictionary. Front. Artif. Intell. Appl. 282, 112–125 (2016). https://doi.org/10.3233/978-1-61499-637-8-112
4. Chekima, K., Alfred, R.: Sentiment analysis of Malay social media text. In: Alfred, R., Iida, H., Ag. Ibrahim, A.A., Lim, Y. (eds.) ICCST 2017. LNEE, vol. 488, pp. 205–219. Springer, Singapore (2018). https://doi.org/10.1007/978-981-10-8276-4_20
5. Sham Awang Abu Bakar, N., Aziehan Rahmat, R., Faruq Othman, U.: Polarity classification tool for sentiment analysis in Malay language. IAES Int. J. Artif. Intell. (IJ-AI) 8(3), 258–263 (2019). https://doi.org/10.11591/ijai.v8.i3.pp258-263
6. Basiri, M.E., Kabiri, A.: Translation is not enough: comparing Lexicon-based methods for sentiment analysis in Persian. In: 18th CSI International Symposium on Computer Architecture and Digital Systems, CSSE 2017 (2017). https://doi.org/10.1109/CSICSSE.2017.8320114
7. Ismail, Z.: Panduan Singkatan Khidmat Pesanan Ringkas (SMS) BM (2008)
8. Bin Rodzman, S.B., Hanif Rashid, M., Ismail, N.K., Abd Rahman, N., Aljunid, S.A., Abd Rahman, H.: Experiment with lexicon based techniques on domain-specific Malay document sentiment analysis. In: ISCAIE 2019 - 2019 IEEE Symposium on Computer Applications and Industrial Electronics, pp. 330–334, April 2019. https://doi.org/10.1109/ISCAIE.2019.8743942
9. Imanina Zabha, N., Ayop, Z., Anawar, S., Hamid, E., Zainal Abidin, Z.: Developing Cross-lingual Sentiment Analysis of Malay Twitter Data Using Lexicon-based Approach (2019). www.ijacsa.thesai.org. Accessed 15 Jan 2021
10. Darwich, M., Azman, S., Noah, M., Omar, N.: Automatically generating a sentiment lexicon for the Malay language. Asia Pac. J. Inf. Technol. Multimedia 5(1), 49–59 (2016). http://www.ftsm.ukm.my/apjitm
11. Hamilton, W.L., Clark, K., Leskovec, J., Jurafsky, D.: Inducing Domain-Specific Sentiment Lexicons from Unlabeled Corpora (2016). http://nlp.stanford.edu/projects/socialsent. Accessed 19 Feb 2021
12. Mudinas, A., Zhang, D., Levene, M.: Bootstrap domain-specific sentiment classifiers from unlabeled corpora. Trans. Assoc. Comput. Linguist. 6, 269–285 (2018). https://doi.org/10.1162/tacl_a_00020

13. Mikolov, T., Chen, K., Corrado, G., Dean, J.: Efficient Estimation of Word Representations in Vector Space (2013). http://ronan.collobert.com/senna/
14. Manning, C.D., Raghavan, P., Schutze, H.: Introduction to Information Retrieval. Cambridge University Press (2008). https://doi.org/10.1108/00242530410565256
15. Liu, B., Hu, M., Cheng, J.: Opinion Observer: Analyzing and Comparing Opinions on the Web. https://dl.acm.org/doi/abs/10.1145/1060745.1060797
16. Baccianella, S., Esuli, A., Sebastiani, F.: SENTIWORDNET 3.0: An Enhanced Lexical Resource for Sentiment Analysis and Opinion Mining (2011). http://wordnetcode.princeton. Accessed 22 Feb 2021
17. Le, Q., Mikolov, T.: Distributed representations of sentences and documents (2014)
18. Ong, J.-Y., Mun'im Ahmad Zabidi, M., Ramli, N., Ullah Sheikh, U.: Sentiment analysis of informal Malay tweets with deep learning. IJ-AI **9**(2), 212–220 (2020). https://doi.org/10.11591/ijai.v9.i2.pp212-220
19. Qaisar, S.M.: Sentiment analysis of IMDb movie reviews using long short-term memory. In: 2020 2nd International Conference on Computer and Information Sciences, ICCIS 2020, October 2020. https://doi.org/10.1109/ICCIS49240.2020.9257657

Agent and Group-Based Systems

Agent-Based Intermodal Behavior
for Urban Toll

Azise Oumar Diallo[1], Guillaume Lozenguez[1(✉)], Arnaud Doniec[1],
and René Mandiau[2]

[1] CERI Systèmes Numériques, Institut Mines-Télécom (IMT) Nord Europe,
Villeneuve d'Ascq, France
{azise.oumar.diallo,guillaume.lozenguez,arnaud.doniec}@imt-nord-europe.fr
[2] LAMIH UMR CNRS 8201, Université Polytechnique Hauts-de-France,
Valenciennes, France
rene.mandiau@uphf.fr

Abstract. To reduce the pollution and noise in the cities, the authorities encourage intermodality, notably through private cars and public
transport combinations. The application of dissuasive measures such as
urban tolls is an increasingly investigated solution. This paper proposes
an agent-based simulation to assess the impact of an urban toll on intermodal trip behaviors (private car + public transport) in a city. The
impact of the urban toll is modeled through a multinomial logit (MNL)
model, which is used to estimate the modal choice for each agent. To
avoid paying the toll tax, people (agents in our simulation) prefer to
combine different modes of transportation, e.g., their private cars and
public transport, by parking their vehicles in the park and ride facilities at the entrance to the city. Our experiments based on the MATSim
platform infer that with 20 euros of toll tax, it is possible to reduce by
20% the use of the private car in the European Metropolis of Lille (MEL,
France).

Keywords: Agent-based modeling · Intermodality · Urban toll ·
MATSim

1 Introduction

One of the political authorities' problems nowadays is reducing negative exter-
nalities (e.g., congestion, CO2 emissions, accidents) of transport systems. Most
of these reforms aim to reduce private car use in favor of environmentally friendly
modes of transportation (e.g., cycling, walking) and public transport (PT). They
generally consist in applying dissuasive measures such as road tolls [4]. Policies
encouraging the use of public transport by adjusting the pricing system are also
used [15]. Finally, to accelerate the energy transition, the political authorities
have significantly developed intermodality through urban transport infrastructure (e.g., park-and-ride facilities, multimodal exchange hubs).

© Springer Nature Switzerland AG 2022
H. Fujita et al. (Eds.): IEA/AIE 2022, LNAI 13343, pp. 397–408, 2022.
https://doi.org/10.1007/978-3-031-08530-7_33

These various measures would modify the travel behavior of users (e.g., the combination of several modes of transportation). This phenomenon affects mobility and leads to a new balance of transport flows. In this context of upheaval in transport modes, intermodality may be defined as the combination of several means of transportation during the same trip [13]. It aims to reduce the use of private cars and, in turn, reduce mobility problems such as congestion, lack of parking, and gas emissions [6]. Among the forms of intermodal travel, the combination of a private car and public transport (car+pt) is particularly emblematic [20]. This combination comprises individuals joining urban public transport services from their home located away from the network. This use is generally done through park and ride facilities, placed at the entrances to public transport (train, tram, metro, bus).

Based on this observation, simulation, and in particular, simulations based on a multi-agent model [2,8] can help to assess the effectiveness of intermodality policies. In this paper, we present an approach allowing the simulation of intermodal trips behavior when faced with the application of an urban toll. The work presented focuses on the combination of private cars and public transport (car+pt) requiring park and ride facilities (P+R). Our approach is based on a modal choice model considering the intermodal alternative and the impact of the urban toll. Later, we show how to combine the model obtained in a simulation tool (in our case *MATSim*) to study the displacement behaviors. The first case of the application of the model is carried out at the perimeter of the European Metropolis of Lille (MEL) in France.

The article is organized as follows. Section 2 presents a state of the art of transport-based simulation intending to include intermodal behaviors. Section 3 describes our methodological approach. Then, Sect. 4 evaluates the results obtained with data from the European Metropolis of Lille (MEL) before ending the paper with our conclusions and perspectives in the last section.

2   Background

To study the impact of their reforms, mobility and transport actors most often use traffic simulation software [21]. In the traditional models aggregated (macroscopic) in four stages (generation, distribution, modal choice, and assignment), the options of individuals' trips are carried out successively in an independent manner (*trip-based approach*) [17].

However, in practice, individuals make their choices by taking into account all the trips to be completed and the activities to be carried out, from leaving home to returning home (*tour-based approach*). This *activity and trip-based* approach of individuals allows taking into account the consistency of the modes of transportation successively chosen (for example, it is not possible to return by private car if the outward journey was made by public transport). It is also possible to consider the impact of the unavailability of a given mode of transportation on the day's trips chain. Such interactions cannot be modeled using a four-step aggregated model [18]. In addition, activity-based models allow, through synthetic population generation techniques [5], to evaluate different policies more directly than in the context of a four-step model. By combining the two models (based on activities with agents), it is then possible to efficiently simulate interactions between individuals of the synthetic population in the realization of their chains of activities [1]. This approach, therefore, allows modeling the decision-making processes of individuals more explicitly than traditional models.

There are different road pricing systems depending on the desired objective. We mainly distinguish the following cases [19]:

- distance: this type of toll consists of charging per distance traveled by car.
- road section: in this case, it is a question of applying a particular charge each time an area (regardless of its length) is crossed (e.g., tunnel and bridge).
- cordon: this consists of creating a cordon around a given zone and applying the toll to each finite perimeter entry and/or exit.
- area: the toll is used as soon as a car drives in the area concerned. The difference with the previous one lies in the payment, performed only once to circulate in the area and not at each passage of the cordon.

In the following, we will consider toll area, which leads the individuals to use public transport to enter the city.

3   Simulation Framework

This section presents our modeling approach to the intermodal behaviors and the road toll in the simulation framework.

3.1   *MATSim* and *eqasim*

The work presented in this paper was carried out using the agent-and-activity-based simulation software *MATSim* (*Multi-Agent Transport Simulation*) [12]. We used more precisely its *eqasim* module, which focuses on the modal choice by integrating a discrete choice model.

Figure 1 describes the process of coupling the discrete choice module and *MATSim* in *eqasim*. The decision-making stage, within particular the evaluation of the agents' plans (*scoring*) in the standard version of *MATSim* is no longer used. It is the discrete choice module (elements in light blue in Fig. 1) that replaces this step for determining the modal choice. The **estimators** calculate the utility of each alternative based on the attributes (e.g., mode cost) from **variables** and **mode parameters**, operations (e.g., calculation of time and cost) defined in the **predictors** and **cost models** of each mode and by integrating the information from the simulation (QSim) such as trips duration, service costs [11]. Then, the discrete choice module (DCM) chooses the mode(s) according to the approach used (*trip-based*, i.e., a single-mode for each trip or *tour-based*, i.e., the chain of modes for the round trip) and the choice model (e.g., multinomial logit model-MNL) from the computed utilities. This choice is also performed by considering certain constraints to avoid generating unrealistic

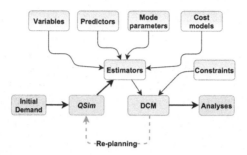

Fig. 1. Schematic representation of discrete choice model and *MATSim* coupling process in *eqasim* [11,12]. (Color figure online)

combinations, such as using the car without a driving license. Therefore, the plan choice and, consequently, the mode of transportation is performed *a posteriori* and not *a priori* as it was initially the case in *MATSim*. In fact, the choice of transport mode is determined according to the information resulting from the simulation (*QSim*). While in the standard version, this choice is performed before simulating without taking this information into account. Therefore, the discrete choice module makes it possible to consider the traffic situation better when choosing the mode of transport. Another advantage of integrating the discrete choice module is obtained from relevant parameter values necessary for rapid simulation calibration. The coefficients of the simulation parameters resulting from the utility functions of the activities and of the trips (in the standard version of *MATSim*) are considered challenging to calculate during the calibration [14].

3.2   Creation of the Intermodal Alternative (car+pt)

Private car and public transport combination (car+pt) must be done through a dedicated infrastructure, particularly park and ride facilities (P+R). To make this combination possible, we have created two new complementary modes of transportation designated by *car_pt* and *pt_car* respectively for outward and return journeys. Figure 2 illustrates our approach to enable intermodal travel combining car and public transport in *eqasim*. The alternative *car_pt* (ditto

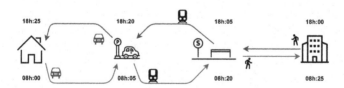

Fig. 2. Illustration of our approach to achieving intermodal travel combining car and public transport in *eqasim*. The red line corresponds to the *car_pt* mode for the outward journey and the blue line for the *pt_car* mode for the return. (Color figure online)

for *pt_car*) is composed of two modes of transportation: private car (*car*) and public transport (*pt*) to correspond to the two journeys of the shift. For the moment, we have focused on intermodal trips made up of two journeys whose origin is home. The destination can be any other activity such as work in the case of the example shown in Fig. 2. In addition, we have adopted an *tour-based* approach (unlike *trip-based*) to be able to take into account the constraints of modal combination. The alternative *car_pt* (shown schematically in red) is used for leaving home with the car as a mode of transport to the relay car park (first trip). The second trip is performed by public transport (e.g., bus, metro, tram, train) from the park and ride to the trip's destination, specifically at the nearest station (individual completes the journey by walking). We do the reverse process for the return using the alternative *pt_car*. The first trip is made by public transport from the previous destination to the same relay car park used on the outward journey. Finally, the individual returned to his/her home using his/her car, which had remained in the P+R. The two modes of transport *car_pt* and *pt_car* used to represent the intermodal alternative of private car and public transport (car+pt), have the same characteristics.

3.3   Considering the Road Toll in the Utility Function of the Car

Discrete choice models are generally based on random utility that an individual i has for an alternative (mode) m (or n) [3]. Let $U_{i,m}$ and $U_{i,n}$ be the utility perceived by i for m and n respectively, i will choose m instead of n if and only if $U_{i,m} > U_{i,n}$. However, the utility is a function based on known characteristics (e.g., age, travel time, and cost) and unknown (error) to the individual. Therefore, the choice of the individual i is not predicted precisely but randomly. The error terms are generally evaluated either by a normal distribution for the Probit models (possible correlation between alternatives) or by Gumbell's law for the Logit models (no correlations assumed between alternatives).

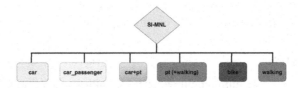

Fig. 3. Structure of the *multinomial logit* (*MNL*) model for modal choice estimation taking into account the intermodal alternative.

To consider the new intermodal alternative in the modal choice, we used a multinomial logit model (MNL) called Simple Intermodal MNL (SI-MNL), whose structure is shown in Fig. 3. The choice of multinomial logit models is justified by the simplicity of their implementation, which existing works can support [16,22]. This model is based on one of the properties of MNL models, namely the independence between alternatives IIA (Independence of Irrelevant Alternatives). We, therefore, assume that these alternatives are independent and that the error terms are not correlated. Subsequently, we have defined the utility function representing the intermodal alternative. The utility functions of the private car, public transit, and car+pt are adapted from [7] and defined as follows (by adding the processing on toll cost).

$$
\begin{aligned}
U_{i,car} = \beta_{ASC,car} &+ \beta_{inVehicleTime,car} \times x_{inVehicleTime,car} \\
&+ \beta_{inVehicleTime,car} \times \theta_{parkingSearchPenalty}
\end{aligned}
$$

$$
+ \beta_{accessEgressWalkTime} \times \theta_{accessEgressWalkTime} + \beta_{cost} \times x_{cost,car} \qquad (1)
$$

$$
+ \beta_{cost} \times toll\_fee \times \begin{cases} 1 & if \quad \text{if the trip is in the toll zone} \\ 0 & else \end{cases}
$$

$$
\begin{aligned}
U_{i,pt} = \beta_{ASC,pt} &+ \beta_{inVehicleTime,pt} \times x_{inVehicleTime,pt} \\
&+ \beta_{accessEgressTime,pt} \times x_{accessEgressTime,pt} \\
&+ \beta_{numberTransfers} \times x_{numberTransfers} \\
&+ \beta_{transferTime,pt} \times x_{transferTime,pt} + \beta_{cost} \times x_{cost,pt}
\end{aligned} \qquad (2)
$$

$$
\begin{aligned}
U_{i,car\_pt} = \beta_{ASC,car\_pt} &+ U_{i,car} - \beta_{inVehicleTime,car} \times \theta_{parkingSearchPenalty} \\
&+ U_{i,pt} - \beta_{ASC,car} - \beta_{ASC,pt}
\end{aligned} \qquad (3)
$$

where, i: individual, β: parameters (coefficients) of the model to be estimated, x: explanatory variables (e.g., travel time and cost) of the choice of mode, *ASC* (*Alternative Specific Constants*): constants specific to the alternative describing the variation of choice not explained by the attributes only, *toll_fee* is urban toll tax, and θ: calibration parameters to be adjusted manually.

4   Case Study

This section provides general information on the study area, details of the implementation of the simulation, and the results obtained.

4.1   Study Area

The MEL comprises 95 municipalities with 1.1 million inhabitants over an area of $672\,km^2$. Figure 4 provides an overview of the spatial distribution of populations.[1] and possession of car (at least one car) per household[2] in the municipalities of the MEL. There is a high concentration of the population in the main cities of the metropolis. In contrast, the population density remains low in the peripheral areas. The opposite phenomenon is observed concerning the number of cars per household. Peripheral regions have a higher concentration of vehicles per inhabitant, unlike large cities where less than 75% of households have a car.

---

[1] Source: https://www.insee.fr/fr/statistiques/3698339#consulter.
[2] Source: https://www.data.gouv.fr/fr/datasets/taux-de-motorisation-des-menages/.

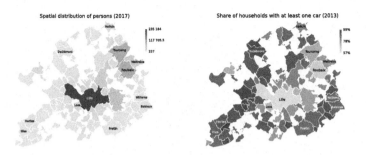

Fig. 4. Spatial distribution of the population (left) and car ownership per household (right) in the municipalities of the MEL.

The MEL household-travel survey (HTS), carried out in 2016, was used in this study. In total, 9,479 persons belonging to 4,539 households (approximately 0.87% of the actual population) were interviewed about their trips performed the day before. 24,629 were listed in this survey. More than half of all trips (all purposes combined) are performed using a private car (57.9%), walking comes second with nearly 30%, followed by public transport (TP) with 6, 7% and the bicycle which has less than 2% of modal share. Only 4% of trips are intermodal (a combination of two more mechanized modes of transportation). A public version of this survey is available on the MEL open data site[3].

4.2   Scenario

The synthetic population *MATSim* generated in this study represents a sample of 20% of the actual population, which corresponds to 225,240 agents with a total of 746,318 trips. This sample is based on recent works about population down-scaling [9] for multi-agent transport simulations. This choice is mainly justified for reasons of computational cost and for obtaining exploitable simulation times. The flow and storage capacities of the car network are multiplied by 0.18 to consider the down-sampling of traffic flow on the existing road network. The number of iterations is set to 100, and the share of agents allowed for re-planning to 20%.

Figure 5 provides information on the area and the geographical location where the urban toll is applied. The application area (hatched in black line) has been defined on the perimeter of the city of Lille, avoiding containing the nearby park and ride lots. The urban toll application configuration is defined as follows:

– Application time: all day.
– Price: variation of different amounts between 2 and 25 euros to significantly reduce the share of the car (10–20%).
– Who: the toll applies to all trips performed by a private car whose origin and/or destination is in the toll area.

---

[3] https://opendata.lillemetropole.fr/explore/dataset/enquete-deplacement-2016/ information/?location=10,50.65641,3.03338&basemap=jawg.streets.

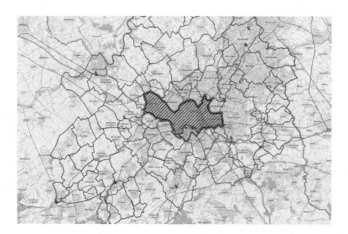

Fig. 5. Urban toll area (hatched in black line) in Lille of the MEL.

– Intermodality: two cases with or without the *car+pt* alternative to evaluate the use of P+Rs.

4.3   Results

The authors [7] described in Table 1, the estimated parameters for this model, which are based on a third of the data from the MEL household travel survey.

The β and θ are the coefficients of the explanatory variables that determine the utility value for each alternative. For example, the β_{cost} is intuitively nega- tive, thus reflecting a certain disutility towards modes of transport whose cost of use is high, such as the car. We want to describe in this paper different experi- ments. They were carried out to study the impact of urban tolls on the reduction of private car use by considering without modal combination (case 1) and the intermodal alternative *car+pt* (case 2). Thus, we performed several simulations by changing the value of the toll tax to obtain a significant reduction (20%) in the use of the private car. We have found that to apply a tax of 20 euros is necessary to achieve this goal.

Figures 6a and b provide the results (modal shares) of the MEL-scale sim- ulation with 20 euros for toll taxes, while Figs. 6c and d give the modal shares within the toll zone (Lille). The reference situation (EMD 2016) is in dotted line, the simulation without road pricing in red and the toll application in blue. We can see the effect of the road pricing on the modal shift by leaving the private car in favor of public transport (10% increase). This reduction even reaches almost 20% for long distances. The modal shift over short distances (less than 2 km) is not visible globally. There is also a slight increase in walking, especially for trips from 2 km. With the possibility of combining the car and public transport (car+pt), the share of the increase in public transport decreases by half in favor of the intermodal alternative (5% increase). Therefore, this trip possibility is more "useful" for users than simply using the PT. The modal shift over short

Table 1. Estimated model parameters.

	Parameters	Value	Rob. Std err
Car	$\beta_{ASC,car}$	-0.816	0.405
	$\beta_{inVehicleTime,car}$	$-0.00146[min^{-1}]$	0.00975
PT	$\beta_{ASC,pt}$	-1.75	0.557
	$\beta_{inVehicleTime,pt}$	$0.0035[min^{-1}]$	0.00186
	$\beta_{accessEgressTime,pt}$	$-0.0186[min^{-1}]$	0.0409
	$\beta_{numberTransfers}$	-0.207	2.21
	$\beta_{transferTime,pt}$	$-0.914[min^{-1}]$	0.403
Bike	$\beta_{ASC,bicycle}$	-1.27	0.44
	$\beta_{travelTime,bicycle}$	$-0.107[min^{-1}]$	0.0157
	$\beta_{highAge,bicycle}$	0.00905	0.00472
Walking	$\beta_{ASC,walk}$	2.16	0.418
	$\beta_{travelTime,walk}$	$-0.135[min^{-1}]$	0.00677
Other	β_{cost}	$-0.298[euro^{-1}]$	0.038
Calibration	$\theta_{parkingSearchPenalty}$	4	$-$
	$\theta_{accessEgressWalkTime}$	4	$-$
	β_{ASC,car_pt}	1.25	$-$
	Number of parameters	13	
	Sample size	5.114	
	Init log likelihood	$-8.668.555$	
	Final log likelihood	$-5.086.149$	

distances (less than 3 km) is not visible globally. A more detailed analysis at the scale of the city of Lille (toll area) allows understanding better this behavior on trips performed by car in the town.

The impact of urban toll on the scale of the city of Lille is much more striking (Figs. 6c and d). We note a drop of 50% in the use of private cars in the city in favor of softer modes (walking, bicycle, and public transport) (Fig. 6c). Walking even reaches a 50% increase for distances less than 3 km. Trips performed by bicycle increase between distances between 2.5 and 4 km. The modal share of public transport (PT) increases to nearly 75% for trips over 5 km in the city. Integrating the intermodal alternative (private car + public transport) offers a more attractive means of travel for agents (users) in urban toll applications. This combination allows users going to Lille to park their car in the P+R lot and then take public transport to enter the city. As a result, they avoid the payment of the toll tax.

This phenomenon is visible through the flow of cars at the P+R at the city's entrance. Figure 7 gives an overview of this situation from 8:30 am. Cars in red are in a congestion situation. They move very slowly. Those in yellow also circulate slower than when the traffic is fluid (cars in green). We can see the impact of the road pricing on the flow of cars within the city of Lille (Fig. 7b).

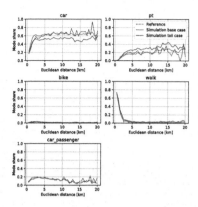

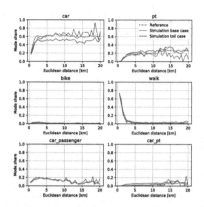

(a) Case 1: Without the intermodal alternative in the MEL.

(b) Case 2: With the intermodal alternative *car+pt* in the MEL.

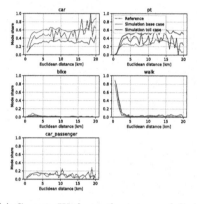

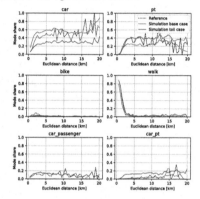

(c) Case 1: Without the intermodal alternative within Lille.

(d) Case 2: With the intermodal alternative *car+pt* within Lille.

Fig. 6. Modal shares with a toll area (20 euros tax).

The number of cars has dropped considerably compared to the situation without an urban toll (Fig. 7a).

4.4   Discussion

The road toll system studied in this work relates to a zonal application on the scale of the city of Lille. This strategy, in simulation, has reduced road traffic in the town with a toll tax of 20 euros. A similar method was applied to the city of Zurich (Switzerland) [10]. Unlike our approach based on the toll area, this study focused on using road pricing through a cordon around the city. In addition, the toll is only applied during peak hours (5:30–9:00 and 17:30–19:00) to enter and exit the city. Several simulations were carried out with different tax amounts to obtain a 20% reduction in automobile traffic in the city.

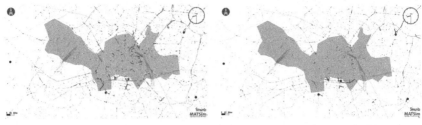

(a) Traffic situation without urban toll.   (b) Traffic situation with urban toll.

Fig. 7. Cars' flow at the P+R located at the entrance to Lille (Color figure online)

The drawback of the zonal toll compared to the cordon is that the flow of cars crossing the area is not considered. However, the main objective of the toll is to reduce or limit these types of trips in the city by making bypass roads. Another point not dealt with in this paper is the acceptability of such measures by individuals depending in particular on their income [15].

5   Conclusion and Perspectives

In this paper, we proposed to study the intermodal travel behavior of the population from the perspective of multi-agent simulation following the application of an urban toll. Compared to traditional four-step models, the agent-based approach is particularly suitable for studying individuals' mobility behaviors (intermodality). To take into account the intermodal alternative (combination of a private car and public transport) in the modal choice decision process, we have proposed a discrete choice model *multinomial logit* (*Simple intermodal MNL* (*SI-MNL*). Later, we created new modes of transport (*car_pt* and *pt_car*) while defining their usage constraints, intermodal routing through relay car parks. Our experiments were carried out on the European Metropolis of Lille (MEL) data. Our study shows that with 20 euros in toll taxes, it was possible to reduce personal car use by 20% in the MEL.

As part of future work, we plan to extend our simulation framework to other modal combinations such as bike or scooter (in particular self-service) and public transport or even the use of taxis or autonomous shuttles.

References

1. Balmer, M.: Travel demand modeling for multi-agent transport simulations: algorithms and systems. Ph.D. thesis, ETH Zurich (2007)
2. Bazzan, A.L., Klügl, F.: A review on agent-based technology for traffic and transportation. Knowl. Eng. Rev. **29**(3), 375–403 (2014)
3. Ben-Akiva, M.E., Lerman, S.R., Lerman, S.R.: Discrete Choice Analysis: Theory and Application to Travel Demand, vol. 9. MIT Press, Cambridge (1985)

4. Cervero, R.: Traffic impacts of variable pricing on the san Francisco-Oakland bay bridge, California. Transp. Res. Rec. **2278**(1), 145–152 (2012)
5. Chapuis, K., Taillandier, P.: A brief review of synthetic population generation practices in agent-based social simulation. In: submitted to SSC2019, Social Simulation Conference (2019)
6. Dacko, S., Spalteholz, C.: Upgrading the city: enabling intermodal travel behaviour. Technol. Forecast. Soc. Chang. **89**, 222–235 (2014)
7. Diallo, A.O., Doniec, A., Lozenguez, G., Mandiau, R.: Agent-based simulation from anonymized data: an application to Lille metropolis. In: Proceedings of the 12th International Conference on Ambient Systems, Networks and Technologies (ANT) (2021)
8. Diallo, A.O., Lozenguez, G., Doniec, A., Mandiau, R.: Comparative evaluation of road traffic simulators based on modeler's specifications: an application to intermodal mobility behaviors. In: Proceedings of the 13th International Conference on Agents and Artificial Intelligence (ICAART), vol. 1, pp. 265–272 (2021)
9. Dor, G.B., Ben-Elia, E., Benenson, I.: Population downscaling in multi-agent transportation simulations: a review and case study. Simul. Model. Pract. Theory, 102233 (2020)
10. de Freitas, L.M., Schuemperlin, O., Balac, M.: Road pricing: an analysis of equity effects with MATSim. In: Proceedings of the 16th Swiss Transport Research Conference, Ascona, Switzerland, pp. 18–20 (2016)
11. Hörl, S., Balać, M., Axhausen, K.W.: Pairing discrete mode choice models and agent-based transport simulation with MATSim. In: TRB Annual Meeting. Transportation Research Board (2019)
12. Horni, A., Nagel, K., Axhausen, K. (eds.): Multi-Agent Transport Simulation MATSim. Ubiquity Press, London (2016). https://doi.org/10.5334/baw
13. Jones, W.B., Cassady, C.R., Bowden, R.O., Jr.: Developing a standard definition of intermodal transportation. Transp. LJ **27**, 345 (2000)
14. Kickhofer, B., Hosse, D., Turnera, K., Tirachinic, A.: Creating an open MATSim scenario from open data: the case of Santiago de Chile. In: Working Paper 16-02, TU Berlin, Transport System Planning and Transport Telematics (2016)
15. Kilani, M., Proost, S., Van der Loo, S.: Road pricing and public transport pricing reform in Paris: complements or substitutes? Econ. Transp. **3**(2), 175–187 (2014)
16. Krajzewicz, D., Heinrichs, M., Beige, S.: Embedding intermodal mobility behavior in an agent-based demand model. Procedia Comput. Sci. **130**, 865–871 (2018)
17. McNally, M.G.: The four step model. Handb. Transp. Model. **1**, 35–41 (2000)
18. Miller, E.J., Roorda, M.J.: Prototype model of household activity-travel scheduling. Transp. Res. Rec. **1831**(1), 114–121 (2003)
19. Nagel, K.: Road pricing. In: Horni, A., Nagel, K., Axhausen, K. (eds.) The Multi-Agent Transport Simulation MATSim, Chap. 15, pp. 97–102. Ubiquity Press, London (2016)
20. Oostendorp, R., Gebhardt, L.: Combining means of transport as a users' strategy to optimize traveling in an urban context: empirical results on intermodal travel behavior from a survey in berlin. J. Transp. Geogr. **71**, 72–83 (2018)
21. Pursula, M.: Simulation of traffic systems - an overview. J. Geogr. Inf. Decis. Anal. **3**(1), 1–8 (1999)
22. Train, K.E.: Discrete Choice Methods with Simulation. Cambridge University Press, Cambridge (2009)

Entropy Based Approach to Measuring Consensus in Group Decision-Making Problems

J. M. Tapia[1]($\boxtimes$) ⓘ, F. Chiclana[2] ⓘ, M. J. del Moral[3] ⓘ, and E. Herrera–Viedma[4] ⓘ

[1] Department of Quantitative Methods in Economic and Business, University of Granada, 18071 Granada, Spain
jmtaga@ugr.es
[2] DIGITS, Department of Informatics, Faculty of Technology, De Montfort University, Leicester LE1 9BH, UK
[3] University of Granada, 18071 Granada, Spain
[4] Department of Computer Science and A.I, University of Granada, 18071 Granada, Spain

Abstract. Entropy is a measure of randomness in a given set of data. An entropy measure could be appropriately used to assess consensus across a set of opinions. A Theil-based index is introduced in this paper to obtain the level of consensus in some problems of Group Decision Making. A comparative analysis reveals that the levels of consensus derived from this index are relatively similar to those obtained by using distance functions when a fuzzy preference relations frame is considered. This behavior suggests that this could be a useful tool in the aforementioned context.

Keywords: Group decision making · Fuzzy preferences · Consensus · Entropy · Theil index · Distance functions

1 Introduction

In Group Decision Making (GDM) problems, a group of experts have to decide a solution among a set of alternatives. In this framework is desirable an agreement among experts about the proposed solution.

This state of agreement among the members of the group is usually known by the term consensus [1]. Although consensus can be understood as a full and unanimous agreement among experts, in most situations such a degree of agreement is not necessary. Several measures can be used to express different levels of consensus. Among them is the one originated by the concept known as soft consensus and is the one which we will deal with in this paper [1, 2]. Using soft consensus measures we can express different levels of agreement among experts. The use of these measures is based on the concept of similarity between the different opinions of the experts -preferences-.

In general, to compute the levels of consensus it is necessary to calculate and aggregate the distance measures employed to represent the proximity among the preferences of each pair of experts on each pair of alternatives [3, 4]. These calculations can be complex depending on the aggregator used and may require the construction of the collective preference relation before these calculus can be performed.

© Springer Nature Switzerland AG 2022
H. Fujita et al. (Eds.): IEA/AIE 2022, LNAI 13343, pp. 409–415, 2022.
https://doi.org/10.1007/978-3-031-08530-7_34

In other contexts, alternative measures based on statistical variability have been used to measure consensus [5, 6]. Inequality measures such as entropy have been used in aggregation methods to determine consensus in GDM with linguistic variables [7] or in conjunction with cluster analysis [8].

Entropy was originally used to measure uncertainty, being Zadeh who introduced it as a measure in a fuzzy sets context [9].

There are several measures of statistical/economic inequality. One of them was proposed by Henri Theil, who defined the Theil index by calculating entropy as a measure of randomness in a given data set.

In this paper we introduce a new index of consensus based on measuring the inequality among the preferences of the experts in a context of GDM problems with fuzzy preference relations by taking into account the aforementioned idea. This index could allow consensus computations without using distance measures in iterative or non-iterative processes and in certain situations. The implementation of this new index could successfully replace more complex processes associated to other measures. This index does not depend on the different aggregation criteria that can be applied when obtaining a collective relation used to find solutions in GDM problems. We compare this new consensus measure with a frequently used approach based on an aggregator and a distance function [4] and acceptable results are obtained.

The structure of this article is as follows: Sect. 2 describes the theoretical framework in which this study is developed: basic concepts and elements of GDM problems. Section 3 introduces the new proposed index. Section 4 is dedicated to the experimental study carried out: the design, the conditions under which it was developed, as well as the results obtained. Finally, the Conclusion section ends this article.

2 GDM Problems and Consensus Measurement

In the context of a fuzzy preference relation, a GDM problem consist of finding the best alternative from a set of alternatives $X = \{x_1, \ldots, x_m\}$ according to the preferences of a group of experts $E = \{e^1, \ldots, e^n\}$ (with $m, n > 1$). These preferences are expressed through fuzzy preference relations [1, 10].

A fuzzy preference relation on a finite set of alternatives X is characterized by a function F, with $F(x_i, x_j) = p_{ij}$ denoting the preference degree of the alternative x_i over x_j given by an expert [3, 5], where 0 represents the minimal (null) preference and 1 represents the maximal (total) preference. This function verifies reciprocity, i.e. $p_{ij} + p_{ji} = 1$, with i, j in $\{1, \ldots, m\}$. These relations are frequently showed by a matrix $P = (p_{ij})$.

To calculate the consensus degree, a consensus matrix is introduced, $CM = (cm_{ij})$. This matrix is obtained by aggregating all the similarity matrices previously calculated using an Ordered Weighted Averaging (OWA) operator with weighting vector W [11, 12].

Some examples of OWA operators are: Maximum ($W = [1, 0, \ldots, 0]$), Minimum ($W = [0, \ldots, 0, 1]$) or Average ($W = [1/n, 1/n, \ldots, 1/n]$). Alternative representations for the concept of fuzzy majority can be found in the literature [12].

The computation of the consensus level among experts uses the measurement of the distance among their preference values [13]. In this computation it is necessary the use of a distance function. Manhattan, Euclidean or Cosine distance functions are frequently used [4, 13]. In Sect. 3 we use Euclidean distance function as a tool to introduce the proposed new index of consensus:

$$Eu(A, B) = \sqrt{\sum_{i=1}^{n} |a_i - b_i|^2} \tag{1}$$

being $A = \{a_1, \ldots, a_n\}$ and $B = \{b_1, \ldots, b_n\}$ two sets of real numbers.

Any of the distance functions could be used to find the similarity between preference values through the similarity function by setting similarity as $s = 1 - d$ [5].

Then, we obtain a similarity matrix for every expert r, $SM^r = \left(sm_{ij}^r\right)$, with $sm_{ij}^r = s\left(p_{ij}^r, p_{ij}\right)$. These matrices provide an evaluation of the proximity among preference values comparing each expert with the rest. This proximity is obtained for each pair of alternatives (x_i, x_j). In this situation, $CM = (cm_{ij})$, with i, j in$\{1, \ldots, m\}$, is obtained as an aggregation of sm_{ij}^k with $k = 1, \ldots, n$. CM shows the consensus degree on each pair of alternatives (x_i, x_j) by means of cm_{ij}.

To calculate the consensus degree on the relation, c_r, i.e. the global agreement among all experts, an aggregation operation of all the consensus degrees at the level of pairs of alternatives is introduced.

3 An Entropic Consensus Measure: Theil-Based Index

There are several measures of statistical/economic inequality. One of them was proposed by Henri Theil [14, 15], the so called Theil index, which is based on the on the Shannon entropy concept [16]. The Theil index is obtained by calculating the entropy of a given set of data, a measure of randomness in this set.

This concept could be used in a GDM context in which fuzzy preference relations are considered, in order to introduce new measures of agreement or disagreement [6].

We introduce the Theil-based index for consensus measurement as follows.

Let $\left\{p_{ij}^1, \ldots, p_{ij}^n\right\}$ be the preferences of n experts on a pair of alternatives (x_i, x_j) with i, j in $\{1, \ldots, m\}$. The Theil-based index on each pair of alternatives (x_i, x_j) (TH) is defined as

$$TH(x_i, x_j) = TH_{ij} = \begin{cases} 0, & for\ h_k = 0; \\ 1 + \dfrac{\sum_{k=1}^{n} (h_k \cdot \log_a h_k)}{\log_a n}, & elsewhere. \end{cases}$$

where h_k is obtained by

$$h_k = \begin{cases} 0, & for\ \sum_{l=1}^{n} p_{ij}^l = 0; \\ \dfrac{p_{ij}^k}{\sum_{l=1}^{n} p_{ij}^l}, & elsewhere. \end{cases}$$

To calculate consensus degree on the relation, the Theil-based consensus index (C_{TH}) is defined as

$$C_{TH} = \frac{n-1}{n} \frac{\displaystyle\sum_{i=1}^{m}\sum_{j>i}^{m} CTH_{ij}}{\displaystyle\sum_{k=1}^{m-1}(m-k)}, \quad CTH_{ij} = 1 - Minimum[TH_{ij}, TH_{ji}] \tag{2}$$

4   Comparative Study

A total of 600 random GDM problems were generated: 100 for each combination of alternatives -3, 4 and 5- and experts -3 and 4-, i.e. 100 GDM problems for 3 alternatives and 3 experts, 100 for 3 alternatives and 4 experts, and so on. The OWA operator used was Average, being the weighting vector $w = [1/n,..., 1/n]$, and the chosen distance function was the Euclidean one (1).

This comparative study attempts to answer a question:

"Can the value obtained by the Theil-based index (C_{TH}) (2) be used to calculate the degree of consensus in a GDM problem instead of Euclidean distance function (Eu) ?"

This question is associated with two other questions:
Is the value obtained by $C_{TH} - Eu$ a monotonic function, i.e. $C_{TH} > Eu$ ($C_{TH} < Eu$) always or almost always or does it fluctuate -sometimes $C_{TH} > Eu$, sometimes $C_{TH} < Eu$, and sometimes $C_{TH} = Eu$-?
and
In case of monotonicity -$C_{TH} > Eu$ or $C_{TH} < Eu$-, how much greater is C_{TH} over Eu or vice versa?
The answer to the first related question is reflected in Table 1. In all cases the value of the Theil-based index is greater than the value obtained by the Euclidean distance function. So, we have a monotonic function and C_{TH} is greater than Eu.

Table 1. Percentage of C_{TH} values greater than Eu values.

Alt/Exp	3	4
3	100%	100%
4	100%	100%
5	100%	100%

Table 2 shows the approximate difference between C_{TH} and Eu values, in percentages. It can be seen that the number of alternatives seems to affect this difference very slightly, while the number of experts seems to influence significantly.

Table 2. Increase in the degree of consensus of C_{TH} over Eu (in percentage).

Alt/Exp	3	4
3	36%	13%
4	39%	14%
5	41%	15%
Average	39%	14%

Figure 1, show the level of consensus (in percentage) achieved in the two cases analyzed. The higher the value of the consensus degrees, the higher the global degree of consensus. The results show the relative positions of the proposed Theil-based index facing Euclidean distance function, and also shown that this index could be used as a measurement of consensus degree in GDM problems with these results in mind.

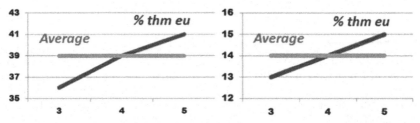

Fig. 1. Increase in the degree of consensus of C_{TH} over Eu for 3 and 4 experts (in percentage).

Figure 1 also shows a possible growth trend of the differences between the measures considered (C_{TH} and Eu) according to the number of alternatives.

Figure 2 clearly shows the difference indicated according to the number of experts that appears in the GDM problem.

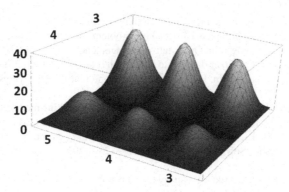

Fig. 2. Increase in the degree of consensus of C_{TH} over Eu for 3 and 4 experts (in percentage).

5 Conclusion

In this study we have proposed an index to measure consensus based on an entropy measure, the Theil-based consensus index. We have compared this new index with the one based on a well-known distance function –Euclidean distance function-. The empirical study carried out shows acceptable -monotonous- results regarding the consensus behavior of the proposed index. The new Theil-based consensus index always exceeds the one associated with the Euclidean distance function, by an average percentage that depends on the number of experts considered and which must be taken into account for the interpretation of this new index.

The possibility of expanding the study with a greater number of experts and alternatives seems to be an interesting question to be addressed in the future.

Acknowledgments. Supported by both the project PID2019-103880RB-I00 funded by MCIN/AEI/https://doi.org/10.13039/501100011033 and the project number P20 00673 funded by the Andalusian Government.

References

1. Herrera-Viedma, E., Cabrerizo, F.J., Kacprzyk, J., Pedrycz, W.: A review of soft consensus models in a fuzzy environment. Inf. Fus. **17**, 4–13 (2014)
2. Del Moral, M.J., Tapia, J.M., Chiclana, F., Al-Hmouz, A., Herrera-Viedma, E.: An analysis of consensus approaches based on different concepts of coincidence. J. Intell. Fuzzy Syst. **34**(4), 2247–2259 (2018)
3. Cabrerizo, F.J., Moreno, J.M., Perez, I.J., Herrera-Viedma, E.: Analyzing consensus approaches in fuzzy group decision making: advantages and drawbacks. Soft. Comput. **14**(5), 451–463 (2010)
4. Del Moral, M.J., Tapia, J.M., Chiclana, F., Herrera-Viedma, E.: Comparing two approaches for consensus computation in group decision making problems. Front. Artif. Intell. Appl. **303**, 312–320 (2018)
5. Akiyama, Y., Nolan, J., Darrah, M., Rahem, M.A., Wang, L.: A method for measuring consensus within groups: an index of disagreement via conditional probability. Inf. Sci. **345**, 116–128 (2016)
6. Del Moral, M.J., Chiclana, F., Tapia, J.M., Herrera-Viedma, E.: An alternative calculation of the consensus degree in group decision making problems. In: Fifth International Conference on Information Technology and Quantitative Management, New Dehli (2017)
7. Lo, C.C., Wang, P.: Using fuzzy distance to evaluate the consensus of group decision-making – an entropy-based approach. In: IEEE International Conference on Fuzzy Systems, Budapest, pp. 1001–1006 (2004)
8. Zhou, X., Zhang, F.M., Hui, X.B., Li, K.W.: Group decision-making method based on entropy and experts cluster analysis. J. Syst. Eng. Electron. **22**(3), 468–472 (2011)
9. Xu, T.-T., Zhang, H., Li, B.-Q.: Axiomatic framework of fuzzy entropy and hesitancy entropy in fuzzy environment. Soft. Comput. **25**(2), 1219–1238 (2020). https://doi.org/10.1007/s00 500-020-05216-9
10. Pedrycz, W., Parreiras, R., Ekel, P.: Fuzzy Multicriteria Decision-Making. Models, Methods and Applications. Wiley, Chichester (2011)
11. Zadeh, L.A.: A computational approach to fuzzy quantifiers in natural languages. Comput. Math. Appl. **9**(1), 149–184 (1983)

12. Yager, R.R.: On ordered weighted averaging aggregation operators in multicriteria decision making. IEEE Trans. Syst. Man Cybern. **18**(1), 183–190 (1988)
13. Deza, M.M., Deza, E.: Encyclopedia of Distances. Springer, Heidelberg (2009). https://doi.org/10.1007/978-3-642-00234-2
14. Hermoso, J.A., Hernandez, A.: Curso básico de Estadística Descriptiva y Probabilidad. Némesis, Granada (2000)
15. Tastle, W.J., Wierman, M.J.: Consensus and dissention: a measure of ordinal dispersion. Int. J. Approx. Reason. **45**(3), 531–545 (2007)
16. Shannon, C.E.: The mathematical theory of communication. Bell Syst. Tech. J. **27**, 379–423 (1948)

Adaptation of HMIs According to Users' Feelings Based on Multi-agent Systems

Alia Maaloul[1(✉)], Houssem Eddine Nouri[2,4], Zied Trifa[3,4], and Olfa Belkahla Driss[2,5]

[1] SMART Laboratory, University of Tunis, Tunis, Tunisia
Maaloul.alia@gmail.com
[2] LARIA Laboratory, University of Manouba, Manouba, Tunisia
olfa.belkahla@esct.uma.tn
[3] MIRACLE Laboratory, University of Sfax, Sfax, Tunisia
[4] Institut Supérieur de Gestion de Gabes, University of Gabes, Gabes, Tunisia
[5] École Supérieure de Commerce de Tunis, University of Manouba, Manouba, Tunisia

Abstract. There is always a component of the field of Human-Machine Interactions (HMI) which constantly tries to provide solutions to the problems relating to adaptation, in order to move towards more ergonomic approaches and more flexible and adaptive tools. It is in this context that this research work is taking place, which will consider an adaptation of User Interfaces (UIs) guided by the results of the analysis of user feelings and this, by adopting a new approach based on Multi-Agent System (MAS) and Deep Learning. To implement this approach, a system admitting a first component as a dynamic Django web application and a second component which corresponds to a SPADE Multi-Agent System, has been produced and tested and has shown effective and interesting experimental results.

Keywords: HMI · HMI adaptation · Multi-Agent System · Sentiment analysis · Deep learning · Python · Django · SPADE

1 Introduction

HMIs have been the subject of several research studies, both in terms of their design and their evaluation and even their adaptation to their context. In the 2000s, Beaudouin-Lafon [1] introduced the term "situated interactions" in order to emphasize the importance of the environment and the characteristics of users in the design and use of HMIs. Indeed, taking into account the context and the situations of use, both in the design and in the execution of the applications, is essential to achieve their main objective: to assist the users as well as possible. This consideration is manifested by the need for dynamic evolution and adaptation of the HMIs to which the applications must respond correctly. In fact, there is always a part of the Human-Computer Interaction (HCI) field that constantly tries to bring solutions to the problems related to adaptation, with the objective of moving towards more ergonomic, flexible and adaptive approaches and tools [2]. It is in this paper that will consider an adaptation of the HCI guided by the results of the analysis of the users' feelings.

© Springer Nature Switzerland AG 2022
H. Fujita et al. (Eds.): IEA/AIE 2022, LNAI 13343, pp. 416–428, 2022.
https://doi.org/10.1007/978-3-031-08530-7_35

In this research work, we will focus on the problem of the adaptation of Human-Computer Interfaces (HCI) to their context of use, characterized by the triplet user, platform, environment [3], and we will put the emphasis precisely on the users through the analysis of their feelings in real time followed by an automatic quasi-immediate adaptation of the HCI. However, even if many works have tackled the problem of adaptive HMIs [4, 5], there are still tracks to be explored such as user reactions and the combination of adaptation rules, especially if the adaptation focuses on the user and not only on the platform and the environment, which requires new approaches and tools such as the combination of the use of Multi-Agent Systems and Deep Learning, which constitutes the contribution of this work.

In this paper, we aim at the following objectives:

- To study the current state of research in HMI adaptation according to the context of use and precisely according to the profiles, characteristics and reactions or even the emotions of the users.
- To propose a new approach to adapting HMIs according to users' feelings based on the use of Multi-Agent Systems (MAS) and Deep Learning.
- To propose an adaptation system based on Multi-Agent Systems and Deep Learning in which the HMI adaptation engine takes into account the analysis of the users' feelings in real time while also considering the platform and the environment.

The organization of this paper reflects the proposed and adopted approach. It is structured around the following parts:

- After the introduction, Sect. 2 describes the state of the art on existing approaches of HMI adaptation in the literature.
- In the Sect. 3, we present the approach proposed to ensure the adaptation of HMIs in the framework of this research work which consists in using jointly Multi-Agent Systems and Deep Learning.
- Sect. 4 and 5 present the details of the implementation and experimentation as well as the validation and evaluation process of the relevance of this implemented approach.

This paper ends with a conclusion and an opening on some perspectives of this research work.

2   State of the Art

In this section, we present a state-of-the-art study on the work done to ensure the adaptation of HMIs to their context in general and to users in particular. This study consists in formulating a synthesis explaining the different approaches and techniques of HMI adaptation based on the research works already done.

2.1   Model-Based Approaches

According to Hachani et al. in 2009 [6], HMI researchers have started to use a model-based approach to ensure HMI adaptation. Indeed, models are essentially formal representations that initially describe user behavior and have subsequently been extended to include adaptive interfaces. As models are developed, they are updated by tracking user actions in the system. The model-based method presented in [6] is based on general and consistent specifications of so-called task models. Its characteristics take into account the similarities and differences between different contexts of simultaneous use of a program. These models are then adjusted to the situation of use by model transformation similar to Model Driven Engineering (MDE) [7]. Model-based approaches express a wide range of variables via model conversion. In particular, Model Driven Engineering provides concepts, languages and tools to create and transform models based on meta-models [7]. The goal is to facilitate compliance with the interactive system's control point code, usage changes, and actions for adaptation. The disadvantage of this approach, however, is that adaptive data is injected into the code, when it is best made available to non-developers. For example, CAMELEON [6] is a benchmark approach that enables context-consistent user interface dynamics. It is based on a process that revolves around four abstract levels: the task model, the abstract user interface, the specific user interface and finally the final user interface. Ultimately, the model-based approach is considered to (i) gather adaptation knowledge with explicit representations without drowning it in code, and also (ii) hide the boundary between system design and execution through the use of so-called live models at runtime. Each model (from the task model to the code) is associated with a perspective of the system [6] that is adapted to the context. Therefore, the adaptation of an interactive system according to this method is an adaptation of its models.

2.2 Approaches Based on Fuzzy Logic

Related to [8], there are many works dealing with the adaptation of HMIs using fuzzy logic. Other existing works deal with the problem of using fuzzy logic to maintain adaptive systems. For example, we can quote FSAM (Fuzzy Services Adaptation Modeling) [9] which is a tool based on fuzzy logic that allows to select the most appropriate service for the user according to the context of use. These works are limited to the choice between predefined services and do not deal with human-computer interaction issues. In [10], a fuzzy logic approach to the development of adaptive mobile applications is proposed, taking into account the variety of mobile operating systems (hardware, operating system, API, etc.) and the capabilities of each of these platforms in terms of performance. Cueva-Fernandez et al. (2016) [11] provides a tool for the development of applications by the driver or passenger in a car via the voice interface. Fuzzy logic is used to calculate the user experience and the level of concentration: beginner, intermediate or advanced. Each level leads to a higher or lower level of options and information provided. It uses 7 fuzzy logic rules and relies on 10 features (number of applications created; number of help questions; number of swear words, etc.) to modify the output result: the level of expertise. Therefore, using this fuzzy logic is not very compatible with the proposed adaptation rules. Nyongesa et al. (2016) [12] provides another fuzzy logic approach for personalizing web pages to improve the user experience. However, it depends on the uses of 7 use cases including: query, search, etc. The data comes from the user's behavior, which limits the size of the user model and therefore possible adaptive rules. It does not allow the combination of adaptation rules which seems to be a central problem. Papatheocharous et al. (2012) [13] proposes an adaptive mechanism for HMI based on fuzzy logic, which takes into account the cognitive characteristics of the users defined by [10], which are their perceptual and imaginative dimensions, their synthesis and analysis dimensions, and their memory capacities. Finally, Soui et al. (2013) [14] presents a fuzzy logic method for HMI personalization based on semantic relationships between HMI components and user preferences to select the most appropriate components. The main drawback of this method is that important aspects such as the size of the application area and the combination of rules are not taken into account.

2.3 Artificial Intelligence-Based Adaptation Approaches

According to Thevenin et al. (2002) [15], study of HMI adaptation is not a new phenomenon. Artificial intelligence (AI), whose main goal is to build intelligent machines, is the first discipline to focus on automating HMI adaptation. In fact, the classical software architecture was consistent with the specification of customization parameters that the user could access via preference menus, but AI tried to customize systems and adapt them. According to Thevenin D. and Al. (2002) [15], we can give the example related to the work of the AID project (Adaptive Intelligent Dialogue) which consists to design and create systems capable of adapting to the characteristics of users such as: their motivations, desires, psychomotor faculties, learning and understanding characteristics. Indeed, in AID, the subject components of adaptation are four in number: help (in accordance with the user's level of understanding), error messages (as a source of comfort and confidence), command language (whose terminology must be in accordance with the user's) and dialogue style. In particular, AID proposes classes of metrics as a thought guide to the design, implementation and evaluation of adaptive HMI. This approach has many advantages such as solid foundation, precise bridge between adaptation and software architecture and gain in precision and clarity. But it has the disadvantage that it is difficult to propose an implementation architecture that covers several levels of adaptation.

2.4 Approaches Based on Machine Learning/Deep Learning and Sentiment Analysis

With the objective of optimizing user interactions and limiting user errors, Ianjafitia (2019) [16] focuses on the use of Machine Learning to ensure the adaptation of HMIs to the context of use, mainly. However, Machine Learning requires the accumulation of a large knowledge base before making changes that actually improve the user experience. Ianjafitia (2019) [16] states that if machine learning is considered a branch of artificial intelligence, then deep learning can be considered a branch of machine learning. The basic ideas of deep learning date back to the late 1980s, with the birth of the first neural networks. However, if the theory was already in place, the means to implement it have only recently appeared. In relation to Sharma et al. (2021) [17], in the context of HMI adaptation, Deep Learning is used in facial expression analysis systems in several domains, especially in educational systems. The advantage of this method is that it implicitly discloses the feelings of the users, but the disadvantage is that the certainty of identifying the feelings varies depending on the algorithm used in relation to a particular library. Galindo et al. (2017) [18] also presents a research work that addresses the problem of adapting the user interface according to the user's emotions (positive, negative and neutral) at runtime. The proposed architecture of the realized system covers three components namely: The inference engine, the adaptation engine and the interactive system. This architecture has been implemented in the form of web pages. From the software point of view, all the components rely on JavaScript and jQuery to perform all the steps of the adaptation process. The interactive system also uses HTML and CSS. This prototype involves real-time adaptation of user interface parameters (color, font size, image size) and variants regarding positive (happiness and joy), negative (anger, disgust, sadness, fear) and neutral feelings. Generic adaptation rules were implemented to adapt the color, font size and image size according to the user's emotions (e.g., image size = large when a negative emotion is evoked). In addition, the generic variable emotionFilter = {positive, negative, neutral} allows to filter the emotions to be taken into account by the adaptation engine. Therefore, the inference engine will not be able to differentiate how positive or negative this change is. According to the proposed approach at this level, it is necessary to identify the relevant adaptations for specific emotions and validate them.

3 Proposed Approach Based on Multi-agent System and Deep Learning

3.1 Presentation of the Proposed Approach

The proposed approach for automatic HMI adaptation system consists of implementing two components which are:

- **Dynamic web application:**

This first component is related to the educational application dedicated to children with learning disabilities and slightly autistic, which consists of a set of interfaces exposing the key concepts of the envisaged learning as well as the proposed exercises with a voice assistance through the different corresponding web pages. This application is designed and scripted according to the specific rules of design for applications for children with learning disabilities on a didactic theme.

- **Multi-Agent System:**

The role of this second component is to offer a set of agents ensuring the visualization of the different interfaces of the educational application while guaranteeing quality pedagogical coaching as well as real-time monitoring of the user's facial expressions in order to analyze their feelings using computer vision and Deep Learning and this, with the objective of automatically making the adaptation of the interfaces of the first component in real time by applying well determined adaptation rules defined in a knowledge base.

3.2 Proposed Architecture

MASs, which constitute an interesting paradigm, promote the exploration of a new path for the design of component-based distributed applications thanks to the various advantages they offer, notably their asynchronous qualities and their moderate use of resources [19]. In addition to the dynamic web application that corresponds to the proposed educational application whose interfaces are the subject of adaptation, the proposed model for the system to be realized is based on the use of MASs while being composed of several autonomous cooperative agents that are able to communicate and cooperate with each other in order to accomplish their tasks during the whole process of exploration of the application.

The MAS component of our system consists of two cooperative agents that work together to present the application and solve the adaptation problem according to the users' feelings, which are:

The Interface Agent. Displays and navigates through the different interfaces of the educational application as well as the set of exercises intended for the user while providing voice assistance. It also provides adaptation.

This agent can also play the role of a virtual assistant who provides pedagogical coaching for the user.

The Sentiment Analyzer Agent. It tracks the user's facial expressions in real time and then analyzes his feelings and asks the first agent to adapt the interface when a negative feeling is detected. Finally, an execution sequence is generated and the interface will be adapted to solve the detected problem.

The Fig. 1 shows the different components of the proposed system architecture:

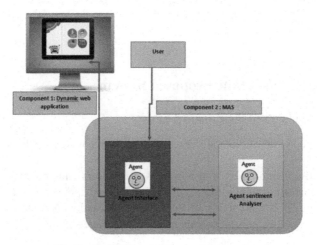

Fig. 1. Architecture of the proposed system

3.3 Parameters and Rules of the GUI Adaptation for our Application

Parameter Tuning. According to the research work on the same theme [17], a set of parameters have been defined, namely:

- Background-color: represents the color of the background of pages
- Font-size, Font-style, Font-color: represents the size, the style and the color of the font used
- Widgets: represents the graphic elements used
- UI layout: represents the general structure of the page or screen
- Picture, Size of picture: represents the images used and their size

Proposal of adaptation rules according to the users' feelings.

a. Inputs/Data:

- We consider these types of sentiments: Neutral (N), Negative (Ng) (sad and angry) and Positive (happy) (Pos)
- Feeling = {N, Ng, Pos} knowing that the initial state is N.

b. Outputs:
The values of the adaptation parameters selected for our application namely: Background-color (white, white), Font-size (Medium, Large), Font-color (black, blue), Picture (Type1, Type2) and Size of picture (Medium, Large).

c. Examples of proposed rules: If sentiment = Neg then Background-color = white, If sentiment = Neg then Font-size = large, If sentiment = Neg then Font-color = blue, If sentiment = Neg then Picture = type2 and If sentiment = Neg then Size of picture = large

4 Implementation of the Proposed System

4.1 The Components of the Proposed System

The implemented system consists of a set of components which are the following:

- **The educational module** which is proposed for users and consists of the various web pages that illustrate the content of the educational application for learning numbers, presented according to a well-studied scenario and whose interfaces have been adapted. In Fig. 2 we present examples of the web pages of our application:

Fig. 2. Examples of pages of the educational module

- **SQLite database** for storing the characteristics of the images and the parameters of the application.
- **Administration module** for online updating of the application data for the creation of the agent interface
- **SPADE Multi-Agent System** that composed of two agents:

✓ **Interface Agent:**

This agent will navigate through the different web pages of the application while offering voice assistance and guidance to the user, it communicates with the second agent of the platform which is the sentiment analyzer agent in order to adapt the interface when a negative sentiment is detected. The implemented tasks related to the Agent Interface are the following:

– Creation of the agent
– Definition of its behavior: Display the application's interfaces and allow the user to explore and do the exercises, Assist the user in progressing through the path, enable communication with the sentiment analyzer Agent and Adapt the interface

✓ **The sentiment analyzer Agent:**

The sentiment analyzer agent works in parallel with the interface agent and communicates with it by requesting the adaptation of the interface when a negative sentiment is detected. The implemented Tasks related to the sentiment analyser agent are the following:

– Creation of the agent
– Definition of its behavior: Launch the camera and capture in video mode and follow in real time the user's facial expressions, analyse the user's feelings, communicate with the agent Interface and request interface adaptation if negative sentiment is detected

We present in Fig. 3 some examples of the results of the execution of the sentiment analyzer agent:

Then the interface agent is launched in parallel with the sentiment analyzer agent also through its supervisor. The interface agent will display the different pages of the educational application with a vocal assistance while listening to the messages coming

Fig. 3. Example of results of execution of the sentiment analyzer agent

from the sentiment analyzer agent which requests an adaptation of the interface when a negative sentiment (angry or sad) is detected. When an adaptation request is detected, the interface agent makes the necessary modifications to the interface in real time, according to the adaptation rules in the knowledge base, and displays the adapted interface to the user. For each agent, we created a supervisor to start and stop it and this is the solution we proposed to ensure the parallel operation of both agents. We were also able to establish direct communication between the two agents using a JSON file.

4.2 Examples of Adaptation

For the home page of the educational application, the initial interface is as follows (Fig. 4):

Fig. 4. Initial interface of the homepage

When negative user sentiment is detected at this level, for the same page, the adapted interface generated in real time is as follows (Fig. 5):

Fig. 5. Adapted interface of the homepage

The automatic adaptation for this example is illustrated by changing the size and color of the font, enlarging the size of the central image and keeping the background color the same (white).

For the page corresponding to number 4, the initial GUI is as follows (Fig. 6):

Fig. 6. Initial interface for the number 4.

For this same page, the adapted interface is as follows (Fig. 7):

Fig. 7. Adapted interface for the number 4

In this adapted page, it is noticeable through this example that the size and color of the font has changed. Also, the second image has changed and is displayed at a larger size than the first image on the original page.

5 Experimental Results and Discussion

5.1 Experimentation

In order to test our approach, we chose to allow children with learning disabilities to use our platform and we found the following values in Table 1:

Table 1. Results of the test of our approach

	Exploration time (numbers part)	NB adapted pages	NB negative feelings	NB positive feelings	Adaptation rate
Kid 1	180 s	5	5	6	45%
Kid 2	170 s	3	3	8	27%
Kid 3	185 s	6	6	5	54%
Kid 4	240 s	10	10	1	90%
Kid 5	175 s	4	4	7	36%

The Fig. 8 highlights the variation of the adaptation percentages according to the feelings of the users who are children with learning disabilities in the context of this experiment.

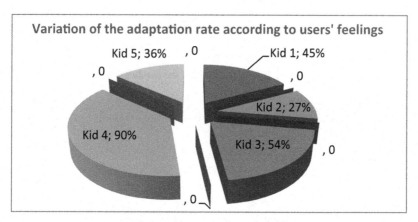

Fig. 8. Variation of the adaptation rate according to users' feelings

5.2 Evaluation and Comparison of the Realized System with Other State-of-the-Art Approaches Based on Sentiment Analysis

We will present in this section comparison between our approach and the other approaches presented in the state of the art. See Table 2.

Table 2. Comparison of our approach with the approaches of the state of the art

Comparison criteria	Our approach	State of the art approaches
Principle	Based on MAS and Deep Learning	Based on machine learning or deep learning
Distribution	Agents used are distributed	Non-distributed(centralised) approaches
Innovation	Innovative	Less innovative
Speed and execution time (exploration and adaptation)	The execution speed is high due to the use of distributed AI and therefore the execution time is reduced	The speed is lower compared to the proposed approach because the adaptation is done offline compared to the sentiment analysis

(continued)

Table 2. (*continued*)

Comparison criteria	Our approach	State of the art approaches
Knowledge base	Our solution integrates a knowledge base	These approaches use a knowledge base
Learning dynamics	The agents used in this project are not learning agents because SPADE does not allow the creation of cognitive agents	The learning of new adaptation rules is provided by some approaches that use a rule base and an inference engine
Ergonomic and usability of the HMI	Ergonomic and useful interface as long as the rules of HMI design and adaptation have been respected	These approaches can also be effective in terms of usability and user-friendliness of the interface since they are implemented at the level of web and mobile applications
Certainty of sentiment identification	Varies depending on the model used and the parameters of the learning environment	Variable depending on the algorithm used in relation to a particular library
Security	The SPADE environment provides multi-level security mechanisms through the use of the XMPP protocol	No special security mechanisms, just the classic ones

6 Conclusion

This research work considers an adaptation of the HMI guided by the results of the analysis of the feelings of the users based on the use of the MAS and Deep Learning. Indeed, in this work, we tried to bring a solution to the problem of the adaptation of the Human Machine Interfaces (HMI) to their users through the analysis of their feelings in real time followed by an automatic quasi-immediate adaptation of the HMI. The various researches and analyses carried out in the framework of the elaboration of this essay have allowed us to understand the specificities of the world of HMI adaptation and the notions related to MAS, sentiment analysis, Deep learning,...

The challenge is to circumvent the difficulties and exploit the opportunities. It is in this perspective that the adaptation approach based on ADM and Deep Learning represents a promising approach. However, even if this contribution has dealt with adaptive HMIs, there are still perspectives such as handling user feedback, learning agents and deducing new adaptation rules. Thus, further research will be necessary to improve the adaptation methods in terms of performance.

References

1. Beaudouin-Lafon, M.: Instrumental interaction: an interaction model for designing post-WIMP user interfaces (2000). https://doi.org/10.1145/332040.332473
2. Weiser, M.: L'ordinateur pour le 21e siècle. Sci. Am. **265**, 94–104 (1991). https://doi.org/10.1038/scientificamerican0991-94
3. Calvary, G., Coutaz, J., Thevenin, D., Limbourg, Q., Bouillon, L., Vanderdonckt, J.: A unifying reference framework for multi-target user interfaces. Interact. Comput. **15**(3), 289–308 (2003)
4. Akiki, P.A., Bandara, A.K., Yu, Y.: Adaptive model-driven user interface development systems. ACM Comput. Surv. **47**(1), 9:1–9:33 (2014)
5. Gupta, A., Anpalagan, A., Khwaja, A-S.: Deep learning for object detection and scene perception in self-driving cars: survey, challenges, and open issues (2016)
6. Hachani, S., Dupuy-Chessa, S., Front, A.: A generic approach for the dynamic adaptation of UIs to the context. In: 21st Francophone Conference on Human-Machine Interaction (IHM 2009), Grenoble, France, pp. 89–96 (2009). ffhal-01002999P
7. Favre, J.-M.: Towards a basic theory to modeldriven engineering. In: 3rd Workshop in Software Model Engineering 2004, Lisboa, Portugal, pp. 9–17 (2004)
8. Giuffrida, T., Dupuy-Chessa, S., Poli, J.-P., Céret, E.: Fuzzy4U: a fuzzy logic system for the adaptation of user interfaces. In: HMI 2018, Brest, France, pp. 23–26 (2018)
9. Cao, J., Xing, N., Chan, A., Feng, Y., Jin, B.: Service adaptation using fuzzy theory in context-aware mobile computing middleware. In: Proceedings of the 11th IEEE International Conference on Embedded and Real-Time Computing Systems and Applications (2005)
10. Desruelle, H., Blomme, D., Gielen, F.: Adaptive mobile web applications: a quantitative evaluation approach. In: International Conference on Web Engineering (2011)
11. Cueva-Fernandez, G., Espada, J.P., García-Díaz, V., Crespo, R.G., GarciaFernandez, N.: Fuzzy system to adapt web voice interfaces dynamically in a vehicle sensor tracking application definition. Soft. Comput. **20**(18), 3321–3334 (2016)
12. Nyongesa, H.O., Shicheng, T., Maleki-Dizaji, S., Huang, S.T., Siddiqi, J.: Adaptive Web interface design using fuzzy logic. In: Proceedings IEEE/WIC International Conference on Web Intelligence (WI 2003), pp. 671–674 (2016)
13. Papatheocharous, E., Belk, M., Germanakos, P., Samaras, G.: Proposing a fuzzy adaptation mechanism based on cognitive factors of users for web personalization. Artif. Intell. Appl. Innov., 135–144 (2012)
14. Soui, M., Abed, M., Ghedira, K.: Fuzzy logic approach for adaptive systems design. In: Kurosu, M. (ed.) HCI 2013. LNCS, vol. 8008, pp. 141–150. Springer, Heidelberg (2013). https://doi.org/10.1007/978-3-642-39342-6_16
15. Thevenin, D., Coutaz, J.: IHM adaptation: taxonomies et software architecture. In: Acts IHM02, pp. 207–210. ACM Press (2002)
16. Ianjafitia, A.: Deep learning-based object detection system in an intelligent store, Master's Thesis, universite d'antananarivo (2019)
17. Sharma, N., Sharma, R., Jindal, N.: Machine learning and deep learning applications-a vision. Glob. Trans. Proc. **2**(1), 24–28 (2021)
18. Galindo, J.A., Dupuy-Chessa, S., Ceret, E.: Toward a UI adaptation approach driven by user emotions. In: ACHI 2017 - The Tenth International Conference on Advances in Computer-Human Interactions, Nice, France, pp. 12–17 (2017). ffhal-01720940
19. Menacer, D.-E.: A mobile Agent-Based Architecture for distributed applications, These Magister, National Institute of Informatics (INI) (2004)

Pattern Recognition

A Generalized Inverted Dirichlet Predictive Model for Activity Recognition Using Small Training Data

Jiaxun Guo[1], Manar Amayri[2], Wentao Fan[3], and Nizar Bouguila[1(✉)]

[1] CIISE, Concordia University, Montreal, Canada
g_jiax@encs.concordia.ca, nizar.bouguila@concordia.ca
[2] G-SCOP Laboratory, Grenoble Institute of Technology, Grenoble, France
manar.amayri@grenoble-inp.fr
[3] Computer Science and Technology, Huaqiao University, Xiamen, China
fwt@hqu.edu.cn

Abstract. In this paper, we develop the predictive distribution of the generalized inverted Dirichlet (GID) mixture model using local variational inference. The main goal is to be able to tackle classification problems involving small training data sets. The two main ingredients of the proposed predictive model are the GID distribution which provides flexibility for the modeling of semi-bounded data that are naturally generated by different sensors outputs and the efficient of variational inference as a deterministic approximation to fully Bayesian approaches. The merits of the proposed model are shown via synthetic data and a real application that concerns activities recognition.

Keywords: Activity recognition · Small sensor data · Mixture models · Variational inference · Predictive distribution

1 Introduction

Activity Recognition has became an emerging research topic over the last two decades and plays a vital role in medicine, surveillance and security, gaming, virtual reality, and even music recommendation [8,13,14,23]. Identifying and detecting a given activity depend on sensors' information processing [12,13]. In the recent years, most of researches have focused on developing activity recognition solutions using inertial signals [18,22,24–27]. This is mainly due to the popularity of smartphones and low cost of wearable sensors. Machine learning techniques have been widely used to model these sensors outputs and then to provide models that can be deployed for the recognition task which is generally reduced to a classification problem. In that context, a challenging necessary task for activity recognition is the collection of labeled data which is a crucial step for supervised learning. However, the step of data labeling is time consuming, error prone and often requires human involvement. In the majority of

© Springer Nature Switzerland AG 2022
H. Fujita et al. (Eds.): IEA/AIE 2022, LNAI 13343, pp. 431–442, 2022.
https://doi.org/10.1007/978-3-031-08530-7_36

real-life scenarios only a small training set is available which prevents using data hungry techniques such as deep learning. Many approaches have been proposed in the past to tackle that problem by mainly increasing the number of labeled data using techniques such as active learning or by deploying training data from other domains using transfer learning. For instance, the authors in [11] propose a cluster based active learning model based on K-Means and the authors in [7] propose a transfer learning approach based on a hierarchical Bayesian method. Unlike previous approaches, we propose a learning technique based on a data-driven predictive distribution by taking advantage of the existing training data at its maximum. The proposed approach is complementary to existing techniques and could be easily integrated within any existing active or transfer learning framework.

The main motivation of considering the predictive distribution, as compared to the commonly used point estimate methods, is that it leads to more reliable results when calculating the predictive likelihood of unseen data when the training data is small. Indeed, it is well-known that when training data is small, point estimate leads to estimated parameters with large variance and then cause uncertainty and unreliability [16]. The predictive density of a new vector $\mathbf{x}$ given the training data $\mathbf{X} = [\mathbf{x}_1, \mathbf{x}_2, ..., \mathbf{x}_N]$ is [2]

$$f(\mathbf{x}|\mathbf{X}) = \int f(\mathbf{x}|\boldsymbol{\theta})f(\boldsymbol{\theta}|\mathbf{X})d\boldsymbol{\theta} \tag{1}$$

where $\boldsymbol{\theta}$, the parameters of both likelihood function $f(\mathbf{x}|\boldsymbol{\theta})$ and posterior distribution $f(\boldsymbol{\theta}|\mathbf{X})$, should be learned from $\mathbf{X}$ via Bayesian inference. An important problem, in predictive modeling, is the choice of the statistical distribution to model the data. In this paper, we consider the generalized inverted Dirichlet mixture model as our distribution. We are mainly motivated by its excellent results when modeling semi-bounded positive vectors (which is our case actually since sensor data are generally described as positive vectors) as compared to other commonly used distributions such as the Gaussian (see, for instance, [1,4,5,19]). The resulting predictive model is validated using a popular activity recognition dataset that contains seven types of activities (Bending1, Bending2, Cycling, Lying, Sitting, Standing, Walking) and obtained from a Wireless Sensor Network (WSN) worn by an actor.

The rest of this paper is organized as follows: Sect. 2 introduces generalized inverted Dirichlet mixture model and our predictive modeling approach. The detailed description and discussion of our experimental results are presented in Sect. 3. Finally, we conclude this paper in Sect. 4.

2   Predictive Model

In this section, we first present briefly the generalized inverted Dirichlet mixture model [3,9,10,17]. Then, its predictive distribution will be approximated by variational inference.

2.1   Generalized Inverted Dirichlet Mixture Model

Let M denotes the number of different components. Assume that a D-dimensional positive vector $\boldsymbol{Y} = (Y_1, \cdots, Y_D)$ follows a finite mixture model of generalized Inverted Dirichlet (GID) Distributions that denoted by a common probability density function $p(\boldsymbol{Y} \mid \boldsymbol{\pi}, \boldsymbol{\alpha}, \boldsymbol{\beta})$ such that

$$p(\boldsymbol{Y} \mid \boldsymbol{\pi}, \boldsymbol{\alpha}, \boldsymbol{\beta}) = \sum_{j=1}^{M} \pi_j \text{GID}\left(\boldsymbol{Y} \mid \boldsymbol{\alpha}_j, \boldsymbol{\beta}_j\right) \tag{2}$$

where $\boldsymbol{\alpha} = \{\boldsymbol{\alpha}_1, \ldots, \boldsymbol{\alpha}_M\}$, $\boldsymbol{\beta} = \{\boldsymbol{\beta}_1, \ldots, \boldsymbol{\beta}_M\}$, $\boldsymbol{\alpha}_j$ and $\boldsymbol{\beta}_j$ are the parameters of the GID distribution representing component j, where $\boldsymbol{\alpha}_j = \{\alpha_{j1}, \cdots, \alpha_{jD}\}$ and $\boldsymbol{\beta}_j = \{\beta_{j1}, \cdots, \beta_{jD}\}$. $\boldsymbol{\pi} = \{\pi_1, \cdots, \pi_M\}$ denotes the mixing weights, such that $\sum_{j=1}^{M} \pi_j = 1$. GID $\left(\boldsymbol{Y} \mid \boldsymbol{\alpha}_j, \boldsymbol{\beta}_j\right)$ is a GID distribution representing component j with parameters $\boldsymbol{\alpha}_j$ and $\boldsymbol{\beta}_j$ and is defined by [4]

$$\text{GID}\left(\boldsymbol{Y} \mid \boldsymbol{\alpha}_j, \boldsymbol{\beta}_j\right) = \prod_{l=1}^{D} \frac{\Gamma\left(\alpha_{jl} + \beta_{jl}\right)}{\Gamma\left(\alpha_{jl}\right)\Gamma\left(\beta_{jl}\right)} \frac{Y_l^{\alpha_{jl}-1}}{\left(1 + \sum_{k=1}^{l} Y_k\right)^{\gamma_{jl}}} \tag{3}$$

where $\alpha_{jl} > 0$, $\beta_{jl} > 0$.

Let $\mathcal{Y} = (\boldsymbol{Y}_1, \ldots, \boldsymbol{Y}_N)$ be a set of N independent identically distributed vectors taken from our mixture model. According to the Bayes' theorem, the probability that the vector $\boldsymbol{Y}_i$ is from component j (also called *responsibilities* of each mixture component j in generating each data sample $\boldsymbol{Y}_i$) can be written as

$$p\left(j \mid \boldsymbol{Y}_i\right) \propto \pi_j GID\left(\boldsymbol{Y}_i \mid \boldsymbol{\alpha}_j, \boldsymbol{\beta}_j\right) \tag{4}$$

We define $\gamma_{jl} = \beta_{jl} + \alpha_{jl+1} - \beta_{jl+1}$ for $l = 1, \cdots, D$, with $\beta_{jD+1} = 0$. After some mathematical manipulations [1], the responsibilities can be factorized as

$$p\left(j \mid \boldsymbol{Y}_i\right) \propto \pi_j \prod_{l=1}^{D} \text{iBeta}\left(X_{il} \mid \alpha_{jl}, \beta_{jl}\right) \tag{5}$$

where $X_{i1} = Y_{i1}$ and $X_{il} = Y_{il}/(1 + \sum_{k=1}^{l-1} Y_{ik})$ for $l > 1$ and iBeta $\left(X_{il} \mid \alpha_{jl}, \beta_{jl}\right)$ is an inverted Beta distribution with parameters $(\alpha_{jl}, \beta_{jl})$:

$$\text{iBeta}\left(X_{il} \mid \alpha_{jl}, \beta_{jl}\right) = \frac{\Gamma(\alpha_{jl} + \beta_{jl})}{\Gamma(\alpha_{jl})\Gamma(\beta_{jl})} X_{il}^{\alpha_{jl}-1}(1 + X_{il})^{-(\alpha_{jl}+\beta_{jl})} \tag{6}$$

where $\Gamma(z) = \int_0^{\infty} x^{z-1}e^{-x}dx$. Thus, the mixture model of the finite GID distribution underlying dataset $\mathcal{Y}$ is the same as that underlying $\mathcal{X} = (\boldsymbol{X}_1, \ldots, \boldsymbol{X}_N)$, where $\boldsymbol{X}_i = (\boldsymbol{X}_{i1}, \ldots, \boldsymbol{X}_{iD}), i = 1, \ldots, N$, using the following clustering structure with conditionally independent features

$$p\left(\boldsymbol{X}_i \mid \boldsymbol{\pi}, \boldsymbol{\alpha}, \boldsymbol{\beta}\right) = \sum_{j=1}^{M} \pi_j \prod_{l=1}^{D} \text{iBeta}\left(X_{il} \mid \alpha_{jl}, \beta_{jl}\right) \tag{7}$$

The formal conjugate prior distribution of the Beta distribution [2] does not have a closed form. Thus, we consider in our work a tractable approximation to the conjugate prior using a product of two independent Gamma distributions as previously used in [1] in the context of a global variational inference (GVI) framework

$$f(\alpha_{jl}, \beta_{jl}) \approx \text{Gam}(\alpha_{jl}; a_0, b_0) \times \text{Gam}(\beta_{jl}; c_0, d_0)$$
$$= \frac{b_0^{a_0}}{\Gamma(a_0)} \alpha_{jl}^{a_0-1} e^{-b_0 \alpha_{jl}} \times \frac{d_0^{c_0}}{\Gamma(c_0)} \beta_{jl}^{c_0-1} e^{-d_0 \beta_{jl}} \tag{8}$$

With available data $\mathcal{X}_l = \{X_{1l}, X_{2l}, \ldots, X_{Nl}\}$, the hyperparameters $(a^*, b^*, c^*$ and $d^*)$ of the posterior distribution can be easily obtained by variational Bayes estimation as detailed in [1]. The posterior distribution could be approximated by a product of two independent Gamma distribution as

$$f(\alpha_{jl}, \beta_{jl} \mid \mathcal{X}_l) \approx \text{Gam}(\alpha_{jl}; a^*, b^*) \times \text{Gam}(\beta_{jl}; c^*, d^*)$$
$$= \frac{b^{*a^*}}{\Gamma(a^*)} \alpha_{jl}^{a^*-1} e^{-b^* \alpha_{jl}} \times \frac{d^{*c^*}}{\Gamma(c^*)} \beta_{jl}^{c^*-1} e^{-d^* \beta_{jl}} \tag{9}$$

2.2 Predictive Distribution of the Mixture Model

The predictive distribution can assess the uncertainty of a new coming observation with respect to the existing dataset. Let Y_i be that new observation independent from the existing $\mathcal{Y}$ which is assumed to be generated from $GID(Y_i \mid \alpha_j, \beta_j)$. Using the transformation presented after Eq. 5, the obtained new observation X_i follows a product of inverted Beta distributions. The predictive distribution of X_i given $\mathcal{X}$ is

$$f(X_i \mid \mathcal{X}) = \int_0^\infty \int_0^\infty \prod_{l=1}^D [\text{iBeta}(X_{il} \mid \alpha_{jl}, \beta_{jl}) f(\alpha_{jl}, \beta_{jl} \mid \mathcal{X}_l)] \, d\alpha_j d\beta_j \tag{10}$$

With the analytically tractable posterior distribution in Eq. 9 and the function of the inverted Beta distribution (Eq. 6), we can extend this predictive distribution as

$$f(X_i \mid \mathcal{X}) \approx \int_0^\infty \int_0^\infty \prod_{l=1}^D \left[\text{iBeta}(X_{il} \mid \alpha_{jl}, \beta_{jl}) \right.$$
$$\times \frac{b^{*a^*}}{\Gamma(a^*)} \alpha_{jl}^{a^*-1} e^{-b^* \alpha_{jl}} \frac{d^{*c^*}}{\Gamma(c^*)} \beta_{jl}^{c^*-1} e^{-d^* \beta_{jl}} \Big] d\alpha_j d\beta_j$$
$$= \prod_{l=1}^D \left[\frac{1}{X_{il}} \frac{b^{*a^*}}{\Gamma(a^*)} \frac{d^{*c^*}}{\Gamma(c^*)} \right] \tag{11}$$
$$\times \int_0^\infty \int_0^\infty \prod_{l=1}^D \left[\frac{\Gamma(\alpha_{jl} + \beta_{jl})}{\Gamma(\alpha_{jl})\Gamma(\beta_{jl})} X_{il}^{\alpha_{jl}} (1 + X_{il})^{-(\alpha_{jl} + \beta_{jl})} \right.$$
$$\alpha_{jl}^{a^*-1} e^{-b^* \alpha_{jl}} \beta_{jl}^{c^*-1} e^{-d^* \beta_{jl}} \Big] d\alpha_j d\beta_j$$

$$f(\boldsymbol{X}_i \mid \mathcal{X}) \le f_{\text{upp}}(\boldsymbol{X}_i \mid \mathcal{X})$$

$$= \prod_{l=1}^{D} \left[\frac{1}{X_{il}} \frac{b^{*a^*}}{\Gamma(a^*)} \frac{d^{*c^*}}{\Gamma(c^*)} \frac{\Gamma(\alpha_0 + \beta_0)}{\Gamma(\alpha_0)\Gamma(\beta_0)} e^{-\alpha_0[\psi(\alpha_0+\beta_0)-\psi(\alpha_0)]-\beta_0[\psi(\alpha_0+\beta_0)-\psi(\beta_0)]} \right]$$

$$\times \int_0^\infty \int_0^\infty \prod_{l=1}^{D} \left[e^{\alpha_{jl}[\psi(\alpha_0+\beta_0)-\psi(\alpha_0)]+\beta_{jl}[\psi(\alpha_0+\beta_0)-\psi(\beta_0)]} \right.$$

$$\left. X_{il}^{\alpha_{jl}} (1+X_{il})^{-\alpha_{jl}} (1+X_{il})^{-\beta_{jl}} \alpha_{jl}^{a^*-1} e^{-b^*\alpha_{jl}} \beta_{jl}^{c^*-1} e^{-d^*\beta_{jl}} \right] d\boldsymbol{\alpha}_j d\boldsymbol{\beta}_j$$

$$= \prod_{l=1}^{D} \left[\frac{1}{X_{il}} \frac{b^{*a^*}}{\Gamma(a^*)} \frac{d^{*c^*}}{\Gamma(c^*)} \frac{\Gamma(\alpha_0 + \beta_0)}{\Gamma(\alpha_0)\Gamma(\beta_0)} e^{-\alpha_0[\psi(\alpha_0+\beta_0)-\psi(\alpha_0)]-\beta_0[\psi(\alpha_0+\beta_0)-\psi(\beta_0)]} \right]$$

$$\times \prod_{l=1}^{D} \left\{ \int_0^\infty e^{\displaystyle -\alpha_{jl} \underbrace{[b^* - \ln X_{il} + \ln(1 + X_{il}) - \psi(\alpha_0 + \beta_0) + \psi(\alpha_0)]}_{g(X_{il},\alpha_0,\beta_0)}} \alpha_{jl}^{a^*-1} d\alpha_{jl} \right\}$$

$$\times \prod_{l=1}^{D} \left\{ \int_0^\infty e^{\displaystyle -\beta_{jl} \underbrace{[d^* + \ln(1 + X_{il}) - \psi(\alpha_0 + \beta_0) + \psi(\beta_0)]}_{h(X_{il},\alpha_0,\beta_0)}} \beta_{jl}^{c^*-1} d\beta_{jl} \right\}$$

$$\tag{12}$$

Equation 11 involves the Inverse Beta function which logarithm has been proved to be concave [6]. Using that concavity property and first order Taylor expansion, we can obtain the following inequality [15]

$$\frac{\Gamma(\alpha + \beta)}{\Gamma(\alpha)\Gamma(\beta)} \le \frac{\Gamma(\alpha_0 + \beta_0)}{\Gamma(\alpha_0)\Gamma(\beta_0)}$$

$$\times e^{[\psi(\alpha_0+\beta_0)-\psi(\alpha_0)](\alpha-\alpha_0)+[\psi(\alpha_0+\beta_0)-\psi(\beta_0)](\beta-\beta_0)} \tag{13}$$

where $\psi(\cdot)$ is the digamma function defined as $\psi(\cdot) = \partial \ln \Gamma(x)/\partial x$. Using Eq. 13 in Eq. 11, we can find an upper bound for the predictive distribution as shown in Eq. 12 by a local variational inference (LVI) method [15]. Compared with the global variational inference [1] which approximates all the model's variables, LVI is considered as a 'local' approach to approximate a subset of variables [2]. In Eq. 12, the integrand in each integration is a Gamma distribution. To simplify the predictive distribution, these integrations can be replaced by

$$\int_0^\infty e^{-\alpha_{jl}g(X_{il},\alpha_0,\beta_0)} \alpha_{jl}^{a^*-1} d\alpha_{jl}$$

$$= \begin{cases} \dfrac{\Gamma(a^*)}{g(X_{il},\alpha_0,\beta_0)^{a^*}} & g(X_{il},\alpha_0,\beta_0) > 0 \\ \infty & g(X_{il},\alpha_0,\beta_0) \le 0 \end{cases}$$

$$\int_0^\infty e^{-\beta_{jl}h(X_{il},\alpha_0,\beta_0)} \beta_{jl}^{c^*-1} d\beta_{jl}$$

$$= \begin{cases} \dfrac{\Gamma(c^*)}{h(X_{il},\alpha_0,\beta_0)^{c^*}} & h(X_{il},\alpha_0,\beta_0) > 0 \\ \infty & h(X_{il},\alpha_0,\beta_0) \le 0 \end{cases} \tag{14}$$

If $g\left(X_{il}, \alpha_0, \beta_0\right) > 0$ and $h\left(X_{il}, \alpha_0, \beta_0\right) > 0$, we obtain a closed-form upper bound for the predictive distribution:

$$
\begin{aligned}
&f_{\text{upp}}(\boldsymbol{X}_i \mid \mathcal{X}) \\
&= \prod_{l=1}^{D} \left[\frac{1}{X_{il}} \left[\frac{b^*}{g\left(X_{il}, \alpha_0, \beta_0\right)} \right]^{a^*} \left[\frac{d^*}{h\left(X_{il}, \alpha_0, \beta_0\right)} \right]^{c^*} \right. \\
&\quad \left. \times \frac{\Gamma\left(\alpha_0 + \beta_0\right)}{\Gamma\left(\alpha_0\right)\Gamma\left(\beta_0\right)} e^{-\alpha_0[\psi(\alpha_0+\beta_0)-\psi(\alpha_0)] - \beta_0[\psi(\alpha_0+\beta_0)-\psi(\beta_0)]} \right]
\end{aligned} \tag{15}
$$

The upper bound is just a function of α_0, β_0 after being given $\mathcal{X}$, which can be rewritten as

$$
f_{\text{upp}}(\boldsymbol{X}_i \mid \mathcal{X}) = \prod_{l=1}^{D} \left[\frac{b^{*a^*} d^{*c^*}}{X_{il}} \times F(X_{il}, \alpha_0, \beta_0) \right] \tag{16}
$$

where $F(X_{il}, \alpha_0, \beta_0)$ can be straightforwardly deduced from Eq. 15. The means $\mathbf{E}(\alpha)$ and $\mathbf{E}(\beta)$ are the most representative values of α_0 and β_0, respectively, which can be taken to approximate the optimal solution (α_0^*, β_0^*). Besides, the means calculated by the observations in $\mathcal{X}$ are independent of X_{il}. To facilitate the calculation, the minimum of the upper bound can be approximated as

$$
\begin{aligned}
&\min_{\alpha_0, \beta_0} f_{\text{upp}}(\boldsymbol{X}_i \mid \mathcal{X}) \\
&= \prod_{l=1}^{D} \left[\frac{b^{*a^*} d^{*c^*}}{X_{il}} \times \min_{\alpha_0, \beta_0} F(X_{il}, \mathbf{E}(\alpha), \mathbf{E}(\beta)) \right] \\
&\approx \prod_{l=1}^{D} \left[\frac{b^{*a^*} d^{*c^*}}{X_{il}} \times F(X_{il}, \mathbf{E}(\alpha), \mathbf{E}(\beta)) \right]
\end{aligned} \tag{17}
$$

Since $\min_{\alpha_0, \beta_0} f_{upp}(\boldsymbol{X}_i \mid \mathcal{X})$ is unnormalized, we need to calculate the normalization factor:

$$
C_{upp} = \int_0^{\infty} \min_{\alpha_0, \beta_0} f_{upp}(\boldsymbol{X}_i \mid \mathcal{X}) d\boldsymbol{X}_i \tag{18}
$$

The approximation to the mixture predictive distribution is finally obtained as

$$
\begin{aligned}
f(\boldsymbol{X}_i \mid \mathcal{X}) &\approx f_{appx}^{LVI}(\boldsymbol{X}_i \mid \mathcal{X}) \\
&= \sum_{j=1}^{M} \frac{\pi_j}{C_{uppj}} \prod_{l=1}^{D} \left[\frac{b_{jl}^{*\,a_{jl}^*} d_{jl}^{*\,c_{jl}^*}}{X_{il}} \times F(X_{il}, \mathbf{E}(\alpha_{jl}), \mathbf{E}(\beta_{jl})) \right]
\end{aligned} \tag{19}
$$

3   Experimental Results

The predictive distribution of GID (Eq. 19) is based on LVI as we mentioned previously. In this section, we first generate different training data to compare the performance of out LVI approach with GVI. GVI is the approximation using the posterior mean as the point estimates:

$$
\begin{aligned}
f(\boldsymbol{X}_i \mid \mathcal{X}) &\approx f_{appx}^{GVI}(\boldsymbol{X}_i \mid \mathcal{X}) \\
&= \sum_{j=1}^{M} \pi_j \prod_{l=1}^{D} \text{iBeta}\left(X_{il} \mid \mathbf{E}(\alpha), \mathbf{E}(\beta)\right)
\end{aligned}
\tag{20}
$$

In the second experiment, we apply our model for activity recognition.

Table 1. Comparison of the KL divergences ($\times 10^2$). The results are averaged over 10 random experiments.

Distribution	KL divergences	$N = 10$	$N = 20$	$N = 50$	$N = 100$	$N = 200$
$iBeta(x; 3, 4)$	$\mathbf{KL}(f\|f_{appx}^{GVI})$	27.59	21.27	2.43	0.80	0.62
	$\mathbf{KL}(f\|f_{appx}^{LVI})$	**17.52**	**13.64**	**2.18**	**0.93**	**0.62**
$iBeta(x; 2, 7)$	$\mathbf{KL}(f\|f_{appx}^{GVI})$	16.78	7.31	2.59	1.42	0.65
	$\mathbf{KL}(f\|f_{appx}^{LVI})$	**12.26**	**6.08**	**2.31**	**1.39**	**0.70**
$iBeta(x; 3, 8) * 0.2$						
$iBeta(x; 5, 5) * 0.2$	$\mathbf{KL}(f\|f_{appx}^{GVI})$	31.63	10.17	3.48	2.34	1.58
$iBeta(x; 4, 6) * 0.6$						
	$\mathbf{KL}(f\|f_{appx}^{LVI})$	**15.63**	**7.98**	**2.74**	**2.03**	**1.62**
$iBeta(x; 1, 6) * 0.1$						
$iBeta(x; 3, 2) * 0.1$	$\mathbf{KL}(f\|f_{appx}^{GVI})$	155.13	204.81	90.39	11.90	6.97
$iBeta(x; 5, 2) * 0.8$						
	$\mathbf{KL}(f\|f_{appx}^{LVI})$	**131.65**	**175.37**	**137.26**	**135.38**	**155.78**
$iBeta(x; 1, 5) * 0.2$						
$iBeta(x; 7, 2) * 0.3$	$\mathbf{KL}(f\|f_{appx}^{GVI})$	40.72	22.07	13.54	9.83	8.08
$iBeta(x; 2, 6) * 0.5$						
	$\mathbf{KL}(f\|f_{appx}^{LVI})$	**36.91**	**21.40**	**13.51**	**10.62**	**9.23**

3.1 Synthetic Data

We generate data from a known GID distribution to test our model in this section. The parameters of GVI and LVI are trained separately under the different training data. And then, all the test data will be used in GVI and LVI frameworks to obtain the predictive distributions. Finally, Kullback-Leibler (KL) divergence is used to judge which obtained predictive distribution is better.

$$
\mathbf{KL}(f_{true}\|f_{appx}^{GVI}) = \int f_{true}(x) \ln \frac{f_{true}(x)}{f_{appx}^{GVI}} dx
\tag{21}
$$

$$
\mathbf{KL}(f_{true}\|f_{appx}^{LVI}) = \int f_{true}(x) \ln \frac{f_{true}(x)}{f_{appx}^{LVI}} dx
\tag{22}
$$

where f_{true} is the true probability distribution, f_{appx}^{GVI} and f_{appx}^{LVI} are the GVI and LVI predictive distributions, respectively. Table 1 summarizes the average results over 10 random experiments. According to this table we can see clearly that our LVI approach provides better results than the previously proposed GVI one.

3.2   Activity Recognition

In this subsection, our statistical framework is applied to tackle the activity recognition problem. First, the considered dataset is described. Some important procedures about data preprocessing are also presented. Second, the recognition results are given and analysed.

Dataset Description and Preprocessing Procedure. The dataset represents a real-life benchmark [21]. Three wireless sensors (IRIS node [20]) are placed on a user to collect Received Signal Strength (RSS) as Fig. 1 shows. In the collected dataset, seven activities are defined: Lying, Sitting, Standing, Walking, two types of Bending (Keeping the legs straight and folded), and Cycling (see Fig. 2). Sensors data were captured as pair (rss12, rss13 and rss 23) with a frequency of 20hz. The final dataset consist of avg_rss12, var_rss12, avg_rss13, var_rss13, avg_rss23, var_rss23 where avg and var are the mean and variance values over 250ms of data, respectively. The normalization for the original data is essential because of the magnitude difference and zero appearance. In this experiment, we proportionally normalized the original data between 0.1 and 1. Finally, 3360 original samples were considered in total with 480 samples in each activity class. Before each experiment, 20 samples are randomly selected from each activity class as test data (140 samples totally). The synthetic data section has indicated that our model has high prediction performance for small training data. Therefore, the amount of training data is incremented dynamically, which is convenient to observe and compare the performance of the prediction model under different amounts of training data. We tested six different amounts of training data namely 5, 10, 20, 50, 100 and 200 samples for each activity class.

Results Analysis. For each of the six different amounts of training data, we have applied our framework 10 times to take into account experimental uncertainty. Random selection of initialization parameters is considered to ensure the consistency of the model. Table 2 displays the average accuracy of our 10 simulations (seed 0–9) considering the seven activities. Figure 3 shows the average accuracy for each activity type as a function of the training data size. In the line chart, x-axis represents the amount of training data for each type of activity. We obtained the results for 5, 10, 20, 50, 100 and 200 training data which are enough to display the general accuracy change for the recognition of the different activities. The y-axis is divided into seven different intervals (one for each activity) meaning 0%–100% in each interval. There are two lines for each type of activity, which are the accuracy fluctuation of our model and the accuracy

Fig. 1. The sensors placement [21]

Fig. 2. The two types of bending activity [21].

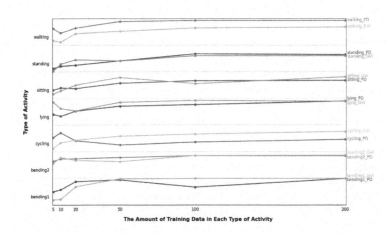

Fig. 3. The accuracy fluctuation when increasing the training data size.

Table 2. Average accuracy of predictive models with different training data sizes. N: The amount of training data in each activity class.

Type of predictive model	$N = 5$	$N = 10$	$N = 20$	$N = 50$	$N = 100$	$N = 200$
PD	0.44	0.480714	0.537143	0.645	0.697857	0.775
GVI	0.292143	0.372143	0.523571	0.674286	0.745	0.799286

fluctuation of GVI. We can clearly observe the changes of two accuracies for each activity. Let us focus on the case when the amount of training data is 5 in Fig. 3. The accuracy of our model is higher than that of GVI in almost all activities except Lying activity. This trend is still being maintained when the amount of training data increases to 10. Only the accuracy of GVI in Bending2 and Standing are slightly better than that of our model. Therefore, the average accuracy of our model is obviously higher than that of GVI when the number of training data equals 5 and 10 as shown in Table 2. Yet, the comparison begins to become difficult when the amount of training data in each type of activity increases to 20. Both average accuracies for $N = 20$ are slightly similar, about 0.53. When the number of training data becomes higher than 20, the average accuracy of GVI tends to be continually higher than that of our model. And the gap between the two average accuracies shows a trend of first expanding and then decreasing. Finally, both accuracies approach 0.8.

4   Conclusion

In this work, we presented and elegant principled statistical framework for predictive modeling based on the GID distribution and a local variational inference. The proposed approach is motivated first by the flexibility of the GID distribution when modeling semi-bounded positive vectors that are naturally generated by several applications involving sensors outputs and second by the efficiency of local variational inference when the amount of training data is limited. Extensive simulations based on synthetic data as well as a challenging real application that concerns activity recognition have shown the merits of our approach. Future works could be devoted to the application and improvement of the proposed framework in other activity recognition contexts such as in the case of smart buildings using nonintrusive sensors. Other potential future works could be devoted to the integration of our model within active or transfer learning frameworks.

Acknowledgement. The completion of this research was made possible thanks to Natural Sciences and Engineering Research Council of Canada (NSERC), the "Nouveaux arrivants Université Grenoble Alpes, Grenoble INP - UGA, G-SCOP" program and the National Natural Science Foundation of China (61876068).

References

1. Bdiri, T., Bouguila, N., Ziou, D.: Variational bayesian inference for infinite generalized inverted Dirichlet mixtures with feature selection and its application to clustering. Appl. Intell. **44**(3), 507–525 (2016)
2. Bishop, C.M.: Pattern Recognition and Machine Learning. Springer (2006)
3. Bouguila, N.: A model-based discriminative framework for sets of positive vectors classification: application to object categorization. In: 2014 1st International Conference on Advanced Technologies for Signal and Image Processing (ATSIP), pp. 277–282 (2014)

4. Bourouis, S., Al Mashrgy, M., Bouguila, N.: Bayesian learning of finite generalized inverted Dirichlet mixtures: application to object classification and forgery detection. Expert Syst. Appl. **41**(5), 2329–2336 (2014)
5. Bourouis, S., Al-Osaimi, F.R., Bouguila, N., Sallay, H., Aldosari, F., Al Mashrgy, M.: Bayesian inference by reversible jump MCMC for clustering based on finite generalized inverted Dirichlet mixtures. Soft. Comput. **23**(14), 5799–5813 (2019)
6. Boyd, S., Boyd, S.P., Vandenberghe, L.: Convex optimization. Cambridge University Press (2004)
7. Diethe, T., Twomey, N., Flach, P.A.: Active transfer learning for activity recognition. In: ESANN (2016)
8. Epaillard, E., Bouguila, N.: Proportional data modeling with hidden Markov models based on generalized Dirichlet and beta-liouville mixtures applied to anomaly detection in public areas. Pattern Recognit. **55**, 125–136 (2016)
9. Fan, W., Bouguila, N.: Nonparametric hierarchical bayesian models for positive data clustering based on inverted Dirichlet-based distributions. IEEE Access **7**, 83600–83614 (2019)
10. Fan, W., Bouguila, N., Liu, X.: A hierarchical Dirichlet process mixture of GID distributions with feature selection for spatio-temporal video modeling and segmentation. In: 2017 IEEE International Conference on Acoustics, Speech and Signal Processing (ICASSP), pp. 2771–2775 (2017)
11. Hossain, H.M.S., Khan, M.A.A.H., Roy, N.: Active learning enabled activity recognition. Pervasive Mob. Comput. **38**, 312–330 (2017). Special Issue IEEE International Conference on Pervasive Computing and Communications (PerCom) 2016
12. Hussain, F., et al.: An efficient machine learning-based elderly fall detection algorithm. arXiv preprint arXiv:1911.11976 (2019)
13. Hussain, Z., Sheng, Q.Z., Zhang, W.E.: A review and categorization of techniques on device-free human activity recognition. J. Netw. Comput. Appl. **167**, 102738 (2020)
14. Kim, H.G., Kim, G.Y., Kim, J.Y.: Music recommendation system using human activity recognition from accelerometer data. IEEE Trans. Consum. Electron. **65**(3), 349–358 (2019)
15. Ma, Z., Leijon, A.: Approximating the predictive distribution of the beta distribution with the local variational method. In: 2011 IEEE International Workshop on Machine Learning for Signal Processing, pp. 1–6. IEEE (2011)
16. Ma, Z., Leijon, A., Tan, Z.H., Gao, S.: Predictive distribution of the Dirichlet mixture model by local variational inference. J. Signal Process. Syst. **74**(3), 359–374 (2014)
17. Mashrgy, M.A., Bdiri, T., Bouguila, N.: Robust simultaneous positive data clustering and unsupervised feature selection using generalized inverted Dirichlet mixture models. Knowl. Based Syst. **59**, 182–195 (2014)
18. Mukherjee, D., Mondal, R., Singh, P.K., Sarkar, R., Bhattacharjee, D.: Ensemconvnet: a deep learning approach for human activity recognition using smartphone sensors for healthcare applications. Multimed. Tools Appl. **79**(41), 31663–31690 (2020)
19. Nasfi, R., Amayri, M., Bouguila, N.: A novel approach for modeling positive vectors with inverted Dirichlet-based hidden Markov models. Knowl.-Based Syst. **192**, 105335 (2020)

20. Palumbo, F., Barsocchi, P., Gallicchio, C., Chessa, S., Micheli, A.: Multisensor data fusion for activity recognition based on reservoir computing. In: Botía, J.A., Álvarez-García, J.A., Fujinami, K., Barsocchi, P., Riedel, T. (eds.) EvAAL 2013. CCIS, vol. 386, pp. 24–35. Springer, Heidelberg (2013). https://doi.org/10.1007/978-3-642-41043-7_3

21. Palumbo, F., Gallicchio, C., Pucci, R., Micheli, A.: Human activity recognition using multisensor data fusion based on reservoir computing. J. Ambient Intell. Smart Environ. **8**(2), 87–107 (2016)

22. Pan, D., Liu, H., Qu, D., Zhang, Z.: Human falling detection algorithm based on multisensor data fusion with SVM. Mobile Information Systems 2020 (2020)

23. Phyo, C.N., Zin, T.T., Tin, P.: Deep learning for recognizing human activities using motions of skeletal joints. IEEE Trans. Consum. Electron. **65**(2), 243–252 (2019)

24. Qi, W., Su, H., Chen, F., Zhou, X., Shi, Y., Ferrigno, G., De Momi, E.: Depth vision guided human activity recognition in surgical procedure using wearable multisensor. In: 2020 5th International Conference on Advanced Robotics and Mechatronics (ICARM), pp. 431–436. IEEE (2020)

25. Sun, S., Folarin, A.A., Ranjan, Y., Rashid, Z., Conde, P., Stewart, C., Cummins, N., Matcham, F., Dalla Costa, G., Simblett, S., et al.: Using smartphones and wearable devices to monitor behavioral changes during covid-19. J. Med. Internet Res. **22**(9), e19992 (2020)

26. Swarnakar, S.K., Agrawal, H., Goel, A.: Smartphone inertial sensors-based human activity detection using support vector machine. In: Sharma, T.K., Ahn, C.W., Verma, O.P., Panigrahi, B.K. (eds.) Soft Computing: Theories and Applications. AISC, vol. 1381, pp. 231–241. Springer, Singapore (2021). https://doi.org/10.1007/978-981-16-1696-9_22

27. Woodstock, T.K.A.: Multisensor Fusion for Occupancy Detection and Activity Recognition in a Smart Room. Rensselaer Polytechnic Institute (2020)

Deepfake Detection Using CNN Trained on Eye Region

David Johnson$^{(\boxtimes)}$ ⓘ, Tony Gwyn ⓘ, Letu Qingge, and Kaushik Roy

North Carolina A&T State University, Greensboro, NC 27411, USA
{dmjohns8,tgwyn}@aggies.ncat.edu, {lqingge,kroy}@ncat.edu

Abstract. In this work, we will develop a simple convolutional neural network to detect deepfakes in videos on a frame-by-frame level, focusing on the region around the eyes. Since deepfakes are increasingly being created using forms of CNN, it should be possible to also detect deepfakes using CNN. OpenCV allows for frame extraction from videos, while also allowing image cropping. The well-developed Multitask Cascade Neural Network (MTCNN) is a stacked neural network for face detection and alignment. MTCNN is used for high accuracy face detection to greatly reduce false positive images in the dataset. Finally, a region around both eyes are cropped, with extra padding, to be used as input to train a CNN, using returned coordinates from MTCNN for the eyes. This research will focus on measuring if the eye region can be a useful area of interest for comparing original videos to deepfake videos.

Keywords: Deep learning · Deepfake · Deepfake detection · CNN · Computer vision · Eye region

1 Introduction

Deepfakes are manipulated images or videos that usually are an attempt to affect an individual's image or reputation, or deepfakes are used to spread disinformation. While they are somewhat easy to distinguish based on the viewer, they are becoming rampant because there are many times where they are difficult to tell if the video is real or fake. The goal of this research is to use deep learning to create a convolutional neural network (CNN) trained in distinguishing between real and fake images to then be able to detect manipulated image sequences in videos. There are a few public datasets that can be used to train a model. For this research, the dataset DeepFakeDetection [1] from the FaceForensics++ collection [1] will be used. DeepFakeDetection [1] consists of one thousand original videos, and three thousand manipulated videos of the originals. While deepfakes can be used to alter both visual and auditorial information, this research will only focus on the visual information from the videos. There exist patterns within deepfakes that can be learned by a CNN [2, 3] to be able to detect potential manipulation, such as the boundary where a fake face is placed over an original.

© Springer Nature Switzerland AG 2022
H. Fujita et al. (Eds.): IEA/AIE 2022, LNAI 13343, pp. 443–451, 2022.
https://doi.org/10.1007/978-3-031-08530-7_37

Fig. 1. An original extracted face (right) and the same face, manipulated at the same frame (left). The left image is an example of a deepfake from the DeepFakeDetection dataset [1, 4].

2   Background

Deepfakes are popularly created using Generative Adversarial Networks (GAN), which creates data after training, and validates against itself to increase realism [5]. In some of the more rudimentary forms, deepfakes present themselves as a false face superimposed on an original, unmanipulated face. Oftentimes, there is some brushing involved to allow better edge blending, this can be seen in Fig. 1. This style of deepfake creation is famously used online with the actor Nicolas Cage's face being superimposed on various people [6], such as other celebrities or world leaders. The increase in hyperrealism from deepfakes also leads to the potential for mis-, or disinformation, to spread. An example of a deepfake created using improved GAN technology would be the video of U.S. President Barack Obama [7] manipulated by the director and comedian Jordan Peele. While these examples show the humor and potential behind deepfakes, oftentimes, they are used as propaganda. Jordan Peele used his deepfake as a way to show how videos on the Internet may not always be original, "good faith" videos, but rather disinformation. Some states have gone to pass laws that make the use of deepfakes for political purposes, although enforcement remains problematic. Deepfakes are created using a type of CNN [5], and typically includes artifacts in the manipulated video that can be detected either by a person or by machine. Below are previous works that have also used CNN detect anomalies and artifacts based on the features from extracted images. The datasets used in the related works are not the same use here, although some use datasets from the same collection, FaceForensics++ [1], others use datasets created from different deepfake methods.

3   Related Works

Deepfakes, in their fundamental nature, are vectors for misinformation. Given their potential, it would be good to have ways to detect various types of deepfakes and their respective patterns in images. The following sections include various works related to the discussed problem.

3.1 Convolutional Neural Networks

Convolutional Neural Networks (CNN) are deep learning models that are commonly used on image data, or data that can be processed as image data. There have been previous works at constructing CNNs to detect deepfakes on different datasets, either by classifying against entire faces [5] or by focusing on specific extracted regions [2, 3]. In building a simple CNN based around the face region on the Celeb-DF dataset, classification close to 70% could be achieved [8]. Other methods of deepfake detection, such as using Scale-Invariant Feature Transform (SIFT) with CNN, which focus on specific regions of the face using SIFT [2], then using those results as input for a CNN, produced classification accuracies closer to 93%. By focusing on a specific region of the extracted face, a CNN can be constructed to detect deepfakes with greater accuracy than training on the entire face.

3.2 Generative Adversarial Networks

Generative adversarial networks (GAN) are deep learning models consisting of paired neural networks, or sub-models, that work against each other to produce realistic images [5]. The paired neural networks, the generator and the discriminator, tend to be CNNs, where the generator creates fake data that looks real, while the discriminator classifies real and fake data. When training with paired sub-models, a generator can create fake images, which are passed to the discriminator, where a classification is predicted. Based on the result, the generator will adjust to try to "fool" the discriminator into classifying fake data as real. This tandem generation and classifying leads to realistic images that are difficult to distinguish from original, unmanipulated images (Fig. 2).

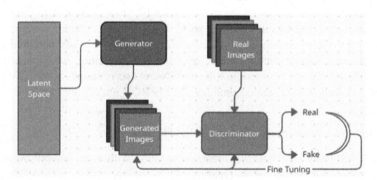

Fig. 2. Deepfake creation using GAN [8].

4   Materials and Methodology

Here, the tools used and methodology will be outlined to classify deepfakes using a CNN.

4.1 Dataset

The dataset used comes from the FaceForensics++ collection of deepfakes [1]. The current collection contains six sets of videos produced in different styles of manipulation. A few of the datasets are Face2Face, DeepFakeDetection, FaceSwap, and Deepfakes. The DeepFakeDetection [1] dataset will be used as the source for videos. The FaceForensics++ collection [1] consists of 1000 original videos that have been manipulated to produced thousands more. In the DeepFakeDetection dataset [1], there are 1000 original videos, and 3000 manipulated videos. To balance original and manipulated videos, every third manipulated video is used for every original video.

4.2 Frame Extraction

Next, using Python [9] and the OpenCV [10] library to load the video frame-by-frame, a frame is extracted every 100 ms and 40 frames are extracted per video. The frames are then used as input to a face detection neural network for high accuracy face detection in each frame. Frames without a high-confidence face are ignored.

4.3 Face and Eye Region Extraction

The face detection neural network is Multitask Cascaded Neural Network (MTCNN) [11], which is a model consisting of three stacked neural networks for bounding box regression, face detection and alignment, and facial landmark location. MTCNN's face detection component is used to detect faces in extracted frames. Faces with a confidence greater than 99% are retained in the extracted faces database, while other images are ignored. MTCNN returns coordinates for facial landmarks, such as the eyes, nose, and mouth. Using the coordinates of the eyes, a bounding box is created to contain both eyes at either end. The bounding boxes width and height are padded with 15 and 10 pixels, respectively, to increase the extracted region to also include potential regions of manipulation and the bounding box edges from face swapping. After extraction, the image database to be used as input to train a CNN consists of shape (40498, 20, 100, 3), where 20,249 RGB images are original and 20,249 RGB images are manipulated.

4.4 Construct Convolutional Neural Network

CNNs are great tools for classifying image data. *Keras*, within the TensorFlow platform [12], is used to build the model. In creating a CNN, the input dataset must be split into training and testing sets. The dataset is split with a ratio of 90% training and 10% testing, or 36,449 training images and 4,049 testing images. Training is set to 200 epochs, using the Adam optimizer with a learning rate of 0.0001. The proposed CNN model consists of six layers: an input layer, three hidden layers, a dense layer, and finally an output layer. To reduce overfitting, dropout is utilized between each layer, with a rate of 0.3. Max pooling layers also proceed every convolutional layer. Accuracies should exceed 80% to be considered statistically relevant (Figs. 3 and 4).

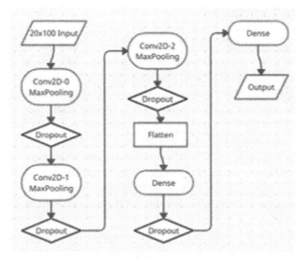

Fig. 3. Architecture of CNN model used

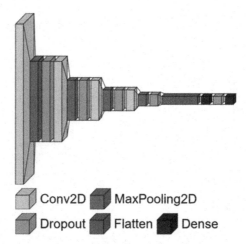

Fig. 4. Visualization of proposed CNN model

4.5 Compare Test Results

It is worth noting that the construction of a CNN model does not have to be limited to the dataset originally designed in mind. Given a collection of datasets, where each dataset is derived from the same set of original videos, in the case of FaceForensics++ [1], it would be fairy simple to retrain the model against similar, but different data. Since manipulated data come from the same source, the original data does not need to be reextracted, just the manipulated data. The new, manipulated data, along with the original data, are used to train the same CNN model-concept to compare metrics.

5   Results and Conclusion

Using the proposed CNN architecture, a model was created on the DeepFakeDetection dataset [1] with accuracy of about 98.3% at epoch 200. At epoch 50, validation accuracy tends to be around 90% but continues to increase, even beyond 200 epochs. Although training could continue beyond 200 epochs, accuracy and loss were not improving enough to warrant the extended training. Considering deepfakes are created using technologies based on CNN, it would seem possible to construct a CNN that could detect deepfakes, and this research shows that it is possible to detect deepfakes with strong confidence. The proposed CNN also provided significant results on other datasets from the FaceForensics++ collection [1], such as Face2Face [1], FaceSwap [1], and NeuralTextures [1]. Some datasets may need additional work on the CNN to increase positive detection rates, such as with detection in the NeuralTextures dataset [1] within FaceForensics++ [1].

This research focused on the creation of a simple CNN, which could be a helpful tool to determine if an image has been manipulated. By restricting images to specific, feature-containing regions, such as around the eyes, a high-accuracy CNN can be created to detect and potentially help mitigate the use of false media for disinformation, while still allowing the creative use of generative adversarial networks (Figs. 5, 6, 7 and Tables 1, 2).

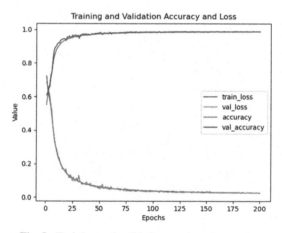

Fig. 5. Training and validation accuracy/loss curves

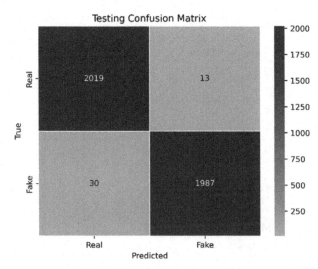

Fig. 6. Resulting confusion matrix

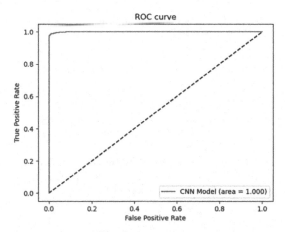

Fig. 7. ROC curve

Table 1. Accuracy results from k-fold validation, k = 10

K-fold accuracies	
0	98.07
1	98.32

(continued)

Table 1. (*continued*)

K-fold accuracies	
2	98.44
3	98.05
4	98.27
5	98.52
6	98.47
7	98.02
8	98.59
9	97.68
Mean	98.24
St. Dev	0.27

Table 2. Accuracy results comparison between datasets on same CNN model architecture

Dataset average accuracies	
DeepFakeDetection	98.3%
FaceSwap	97.8%
Face2Face	86.5%
NeuralTextures	67.2%

Acknowledgements. This research is supported by National Science Foundation (NSF). Any opinions, findings, and conclusions or recommendations expressed in this material are those of the author(s) and do not necessarily reflect the views of NSF.

References

1. Rössler, A., et al.: FaceForensics++: learning to detect manipulated facial images. In:International Conference on Computer Vision (ICCV) (2019)
2. Burroughs, S., Roy, K., Gokaraju, B., Luu, K.: Detection analysis of deepfake technology by reverse engineering approach (DREA) of feature matching. In: Agrawal, S., Kumar Gupta, K., H. Chan, J., Agrawal, J., Gupta, M. (eds.) Machine Intelligence and Smart Systems. AIS, pp. 421–430. Springer, Singapore (2021). https://doi.org/10.1007/978-981-33-4893-6_36
3. Wodajo, D., Atnafu, S.: Deepfake video detection using convolutional vision transformer, 11 March 2021. https://arxiv.org/pdf/2102.11126.pdf
4. Dufour, N., Gully, A., Jigsaw: Contributing data to deepfake detection research. Google Blog. Google, 24 September 2019. https://ai.googleblog.com/2019/09/contributing-data-to-deepfake-detection.html

5. Mo, H., Chen, B., Luo, W.: Fake faces identification via convolutional neural network. In: Proceedings of the 6th ACM Workshop on Information Hiding and Multimedia Security, IH&MMSec 2018, pp. 43–47. Association for Computing Machinery, Innsbruck (2018)
6. derpfakes. Nicolas Cage | Mega Mix Two | Derpfakes. YouTube. 02 February 2019. https://www.youtube.com/watch?v=_Kuf1DLcXeo
7. Buzzfeed Video, Jordan Peele. You Won't Believe What Obama Says in This Video! YouTube, 17 April 2018. https://www.youtube.com/watch?v=cQ54GDm1eL0
8. Karandikar, A., Deshpande, V., Singh, S., Nagbhidkar, S., Agrawal, S.: Deepfake video detection using convolutional neural network. In: International Journal of Advanced Trends in Computer Science and Engineering (2020)
9. Python. https://www.python.org/
10. OpenCV. https://opencv.org/
11. Zhang, K., Zhang, Z., Li, Z., Qiao, Y.: Joint face detection and alignment using multi-task cascaded convolutional networks (2016)
12. Tensorflow. https://www.tensorflow.org/
13. Rossler, A., et al.: FaceForensics++: learning to detect manipulated facial images. In: The CVF (2019). https://openaccess.thecvf.com/content_ICCV_2019/papers/Rossler_FaceForensics_Learning_to_Detect_Manipulated_Facial_Images_ICCV_2019_paper.pdf
14. Dataset, FaceForensics++. GitHub. https://github.com/ondyari/FaceForensics

Face Authentication from Masked Face Images Using Deep Learning on Periocular Biometrics

Jeffrey J. Hernandez V.$^{(\boxtimes)}$, Rodney Dejournett, Udayasri Nannuri, Tony Gwyn, Xiaohong Yuan, and Kaushik Roy

North Carolina Agricultural and Technical State University, Greensboro, NC 27407, USA
{jjhernandezvillarreal,r1dejournett,unannuri,
tgwyn}@aggies.ncat.edu, {xhyuan,kroy}@ncat.edu

Abstract. Nowadays, identity theft is an alarming issue with the growth of e-commerce and online services. Moreover, due to the Covid-19 pandemic, society has been pushed towards the usage of masks for people to safely interact with one another. It is hard to recognize a person if the face is mostly covered, even more so to artificial intelligence who have more difficulty identifying a masked individual. To further protect personal information and to develop a secure information system, more comprehensive bio-metric approaches are required. The currently used facial recognition systems are using biometrics such as periocular regions, iris, face, skin tone and racial information etc. In this paper, we apply a deep learning-based authentication approach using periocular biometric information to enhance the performance of the facial recognition system. We used the Real-World Masked Face Dataset (RMFD) and other datasets to develop our system. We implemented some experiments using CNN model on the periocular region information of the images. Hence, we developed a system that can recognize a person from only using a small region of face, which in this case is the periocular information including both eyes and eyebrows region. There is only a focus on the periocular region with our model in the view of the fact that the periocular region of the face is the main reliable source of information we can get while a person is wearing a face mask.

Keywords: Biometrics · Periocular recognition · Overfitting · Facial landmarks · CNN model · Augmentation · Authentication

1 Introduction

Authentication has become a fundamental issue to any computing system. Moreover, it is also a crucial part in any security-based computing system. Authentication allows only legitimate users to access system resources. Hence implementation of the authentication system is difficult. There are many ways to authenticate a person and give them access into the system. However, memorizing passwords for multiple systems and managing several smart cards are inconvenient. There is always a chance that the means of self-authentication can be stolen or lost. This results in password-based and card-based authentication being less than a reliable way of securing a system or files. Furthermore,

© Springer Nature Switzerland AG 2022
H. Fujita et al. (Eds.): IEA/AIE 2022, LNAI 13343, pp. 452–459, 2022.
https://doi.org/10.1007/978-3-031-08530-7_38

the subject of a biometrics-based authentication system is to be considered when trying to cover someone's identity or role. But, not a lot of focus is brought upon this subject as a means of authenticating someone into the system. When it comes to these forms of authentication, biometric-based authentication is a more secure way of implementing into the system, but it does require a bit more set-up.

Biometrics mostly refer to a part of the human body being utilized for something, which can include the face, iris, fingerprint, and skin tone. One way of using these biometrics is to identify or verify a person. A biometric based authentication system consists of two phases: Feature extraction and verification. During the feature extraction phase, a set of biometric features are extracted from the image dataset that has been collected. From there, a collection of the features gathered from the biometric data is made and stored as a template. In the verification phase, the biometric feature data is applied in the algorithm to verify/authenticate the label with the legitimate person. Biometric features can be different for the same person due to some factors such as variations in scale, pose, lighting and occlusions. So, more images of a subject/person are needed to prepare a sophisticated biometric feature dataset. A few images of a subject may not provide most biometric information. Due to this reason, a large image dataset is preferable in developing any biometric based authentication system.

With this all-in mind, this project aims to replicate facial recognition with the focus of only the periocular region of the face using a CNN model. The periocular region is the most reliable aspect of the face that can be viewed and identified especially in the case of the individual has their face covered up. After being detected, the periocular region will then be used to detect if the person is part of the database of authenticated people, giving a pass or fail to the tested image(s). The dataset will be split to fit the CNN model and made up entirely of masked individuals. The expected result should be the CNN model being able to present a high accuracy from the dataset being tested. As of now, the CNN model is still being tweaked to get the most accurate results possible.

2 Related Work

Biometric data is becoming increasingly used as a means of establishing a secure verification process. In [1], the author proposed a Deep Convolutional Neural Networks for the iris and face based Presentation Attack Mitigation by using machine learning techniques and implementing a feature extraction tool like the discrete wavelet transform (DWT). From there, the author developed a multiple CNN channel model like modified AlexNet, modified-SpoofNet, and modified-VGGNet, tested on the video dataset and get an accuracy for each channel: channel 1 being 99.90% accurate, Channel 2 being 99.83% accurate, and Channel 3 being 99.68%. In [2], Wang et al. described what type of dataset they had collected to test for the peculiar region with masked images, here they used a Masked Face Detection Dataset (MFDD), Real World Masked Face Recognition (RWMFR), and Simulated Masked Face Recognition Dataset (SMFRD). From there, they trained the dataset with a face eye-based multi granularity Recognition model. By testing these datasets, they got an accuracy of 95%. So, in our project we utilized a masked dataset.

Authors in [3] discussed whenever the periocular region is brought up, the report describes mainly facial features like eyelids, eye shape, eyebrows, eyelashes, the top of the nose, and skin texture. The paper provides a detailed survey of periocular biometrics and a deep knowledge of various aspects, like ROI Extraction, and functionality of periocular region stand-alone methods, like LBP, LPQ, PIGP, and a combination with an eye. It applied itself to many applications, such as smartphone authentication, to discover the role of the periocular region in the soft biometric categorization of the facial region [3]. Thus, the importance of the periocular region of the face whenever it comes to our experiment.

Chandana et al. [4] describes about every time they cannot capture an image or video within high resolution. So, to identify these high-definition images, they require these features, like irises, eyebrows, eyelids, and skin texture. These features can get the identification of a person within an image or video surveillance. By using an FGNET dataset and applying machine learning algorithms like logistic regression and naive bayes algorithms, the project was able to score a high accuracy of 96% by using only the periocular region.

With all these reports, we can take in their concepts of facial recognition and the periocular region and apply them into our working project. The result of this leaves us with our own CNN model, a baseline model to compare to, and a large periocular region image dataset made up of masked faces.

3 Methodology

In this research, we begin with gathering up and processing the dataset that is planned to be used the CNN model's testing and training sets. After data collection, the project will then focus on the application of the dataset to the CNN model and achieve the highest accuracy as possible when testing for a masked individual on whether they are an authenticated user or not.

We preprocessed the dataset to become a large dataset of clear images made of masked individuals' faces. Initially, we used the RFMD Dataset [5, 6] that was already separated into individual subjects with multiple images in each subject folder. The images in the dataset were then all adjusted to be the size of 400 pixels by 400 pixels to ensure that all the images were of the same size and prepared to grab the periocular region from each of the faces in the images. To accomplish this, an algorithm was developed, utilizing a combination of facial landmarks, OpenCV, and a library called lib. Once finished, this program would scan the image of the masked subject and grab a particular region of their face.

Using a CNN face detection model and shape predictor from within dlib, the program would detect facial landmarks within the masked images. We wanted to make sure that the new images obtained would contain both eyes and eyebrows, so we did not need to detect all the faces. But to make sure that the landmarks were accurate, we still went ahead and detected (or predicted as much as the program can) all 68 facial landmarks in each face. After that, we modified it in such a way that instead of completely displaying all 68 facial landmarks, the program would focus on just the leftmost part of the left eyebrow through the bottom part of the right eyeball (facial landmarks 18 through 27, landmarks 28–30 were added later to be able to capture the periocular region much better) [7]. Once all the necessary facial landmarks have been detected on the actual face within the image, the program would go through and confirm what parts of the face it managed to grab that relate to the periocular region. This would allow the program to crop that part of the face in the images and save it as a separate image. Figure 1 showcases these facial landmarks and then the extraction process.

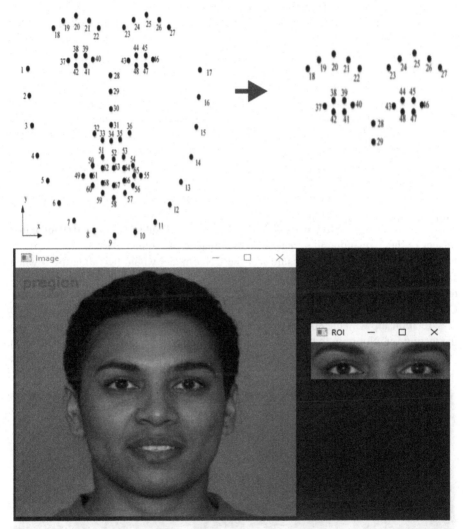

Fig. 1. Facial Landmarks being broken down to only needing landmarks located in the periocular region and the program extracts the region into a new image. Image is taken from [5, 6].

Unfortunately, within the limits of the RMFD dataset, many of the subjects' faces were not detected and their periocular region could not be extracted. This would lead to adding more images of masked faces to the original masked dataset, resulting in new periocular masked faces being placed and more subjects to augment. With this in mind, we proceeded with a Real and Fake Face Detection dataset to add into our periocular dataset [5, 6]. Despite the faces not being masked, the extracted periocular region images will be added into the training set to allow the CNN model to better distinguish and understand what parts of the face it should focus on.

With the facial landmarks extracted, we were able to create a new dataset with periocular images, made from the masked dataset we have. We then started to prepare our data to be used for deep learning to identify if the image matches an authenticated face within the database, all using a CNN binary image classifier. To further establish the dataset, the next focus would have to be image augmentation. Image augmentation is necessary within this project since it will provide training and exposure to our CNN model through allowing multiple different variations of the subject images into the training and testing sets. Unfortunately, a question came up of what style of the original dataset should we go for. For each subject in the masked dataset, there were multiple images of them. The question of style was if the placement of the subjects or where they are distributed really matter enough for the CNN model to operate successfully. From there, we only focused on one format of the dataset, but that same question would be brought into place later.

Then we augmented the images through another program we implemented that replicates the images into different versions and aspects, as shown in Fig. 2. The large, combined dataset soon grew to have 2,005 subjects, with each subject getting 10 augmented images of them, resulting in a total of 20,045 images. We aimed to overcome the overfitting issue of our dataset to ensure that the CNN model would be validating it when calculating its accuracy. Once the augmentation has been completed, the dataset needed to be split into a training set and testing set. These sets would then be utilized by our CNN model. To achieve a high accuracy with our model, we decided to get an 80% to 20% relationship with the 80% of images going into the training set and 20% going into the testing set. Not only that, the 80% that the training set will get must also have an image from each subject to maintain consistency. Once the images have been split, the

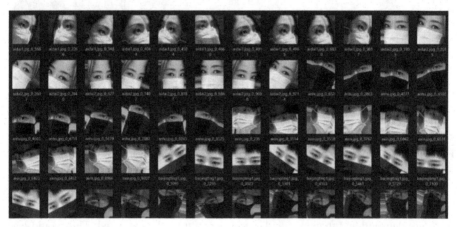

Fig. 2. A preview of how the augmented dataset looked before the images were extracted of their periocular region [5, 6].

CNN model that we developed requires both sets to have authentic and unauthentic sets, so it can try to distinguish. We originally had it named as real and fake sets but, after discussion over the concept, we decided to rename them to authentic and unauthentic sets to better fit what their purpose within the model, For this case, we split each set into 90% of the images going into an unauthentic set and the rest (10%) would go into an authentic set within the folder. With the dataset being as prepared and fitted as possible, we bring out focus to our CNN model and its results.

4 Results

In this section, we provide our experimental results. The CNN model that was developed and used came to be from integrating with Keras Image Classification and following a base model of a classification matrix and its many layers [9]. The model itself is a binary image classifier, meaning that it will identify the images placed in the dataset as one class or the other. In this experiment, the model will take in the training set and testing set as its own classes and focus on it. It operates with layers and an optimizer that helps with the minimizing of classifying the images with the data. As a result, the CNN model produces a report showing the accuracy, precision, recall and two graphs that show the overall accuracy and loss as the model works with the dataset, as shown in Fig. 3. Using the RFMD dataset we used images to fit for binary image classification for authenticated subjects vs unauthenticated subjects. We experimented with this data on our benchmark model to see if we could better results with subjects using their full face. We used the formula, steps_per_epoch * batch_size = total # of images to determine where the new number of epochs or batch sizes should be. On our first attempt to train the model on the data we ended up with 33% loss and 85% accuracy in training and 78% loss and 62% in validation. Total time taken was about 47 min for 15 epochs. Our confusion matrix resulted in 366 true positives, 70 false positives, 370 false negatives, and 68 true negatives. For the second run we lowered the batch size for the testing dataset and increased number of epochs to 18. The training for the model ran for 54 min and resulted in 39% loss and 81% accuracy for training and 58% loss and 70% accuracy for validation. The predictions from the confusion matrix were 333 true positives, 103 false positives, 314 false negatives, and 125 true negatives.

```
ROC Curve- .52
Confusion Matrix
[[333 103]
 [313 125]]

Classification Report
                precision    recall   f1-score   support

     authentic      0.52       0.76      0.62       436
   unauthentic      0.55       0.29      0.38       438

      accuracy                           0.52       874
     macro avg      0.53       0.52      0.50       874
  weighted avg      0.53       0.52      0.50       874
```

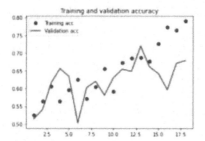 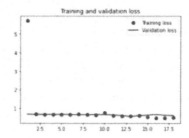

Fig. 3. The classification report that our CNN model provided with the datasets alongside two graphs that show the accuracy and loss of the data

With the results, we see that the accuracy when it comes to the CNN binary image classifier model ranges around 53%.

5   Conclusions

The project continues to develop as the focus of trying to achieve a high accuracy when it comes to the authentication of masked faces is still relevant. Theories have been made and new approaches are currently being taken to get the results we desire. For now, a baseline facial recognition model using binary image classifier has been taken into consideration to compare the results it gives to our own model [10]. Even though it utilized TensorFlow instead of what we used, the model still could be used as a baseline since it gives up accurate results with so little data needed. This model allows us to configure the parameters. This will allow us to break down what will work as a means for better results: an update to the periocular region dataset or reworking the CNN model. At the current iteration of the dataset, we switched into full, uncovered faces instead of the periocular region of masked faces. An idea was brought up to see if the CNN model's facial recognition would operate better with a full-face dataset rather than a dataset with a part of someone's face. Future tests will now experience the full face rather than the periocular region to see if better results appear. This will allow us to improve on the CNN model and promise better results in the future.

Acknowledgement. This research is supported by National Science Foundation (NSF). Any opinions, findings, and conclusions or recommendations expressed in this material are those of the author(s) and do not necessarily reflect the views of NSF.

References

1. Chatterjee, P.: Deep Convolutional Neural Networks for the iris and face based Presentation Attack Mitigation, pp. 1–101. The Graduate School of NCAT, Greensboro, NC (2020)
2. Wang, Z., Wang, G., Huang, B., et. al.: Masked face recognition dataset and application, pp. 1–3 (2020). https://arxiv.org/abs/2003.09093
3. Kumari, P., Seeja, K.R.: Periocular biometrics: a survey.. J. King Saud Univ., 1–12 (2019)
4. Chandana, C.S., Rao, K.D., Sahoo, P.K.: Face recognition through machine learning of periocular region. Int. J. Eng. Res. Technol. (IJERT), **9**(3), 1–5 (2020)
5. Huang, B.: Real-World Masked Face Dataset (RMFD). GitHub, 13 February 2020. https://git hub.com/X-zhangyang/Real-World-Masked-Face-Dataset#real-world-masked-face-datase trmfd
6. Huang, B., et al.: Masked face recognition datasets and validation. In: 2021 IEEE/CVF International Conference on Computer Vision Workshops (ICCVW), pp. 1487–1491 (2021). https://doi.org/10.1109/ICCVW54120.2021.00172. https://ieeexplore.ieee.org/document/9607619
7. Rosebrock, A.: Facial Landmarks with dlib, opencv, and python. PyImageSearch, 13 April 2017. https://www.pyimagesearch.com/2017/04/03/facial-landmarks-dlib-opencv-python/
8. Computational Intelligence and Photography Lab, Yonsei University, 14 January 2019. Real and Fake Face Detection. Kaggle. https://www.kaggle.com/ciplab/real-and-fake-face-detection
9. Brownlee, J.: How to perform face detection with deep learning. Mach. Learn. Mastery (2019). https://machinelearningmastery.com/how-to-perform-face-detection-with-classi cal-and-deep-learning-methods-in-python-with-keras/
10. Phan, B.: 10 minutes to building a fully-connected binary image classifier in tensorflow. Towards Data Sci. (2020). https://towardsdatascience.com/10-minutes-to-building-a-fully-connected-binary-image-classifier-in-tensorflow-d88062e1247f. Accessed 13 Dec 2021

An Optimization Algorithm for Extractive Multi-document Summarization Based on Association of Sentences

Chun-Hao Chen[1(✉)], Yi-Chen Yang[1], and Jerry Chun-Wei Lin[2]

[1] Department of Information and Finance Management, National Taipei University of
Technology, Taipei, Taiwan
chchen@ntut.edu.tw, t109AB8007@ntut.org.tw
[2] Department of Computer Science, Electrical Engineering, and Mathematical Sciences,
Western Norway University of Applied Sciences, Bergen, Norway
jerrylin@ieee.org

Abstract. Designing of automatic summary extraction technology becomes more and more important as the number of documents increasing rapidly. At present, the indicators that are often used for summary evaluation including the coverage, redundancy, and relevance. However, the summary cannot be completely extracted using only these three evaluation indicators. Therefore, with the concept of centrality, we design an association-based centrality criterion firstly, which can be used to evaluate associations of sentences to reinterpret the centrality of summary. Based on the three commonly used and the designed centrality factors, an optimization algorithm is then proposed for obtaining the summary from the given multiple documents. Experiments were also made on the dataset from the Document Understanding Conference 2002 (DUC2002) to show the effectiveness of the proposed algorithm in terms of ROUGE scores.

1 Introduction

Documents summarization is an important topic because a large amount articles has been published rapidly in recent years, which can be utilized to find important events and provide user an efficient way to get information from them. In the literature, summarization approaches can be divided into two types: abstractive and extractive summarization, and most of them are related to extractive summarization [17, 18]. The existing algorithms can be roughly divided into two groups: (1) the use of optimization technology, and (2) non-optimization technology summary extraction methods.

For instance, in the use of optimization technology, Sanchez-Gomez et al. used the multi-objective artificial bee colony (MOABC) algorithm to discuss the effectiveness of the combinations of summary evaluation criteria [2, 3]. Due to multi-objective functions could be used to find summarization, Sanchez-Gomez et al. also proposed four methods for solving multi-objective optimization task [9]. Besides, Alqaisi et al. proposed an Arabic extractive multi-document summarization method based on clustering and evolutionary multi-objective optimization approach [7]. Other meta-heuristic algorithms have also been proposed, including particle swarm optimization (PSO) [8, 10],

© Springer Nature Switzerland AG 2022
H. Fujita et al. (Eds.): IEA/AIE 2022, LNAI 13343, pp. 460–469, 2022.
https://doi.org/10.1007/978-3-031-08530-7_39

Firefly Algorithm (FA) [15], and others [1, 5, 11]. For the non-optimized technology, Li et al. proposed an approach to establish a semantic link network to produce abstractive summarization from multiple documents [4]. Besides, Uçkan et al. proposed the features vector centrality to generate extractive summarization from given documents [14].

To extract a high-quality summarization, many criteria or indicators should be considered. According to the literature, previous scholars commonly used coverage, redundancy, and relevance indicators to measure the quality of generated summarization [1–3, 7–9, 11, 13, 16]. Although these three indicators are currently used by the most scholars, there is still have space for improvement. For example, the centrality criterion is introduced in [14].

Therefore, firstly, we design a new summarization evaluation criterion, named association-based centrality, for measuring the centrality of the extracted key sentences in the summary. From the given documents, the key sentences are first obtained. Then, a set of keywords is derived from a key sentence. After all sentences are processed, sets of keywords are collected and used as transactions for discovering association rules. Then, the derived rules are used to find frequent word combinations for calculating the centrality of sentences. Because the multi-documents summarization is an optimization task, we then propose an optimization approach for extracting summarization using genetic algorithm (GA). The proposed approach consists of four phases that are: (1) The key sentence extraction, (2) Key sentence preprocessing, (3) Association-based centrality calculation, and (4) The summarization optimization algorithm. The first phase is trying to get key sentences from multiple documents. Then, in the second phase, the data preprocessing is executed to generate the attributes needed for next phase. In addition, in order to add other synonyms that may be commonly used in addition to keywords, the Wordnet is utilized to find synonyms that can be employed to effectively include sentences with high centrality values in the summarization. The association-based centrality values of all key sentences are calculated. In the last phase, the genetic algorithm is employed to find the summarization from the given multiple documents. It first encodes key sentences into a chromosome by a bit string. The fitness value of a chromosome is evaluated by not only the coverage, redundancy, and relevance indicators but also the designed association-based centrality indicator. The evolution process is repeated until reaching termination conditions. Experiments were also conducted on the Document Understanding Conference dataset (DUC2002) to show the effectiveness of the proposed approach in terms of ROUGE scores.

2 Details of Proposed Approach

2.1 Encoding Scheme

The goal of the proposed approach is to obtain an appropriate summarization from the given multiple documents. After finding n key sentences from given documents using Word2Vec and TextRank, those key sentences are encoded into a chromosome based on their appearance in the original documents. Assume there are n key sentences, the encoding scheme is shown in Fig. 1.

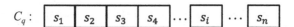

C_q : | s_1 | s_2 | s_3 | s_4 | ... | s_i | ... | s_n |

Fig. 1. The encoding scheme of a summarization.

In Fig. 1, every sentence can be regarded as a gene, and a binary code is used to determine whether the sentence is added to the summary. For example, let n is 10, it means a total of 10 sentences at most will appear in a summary. When the binary string is "1000100001", it means that sentences s_1, s_5 and s_{10} are collected as the summarization.

2.2 Fitness Function

To calculate the fitness value of each chromosome, the survival value of each chromosome can be judged. The quality of a summary cannot use a single criterion to accurately extract the best summary. In this paper, the fitness function was established based on four summary evaluation indicators, namely the coverage, redundancy reduction, relevance, and association-based centrality factors. The designed fitness function is shown in the following formula:

$$f(C_q) = Cov(C_q) * Redund_{Red}(C_q) * Rel(C_q) * assCentral(C_q), \qquad (1)$$

where the purpose of coverage of a chromosome $Cov(C_q)$ is to allow the summary to cover all the topics of the original article, so that each sentence of the summary can represent the meaning of the topic. The formula of $Cov(C_q)$ is given as follows:

$$Cov(C_q) = \sum_{i=1}^{n} sim(s_i, M) \cdot x_i, \qquad (2)$$

where n is the number of key sentences, s_i represents the ith sentence in the key sentence set, M is the mean vector of the overall documents $M = (m_1, m_2, \ldots, m_k)$, and x_i represents whether the ith sentence is one of the sentences in summary. If x_i is 1, the ith sentence is included in the summary, otherwise the it is not in the summary. The cosine similarity is employed to calculate similarity of sentences. Take two sentences as an example. After calculating the weights of words by TFIDF, the vectors of the sentences can be expressed as $s_i = (w_{i1}, w_{i2}, \ldots, w_{ik})$ and $s_j = (w_{j1}, w_{j2}, \ldots w_{jk})$, then the similarity of them can be calculated using the following formula:

$$sim(s_i, s_j) = \frac{\sum_{k=1}^{n} w_{ik} \cdot w_{jk}}{\sqrt{\sum_{k=1}^{n} w^2_{ik}} \sqrt{\sum_{k=1}^{n} w^2_{jk}}}. \qquad (3)$$

As to the redundancy reduction of a chromosome $Redund_{Red}(C_q)$, because summary should be short and must contain the general meaning of the original article, it is necessary to avoid sentences with similar meanings to appear repeatedly in the summary. As a result, the redundancy reduction is given in the following formula:

$$Redund_{Red}(C_q) = \frac{1}{\log[\left(\sum_{i=1}^{n} \sum_{j=i+1}^{n} sim(s_i, s_j) * r_{ij}\right) * \sum_{i=1}^{n} x_i]}, \qquad (4)$$

where $sim(s_i, s_i)$ means the similarity between two sentences, and the decision variable r_{ij} is used to determine whether two sentences are appeared in the summary at the same time. If $r_{ij} = 1$, it indicates that both sentences appearing in the summary. Otherwise, the two sentences do not appear at the same time.

Due to the content in the original text may cover a few small topics, in order to emphasize the main topic and avoid the summary from being affected by the small topics, the relevance of a chromosome $Rel(C_q)$ is to make every sentence of the summary related to the original content as much as possible, and don't deviate from the main topic to cause the problem of irrelevance between the summary and the original content. The $Rel(C_q)$ is defined as follows:

$$Rel(C_q) = sim(S, M) * \left(\sum_{i=1}^{n} sim(S, s_i) * x_i \right), \tag{5}$$

where $sim(S, M)$ is the calculation of the similarity between the summary vector $S = (sw_1, sw_2, \ldots, sw_k)$ which is composed of multiple word weights, and the mean vector M is a vector of words calculated through TFIDF, and $sim(S, s_i)$ is the calculation of the similarity score between the summary vector S and the sentence s_i. When x_i is 0 and the ith sentence is not in the summary, the weight of the word w_{ik} in the sentence s_i will be changed to 0. The weights of all words are summed up and average to form the summary word vector sw_k. The summary word vector is calculated using the following formula:

$$sw_k = \frac{1}{n} \sum_{i=1}^{n} w_{ik}, \text{ if } w_{ik} \notin S \rightarrow w_{ik} = 0. \tag{6}$$

Article is composed of multiple sentences. Some of the sentences are not important but some of them are the meaning of the core theme. In other words, the sentences that have high related to other sentences are the core sentences and could be the more important sentences for the article. In order to find those sentences that have high associations to other sentences, association rule mining approach is executed on the key words that extracted from sentences for discovering the frequent keyword combinations. Besides, to avoid similar sentences are misjudged due to the different using of words, we employ Wordnet to find more synonyms as keywords for finding important sentences. The association-based centrality of a chromosome $assCentral(C_q)$ is thus defined as follows:

$$assCentral(C_q) = \frac{1}{n} \sum_{i=1}^{n} c_i \cdot x_i. \tag{7}$$

where c_i is the centrality score of the ith sentence, where the centrality score is calculated based on the frequent keyword combinations from the extracted association conditions.

2.3 Genetic Operations

In order to retain the good genes in the population can be survived to next generation, we choose the elitist strategy to perform the selection operation. That is, the selection operation is regarded as the survival of the fittest in the survival rule. The scores calculated according to the fitness function are sorted in descending order, and choose the chromosomes with higher fitness value in the top $n\%$ as the candidate chromosomes for generating next generation. By retaining the elite chromosomes, it can be ensured that the gene combination completed in the final iteration is the chromosome with the highest fitness value. Crossover is an important part in the evolution of nature. After the selection operation, the retained elite chromosomes generate new chromosomes through the action of mating and recombination. Two chromosomes are selected from the elite chromosomes as the parents. Then, a crossover point is randomly selected. According to the cut point, the two selected chromosomes are exchange genes, and finally two new chromosomes are produced. Then, in order to increase the diversity of the population, we randomly select a gene from the offspring chromosomes as a mutation point, and change its from 0 to 1, or 1 to 0. This operation not only allows the population to be diversified, but also prevents the evolutionary process falling into the local optimal.

3   Pseudo Code of Proposed Approach

Table 1. Pseudo code of the proposed approach.

Algorithm 1: Association-based Extractive Multiple Documents Summary Optimization Algorithm.
Input: Multiple News Documents $newsDoc = \{ d_1, d_2, ..., d_n \}$.
Parameter: Number of top key sentences T, iterations $iter$, population size $pSize$, retain rate $rRate$, crossover rate $cRate$, and mutation rate $mRate$.
Output: The best chromosome as the final summarization.
Procedure:
1. $S = \{s_1, s_2, ..., s_T\} \leftarrow$ keySentencesExtraction($newsDoc$);
2. $centralitySet = \{c_1, c_2, ..., c_T\} \leftarrow$ calculateAssoCentrality(S);
3. $MV \leftarrow$ calculateMeanVector(S);
4.   GA_Procedure(){
5. $population \leftarrow$ initialChro($pSize$, S);
6.   **FOR** $i = 0$ to $iter$ **DO:**
7.   **FOR** $j = 0$ to pSize **DO:**
8. $coverage_j \leftarrow$ calculateCoverage(S, MV);
9. $redun\_red_j \leftarrow$ calculateRedundancyReduction(S);
10. $SV \leftarrow$ calculateSummaryWordVector(S);
11. $relevance_j \leftarrow$ calculateRelevance(S, SV);
12. $centrality_j \leftarrow$ calculateCentrality(S, $centralitySet$);
13. $fitness_j \leftarrow$ calculateFitness($coverage_j$, $redun\_red_j$, $relevance_j$, $centrality_j$);
14.   **END** j **FOR LOOP**
15. $population \leftarrow$ executeSelection($population$, $pSize$, $rRate$);
16. $population' \leftarrow$ executeCrossover($population$, $cRate$);
17. $population'' \leftarrow$ executeMutation($population'$, $mRate$);
18.   **END** i **FOR LOOP**
19. $bestChrom \leftarrow$ findtheBestChrom($population''$);
20.   **RETURN** $bestChrom$;
21. }

In this section, the pseudo code of the proposed association-based extractive multiple documents summary optimization algorithm is given in Table 1.

From Table 1, the data preprocessing is done firstly, including extracting key sentences, calculating association-based centrality values and mean vectors, to get S, *centralitySet*, and MV. The evolutionary process is then executed (Lines 4 to 20). According to the population size and the extracted key sentences S, the initial population is then generated (Line 5). The four factors, coverage, redundancy reduction, relevance and centrality, are calculated (Lines 8 to 12). After that, the fitness value of a chromosome is set using formula (1) (Line 13). Next, the genetic operators are performed on the population to generate new offspring. In selection operator, based on the given retain rate, the candidate chromosomes are generated. The *pSize* chromosomes are then selected from the candidate chromosomes to form the next population (Line 15). The crossover and mutation operators are then executed to obtain new chromosomes (Lines 16 to 17). At last, if the termination condition is reached, the chromosome with the highest fitness value is outputted. Otherwise, it will go to next iteration.

In the following, the pseudo code for the association-based centrality calculation is described. Based on the sorted key sentences, the association-based centrality values of the extracted key sentences are calculated using the association-based centrality calculation algorithm which is shown in Table 2.

From Table 2, the data preprocessing is needed for the sentences, including splitting sentence to words (Line 3), getting stem words (Line 4), and removing stopwords (Line 5), to extract key words set *sentenceKeyWord*. The *sentenceKeyWord* is then used as transactions along with the given minimum support and minimum confidence for generating association rules (Line 8). Then, the itemsets in left- and right-hand side of the rules are gathered to form the association word set *assWordSet* (Line 9). The *assWordSet* is then employed to find synonyms (Line 10). The association-based centrality values of all sentences are calculated (Lines 11 to 23). Take $sentence_i$ as an example. For any compared sentence, said $sentence_j$, when they have the same keyword or synonym, then the centrality value of $sentence_i$ will plus one (Lines 16 and 20). After that, the association-based centrality set ass*CentralitySet* is returned.

Table 2. Pseudo code of the association-based centrality calculation.

Algorithm 2: Association-based Centrality Calculation Algorithm
Input: Sorted key sentence set $S = \{s_1, s_2, \ldots, s_T\}$.
Parameter: Minimum support $minSup$, and minimum confidence $minConf$.
Output: Association-based centrality values for all key sentences $assCentralitySet$.
Procedure:
1. $sentenceKeyWord = []$;
2.   **FOR** $sentence$ in S **DO**:
3. $word \leftarrow$ splitSentenceToWords($sentence$);
4. $word' \leftarrow$ stemWords($word$);
5. $word'' \leftarrow$ filterStopwords($word'$);
6. $sentenceKeyWord$.append($word''$);
7.   **END FOR**
8. $assRuleSet \leftarrow$ generateAssociationRules($sentenceKeyWord$, $minSup$, $minConf$);
9. $assWordSet \leftarrow$ findWordInAssociationRules($assRuleSet$);
10. $syno \leftarrow$ wordnetToFindSynonym($assWordSet$);
11.   **FOR** $i = 0$ to $|S|$ **DO**:
12.   For $j = 0$ to $|S|$ Do:
13.   **IF** $sentence_i == sentence_j$ **THEN**
14.   continue;
15.   **ELSE IF** $sentence_i \cap sentence_j \in assRuleSet$ **THEN**
16. $centralityScore_i$ ++;
17.   **ELSE**
18. $newAssWordSet \leftarrow$ replaceWithSynonym($assWords$, $syno$);
19.   **IF** $sentence_i \cap sentence_j \in newAssociateRuleSet$ **THEN**
20. $centralityScore_i$ ++;
21.   **END FOR**
22. $assCentralitySet \leftarrow$ addScore($centralityScore_i$);
23.   **END FOR**
24.   **RETURN** $assCentralitySet$;

4 Experimental Result

In order to effectively evaluate the proposed multi-document summarization optimization approach, experiments were conducted on the Document Understanding Conference (DUC2002) dataset. The DUC2002 dataset contains 59 topics. Each topic contains an average of 5 to 15 related documents, and there are 567 documents in total. DUC also provides extractive and abstractive reference summary for each topic. Not only that, but the extractive summarization is also divided into 200-words and 400-words summaries, so that everyone can choose appropriate reference summary according to their respective summarization algorithms to facilitate follow-up summary evaluation. In this paper, the 400-words summary is used for experimental evaluation.

All the experiments in this study were implemented on the macOS version 11.6 system which is equipped with an Intel i5 1.4 GHz quad-core processor and 8GB 2133 MHz LPDDR3 memory. The development platform is developed using the Spyder IDE in the Anaconda 2020.07 version, and the python version is 3.8. The parameter setting is stated as follows: population size $pSize$ is 100, the length of a chromosome T is 50, number of iterations $iter$ is 150, retain rate $rRate$ is 0.5, crossover rate $cRate$ is 1, mutation rate *mRate* is 0.1. That is, 50 key sentences are used to generate a summarization by the proposed approach. Additionally, the ROUGE-1, ROUGE-2 and ROUGE-L are employed to evaluate the proposed and compared approaches.

Firstly, using the derived chromosome, comparing to the standard summary, the recall, precision and F-score of all ROUGE types are shown in Table 3.

Table 3. Recall, precision and F-score of the proposed approach.

Rouge-type	Recall	Precision	F-score
Rouge-1	0.6580	0.4306	0.5205
Rouge-2	0.4360	0.2976	0.3538
Rouge-L	0.6060	0.3966	0.4795

From Table 3, we can observe that the proposed approach is good to get better recall values that are 0.658, 0.436, and 0.606 than precision values for Rouge-1, Rouge-2, and Rouge-L. The F-score values are 0.5205, 0.3538, and 0.4795. In order to show the merits of the proposed approach, the F-score values are used to compare with other existing approaches in next experiments.

Because ROUGE is a commonly used evaluation criterion for summary evaluation, in order to effectively measure the performance of the proposed approach, we use ROUGE-1, ROUGE-2, and ROUGE-L to compare with other optimized and non-optimized summarization algorithms, including latent semantic analysis (LSA) [19, 20], the graph-based ranking approach for summarization [21, 22], the SumBasic approach [6], the generative probabilistic models for multi-document summarization [12], the eigenvector centrality to generate extractive summarization [14], and the evolutionary multi-objective optimization approach [7]. The length of summary for comparison is the 400-words summary. Table 4 shows the comparison results of the proposed and other six approaches.

From Table 4, according to the performance of ROUGE scores, we can discover that the proposed approach is ranked third on ROUGE-1 and ROUGE-L, with 0.5205 and 0.4795 respectively. It is worth mentioning that in the performance of the ROUGE-2 score which is a more representative summary evaluation criterion, the proposed approach got 0.3538 which is completely higher than the other six approaches. From the experimental results, we can conclude that the proposed approach is effective in terms of ROUGE-2, and reach similar results to other approaches.

Table 4. Comparisons of proposed and other existing approaches on DUC2002.

Approach	Category	Fitness function	Rouge-1	Rouge-2	Rouge-L
Landauer [19, 20]	LSA	N/A	0.4680	0.1614	0.4269
Mihalcea [21, 22]	TextRank	N/A	0.5517	0.2444	0.5393
Vanderwende [6]	SumBasic	N/A	0.5146	0.1854	0.4599
Haghighi [12]	KLSum	N/A	0.4349	0.1613	0.3992
Uçkan [14]	Eigenvector Centrality	N/A	**0.5806**	0.3182	**0.5525**
Alqaisi [7]	NSGA-II	Cov, Div, Rel	0.4705	0.237	0.471
Proposed approach	GA	Cov, Redun_red, Rel and Central	0.5205	**0.3538**	0.4795

5   Conclusions and Future Work

With the increase in the number of news, to design an algorithm for generating summary of news for users to read for saving more time and expenditure is needed. Therefore, extractive summarization is the problem to be explored in this paper. In this research, we propose a multi-document extractive summarization model based on genetic algorithms. Unlike the past only using common summary evaluation indicators, such as coverage, redundancy, and relevance. We hope to find out the core sentences from the original text based on the characteristics of centrality. Therefore, the association-based centrality factor is designed firstly. More than that, in order to avoid using different words but similar keywords between sentences, we also use Wordnet to find synonyms of keywords to improve the centrality score calculation. The results show that, based on the four summary evaluation indicators, including the coverage, redundancy reduction, relevance, and association-based centrality factors as fitness function, the proposed optimization approach is better than the previous approaches in terms of ROUGE-2 score on DUC2002 dataset. In the future, we will continue to improve the extractive optimization summarization model, hoping to build an automatic extractive summarization similar to the manual extractive summarization, e.g., trying different parameter settings, the type of use of genetic operators in genetic algorithm selection, and even the combination of multi-objective summary evaluation indicators.

References

1. Ghodratnama, S., Beheshti, A., Zakershahrak, M., Sobhanmanesh, F.: Extractive document summarization based on dynamic feature space mapping. IEEE Access **8**, 139084–139095 (2020)
2. Sanchez-Gomez, J.M., Vega-Rodríguez, M.A., Pérez, C.J.: Extractive multi-document text summarization using a multi-objective artificial bee colony optimization approach. Knowl. Based Syst. **159**, 1–8 (2018)

3. Sanchez-Gomez, J.M., Vega-Rodriguez, M.A., Carlos, J.P.: Experimental analysis of multiple criteria for extractive multi-document text summarization. Expert Syst. Appl. **140**, 112904 (2020)
4. Li, W., Zhuge, H.: Abstractive multi-document summarization based on semantic link network. IEEE Trans. Knowl. Data Eng. **33**(1), 43–54 (2021)
5. Liu, W., Gao, Y., Li, J., Yang, Y.: A combined extractive with abstractive model for summarization. IEEE Access **9**, 43970–43980 (2021)
6. Vanderwende, L., Suzuki, H., Brockett, C., Nenkova, A.: Beyond SumBasic: task-focused summarization with sentence simplification and lexical expansion. Inf. Process. Manag. **43**(6), 1606–1618 (2007)
7. Alqaisi, R., Ghanem, W., Qaroush, A.: Extractive multi-document Arabic text summarization using evolutionary multi-objective optimization with k-medoid clustering. IEEE Access **8**, 228206–228224 (2020)
8. Rautray, R., Balabantaray, R.C.: An evolutionary framework for multi document summarization using cuckoo search approach: MDSCSA. Appl. Comput. Inform. **14**(2), 134–144 (2018)
9. Sanchez-Gomez, J.M., Vega-Rodríguez, M.A., Pérez, C.J.: Comparison of automatic methods for reducing the Pareto front to a single solution applied to multi-document text summarization. Knowl. Based Syst. **174**, 123–136 (2019)
10. Al-Abdallah, R.Z., Al-Taani, A.T.: Arabic single-document text summarization using particle swarm optimization algorithm. Procedia Comput. Sci. **117**, 30–37 (2017)
11. Saini, N., Saha, S., Bhattacharyya, P.: Multiobjective-based approach for microblog summarization. IEEE Trans. Comput. Soc. Syst. **6**(6), 1219–1231 (2019)
12. Haghighi, A., Vanderwende, L.: Exploring content models for multi-document summarization. In: Proceedings of Human Language Technologies, pp. 362–370 (2009)
13. Lamsiyah, S., El Mahdaouy, A., Espinasse, B., Ouatik, S.E.A.: An unsupervised method for extractive multi-document summarization based on centroid approach and sentence embeddings. Expert Syst. Appl. **167**, 114152–114167 (2021)
14. Uçkan, T., Karcı, A.: Extractive multi-document text summarization based on graph independent sets. Egypt. Inform. J. **21**(3), 145–157 (2020)
15. Tomer, M., Kumar, M.: Multi-document extractive text summarization based on firefly algorithm. J. King Saud Univ. Comput. Inf. Sci. (2021)
16. Verma, P., Om, H.: Collaborative ranking-based text summarization using a metaheuristic approach. In: Abraham, A., Dutta, P., Mandal, J., Bhattacharya, A., Dutta, S. (eds.) Emerging Technologies in Data Mining and Information Security. AISC, vol. 814, pp. 417–426. Springer, Singapore (2019). https://doi.org/10.1007/978-981-13-1501-5_36
17. El-Kassas, W.S., Salama, C.R., Rafea, A.A., Mohamed, H.K.: Automatic text summarization: a comprehensive survey. Expert Syst. Appl. **165**, 113679 (2021)
18. Nasar, Z., Jaffry, S.W., Malik, M.K.: Textual keyword extraction and summarization: state-of-the-art. Inf. Process. Manag. **56**(6), 102088 (2019)
19. Landauer, T.K., Foltz, P.W., Laham, D.: An introduction to latent semantic analysis. Discourse Process **25**(2–3), 259–284 (1998)
20. Landauer, T.K., Dutnais, S.T.: A solution to Plato's problem: the latent semantic analysis theory of acquisition, induction, and representation of knowledge. Psychol. Rev. **104**(2), 211–240 (1997)
21. Mihalcea, R.: Language independent extractive summarization. In: Proceedings of the ACL on Interactive Poster and Demonstration Sessions, vol. 5, pp. 49–52 (2005)
22. Mihalcea, R., Tarau, P.: TextRank: bringing order into texts. In: Proceedings of the ACL on Interactive Poster and Demonstration Sessions, pp. 404–411 (2004)

A Spatiotemporal Image Fusion Method for Predicting High-Resolution Satellite Images

Vipul Chhabra[1], R. Uday Kiran[2,3(✉)], Juan Xiao[4], P. Krishna Reddy[1],
and Ram Avtar[4]

[1] IIIT-Hyderabad, Hyderabad, Telangana, India
`vipul.chhabra@research.iiit.ac.in, pkreddy@iiit.ac.in`
[2] University of Tokyo, Tokyo, Japan
`uday_rage@tkl.iis.u-tokyo.ac.jp`
[3] University of AIZU, Fukushima, Japan
[4] Graduate School of Environmental Science, Hokkaido University,
Sapporo 060-0810, Japan
`xiao@eis.hokudai.ac.jp, ram@ees.hokudai.ac.jp`

Abstract. Given a coarse satellite image and a fine satellite image of a particular location taken at the same time, the high-resolution spatiotemporal image fusion technique involves understanding the spatial correlation between the pixels of both images and using it to generate a finer image for a given coarse (or test) image taken at a later time. This technique is extensively used for monitoring agricultural land cover, forest cover, etc. The two key issues in this technique are: (i) handling missing pixel data and (ii) improving the prediction accuracy of the fine image generated from the given test coarse image. This paper tackles these two issues by proposing an efficient method consisting of the following three basic steps: (i) imputation of missing pixels using neighborhood information, (ii) cross-scale matching to adjust both the Point Spread Functions Effect (PSF) and geo-registration errors between the course and high-resolution images, and (iii) error-based modulation, which uses pixel-based multiplicative factors and residuals to fix the error caused due to modulation of temporal changes. The experimental results on the real-world satellite imagery datasets demonstrate that the proposed model outperforms the state-of-art by accurately producing the high-resolution satellite images closer to the ground truth.

Keywords: Satellite images · Monitoring · Land cover · Image fusion

1 Introduction

A satellite image represents the data captured by an artificial satellite for a particular area at a timestamp. Government organizations and industries use the

---

V. Chhabra and R. Uday Kiran—Contributed 90% to the paper equally.

© Springer Nature Switzerland AG 2022
H. Fujita et al. (Eds.): IEA/AIE 2022, LNAI 13343, pp. 470–481, 2022.
https://doi.org/10.1007/978-3-031-08530-7_40

information generated from these images for efficient planning and monitoring purposes. Depending on the surface area covered by a pixel in an image (also called the spatial resolution of an image), we can classify the satellite imagery data into the following two types: (i) high-resolution images and (ii) coarse-resolution images. High-resolution means that pixel sizes are smaller, providing more detail. For example, a 1-m resolution satellite imagery can capture details on the ground greater than or equal to 1-m by 1-m. Anything on the ground that is less than 1-m will be blended with the surrounding area to make a 1-m by 1-m^2. Coarse-resolution means that pixel sizes are large, thus capturing only large features. For example, a 1-km resolution satellite imagery can capture details on the ground greater than or equal to 1-km by 1-km.

With increased automation demands in agriculture [1,3], intelligent systems are in need of frequent high-resolution images [2] to help efficient planning. However, high-resolution images are not abundantly available due high operational cost and low satellite revisit frequency [5,8,13]. Coarse-resolution images are highly available due to low operational cost and high satellite revisit frequency. However, they do not satisfy the data needs of intelligent systems. When confronted with this problem in real-world applications, researchers introduced a **high-resolution spatiotemporal image fusion approach** that provides a feasible alternative for generating high-resolution images by blending the information of the course-resolution image, which is widely available. This approach involves understanding the temporal correlation between the pixels in the coarse and finer images at a particular time and later building a finer image from a given coarse (or test) image taken at a later time.

Example 1. Figure 1(a) shows the coarse image of a location taken by a satellite at a particular timestamp, say t_0. Figure 1(b) shows the fine image of the same location taken by a different satellite almost at the same timestamp, say t_0. High-resolution spatiotemporal image fusion technique involves understanding the correlation between the pixels in Fig. 1(a) and (b), and later predicting the finer image from the coarse image taken for the same location at a different timestamp say t_1. The intelligent systems use the generated finer image for various decision making purposes, such as fertilizer application and harvesting.

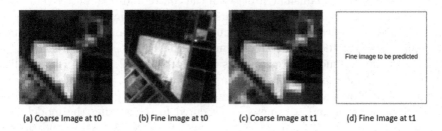

(a) Coarse Image at t0   (b) Fine Image at t0   (c) Coarse Image at t1   (d) Fine Image at t1

Fig. 1. Illustration of high resolution spatiotemporal image fusion technique

Numerous image-fusion techniques, such as HISTIF [7], STARFM [4] and ESTARFM [19], have been described in the literature. These techniques predict high-resolution satellite data from low-resolution data for a given timestamp based on the temporal and spatial features extracted from the set of pair of input images of the previous timestamp. The fusion techniques vary from the fusion level depending upon the use case. However, some existing image fusion methods are used explicitly to capture the land's phenological changes, such as the spatial and temporal adaptive reflectance model (STARFM) [4] and Enhanced STARFM (ESTARFM) [19]. These algorithms are entirely dependent on classification accuracy and are adversely affected by misclassification errors. The HISTIF algorithm also captures the phenological changes considering the fusion error and cross-scale spatial mismatch caused by the geo-registration error and point spread function (PSF) errors [18].

The above approaches suffer from the following two problems.

1. **(Geo-registration errors).** Since most of these fusion techniques operate at the pixel level [11], they are susceptible to geo-registration errors, which include cross-scale spatial errors and Point Spread Function (PSF) effect [18] of the sensors.
2. **(Non-uniformity in pixels).** The temporal changes imposed by the multiplicative modulation may not transfer the temporal aspects effectively, leading to sudden pixel changes between the adjacent pixels especially for heterogeneous landscapes.
3. **(Missing pixel data).** A common problem while dealing with the satellite data is missing pixels, which is completely ignored by the existing image fusion techniques. The temporal changes obtained from the training dataset cannot be applied directly to missing pixels.

Considering the limitations mentioned above, this study proposes a high-resolution spatiotemporal image fusion model. The aim is to predict the high-resolution satellite image from corresponding course-resolution satellite data by the fusion of features from the training data. The algorithm consists of three primary stages: estimation of missing pixels, filtering for cross-scale spatial matching, and modulation of the temporal changes. The missing pixels are fixed in both training and test data based on the neighboring pixel. The next step tries to reduce the spatial scale mismatch between the coarse and fine images. Further, it reconstructs the image from the coarse by accomodating temporal changes pixel by pixel at fine resolution. We evaluated its performance against the three different datasets, which reveal that it outperforms the existing approaches.

The remaining study is organized as follows. In Sect. 2, we describe the literature on spatiotemporal image fusion. Further in Sect. 3, we introduce our proposed model of spatiotemporal image fusion method. In Sect. 4, we describe the experimentation and parameters considered for evaluating the model and present the experimental results. Finally, in Sect. 6, we conclude this study along with future research directions.

2   Related Work

Several spatiotemporal fusion approaches have been discussed in the literature to cater various real world applications. Depending upon the level of operation, these fusion approaches can be broadly divided into three types [12]: (i) the pixel level, (ii) feature level, and (iii) decision level. The crops within the fields exhibit different growth due to variations in sowing, moisture, specimen, and other management measures. Since it is better to use the approaches with more specific use-case than the generic ones, we discuss the methods focusing primarily on the sub-field level changes. The most widely used methods for phenological changes are the spatial and temporal adaptive reflectance fusion model (STARFM) and Enhanced STARFM (ESTARFM). However, one primary problem with these methods is that they don't consider sub-field variability with particular attention, and the results are entirely dependent upon the classification accuracy. Similarly, there are multiple other STARFM-like methods [14–17] and some spatial unmixing-based methods [20,21] which depend upon the classification. Besides the problem of sub-field variability, they don't consider the fusion error because of the cross-scale spatial mismatch and point spread function (PSF) of sensors [18,20]. Since most of these methods operate at the pixel level, making them highly sensitive to the geo-registration error [11]. The algorithms that work at the feature level, such as the sparse-representation-based spatiotemporal reflectance fusion model (SPSTFM) [6], are less susceptible to geo-registration errors but focus more on shape than spatial details. A recently proposed method, HISTIF, considered the above limitations and resolved the issue of fusion errors because of cross-scale spatial mismatch and PSF and used Multiplicative modulation for Temporal Change (MMTC) for phenological changes. However, the study didn't consider the bais because of fusion error by Multiplicate Modulation, which adversely affects the results for heterogeneous landscapes. Most of the algorithms failed to consider the practical issue of missing pixels in satellite data. In contrast, The proposed model overcomes all the limitations mentioned above and demonstrates higher accuracy in prediction than HISTIF.

3   Proposed Model

3.1   Preprocessing

The proposed models require one image pair at the reference time t_0 and one coarse image at time t_1 as the input for predicting the fine image at time t_1. Before using the proposed model, we need to ensure that all input images are calibrated to the surface reflectance for monitoring temporal changes at the sub-field level. They also need to be reprojected to the same coordinate system. To preserve the spectral integrity of the original images, we may use the nearest neighbor interpolation to resample the coarse images to fine resolution. After the initial processing of the data, Let $C(b_n, t_0)$, $F(b_n, t_0)$, $C(b_n, t_1)$ be the coarse image at time t_0, fine image at time t_0, coarse image at time t_1 where (n = 1, 2, 3, 4) and bands are blue, green, red, and NIR, respectively. The model contains three basic steps: EMP, FCSM, and MTC.

3.2   Estimation of Missing Pixels (EMP)

The satellite data is prone to missing pixels because of the sudden variation in atmospheric conditions or electronic failure. The missing pixels in input data can lead to an increase in geospatial error. Also, if there are missing pixels in any band of $C(t_1, b_n)$, it will not be able to modulate temporal aspects on those pixels leading to distortion in the predicted image. Therefore, we proposed the EMP to estimate the missing pixels, enabling it to reduce the geo-registration and provide distortion-free predictions (Fig. 2).

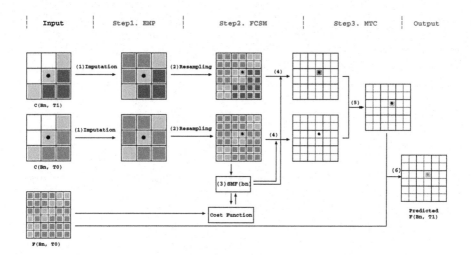

Fig. 2. Illustration of the overall workflow

The pixels in images possess some relation with the neighboring pixels, so it can be estimated that the missing pixel would be the weighted average of the nearest available pixels. Since the pixels could be missing from the multiple positions in the image, so to efficiently query the closest available pixels, we use the KD-Tree to store all available pixels with their locations. The number of pixels taken into consideration is a user-defined parameter, and the pixel value is calculated using the formula

$$\sum_{i=1}^{n} w_i x_i \quad where \sum_{i=1}^{n} w_i = 1 \tag{1}$$

where w_i is the weight assigned to the neighboring pixel, which is proportional to the distance from the missing pixel, and x_i is the neighboring pixel value.

3.3   Filtering for Cross-scale Spatial Mismatch (FCSM)

Since the traditional filtering is based on the assumption of regular within-class change occurring from t_0 to t_1 and is susceptible to classification accuracy [9] and the geo-registration error. Therefore, we use FCSM to overcome traditional filtering limitations and account for the PSF effect of the sensor. Since the sensor, PSF could comprise multiple components. We consider the two critical components (optical and detector PSF), which can be modeled using the 2-D gaussian described as follows [10]

$$PSF(x, y) = \frac{G(x) \cdot G(y)}{\sum G(x) \cdot G(y)} \tag{2}$$

where the gaussian functions used are

$$G(x) = e^{\frac{-X^2}{2\sigma_x^2}} \ and \ G(y) = e^{\frac{-Y^2}{2\sigma_y^2}} \tag{3}$$

Since there could be a difference in the observation angle by the sensors hence for the reduction in the error, the filter can be built by multiplying the PSF by rotation matrix as follows

$$SMF(FWHM_x, FWHM_y, \theta) = PSF \cdot \begin{bmatrix} \cos\theta & \sin\theta & 0 \\ -\sin\theta & \cos\theta & 0 \\ 0 & 0 & 1 \end{bmatrix} \tag{4}$$

To reduce the geo-registration error, the SMF is convolved with the coarse image as follows

$$\hat{C}_{SMF}(b_n, t_0) = C(b_n, t_0) \quad \otimes SMF(b_n) \tag{5}$$

The convolved version could be shifted in both East-West and North-South dimensions as measured by $shift_x$ and $shift_y$, resulting in the removal of the blocky artifacts in $C(b_n, t_0)$ and reduction in a cross-scale spatial mismatch between the image pairs. Since the size, rotation, and shift are the three sets of parameters useful in determining the SMF hence for automating the error optimization process due to the cross-scale spatial mismatch and the removal of blocky artifacts, we use root mean square error (RMSE) as the cost function.

$$RMSE = \sqrt{\frac{\sum(\hat{C}_{SMF}(b_n, t_0) - F(b_n, t_0))^2}{N}} \tag{6}$$

We use particle swarm optimization (PSO) for the optimization of the five parameters used in the equation. The primary reason for choosing particle swarm optimization is its computational efficiency, and also it doesn't require the optimization problem to be differentiable. Apart from its primary advantage, it is simple and easy to implement compared to other optimization algorithms.

In cases where the PSF effect and geometry are constant, mainly when the same sensors are used for the observation at t_0 and t_1. The coarse image at time t_1 can be convolved with the same SMF generated from the image pair at time t_0 to produce the filtered image at time t_1

$$\hat{C}_{SMF}(b_n, t_1) = C(b_n, t_1) \quad \otimes SMF(b_n) \tag{7}$$

where $\hat{C}_{SMF}(b_n, t_1)$ is the filtered coarse image in band n at time t_1.

3.4   Modulation of Temporal Changes (MTC)

To accommodate the temporal changes which occurred at the sub-field level with the time, we quantify the changes at the pixel level rather than per class. The pixel-level changes which occurred in the coarse image at t_0 to the coarse image at t_1 can be represented using the two factors.

$$MMF(b_n) = \frac{\hat{C}_{SMF}(b_n, t_1)}{\hat{C}_{SMF}(b_n, t_0)} \tag{8}$$

$$\xi(b_n) = F(b_n, t_0) - \frac{\hat{C}'_{SMF}(b_n, t_0)}{\hat{C}_{SMF}(b_n, t_0)} \times F(b_n, t_0) \tag{9}$$

where the $\hat{C}_{SMF}(b_n, t_0)$ is the coarse image at t_0 convoluted with the different SMF.

The primary reason for representing it as two factors is only using multiplicative modulation leads to distortions in the adjacent pixels, especially for the heterogeneous landscapes. Hence to reduce the distortion and remove the errors in multiplicative modulation, an additional term for the bias is added at the pixel level. This way, the temporal changes between the coarse image at two timestamps could be efficiently transferred in predicting the fine image at t_1. So after combining all the intermediate equations, the fine image at t_1 can be predicted from the coarse image at t_1 as follows

$$F(b_n, t_1) = \frac{\hat{C}(b_n, t_1)}{\hat{C}(b_n, t_0)} \times F(b_n, t_0) + \xi(b_n) \tag{10}$$

4   Experimentation and Results

The experimentation of the proposed model was carried out using the real dataset to address the complicated factors and feasibility of its usage in real-life application compared to the existing approach HISTIF.

The dataset included the fine images from the Planet dataset, which were captured at 3m resolution. In contrast, the coarse images were obtained from the Landsat dataset, captured at 30m resolution. The fine images were obtained on 2021/07/28, and 2021/08/07 and coarse images were obtained on 2021/07/28 and 2021/08/06 for t_0 and t_1, respectively.

For the statistical assessment of the band wise predictions with the ground truth, we used correlation coefficient (CC), Root Mean Square Error (RMSE), and the mean absolute difference (MAD)

$$CC = \frac{\sum(x - \bar{x}) \cdot (y - \bar{y})}{\sqrt{\sum(x - \bar{x})^2 \cdot \sum(y - \bar{y})^2}} \tag{11}$$

$$RMSE = \sqrt{\frac{\sum(x - y)^2}{n}} \tag{12}$$

$$MAD = \frac{\sum|x - y|}{n} \tag{13}$$

4.1   Comparison with HISTIF

For comparison, we extracted three different patches containing different levels of changes from the dataset named D1, D2, and D3 with comparison results in Figs. 3, 4 and 5 respectively.

In comparison with HISTIF, we can see using the statistical assessment that the proposed model could reduce the errors and possess a higher correlation coefficient for the given set of hyper-parameters for most cases. Upon the visual inspection, we can see that the pixels are more uniform and capture the changes more accurately than the original version (Tables 1, 2 and 3).

Table 1. Statistical evaluation using Correlation Coefficient (CC)

Band-wise evaluation using the Correlation Coefficient (CC)			
Dataset	Band	HISTIF	IHISTIF
D1	Blue	0.870	**0.891**
	Green	0.805	**0.857**
	Red	0.863	**0.903**
	NIR	0.600	**0.647**
D2	Blue	0.962	**0.967**
	Green	0.947	**0.961**
	Red	0.956	**0.981**
	NIR	0.882	**0.936**
D3	Blue	0.912	**0.927**
	Green	**0.919**	0.901
	Red	0.926	**0.946**
	NIR	0.824	**0.827**

Table 2. Statistical evaluation using Root Means Square Error (RMSE)

Band-wise evaluation using the Root Mean Square Error (RMSE)			
Dataset	Band	HISTIF	IHISTIF
D1	Blue	208.003	**175.630**
	Green	270.352	**203.504**
	Red	447.369	**343.739**
	NIR	1149.977	**961.840**
D2	Blue	102.525	**97.509**
	Green	203.591	**202.690**
	Red	236.574	**215.020**
	NIR	409.106	**279.090**
D3	Blue	101.568	**88.400**
	Green	**197.163**	198.68
	Red	254.627	**251.000**
	NIR	614.455	**610.438**

Table 3. Statistical evaluation using Mean Absolute Difference (MAD)

Band-wise evaluation using the Mean Absolute Difference (MAD)			
Dataset	Band	HISTIF	IHISTIF
D1	Blue	135.221	**118.100**
	Green	187.475	**153.112**
	Red	262.954	**222.254**
	NIR	806.7187	**672.840**
D2	Blue	84.893	**81.044**
	Green	**187.003**	189.321
	Red	209.562	**197.730**
	NIR	300.911	**213.629**
D3	Blue	74.115	**68.325**
	Green	178.129	**177.296**
	Red	210.360	**208.657**
	NIR	352.776	**347.557**

The dataset D1 comprises a few abrupt changes and a few sub-field level changes with time. The proposed model captured both the changes more accurately than the HISTIF.

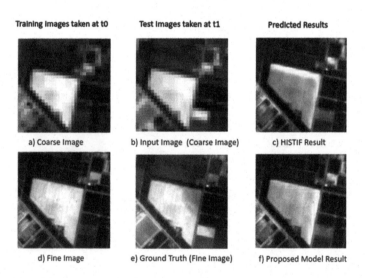

Fig. 3. Comparison of the results generated by HISTIF and proposed model against ground truth for D1 dataset

While the dataset D2 comprises only sub-field level changes with time. The proposed model predicted most of the changes accurately and showed higher correlation and lesser error than the HISTIF w.r.t to ground truth.

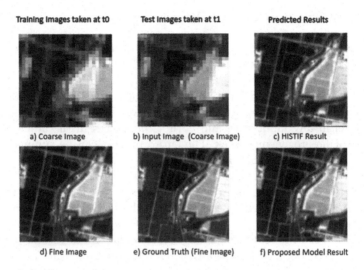

Fig. 4. Comparison of the results generated by HISTIF and proposed model against ground truth for D2 dataset

The dataset D3 comprises majorly abrupt changes and minor sub-field level changes. The proposed model could accurately predict sub-field changes while

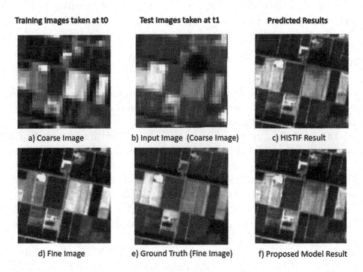

Fig. 5. Comparison of the results generated by HISTIF and proposed model against ground truth for D3 dataset

the abrupt changes were partially transmitted to the predicted image because the model is suited primarily for sub-field level changes. Hence, the prediction of the proposed model had a higher correlation coefficient than HISTIF.

5   Conclusion and Future Work

For improving the monitoring at the sub-field level, this article presents an improved variant of the spatiotemporal fusion algorithm HISTIF, which predicts the fine resolution satellite data from the coarse resolution satellite data at time t_1, given the pair of fine and its corresponding coarse image at time t_0 by the fusion of temporal features. It addresses two challenging issues: the missing pixel in any input data and abrupt differences in the adjacent pixels, especially for heterogeneous areas. The entire process can be summarized into three basic steps EMP, FCSM, and MTC. It was tested against the real dataset obtained from Planet and Landsat dataset, considering HISTIF as the baseline. The missing pixels estimation using the weighted average helped in improving the predictions. The bias consideration at pixel level provided more accurate predictions with homogeneity. The visual assessment demonstrated that the colors were more accurate and were homogenous similar to the ground truth. This method has great potential, especially for applications requiring monitoring at the sub-field level with low resources such as crop monitoring, precision agriculture sector, etc., which could predict with high stability and great computational efficiency.

As part of future work, we would like to extend the current model by using different ways of applying the Modulation Factor (MF). For example, Kwan et al. [8] observed that the prediction could be more accurate if we divided the image into multiple patches. Also, the proposed model assumes that the temporal changes on different scales don't have much significant difference. So further work can be done to remove this assumption to a certain extent.

References

1. Cai, Y., et al.: A high-performance and in-season classification system of field-level crop types using time-series Landsat data and a machine learning approach. Remote Sens. Environ. (2018)
2. Campos-Taberner, M., et al.: Multitemporal and multiresolution leaf area index retrieval for operational local rice crop monitoring. Remote Sens. Environ. 102–118 (2016)
3. Gao, F., et al.: Toward mapping crop progress at field scales through fusion of Landsat and MODIS imagery. Remote Sens. Environ. 9–25 (2017)
4. Gao, F., Masek, J., Schwaller, M., Hall, F.: On the blending of the Landsat and MODIS surface reflectance: predicting daily Landsat surface reflectance. IEEE Trans. Geosci. Remote Sens. 2207–2218 (2006)
5. Gevaert, C., Garcia-Haro, F.: A comparison of STARFM and an unmixing-based algorithm for Landsat and MODIS data fusion. Remote Sens. Environ. 34–44 (2015)

6. Huang, B.: Spatiotemporal reflectance fusion via sparse representation. IEEE Trans. Geosci. Remote Sens. 3707–3716 (10 2012)
7. Jiang, J., et al.: HISTIF: a new spatiotemporal image fusion method for high-resolution monitoring of crops at the subfield level. IEEE J. Sel. Top. Appl. Earth Observ. Remote Sens. 4607–4626 (2020)
8. Kwan, C., Budavari, B., Gao, F., Zhu, X.: A hybrid color mapping approach to fusing MODIS and Landsat images for forward prediction. Remote Sens. (2018)
9. Liu, M., Liu, X., Wu, L., Zou, X., Jiang, T., Zhao, B.: A modified spatiotemporal fusion algorithm using phenological information for predicting reflectance of paddy rice in Southern China. Remote Sens. 772 (2018)
10. Mira, M., et al.: The MODIS (collection V006) BRDF/albedo product MCD43D: temporal course evaluated over agricultural landscape. Remote Sens. Environ. 216–228 (2015)
11. Pohl, C., Genderen, J.L.V.: Review article multisensor image fusion in remote sensing: concepts, methods and applications. Int. J. Remote Sens. 823–854 (1998)
12. Pohl, C., Van Genderen, J.L.: Review article multisensor image fusion in remote sensing: concepts, methods and applications. Int. J. Remote Sens. 823–854 (1998)
13. Song, H., Huang, B.: Spatiotemporal satellite image fusion through one-pair image learning. IEEE Trans. Geosci. Remote Sens. 1883–1896 (2013)
14. Wang, P., Gao, F., Masek, J.: Operational data fusion framework for building frequent Landsat-like imagery. IEEE Trans. Geosci. Remote Sens. 7353–7365 (2014)
15. Wang, Q., Zhang, Y., Onojeghuo, A., Zhu, X., Atkinson, P.: Enhancing spatiotemporal fusion of MODIS and Landsat data by incorporating 250 m MODIS data. IEEE J. Sel. Top. in Appl. Earth Observ. Remote Sens. 4116–4123 (2017)
16. Weng, Q., Fu, P., Gao, F.: Generating daily land surface temperature at Landsat resolution by fusing Landsat and MODIS data. Remote Sens. Environ. 55–67 (2014)
17. Wu, B., Huang, B., Cao, K., Zhuo, G.: Improving spatiotemporal reflectance fusion using image inpainting and steering kernel regression techniques. Int. J. Remote Sens. (2017)
18. Zhu, X., Cai, F., Tian, J., Williams, T.K.A.: Spatiotemporal fusion of multisource remote sensing data: literature survey, taxonomy, principles, applications, and future directions. Remote Sens. (2018)
19. Zhu, X., Chen, J., Gao, F., Chen, X., Masek, J.: An enhanced spatial and temporal adaptive reflectance fusion model for complex heterogeneous regions. Remote Sens. Environ. 2610–2623 (2010)
20. Zhukov, B., Oertel, D., Lanzl, F., Reinhackel, G.: Unmixing-based multisensor multiresolution image fusion. IEEE Trans. Geosci. Remote Sens. 1212–1226 (1999)
21. Zurita-Milla, R., Clevers, J.G.P.W., Schaepman, M.E.: Unmixing-based Landsat tm and MERIS FR data fusion. IEEE Geosci. Remote Sens. Lett. 453–457 (2008)

Security

WHTE: Weighted Hoeffding Tree Ensemble for Network Attack Detection at Fog-IoMT

Shilan S. Hameed[1,2] (ID), Ali Selamat[1,3,4,5]([✉]), Liza Abdul Latiff[6], Shukor A. Razak[3], and Ondrej Krejcar[1,5] (ID)

[1] Malaysia-Japan International Institute of Technology (MJIIT), University Teknologi Malaysia, 54100 Kuala Lumpur, Malaysia
hameed.s@graduate.utm.my, aselamat@utm.my
[2] Directorate of Information Technology, Koya University, 44023 Koya, Iraq
[3] School of Computing, Faculty of Engineering, Universiti Teknologi Malaysia, Skudai, Johor Bahru 81310, Johor, Malaysia
shukorar@utm.my
[4] Media and Games Center of Excellence (MagicX), Universiti Teknologi Malaysia, Skudai 81310, Johor Bahru, Malaysia
[5] Center for Basic and Applied Research, Faculty of Informatics and Management, University of Hradec Kralove, Rokitanskeho 62, Hradec Kralove 50003, Czech Republic
ondrej.krejcar@uhk.cz
[6] Razak Faculty of Technology and Informatics, Universiti Teknologi Malaysia, 54100 Kuala Lumpur, Malaysia
liza.kl@utm.my

Abstract. The fog-based attack detection systems can surpass cloud-based detection models due to their fast response and closeness to IoT devices. However, current fog-based detection systems are not lightweight to be compatible with ever-increasing IoMT big data and fog devices. To this end, a lightweight fog-based attack detection system is proposed in this study. Initially, a fog-based architecture is proposed for an IoMT system. Then the detection system is proposed which uses incremental ensemble learning, namely Weighted Hoeffding Tree Ensemble (WHTE), to detect multiple attacks in the network traffic of industrial IoMT system. The proposed model is compared to six incremental learning classifiers. Results of binary and multi-class classifications showed that the proposed system is lightweight enough to be used for the edge and fog devices in the IoMT system. The ensemble WHTE took trade-off between high accuracy and low complexity while maintained a high accuracy, low CPU time, and low memory usage.

Keywords: Intrusion detection · Machine learning · Incremental ensemble classifier. Fog-computing · Attack detection

1 Introduction

The Internet of Things (IoT) is a rapidly evolving technology that uses networking to connect infrastructure, digital devices, physical objects, applications, and persons [1].

© Springer Nature Switzerland AG 2022
H. Fujita et al. (Eds.): IEA/AIE 2022, LNAI 13343, pp. 485–496, 2022.
https://doi.org/10.1007/978-3-031-08530-7_41

The Internet of Medical Things (IoMT) is a use of the Internet of Things (IoT) in the health care sector [2, 3]. It is undeniable that smart medical gadgets have made life simpler and healthier for many people. However, security and privacy issues in the IoMT system remain unsolved issue [4, 5]. Hence, cyber-attack detection systems are considered as a defensive layer for the IoMT devices and networks. Machine learning and deep learning have been employed for intrusion and cyber-attack detection for the IoMT system. Solutions include on gadgets embedded models to cloud based systems. However, the chips and gadgets are not much efficient to hold the models and the IoMT network data. Additionally, cloud-based systems are centralized, and their detection is associated with delay. Hence, new approaches of network cyber-attack detection is required to overcome those limitations.

Fog computing is a novel concept that was developed to address the cloud's latency, centralization, and privacy problems [6]. Some cloud computing responsibilities will be moved closer to the smart devices in fog oriented IoT [7]. Moreover, a fog node might serve as the initial defense line for small devices that lack security features [8]. Fog-based attack detection is not widely used, especially for the IoMT system. Few studies have proposed a fog-based detection system. The authors in [9] presented a distributed Intrusion Detection System (IDS) which works based on fog-computing principle. Their system is designed for smart medical system that uses an online method specifically sequential extreme learning machine (EOS-ELM). They demonstrated that their proposed system is superior to cloud-based systems regarding detection time and true positive rate. In another study [10], the authors have used an ensemble learning for binary network cyber-attack detection using ensemble of (Decision Tree, Naive Bayes, and Random Forest) and XGBoost classifiers in the IoMT system following fog-cloud architecture. Because their system is too heavy for fog devices, they recommend using cloud computing for training and fog computing for testing. Another study [11] employs an ensemble incremental learning technique for fog devices for network intrusion detection in medical IoT networks. However, the dataset utilized isn't a recent IoT. Then, based on the current research gaps, a fog-based attack detection system is proposed using incremental ensemble learning for the IoMT system. The proposed system has two-folds advantages; firstly, the cyber-attacks will be detected accurately and soon they appear; secondly, the system is lightweight and does not use many resources.

2 Methodology

2.1 Datasets

We used two datasets of NSL-KDD and ToN-IoT. The NSL-KDD is a well-known bench making dataset, which was originally developed for conventional network and used by many researchers [12]. Hence, we have included this for comparison purpose. The dataset has 41 features and total of 148,517 samples in both train and test samples. The dataset contains Normal, Denial of Service (DoS), Probe, User to Root (u2r), and Root to Local (r2l)) samples. The detail of each class count is shown in Fig. 1. The second dataset is a new cyber-attack dataset which was developed for IoT and IIoT systems. The dataset was built in a real-world IoT network context, using seven different sensors and telemetry services. As a result, the dataset exemplifies the IoT system's diversity. The dataset is the

IoT system's network traffic, converted into NetFlow files [13]. NetFlow format is lighter than payload data as it only uses metadata instead of the packet contents. Additionally, since it does not use the packets holding the patient's data, it does not violate the privacy rules, making the approach more compatible with an IoMT system. This version of data was curated and most informative records and features are selected to achieve a high performance [13]. The number of data records is 1,379,274, while the feature count is 13. There are multiple attacks available in the dataset as their sample counts are shown in Fig. 1.

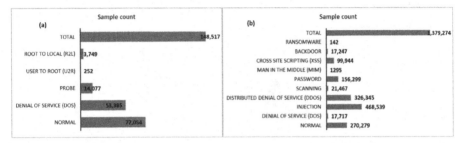

Fig. 1. The sample count of the attacks and normal class in the (a) NSL KDD and (b) ToN-IoT_NetFlow datasets.

2.2 A Lightweight Network Attack Detection System

The proposed system in this study follows the guidelines of fog-computing architecture by IEEE [8]. Figure 2 illustrates the proposed system.

The absent security measures are shown by the red alert symbol. As a result of security absence, the medical devices and their network communication at edge-fog layers are exposed to various attacks. The amount of network data arriving through fog devices will grow with time, resulting in massive data, but fog devices are inefficient at storing it. As a result, training the data in stages would be preferable to retrain the whole data every time they aggregate. We have used a sliding window setting to train the classifiers incrementally. Unlike batch learning which uses cross-validations and hold-out, in our online learning a prequential evaluation was utilized, which uses the samples incrementally to train and test each sample record at a time. In this experiment, the maximum memory was set to 5 thousand samples, while the sliding window was set to 1000 samples at a time.

Compared Incremental Classifiers. In this study, a collection of single incremental classifiers and Bagging Hoeffding Tree ensemble was utilized to be compared to the proposed WHTE ensemble model. Each of them was deployed with their best tuned parameters using the same experimental environment. The following list is the utilized single classifiers:

- Incremental K-Nearest Neighbor (IKNN) [6].

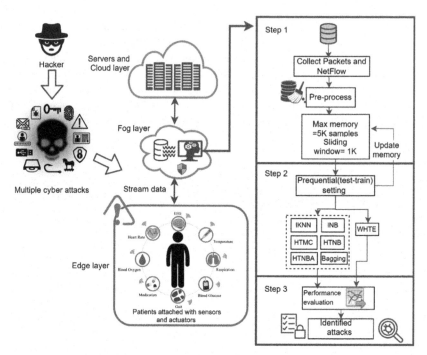

Fig. 2. The proposed fog-based network attack detection system and its architecture.

- Incremental Naïve Bayes (INB) [14].
- Hoeffding Tree-based Majority Class (HTMC) [15].
- Hoeffding Tree Naïve Bayes (HTNB) [15].
- Hoeffding Tree Naïve Bayes Adaptive (HTNBA) [16].
- Bagging Hoeffding Tree [17]

Weighted Majority Hoeffding Tree Ensemble (WHTE). The previously mentioned single classifiers may not produce high performance when the data is heterogeneous such as the IoT data. Hence, the ensemble of the single classifiers could maximize their performance and minimize their weakness. As a result, we propose an ensemble strategy in which a collection of single classifiers, particularly distinct types of Hoeffding Tree classifiers, are combined (HTMC, HTNB, HTNBA). Figure 3 depicts a summary of our ensemble technique flowchart. The ensemble is called Weighted Hoeffding Tree Ensemble (WHTE), which uses a weighted majority approach. It considers all of the classifiers' decisions equally at the beginning [18]. It will, however, penalize a classifier if they make a wrong decision by not treating their decisions as significant as they formerly were [19]. The overall performance of the ensemble is the maximum because the errors created by the entire algorithm will essentially be the same as a constant error made by the best approach. When the expert makes a mistake in the initial weighted majority algorithm, the weighted value is doubled by ½. As a result, the error bound equation is as follows:

$$M \leq 2.41(m + \log N) \tag{1}$$

where, m is the total of mistakes of the best classifier, M is the total of mistakes of the ensemble, and N is the total number of single classifiers. A randomized form of weighted majority algorithm can be used to reduce the error value to a minimum, which reduces the error equation's constant value to close to one by adding (Beta β) to the equation. Hence, for the WHTE ensemble, the error equation can be defined as follow:

$$M \leq \frac{m\ In\ (1/\beta) + In\ N}{1 - \beta} \tag{2}$$

The value of β is set to be 0.5. Hence, the value of M for each iteration or a sample at a time will be counted as follow:

$$M \leq 1.39m + 2In\ N \tag{3}$$

Performance Metrics. Multiple metrics are used to evaluate the proposed method in the current study. The detail of each metric is given in Table 1.

Table 1. The utilized evaluation metrics for evaluating the proposed method.

Metric	Discretion
Average accuracy	It is the average of all the sliding windows' accuracies. Its equation is given below Average accuracy $= \frac{\sum_i acc}{N}$ .... (4) While acc is the accuracy per each i sliding window over N total of the sliding windows
Average time (s)	The cumulative learning method's average CPU time for all sliding windows
Average memory (MiB)	It is the average memory usage taken by each method for all datasets while considering the device's memory

3   Results and Discussion

First, the proposed methods were evaluated on NSL-KDD dataset for the purpose of comparison with literature. As shown in Table 2, the WHTE ensemble outperformed the other single and ensemble classifiers with a high accuracy of 98.0%. Also, it is an ensemble method which recorded lower memory usage and CPU time compared to the Bagging ensemble.

After that the proposed model was evaluated on the ToN-IoT_NetFlow dataset using binary and multi-class classification. In the binary classification, the results were much better than multi-class classification, as expected. This is because multi-classification of 10 (refer to Fig. 1) classes in incremental fashion reduces the accuracy. Table 3 shows

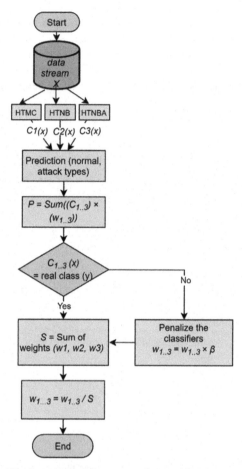

Fig. 3. The flowchart of the proposed ensemble WHTE method.

Table 2. An average performance of the WHTE classifier compared to the other single and ensemble classifiers for the NSL-KDD dataset.

Method	Average accuracy (%)	Average time (s)	Average memory (MiB)
IKNN	96.94	61.88	2.46
INB	87.83	1.83	0.03
HTMC	94.12	3.94	1.71
HTNB	97.0	3.29	1.71
HTNBA	97.50	3.19	1.71
Bagging	95.0	20.97	16.40
WHTE	98.0	8.94	5.19

that the model's average accuracy for the ensemble WHTE was 100%. In addition, Bagging had 99.40% average accuracy and took the second place. The average memory use for the WHTE technique was 0.37 MiB and Bagging recorded the highest of 8.63 MiB while the HTNBA, HTNB, and HTMC methods used 0.08 MiB each. The average CPU time required to identify all intrusions was just 12.89 s for the WHTE technique, while Bagging needed 77.49 s. The IKNN approach, on the other hand, has highest complexity. In the multiclass classification, WHTE again took tradeoff between accuracy and complexity. Table 3 shows that the proposed ensemble had higher accuracy than single classifiers. Although its accuracy was slightly better than Bagging, its time and memory complexity were much lower. This is what we need for the lightweight devices.

Table 3. An average performance of the WHTE classifier compared to the other single and ensemble classifiers for ToNIoT-Netflow dataset using binary and multiclass classification

Method	Average accuracy (%)	Average time (s)	Average memory (MiB)	Average accuracy (%)	Average time (s)	Average memory (MiB)
Binary classification				Multiclass classification		
IKNN	98.79	184.69	1.15	70.50	255.79	1.25
INB	97.62	8.47	0.22	60.03	15.08	0.24
HTMC	99.01	3.90	0.08	70.82	24.92	9.85
HTNB	98.94	5.75	0.08	69.16	26.84	9.85
HTNBA	99.01	5.02	0.08	70.07	27.82	9.87
Bagging	99.40	77.49	8.63	71.08	299.89	86.52
WHTE	100.00	12.89	0.37	72.01	115.85	29.78

For the rest of the analysis, we have chosen the results of binary classification due to avoiding multiple and duplicate figures. To see the effect of concept drift on each classifier, we ha each technique's incremental accuracy per five thousand records is conceptualized. From Fig. 4, it can be observed that the INB classifier was sensitive to the concept drift, and it had instability in its accuracy. Comparably, the rest of the classifiers looked much more stable due to the figure's high variance in INB accuracy.

Hence, to see the other classifiers' performance, INB is removed from the illustration presented in Fig. 5. It was seen that the HTNB and IKNN were more sensitive to the changes in the data, and their accuracy was constantly changing. Notably, the WHTE classifier showed a stable accuracy of 100% for each frequent sample. Moreover, Bagging, HTMC and HTNBA performed better instability than the rest of the classifiers.

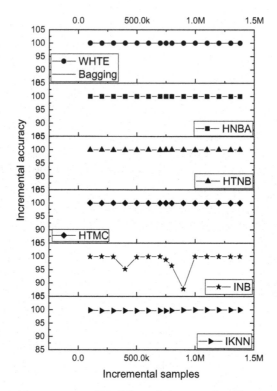

Fig. 4. The incremental accuracy per 5K sliding window samples. The accuracy was averaged for every 100K samples for clear visualization.

In terms of total CPU time per each sample frequencies, a 3D waterfall color surface was drawn, as demonstrated in Fig. 6. It is obvious that IKNN's and Bagging's CPU time were significantly impacted by the rising arrived samples to the system, in which the surface color rises from red to dark blue. Though, for other classifiers the CPU time was risen linearly with the increased samples.

A comparison has been made between the current work and related studies, as shown in Table 4. The proposed system outperformed the previous studies. Additionally, the current system is lightweight, and the system's complexity is comprehensively analyzed, while previous studies were not lightweight nor considered these metrics for their evaluation.

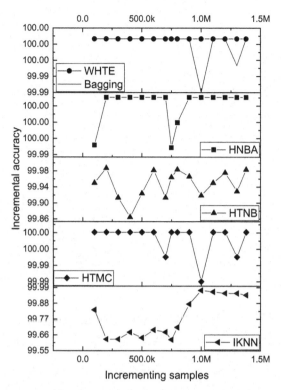

Fig. 5. Except for INB, the incremental accuracy of the utilized classifiers per 5K sliding window samples. The accuracy was averaged for every 100K samples for clear visualization.

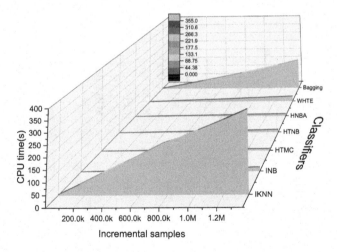

Fig. 6. The incremental CPU time per each subset of data samples for each classifier represented by a 3D waterfall colormap surface

Table 4. A comparison between the proposed system and related fog-based attack detection systems.

Ref	Architecture	Lightweight	Device specs	Best testing accuracy (%)	Type/Name of dataset	Type of learning	Complexity metrics	Splitting method
[10]	Cloud-Fog	No	CPU 2.20 GHz (10 cores, 13.75 MB L3 Cache), and 128 GB RAM	96.35	Network packet/ToNIoT	Batch	Not considered	Holdout Train-test (80:20)
[9]	Fog	No	Intel core i7 CPU processor and 16 GB RAM	98.19	Network packet/NSL-KDD	Batch	Not considered	Holdout Train-test (80:20)
[11]	Fog	No	CPUs ≈2.2GHz and 8GB RAM	98.95	Network packet/NSL-KDD	Incremental batch	Not considered	Cross-validation
This work	Edge-Fog	Yes	CPUs ≈2.2GHz (4 cores, 3 MB L3 Cache), and 8GB RAM	100.00	Network packet and NetFlow/ NSL-KDD and ToNIoT	Incremental	Considered	Prequential sliding window evaluation

4   Conclusions

In this study, a lightweight network attack detection was proposed for the fog devices of the IoMT system. For this purpose, we have proposed a fog-based architecture and an incremental ensemble called WHTE which its performance was compared to another six incremental learning methods. It was seen that the system detects attacks with high accuracy of 100.0. In addition to that, the model is considered lightweight as it uses less low memory and CPU time. As a result, the proposed approach surpassed the earlier conducted solutions.

Acknowledgments. The authors sincerely thank Universiti Teknologi Malaysia (UTM) under Malaysia Research University Network (MRUN) Vot 4L876, for the completion of the research. This work was also partially supported/funded by the Ministry of Higher Education under the Fundamental Research Grant Scheme (FRGS/1/2018/ICT04/UTM/01/1) and Universiti Tenaga Nasional (UNITEN). The work and the contribution were also supported by the SPEV project "Smart Solutions in Ubiquitous Computing Environments", University of Hradec Kralove, Faculty of Informatics and Management, Czech Republic (under ID: UHK-FIM-SPEV-2022–2102). We are also grateful for the support of student Michal Dobrovolny in consultations regarding application aspects.

References

1. Farahani, B., Firouzi, F., Chang, V., Badaroglu, M., Constant, N., Mankodiya, K.: Towards fog-driven IoT eHealth: promises and challenges of IoT in medicine and healthcare. Futur. Gener. Comput. Syst. **78**, 659–676 (2018)
2. Alsubaei, F., Abuhussein, A., Shiva, S.: A framework for ranking IoMT solutions based on measuring security and privacy. In: Arai, K., Bhatia, R., Kapoor, S. (eds.) FTC 2018. AISC, vol. 880, pp. 205–224. Springer, Cham (2019). https://doi.org/10.1007/978-3-030-02686-8_17
3. He, D., Ye, R., Chan, S., Guizani, M., Xu, Y.: Privacy in the Internet of Things for smart healthcare. IEEE Commun. Mag. **56**(4), 38–44 (2018)
4. Yaacoub, J.-P.A., et al.: Securing internet of medical things systems: limitations, issues and recommendations. Futur. Gener. Comput. Syst. **105**, 581–606 (2020). https://doi.org/10.1016/j.future.2019.12.02812
5. Rathore, H., Al-Ali, A.K., Mohamed, A., Du, X., Guizani, M.: A novel deep learning strategy for classifying different attack patterns for deep brain implants. IEEE Access **7**, 24154–24164 (2019)
6. Hameed, S.S., et al.: A hybrid lightweight system for early attack detection in the IoMT fog. Sensors **21**(24), 8289 (2021)
7. Cisco, C.: Fog computing and the Internet of Things: extend the cloud to where the things are. Электронный ресурс]. https://www.cisco.com/c/dam/en_us/solutions/trends/iot/docs/computing-overview.pdf(дата обращения: 10.03. 2019) (2015)
8. Group, O.C.A.W.: OpenFog reference architecture for fog computing. OPFRA001 **20817**, 162 (2017)
9. Alrashdi, I., Alqazzaz, A., Alharthi, R., Aloufi, E., Zohdy, M.A., Ming, H.: FBAD: fog-based attack detection for IoT healthcare in smart cities. In: 2019 IEEE 10th Annual Ubiquitous Computing, Electronics & Mobile Communication Conference (UEMCON). IEEE, pp. 0515–0522 (2019)

10. Kumar, P., Gupta, G.P., Tripathi, R.: An ensemble learning and fog-cloud architecture-driven cyber-attack detection framework for IoMT networks. Comput. Commun. **166**, 110–124 (2021). https://doi.org/10.1016/j.comcom.2020.12.003

11. Hameed, S.S., Hassan, W.H., Latiff, L.A.: An efficient fog-based attack detection using ensemble of MOA-WMA for Internet of Medical Things. In: Saeed, F., Mohammed, F., Al-Nahari, A. (eds.) IRICT 2020. LNDECT, vol. 72, pp. 774–785. Springer, Cham (2021). https://doi.org/10.1007/978-3-030-70713-2_70

12. Tavallaee, M., Bagheri, E., Lu, W., Ghorbani, A.A.: A detailed analysis of the KDD CUP 99 data set. In: 2009 IEEE symposium on computational intelligence for security and defense applications, pp. 1–6. IEEE (2009)

13. Sarhan, M., Layeghy, S., Moustafa, N., Portmann, M.: Netflow datasets for machine learning-based network intrusion detection systems. arXiv preprint arXiv:2011.09144 (2020)

14. Alaei, P., Noorbehbahani, F.: Incremental anomaly-based intrusion detection system using limited labeled data. In: 2017 3th International Conference on Web Research (ICWR), pp. 178–184. IEEE (2017)

15. Gama, J., Medas, P., Rodrigues, P.: Learning decision trees from dynamic data streams. In: Proceedings of the 2005 ACM Symposium on Applied computing, pp. 573–577 (2005)

16. Holmes, G., Kirkby, R., Pfahringer, B.: Stress-testing hoeffding trees. In: Jorge, A.M., Torgo, L., Brazdil, P., Camacho, R., Gama, J. (eds) European Conference on Principles of Data Mining and Knowledge Discovery, PKDD 2005. Lecture Notes in Computer Science, vol. 3721. Springer, Berlin, Heidelberg (2005). https://doi.org/10.1007/11564126_50

17. Oza, N.C., Russell, S.J.: Online bagging and boosting. In: International Workshop on Artificial Intelligence and Statistics, pp. 229–236. PMLR (2001)

18. Kolter, J.Z., Maloof, M.A.: Dynamic weighted majority: an ensemble method for drifting concepts. J. Mach. Learn. Res. **8**, 2755–2790 (2007)

19. Littlestone, N., Warmuth, M.K.: The weighted majority algorithm. Inf. Comput. **108**(2), 212–261 (1994)

An Improved Ensemble Deep Learning Model Based on CNN for Malicious Website Detection

Nguyet Quang Do[1], Ali Selamat[1,2,3]([✉]), Kok Cheng Lim[4], and Ondrej Krejcar[1,3]

[1] Malaysia-Japan International Institute of Technology (MJIIT), Universiti Teknologi Malaysia, Kuala Lumpur, Malaysia
doquang@graduate.utm.my, aselamat@utm.my
[2] School of Computing, Faculty of Engineering, Universiti Teknologi Malaysia, Johor Bahru, Malaysia and Media and Games Center of Excellence (MagicX), Universiti Teknologi Malaysia, Johor Bahru, Malaysia
[3] Center for Basic and Applied Research, Faculty of Informatics and Management, University of Hradec Kralove, Rokitanskeho 62, 500 03 Hradec Kralove, Czech Republic
ondrej.krejcar@uhk.cz
[4] College of Computing and Informatics (CCI), Universiti Tenaga Nasional (UNITEN), Kajang, Malaysia
kokcheng@uniten.edu.my

Abstract. A malicious website, also known as a phishing website, remains one of the major concerns in the cybersecurity domain. Among numerous deep learning-based solutions for phishing website detection, a Convolutional Neural Network (CNN) is one of the most popular techniques. However, when used as a standalone classifier, CNN still suffers from an accuracy deficiency issue. Therefore, the main objective of this paper is to explore the hybridization of CNN with another deep learning algorithm to address this problem. In this study, CNN was combined with Bidirectional Gated Recurrent Unit (BiGRU) to construct an ensemble model for malicious webpage classification. The performance of the proposed CNN-BiGRU model was evaluated against several deep learning approaches using the same dataset. The results indicated that the proposed CNN-BiGRU is a promising solution for malicious website detection. In addition, ensemble architectures outperformed single models as they joined the advantages and cured the disadvantages of individual deep learning algorithms.

Keyword: Cybersecurity · Malicious website · Phishing detection · Deep learning (DL) · Convolutional Neural Network (CNN) · Bidirectional Gated Recurrent Unit (BiGRU)

1 Introduction

There have been numerous attempts to mitigate phishing attacks by applying deep learning (DL) techniques [1, 2]. Malicious website detection based on CNN is one of the most common approaches to combat phishing activities [3, 4]. Although CNN can extract features from raw input efficiently, it still suffers from two major limitations. First, CNN

© Springer Nature Switzerland AG 2022
H. Fujita et al. (Eds.): IEA/AIE 2022, LNAI 13343, pp. 497–504, 2022.
https://doi.org/10.1007/978-3-031-08530-7_42

performs better on multi-dimensional data (e.g., videos, images, speech signals, etc.) than textual data [5]. Since webpages contain different types of information, such as text, frame, and images, using CNN as a single feature extractor or a classifier cannot guarantee maximum classification accuracy. Secondly, CNN is best fitted to extract spatial feature representation, and not temporal feature representation of URLs or webpage content [6]. It lacks the capability to learn contextual information, resulting in limited detection accuracy [2]. Therefore, CNN can be integrated with other DL algorithms in an ensemble model to combine their strengths, resolve their weaknesses, and improve the overall performance accuracy. Motivated from the above, this study examines an ensemble DL model based on CNN to classify legitimate and malicious websites effectively. CNN was used as the first classifier, while BiGRU was utilized as the second classifier in a CNN-BiGRU architecture. The outcome of this research is a phishing detection model based on ensemble DL technique with improved performance accuracy.

The existing DL-based phishing detection models can be categorized into two groups: single DL and ensemble DL. In single DL models, only one DL technique is used as a feature extractor and/or classifier to detect malicious websites. Meanwhile, in ensemble DL models, several DL algorithms are combined where the output of one classifier is used as an input to the next classifier. In this paper, the term ensemble refers to the combination of two DL algorithms, namely CNN and BiGRU. The major contributions of this research are:

- A novel ensemble DL model was proposed by combining CNN with BiGRU for effective classification of phishing websites
- A comparative analysis was performed among several ensemble architectures, and the proposed CNN-BiGRU was benchmarked with single DL techniques to highlight the advantages of hybrid models over individual algorithms.

The remaining of this paper is organized as follows. Section 2 provides research background on ensemble DL models based on CNN. Section 3 explains the methodology of the proposed CNN-BiGRU model. Section 4 describes the experiments and discusses the obtained results. Section 5 summarizes the paper and recommends future research directions.

2 Literature Review

CNN has been applied in various domains within cybersecurity, including phishing detection. However, it was primarily used for malicious URL (Uniform Resource Locator) recognition and prediction [1, 3]. Besides, CNN has its limitation when used as a stand-alone classifier. Consequently, it can be combined with other DL techniques in an ensemble model to improve the overall detection accuracy. Al-Ahmadi and Alharbi [7] proposed an ensemble model consisting of two CNNs to distinguish malicious and benign websites. The first CNN extracted the text features from the webpage URLs, while the second CNN extracted the visual features from the webpage design and layout. The two CNNs received the webpage's URLs and screenshots as inputs simultaneously. Their outputs were combined, and the webpage was classified as either legitimate or phishing.

In this study, experiments were conducted on a dataset containing 2,000 screenshots and URLs, and an accuracy of 99.67% was achieved.

CNN-LSTM is one of the most common pairs among the ensemble DL models for detecting phishing websites. Previous authors have widely used it in numerous studies to achieve more accurate detection results [4, 8]. For example, an Intelligent Phishing Detection System (IPDS) was constructed in [4] to extract features from the website's URL and content. The model was built, compiled, and evaluated on a dataset comprising one million URLs and 10,000 images. The classification results of the proposed IPDS showed a higher level of performance accuracy (93.28%) than each DL algorithm (CNN and LSTM, separately). Another model based on CNN and LSTM, named Deep-URLDetect (DUD), was suggested in [8] to leverage the advantages of applying DL with character-level embedding in detecting phishing URLs. The model was tested on different datasets to examine the generalization of the proposed method, and the obtained accuracies were in a range of 93–98%.

Similarly, Rasymas and Dovydaitis [9] combined CNN and GRU in an ensemble model to achieve high-accuracy classification results. The model consists of two convolutional layers, one GRU layer, and several dense layers. In this study, different URL features were examined, including lexical features, character-level, and word-level embeddings, to find an approach that yields the highest accuracy. More than 2.5 million samples were collected for the experiment, and the final results showed that a combination of character and word-level embeddings produced the best performance accuracy (94.40%).

Although both LSTM and GRU are variants of Recuccurent Neural Network (RNN), GRU is considered as a light-weitght version of LSTM due to its simple internal architecture. The performance of GRU is similar to or in some cases, even better than that of LSTM when training on small datasets [10]. Nevertheless, one of their major disadvantages is that LSTM and GRU only consider the forward connection and ignore the backward connection between time steps [11]. Therefore, bidirectional LSTM and bidirectional GRU can be used to address this problem. Despite the benefits that BiGRU might offer, this DL algorithm has not been widely used in comination with CNN as compared to BiLSTM. Specifically, none of the related studies in the literature integrated CNN with BiGRU for malicious website detection. To the best of our knowledge, this is the first paper that adopted an ensemble model based on CNN-BiGRU to classify phishing and legitimate websites. This paper is also a continuation of our previous studies [12, 13] to further explore the application of deep learning in phishing website detection.

3   The Proposed CNN-BiGRU Model

CNN is generally comprised of three layers: convolutional layer, pooling layer, and fully connected layer. Considering a website corpus S with s_i be the i-th website in S ($s_i \in S$). The first step is to convert the website s_i into a feature vector $s_i \rightarrow x_i \in R^K$ so that it can be used as input to the convolutional layer. After conversion, the matrix representation $s \rightarrow x \in R^{LK}$ is obtained, where x is a set of adjoining components x_i (i = 1, 2, ..., L), L is the total number of websites, and K is the number of features in the dataset S. Next,

CNN performs a convolution operation $\otimes$ over $x \in R^{LK}$ to produce a new feature:

$$c_i = f(W \otimes X_{i:i+h-1} + b_i) \tag{1}$$

In the above equation, W denotes the convolution kernel or convolving filter (W $\in R^{Kh}$), h is the kernel size of the convolutional layer, b represents bias, and f(.) is a non-linear activation function.

In the second layer, max pooling is used to reduce the feature dimension as the result of the convolution operation:

$$p_i = \max(c_i) \tag{2}$$

Finally, the output of a fully connected layer is passed through a sigmoid function so that it is in a range of (0,1):

$$\text{Sigmoid}(x) = \frac{1}{1 + e^x} \tag{3}$$

There is no relation between the previous and next input in the CNN architecture. As a result, CNN is unable to handle time-series or sequential information and does not perform well when it comes to non-spatial data.

Meanwhile, the architecture of RNN is superior than that of a conventional neural network in terms of internal feedback, where there is a feedforward connection between the nodes in each layer. Moreover, GRU solves the problems of the standard RNN by introducing memory unit with gate structure: update gate and reset gate. The update gate, reset gate, hidden state, and output are defined as follows:

$$z_t = \sigma(w_z h_t - 1 + u_z x_t + b_z) \tag{4}$$

$$r_t = \sigma(w_r h_t - 1 + u_r x_t + b_r) \tag{5}$$

$$\tilde{h}_t = \tanh(w_h h_t - 1 + z_t u_z x_t + b_h) \tag{6}$$

$$h_t = u_t \tilde{h}_t + (1 - z_t)h_{t-1} \tag{7}$$

Although GRU demonstrate powerful capabilities in dealing with sequential data, their hidden state can only capture information from one direction. In contrast, BiGRU can attain past information and predict future information. Initially, the forward and backward GRU are used separately and then combined to form the final output $H_t = [\overrightarrow{h_t}, \overleftarrow{h_t}]$, where $\overrightarrow{h_t}$ denotes the forward and $\overleftarrow{h_t}$ represents the backward GRU, to obtain the long-term dependent features in both directions:

$$H_t = \overrightarrow{h_t} \oplus \overleftarrow{h_t} \tag{8}$$

With the help of memory cells, BiGRU is capable of handling sequential information which overcome the drawbacks of CNN technique. Consequently, CNN and BiGRU are combined together in this study to build an ensemble DL model, where the output vector of CNN is used as input of BiGRU to detect phishing or malicious website.

The proposed CNN-BiGRU model was trained on a publicly available data source, named University of California Irvine Machine Learning Repository (UCI). This dataset contains 11,055 URLs and is considered a balanced dataset, since the number of phishing and legitimate websites are 4898 and 6157, respectively. The architectures and pseu-do code of the proposed CNN-BiGRU models are shown in Fig. 1.

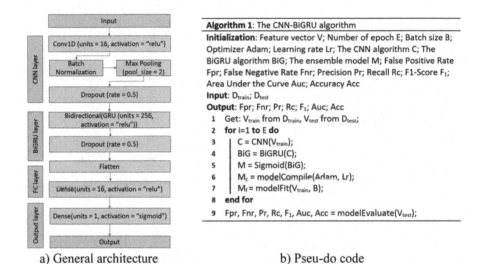

a) General architecture   b) Pseu-do code

Fig. 1. The architecture and pseu-do code of the proposed CNN-BiGRU model

First, in the data pre-processing and feature engineering stage, URL features were converted into feature vectors (V, V_{train}, V_{test}). Data cleaning and data transformation were also performed and a hold-out cross validation method was used by splitting the dataset into two parts: 80% for training (D_{train} = 8844) and 20% for testing (D_{test} = 2211). Next, in the classification stage, CNN (C) was combined with BiGRU (BiG) to form an ensemble DL architecture (M). The pseudo-code of this ensemble DL model is presented in Algorithm 1. Finally, in the output stage, the website URL is classified as either legitimate or phishing. Several metrics, such as False Positive Rate (Fpr), False Negative Rate (Fnr), Precision (Pr), Recall (Rc), F1-Score (F1), Area Under the Curve (Auc), and Accuracy (Acc) are measured to assess the performance of the underlying ensemble DL model.

4   Results and Discussion

The proposed CNN-BiGRU was compared with other ensemble DL models, including CNN-GRU, CNN-LSTM, and CNN-BiLSTM. The architecture of these models was similar to that of the proposed CNN-BiGRU. The BiGRU layer in the original model was replaced by the GRU, LSTM, and BiLSTM layers with the same parameter settings (e.g., number of layers, activation function, dropout rate). The number of neurons in the GRU and LSTM layers were set to 128, which was halved those of BiGRU and BiLSTM models. Table 1 and Fig. 2 displays several metrics used to evaluate the performance of these ensemble architectures. It is observed that CNN-BiGRU achieved better results since it obtained the highest scores in four out of five performance metrics. Throughout the experiments, parameter optimization, also known as fine-tuning, was performed to find the optimal set of parameters that produce the highest detection accuracy. The detection accuracies of four ensemble DL models were presented in Table 1.

Table 1. Evaluation metrics of various ensemble DL models

Ensemble DL model	Precision (%)	Recall (%)	F1-Score (%)	AUC (%)	Accuracy (%)
CNN-LSTM	96.91	97.79	97.35	99.61	97.06
CNN-BiLSTM	96.47	97.84	97.15	99.56	96.88
CNN-GRU	**97.44**	97.91	97.68	99.34	97.38
The proposed CNN-BiGRU	97.28	**98.46**	**97.86**	**99.64**	**97.60**

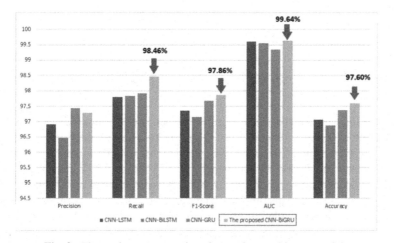

Fig. 2. The performance metrics of several ensemble DL models

The performance of CNN-BiGRU was also compared with two baseline classifiers (CNN and BiGRU, individually) used to construct the hybrid model. The results demonstrated that the proposed ensemble model achieved higher performance accuracy in detecting phishing websites compared to the individual algorithms, since ensemble models join the strengths and cure the weaknesses of single DL classifiers. Similar to ensemble models, single DL techniques also went through a fine-tuning process to optimize the hyper-paramters, and the highest accuracies achieved were displayed in Table 2.

Table 2. Benchmarking with single DL algorithms

DL model	Precision (%)	Recall (%)	F1-Score (%)	AUC (%)	Accuracy (%)
CNN	94.47	97.31	95.87	99.29	95.34
BiGRU	93.18	97.46	95.27	98.94	94.66
The proposed CNN-BiGRU	**97.28**	**98.46**	**97.86**	**99.64**	**97.60**

5   Conclusion and Future Work

To conclude, this study proposed an ensemble DL model based on CNN and BiGRU to improve the detection accuracy and solve the problem of the existing CNN architecture. The proposed CNN-BiGRU demonstrated the best results when compared with other three ensemble DL models, since it achieved the highest measures in Recall, F1-Score, AUC and Accuracy. For future work, we will implement another variant of CNN, called Graph Convolutional Network (GCN), to examine its performance in detecting malicious websites. In addition, to enhance the feature selection process, other technique such as attention-based mechanism can be used to strengthen the influence of important features and improve the detection ability of the model.

Acknowledgement. The authors sincerely thank Universiti Teknologi Malaysia (UTM) under Malaysia Research University Network (MRUN) Vot 4L876, for the completion of the research. This work was also partially supported/funded by the Ministry of Higher Education under the Fundamental Research Grant Scheme (FRGS/1/2018/ICT04/UTM/01/1) and Universiti Tenaga Nasional (UNITEN). The work and the contribution were also supported by the SPEV project "Smart Solutions in Ubiquitous Computing Environments", University of Hradec Kralove, Faculty of Informatics and Management, Czech Republic (under ID: UHK-FIM-SPEV-2022–2102). We are also grateful for the support of student Michal Dobrovolny in consultations regarding application aspects.

References

1. Wei, W., Ke, Q., Nowak, J., Korytkowski, M., Scherer, R., Woźniak, M.: Accurate and fast URL phishing detector: a convolutional neural network approach. Comput. Netw. **178** (2020). https://doi.org/10.1016/j.comnet.2020.107275
2. Feng, J., Zou, L., Yang, Y., Han, O., Zhou, J.: Web2Vec: phishing webpage detection method based on multidimensional features driven by deep learning. IEEE Access. **8**, (2020). https://doi.org/10.1109/ACCESS.2020.3043188
3. Xiao, X., Zhang, D., Hu, G., Jiang, Y., Xia, S.: CNN–MHSA: a Convolutional Neural Network and multi-head self-attention combined approach for detecting phishing websites. Neural Netw. **125**, 303–312 (2020). https://doi.org/10.1016/j.neunet.2020.02.013
4. Adebowale, M.A., Lwin, K.T., Hossain, M.A.: Intelligent phishing detection scheme using deep learning algorithms. J. Enterp. Inf. Manag. (2020). https://doi.org/10.1108/JEIM-01-2020-0036

504 N. Q. Do et al.

5. Liu, D., Lee, J., Wang, W., Wang, Y.: Malicious Websites Detection via CNN based Screenshot Recognition*. 115–119 (2019)
6. Huang, Y., Yang, Q., Qin, J., Wen, W.: Phishing URL detection via CNN and attention-based hierarchical RNN. Proc. - 2019 18th IEEE Int. Conf. Trust. Secur. Priv. Comput. Commun. IEEE Int. Conf. Big Data Sci. Eng. Trust. 112–119 (2019). https://doi.org/10.1109/TrustCom/BigDataSE.2019.00024
7. Al-Ahmadi, S., Alharbi, Y.: A deep learning technique for web phishing detection combined URL features and visual similarity. Int. J. Comput. Netw. Commun. **12**, 41–54 (2020). https://doi.org/10.5121/ijcnc.2020.12503
8. Srinivasan, S., Vidyapeetham, A.V., Ravi, V., Arunachalam, A., Universitet, O., Alazab, M.: Malware analysis using artificial intelligence and deep learning. Malware Anal. Using Artif. Intell. Deep Learn. (2021). https://doi.org/10.1007/978-3-030-62582-5
9. Rasymas, T., Dovydaitis, L.: Detection of phishing URLs by using deep learning approach and multiple features combinations. Balt. J. Mod. Comput. **8**, 471–483 (2020). https://doi.org/10.22364/BJMC.2020.8.3.06
10. Yuan, L., Zeng, Z., Lu, Y., Ou, X., Feng, T.: A character-level bigru-attention for phishing classification. In: Zhou, J., Luo, X., Shen, Q., Xu, Z. (eds.) ICICS 2019. LNCS, vol. 11999, pp. 746–762. Springer, Cham (2020). https://doi.org/10.1007/978-3-030-41579-2_43
11. Ozcan, A., Catal, C., Donmez, E., Senturk, B.: A hybrid DNN–LSTM model for detecting phishing URLs. Neural Comput. Appl. (2021)https://doi.org/10.1007/s00521-021-06401-z
12. Quang, D.N., Selamat, A., Krejcar, O.: Recent research on phishing detection through machine learning algorithm. In: Fujita, H., Selamat, A., Lin, J.-W., Ali, M. (eds.) IEA/AIE 2021. LNCS (LNAI), vol. 12798, pp. 495–508. Springer, Cham (2021). https://doi.org/10.1007/978-3-030-79457-6_42
13. Do, N.Q., Selamat, A., Krejcar, O., Yokoi, T., Fujita, H.: Phishing webpage classification via deep learning-based algorithms: an empirical study. Appl. Sci. 11 (2021). https://doi.org/10.3390/app11199210

Intrusion-Based Attack Detection Using Machine Learning Techniques for Connected Autonomous Vehicle

Mansi Bhavsar$^{(\boxtimes)}$, Kaushik Roy, Zhipeng Liu, John Kelly, and Balakrishna Gokaraju

North Carolina A&T State University, Greensboro, NC, USA
{mhbhavsar,zilu2}@aggies.ncat.edu, {kroy,jck,
bgokalraju}@ncat.edu

Abstract. With advancements in technology, an important issue is ensuring the security of self-driving cars. Unfortunately, hackers have been developing increasingly complex and harmful cyberattacks, making them difficult to detect. Furthermore, due to the diversity of the data exchanged amongst these vehicles, traditional algorithms face difficulty detecting such threats. Therefore, a network intrusion detection system is essential in a connected autonomous vehicle's communication infrastructure. The IDS (intrusion detection system) aims to secure the network by identifying malicious and abnormal traffic in real-time. This paper focuses on the data preprocessing, feature extraction, attack detection for such a system.

Additionally, it will compare the performance of this proposed IDS when operating in different machine learning models. We apply Linear Regression (LR), Linear Discriminant Analysis (LDA), K Nearest Neighbors (KNN), Classification and Regression Tree (CART), and Support Vector Machine (SVM) to classify the NSL-KDD dataset. The dataset was classified using binary and multiclass classification to train and test files. This data resulted in 94% and 98% accuracy for the train and test files, respectively, with KNN and CART algorithms.

Keywords: Machine learning · Autonomous vehicle · Cyberattacks · Intrusion · Data preprocessing · Feature engineering · ML model · Accuracy

1 Introduction

As advances in machine learning (ML) and deep neural networks (DNN) bring colossal potential to search for and develop self-driving cars a reality. Technology advancements in both the software and hardware side open new doors for huge applications in different domains. Many companies are in the race to develop safe and secure autonomous cars. (Such as Ford, Toyota, NVIDIA, NCA&T, and many more). Due to its communication system, more chances of threat surface access to exploit system vulnerabilities for malicious hackers. Connected autonomous vehicles (CAV) is a transformative technology that has increased potential in the research area. It helps reduce traffic congestion and accidents, improve efficiency, and the improved quality of vehicular systems. Moreover, using developed technologies such as ML, big data, IoT, and sharing economy extensively benefit intelligent cities [1].

© Springer Nature Switzerland AG 2022
H. Fujita et al. (Eds.): IEA/AIE 2022, LNAI 13343, pp. 505–515, 2022.
https://doi.org/10.1007/978-3-031-08530-7_43

Autonomous cars are vulnerable to different security threats. Network security is an important topic, and an intrusion detection system (IDS) can help us mitigate network threats without disrupting the safety and security of the host and the network. An intrusion detection system (IDS) is being implemented by applying ML techniques. It may be grouped by Host-based IDS and Network-based IDS, described by its placement over the network system [2]. There have been two types of detection: Misuse and anomaly detection. The misuse detects the known attacks, whereas anomaly detects the abnormal behavior. ML models are used to build anomaly-based detection systems. ML models also assist in feature engineering. This paper uses an existing labeled dataset (NSL-KDD) [3] to evaluate an anomaly-based intrusion detection system (IDS) to mitigate the threats and attacks. Dataset has made researchers compare different IDS methods and build an IDS system, either host or network-based. We apply different ML algorithms and present a comparative analysis.

The rest of the paper is organized as follows. Section 2 covers the related work. Section 3 includes the methodology, including data description, data preprocessing, and feature engineering. Section 4 presents results and performance metrics. Finally, Sect. 5 is the discussion and conclusion.

2   Literature Review

Cyber threats have become an essential issue with the emergence of self-driving vehicles and require the system to provide safe and secure connected vehicles. NSL-KDD [3] dataset is a refined version of the KDDcup99 data set [4]. Many analyses have been taken place by applying different techniques and tools to develop an effective intrusion detection system. The detailed implementation of various machine learning techniques with the WEKA tool [5]. A detailed description of the dataset is given in [6].

The problem of redundancy gets biased while learning, which might be one reason why a specific classifier shows an accuracy of above 95% [7, 8]. In [9], results show that the machine learning algorithm does not produce good results in the case of detection of misuse. In [10], the author compared the supervised ML classifiers for intrusion detection in a network. The efficient algorithm has been selected via performance matrices and concluded that Random Forest performs better than other classifiers. Authors in [11] proposed a lightweight IDS method that focuses on data-preprocessing to use essential features. They removed the redundant data from the dataset, which helps them get unbiased results with machine learning models. Different feature selection techniques have been used, such as wrapper or embedded feature selection, to improve the results [12]. Correlation-based feature selection filter methods have been used, which verify the model's effectiveness in terms of the detection rate as keeping a low false-positive rate with the use of a full attack scenario [13]. The IDS uses supervised machine learning models to detect normal and abnormal attacks [14]. The proposed method only classifies the Denial of Service (DOS) and probe attacks, but the remaining episodes are not considered.

The [15] proposed anomaly intrusion detection using an improved self-Adaptive Bayesian algorithm to process a large amount of data. The paper proposed an intrusion detection method using a support vector machine [16]. They used the feature removal method to improve the efficiency of the algorithm.

3 Methodology

3.1 Dataset Used

The project used the NSL-KDD [3] dataset with 42 attributes. Data is an improved version of the KDD99 dataset [17], a standard dataset for intrusion detection. The dataset has several versions available, from which the KDDTrain+ and KDDTest+ (training and testing data, respectively considered, which have a total of 125912 and 22544 instances.

The dataset contains network attacks related to the autonomous vehicle, including the 24 training attack types with 14 classes in the test file. Therefore, the dataset has KDD_train.csv and KDD_test.csv, which are not recorded from the same probability distribution, making it more realistic. Moreover, some intrusion experts believe that most novel attacks are variants of known attacks, and those can be sufficient to catch the novel variants.

This is a classification problem. The dataset description is given in Fig. 1, which provides the total instances of both files. The measure of different attacks and features of a dataset is shown in Table 1.

Table 1. Detailed instances in the dataset

NSL-KDD	Total instances	Normal	Dos	Probe	R2L	U2R
KDD_train	125973	67343	45927	11656	995	52
KDD_test	22544	9711	7460	2885	2421	67

The Normal traffic shows no attack recorded, and the other four subtypes show the documented Distributed Denial of service attack (DDoS). DDoS is a malicious attempt to disrupt the traffic of a targeted server [18]. The four sub attacks description is given below:

- DOS: (Denial of Service) is recorded when overloading the server with too many requests to be handled.
- Probe: the hacker scans the network to misuse a known vulnerability.
- R2L: (remote-to-local) attacks in which the attacker tries to gain local access to unauthorized information through sending packets to the victim machine.
- U2R: (user to root) attack where an attacker gets the core access of the system using his standard account to exploit the system vulnerabilities.

3.2 Data Preprocessing

It is vital to preprocess the dataset to apply the ML techniques to any given dataset. The less essential attributes in the dataset do not affect the accuracy of the classifier we want to use. This report aims to provide the complete preprocessing steps of the two files of the

NSL-KDD [3] dataset. To preprocess the dataset, python-Anaconda-navigator (Jupyter notebook) was used. The same methods were used [19] to preprocess the dataset of both files.

Preprocessing contains:

- load the dataset and analyze the statistics of the dataset.
- Change sub attach labels to their respective class.
- Check the missing value in the dataset.
- Check the outliers.

After performing the above steps, the dataset contains no missing values but outliers. The results of paired boxplot Fig. 1 with different ranges (a b). So, the IQR (Interquartile range) method was used to remove outliers. However, the box and whisker plot (in Fig. 2) provide removed outliers scenario between attributes. The shape of the cleaned data (which is (40118,42)) is not suitable as the dataset losses more than half of the information, which is not acceptable because if we keep the cleaned dataset, then the model overfits. (Which checked with applying the spot-check algorithm and getting 0.99 accuracy results).

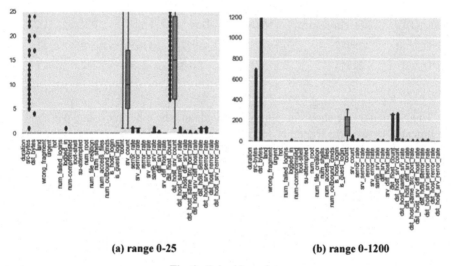

(a) range 0-25   (b) range 0-1200

Fig. 1. Paired box plot

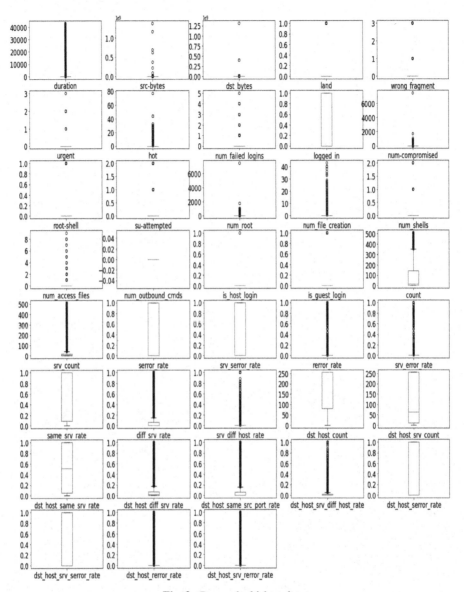

Fig. 2. Box and whisker plot

The histogram result (shown in Fig. 3) helped consider the data itself. This is because many attributes such as (duration, host_srv_rate, serror_rate, …) have many values in one class, whereas the difference between the two classes is not distinct. (The difference is too significant, and it considered the outlier as the highest value comparedto that one class which means it dropped the valuable figures from the dataset.)

It is not worth removing the outliers from the dataset and keeping the original data for our use case.

Fig. 3. Histogram plot

Preprocessing steps continue…

- below are the converted labels into five classes for multiclass classification for training and testing.
- Used binary classification for changing attack labels into two categories: normal and abnormal attacks, with the help of a label encoder.
- Used multiclass classification for changing attack labels into five usual categories, R2L, Probe, U2R, and Dos, respectively, with the help of a Label encoder.

3.3 Feature Engineering

Feature engineering is a crucial step to improve the performance of ML techniques. The feature selection is made using the Pearson correlation method [19]. It is a standard method used for filtering the essential features from the dataset. The paper research [20] concluded that a correlation coefficient value below 0.2 is considered a negligible correlation. We selected the threshold value of 0.5 to extract the feature with moderate to high correlation. A correlation matrix completes the filtering for more than 0.5 correlation attributes with encoded attack label features selected for binary and multiclass classification.

```
corr= numeric_bin.corr()
corr_y = abs(corr['intrusion'])
highest_corr = corr_y[corr_y >0.5]
highest_corr.sort_values(ascending=True)

count                          0.576444
srv_serror_rate                0.648289
serror_rate                    0.650652
dst_host_serror_rate           0.651842
dst_host_srv_serror_rate       0.654985
logged_in                      0.690171
dst_host_same_srv_rate         0.693803
dst_host_srv_count             0.722535
same_srv_rate                  0.751913
intrusion                      1.000000
Name: intrusion, dtype: float64
```

```
corr= numeric_bin.corr()
corr_y = abs(corr['intrusion'])
highest_corr = corr_y[corr_y >0.5]
highest_corr.sort_values(ascending=True)

same_srv_rate              0.510634
logged_in                  0.551159
dst_host_same_srv_rate     0.575526
dst_host_srv_count         0.593344
intrusion                  1.000000
Name: intrusion, dtype: float64
```

Fig. 4. Feature selection with the correlation method for train and test file

Figure 4 explains the features selected with the highest correlation greater than 0.5 and selected that attribute for binary and multiclass classification. The same procedure was followed for both train and test files.

4 Results and Discussion

We applied logistic regression (LR), support vector machine (SVM), K-nearest neighbor (KNN), classification and regression tree (CART), and linear discriminant analysis (LDA) to the modified version of the NSL-KDD dataset. The above models are standard for machine learning models with their respective advantages and disadvantages, making them unique. This algorithm works efficiently according to the user's use case and data type. The results of the model's accuracy are shown below in Tables 2 and 3 and model comparison in Figs. 5 and 6, respectively, for the train and test files.

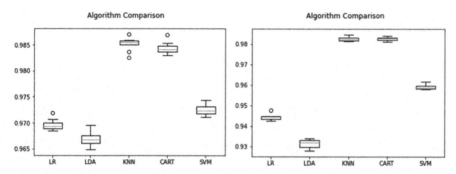

Fig. 5. Train data box plot for model comparison of binary and multiclass classification

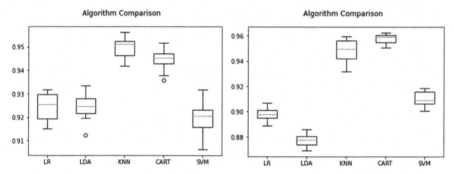

Fig. 6. Test data box plot for model accuracy of binary and multiclass classification

Table 2. Train accuracy for binary and multiclass classification

Accuracy (%)	Binary	Multiclass
LR	96.95	94.43
LDA	96.68	93.14
KNN	98.52	98.25
CART	98.45	98.25
SVM	97.24	95.91

The train Table 2 gives the likely results for binary and multiclass classification. It is a classification problem; therefore, the CART algorithm and KNN give higher accuracy results than LR and LDA. Boxplot Figs. 5 and 6 describe the model accurately predicting binary and multiclass classification data for algorithm comparison. The accuracy of multiclass classification for test files the LR and SVM algorithm accuracies were increased by tuning the parameters. Tolerance and different solvers were used for LDA from the Sklearn documentation, but it didn't help improve the accuracy.

Test Table 3 gives the acceptable results for binary and multiclass classification. Since it is a classification problem, the CART algorithm and KNN give higher accuracy results

Table 3. Test accuracy for binary and multiclass classification

Accuracy (%)	Binary	Multiclass
LR	92.45	90.16
LDA	92.45	87.76
KNN	94.98	94.84
CART	94.47	95.78
SVM	92.00	91.02

than LR and LDA. Therefore, it can be said that the model is accurate and predicted with 94% overall accuracy of the suitable attack classes with the CART and KNN algorithm.

Performance matrices check the model's performance, behavior, and activities. It describes whether the model is performing well or not, predicting as per the test data. Different parameters of performance matrices were checked, and the results for binary and multiclass classification of test and train data files are given in Tables 4 and 5, respectively.

Table 4. Performance metrics for binary classification

Performance metrics for binary classification		
Accuracy	KDD_test	KDD_train
Precision	0.875	0.967
Recall	0.942	0.973
Accuracy	0.917	0.967
False alarm rate	0.05803	0.0270

Table 5. Performance metrics for multiclass classification

Performance metrics for multiclass classification		
Accuracy	KDD_test	KDD_train
Precision	0.898	0.952
Recall	0.900	0.961
Accuracy	0.900	0.961
False alarm rate	0.1572	0.0933

It is evident from Tables (4 and 5) that our model outperforms with high accuracy of 91% for the test, 97% for train binary classification, a recall of 94% for the test, and

97% for train binary classification to classify the attacks. The CART, KNN, and SVM perform much better as its mostly preferred classifiers for the classification problem. Each sort achieves higher precision because they have a relatively low false alarm rate, which is consistent. The results are similar in pattern for multiclass classification.

5   Conclusion

This paper briefly explains how the ML algorithms are applied to the CAV dataset step by step. The procedure starts with data massaging, feature extraction, and machine learning classifiers. Next, the dataset predicts intrusion-based attacks on a self-driving car. Finally, the results were studied to predict the DDoS attack with binary and multiclass classification. The accuracy of models is comparatively similar for multiclass and binary (approximately 94 to 98%).

Furthermore, the feature selection helps to reduce the training and testing times. Testing is performed with the help of 10-fold cross-validation, where each fold is used once for testing and nine times for training. Results showed that the proposed preprocessing and feature selection method delivers excellent accuracy in the model.

6   Future Scope

The model results are reasonable for both cases. However, the gap here is in the class attack dataset instances. When looking back into the attack labels values per class, there has been an imbalance of the dataset in the U2R attack (class 3- the value is 52 instances), which is significantly less compared to the other courses, which may be one of the reasons behind the varying accuracy or dropping the accuracy in multiclass classification than the binary classification. Furthermore, it can be limited to see the difference while removing the third class (U2R), with fewer attributes. To add to it, as the details do not have effective differences amongst them, maybe removing the third class itself will not improve the results by a reasonable amount. Unlikely, based on the mechanisms of attack on IoT, exterior features are being evaluated and need to be considered. A data spike can be retrieved in the simple anomaly technique as future attacks exist. Thus, we believe that our model is capable of such intrusions. In the future, we will improve dataset techniques (such as interpolation) to check whether that helps with the prediction model or improves the data redundancy and vulnerability of the novel attacks. We will also consider various outlier handling techniques.

Acknowledgment. This research is supported by Palo Alto Networks. Any opinions, findings, conclusions, or recommendations expressed in this material are those of the author(s) and do not necessarily reflect the views of Palo Alto Networks.

References

1. Bimbraw, K.: Autonomous cars: past, present, and future. A review of the developments in the last century, the present scenario, and the expected future of au-tonomous vehicle technology. In: In Proceedings of the 12th International Conference on Informatics in Control, Automation and Robotics, pp. 191–198, ICINCO (2015)

2. Ieracitano, C., Adeel, A., Morabito, F.C., Hussain, A.: A novel statistical analysis and autoencoder driven intelligent intrusion detection approach. Neurocomputing, 387, 51–62 (2020). https://doi.org/10.1016/j.neucom.2019.11.016

3. Tavallaee, M., Bagheri, E., Lu, W., Ghorbani, A.: A Detailed Analysis of the KDD CUP 99 Data Set, Submitted to Second IEEE Symposium on Computational Intelligence for Security and Defense Applications, CISDA (2009)

4. Tavallaee, M., Bagheri, E., Lu, W., Ghorbani, A.: A detailed analysis of the KDD CUP 99 data set. In: Proceedings of the 2009 IEEE Symposium on Computational Intelligence in Security and Defense Applications, CISDA (2009)

5. Revathi, Dr, S., Malathi, A.: A detailed analysis on NSL-KDD dataset using various machine learning techniques for intrusion detection. Int. J. Eng. Res. Technol. (IJERT), 2, (2013), ISSN: 2278–0181

6. Dhanabal1, Dr. L., Shantharajah, S.P.: A study on NSL-KDD dataset for intrusion detection system based on classification algorithms. Int. J. Adv. Res. Comput. Commun. Eng. 4(6) (2015)

7. Revathi, Dr. S., Malathi, A.: A detailed analysis of KDD cup99 Dataset for IDS. Int. J. Eng. Res. Technol. (IJERT), 2(12) (2013)

8. Lippmann, R.P., Fried, D.J., Graf, I.: Evaluating intrusion detection systems: the 1998 DARPA off-line intrusion detection evaluation. In: Proceedings of the 2000 DARPA Information Survivability Conference and Exposition, DISCEX 2000 (2000)

9. Sabhnani, M., Serpen, G.: Why machine learning algorithms fail in misuse detection on kdd intrusion detection dataset. ACM Trans. Intell. Data Anal. 403–415 (2004)

10. Hamid, Y., Sugumaran, M., Balasaraswathi, V.R.: IDS using machine learning - current state of art and future directions. Curr. J. Appl. Sci. Technol. 15(3), 1–22 (2016). https://doi.org/10.9734/BJAST/2016/23668

11. Manjula, J., Belavagi, C., Muniyal, B.: Performance evaluation of supervised machine learning algorithms for intrusion detection. In: 25th International Multi-Conference on Information Processing (2016)

12. Wahba, Y., Elsalamouny, E., Eltaweel, G.: Improving the performance of multiclass intrusion detection systems using feature reduction, Research gate, article, June 2015

13. Sharafaldin, I., Lashkari, A.H., Ghorbani, A.A.: Toward generating a new intrusion detection dataset and intrusion traffic characterization. In: 4th International Conference on Information Systems Security and Privacy, ICISSP, Portugal, January 2018

14. Sangkatsanee, P., Wattanapongsakorn, N., Charnsripinyo, C.: Practical real-time intrusion detection using machine learning approaches. Comput. Commun. 34(18), 2227–2235 (2011)

15. Farid, D.M., Rahman, M.Z.: Anomaly network intrusion detection based on improved self-adaptive Bayesian algorithm. J. Comput. 5(1), 23–31 (2010)

16. Li, Y., Xia, J., Zhang, S., Yan, J., Ai, X., Dai, K.: An Efficient intrusion detection system based on support vector machines and gradually feature removal method. Expert Syst. Appl. 39(1), 424430 (2012)

17. The NSL-KDD dataset from the Canadian Institute for Cybersecurity (an updated version of the original KDD Cup 1999 Data (KDD99). https://www.unb.ca/cic/da-tasets/nsl.html

18. Sharafaldin, I., Lashkari, A.H., Hakak, S., Ghorbani, A.A.: Developing realistic distributed denial of service (DDOS) attack dataset and taxonomy. In: IEEE 53rd International Carnahan Conference on Security Technology, Chennai, India (2019)

19. https://github.com/abhinav-bhardwaj/Network-Intrusion-Detection-Using-Ma-chine-Lea rning/blob/master/README.md

20. Mukaka, M.M.: Statistics corner: a guide to the appropriate use of correlation coefficient in medical research. Malawi Med. J. 24(September), 69–71 (2012)

Detection of Anti-forensics and Malware Applications in Volatile Memory Acquisition

Chandlor Ratcliffe, Biodoumoye George Bokolo, Damilola Oladimeji, and Bing Zhou[✉]

Sam Houston State University, Huntsville, TX 77340, USA
{ctr019,bgb023,dko011,bxz003}@shsu.edu

Abstract. Malicious software operating on a target system, whether malware or anti-forensic, can impede data collecting, processing, and testing in digital and cyber forensic research. VolMemLyzer was developed by Lashkari et al. to identify malware executing in memory dumps using machine learning techniques. The usage of VolMemLyzer to detect the presence of Malware or Anti-Forensic software using characteristics retrieved from a memory dump was expanded in this research. We also implemented the Multi-layer Perceptron, Random Forest, K-Nearest Neighbors, adaBoost, and Decision Tree machine learning models. The results demonstrated that the Multi-layer Perceptron can compete with Random Forest and K-Nearest Neighbors. We were also able to perform multi-classification to detect numerous, overlapping application types, and we added features to VolMemLyzer to expand its applicability to any profile supported by Volatility 2.

Keywords: Memory analysis · Memory acquisition · Anti-forensic application · Malware · Machine learning · Ada-boost · Multi-layer perceptron · K-nearest neighbours · Decision tree · Random forest · Memory forensics

1 Introduction

In digital and cyber forensic science, we are trained in a variety of tools and techniques for engaging with systems that have data which we must recover, analyze, organize, and present. Generally, this is done one of two ways: live or static acquisition. Live acquisition is the process of collecting data from a system that is currently running, and static acquisition is collecting data from an offline system. The information from various components is collected in this process which may contain something of use to the investigator. Applications running on a system at the time of collection are of interest in this paper because it is possible that anti-forensic software or malware is running on the system that has not yet been identified by other means and may interfere with the process,

© Springer Nature Switzerland AG 2022
H. Fujita et al. (Eds.): IEA/AIE 2022, LNAI 13343, pp. 516–527, 2022.
https://doi.org/10.1007/978-3-031-08530-7_44

be relevant to the case, or be of other interest to the researcher or technician conducting the analysis.

When an application is running, it is loaded into live memory and iterated through by the processor. While some applications leave footprints on the hard drive in the form of logs, recently modified files, and events, some applications run silently. Detecting these applications is possible through an inspection of the live system memory.

System memory refers to random access memory for the purposes of this paper. In other contexts, it also refers to the cache, swap spaces, page files, long-term storage media, read-only memory, and more. When performing a live acquisition, tools can be used to generate and collect volatile data from the system memory; this is known as a memory dump. Live memory acquisition and analysis is a difficult task. Several approaches exist from memory dump sampling [8,14] and examining hardware interactions directly [5] to completely lifting the existing operating system into a virtual environment without disruption [9].

Many resources cite two major tools [8,14] in memory analysis and acquisition, namely Rekall[1], and Volatility[2]. VolMemLyzer [14] is an application which wraps Volatility 2 to extract specific measures of a memory dump. Those measures are then fed to machine learning models in an attempt to identify whether or not malware was actively loaded into memory at the time of the memory dump. The initial paper explored a few specific models outlined later, and it presented a breakdown of a variety of features that they found particularly useful for those models.

Machine Learning is a subfield of artificial intelligence dedicated to developing agents that improve over time for a certain problem-set without requiring a complete and exhaustive exposition of the issue space at the time of development. They typically train by exposing themselves to data-sets, and trained models may be applied to fresh data without incurring the cost associated with training.

The main objective of this research is to utilize machine learning techniques to detect malware, benign applications and anti-forensic applications running on a system using the features extracted by VolMemLyzer from memory dumps of live systems. Based on the result of this research, we established that it is feasible to detect the behavior of malware and apps operating on a Windows 7 and Windows 8 system using VolMemlyzer. We extended their implementation to support any memory dump for which Volatility 2 has a compatible profile. We reproduced the comparison of the models that they chose with the addition of one which they did not, and we concluded that multi-layer perceptron, K-Nearest neighbors, and Random Forest performed competitively in classifying the memory dumps as containing malware, benign applications, anti-forensic applications, or some combination of the three according to the features selected by VolMemLyzer. We were also able to identify from the features which applications from the training data were likely to be running at the time of the collection of the memory dump.

The rest of this paper is structured as follows. Section 2 highlights the relevant work carried out by others in the research area. Section 3 gives a detailed

---

[1] http://www.rekall-forensic.com/.
[2] https://github.com/volatilityfoundation/volatility.

description of the methodology employed to set-up, carry out, and analyze the results of our experiments. Section 4 discusses the results of the experiments. Finally, Sect. 5 gives the conclusion, limitations and future works.

2   Literature Review

Several comparative analyses discuss the value of memory analysis in identifying what is taking place on a system at the time of acquisition [2,7,21], and each proposes different directions. Case and Richard propose that memory acquisition would benefit from a better understanding of how malware and tools utilize system features [7], and Lashkari et al. applied machine learning to the task of extracting and correlating hardware activity features and processes to classify memory dumps based on different types of running applications [14].

Malware detection can be categorized into static detection techniques, dynamic detection techniques, and hybrid detection techniques [11,12,25,26]. It is noteworthy to cite that each of the papers used different approaches with [26] stating the use of the hybrid techniques was to circumvent the challenges that both the static [25] and dynamic techniques [11] may pose, while a few others referred to the utilization of sandboxed execution traces of malware in the memory [13,16,17].

Case et al. proposed new memory forensics capabilities, which encompasses hooktracer_messagehooks, a Volatility plugin that uses Hooktracer to determine whether hooks in memory are created by a keylogger or by benign software [6].

Yunus et al. [26] employed a different approach by combing hybrid analysis techniques with memory analysis techniques to detect, analyse and classify malware. However, this paper aims to use a machine learning approach to detect both malware and anti-forensics applications running on a system by carrying out feature analysis on memory dumps.

Panker et al. [20] presented a framework that utilized machine learning to identify unknown malware in Linux VM cloud environments. By asking the hyper-visor in a trustworthy manner, the framework obtains volatile memory dumps from the inspected VM which overcomes the malware's ability to identify the security mechanism and elude detection. While this research focused on Linux OS, Handaya et al., opted to use machine learning to strictly detect file-less crypto-currency mining malware [10], which is usually hidden in the memory by attackers and may be camouflaged as part of a legitimate business process to prevent being discovered [12]. Our work also took advantage of memory dumps obtained from a VM.

Another important aspect of this research is the detection of anti-forensic tools in the memory of a system used to carry out malevolent activities. Palutke et al. [19] reviewed various approaches to detecting anti-forensics techniques, as well as implemented three Rekall plugins that automate the detection of hidden memory in the shared memory scenario. Block and Dewald utilized a plugin from the memory forensics tool Rekall to immediately report segments of the memory with executable files as well as review them against own implementation of anti-forensics hiding tools and malware samples [4]. Similar plugins exist for Volatility, and VolMemLyzer utilized those to similar effect [14]. We chose to extend VolMemlyzer.

Anti-forensics tries to obstruct, or corrupt the forensic process of acquiring evidence, analyzing it, and/or determining admissibility [1,3]. To accomplish this purpose, anti-forensic tools either conceal or modify digital evidence so that it is inadmissible in court, or cause the forensic process of retrieving and analyzing digital evidence to become so expensive and time consuming that the investigation is deemed inconclusive or cost prohibitive [24]. It is for this reason that we chose to create a category specifically for anti-forensic application detection in our extension of VolMemLyzer.

Lashkari et al. [14] introduced a python tool for memory analysis called VolMemlyzer that extracts the most important characterisation features from memory dumps captured during a live malware infection. The aim of this research is to expand this work.

Most researchers focused on either the detection of malware or the detection of anti-forensic tools from the memory using mainly volatility or other software suites. We consolidate the detection of malware and anti-forensic tools. We extended the capabilities of the VolMemLyzer tool by expanding the domain of memory dump profiles the tool can handle, by extending the machine learning models it currently uses, and by adding a distinction to the training that affords recognition of anti-forensic applications as a category in addition to the benign and malicious categories they showed.

3   Methodology

3.1   Lab Setup

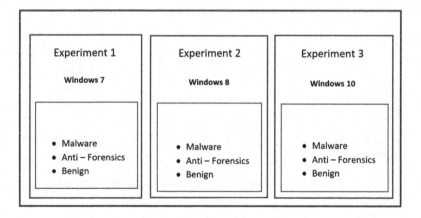

Fig. 1. Experiment virtual machine layout

For our experiment, a machine running Windows 10 was configured to host three virtual machines using Oracle VirtualBox, namely Windows 7, 8, and 10 as outlined in Fig. 1. This acted as our base, and was imaged and used on three other hosts in order to permit the testing of a wide variety of configurations of software and strains of malware without significant variability in operating system configuration. See our index Table 1 for specific applications used.

On each virtual system, a randomly selected combination of benign, malicious, and anti-forensic applications were deployed and run. Then, several memory dumps were collected from each virtual system as .elf files using Virtual Box's VBoxManage utility. These were done several times in order to collect memory in different states: Once in the beginning to try and catch near-startup behavior, once after a few minutes of running, and once again a little while later. The memory dumps were moved to a different workstation for analysis, and their features were extracted from the .elf files using VolMemLyzer. The extracted files generated by VolMemLyzer were consolidated into a single compressed file, and additional columns were added to aid in the generation of effective labels for classification. A more detailed look at how this data was labeled and used will be discussed in Sect. 3.2.

The first category in Table 1 below is the malware identified by [14]. The second is a list of anti-forensics tools. The third is a list of benign applications.

Table 1. Software targets

Anti-Forensics Tools	Malware Application	Benign Applications
Wipe		Microsoft Office [e]
ConfidentSend 3.0		Mozilla Firefox
Cryptapix	Zeus[d]	Mozilla Thunderbird
Gilsoft File lock pro[a]	VmZeus	Google Chrome
Puffer	MSI true Colors	Libre Office Calc
USB Stick Encryption	Sphinx	Adobe Reader
Wbstego4	Panda Debunker	Notepad++
BestCrypt Data Shelter Dban[b]		Calculator
BCArchive		
BCTextEncoder [c]		
BestCrypt Traveller		

[a]http://gilisoft.com
[b]https://dban.org/
[c]http://www.briggsoft.com
[d]https://bazaar.abuse.ch/sample/d03339104a85a196ea98e3caebda458931f2af281e5e d93f867d8caf1b157726/
[e]https://filehippo.com/search/?q=microsoft

3.2   Analysis/Implementation

This phase consisted of several critical stages outlined below. VolMemLyzer was extended to support additional profiles. It was used to extract features from the memory dumps to use as samples. The samples were annotated, amplified, and divided into training and test sets. The machine learning models configured and trained for the analysis are AdaBoost, k-Nearest Neighbor, Decision Tree, Random Forest, Multi-layer Perceptron.

Accuracy of the models was scored against the test data. This process was repeated 200 times. Visualizations and additional discussion are presented in Sect. 4.

Modifications to VolMemLyzer. VolMemLyzer required to be extended to accommodate analysis of certain Volatility 2 profiles in order to accomplish the feature extraction. To allow for this, a new optional command-line parameter was introduced and threaded through.

Feature Extraction with VolMemLyzer. Thirteen memory images were collected, processed through VolMemLyzer to create a table structured format (csv) of critical features, and consolidated into a single file. Additional columns were added to indicate software that was actively running at the time that the memory dump was collected, and separate columns were added to flag whether malicious, anti-forensic, or benign software were among those running. This became the starting point for amplification.

Sample Amplification. 200 runs were executed on the extracted and annotated seed data. The seed was amplified to a total of 6,656 samples in each iteration, with random fluctuation in the range -100 to 100 for each feature across the samples and clamped to non-negative values. Every run through the sample quadrupled its size. This was repeated until the run size of 6,656 samples was achieved. 20% of the data including the original seed data was set aside for validation. The models were each fit to the remaining 80% of the data.

Model Configuration Details. All five of the models employed were imported from SKlearn [22]. For adaBOOST, the AdaBoostClassifier was initialized with a DecisionTreeClassifier of maximum depth 2, 600 estimators, and a learning rate of 1. For Decision Tree model, the DecisionTreeClassifier was initialized to an initial maximum depth of 3. For Random Forest, the RandomForestClassifer was initialized with 500 estimators.

For K-Nearest Neighbors, the K-NeighborsClassifier was initialized with 5 neighbors. For the Multi-layer Perceptron, The MLPClassifier was initialized to use the lbfgs solver with $\alpha = 10^{-5}$, 15 hidden layers, random states, and a maximum iteration depth of 1,500. Importantly, it was not allowed to start warm because information from the previous runs would invalidate the statistics collected for comparison.

Each of these models were configured to explicitly have 31 classes named after the 31 columns added to the CSVs. The models were fit to the features with respect to compound labels formed from those columns.

Execution of Runs. To train the models, the divided data was further split into features and labels. Features were stored in a two-dimensional array such that every row corresponded to a sample, and every column corresponded to a feature. The expected results of the model were provided as an array of labels.

For each sample, the annotation columns were used to generate a compound label indicating which columns were marked in feature extraction with VolMem-lyzer stage explained in Sect. 3.2. An example label from our data is "antiforensic, ashisoft, crypta pix, puffers, benign, thunderbird". This indicated that the antiforensic applications ashisoft, crypta pix, and puffers, as well as the benign application thunderbird were running at the time of the collection of that sample or the sample from which it was amplified.

Analysis of Runs. All of sample data for each run was saved in separate files, and the scores from each run were consolidated. The averages and standard deviations from these scores formed the basis of our conclusions about the relative performance of the different models discussed in Sect. 4.

4   Results

Before diving too deeply, it's worth noting that the word "score" in this section acts as an abbreviation of "accuracy score" for both brevity and consistency with the implementation. It is a quantitative analysis of the number of test cases that a model could correctly reproduce after training as a percentage. For each of the models imported from [22], a score function was provided to make collecting this data easier. The scores were obtained by calling that function on the test data in each run.

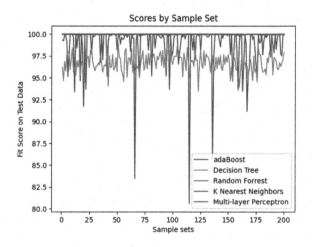

Fig. 2. Scores by sample set

Figure 2 is a direct plot of the scores of each of the models on each of the sample sets. First, Decision Tree performed consistently worse than the other models across the tests. Second, ada-Boost was very inconsistent, occasionally performing worse than Decision Tree. Figures 3 and 4 depict the results listed in Table 2, and further make clear the scale of the difference between the three competitive models and the two trailing them. In Fig. 3, higher is better, while in Fig. 4, lower is better.

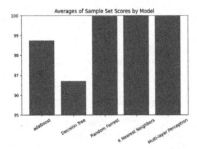

Fig. 3. Averages of sample set scores by model

Fig. 4. Standard deviations of sample set

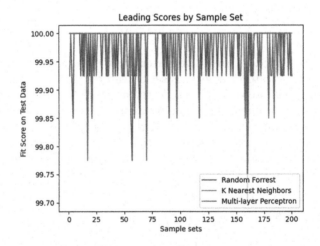

Fig. 5. Leading scores by sample set

To examine the leaders a bit more closely, Figs. 5, 6, and 7 gives a more information-dense and an expanded view of their relative performance. While all three had outliers, K-Nearest Neighbors appears to have had the least. The multi-layer perceptron has greater variability and more outliers, but is still extremely competitive with the other two. Of note is that the multi-layer perceptron was only allowed to run its training once on the same data as the other models. The documentation for the classifier suggests that if it were allowed to run warm multiple times, it would converge on a stronger result.

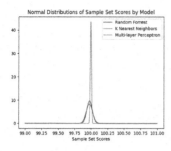
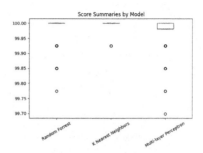

Fig. 6. Normal distributions of sample set

Fig. 7. Score summaries by model

Table 2. Statistical comparison of different models

	Averages	Standard deviations
AdaBoost	98.747	2.579
Decision tree	96.715	1.031
Random forest	99.979	0.041
K-nearest neighbors	99.999	0.001
Multi-layer perceptron	99.976	0.048

These scores are indications of how accurately each model could reproduce a compound label of categories given a set of features. Again, an example label from our data is "antiforensic, ashisoft, crypta pix, puffers, benign, thunderbird" indicating that the antiforensic applications ashisoft, crypta pix, and puffers were detected, and the benign application thunderbird was detected. The categories of antiforensic and benign in this case tell us that the model predicts that software in those categories is likely to be running on the system based on the features extracted by VolMemLyzer.

5   Conclusions

From this work, we have concluded that VolMemLyzer can be extended to identify the presence of anti-forensic software patterns on Windows 7 and Windows 8 systems, and that the multi-layer perceptron is a competitive model with K-Nearest Neighbors and Random Forest in classifying and categorizing the features extracted from memory dumps.

5.1   Limitations

Several limitations arose throughout the process. The most critical of these is that VolMemLyzer relies on Volatility 2's modules to extract data about what was running on the system at the time of the dump, and it does not currently have profiles to support the most recent version of Windows 10. Consequently, the bulk of our analysis is on the Windows 7 and Windows 8 memory dumps. Volatility 3 was created to resolve this and many other scalability issues in Volatility 2, so it would be productive to extend VolMemLyzer to utilize Volatility 3's modules.

VirtualBox was used to host the virtual machines used in sandboxing. The version used was 6.1.28, but this has several bugs that made getting data into and out of the virtual machine somewhat cumbersome. In the future, version 6.1.26 may be a more productive base until a more stable release of VirtualBox is produced.

The creation and collection of seed data to be amplified was chunked. That is, there are no idle memory dumps or memory dumps with standalone applications in the data-set. It would bolster the findings of this paper to more exhaustively collect images with each application running standalone, to have at least one with no applications, and to have a larger selection of combinations.

The space requirement for memory dumps is significant. The .elf files ranged from 2.02 GB to 17.9 GB depending on the operating system and specifications of the virtual machine. This significantly impacts analysis time.

Finally, due to the extremely high performance of K-Nearest Neighbors, it is possible that it has been overfit. A more diverse initial dataset may reduce this potential issue.

5.2   Contribution

Based on the result of this research, we established that it is feasible to detect the behavior of benign, malicious, and anti-forensic software running on a Windows 7 and Windows 8 system using VolMemlyzer. We extended their implementation to support any memory dump for which Volatility 2 has a compatible profile. We reproduced the comparison of the models that they chose with the addition of one which they did not, and we concluded that multi-layer perceptron, K-Nearest neighbors, and Random Forest performed competitively in classifying the memory dumps as containing malware, benign applications, anti-forensic applications, or some combination of the three according to the features selected by VolMemLyzer. We moved from the binary classification of malign and benign to multi-classification. We were also able to identify from the features which applications from the training data were likely to be running at the time of the collection of the memory dump.

5.3   Future Work

Volatile memory forensics is an analysis technique for extracting relevant information from a memory dump file and detecting the existence and behavior of malware. When malware infects a personal computer, it can detect if the user has been using it for a long period because it leaves traces on the system. However, analysis environments are regularly wiped clean to allow "a fresh start" between malware examinations. Malware can leverage the lack of evidence of users' activity to detect an analysis environment [18]. Malware developers conceal their presence on the victim's hard drive by executing and operating malware in RAM; forensic memory analysis is essential in the investigation of these malware attacks. [23].

Malware modifies data structures by overwriting non-essential fields used to fingerprint the memory location, it may be undetectable, which sometimes makes Memory Forensic analysis more onerous [15].

In the future, overcoming any of the limitations expressed in the previous section would be progress. Support for other memory analysis tools like Rekall would increase the portability of VolMemLyzer to environments where Volatility isn't used. Finding a way to directly measure the target features on a live system rather than extracting them from memory images would be a significant step forward. Finally, it would be worthwhile to compare the performance of a better configured multi-layer perceptron permitted to retain its training between runs with the current models under consideration.

References

1. AlHarbi, R., AlZahrani, A., Bhat, W.A.: Forensic analysis of anti-forensic file-wiping tools on windows. J. Forensic Sci. (2021)
2. Aljaedi, A., Lindskog, D., Zavarsky, P., Ruhl, R., Almari, F.: Comparative analysis of volatile memory forensics: live response vs. memory imaging. In: 2011 IEEE Third International Conference on Privacy, Security, Risk and Trust and 2011 IEEE Third International Conference on Social Computing, pp. 1253–1258 (2011). https://doi.org/10.1109/PASSAT/SocialCom.2011.68
3. Bhat, W.A., AlZahrani, A., Wani, M.A.: Can computer forensic tools be trusted in digital investigations? Sci. Justice **61**(2), 198–203 (2021)
4. Block, F., Dewald, A.: Windows memory forensics: detecting (un)intentionally hidden injected code by examining page table entries. Digit. Investig. **29**, S3–S12 (2019)
5. Botacin, M., Grégio, A., Alves, M.A.Z.: Near-memory & in-memory detection of fileless malware. In: The International Symposium on Memory Systems. MEMSYS 2020, pp. 23–38. Association for Computing Machinery, New York (2020). https://doi.org/10.1145/3422575.3422775
6. Case, A., et al.: HookTracer: automatic detection and analysis of keystroke loggers using memory forensics. Comput. Secur. **96**, 101872 (2020)
7. Case, A., Richard, G.G.: Memory forensics: the path forward. Digit. Investig. **20**, 23–33 (2017). https://doi.org/10.1016/j.diin.2016.12.004, https://www.sciencedirect.com/science/article/pii/S1742287616301529, special Issue on Volatile Memory Analysis
8. Chan, E., Venkataraman, S., David, F., Chaugule, A., Campbell, R.: ForenScope: a framework for live forensics. In: Proceedings of the 26th Annual Computer Security Applications Conference. ACSAC 2010, pp. 307–316. Association for Computing Machinery, New York (2010). https://doi.org/10.1145/1920261.1920307
9. Cheng, Y., Fu, X., Du, X., Luo, B., Guizani, M.: A lightweight live memory forensic approach based on hardware virtualization. Inf. Sci. **379**, 23–41 (2017). https://doi.org/10.1016/j.ins.2016.07.019, https://www.sciencedirect.com/science/article/pii/S0020025516305011

10. Handaya, W., Yusoff, M., Jantan, A.: Machine learning approach for detection of fileless cryptocurrency mining malware. J. Phys. Conf. Ser. **1450**, 012075. IOP Publishing (2020)
11. Jeon, J., Park, J.H., Jeong, Y.S.: Dynamic analysis for IoT malware detection with convolution neural network model. IEEE Access **8**, 96899–96911 (2020)
12. Jerbi, M., Dagdia, Z.C., Bechikh, S., Said, L.B.: On the use of artificial malicious patterns for android malware detection. Comput. Secur. **92**, 101743 (2020)
13. Kawaguchi, N., Omote, K.: Malware function classification using APIs in initial behavior. In: 2015 10th Asia Joint Conference on Information Security, pp. 138–144. IEEE (2015)
14. Lashkari, A.H., Li, B., Carrier, T.L., Kaur, G.: Volmemlyzer: volatile memory analyzer for malware classification using feature engineering. In: 2021 Reconciling Data Analytics, Automation, Privacy, and Security: A Big Data Challenge (RDAAPS), pp. 1–8 (2021). https://doi.org/10.1109/RDAAPS48126.2021.9452028
15. Lengyel, T.K., Neumann, J., Maresca, S., Payne, B.D., Kiayias, A.: Virtual machine introspection in a hybrid honeypot architecture. In: CSET (2012)
16. Liang, G., Pang, J., Dai, C.: A behavior-based malware variant classification technique. Int. J. Inf. Educ. Technol. **6**(4), 291 (2016)
17. Lin, C.T., Wang, N.J., Xiao, H., Eckert, C.: Feature selection and extraction for malware classification. J. Inf. Sci. Eng. **31**(3), 965–992 (2015)
18. Or-Meir, O., Nissim, N., Elovici, Y., Rokach, L.: Dynamic malware analysis in the modern era-a state of the art survey. ACM Comput. Surv. **52**(5) (2019). https://doi.org/10.1145/3329786
19. Palutke, R., Block, F., Reichenberger, P., Stripeika, D.: Hiding process memory via anti-forensic techniques. Forensic Sci. Int. Digit. Investig.' **33**, 301012 (2020)
20. Panker, T., Nissim, N.: Leveraging malicious behavior traces from volatile memory using machine learning methods for trusted unknown malware detection in Linux cloud environments. Knowl.-Based Syst. **226**, 107095 (2021)
21. Patil, D.N., Meshram, B.B.: Extraction of forensic evidences from windows volatile memory. In: 2017 2nd International Conference for Convergence in Technology (I2CT), pp. 421–425 (2017). https://doi.org/10.1109/I2CT.2017.8226164
22. Pedregosa, F., et al.: Scikit-learn: machine learning in Python. J. Mach. Learn. Res. **12**, 2825–2830 (2011)
23. Rathnayaka, C., Jamdagni, A.: An efficient approach for advanced malware analysis using memory forensic technique. In: 2017 IEEE Trustcom/BigDataSE/ICESS, pp. 1145–1150. IEEE (2017)
24. Wani, M.A., AlZahrani, A., Bhat, W.A.: File system anti-forensics-types, techniques and tools. Comput. Fraud Secur. **2020**(3), 14–19 (2020)
25. Wüchner, T., Ochoa, M., Pretschner, A.: Robust and effective malware detection through quantitative data flow graph metrics. In: Almgren, M., Gulisano, V., Maggi, F. (eds.) DIMVA 2015. LNCS, vol. 9148, pp. 98–118. Springer, Cham (2015). https://doi.org/10.1007/978-3-319-20550-2_6
26. Yunus, Y.K.B.M., Ngah, S.B.: Review of hybrid analysis technique for malware detection. In: IOP Conference Series: Materials Science and Engineering, vol. 769, p. 012075. IOP Publishing (2020)

Malware Classification Based on Graph Convolutional Neural Networks and Static Call Graph Features

Attila Mester[1,2]([⊠]) [iD] and Zalán Bodó[1] [iD]

[1] Faculty of Mathematics and Computer Science, Babeş–Bolyai University,
Cluj-Napoca, Romania
[2] Bitdefender, Cluj-Napoca, Romania
{attila.mester,zalan.bodo}@ubbcluj.ro

Abstract. Advanced Persistent Threats (APT) are targeted, high level cybersecurity risk factors facing governments, financial units and other organizations. The attribution of APTs – gathering information about the origin of an attack – is an important key in the process of securing an organisation's infrastructure, prioritizing the measures to be taken depending on the actor(s) targeting the organisation. In practice, an elementary step in the process of attribution is determining the family and/or author of a sample, based on the binary file and/or its dynamic analysis – i.e. a multi-class classification problem regarding the family/author label. There are numerous methods in the literature aimed to label a sample based on its control flow graph or API sequence graph. We aim to summarize the literature on these methods, and offer another method to classify malware families leveraging the static call graph of a PE executable, as well as the functions' instruction lists, using a locality-sensitive hashing method to obtain the node feature vectors. Our results are compared to recent publications in the field.

Keywords: Static analysis · Static call graph · Graph convolutional networks · Locality-sensitive hashing

1 Introduction

One of the most relevant and valuable label a malicious sample can be assigned is the family and author information. Not only do these help an analyst predict the behaviour of the sample based on an internal or public family database, but they also offer attribution information, a key aspect of antivirus (AV) products. Due to this motive, being able to predict a sample's family/author with high confidence is extremely valuable. The main difficulty of this task lies in the noisiness of the ground truth labels (similarly to other machine learning problems), but also the nature of this multi-class classification problem: there are thousands of malware families, their naming can vary between different AV products – so that there may be cases where different labels from different vendors in fact cover the same

© Springer Nature Switzerland AG 2022
H. Fujita et al. (Eds.): IEA/AIE 2022, LNAI 13343, pp. 528–539, 2022.
https://doi.org/10.1007/978-3-031-08530-7_45

family – and also collecting ground truth labels is a lengthy process, involving manual analysis and documentation, and may result in uncertain labeling.

One method to capture a malware family is to analyze the sample's static call graph, a highly popular technique in this domain. While there are various methods in the literature, we offer a new approach leveraging both the structural features of the static call graph, but also node-level information obtained from its instruction n-gram distribution, projecting the high-dimensional vectors to a lower representation applying locality-sensitive hashing (LSH) [21]. The novelty in this method is to combine both of these features to improve classification accuracy. It is important to mention that we use the static call graph based features, avoiding the potentially slow runtime implied by dynamic analysis, which would lead to a harder scaling of the method in real-time usecase.

This rest of the paper is structured as follows. In Sect. 2 the work related to the topic of the paper is presented. The main part, Sect. 3 introduces graph convolutional neural networks and presents our malware classification approach: the features extracted from the call graph and the LSH codes assigned to each function, as well as the architecture of the GCN used in the experiments. In Sect. 4 the performed experiments are detailed, evaluating the proposed model and comparing it to other solutions, while Sect. 5 concludes the paper and discusses possible future directions.

2 Related Work

We analyzed the literature regarding API call features, embedding methods and neural network models in malware analysis in order to be able to bring value to the field. The short summary of this survey process is that API calls (dynamic as well as static) are by far one of the most extensively used features in the literature for malware classification. Another conclusion is that graph convolutional neural networks (GCN), deep GCN models (DGCNN) show an increasing popularity in cyber threat intelligence, likely due to the novelty as well as the benefits of these approaches. Furthermore, we can also conclude that malware detection methods are in general validated using only around 10 classes.

A more detailed overview on the literature is as follows. API calls is a feature leveraged both when analyzing Android malware (.apk) as well as x86 Portable Executables (PE). In the case of Android malware classification we can find methods using dynamic analysis based on API system calls, using embedding techniques or other classifiers to determine families [2,15].

PE analysis contains an exhaustive analysis of API calls and embedding methods. API calls and functions obtained from static analysis are used for malware detection and family classification in [7,12,13,23]. Similarly, these features are used in node and graph embedding techniques [13,14,26] and GCN methods [18]. Dynamic analysis of malware, based on API and system call features is also applied with promising results. Several papers mention the use of dynamic sys call sequences for classification [12,13,19,24]; these features are also used in embedding and GCN methods [13,22]. However, the limitations of such analysis cannot be ignored: the authors of [22] report 3000 hours of runtime to process 40 thousand samples and obtaining 1.5 TB data of JSON reports describing the dynamic API call sequences, making it virtually impossible for real-time use cases.

Based on the literature analyzed by us, it is important to mention the size of the datasets, and the number of classes used in the classification process. Simple binary classification (malicious/benign) is applied in [7,14,22,23], [12] mentions 7 classes of authors, [13] classifies samples into 6 families, while [24] into 9 families, and [26] mentions 12 families.

In this paper we analyze the static API call sequences of a PE executable, assigning node-level features in the call graph, obtained by a LSH method on the function's instruction n-gram distribution – thus grasping its coding style and functionalities (based on our previous work [21]) – and categorize the sample into one of the 223 families, a number significantly larger – and much more realistic – than the families used in the literature.

3   Graph Convolutional Networks in Malware Analysis

3.1   Graph Convolutional Networks

Convolutional neural networks (CNNs) are shift-invariant neural networks extracting local features from signals using the convolution operation, repeating this procedure in a stacked manner to obtain a better representation and "understanding" of the input. The architecture of CNNs is inspired by biological phenomena, copying the activation procedure in the visual cortex of animals. While CNNs are best known for their successful applications in computer vision, they can be employed for all kinds of sequential data, e.g. natural language texts [27]. Currently, CoAtNet [6], a novel type of CNN offers the best evaluation scores on the ImageNet dataset [9].

Graph convolutional networks (GCN) generalize CNNs to graph structures using convolutions from spectral graph theory. The two types of GNCs that can be found in the literature are called spatial and spectral GCNs. Spatial GNCs (e.g. [16,25]) utilize the neighborhood information in order to construct a graph embedding model, while spectral GCNs (e.g. [1,11]) build on the eigenvectors of the graph Laplacian [5]. Since determining the eigenvectors is computationally prohibitive in case of large graphs, it is not uncommon to approximate these using Chebyshev polynomials [8].

In this research we use the GCN model introduced in [16] based on Laplacian smoothing, i.e. averaging every point over its neighbors in the graph. This operation makes the vertices of the same cluster similar (see the cluster assumption of semi-supervised learning [3]), however, deep architectures of such networks run into the problem of over-smoothing [17]. The propagation rule of this spatial GCN is the following,

$$\mathbf{H}^{(i+1)} = \sigma\left(\tilde{\mathbf{A}}\mathbf{H}^{(i)}\mathbf{W}^{(i)}\right) \tag{1}$$

where $\mathbf{H}$ denotes the (embedded) data representation, which initially contains the input features, i.e. $\mathbf{H}^{(0)} = \mathbf{X}$, $\tilde{\mathbf{A}}$ is the symmetrically normalized adjacency matrix, $\mathbf{D}^{-1/2}\mathbf{A}\mathbf{D}^{-1/2}$, containing self-loops, and $\mathbf{D} = \mathrm{diag}(\mathbf{A}\cdot\mathbf{1})$ is the diagonal degree matrix. The matrix $\mathbf{W}$ denotes the weights of the neural network, and σ is a non-linear activation function, usually ReLU [10]. By stacking together such layers a graph embedding of the data is arrived at.

In graph learning the following three types of tasks are considered [28]: (i) *node-level* tasks focusing on node-level predictions, i.e. node classification, node clustering, (ii) *edge-level* tasks such as edge classification and link prediction, and (iii) *graph-level* tasks corresponding to graph classification, graph regression and graph matching. The above GCN model produces an embedding vector for each node, therefore these have to be summarized for the entire graph, for which a common choice is to use average pooling, that is to represent the graph as the average of its node vectors.

3.2   Malware Classification Using GCNs

The aim of using graph convolutional neural network is in fact to use a special feedforward neural network model which can be fed with the static call graph of a binary executable, and will output a probability distribution (using softmax activation function on the last layer) of the desired classes, i.e. malware families in this case. This is especially useful in the industry, since most of the time a sample may reflect a mixture of multiple families. In such cases, offering an analyst a probability over the families is of great value.

The idea is a continuation of the analysis of the static call graph from [21], where we assigned a LSH codeword for each function based on its instruction n-gram distribution (which in fact is a $\approx 600\,000$ long sparse vector, accounting for mnemonic unigrams, bigrams and trigrams). In this work we proved that the signatures extracted from the samples can successfully group malware families, but other than extracting bigrams from the call graph (i.e. edges, with LSH codewords as endpoints), we could not grasp the high level structural, topological similarities of the call graphs of two samples. A GCN model however exploits both the node-level feature vectors (in our case, the 8-dimensional integer LSH codeword), and the adjacency matrix. In each GCN layer, the nodes obtain an embedded vector which is influenced by its neighbourhood. Thus, applying multiple GCN layers we can make use of a higher level abstraction of the entire call graph.

Input of the GCN: Call Graph and Node-Level Features. The call graph is in fact the combination of the sample's control flow graph (CFG, offered by IDA Pro's *GenFuncGdl*) and its function call graph (offered by *GenCallGdl*). In this paper, by the term *call graph* we mean a unified graph, which is the result of a DFS traversal on the control flow graph, merged with the function call graph – details in [20,21]. There is a significant difference between using such a call graph, than using the control flow graph of the sample (i.e. the output

```
ModuleList(
    (0): GCNConv(8, 128)
    (1): ReLU()
    (2): Dropout(p=0.5)
    (3): GCNConv(128, 128)
    (4): ReLU()
    (5): Dropout(p=0.5)
    (6): GCNConv(128, 128)
    (7): ReLU()
    (8): Dropout(p=0.5)
    (9): GCNConv(128, 128)
    (10): Dropout(p=0.5)
)
(f): Linear(in_features=128, out_features=223, bias=True))
```

Fig. 1. Architecture of the GCN model used in the experiments.

of *GenFuncGdl*), since the latter has far more vertices than the call graph (due to the simple fact that in the call graph, instruction blocks are merged together based on DFS). The significant differences are reflected by the number of 150 thousand vertices (instruction blocks of a CFG) reported by [23], in contrast to the avg. 10 thousand nodes (functions in the call graph) in our dataset – as shown in Fig. 4.

Node-level features are the LSH codewords used in [21] – each node will have a 8-dimensional vector, representing its 600-thousand-long instruction distribution vector's projection onto 8 random hyperplanes [4].

Structure of the GCN. We built a GCN using the PyTorch library, specifically, the geometric package[1]. Various experiments were carried out to test different hyperparameter combinations, which we detail in Sect. 4. The model resulting the best F_1-scores while keeping low variance is as follows (using the official *torch* layer names): three *GCNConv* layers, each followed by *ReLU*, then *Dropout*, then a *GCNConv* followed by *Dropout* and a *global_mean_pool* layer, finally, a fully connected layer, serving the output probabilities. For model optimization, we use the *CrossEntropyLoss* loss, and the *Adam* optimizer, with a learning rate of 0.01. This structure is shown in Fig. 1.

3.3   Comparison: Node-Level Features vs. Topological Features

After selecting the best possible model according to various aspects (F_1-score, balanced accuracy, loss trends of training and validation set – to exclude the problem of overfitting), we continued to test the model by introducing four new experiments.

---

[1] https://www.pyg.org/.

```
(stack): Sequential(
    (0): Linear(in_features=8, out_features=128, bias=True)
    (1): ReLU()
    (2): Dropout(p=0.5)
    (3): Linear(in_features=128, out_features=128, bias=True)
    (4): ReLU()
    (5): Dropout(p=0.5)
    (6): Linear(in_features=128, out_features=223, bias=True)
)
```

Fig. 2. Architecture of the MLP model used for learning only on node-level features.

Only Topological Features. By training the same GCN model without node-level features (i.e. based only on the call graphs' adjacency matrix), we may answer the question whether the LSH codewords are truly useful when classifying families.

Only Node-Level Features. This experiment in fact meant the elimination of the GCN model, and training instead a multilayer perceptron (MLP) model on the node-level features, the LSH codewords. By doing so, we could examine whether these codewords can classify malware families with the same success (i.e. F_1-score) as the full GCN model – in other words, whether the topological features represented by the adjacency matrices bring valuable information for the model to learn on. The structure of the MLP model is shown in Fig. 2. Similarly to the previously mentioned GCN model, for this MLP we used *CrossEntropyLoss* loss function, and *Adam* optimizer with a learning rate of 0.01.

Other Node-Level Features from the Literature. This experiment refers to the training of the above GCN and MLP models, but with other node-level features, described in [26]. Similarly to the previous section, in this experiment we trained the same MLP model as before (only changing the input layer's size from 8 to 14 neurons), but using other set of node-level features, specifically, the ones suggested in [26]. In this state-of-the-art paper, the authors train a GCN model on the CFG of the samples, using as node-level features a 11-long vector, representing the number of different kind of instructions (e.g. transfer, call, arithmetic, compare, mov, etc.). In our experiments we simulate the method described in [26] by assigning each function in the call graph a vector containing 14 numbers, according to the number of occurrences of the instruction mnemonics, divided into 14 categories[2]: data transfer, *mov*, control transfer, *call*, arithmetic, compare, flag, bit and byte, shift and rotate, logical, string, I/O instructions, and in- and out-degree of the respective function.

Likewise, the above mentioned GCN is trained using these node-level feature vectors.

---

[2] x86 Assembly Language Reference Manual, https://docs.oracle.com/cd/E36784_01/html/E36859/enmzx.html.

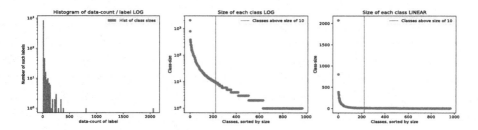

Fig. 3. Distribution of family sizes within the dataset of 15k samples.

4 Experimental Results

During the experiments the following languages, tools, frameworks were used: Python3, IDA Pro 6, GraphViz, PyTorch 1.10.0, Pytorch Geometric (pyg) 2.0.2, Tensorboard. The training of the GCN model was carried out on a system with Intel Xeon E5-2697A v4, 64 GB RAM, and a GeForce RTX 2080 Ti video card.

As mentioned in Sect. 3.2, different combinations of hyperparameters were tested in order to select the best model: number of hidden layers: $1-4$, size of hidden GCN layers: 64, 128 or 256, dropout probability: 0.2, 0.4 or 0.5, dropout only after the last GCN layer or after each of them.

4.1 Dataset Used

The private Bitdefender dataset contains 15 375 samples from 967 families. Due to the highly imbalanced nature of this dataset, filters were applied to select only those families to which at least 10 samples belong to. In this way, a set of 8620 samples from 223 families were obtained. This selection is illustrated in Fig. 3. On the leftmost plot we show the histogram of family class sizes regarding the whole dataset (plotted in logarithmic scale). The other two plots show the size of each family, plotted in logarithmic and linear scale, showing the thresholding line as well. All experiments were carried out on this filtered dataset (i.e. the classification task includes 223 classes).

It is important to mention that this dataset is probably still noisy, containing mislabelled samples (e.g. samples having other kind of internal label, due to the lack of its proper family label), although we tried to collect samples having confident ground truth information. As we can see on Fig. 3, after family-size filtering the dataset is still imbalanced – we did not aim to address this issue in this work, no data-balancing methods were applied.

4.2 Results and Discussion

In this section, we include the results according to 5 different models – as described in Sects. 3.2 and 3.3. Figure 6 contains multiple Tensorboard plots regarding the models' F_1-score, at different epochs. This score is evaluated on

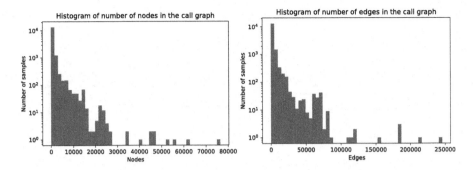

Fig. 4. Histogram of number of nodes and edges in a call graph.

the train and validation dataset (65% train, 15% validation, 20% test set). Splitting the dataset was carried out in a way to account for label imbalance. We trained the model for 300 epoch, as this was observed as an optimal number of iterations over the training dataset. The test datasets' F_1-scores are shown in Table 1.

Due to the imbalanced nature of the labels, the F_1-score was an appropriate evaluation method, since this encapsulates both the precision and recall of the model, being defined by their harmonic mean: $F_1 = 2 \cdot PR/(P+R)$. The F_1-scores are *micro* and *macro*-averaged (resulting in 4 F_1-trends), the former calculating the global false positives and false negatives, while the latter calculating it for each label, yielding their unweighted mean – ignoring label imbalance. Thus, the macro scores are always lower in our plot than the micro scores. Also, this score was evaluated for each family class, separately, and the following conclusion could be drawn. Although in general a larger class infers better classification, in this case, the size of the family label (i.e. how many samples belong to it) does not seem to affect the F_1-score of the class (Fig. 5) – a possible explanation could be the noisiness of the ground truth labeling and the differences across family labels regarding the samples' code-level features, and also treating distinct classes as one family due to their static similarities. An example could be the case of metamorphic viruses, whose code will contain a lot of register usage, data transfer operations, circular calls (reflected in the adjacency matrix of the call graph), etc.

In Fig. 6 a. we display the GCN model's metrics, according to the best configuration: 4 GCN layers, layer size of 128, dropout of 0.5. The b. plot shows the results of the same GCN model, but using other node-level features [26]. We can observe that these models have the best F_1-scores, meaning that using both topological features of the static call graph and the feature vectors of the local functions yields the best results. The F_1-score of each family is shown in Fig. 5, according to these two models. It can be seen that there are many classes having high scores, but due to the large number of families, the overall F_1-scores are affected in a negative way. Also, the features described in [26] result in higher F_1-metrics.

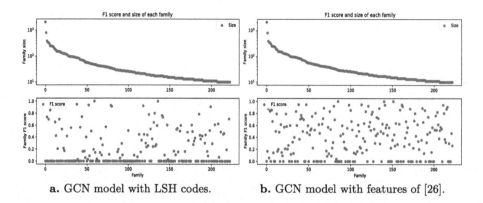

a. GCN model with LSH codes.   **b.** GCN model with features of [26].

Fig. 5. F_1-score of each class, plotted against the size of the family.

Table 1. F_1-scores of each model on the test dataset.

Model	Micro-F_1	Macro-F_1
GCN model with LSH codes	0.360	0.151
GCN model with features of [26]	0.576	0.361
GCN model without node-level features	0.251	0.028
MLP model with LSH codes	0.313	0.050
MLP model with features of [26]	0.242	0.020

In Fig. 6 **c.** and **d.** we display two MLP models: the one trained on LSH codewords, and the other using features of [26]. We can observe that these models have similar performance, and significantly lower F_1-scores compared to the GCN models. A conclusion can be that both of these node-level features bring some value to the problem of classifying 223 families, but does not have such high F_1-scores as the GCN model using both topological and node-level features.

Figure 6 **e.** displays the results regarding the GCN model using only topological features (i.e. ignoring node-level features). As shown in Table 1 as well, this model performs the worst out of all the other ones – suggesting that although topology of a call graph is important, the attributes of the functions are essential as well.

The arguably low F_1-scores can be attributed to the nature of the classification problem, having 223 classes, and to the imbalanced dataset. Examining the F_1-scores of each family label (Fig. 5) we can observe that many families have low scores, close to 0, which results in an overall lower F_1-score, although there are many families whose score is close to 1, even though the family contains only 10–20 samples. It is worth to analyze these cases in the future, as we can offer a relevance score for the family, a numeric measurement to describe how coherent, consistent the family is regarding its samples' static features – similarly, to filter the erroneous family labels.

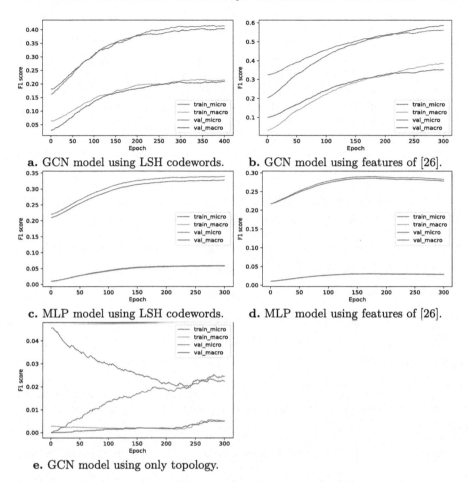

a. GCN model using LSH codewords.   b. GCN model using features of [26].

c. MLP model using LSH codewords.   d. MLP model using features of [26].

e. GCN model using only topology.

Fig. 6. F_1-score of the GCN and MLP models using various features: topology and node-level features – LSH codewords and features according to [26].

We would like to highlight the comparison of the models, both GCN and MLP, using two different node-level feature vectors. The one described in [26] has better detection accuracy, as reflected by the higher F_1-scores (Fig. 5, Table 1). This is probably due to the fact that in our implementation we chose a 14-dimensional feature vector to reproduce the results described in [26], whereas the LSH codewods are just 8-dimensional embeddings of a large feature vector [21]. Increasing the dimensionality of these LSH codewords will arguably result in a denser separation of the input vectors, which could increase the model's accuracy – further experiments are needed to prove these claims.

5   Conclusions and Future Work

We presented a GCN-based approach to classify malware, using static features based on the call graph of a PE. The node-level features are LSH codewords obtained from instruction n-gram frequencies from the code [21]. The GCN helps to aggregate topological information and use it in the classification process. Furthermore, we compare different models, using topological features and node-level features from the literature as well.

We mention that averaging the node embeddings by using a *global_mean_pool* layer in order to obtain the final embedding of the call graph (see Sect. 3.2) is probably not an adequate choice, future work is needed to examine other possible aggregation methods. Other future ideas include and are not limited to experiments on graph aggregation operators (e.g. concatenation), deep graph NN architectures, improving the node-level features by taking into account other instruction-level information such as parameters, and using a smaller curated dataset to observe the effects of noisy labels.

Acknowledgements. This project was supported by Bitdefender, offering the malware dataset and the infrastructure for training the models. I am grateful for the academic guidance of my supervisor, prof. dr. A. Andreica.

References

1. Bruna, J., Zaremba, W., Szlam, A., LeCun, Y.: Spectral networks and deep locally connected networks on graphs. In: International Conference on Learning Representations (ICLR) (2014)
2. Cai, M., Jiang, Y., Gao, C., Li, H., Yuan, W.: Learning features from enhanced function call graphs for Android malware detection. Neurocomputing **423**, 301–307 (2021)
3. Chapelle, O., Schölkopf, B., Zien, A. (eds.): Semi-supervised learning. MIT Press (2006)
4. Charikar, M.S.: Similarity estimation techniques from rounding algorithms. In: STOC, pp. 380–388. ACM, Montréal, Québec, Canada (2002)
5. Chung, F.R.: Spectral Graph Theory. No. 92 in Regional Conference Series in Mathematics, American Mathematical Society (1997)
6. Dai, Z., Liu, H., Le, Q.V., Tan, M.: CoAtNet: Marrying Convolution and Attention for All Data Sizes. arXiv preprint arXiv:2106.04803 (2021)
7. Dam, K.H.T., Touili, T.: Malware detection based on graph classification. In: Proceedings of the 3rd International Conference on Information Systems Security and Privacy, SCITEPRESS-Science and Technology Publications (2017)
8. Defferrard, M., Bresson, X., Vandergheynst, P.: Convolutional neural networks on graphs with fast localized spectral filtering. Adv. Neural. Inf. Process. Syst. **29**, 3844–3852 (2016)
9. Deng, J., Dong, W., Socher, R., Li, L.J., Li, K., Fei-Fei, L.: ImageNet: a large-scale hierarchical image database. In: CVPR, pp. 248–255. IEEE (2009)
10. Goodfellow, I., Bengio, Y., Courville, A.: Deep Learning. MIT Press (2016)
11. Henaff, M., Bruna, J., LeCun, Y.: Deep convolutional networks on graph-structured data. arXiv preprint arXiv:1506.05163 (2015)

12. Hong, J., Park, S., Kim, S.W., Kim, D., Kim, W.: Classifying malwares for identification of author groups. Concurrency Comput. Practice Exp. **30**(3), e4197 (2018)
13. Hong, J., et al.: Malware classification for identifying author groups: a graph-based approach. In: Proceedings of the Conference on Research in Adaptive and Convergent Systems, pp. 169–174 (2019)
14. Jiang, H., Turki, T., Wang, J.T.: DLGraph: malware detection using deep learning and graph embedding. In: 2018 17th IEEE International Conference on Machine Learning and Applications (ICMLA), pp. 1029–1033. IEEE (2018)
15. John, T.S., Thomas, T., Emmanuel, S.: Graph convolutional networks for Android malware detection with system call graphs. In: Third ISEA Conference on Security and Privacy (ISEA-ISAP), pp. 162–170. IEEE (2020)
16. Kipf, T.N., Welling, M.: Semi-supervised classification with graph convolutional networks. arXiv preprint arXiv:1609.02907 (2016)
17. Li, Q., Han, Z., Wu, X.M.: Deeper insights into graph convolutional networks for semi-supervised learning. In: Thirty-Second AAAI Conference on Artificial Intelligence (2018)
18. Li, S., Zhou, Q., Zhou, R., Lv, Q.: Intelligent malware detection based on graph convolutional network. J. Supercomput. **78**, 1–17 (2021)
19. Luh, R., Schrittwieser, S.: Advanced threat intelligence: detection and classification of anomalous behavior in system processes. e & i Elektrotechnik und Informationstechnik **137**(1), 38–44 (2020)
20. Mester, A.: Scalable, real-time malware clustering based on signatures of static call graph features. Master's thesis, Babeş-Bolyai University, Faculty of Mathematics and Computer Science, Cluj-Napoca, Romania (2020)
21. Mester, A., Bodó, Z.: Validating static call graph-based malware signatures using community detection methods. In: Proceedings of ESANN (2021)
22. de Oliveira, A.S., Sassi, R.J.: Behavioral malware detection using deep graph convolutional neural networks. Int. J. Comput. Appl. **174** (2021)
23. Phan, A.V., Le Nguyen, M., Nguyen, Y.L.H., Bui, L.T.: DGCNN: a convolutional neural network over large-scale labeled graphs. Neural Netw. **108**, 533–543 (2018)
24. Tang, M., Qian, Q.: Dynamic API call sequence visualisation for malware classification. IET Inf. Secur. **13**(4), 367–377 (2019)
25. Wu, F., Souza, A., Zhang, T., Fifty, C., Yu, T., Weinberger, K.: Simplifying graph convolutional networks. In: International Conference on Machine Learning, pp. 6861–6871. PMLR (2019)
26. Yan, J., Yan, G., Jin, D.: Classifying malware represented as control flow graphs using deep graph convolutional neural network. In: 49th Annual IEEE/IFIP International Conference on Dependable Systems and Networks (DSN), pp. 52–63. IEEE (2019)
27. Zhang, X., Zhao, J., LeCun, Y.: Character-level convolutional networks for text classification. In: Advances in Neural Information Processing Systems, vol. 28, pp. 649–657 (2015)
28. Zhou, J., et al.: Graph neural networks: a review of methods and applications. AI Open **1**, 57–81 (2020)

Modelling and Diagnosis

The Java2CSP Debugging Tool Utilizing Constraint Solving and Model-Based Diagnosis Principles

Franz Wotawa[(✉)] and Vlad Andrei Dumitru

Institute for Software Technology, TU Graz, 8010 Graz, Austria
wotawa@ist.tugraz.at, dumitru@tugraz.at

Abstract. Localizing faults in programs and repairing them is considered a difficult, time-consuming, but necessary activity of software engineering to assure programs fulfilling their expected behavior during operation. In this paper, we introduce the Java2CSP debugging tool implementing the principles of model-based diagnosis for fault localization, which can be accessed over the internet using an ordinary web browser. Java2CSP makes use of a constraint representation of a program together with a failing test case for reporting debugging candidates. The tool supports a non-object-oriented subset of the programming language Java. Java2CSP is not supposed to be used in any production environment. Instead, the tool has been developed for providing a prototypical implementation of a debugger using constraints. We present the underlying foundations behind Java2CSP, discuss some preliminary results, and show how the tool can also be used for test case generation and other applications.

Keywords: Automated debugging · Application of constraint solving · Debugging research web tool

1 Introduction

Debugging, i.e., finding, locating, understanding, and repairing bugs in software is an inevitable activity of software debugging and accounts for 40–50% of the total costs [2,3]. Even in the case that maybe half or more of these costs are related to bug finding, activities related to localizing and finally repairing bugs are still costly and (at least partially) automation is highly welcomed. Since the early 80s of the last century, a lot of fault localization research has been carried out. In a more recent survey, Wong et al. [10] categorized the debugging research field indicating spectrum-based fault localization (SFL) as the one having the largest share on publications, followed by model-based debugging (MBD). Whereas there are tools available for SFL, there is a lack of availability of MBD tools.

© Springer Nature Switzerland AG 2022
H. Fujita et al. (Eds.): IEA/AIE 2022, LNAI 13343, pp. 543–554, 2022.
https://doi.org/10.1007/978-3-031-08530-7_46

To close this gap, we developed the tool Java2CSP[1] that takes Java-like imperative programs and converts them into a constraint satisfaction problem that formally represents the corresponding debugging problem. It is supposed to be used for teaching and research. In addition, the tool also provides a user interface guiding the user through the debugging process, i.e., specifying input values of methods to be debugged and their expected outcome, as well as an interface to a constraint solver. In this paper, we introduce the underlying foundations, describe the tool, and provide an initial experimental evaluation. Furthermore, we briefly discuss how the constraint representation can be used for other purposes like test case generation and, therefore, provides a valuable contribution for researchers from several areas of software engineering.

```
1. public class Foo {
2.    public int behav(int a, int b,
                       boolean sar) {
3.        int res;
4.        int i;
5.        if (sar) {
6.          res = a * b ;
7.        } else {
8.          i = a + b;
9.          res = 2 * i; }
10.       return res; } }
```

Fig. 1. A simple program computing either the area or the circumference of a rectangle with given values a and b.

In the following, we discuss the debugging problem making use of a small example program depicted in Fig. 1. The program comprises 10 lines of code implementing a class Foo. The method behav takes three inputs and computes depending on the value of the input variable sar the surface or the circumference of a rectangle specified using the inputs a and b holding the width and height of a rectangle respectively. Obviously, Foo.behav implements the proposed function correctly. However, let us now assume that Line 9 is int i = a * b; resulting in a method Foo.behav$_F$ Such a bug may occur when copying and pasting a part of the program like Line 6 during programming.

Such a bug will be detected during testing the program, i.e., calling behav$_F$ on a=2, b=3, and sar=$false$. The expected result stored in res should be 10, but Foo.behav$_F$ would return 12 instead. Hence, we would be interested in identifying the fault, i.e., localizing faulty statements leading to the observed

---

[1] Access the web interface of Java2CSP using the following link: http://modelinho. ist.tugraz.at. Note that there are currently two frontends available. The one we are talking about in this paper, and an early release of a new improved interface that is currently under development. To access the version of the tool we are describing in this paper follow the first version link given on the mentioned web page.

wrong output value. When using Java2CSP we would obtain two potential single fault diagnoses for Foo.behav$_F$, i.e., statement i = a * b in Line 8 and res = 2 * i in Line 9. Both statements can be changed in order to cope with the problem. However, changing * to + would be the easiest correction.

The debugging problem, as considered in this paper, comprises a program, and a failing test case, i.e., a test case where the program computes an output value that is different from the expected output. This view on debugging is similar to the view taken when considering program slicing [7,9] but different to other approaches like SFL [1] where we require a set of passing test cases as well.

We structure the paper as follows: We first discuss the basic foundations behind model-based diagnosis and its application to debugging. Afterward, we briefly explain how the tool can be used followed by a section where we introduce obtained initial experimental results. Finally, we discuss potential application scenarios of the tool, summarize the paper and discuss future research.

2 Basic Definitions

The foundations behind Java2CSP originate from model-based diagnosis [8,11] where the underlying idea has been to use a model of a system directly together with observations for identifying faults in the system. The application of model-based diagnosis to software debugging relying on constraints has been investigated already [15]. However, to be self-contained, we briefly discuss the underlying definitions focusing on software debugging and its formalization, which forms the basis behind our Java2CSP tool.

We start formulating and formalizing the debugging problem. We assume that we have a program Π, and a test case (I, O), where the inputs I and outputs O are both sets comprising pairs of variables (from $VARS$) and their (expected) values (from a set $VALS$), i.e., $I, O \subseteq \{(x, v) | x \in VARS, v \in VALS\}$. Note that we further restrict I and O to store values for different variables, and to have only one value specified for any variable. The *debugging problem* can be formulated as follows: Given a program Π and a test case (I, O) where at least one value of an output stored in O is different from the value obtained when executing Π using the input I, search for the root cause, i.e., a statement, that is responsible for the detected discrepancy.

Model-based diagnosis provides means for solving the debugging problem. For this purpose we have to come up with a model of a program Π, i.e., $M(\Pi)$, fulfilling the following requirements: (i) $M(\Pi)$ comprises a representation of statements (from $STMNTS$) and their interconnection, and (ii) a model for each statement $s \in STMNTS$ that capture the semantics if and only if s is assumed to behave not abnormal. The latter requirement can be realized as a constraint $ab_s \lor behav_s$ where ab_s states that s is faulty (or abnormal), and $behav_s$ the statement's behavior. The latter can be extracted from the source code of Π. For example, the assignment statement res = a * b ; of Line 6 of program Foo.behav from Fig. 1 can be represented as equation/constraint

$res = a \cdot b$. Taking these considerations into account, we can define a *diagnosis*, i.e., a solution of the diagnosis problem, as follows:

Definition 1. *Given* $(M(\Pi), M(I, O), STMNTS)$ *where* $M(\Pi)$ *is the model of program* Π, $M(I, O)$ *is a set of equations representing the values of variables from the test case* (I, O), *and* $STMNTS$ *the set of statements in* Π. *A set* $\Delta \subseteq STMNTS$ *is a diagnosis if and only if* $M(\Pi) \cup M(I, O) \cup \{ab_s = \bot | s \in STMNTS \setminus \Delta\} \cup \{ab_s = \top | s \in \Delta\}$ *is consistent.*

Consistency can be checked using a constraint solver. We are using the Z3 Theorem Prover [5] in our **Java2CSP** tool for this purpose. Note that diagnosis as defined only requires checking assumptions about the correctness of statements. In the following, we briefly discuss how models for programs can be automatically derived. For more information, we refer to [15].

As already mentioned, assignment statements can be easily represented as equations, but we have to take care of the fact that variables may be defined more than once in a program. Hence, only referring to the variable name would lead to a model that does not capture the semantics of a program. The solution to this problem relies on the use of the static single assignment form (SSA) [4] where each variable has an additional index assigned, starting with 0 at the beginning. Each time a variable is redefined, its index is increased by 1. If a variable is used in an expression, the index of its last definition is used for reference.

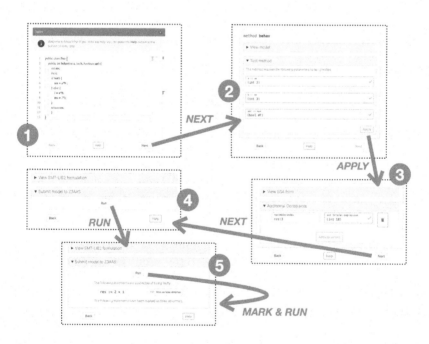

Fig. 2. Walk through the **Java2CSP** tool based on the **Foo.behav**$_F$ method.

Conditional statements are handled as follows. Each branch is converted separately using different indices for the variable definitions. Afterward, if a variable x is defined in one or both branches, we obtain two distinct indices for this variable and each branch. Note that if a variable is not used in one branch this index is the index we used before converting the conditional statement. Let us now assume that we have the SSA variables x_T and x_E from the then and the else branch respectively. We introduce a function Φ taking three parameters. One is the (converted) condition C of the conditional statements itself, the others are x_T, and x_E. The converted statement for the conditional is $x_{\max(T,E)} = \Phi(C, x_T, x_E)$.

Loop statements (and also in principle recursive statements) are converted into nested conditional statements considering a maximum nesting depth that depends on the number of iterations to be considered. Afterward, the nested conditional statements are converted into a model as already described.

It is worth noting that during conversion we also consider the ab variables and combine them with the converted statements using a logical disjunction. For the faulty Foo.behav program from Fig. 2 the model M looks like[2]:

$$ab_5 \vee tmp_0 = \mathsf{sar}_0$$
$$ab_6 \vee \mathsf{res}_1 = \mathsf{a}_0 * \mathsf{b}_0$$
$$ab_8 \vee \mathsf{i}_1 = \mathsf{a}_0 * \mathsf{b}_0$$
$$ab_9 \vee \mathsf{res}_2 = 2 * \mathsf{i}_1$$
$$\mathsf{res}_3 = \Phi(tmp_0, \mathsf{res}_1, \mathsf{res}_2)$$

The whole conversion process has to terminate because of considering a finite number of statements, and a maximum nesting depth when unrolling loop statements or recursive functions. The computational complexity depends on the nesting depth and the number of loops. For smaller example programs, which we consider in this work, conversion can be done within a fraction of a second when using today's standard computer. The Java2CSP tool implements the described debugging approach and delivers besides diagnosis results also a constraint representation for the Z3 constraint solver. In the next section, we discuss the tool and its usage in more detail.

3 Java2CSP tool

The Java2CSP tool implements the foundations behind model-based diagnosis for computing those statements that are capable of explaining the faulty behavior revealed in the given test case. The tool is not supposed to be used within a software development process. Instead, the tool is intended to be used for doing hands-on experiments in the area of automated debugging assuming programs that are smaller in size, and not considering all possibilities a today's programming language like Java offers. In particular, we restrict variables to be of type

---

[2] The model used in Java2CSP and especially the given indices might vary but follows the same idea behind conversion.

integer and Boolean. We debug at the level of methods not comprising method calls. Support for data types and structures like arrays is currently not fully implemented. However, we plan to make adaptations and improve functionality.

We depict the graphical user interface of Java2CSP in Fig. 2. We first have to add a class with at least one method. After pressing the Next button, we arrive at the second window where we are able to select a method for debugging and to add the values of input variables. After pressing the Apply button, we may also enter the expected output variables in the third window. Note that because of using the static single assignment form, we have to select a variable with a particular index. The highest index is used for the last definition of a variable. After pressing Next, we arrive at the last window where we can have a look at the constraint representation (in SMT-LIB format), or we are able to start debugging when pressing Run. The debugger returns the first statement, which can be marked as being falsely positive. In this case, we are able to re-run the diagnosis process and may receive another candidate statement. This process can be repeated until either there are no new diagnoses, or we arrive at a stage where we have a diagnosis of higher cardinality.

Besides debugging, Java2CSP offers the constraint representation of the method given including the input and expected output values. For example, the SMT-LIB representation of the program Foo.behav with the fault and input values provided in the introduction is:

```
(declare-const res_1 Int)
...
(declare-const ab_1 Bool)
(assert (= a_1 2))
(assert (= b_1 3))
(assert (= sar_1 false))
(assert (or ab_0 (= tm_0_0 sar_1)))
(assert (or ab_1 (= res_1 (* a_1 b_1))))
(assert (or ab_2 (= i_1 (* a_1 b_1))))
(assert (or ab_3 (= res_2 (* 2 i_1))))
(assert (= i_2 (ite tm_0_0 i_0 i_1)))
(assert (= res_3 (ite tm_0_0 res_1 res_2)))
(assert (= objective (+ (ite ab_0 1 0) ... (ite ab_3 1 0))))
```

The first part of the SMT-LIB program comprises all declaration of variables. Afterward, there are three **assert** statements capturing the input values. Afterward, there are the constraints representing the program. In the last line, there is an objective function used to start searching for the smallest diagnoses first. The SMT-LIB program can be also used for other purposes like test case generation. For this purpose, we have to remove the statements representing input and output values as well as the objective function. In addition, we have to add assert statements setting the ab_i values to false. When using a constraint solver like Z3 together with the modified constraint representation, we are able to compute input and output values (or input values for specified output values). See, for example, [14] or [6] on how to use constraints for testing and in particular test data generation, e.g., the generation of inputs for given outputs or any input/output combination.

For the purpose of test data generation based on constraints, we only need to take the part of the program that is used for representing the behavior and set the ab_i values to false using the `assert statement`. For `Foo.behav` we may come up with the following constraints and ask Z3 for solutions that make the constraints valid, which represent test data:

```
(declare-const res_1 Int) ...
(assert (or ab_0 (= tm_0_0 sar_1)))
(assert (or ab_1 (= res_1 (* a_1 b_1))))
(assert (or ab_2 (= i_1 (* a_1 b_1))))
(assert (or ab_3 (= res_2 (* 2 i_1))))
(assert (= ab_0 false)) ...
        ... (assert (= ab_3 false))
(assert (= i_2 (ite tm_0_0 i_0 i_1)))
(assert (= res_3 (ite tm_0_0 res_1 res_2)))
```

In summary, the Java2CSP tool can handle the following parts of Java:

- Compiling one class that comprises methods, where we assume that there is no method call used. Methods can take an arbitrary number of parameters.
- Conditional statements (if-then-else statements) and while loops are supported. The number of expected iterations can be set but there is also a default value of 5. To set the number of iterations n use /*== unroll-depth=n /* immediately before the `while(...){...}` statements begins.
- The basic data types Boolean, Integer, and Float are supported and compiled accordingly to their SMT-LIB representation. The set a value *val* for a particular variable in the Java2CSP tool we have to use the following format:
  - Boolean: (`bool` *val*), where *val* can be #T or #F for true or false respectively.
  - Integer: (`int` *val*)
  - Float: (`float` *val*)
- Arrays of basic data types are supported by our tool. Use (`array` *type* ((*type* *val*$_1$) (*type* *val*$_2$) ...)) to set the values of the 1st, 2nd, ...element of the array.
- It is possible to state the parameter values as well as an arbitrary number of variable values expected to be taken during execution.

Object-oriented concepts like inheritance are currently not supported. We also do not support method calls (neither non-recursive nor recursive calls). In the future method calls as well as other data types may be supported. However, the main purpose of the tool is to provide an initial implementation of the application of model-based diagnosis to software debugging not requiring the installation of tools on a computer, which is good enough for carrying out initial experiments.

Table 1. Experimental results using Java2CSP on small example programs. Diagnoses in bold face are indicating the real fault introduced in the programs.

Program	Method	Parameters	Observations	Diagnoses
D74	behaviorF1	a=2, b=3, c=3, d=2, e=2	f_1=12, g_1=12	1
D74	behaviorF2	a=2, b=3, c=3, d=2, e=2	f_1=12, g_1=12	0, **3**
D74	behaviorF3	a=2, b=3, c=3, d=2, e=2	f_1=12, g_1=12	0, 3
HalfAdder	behaviorF1	a=true, b=false	c_2=false, s_2=true	0, 1
HalfAdder	behaviorF2	a=true, b=false	c_2=false, s_2=true	8, **9**
HalfAdder	behaviorF3	a=false, b=false	c_2=false, s_2=false	7, 4, **5**, 6
HalfAdder	behaviorF3	a=false, b=true	c_2=false, s_2=true	6, **5**, 2, 3, 7, 4
IfTest	test1	a=1, b=5	i_3=10	0, **2**
IfTest	test2	a=-1, b=1	i_3=4	**2**
IfTest	test3	a=-1, b=1	i_7=10	**5**
IfTest	test4	a=0, b=1, c=1	$field2_1$=4	**2**
IfTest	test5	a=0, b=1, c=-1	$field2_3$=10	**4**
IfTest	test6	a=5, b=1, c=-1, d=1	$field2_6$=1	0, **6**
WhileTest	test1_f1		a_{21}=4, b_{21}=4, c_{21}=4	**4**, 8
WhileTest	test1_f2		a_{21}=4, b_{21}=4, c_{21}=4	0, 5, 6, 3, 9, 7
WhileTest	test1_f3		a_{21}=4, b_{21}=4, c_{21}=4	7, **8**, 3
WhileTest2	test2_f1		a_{13}=5, b_{63}=7	0, 10, **2**, 11, 4
WhileTest2	test2_f2		a_{13}=5, b_{63}=7	2, 0, 11, 10, **4**
WhileTest2	test2_f3		a_{13}=5, b_{63}=7	**8**, 10
TrafficLight	test_f1	state=3	$nextState_7$=0	**6**
TrafficLight	test_f2	state=0	$nextState_7$=1	**1**

4   Initial Evaluation

We provided an initial evaluation of our Java2CSP tool making use of 6 example classes comprising different methods. In Fig. 3 all the programs including comments indicating considered manually seeded faults are depicted. The objective behind the evaluation was to show that the tool is capable of delivering correct solutions and to have a first look at the precision of debugging results. For the latter, we compare the number of generated diagnosis candidates with the total number of statements. If both numbers are the same, a diagnosis does not allow to distinguish correct from incorrect statements and precision is considered low.

The programs used for evaluation comprise only a few lines of code for each method varying between 5 and 13. For an evaluation of a general debugging methodology to be used in practice, this would not suffice. However, Java2CSP has been developed for showing the basic capabilities behind the model-based diagnosis approach and has not been intended to be used within a development environment, where run-time considerations are important.

We depict the obtained results of the evaluation in Table 1 considering single faults only. Note that the tool starts enumerating statements with 0. Hence, statement 0 is the first statement in a method. We see that in all diagnosis runs the real bug has been found. The quality of diagnosis, i.e., its precision, varies depending on the test case and also on the program. In none of the cases Java2CSP was returning all statements as the potential candidates. We further

```
public class HalfAdder {
    public void behavior(boolean a, boolean b) {
        boolean n1 = a && b;
        n1 = !n1; // n1=n1; (F1)
        boolean n2 = a && n1;
        n2 = !n2;
        boolean n3 = n1 && b;
        n3 = !n3; // n3=n3 (F3)
        boolean s = n2 && n3;
        s = !s;
        boolean c = n1 && n1;
        c = !c; }} // c=c; (F2)
```

HalfAdder

```
class WhileTest {
    static public int test1() {
        int a,b,c,d,i;
        i = 1; // i=8; (f2)
        a = 1;
        b = 2;
        c = 3;
        d = 4; // d=5; (f1)
        while (i < 10) {
            a = b;
            b = c;
            c = d; // c=b; (f3)
            i++; }
        return a; }}
```

WhileTest

```
class WhileTest2 {
    static public void test2() {
        int a,b,i,j;
        i = 0;
        j = 0;
        a = 0; // a=1; (f1)
        b = 0;
        while (i < 5) { // i<4 (f2)
            j = 0;
            b = a;
            while (j < 3) {
                b = b+1; // b=a+1 (f3)
                j++;
            }
            a = a + 1;
            i++; }}}
```

WhileTest2

```
public class TrafficLight {
    public static void test(int state) {
        int nextState;
        if (state == 0) {
            nextState = 1; // nextState=0; (f2)
        } else {
            if (state == 1) {
                nextState = 2;
            } else {
                if (state == 2) {
                    nextState = 3;
                } else {
                    nextState = 0; // nextState=3; (f1)
                }}}
        return nextState; }
```

TrafficLight

```
public class D74Circuit {
    public void behavior(int a, int b, int c, int d, int e
        ) {
        int s1=a*c; // s1=a+c; (F3)
        int s2=b*d; // s2=b+d; (F1)
        int s3=c*e;
        int f=s1+s2; // f=s1+s2-2; (F2)
        int g=s2+s3; }}
```

D74Circuit

```
class IfTest {
    public static int test1(int a, int b) {
        int i;
        i = 0;
        if (a>0) {
            i = 2*i; // should be i = 2*b; }
        return i; }
    public static int test2(int a, int b) {
        int i;
        if(a>0) {
            i = 2*b;
        } else {
            i = 3*b; // should be i = 4*b; }
        return i; }
    public static int test3(int a, int b) {
        int i;
        if (a>0) {
            if (b>0) {
                i = 1;
            } else {
                i = 2;
            }} else {
            if (b>0) {
                i = 3; // should be i = 10;
            } else {
                i = 4; }}
        return i; }
    public void test4(int a, int b, int c) {
        int field1;
        int field2
        i = new O(a,b);
        if (c>0) {
            field1 = a*2;
            field2 = b*3; // should be field2 = b*4;}}
    public void test5(int a, int b, int c) {
        int field1;
        int field2;
        i = new O();
        if (c>0) {
            field1 = a*2;
            field2 = b*2;
        } else {
            field1 = a*3;
            field2 = b*3; //should be field2 = 10;}}
    public void test6(int a, int b, int c , int d) {
        int field1;
        int field2;
        i = new O();
        if (c>0) {
            if (d>0) {
                field1 = a;
                field2 = b;
            } else {
                field1 = a;
            }} else {
            if (d>0) {
                field2 = a; // should be i.field2 = b;
            } else {
                field1 = -1;
                field2 = -1; }}}}
```

IfTest programs

Fig. 3. Programs used for the initial evaluation of Java2CSP.

had a look at whether the obtained results are feasible, which was the case. Diagnosis time for all examples was always a fraction of a second and could be neglected.

5   Application Scenarios

The original objective behind Java2CSP was to come up with a debugging tool utilizing model-based diagnosis principles together with an easy to access web-interface allowing others to play with such a tool and to further gain experience on its capabilities, without requiring additional installation effort. Hence, the original application scenario has been debugging at the method level of Java-like programs focusing on the non-object-oriented part only. Moreover, when describing models of systems using a program Java2CSP can also be used to provide means for diagnosing systems as well. This scenario has already been discussed in our previous work [12] where we introduced an initial version of Java2CSP.

Besides the more obvious application scenarios of Java2CSP for debugging and diagnosis, we already mentioned the use constraints for testing. In particular, [6] showed how a constraint representation of a program can be used to obtain inputs and outputs. Because Java2CSP allows for accessing an SMT-LIB presentation of a program used by Z3 for constraint solving, the presentation can also be used for testing. The only changes that are necessary are to add constraints setting the ab_i variables to false as already described in a previous section. With additional constraints test cases for given outputs as well as test cases fulfilling certain properties can be generated.

The last application scenario we have in mind is to use Java2CSP for obtaining constraints representing system invariants that are checked during the operation of systems. Such monitoring functionality can be used at runtime to identify critical scenarios that have to be avoided. In the context of autonomous driving, [13] outlined ideas behind a monitoring system that raises a warning or error message in case the observed behavior contradicts given properties. When using Java2CSP, properties can be expressed using the syntax of Java instead of other specialized programming languages like SMT-LIB. Because Java is more well-known its use might be more appropriate for practical applications.

To make use of Java2CSP for monitoring we may represent any property p as a Boolean variable v where its value given using the property. For example, if we want to state that an autonomous car in a town should never exceed the speed limit of 50 km/h and on a highway of 130 km/h, we can express this easily using the following Java code fragment:

```
if (inCity) { v = (velocity <= 50) }
else {
    if (onHighWay) { v = (velocity <= 130) }
    else { v = true } }
```

Java2CSP converts such code into a constraint representation that can be checked using Z3 during operation. We only need to specify that v should be true at the output as an observation. Any violation of the property would lead to setting v to false contradicting the observation. Besides being able to identify the violation we can further indicate the reason behind the violation by making use of diagnosis.

6   Conclusions

In this paper, we introduce the debugging tool Java2CSP that is publicly available over the internet. The only prerequisites are the availability of a web browser. We discuss the underlying foundations, how to use the tool, and the results obtained when carrying out an initial evaluation. The tool has not been developed to be used within any ordinary software development environment. Instead, we focused on providing a tool for demonstrating the capabilities of model-based reasoning when applied to debugging students and researchers can easily access.

Java2CSP supports only an imperative subset of Java, which includes ordinary if-then-else statements and while loops. Basic data types and basic arrays are supported as well. In the future, we plan to (1) improve the user interface. Note that a new version of the interface is available on the same web page for testing purposes. (2) to extend the subset supported for debugging including non-recursive method calls, as well as other statements. (3) improve the evaluation by considering more and larger examples. It is planned to make all the programs and inputs/outputs used for evaluation public available. (4) Provide an installer for loading and installing the Java2CSP tool locally on a computer to allow its integration into development environments. Furthermore, (5) we plan to make use of the tool for integrating property checking into system monitoring, allowing users to specify properties in a Java-like syntax.

Acknowledgements. ArchitectECA2030 receives funding within the Electronic Components and Systems For European Leadership Joint Undertaking (ESCEL JU) in collaboration with the European Union's Horizon2020 Framework Programme and National Authorities, under grant agreement number 877539. All ArchitectECA2030 related communication reflects only the author's view and the Agency and the Commission are not responsible for any use that may be made of the information it contains. The work was partially funded by the Austrian Federal Ministry of Climate Action, Environment, Energy, Mobility, Innovation and Technology (BMK) under the program "ICT of the Future" project 877587.

References

1. Abreu, R., Zoeteweij, P., Golsteijn, R., van Gemund, A.J.C.: A practical evaluation of spectrum-based fault localization. J. Syst. Software **82**(11), 1780–1792 (2009). https://doi.org/10.1016/j.jss.2009.06.035
2. Beizer, B.: Software Testing Techniques. Van Nostrand Reinhold, New York (1990)
3. Boehm, B.W.: Improving software productivity. Computer **20**, 43–57 (1987)
4. Brandis, M.M., Mössenböck, H.: Single-pass generation of static assignment form for structured languages. ACM TOPLAS **16**(6), 1684–1698 (1994)

5. de Moura, L., Bjørner, N.: Z3: an efficient SMT solver. In: Ramakrishnan, C.R., Rehof, J. (eds.) TACAS 2008. LNCS, vol. 4963, pp. 337–340. Springer, Heidelberg (2008). https://doi.org/10.1007/978-3-540-78800-3_24

6. Gotlieb, A., Botella, B., Rueher, M.: Automatic test data generation using constraint solving techniques. In: Proceedings of the 1998 ACM SIGSOFT International Symposium on Software Testing and Analysis. ISSTA 1998, New York, NY, USA, pp. 53–62. Association for Computing Machinery (1998). https://doi.org/10.1145/271771.271790

7. Krinke, J.: Effects of contex on program slicing. J. Syst. Softw. **79**, 1249–1260 (2006)

8. Reiter, R.: A theory of diagnosis from first principles. Artif. Intell. **32**(1), 57–95 (1987)

9. Weiser, M.: Program slicing. IEEE Trans. Software Eng. **10**(4), 352–357 (1984)

10. Wong, W.E., Gao, R., Li, Y., Abreu, R., Wotawa, F.: A survey on software fault localization. IEEE Trans. Software Eng. **42**(8), 707–740 (2016). https://doi.org/10.1109/TSE.2016.2521368

11. Lughofer, E., Sayed-Mouchaweh, M. (eds.): Predictive Maintenance in Dynamic Systems. Springer, Cham (2019). https://doi.org/10.1007/978-3-030-05645-2

12. Wotawa, F., Dumitru, V.A.: Java2csp - a model-based diagnosis tool not only for software debugging. In: Intelligent Decision Technologies 2021 - Proceedings of the 13th KES International Conference on Intelligent Decision Technologies (KES-IDT-21). Smart Innovation, Systems and Technologies, Springer Nature Singapore Pte Ltd. (2021)

13. Wotawa, F., Lewitschnig, H.: Monitoring hierarchical systems for safety assurance. In: Intelligent Distributed Computing XIV. Studies in Computational Intelligence. Springer (2022)

14. Wotawa, F., Nica, M., Aichernig, B.K.: Generating distinguishing tests using the minion constraint solver. In: ICST Workshops. pp. 325–330. IEEE Computer Society (2010)

15. Wotawa, F., Nica, M., Moraru, I.: Automated debugging based on a constraint model of the program and a test case. J. Log. Algebraic Methods Program. **81**(4), 390–407 (2012). https://doi.org/10.1016/j.jlap.2012.03.002

Formal Modelling and Security Analysis of Inter-Operable Systems

Abdelhakim Baouya[1(✉)], Samir Ouchani[2], and Saddek Bensalem[1]

[1] VERIMAG, Université Grenoble Alpes, Saint-Martin-d'Hères, France
{abdelhakim.baouya,saddek.bensalem}@univ-grenoble-alpes.fr
[2] LINEACT CESI, Aix-en-Provence, France
souchani@cesi.fr

Abstract. Emerging technologies utilised in building modern systems make them inter-operable but potentially exposed to security threats. Thus, engineers need to consider the system structure and behaviour at the design level. This paper addresses the security risk assessment of inter-operable IoT systems designed in BIP (Behaviour-Interaction-Priority). For this purpose, we model different attacks scenarios from Microsoft STRIDE threats catalogue and identify the threats entry points in the component-port-connector architecture. Using standards communication styles such as message passing, we compose architectural components to model data flow between communication entities. We use BIP statistical model checking to assess the architecture conformance regarding security properties expressed in temporal logic.

Keywords: IoT · Component & Connectors · Threats · Automata · Statistical model checking

1 Introduction

Security is a pillar system requirement and its need to be considered at the design level leads to the adoption of *security-by-design* paradigm [4]. Yet, a more ambitious challenge is to specify precisely, express accurately, and measure correctly the system's security level based on its design structures. Security analysis at the architectural level is essential as it helps prevent flaws from being propagated to the implementation and then the deployment. G. McGraws [12] defines a set of software security activities at different software development lifecycle commonly considered by Microsoft's Security Development Lifecycle[1] and Adobe's Secure Product Lifecycle[2] Gary McGraws focuses on architectural level security since 50% of reported flaws constitute the detected vulnerabilities in the implementation.

From a practical side, emerging platforms under the IoT -Internet of Things-moniker involves lightweight communication protocols (i.e., MQTT[3] Bluetooth,

---

[1] SDL: www.microsoft.com/security/sdl.

[2] SPLC: www.adobe.com/security/splc.

[3] MQTT: http://docs.oasis-open.org/mqtt/mqtt/v3.1.1/mqtt-v3.1.1.html.

© Springer Nature Switzerland AG 2022
H. Fujita et al. (Eds.): IEA/AIE 2022, LNAI 13343, pp. 555–567, 2022.
https://doi.org/10.1007/978-3-031-08530-7_47

CoAP[4]), making them vulnerable to attacks since devices are constrained in terms of resources. IoT Communication protocols rely on common existing state-of-the-art communication styles with well-known behaviour such as message passing implementing FIFO, buffering, Remote procedure call, and distributed shared memory.

In this context, we rely on formal methods to precisely model IoT systems in a component-based fashion using BIP[5] (Behaviour-Interaction-Priority). Following the characteristics of the communication protocol (Client-Server), we address security issues by composing attack scenarios (i.e., threats) based on Microsoft STRIDE [13] threats classification to the classical communication styles using BIP formalism (Component-Port-Connector [7]). Then, logical specifications of security properties are expressed in LTL [2] to perform Statistical Model Checking using SMC-BIP [14]. The contributions of the paper are: (1) Formal modelling using BIP of inter-operable systems is proposed to describe components, structures, behaviours, and composition, (2) Specifying a set of formal descriptions of threats that use connectors entry points, (3) Performing statistical model checking to verify the security vulnerability at the architectural level.

2   Related Work

Threats formalization for verification and validation of systems architecture has been addressed recently. Existing formalization attempts for threats modelling include the work in [8] using VDM++ to specify the threats core components based on STRIDE and DREAD catalogues in addition to confidentiality, integrity, availability, authorization and non-repudiation security mechanisms. The authors advocate the adoption of formal methods and do not propose the framework that performs the verification. Authors of [15] model the software systems and CAPEC [19] threats in SysML. They propose a Model-To-Model transformation where the specification is converted into PRISM language to perform probabilistic model checking. The specified probabilities in SysML depend on the threats load, and its correspondence to Kent's Words of Estimative [9] Probability (intelligence analysts use i.e., Kent's Words of Estimative). In [16], the authors model a set of attack scenarios (i.e., Flooding, Session Hijacking and TEARDOWN). They are composed with a communication protocol to check the validity of the composed systems written in PRISM for probabilistic verification. Modelling threats composed with the software behaviour in PRISM is not advocated due to (1) The state-space explosion problem and (2) The lack of modelling systems in Component-Port-Connector formalism and thus make the model abstracted. In [18], authors propose FATHoM framework (FormAlizing THreat Models) to define threat models. A set of relations and rules are used to define how to protect the derived vulnerable components. FATHoM compiles the threat model (derived from the template and the user choices) and the model's

---

[4] CoAP: https://www.ietf.org/archive/id/draft-shelby-core-coap-01.txt.

[5] BIP: https://www-verimag.imag.fr/TOOLS/DCS/bip/doc/latest/html/index.html.

rules into an executable program for XSB[6]. The tool is only used to observe the derivation in action using predefined relations and rules. The work presented in [6] propose a new metamodel to model software architecture in Component-Port-Connector formalism. The architecture behaviour uses Wright [5] where the connector type is defined as communicating Sequential Process (CSP) [10]. The architecture is checked against a set of LTL properties using PAT model checker [11]. Also, the model is animated against different attack scenarios. Unfortunately, Wright is not able to model complex systems with different data types. To show the expressiveness of BIP language, [3] highlights the efficiency of modelling Blockchain architecture for IoT systems in BIP following the Component-Port-Connector formalism and then protecting IoT platforms against classical threats such as tampering and repudiation. The modelled system is then checked using SMC-BIP [14] against liveness properties (i.e., locating safe states). The work done in [17] propose a metamodel to instantiate system architecture in Component-Port-Connector formalism that is converted to Alloy. The authors use modal logic to express threats and then permits security threats analysis in the Alloy analyzer. However, the approach cannot model complex systems.

3 Systems Composition Semantics

To develop a precise system composed from different components where each executes a local task, the composition operators should be well defined to aggregate local tasks into a global task. We rely on BIP to model them precisely where a system is equivalent to a BIP model, sub-systems are atomic components in BIP that connect through ports to compose an inter-operable global system. Atomic components are elementary building blocks for modelling a sub-system in BIP. They are described as labelled transition systems extended with *variables.* To deal with the system's inter-operability, BIP formalism provides connection operators that allow composing and coordinating atomic components called *glue.* The latter provides mechanisms for harmonizing and coordinating components behaviours, namely priorities and interactions. In the following, we focus on the inter-operability between the system's components to analyze the security of a composed system using message passing and remote procedure calls that are known as attack surfaces.

3.1 Message Passing

In message-passing styles (MPS), a channel is responsible to rout data between components. The message is transmitted by following two communication policies: FIFO ordering or buffering. However, the buffered message passing involves three parameters: `size` is the length of the current buffer, `buffer` is the memory space (stack) where messages arrive, and `iterator` to point the current message address. For a simple client-server interaction, MPS semantics is an automata of three states Init, S1, and S2 as portrayed in Fig. 1.

---

[6] http://xsb.sourceforge.net/.

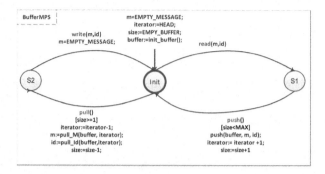

Fig. 1. Automata for buffer based MPS.

For our evaluation, we express the whole automata in BIP. For example Listing 1.1 expresses in BIP modeling language of the state machine presented in Fig. 1. The BIP states are defined in line 8 by the preceded keyword **place** . We identify in the model two kinds of ports: export and internal ports. The export ports are represented by a red colour such as `write` and `read` . Both ports are used to write and read communicated parameters from clients or servers. The used parameters in this model for the export port are the message m and the identifier of the sender id. In addition, the ports **pull** and **push** are internal, silent, and do not synchronize with external components. The BIP definition of ports is mentioned in lines 15–18, where the exported ports are preceded by the keyword **export** . The used ports are instantiated from the ports definition in lines 6–7.

Listing 1.1: Buffered message passing component in BIP

```
1   const data int MAX = 10
2   const data int EMPTY_BUFFER = 0
3   const data int EMPTY_MESSAGE = -1
4   const data int HEAD =0
5   extern data type Buffer
6   port type port_type_mess (int m, int id)
7   port type port_type_Silent ()
8   atom type BufferMPS ()
9       data int iterator
10      data int size
11      data int m
12      data Buffer buffer
13      data int id
14
15      export port port_type_mess write(m, id)
16      export port port_type_mess read(m, id)
17      port port_type_Silent push()
18      port port_type_Silent pull()
19
20      place Init, S1, S2      //Transitions definition
21  end
```

Running the BIP automata triggers the initialization that performs BIP actions as portrayed in Fig. 1 which is represented by BIP block preceded by the keyword **initial to** as follow.

```
initial to Init
do {          size = EMPTY_BUFFER;
              m = EMPTY_MESSAGE;
              iterator = HEAD;
              buffer= init_buffer(); }
```

The component variables are initialized using constant data defined in Listing 1.1 in lines 1–4 and preceded by the keyword **const data** . Four transitions are identified and labelled by internal and export ports. For instance, the synchronized transition labelled with port read enables the synchronized actions with the sender Init → S1 by collecting the message m and the sender identified by id, whereas, the synchronized transition S2 → Init labelled with port **write** resets the received message while it writes the message m and the identifier of the sender to the server. Both transitions are modelled as follows.

```
         on read from Init to S1
on write from S2 to Init do { m = EMPTY_MESSAGE; }
```

The following Internal transitions are labelled with silent ports. For instance, the transition S1 → Init is labelled with port **push** performing three actions; (1) it pushes to the buffer the received message and the identifier of the sender using push(), (2) it increments the iterator, and (3) it increments the size of the buffer.

```
on push from S1 to Init provided (size <=MAX) do {
m=push(buffer,m, id iterator); iterator =iterator +1; size=size+1;}
```

The transition described below Init → S2 labelled with port **pull** (1) retrieves the message m and the identifier of the sender id using the functions pull_M() and pull_Id(), and (2) decrements the iterator, then (3) decrements the buffer size.

```
on pull from Init to S2 provided (size >=1) do {
iterator =iterator -1;
m=pull_M(buffer, iterator); id=pull_Id(buffer, iterator); size=size-1;}
```

For the inter-operability, **Rendez-vous** connector is used to accomplish the synchronization between the communicating entities using the MPS connector. The connector is endowed with two port parameters (i.e., **Entry points**) of type port_type_mess defined in Listing 1.1, line 6. The behaviour of the connector is defined in Listing 1.2. The interaction points are defined in line 3 and the behaviour is defined in lines 4–5 to perform a transfer of the message m and the identifier id from port a to port b.

Listing 1.2: Rendez-vous connector definition in BIP

```
1   connector type Connector_Type_Client_Server
2                   (port_type_mess a, port_type_mess b)
3       define a b
4       on a b down { b.m = a.m;
5                     b.id= a.id;}
6   end
```

3.2 Remote Procedure Call

In the Remote Procedure Call (RPC) communication style, a channel is used for sending invocation (request) messages from client to server and for receiving acknowledgement (reply) messages from a server to a client. The buffer is also used to model RPC communication since it implements the same functions as in MPS.

Listing 1.3: Client and server components in BIP

```
1   atom type Client (int id)
2       data int m
3       export port port_type_mess write(m, id)
4       export port port_type_mess read(m, id)
5       port port_type_silent produce( )
6       place Init, S1, S2
7       initial to Init { m=EMPTY_MESSAGE;}
8       on produce from Init to S1 do { m= produce_message();}
9       on write from S1 to S2
10      on read from S2 to Init do {m=EMPTY_MESSAGE;}
11  end
12  atom type Server ()
13      data int m
14      data int id
15      export port port_type_mess write(m, id)
16      export port port_type_mess read(m, id)
17      port port_type_silent consume( )
18      place Init, S1, S2
19      initial to Init { m=EMPTY_MESSAGE;}
20      on read from Init to S1
21      on consume from S1 to S2
22      on write from S2 to Init do {m=EMPTY_MESSAGE;}
23  end
```

The BIP code of the client and server components is enhanced with new ports and also implies behaviour modification in Listing 1.3. The state S2 is necessary for both atomic components (i.e., client and server). For instance, the client behaviour is developed with three transitions in lines 8–10: (1) The transition Init → S1 labelled with port **produce** produces the data, (2) S1 → S2 labelled with port write writes data to the server, and (3) transition S2 → S1 labelled with port read reads the reply from the server. Moreover, the server component holds three transitions in lines 21–23: (1) The transition Init → S1 labelled with port read receives data from the client, (2) S1 → S2 labelled with port **consume** consumes the received data from the client, and (3) transition S2 → S1 labelled with port write sends a reply message to the client.

4 Security Threats Analysis Using SMC-BIP

In this section, we specify representative properties for each STRIDE category (spoofing, Tampering, Repudiation, Information Disclosure, Denial of Service, Elevation of privilege) such that the violation of the specified property expressed in LTL indicates the presence of the threat. The BIP codes for each threat are available in [1].

4.1 Spoofing

Spoofing refers to the impersonation of a component in the system to mislead other system entities into falsely believing that the attacker is legitimate. The spoofing property φ_1 is expressed in LTL as follow.

$$\varphi_1 : P_{=?}[(\text{C1.m} = \text{x \& C1.id} = 1) \ \cup \ (\text{C2.m} = \text{x \& C2.id} = 1)] \tag{1}$$

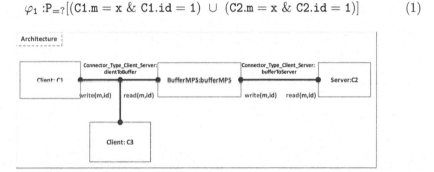

Fig. 2. Architecture of client-server including malicious component C3.

The property 1 shows that every message that is received by component C2 makes it believes it was sent only by component C1. SMC-BIP will explore the traces generated and then check to outcome the verdicts. Also, to model the threat, we augment the model using MPS communication style with a new sender that plays a role of a malicious component called C3 with id equals to 3 as portrayed in Fig. 2.

Figure 3 shows that the message can be received at component C2 with a different id since the probability to collect the message $\text{m} \in \{5,6\}$ during the simulation using MPS style is equal to 0% whereas the probability to collect the message $\text{m} \in \{7,8,9\}$ using RPC style is also equal to 0%. SMC BIP employs a random execution with all possible executions that may happen. The results show that the server can receive a message with falsely id in the case where no authentication is employed.

4.2 Tampering

Tampering threats involve the modification of data in transit. As expressed in LTL by φ_2, tampering refers to unauthorized modification of data.

$$\varphi_2 : P_{=?}[(\text{C1.m} = \text{x \& C1.id} = 1) \ \cup \ (\text{C2.status} = 1 \ \& \ \text{C2.id} = 1)] \tag{2}$$

For the property φ_2, the BIP code is rewritten as follows to add the payload variable at the client and server atomic components. The port type is also redefined regarding Listing 1.1 where d is the payload.

```
port type port_type_mess ( int m, int id, int d)
```

The attacker atomic component that we integrate with a component shall intercept the message when the communication component (MPS or RPC) are synchnized with the client component and then send a fake payload. The redefinition of the connector is described as in Listing 1.4 using a **broadcast** style. The connector ports a, b, and c are used by a client, server, and attacker respectively. When the attacker synchronizes with the client and the server, it sent the altered message m (line 8) and the wrong payload d (line 10).

```
Listing 1.4: Broadcast connector definition for tampering illustration in BIP
1    connector type Connector_Type_Client_Server_Attacker
2                   (port_type_mess a, port_type_mess b, port_type_mess c)
3       define (a b)' c
4       on a b down {  b.m = a.m;
5                      b.id= a.id;
6                      b.d= a.d;}
7       on a b c down {  b.m = c.m;
8                        b.id= c.id;
9                        b.d= a.d;}
10   end
```

When the server receives the message, it computes the message payload when its consumed. Then, actions related to the transition S1 → Init labelled with **consume** are updated. The server evaluates the payload of the message m and compares it against the payload d using if-then-fi structure supported by BIP. For the case where MPS style is used, the server code is updated as follows.

```
on consume from S1 to Init do {status=0; if (payload(m)==d) then status =1; fi }
```

Property 2 is checked using SMC-BIP and the results are retrieved from the traces analysis portrayed in Fig. 5. We observe that the message is altered during the transfer when the message $m \in \{5, 9\}$ use MPS communication style. Also, the message is altered when the message $m \in \{6, 8\}$ use RPC communication style (Fig. 4).

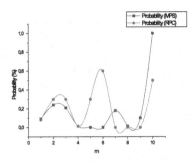

Fig. 3. Spoofing checking for φ_1.

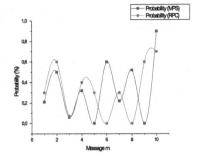

Fig. 4. Tampering checking for φ_2.

4.3   Repudiation

Repudiation refers to a component claiming to have not performed an action that was performed, in reality. The sender generates the message hash, encrypts it (i.e., signature), and transmits it. When the server receives the message, it computes the hash message and compares it with the decrypted received signature. In that case, the repudiation property φ_3 is expressed in LTL as follows.

$$\varphi\_3 : P_{=?}[(C1.m = x) \ \cup \ (C2.status = 0)] \tag{3}$$

The status of the computed value is the comparison the the hash message and the decrypted received signature at the server C2. Also, the BIP code is rewritten to add the signature variable `sig` at the client and the server atomic component. The port type is redefined regarding Listing 1.1 where `sig` is the signature described below.

port type port_type_mess (**int** m, **int** sig)

The connector between the clients and the component handling the communication operator is updated to make the malicious attacker sending messages with wrong signature. The defined connector in Listing 1.5 where the port b triggers the communication when the client using port a or attacker using port c are available.

```
Listing 1.5: Repudiation illustration in connector definition
1    connector type Connector_Type_Client_Server_Attacker
2                   (port_type_mess a, port_type_mess b, port_type_mess c)
3        define b' a c
4        on b a down {   b.m = a.m;
5                        b.sig = a.sig;}
6        on b c down {   b.m = c.m;
7                        b.sig = c.sig;}
8    end
```

The following transition S1 $\rightarrow$ Init expresses when the server receives the signature, and performs the hashing function and decryption.

```
on consume from S1 to Init do {
status=0;
if (decrypt(sig)==Hash(m)) then status =1; fi }
```

Property 3 is checked using SMC-BIP and the obtained results are plotted in Fig. 5. We observe that the message is correctly received by the server using RPC communication style when $m \in \{1, 2, 3, 8\}$. Moreover, m are correctly received by the server using MPS style when $m \in \{3, 6, 7\}$. Exceptions arise when some explored traces do not highlight the message correctness at the server.

4.4 Information Disclosure (IDS)

Information disclosure refers to the unauthorized exposure of information to a component for which it is not intended. It occurs when components other than those for which a message was intended are able to receive the sent message. Property φ_4 expresses whether the message x is received by the malicious component C3 placed between the server and the communication component (RPC and MPS). The message shall be received with the sender identifier id ==1.

$$\varphi_4 : P_=?[(C1.m == x) \cup (C3.m == x \text{ \& } C3.id == 1)] \tag{4}$$

The checking results of Property 4 using SMC-BIP are portrayed in Fig. 6. We observe that φ_4 is satisfied for almost all messages in RPC and MPS communication modes. It proves that the probability of intercepting the messages by a malicious component playing a server role is high.

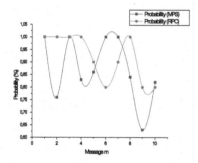

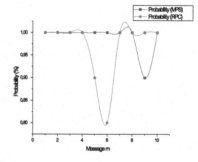

Fig. 5. Repudiation checking for φ_3.

Fig. 6. IDS checking for φ_4.

4.5 Denial of Service (DoS)

In the context of message passing communication, denial of service threats may involve blocking the transmission of messages from senders to receivers. Thus, the receiver never receives any of the messages before their freshness expire. Property

φ_5 expresses whether threat can be identified by verifying if sent messages are delayed, destroyed, or deleted in transit.

$$\varphi_5 : P_{=?}[F \ (C1.m == x) \ \& \ (C2.status == 0 \ || \ C2.i > 0)] \tag{5}$$

Property 5 states that every message m sent by a component C1 intended for a component C2 and that contains a payload d with a freshness interval i has the opportunity to be received at some point by the intended receiver C2 in the interval i. The server configuration will be the same for tampering modeling in BIP, except that the client is modified to add the interval i that is incremented if the synchronization with the communication channel does not occur and performed on client transition labelled with **produce** if the message has not been consumed (i.e., sent). The results are portrayed in Fig. 7. Regarding the results, a high probability is observed in both communication modes (RPC and MPS) caused by interval timeout or payload failure checking (i.e. status=0).

4.6   Elevation of Privilege

Elevation of privilege refers to the ability of a component to gain capabilities without proper authorization to have such capabilities. The BIP model is enhanced with a new connector definition that asserts that the channel is used only if the source identifier is greater or equal to authorized one. The properties φ_6 and φ_7 express whether the messages sent by C1 and C2 are received at the server C3.

$$\varphi_6 : P_{=?}[(C1.m == x) \ \cup \ (C3.m == x \ \& \ C3.id == 1)] \tag{6}$$

$$\varphi_7 : P_{=?}[(C2.m == x) \ \cup \ (C3.m == x \ \& \ C3.id == 1)] \tag{7}$$

The connector redefinition is portrayed in Listing 1.6. Two entry points for the client and the tester are identified (a and c). The constraints are set using the provided keyword. Only client with id=1 are allowed to send the messages.

```
    Listing 1.6: Elevation of privilege in connector definition
1   connector type Connector_Type_Client_Server_Attacker
2                  (port_type_mess a, port_type_mess b, port_type_mess c)
3       define b' a c
4       on b a provided ( a.id == 1 )
5              down { b.m = a.m;
6                     b.id= a.id;}
7       on b c provided ( c.id==1)
8              down { b.m = c.m;
9                     b.id = c.id;}
10  end
```

The results portrayed in Fig. 8 show that the server will not receive the data from the tester since it has not the authorization defined in the connector. Also, the constraints can be defined at the server level while processing the messages and the identifiers consumed after buffer and server synchronization.

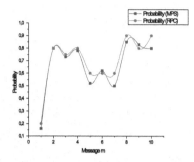

Fig. 7. DoS checking for φ_5.

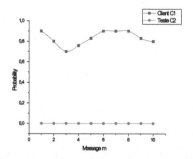

Fig. 8. Privilege checking for φ_6, φ_7.

5   Conclusion

In this paper, we have demonstrated the practical application of BIP modeling language to design complex systems in Component-Port-Connector formalism with behaviour expressed in automata fashion. We adopt the BIP language due to itse high expressiveness compared to the existing ones. Microsoft's STRIDE catalogue is used to model classical threats observed in IoT systems. The modelled IoT system is composed with threats and then checked using SMC-BIP against properties that express the validity of the failure caused by malicious components. In our future work, we intend to reuse the defined communication styles as BIP libraries to model complex protocols composed with CAPEC attacks scenarios and analyse the outcoming results. In addition, we want to model the software composition with hardware failures to check the system's safety.

Acknowledgement. The research leading to the presented results has been undertaken within the research profile **CPS4EU**, funded by the European Union, grant number: 826276.

References

1. Stride threats in BIP. https://github.com/hakimuga/Threat_Modeling_BIP
2. Amir, P.: The temporal logic of programs. In: 2013 IEEE 54th Annual Symposium on Foundations of Computer Science, pp. 46–57. IEEE Computer Society, October 1977
3. Baouya, A., Chehida, S., Bensalem, S., Bozga, M.: Formal modeling and verification of blockchain consensus protocol for IoT systems. In: Proceedings of the 19th International Conference on New Trends in Intelligent Software Methodologies, Tools and Techniques, SoMeT 2020, 22–24 September 2020, vol. 327, pp. 330–342. IOS Press (2020)
4. van den Berghe, A., Yskout, K., Scandariato, R., Joosen, W.: A lingua franca for security by design. In: 2018 IEEE Cybersecurity Development (SecDev), pp. 69–76 (2018)

5. Chondamrongkul, N., Sun, J., Warren, I.: PAT approach to architecture behavioural verification, pp. 187–192, July 2019
6. Chondamrongkul, N., Sun, J., Warren, I.: Formal security analysis for software architecture design: an expressive framework to emerging architectural styles. Sci. Comput. Program. **206**, 102631 (2021)
7. Crnkovic, I.: Component-based software engineering for embedded systems. In: ICSE 2005, pp. 712–713 (2005)
8. Hussain, S., Erwin, H., Dunne, P.: Threat modeling using formal methods: a new approach to develop secure web applications. In: 2011 7th International Conference on Emerging Technologies, ICET 2011, September 2011
9. Kent, S.: Sherman Kent and the profession of intelligence analysis, center for the study of intelligence, central intelligence agency, p. 55, November 2002. https://www.cia.gov/library/kent-center-occasional-papers/vol1no5.htm
10. Kruchten, P.: The 4+1 view model of architecture. IEEE Softw. **12**(6), 42–50 (1995)
11. Liu, Y., Sun, J., Dong, J.S.: PAT 3: An extensible architecture for building multi-domain model checkers. In: 22nd ISSRE 2011, pp. 190–199. IEEE (2011)
12. McGraw, G.: Software Security: Building Security in. Addison-Wesley Professional Computing Series. Addison-Wesley, United States (2006)
13. Microsoft, the STRIDE Threat Model: Microsoft corporation (2009). https://docs.microsoft.com/en-us/previous-versions/commerce-server/ee823878(v=cs.20)?redirectedfrom=MSDN
14. Nouri, A., Mediouni, B.L., Bozga, M., Combaz, J., Bensalem, S., Legay, A.: Performance evaluation of stochastic real-time systems with the SBIP framework. Int. J. Crit. Comput. Based Syst. **8**, 1–33 (2018)
15. Ouchani, S., Jarraya, Y., Ait Mohamed, O., Debbabi, M.: Probabilistic attack scenarios to evaluate policies over communication protocols. J. Softw. **7**, 1488–1495 (2012)
16. Ouchani, S., Mohamed, O.A., Debbabi, M.: A security risk assessment framework for SysML activity diagrams. In: IEEE 7th International Conference on Software Security and Reliability, pp. 227–236 (2013)
17. Rouland, Q., Hamid, B., Jaskolka, J.: Specification, detection, and treatment of stride threats for software components: modeling, formal methods, and tool support. J. Syst. Archit. **117**, 102073 (2021)
18. Sgandurra, D., Karafili, E., Lupu, E.: Formalizing threat models for virtualized systems. In: Ranise, S., Swarup, V. (eds.) DBSec 2016. LNCS, vol. 9766, pp. 251–267. Springer, Cham (2016). https://doi.org/10.1007/978-3-319-41483-6_18
19. U.S. Department of Homeland Security: Common attack pattern enumeration and classification, November 2002. http://capec.mitre.org

Social Network Analysis

Content-Context-Based Graph Convolutional Network for Fake News Detection

Huyen Trang Phan[1] ORCID, Ngoc Thanh Nguyen[2] ORCID, and Dosam Hwang[1]([⊠]) ORCID

[1] Department of Computer Engineering, Yeungnam University, Gyeongsan, Republic of Korea
dosamhwang@gmail.com
[2] Department of Applied Informatics, Wroclaw University of Science and Technology, Wroclaw, Poland
Ngoc-Thanh.Nguyen@pwr.edu.pl

Abstract. The development of social networks is increasing, and such networks are gradually becoming critical news sources for many people. However, not all news sources on social media are trustworthy. Numerous news stories containing false information have appeared and have spread on social networks, fulfilling the specific aims of certain individuals. Such misinformation is known as fake news, which is becoming increasingly sophisticated, making it difficult to immediately distinguish from real news. Therefore, fake news remains an entirely unresolved problem in social networks and is an attractive subject to many researchers. We propose a content-context-based graph convolutional network (C&C-GCN) as a novel method for representing and learning social content and context for fake news detection. The proposed method integrates information in terms of (i) the content of the news, and (ii) the social context of the news, such as the user who created or shared the news and the source that published the news. Unlike previous methods, C&C-GCN is better at simultaneously capturing the content and context of news. Experiment results when applying benchmark datasets show that this model can extract more information from shared news into a better representation of the graph structure and learn better features represented on a graph. Furthermore, it considers user sentiment to be a significant feature of news content.

Keywords: Fake news detection · GCNs · C&C-GCN · Social networks

1 Introduction

Social networks have contributed to an information explosion and have become the main communication channel for people worldwide [12]. However, the authenticity of news posted on social networks often cannot be determined. Thus, there are both pros and cons to the use of social networks: If the news received from

© Springer Nature Switzerland AG 2022
H. Fujita et al. (Eds.): IEA/AIE 2022, LNAI 13343, pp. 571–582, 2022.
https://doi.org/10.1007/978-3-031-08530-7_48

a social network is real, it will be of benefit to users, whereas if the news is fake, it can have significantly harmful consequences, and the extent of the damage incurred when fake news is widely spread can be incalculable. Fake news comprises completely fabricated information, deceptive content, or completely distorted factual reports [1]. Many examples of fake news have occurred, mainly since the emergence of the COVID-19 pandemic at the end of 2019. With the implementation of lockdowns and regulations restricting people from leaving their homes when being unnecessarily in certain countries, much fake news has been published and spread on social media. In particular, when governments began vaccination campaigns, fake news exploded. These examples illustrate that the spread of fake news on social media has significantly impacted many sectors. Thus, the timely detection and prevention of fake news before widespread spreading are urgent tasks. Therefore, many methods to detect and limit fake news dissemination have been proposed and grown in the past decade. Graph convolutional networks (GCNs) based approach is among the most recent.

GCNs are attracting increasing interest as a way to extend deep learning approaches to graph data [16]. Before their application in fake news detection systems, GCNs were applied in many natural language processing tasks with satisfaction performance, such as object detection [13], machine translation [7], and sentiment analysis [17]. The appearance of more and more social networks leads to increased users, the posted news, and the users' interactions. Although some promising results have been achieved, using the information related to the users, the posted news, and the users' interactions simultaneously for fake news detection is still a challenge for previous methods. Solving this problem is the primary motivation for GCN-based fake news detection methods that are proposed and grown (the first justification).

Many algorithms that use GCNs for fake news detection have recently been proposed. However, most GCN-based fake news detection approaches focus on the use of propagation, content, or context features independently for classification. Although a significant number of methods have relied on the combination of two out of three features simultaneously, there are no approaches that use a hybrid of content and context features simultaneously in a single model. Therefore, based on our assessment, this is also a challenge currently faced by fake news detection methods (the second justification).

The two justifications above motivated us to propose a content-context-based graph convolutional network (C&C-GCN), which is a novel method for representing and learning the social content and context for fake news detection. The proposed method integrates information in terms of (i) the content of the news, and (ii) the social context of the news, such as the user who created or shared the news, and the source that contained the news. Experiment results based on some available datasets have shown that the C&C-GCN model is more effective than recent GCN-based algorithms, such as a GCN-based method [5], and CNN-based text classification (TextCNN) [18]. Our contribution is briefly as follows: (i) We propose a C&C graph data structure to efficiently represent the content and context of the news. (ii) We propose a C&C-GCN algorithm for fake news detection in benchmark dataset. (iii) We evaluate the proposed algorithm by comparing it to baseline methods using the test dataset.

The remainder of this paper is organized as follows. Section 2 reviews previous related works. Section 3 introduces the research problem of the proposed method. Section 4 describes a mathematical model for solving the research problem. A brief introduction to the datasets and experiment results of the proposed approach comparison to other well-known methods are provided in Sect. 5. Some concluding remarks and areas of future work are presented in Sect. 6.

2 Related Works

The GCN-based approach is a category of methods that is mostly used for fake news detection relying on GNNs. GCNs are an extension of GNNs that take a graph structure and integrate node information from neighborhoods based on a convolutional function. GCNs can represent graphs and achieve a state-of-art performance in various applications and tasks, including fake-news detection.

Nguyen et al. [15] presented two textual- and graph-based methods for predicting false news related to the Covid-19 pandemic and the 5G networking. The first method captured textual and metadata features based on a pre-trained BERT method and a multilayer perceptron method. The second method extracted nine content features, including the page rank, hub, and authority, to construct a GCN-based fake news detection model. After that, the first method's performance is better than the second because metadata significantly improves the efficacy. For the same topic regarding Covid-19 and 5G conspiracies, unlike other GCN-based methods, in [9], two methods, GCN and DGCN, are used to capture the temporal features of the social networks without considering any of the textual features. Li et al. [6] proposed a political perspective determination method which is a GCN-based approach by extracting the propagation of social contextual information. Meyers et al. [8] proposed a feature propagation-based model. First, a propagation graph is built to extract the critical information. Then, a random forest model is used over the propagation graph for graph representations. Finally, the GCN model is employed for fake news prediction. Different from other graphs, the propagation graph in [8] is constructed from a set of sub-graphs, where each sub-graph consists of a tweet node and its retweet nodes. Therefore, its depth is never more than 1.

The above studies prove that GCN-based methods have been applied to fake news detection, achieving quite a good performance, and are still an area of interest to researchers. However, most of these methods construct graphs by representing features such as content and context separately. We wondered whether using content and context features simultaneously when building graphs improves the performance of fake news detection methods. In other words, the development of a content- and context-based GCN for fake news detection is still a challenge, which we hope to resolve through this study.

3 Problem Definition and Research Questions

Given a finite set of p news, for a specific news n_i, let c_i be a set of features related to news content (such as semantic, lexical, syntactic) and t_i be a set of features regarding news context (such as news sources, news publishers, and news interactions). The objective of this proposal is to construct a C&C-GCN for fake-news detection. This objective can be formalized by finding a mapping function $F : (c_i \cup t_i) \rightarrow \{0, 1\}$ such that:

$$F(c_i \cup t_i) = \begin{cases} 1, & if\ n_i\ is\ a\ fake\ news, \\ 0, & otherwise \end{cases} \tag{1}$$

The main aim of this study is to propose a novel method for fake news detection, called C&C-GCN. Therefore, we attempted to answer the following research questions:

- How can a C&C graph structure be constructed to represent the content and context of news?
- How can a C&C-GCN be constructed to create node representations?
- How can a C&C-GCN be used to improve the fake news detection performance?

4 Proposed Method

In this section, we describe the concept and flow of our proposed C&C-GCN method. A C&C-GCN is a content-context-based graph convolutional network for fake news detection. The proposed method is illustrated in Fig. 1.

Given a set of p news samples, $N = \{n_1, n_2, \ldots, n_p\}$. For $n \in N$, let $M = \{m_1, m_2, \ldots, m_k\}$ be a set of words of news sample n. For $n \in N$, let s, u, and r be source, user, and relation, respectively, regarding news n. Hence, we have $S = \cup s = \{s_1, s_2, \ldots, s_{p_2}\}$, $U = \cup u = \{u_1, u_2, \ldots, u_{p_1}\}$, and $R = \cup r = \{follow, like, post, reply\}$ which are sets of sources, users, and relations, respectively. The C&C-GCN includes the following steps:

4.1 Content and Context Feature Representations

Content feature: The content of the news is considered the content feature. For news $n \in N$, we converted news n to a content vector, denoted by vc, based on contextualized word representations of news.

Context features: News sources, news users, and interactions are the context features focused on in this study.

- The news source is a platform (website) that contains the news. For news $n \in N$, using the news source, we created a news source vector, denoted by vs, based on the contextualized word representations of the texts found on the website.

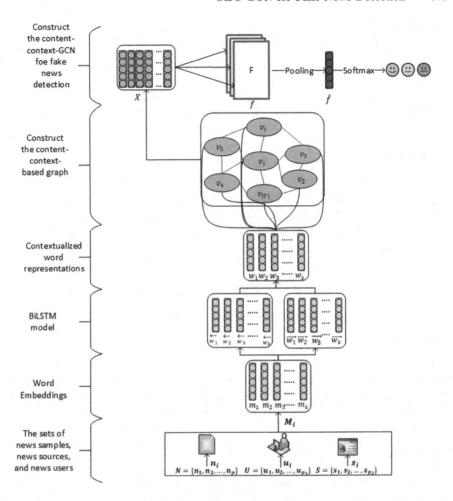

Construct
the content-
context-GCN
foe fake
news
detection

Construct
the content-
context-
based graph

Contextualized
word
representations

BiLSTM
model

Word
Embeddings

The sets of
news samples,
news sources,
and news users

Fig. 1. Overall framework for proposed method

- The news user is the textual description of the user profile. We converted the user profiles text description into a news user vector, denoted by vu, based on the contextualized word representations of the text description.
- An interaction is the relationship between two entities (source, user, and news). The relation may be a follow, like, post, or reply.

Contextualized word representation: The BiLSTM model [3] was used for a contextualized word representation. Unlike basic grammar-based methods, only statistical characteristics that disregard the real meaning between words are considered based on the context information. BiLSTM can learn contextual information that is suitable for the logic of human language [2]. The BiLSTM model was built according to the following steps [11]:

Word Embedding Layer: For $m_i \in M$, $i = [1, k]$, let $X \in R^{k \times d_m}$ is a word embeddings matrix, where d_m is the size of the word vector and each row $x_i \in R^{d_m}$ is identified as follows:

$$x_i = GloVe(m_i) \tag{2}$$

where $GloVe(m_i)$ is the corresponding vector of word m_i created based on GloVe embeddings [10].

BiLSTM Layer: This layer integrates contextual information from both directions for the words. A forward LSTM $(\overrightarrow{lstm})$ encodes the sentence via the direction of left to right, and a backward LSTM $(\overleftarrow{lstm})$ the sentence in the opposite direction of $(\overrightarrow{lstm})$. Therefore, in this layer, each word vector x_i is mapped to a pair of hidden vectors $\overrightarrow{w}_i$ and $\overleftarrow{w}_i$. This layer is formulated as follows:

$$\overrightarrow{w_i} = \overrightarrow{lstm}(x_i) \in R^{d_w}, i = [1, k] \tag{3}$$

$$\overleftarrow{w_i} = \overleftarrow{lstm}(x_i) \in R^{d_w}, i = [k, 1] \tag{4}$$

$$w_i = [\overrightarrow{w_i}, \overleftarrow{w_i}] \tag{5}$$

where w_i is the contextualized representation of word m_i; and d_w is the size of the hidden vectors.

Output Layer: From the word embeddings matrix $X \in R^{k \times d_m}$, we obtain the contextualized word representation matrix $W = (w_1, w_2, ..., w_k) \in R^{k \times d_w}$, where d_w is the size of the contextualized word vector.

4.2   Construct the C&C Graph from Content and Context Features

For $N = \{n_1, n_2, \ldots, n_p\}$, $S = \{s_1, s_2, \ldots, s_{p_2}\}$ and $U = \{u_1, u_2, \ldots, u_{p_1}\}$, let W^N, W^S, and W^U are a news feature matrix, a source feature matrix, and a user feature matrix, respectively, where $W^N = [vn_1 \oplus vn_2 \oplus ... \oplus vn_p]$, $W^S = [vs_1 \oplus vs_2 \oplus ... \oplus vs_{p_2}]$, and $W^U = [vu_1 \oplus vu_2 \oplus ... \oplus vu_{p_1}]$.

A C&C graph $G = (V, E, A)$ includes a set of nodes V, a set of edges E, and an adjacency matrix A. Nodes represent the news, sources, and users regarding the news. Edges contain all pairs of nodes. The adjacency matrix $A \in R^{|V| \times |V|}$ represents the relations between nodes. In addition, G has a node feature matrix $Q = [W^N \oplus W^S \oplus W^U] \in R^{|V| \times d_w}$, where each row Q_i is the feature vector of the news, source, or user node $v_i \in V$. The adjacency matrix A is defined as follows:

$$A_{ij} = \begin{cases} 1, & \text{if } u_i \text{ follows } u_j \text{ and } i \neq j, \\ 1, & \text{if } u_i \text{ likes } s_j, n_j \text{ and } i \neq j, \\ 1, & \text{if } u_i \text{ post } s_j, n_j \text{ and } i \neq j, \\ 1, & \text{if } u_i \text{ reply } s_j, n_j \text{ and } i \neq j, \\ 1, & \text{if } v_i = v_j, \\ 0, & \text{otherwise} \end{cases} \tag{6}$$

4.3   Build the C&C-GCN to Create Node Representations

In this study, we applied two graph convolution layers over the C&C graph for node representations as follows:

After building the C&C graph, we fed matrices Q and A into a simple two-layer GCN proposed by Kipf et al. [5] as follows:

$$Z^1 = \alpha(K \cdot Z^0 \cdot H^1 + b^1) \tag{7}$$

Hence,

$$X = Z^2 = \alpha(\alpha(K \cdot Z^0 \cdot H^1 + b^1) \cdot H^2 + b^2)) \tag{8}$$

where $X \in R^{|V| \times d_w}$; $Z^0 = Q$. α is a non-linear activation function, such as ReLU. $H^1 \in R^{d_w \times |V|}$ and $H^2 \in R^{|V| \times d_w}$ are the weight matrices of the i-th layer. b^1 and b^2 are the biases of two layers, respectively.

$$K = B^{-0.5} A B^{-0.5} \tag{9}$$

K is the normalized symmetric adjacency matrix of A; B is the degree matrix of A, where:

$$B_{ii} = \sum_j A_{ij} \tag{10}$$

4.4   Construct the C&C-GCN-Based Fake New Detection Model

In this step, we used the CNN model which is presented in our previous paper [11] over a node representation matrix X to convert elements in matrix H into the low-dimensional matrix in the following phases:

Input Layer: The node representations matrix $X \in R^{|V| \times d_w}$ is used as the input of the classifier.

Convolutional Layer: This layer is extract important information via a feature map f from the C&C node representations using a filter $F \in R^{q \times d_w}$ of length q from i to $i + q - 1$ to slide over the matrix X as follows:

$$f = [f_1, f_2, .., f_{|V|}] \tag{11}$$

where f_i, $1 \le i \le |V|$, is defined as

$$f_i = \alpha(F \ominus X_{i:i+|V|-1} + b) \tag{12}$$

where $\ominus$ is the convolution operator; α is a ReLU activation function. b is a bias.

Max-Pooling Layer: Because the size of the feature vectors $f_i \in f$ is different if the news length and the size of the filter exist any changes. Therefore, this layer is to convert a feature map f into a new feature map $\hat{f}$ including vectors with the same size by extracting the maximum number from each f_i vector as follows:

$$\hat{f} = [\hat{f}_1, \hat{f}_2, .., \hat{f}_{|V|}] \tag{13}$$

where $\hat{f}_i = Max(f_i)$.

Fully Connected Layer: This layer is to choice the characteristics related to the fake news from the output of the max-pooling layer and detect the news as the fake or real as follows:

$$\hat{y} = \delta(E \cdot \hat{f} + b) \tag{14}$$

where δ is an activation function, such as Softmax. $E \in R^{l \times |V|}$ and $b \in R^l$ are a weight matrix and a bias of the activation function, respectively. l is the number of classifications.

Model Training: The C&C-GCN model is trained by minimizing the cross-entropy error of the predicted and true label distributions as the following equation:

$$L = -\sum_i^l y_i \log(\hat{y}_i) + \lambda \|\theta\|^2 \tag{15}$$

where y_i represents the i-th real distribution of class, and $\hat{y}_i$ indicates the i-th predicted distribution of class. λ is the coefficient of L_2 regulation. θ is the parameter set from the previous layers.

5   Experimental Evaluation

5.1   Dataset and Baselines

In this study, to demonstrate the performance of our model and to ensure a fair comparison of the proposed method with other methods, we used the benchmark dataset named PolitiFact[1] in FakeNewsNet[2] [14]. The details information of dataset is shown in Table 1.

Table 1. Databases used in experiments

Features	Fake	Real
#News with text	420	528
#Users posting tweets	95,553	249,887
#Users involved in likes	113,473	401,363
#Users involved in retweets	106,195	346,459
#Tweets posting news	164,892	399,237
#Tweets with replies	11,975	41,852
#Tweets with likes	31,692	93,839
#Tweets with retweets	23,489	67,035
#Followers	405,509,460	1,012,218,640
#Followees	449,463,557	1,071,492,603

To prove that the performance of our model is better than that of other models, we deployed three different methods, including our proposal and two baselines, on the FakeNewsNet dataset.

---

[1] https://www.politifact.com/.
[2] https://github.com/KaiDMML/FakeNewsNet/tree/master/dataset.

- Text-CNN: This is a convolutional neural network-based text-classification model requiring only features related to textual information (tweet content), and is implemented using the following steps: (i) The content of a tweet is converted into a text vector, and (ii) this vector is fed into the CNN model for text classification.
- GCN: This is a graph convolutional network-based fake news detection approach. This model is implemented through the following steps: (i) Social interactions are represented as adjacency matrices, (ii) the output of the GCN [5] is then concatenated with adjacency matrices to represent nodes, and (iii) node representations are passed through a linear layer for classification.
- Content-based GCN: This is our proposed content-based graph convolutional network model using only content features.
- Context-based GCN: This is our proposed context-based graph convolutional network model using only context features.

5.2 Experimental Setup

We used 70%, 20%, and 10% of the total news as training, validation, and testing datasets, respectively. In all experiments, we used word embeddings by the pretrained GloVe with the dimension by 64. The dimension of the hidden vectors in the BiLSTM model was set to 64. The Adam optimizer [4] was used with a learning rate of 0.001 and the dropout rate was set to 0.5 to avoid the overfitting error. The filter sizes were (3, 4). All models were run five times, and the average performance for the test dataset was statistics. In addition, the hyperparameters were fine-tuned for all baselines using a cross-validation technique. Finally, an L_2 norm regularization was set to 10^{-5}.

5.3 Results and Discussion

The performance of the proposed method is presented in Table 2. The comparison results between the proposed method and the baselines are presented in Table 3.

Table 2. Performance comparison of our proposal trained with/without interactions

Features			Precision	Recall	F_1 score
Content features		With interactions	0.806	0.801	0.803
		Without interactions	0.561	0.635	0.596
Context features	News Source	With interactions	0.798	0.803	0.800
		Without interactions	0.581	0.557	0.569
	News User	With interactions	0.816	0.792	0.804
		Without interactions	0.607	0.621	0.614
	Both news source and news user	With interactions	0.835	0.829	0.832
		Without interactions	0.694	0.706	0.700
Content-context features		Without interactions	0.871	0.862	0.866
		With interactions	0.751	0.767	0.759

Table 2 indicates the contribution of each C&C-GCN component by removing the components individually from the complete model. We can see that each component indeed makes a significant contribution to the performance of the C&C-GCN model. First, C&C-GCN outperforms its variants without content or context features. This indicates that the source posts play an important role in fake news detection. Second, content- and context-based GCNs cannot always achieve better results than C&C-GCN. This implies the importance of simultaneously considering both the content and context features. Finally, all variants having interactions achieve a better performance than those variants without interactions.

To confirm the performance of the proposed method, we implemented two additional baseline methods for comparison in terms the *precision, recall*, and F_1. In addition, we analyzed the effect of each variant of the C&C-GCN by comparing the proposed method with content- and context-based GCNs, which are the proposed variants without context and content features, respectively. Table 3 presents a performance comparison of the methods.

Table 3. Performance comparison of models

Method	Class	*Precision*	*Recall*	$F_1 score$
Text-CNN	Fake	0.695	0.724	0.709
	Real	0.683	0.775	0.726
GCN	Fake	0.762	0.745	0.753
	Real	0.748	0.751	0.749
Context-based GCN	Fake	0.842	0.850	0.846
	Real	0.833	0.825	0.829
Content-based GCN	Fake	0.794	0.786	0.790
	Real	0.814	0.806	0.810
Proposed method	Fake	0.867	0.859	0.863
	Real	0.873	0.865	0.869

This can be seen in the following discussion. First, among the baselines, the GCN method performs significantly better than the Text-CNN method. This is not surprising because the GCN method can learn high-level representations of fake news for capturing valid features. This demonstrates the importance and necessity of using GCNs for fake-news detection. Second, the proposed method outperformed the GCN method in terms of all performance measures, which indicates the effectiveness of using a CNN model over a GCN structure for fake news detection. Because a Text-CNN cannot process data with a graph structure, a GCN ignores the important content, context, and interaction features of spreading fake news, preventing it from obtaining efficient high-level representations of fake news and resulting in a worse fake news detection performance. Finally, C&C-GCN is significantly superior to its variants, including the content- and context-based GCN methods. Content- and context-based GCNs use only the hidden feature vector of all leaf nodes and thus are heavily impacted by the information of the latest posts. However, the latest posts always lack information, such as comments, and simply follow the former posts. Unlike a content-based GCN, the root feature enhancement allows the proposed method to focus more on the information of the source posts, which helps improve our models.

6   Conclusions and Future Works

In this study, we introduce a fake news detection method, i.e., C&C-GCN, which can predict whether a short-text is fake. The proposed method integrates information in terms of (i) the content of the news, and (ii) the social context of the news, such as the user who created or shared the news, and the source that contained the news. The evaluation results show that the C&C-GCN can obtain more effective than the baseline methods. We believe that the C&C-GCN model is used not only for fake news detection but also for other tasks on social media related to the text classification, such as sentiment analysis and hate speech detection. However, this study only focuses on analyzing the news texts without considering the news images or videos. We will research these orientations in a future study.

References

1. Bovet, A., Makse, H.A.: Influence of fake news in Twitter during the 2016 US presidential election. Nat. Commun. **10**(1), 1–14 (2019)
2. Cai, R., et al.: Sentiment analysis about investors and consumers in energy market based on BERT-BILSTM. IEEE Access **8**, 171408–171415 (2020)
3. Huang, Z., Xu, W., Yu, K.: Bidirectional LSTM-CRF models for sequence tagging. arXiv preprint arXiv:1508.01991 (2015)
4. Kingma, D.P., Ba, J.: Adam: a method for stochastic optimization. arXiv preprint arXiv:1412.6980 (2014)
5. Kipf, T.N., Welling, M.: Semi-supervised classification with graph convolutional networks. arXiv preprint arXiv:1609.02907 (2016)
6. Li, C., Goldwasser, D.: Encoding social information with graph convolutional networks for political perspective detection in news media. In: Proceedings of the 57th Annual Meeting of the Association for Computational Linguistics, pp. 2594–2604 (2019)
7. Marcheggiani, D., Bastings, J., Titov, I.: Exploiting semantics in neural machine translation with graph convolutional networks. arXiv preprint arXiv:1804.08313 (2018)
8. Meyers, M., Weiss, G., Spanakis, G.: Fake news detection on Twitter using propagation structures. In: van Duijn, M., Preuss, M., Spaiser, V., Takes, F., Verberne, S. (eds.) MISDOOM 2020. LNCS, vol. 12259, pp. 138–158. Springer, Cham (2020). https://doi.org/10.1007/978-3-030-61841-4_10
9. Pehlivan, Z.: On the pursuit of fake news: from graph convolutional networks to time series (2020)

10. Pennington, J., Socher, R., Manning, C.D.: Glove: global vectors for word representation. In: Proceedings of the 2014 Conference on Empirical Methods in Natural Language Processing (EMNLP), pp. 1532–1543 (2014)
11. Phan, H.T., Nguyen, N.T., Hwang, D.: Convolutional attention neural network over graph structures for improving the performance of aspect-level sentiment analysis. Inf. Sci. (2022)
12. Phan, H.T., Nguyen, N.T., Tran, V.C., Hwang, D.: A sentiment analysis method of objects by integrating sentiments from tweets. J. Intell. Fuzzy Syst. **37**(6), 7251–7263 (2019)
13. Shi, W., Rajkumar, R.: Point-GNN: graph neural network for 3D object detection in a point cloud. In: Proceedings of the IEEE/CVF Conference on Computer Vision and Pattern Recognition, pp. 1711–1719 (2020)
14. Shu, K., Mahudeswaran, D., Wang, S., Lee, D., Liu, H.: FakeNewsNet: a data repository with news content, social context, and spatiotemporal information for studying fake news on social media. Big data **8**(3), 171–188 (2020)
15. Tuan, N.M.D., Minh, P.Q.N.: FakeNews detection using pre-trained language models and graph convolutional networks (2020)
16. Wu, Z., Pan, S., Chen, F., Long, G., Zhang, C., Philip, S.Y.: A comprehensive survey on graph neural networks. IEEE Trans. Neural Netw. Learn. Syst. **32**(1), 4–24 (2020)
17. Zhang, C., Li, Q., Song, D.: Aspect-based sentiment classification with aspect-specific graph convolutional networks. arXiv preprint arXiv:1909.03477 (2019)
18. Zhang, Y., Wallace, B.: A sensitivity analysis of (and practitioners' guide to) convolutional neural networks for sentence classification. arXiv preprint arXiv:1510.03820 (2015)

Multi-class Sentiment Classification for Customers' Reviews

Cuong T. V. Nguyen[2,3], Anh M. Tran[1,2,3], Thao Nguyen[2,3], Trung T. Nguyen[4], and Binh T. Nguyen[1,2,3(✉)]

[1] AISIA Research Lab, Ho Chi Minh City, Vietnam
[2] University of Science, Ho Chi Minh City, Vietnam
ngtbinh@hcmus.edu.vn
[3] Vietnam National University, Ho Chi Minh City, Vietnam
[4] Hong Bang International University, Ho Chi Minh City, Vietnam

Abstract. The rise of e-commerce due to the Covid-19 situation is becoming more significant in 2021. It could lead to great demands to understand customers' opinions usually shown in their reviews. An e-commerce platform with the ability to be aware of its users' viewpoint can have a higher possibility of meeting customer expectations, attracting new users, and increasing sales. With the tremendous data in e-commerce platforms presently, sentiment analysis is a powerful tool to understand users. However, the sentiment in reviews data may contain more than two states, positive and negative, and then a binary sentiment classifier may not be helpful in practice. According to our knowledge, research on this subject is often restricted access. Therefore, this paper presents a multi-class sentiment analysis for Vietnamese reviews on a large-scale dataset, including 480,702 reviews. We collected these reviews from popular Vietnamese e-commerce websites and manually did the labeling process with three classes of sentiments (positive, negative, and neutral). To build a suitable classification model for the main problem, we propose a deep learning approach using different architectures (LSMT, GRU, TextCNN, LSTM + CNN, and GRU+CNN) and compare the performance among other ensemble techniques. The experimental results show the outperformance of the ensemble techniques on the multi-class sentiment classification problem, and the combination of chosen architectures using the attention mechanism could obtain the best F-1 score of 73.64%.

Keywords: Sentiment classification · Ensemble methods · FDA

1 Introduction

Nowadays, natural language processing (NLP) is one of the most researched fields with some duty to be assessed, especially sentiment analysis tasks. While the Covid-19 situation is still taking place very complicatedly in many countries around the world, the application of the sentiment analysis in social media, e-commerce, customer service, and market research becomes ever more critical than

© Springer Nature Switzerland AG 2022
H. Fujita et al. (Eds.): IEA/AIE 2022, LNAI 13343, pp. 583–593, 2022.
https://doi.org/10.1007/978-3-031-08530-7_49

ever [3,14]. Furthermore, many large companies and corporations have continuously collected user feedback and suggestions on various products and developed customer support teams. Therefore, it can be considered an opportunity for businesses to understand customer psychology and make more accurate decisions than competitors. Also, one can extract valuable information from user comments, discovering all suggestions regarding current products (e.g., new features, new designs, potential bugs, security concerns, and customer service improvements).

This work focuses on studying the Vietnamese sentiment analysis on a large-scale dataset crawled from many major e-commerce platforms in Vietnam. This dataset has 480,702 customer reviews, and we labeled each review with three classes of sentiment (negative, positive, and neutral). We first study five deep learning-based models: TextCNN, LSTM, GRU, LSTM + CNN, and GRU + CNN. After that, we deep dive with different ensemble schemes based on the trained models: Gating Network, Squeeze-Excitation Network, Attention Network, Uniform Weighting, and Linear Ensembling. The experimental results show that using the Attention Network technique can help us achieve performance dominance compared with other approaches in terms of the F1-score as 73.64%.

This paper focuses on introducing related information about credit scoring problems in the our dataset. In Sect. 2, we represent some related works. We describe the processing of the dataset in detail in Sect. 3. In Sect. 4, the solutions and models are represented. Our experiments and results are presented in Sect. 5. Finally, we conclude the paper in Sect. 6.

2   Related Works

Zhang et al. had great success in sentiment analysis using deep learning for popular languages like English and Chinese [16]. A popular approach is using well-known word embedding models like GloVe [11], or Word2Vec [7] for word representation. It feeds them to a deep learning architecture with a Softmax function to predict class probability. Two main kinds of architectures to be used are convolutional neural network (CNN) and recurrent neural network (RNN) [15]. Another approach is applying transfer learning with state-of-the-art pre-trained models and fine-tune for the main problem. One of the most well-known pre-trained models presently is BERT [4].

Most of the researches and points of reference specialize in the English language. For Vietnam, most studies on e-commerce data are primarily built-in binary classification (negative or positive). For the Vietnamese sentiment analysis, Hung et al. [13] explored the topic classification and sentiment analysis for a Vietnamese education survey system using the Bag-of-Structure technique and traditional machine learning methods. Cuong et al. [8] studied sentiment analysis for Vietnamese language reviews using deep learning methods. But not all sentences have a clear sentiment towards polarities, and one should not ignore them in learning polarity. Besides, one could not regard them as a state between negative and positive sentiments but as a class without sentiment. Therefore, training models based on negative and positive alone will not guarantee the prediction accuracy of those cases. Other works can be found at [10].

3 Datasets

To review the effectiveness of the proposed approaches, we employ our large-scale dataset, which was crawled from the Vietnamese e-commerce websites and labeled by ourselves. Our dataset collected contains more than 480K user reviews, including 328k positive reviews, 54k neutral reviews, and 98k negative reviews. In our experiments, we randomly allocate our dataset into the training set, the validation set, and the testing set as depicted in Table 1.

Table 1. The dataset collected and used in our work.

Our Dataset	Positive	Neutral	Negative	Total
Train	183913	30562	54717	269192
Validation	45978	7641	13679	67298
Test	98515	16241	29456	144212

4 Our Proposed Methods

This section will present the data processing step, selected learning models, and how we construct the proper ensemble method for the main problem.

4.1 Data Processing

We apply various processing steps for the raw text data and a script automatically does the processing before training to eliminate the huge effort in manually checking by humans. No text augmentation is applied in the two datasets.

(a) Correct elongate words such as "đẹpppppp quá" to "đẹp quá" (so beautiful).
(b) Standard accents in Vietnamese such as "a'" to "à", "y~" to "ỹ".
(c) Standard a lot of sentiment words/English words.
(d) Remove elements which likely do not contribute to the sentiment of a sentence:
    + URLs, HTML, website, email, username, #hashtag.
    + Punctuations, special characters, icons such as !" # $ % & ' () * + , - .
      / : ; < = > ? [] _ ` {|} ~
    + Numbers, Arabic numbers, or Roman numbers.
(e) Even though this text data was collected from Vietnam e-commerce websites, there are a few reviews in other languages such as Korean, Chinese, etc. These reviews are removed.
(f) Especially, we observe that there exist many freestyle letters and Acronyms in Vietnamese reviews, and these letters represent the words heavily contributing to the sentiment of a sentence. Thus, it is important to replace these letters with the correct words such as "kp" to "không phi" (not), "ô kê" to "ok", etc.
(g) Remove Vietnamese stop words in a sentence and extra white spaces.

4.2 Learning Models

Deep neural networks have proven potential applications in different machine learning tasks, including NLP problems, and have shown remarkable outperformance compared to other techniques. This paper investigates a suitable deep learning approach for the multi-class sentiment classification problem in the dataset collected.

There are two main components in the deep learning architecture we aim to construct for this problem. The first component is the word embedding for expressing words with learned representation. To choose the most suitable embedding method, we consider the following pre-trained models: Word2vec [7], GloVe [11], and Fasttext [1]. For training an appropriate model for the man problem, we utilize five different architectures of deep neural networks, including TextCNN [5], Bi-LSTM [12], Bi-GRU [2], LSTM + CNN, and GRU + CNN [6,9] and five different ensembling approaches. One can see more details in our proposed architectures in Fig. 1.

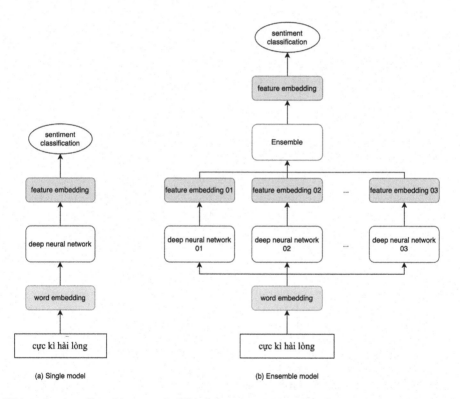

Fig. 1. An overall architecture for Vietnamese sentiment classification, where the input review is "Cc kì hài lòng" ("extremely satisfied"). (a) A single model using deep neural network. (b) An ensemble model which combines different feature embeddings from multiple models.

In what follows, we aim to describe our chosen architectures for a single model approach before considering the ensemble techniques for selected single classifiers. As mentioned before, they are TextCNN, Bi-LSTM, Bi-GRU, Bi-LSTM + CNN, and Bi-GRU + CNN. One can see more details of these methods in the following sections.

Long Short-Term Memory (LSTM) [12] and Gated Recurrent Unit (GRU) [2] are two particular types of RNN. They are designed to address the problem of sequence input and can learn long-term dependencies in sequences. Remarkably, they use gating units to control the flow of information and decide which part to be forgotten or updated. Our work applies LSTM and GRU with bi-directional architectures that can capture both backward and forward sequence information at every time step as the baseline models.

4.2.1   Bi-LSTM/GRU + CNN

One of the chosen methods in this paper is the following combinations, Bi-LSTM + CNN and Bi-GRU + CNN, between RNN architectures Bi-LSTM/GRU and CNN. When considering these combinations, one of the main ideas is that RNN models can capture long-term dependencies and contextual information but are biased towards last words. In contrast, CNN can extract local and position-invariant features. Therefore, this study utilizes Bi-LSTM/GRU as the RNN baselines to extract hidden vectors. Then, at each time t, one can define a new representation of word w_t as the concatenation of the forward context, the word embedding, and the backward context $[fh_t, x_t, bh_t]$. The newly obtained representation of the word w_t can be fed to two linear layers; the first linear layer aims to reduce the dimension of the new representation embedding followed by a Max Pooling 1D, and the last linear layer delivers sentiment output. We depict all details of this architecture in Fig. 2.

4.2.2   TextCNN

With the ability to detect local spatial patterns, CNN architectures are ubiquitously used in text classification problems. In our work, we use TextCNN [5] as a baseline model, which defines multiple convolution kernels uses them to perform convolution calculations on the inputs. Then it performs max pooling on all output channels, concatenates the pooling output in a vector fed to two linear layers, and outputs the sentiment.

4.3   Ensemble Models

Besides using different individual models above, we consider ensemble techniques for combining multiple embedded features and measuring the corresponding impact on the performance of the ensemble methods into the final problem. Let us denote $E_1, E_2, \ldots, E_m$ as a set of embedded features from m different models. There are several ways to combine them appropriately to enhance the new model's performance. In this work, we choose five popular methods, and we

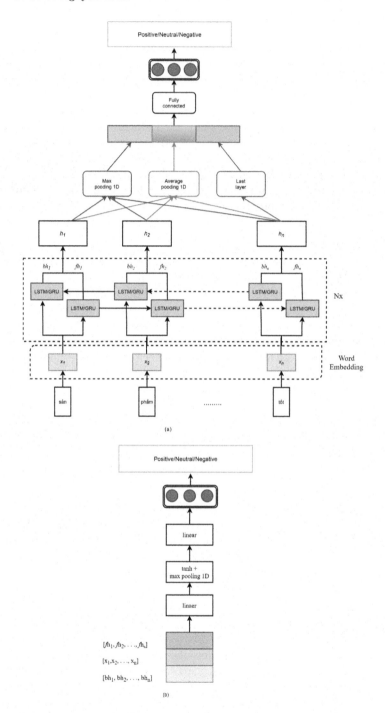

Fig. 2. Our proposed architectures using Bi-LSTM/GRU for Vietnamese review sentiment. (a) The detail of our proposed Bi-LSTM/GRU architecture. (b) The detail of our proposed Bi-LSTM/GRU + CNN architecture.

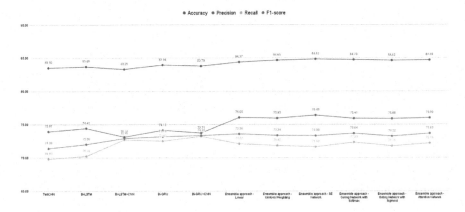

Fig. 3. The performance of different methods in the multi-class sentiment classification in our work.

can illustrate the proposed ensemble networks for Vietnamese sentiment multi-classification in Fig. 1.

First of all, one of the standard methods is the **uniform weighting** approach in which the overall embedding values are calculated as the average values of individual ones:

$$E = [\frac{\sum_{i=1}^{m} E_i}{m}]$$

The next one is the **linear ensemble** approach which is simply a concatenation of input embeddings.

$$E = [E_1, E_2, ..., E_m]$$

Although different features impact the output, the above methods are not optimized to learn with feature importance. Therefore, our study analyzes three more ensemble techniques that dynamically learn the feature importance, leverage multi-feature embeddings, and boost performance more efficiently Fig. 3.

In recent years, the Attention mechanism has risen as one of the most successful state-of-the-art methods in natural language processing [8]. For this reason, we also use the **Attention technique** to determine the importance coefficients for each embedding feature.

Two linear layers and a SiLU activation are used as an encoder and a decoder to convert the concatenation of feature embeddings to a weights vector with the dimension $m \times K$, which m as the number of individual models and K as the dimension of each feature embedding. Thus, we have the weights vector $Z = [z_1, z_2, ..., z_{m \times K}]$ can be calculated as follows:

$$Z = \text{Sigmoid}(W_2 SiLU(W_1 S)),$$

where W_1 and W_2 are the linear layers' parameters matrices. The last feature embedding from the Attention network can be calculated as the feature-wise multiplication between feature embeddings and the weights vector. With $i \in \{1, 2, ..., m\}$ and $j \in \{1, 2, ..., K\}$, we have the equation

$$E = [e_{ij} \cdot ij]$$

Another method we use in this work is the **Gating network** which can learn the importance coefficients of feature embeddings [8]. In practice, we use one linear layer and an activation function to transform the concatenation of all feature embeddings to an m-dimensional weights vector $Z = [z_1, z_2, ..., z_m]$. For choosing the activation function, we consider between Softmax and Sigmoid functions. Later, the final feature embedded from the Gating network can be calculated as the feature embedding-wise multiplication between feature embeddings and the weights vector.

$$Z = \sigma(W([E_1, E_2, ..., E_m],$$

$$E = [E_1 \cdot z_1, E_2 \cdot z_2, ..., E_m \cdot z_m]$$

Lastly, we utilize the **Squeeze-Excitation mechanism** to learn better the importance of each feature embedding in the sentiment classification problem. The SE Network has two main modules as follows:

Squeeze. This module employs 1D-pooling techniques, such as Max or Average, to squeeze a K-dimensional feature embedding into a scalar value representing the global information represented for that feature embedding. Feature embeddings are squeezed into an m-dimensional statistic vector $S = [s_1, s_2, ..., s_m]$, where:

$$s_i = \frac{\sum_{j=1}^{K} e_{ij}}{K}, i \in \{1, 2, ..., m\}$$

Excitation. This module is applied to learn the weights for each feature embedding based on vector S above. We use two linear layers to learn the corresponding weights of each embedding feature. The first linear layer, working as an encoder, transforms the statistic vector into a latent vector with a reduction factor. The second linear layer, working as a decoder, transforms the latent vector into a weights vector. We put an activation function between these two layers, and after the second layer, another activation function is placed. The m-dimensional weights vector $Z = [z_1, z_2, ..., z_m]$ can be calculated as:

$$Z = \text{Sigmoid}(W_2 SiLU(W_1 S)),$$

where W_1 and W_2 are the parameter matrices of two linear layers. A feature embedding-wise multiplication between feature embeddings and the weights vector learned above is processed at the network's end to amplify the most important features:

$$E = [E_1 \cdot z_1, E_2 \cdot z_2, ..., E_m \cdot z_m]$$

where $i \in \{1, 2, ..., m\}$ and $j \in \{1, 2, ..., K\}$.

5   Experiments

5.1   Data Preparation

We implement data preprocessing, as mentioned in Sect. 4. Subsequently, the dataset is split into training and test sets. We also take advantage of the Cross-validation method on the training set while training the models. This technical method helps us train the models on some subsets from the training set to directly evaluate the models in the training phase to find out the best models to apply to the testing set.

5.2   Settings

We use Pytorch to run all the experiments. The Vietnamese version of Fasttext, which has an embedding dimension of 300, is used as a word embedding layer. We train single models with a batch size of 256, a learning rate of 1e-3, weight decay of 1e-5, maximum epochs of 20, and the dimensionality of each feature embedding is 256.

We use similar hyper-parameter values for ensemble models, except for the learning rate of 1e-5 and maximum epochs of 16. The single models are trained independently. Models with the best results on the validation set are saved and weights-frozen and used as feature extractors for the ensemble approaches. Accuracy, F1-Score, Precision, and Recall are used to measure models' performance.

5.3   Results

Table 2 presents the effectiveness of the proposed individual and ensemble methods in enhancing Vietnamese multi-class sentiment analysis performance.

Among five individual methods (TextCNN, Bi-LSTM, Bi-GRU, Bi-LSTM+CNN, and Bi-GRU+CNN), the Bi-GRU outperforms other techniques in the metrics of accuracy (83.94%). For precision, using the Bi-LSTM approach can perform better than other approaches (74.41%). The Bi-GRU+CNN model delivers outstanding performances in terms of recall and the F1-score, where this method can achieve a recall of 73.20% and an F1-score of 73.34%.

Using five different ensemble techniques for these five individual methods can show the outperformance compared to any single model in terms of all considered metrics. Generally, SE Network delivers the best performance in the metrics of accuracy (84.82%), and precision (76.40%). For F1-score, using the gating network with the Softmax approach can obtain the best performance of 73.64%. It is worth noting that one should focus on F1-score as the most critical metric in an imbalanced dataset like the current dataset.

Besides the advantages of ensemble techniques, they also have some disadvantages. For example, no ensemble method performs better than the individual model (Bi-GRU+CNN) in all terms; especially, this model gives the best recall performance in our work. This can show that we need a better ensemble method to combine and take advantage of individual models. However, the limitations motivate us to research new approaches that have the best performances in the future.

Table 2. The experimental results of both ensemble networks and other baseline methods.

Our Model	Accuracy	Precision	Recall	F1-Score
TextCNN	0.8350	0.7391	0.6980	0.7138
Bi-LSTM	0.8369	0.7441	0.7023	0.7200
Bi-LSTM+CNN	0.8329	0.7310	0.7270	0.7290
Bi-GRU	0.8394	0.7412	0.7250	0.7316
Bi-GRU+CNN	0.8378	0.7371	**0.7320**	0.7334
Ensemble approach - Linear	0.8437	0.7605	0.7207	0.7356
Ensemble approach - Uniform Weighting	0.8465	0.7595	0.7182	0.7334
Ensemble approach - SE Network	**0.8482**	**0.7640**	0.7163	0.7330
Ensemble approach - Gating Network with Softmax	0.8473	0.7591	0.7230	**0.7364**
Ensemble approach - Gating Network with Sigmoid	0.8462	0.7588	0.7176	0.7322
Ensemble approach - Attention Network	0.8468	0.7600	0.7216	0.7363

6   Conclusion and Future Works

This paper has investigated the Vietnamese multi-class sentiment analysis by exploring the impact of different approaches: individual models such as TextCNN, LSTM, GRU, LSTM + CNN, and GRU + CNN and ensemble models such as Gating Network, SE Network, Attention Network, Uniform Weighting, and Linear Ensemble. We ran all experiments on one large-scale crawled dataset with four standard metrics of Accuracy, Precision, Recall, and F1-score to compare these methods' performance. The experimental results show the enhanced performance of the proposed ensemble networks compared with other methods in terms of Accuracy, Precision, and F1 score.

It is worth noting that the data collected from e-commercial is not large enough to generalize all Vietnamese sentiment situations as every person in the different regions in Vietnam has other ways to express their sentiment for a case. As a result, our dataset seems not good enough to train models from scratch. To solve this problem, using pre-trained models is the best choice.

In the scope of this work, we still did not involve pre-trained transformer models such as multilingual BERT, PhoBERT, or BARTpho, which are dominating in some natural language processing tasks. However, we can focus on them and use treatments to deal with imbalanced problems for future work.

Acknowledgments. We want to thank the University of Science, Vietnam National University in Ho Chi Minh City, and AISIA Research Lab in Vietnam for supporting us throughout this paper.

References

1. Bojanowski, P., Grave, E., Joulin, A., Mikolov, T.: Enriching word vectors with subword information. Trans. Assoc. Comput. Linguist. **5**, 135–146 (2017)

2. Cho, K., et al.: Learning phrase representations using RNN encoder-decoder for statistical machine translation (2014)
3. Contratres, F.G., Alves-Souza, S.N., Filgueiras, L.V.L., DeSouza, L.S.: Sentiment analysis of social network data for cold-start relief in recommender systems. In: Rocha, Á., Adeli, H., Reis, L.P., Costanzo, S. (eds.) WorldCIST'18 2018. AISC, vol. 746, pp. 122–132. Springer, Cham (2018). https://doi.org/10.1007/978-3-319-77712-2_12
4. Devlin, J., Chang, M.-W., Lee, K.: Bert: pretraining of deep bidirectional transformers for language understanding (2019)
5. Kim, Y.: Convolutional neural networks for sentence classification. In: Proceedings of the 2014 Conference on Empirical Methods in Natural Language Processing (EMNLP), pp. 1746–1751. Association for Computational Linguistics, Doha, Qatar (2014)
6. Lai, S., Liheng, X., Liu, K., Zhao, J.: Recurrent convolutional neural networks for text classification. In: Bonet, B., Koenig, S. (eds.) AAAI, vol. 333, pp. 2267–2273 (2015)
7. Mikolov, T., Sutskever, I., Chen, K., Corrado, G., Dean, J.: Distributed representations of words and phrases and their compositionality. In: Proceedings of the 26th International Conference on Neural Information Processing Systems, NIPS'13, Red Hook, NY, USA, vol.2, pp. 3111–3119. Curran Associates Inc (2013)
8. Nguyen, C.V., Le, K.H., Nguyen, B.T.: A novel approach for enhancing vietnamese sentiment classification. In: Fujita, H., Selamat, A., Lin, J.C.-W., Ali, M. (eds.) IEA/AIE 2021. LNCS (LNAI), vol. 12799, pp. 99–111. Springer, Cham (2021). https://doi.org/10.1007/978-3-030-79463-7_9
9. Nguyen, V.C., Le, K.H., Nguyen, B.T.: An efficient framework for Vietnamese sentiment analysis. In: Proceedings of The 18th International Conference on Intelligent Software Methodologies, Tools, and Techniques (SoMeT) (2020)
10. Nguyen, H., et al.: Multi-level sentiment analysis of product reviews based on grammar rules. In: New Trends in Intelligent Software Methodologies, Tools and Techniques: Proceedings of the 20th International Conference on New Trends in Intelligent Software Methodologies, Tools and Techniques (SoMeT_21), vol. 337, pp. 444–456. IOS Press (2021)
11. Pennington, E., Socher, R., Manning, C.: Glove: global vectors for word representation. In: Proceedings of the 2014 Conference on Empirical Methods in Natural Language Processing (EMNLP)s, pp. 1532–1543 (2014)
12. Hochreiter, S., Schmidhuber, J.: Long short-term memory. Neural Computing **9**, 1753–1780 (1997)
13. Vo, H., Lam, H., Nguyen, D.D., Tuong, N.: Topic classification and sentiment analysis for Vietnamese education survey system. Asian J. Comput. Sci. Inf. Technol. **6**, 27–34 (2016)
14. Wang, G., Sun, J., Ma, J., Kaiquan, X., Jibao, G.: Sentiment classification: the contribution of ensemble learning. Decis. Support Syst. **57**, 77–93 (2014)
15. Wilcox, E., Levy, R., Morita, T., Futrell, R.: What do RNN language models learn about filler-gap dependencies?. In: Proceedings of the 2018 EMNLP Workshop BlackboxNLP: Analyzing and Interpreting Neural Networks for NLP, pp. 211–221 (2018)
16. Zhang, L., Wang, S., Liu, B.: Deep learning for sentiment analysis: a survey. Wiley Interdisc. Rev. Data Min. Knowl. Disc. **8**(4), e1253 (2018)

Transportation and Urban Applications

MM-AQI: A Novel Framework to Understand the Associations Between Urban Traffic, Visual Pollution, and Air Pollution

Kazuki Tejima[1] , Minh-Son Dao[2]([✉]) , and Koji Zettsu[2]

[1] The University of Aizu, Aizu-Wakamatsu, Fukushima, Japan
s1270245@u-aizu.ac.jp
[2] National Institute of Information and Communications Technology, Koganei, Japan
{dao,zettsu}@nict.go.jp

Abstract. Understanding the associations between different traffic factors (e.g., time, vehicles, trees, and people) and the air pollution in a particular region is a challenging problem of great concern in Intelligent Transportation Systems. Most previous works primarily focused on efficient prediction of air pollution levels the given traffic imagery data. To the best of our knowledge, there exists no study that tries to discover hidden associations (or correlation) that exist between the traffic factors and the air pollution towards predicting PM2.5 levels within a certain period of time. With this motivation, this paper proposes a novel framework that aims to discover hidden associations that exist between the traffic factors and the air pollution towards predicting air pollution level in short- and medium-term time. Our framework has the following six steps: (i) Extract features from the traffic images using any machine learning algorithm, (ii) generate a new dataset by joining the extracted features dataset and air pollution dataset using time, (iii) transform this new dataset into an uncertain temporal database using fuzzy rules, (iv) apply uncertain periodic-frequent pattern mining techniques to discover hidden associations between various traffic factors and air pollution, (v) estimate air pollution level from a given image using transfer learning on a pre-trained model, and (vi) predict air pollution level using estimated air pollution level and mined patterns dataset. Experimental results show that our method can estimate and predict air pollution level with high accuracy (from 77% to 98%).

Keywords: Air pollution · Transfer learning · Uncertain temporal transaction · FPGrowth · First-view images · Fuzzy logic

1 Introduction

Improving the quality of human life in smart cities is an important objective that is directly linked to several United Nations Sustainable Developmental Goals,

---

Kazuki was the first to introduce the idea presented in this paper.

© Springer Nature Switzerland AG 2022
H. Fujita et al. (Eds.): IEA/AIE 2022, LNAI 13343, pp. 597–608, 2022.
https://doi.org/10.1007/978-3-031-08530-7_50

such as climate action and life on land. Urban transportation and air pollutant are two key factors that affect the quality of life in smart cities. Thus, most previous studies focused on determining traffic factors influencing air pollution using statistical [17] and deep-learning [14] techniques. As statistical techniques suffer from scalability problem, these studies confined their field of study to small areas. Deep-learning techniques were able to address the scalability problem encountered by the statistical techniques; however, their black-box nature has limited the users in understanding the correlation between the various traffic factors and air pollution in a locality. Most of the previous research mostly focused on efficient prediction of air pollution values given the traffic imagery data generated by satellites, CCTVs, and drones [18]. To the best of our knowledge, there exists no literature that aims to understand the correlation between the various traffic factors and air pollution in traffic imagery data. This paper aims to develop a novel and generic framework that can discover correlation (or association) between various traffic factors ad air pollution in a locality.

It has to be noted that developing a framework to discover hidden associations between various traffic factors and air pollution levels is a non-trivial and challenging task due to the following reason: "*Instance segmentation* [10] *and semantic segmentation* [5] *techniques exist to identify different features in an image. Unfortunately, these techniques do not discuss how to transform the features discovered in multiple images into a database (particularly, uncertain temporal database).*" This paper tackles this challenge in order to discover (temporal) associations between the various traffic factors and air pollution levels.

The contributions of this paper are as follows. This paper proposes a novel fuzzy framework that extracts (temporal) associations between different traffic factors and air pollution in a database. Our framework has the followingsteps: (i) extract features from the input images using any segmentation technique, apply fuzzy logic on the generated features, and produce an uncertain traffic temporal database. (ii) construct uncertain pollution temporal database by converting the air pollution recordings using fuzzy-logic, (iii) generate an uncertain traffic-pollution temporal database by joining uncertain traffic temporal database and uncertain pollution temporal databases with respect to time, and (iv) apply periodic-frequent pattern mining on the uncertain traffic-pollution database to understand the correlation between the various traffic factors and pollution levels (in a locality or area of interest), (v) estimate PM2.5 level at the given time ti_c using the input image im_c, and (vi) use this predicted PM2.5 level and the periodic-frequent pattern database to predict the PM2.5 level in short- and medium-term time. Experimental results on real-world databases demonstrate that our model can discover useful information.

The rest of the paper is organized as follows. Section 1 introduced the research problem and the motivation, Sect. 2 discusses related works, Sect. 3 introduces the proposed method, Sect. 4 explains the experimental results, and Sect. 5 concludes the paper and sketches out the future works.

2   Related Work

Several efficient and effective methods have been described in the literature to predict air pollution using single- [14] or multi-modality [2] data. Among the methods, Image-based air pollution estimation methods are gaining popularity due to their practicability and low investment cost. Several deep learning methods have been developed to create classifying/regressive models to estimate air pollution using only images, such as Hybrid Convolutional Neural Network [14], Ensemble of Deep Neural Networks [20], CNN and LSTM [13]. Other researchers have integrated images with other data types to estimate air pollution values/levels, such as weather data [4] and traffic density [2].

Instead of estimating exact air pollution values (e.g., PM2.5, PM10), people apply fuzzy logic to enhance the tolerance of features and classifiers [9]. This approach might be suitable for daily applications when citizens pay attention to air pollution levels (e.g., good, bad, moderate) than exact values.

Understanding the association between high-semantic visual cues and air pollution is another approach to building necessary knowledge to enhance prediction accuracy. In [7], the authors try to discover the correlation between traffic and air pollution by analyzing built environment aerial images. In [12,23], the greenness of the city, the air pollution, and temperature are proved that they have a strong correlation.

3   Proposed Framework

This section describes our proposed method in detail, including problem statement, uncertain temporal dataset generation, periodic frequent-pattern mining, and PM2.5 estimation and prediction. The significant difference between our method from other image-based PM2.5 prediction methods is that our method does not need to define the fixed-size time window for predicting. We inherit the framework introduced in [11] and implement it by redefining the problem statement, considering the haze level estimated from images, applying different visual feature extraction models and fusion schema, and testing the methods on more datasets collected from other countries. The experimental results show that our implementation significantly improves the accuracy and captures more cognitive factors in the patterns. We also discuss the role of deep-learning-based features and high-semantic domain-based handcraft features in the accuracy of PM2.5 prediction. Figure 1 illustrates the operational flow and significant components of the system.

3.1   Problem Statements

Let $IM_{db} = \{(ti_k, im_k)\}_{k=1..N}, N \geq 1$, be a traffic imagery database constituting of a set of first-view images and their timestamps. Let $FE(im_k)$ be a feature extraction algorithm, say an instance segmentation, that extracts a set of features and their associated probabilities from an item im_k. That is,

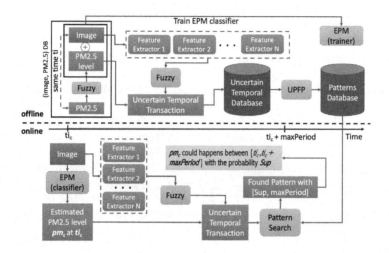

Fig. 1. Framework overview

$FE(im_k) = \{(f_x, p_x)\}_{x=1..M}$, where f_x is a xth-feature and $p_x \in [0,1]$ is the probability of the respective feature in the image. Let $\widehat{FE(im_k)} \subseteq FE(im_k)$ be a set of interesting features that have probability values no less than the user-specified threshold value γ, $0 \le \gamma \le 1$. Let $UTrafficTDB = \cup_k = 1^N(ts_k, \widehat{FE(im_k)})$ be an uncertain traffic temporal database generated from IM_{db}.

Let $PM_{db} = \{(ti_k, pm_k)\}_{k=1..N}, N \ge 1$, be an air pollution database constituting of air pollutant values and their timestamps. Let $FU(pm_k)$ be a fuzzy function that transforms the given air pollutant value pm_k into a fuzzy value, say fpm_k. That is, $FU(pm_k) = fpm_k$. Let $UncertainPollutionTDB = \{(ti_k, pm_k)\}_{k=1..N}$ be an uncertain pollution temporal database generated from PM_{db}.

The uncertain traffic-pollution database, denoted as $UTDB$, is generated by joining $UTrafficTDB$ and $UPollutionTDB$ with respect to time. That is, $UTDB = UTrafficTDB \bowtie UPollutionTDB$.

Given a new first-view traffic image im_c associated with a timestamp ti_c, the problems here are to (i) estimate a pm_c at ti_c, (ii) predict a PM2.5 within a time window $[ti_c, ti_{c+\Delta t}]$, and (iii) explain the association of urban traffic, visual pollution, and air pollution within the time window.

3.2 Uncertain Temporal Dataset Generation

Many researchers have pointed out the association between transportation, factories, and air pollution [6,8,12,21]. In [6], we also confirmed our hypothesis of the correlation between urban nature, moving objects, and air pollution and pointed out some significant high-semantic features associated with PM2.5 levels.

In light of these observations, we construct the function to extract the ratio of these high-semantic handcraft features (e.g., urban nature segments, moving objects, and haze degree) from images.

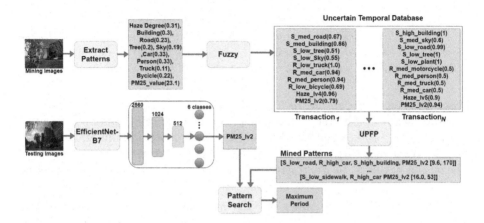

Fig. 2. Example of generating UPFP, predicting PM2.5 at the current time, and predicting PM2.5 in the future

Urban Nature Ratio. In [5], the author mentioned semantic segmentation as one of the most popular techniques for clustering semantic areas. Hence, we utilize this technique to extract urban nature segments as one of our high-semantic handcraft features.

Let $FE_{y=1}(im_k)$ be a semantic segment extractor. Hence we convert image im_k into the urban nature segment ratio set $UNR_k = \{(f_x^y, p_x^y)\}_{x=1..P_y}$. Here, the f_x^y indicates the name of xth semantic segment (e.g., tree, sky, road), and

$$p_x^y = pixel(f_x^y) / \sum_{x=1}^{P_y}(pixel(f_x^y)) \qquad (1)$$

where $pixel(f_x^y)$ returns the number of pixels occupied by f_x^y.

Figure 2 illustrates an example of this set. In this example, UNR_k is $\{Sky(0.19),\ Building(0.3), Road(0.23), Tree(0.2)\}$. We define the pre-defined threshold by which some small segments are ignored.

Moving Object Ratio. In [10], the authors introduced the instance segmentation technique to extract interesting objects from images. In our case, these interesting objects are moving objects.

Let $FE_{y=2}(im_k)$ be a instance segment extractor. Hence we convert image im_k into the moving object ratio set $MOR_k = \{(f_x^y, p_x^y)\}_{x=1..P_y}$. Here, the f_x^y

indicates the name of xth moving object (e.g., car, pedestrian, motorbike), and

$$p_x^y = number(f_x^y)/\sum_{x=1}^{P_y}(number(f_x^y)) \tag{2}$$

where $number(f_x^y)$ returns the number of detected moving object f_x^y.

Figure 2 illustrates an example of this set. In this example, MOR_k is $\{Car(0.33),\ Person(0.33),\ Truck(0.11),\ Bicycle(0.22)\}$. We define the pre-defined threshold by which some objects with small volumes are ignored.

Haze Degree. Haze, smoke, or fog could be related to air pollution. Hence, we integrate haze degree extracted from images as one of the air-pollution-correlated features. In our work, we utilize the haze degree detection introduced by Mao et al. [15]. We define $FE_{y=3}(im_k)$ as the haze feature extractor running on im_k. The output of this feature extractor includes six haze levels and one foggy level.

Fuzzy Negation. The accuracy of $FE_y(im_k)$ outputs depends on the models utilized to extract features. That leads to the fact that feature qualification depends on such models. Hence, we decide to use the scaling space (i.e., category) to present the feature's value instead of the numeric space (i.e., regression) to decrease the impact of uncertainty towards approximate reasoning.

For example, we cluster each feature of UNR_k into the scaling space (LOW, MED, HIGH) where LOW, MED, and HIGH are lowerbounded and upperbounded by the pre-defined thresholds. We apply the fuzzy negation method [3] to approximate the uncertainty of our features UNR_k. Hence, the UNR_k = $\{Road(0.23),\ Tree(0.2),\ Sky(0.19), Building(0.3)\}$ of image im_k is converted into $\{S\_med\_road(0.67),\ S\_low\_tree(0.51),\ S\_low\_sky(0.55), S\_med\_building (0.86)\}$.

We apply fuzzy negation for urban nature ratio, moving object ratio, haze degree and PM2.5 level using the parameters illustrated in Fig. 3.

Category	Range Value		Fuzzy Label	Category	Range Value		Fuzzy Label
Urban Nature Ratio	0.05	0.1	S_ Low(1)	Moving Object Ratio	0	0.15	R_ Low(1)
	0.1	0.275	S_ Low(1-0)—S_Medium(0-1)		0.15	0.45	R_ Low(1-0)—R_Medium(0-1)
	0.275	0.425	S_ Medium(1-0)—S_High(0-1)		0.45	0.75	R_ Medium(1-0)—R_High(0-1)
	0.425	0.5	S_ High(1)		0.75	0.9	R_ High(1)
PM25	0	6	PM25_lv1(1)	Haze Degree	0	0.04	Haze_lv0(1)
	6	23.8	PM25_lv1(1-0)—PM25_lv2(0-1)		0.04	0.12	Haze_lv0(1-0)—Haze_lv1(0-1)
	23.8	45.5	PM25_lv2(1-0)—PM25_lv3(0-1)		0.12	0.2	Haze_lv1(1-0)—Haze_lv2(0-1)
	45.5	103	PM25_lv3(1-0)—PM25_lv4(0-1)		0.2	0.28	Haze_lv2(1-0)—Haze_lv3(0-1)
	103	200.5	PM25_lv4(1-0)—PM25_lv5(0-1)		0.28	0.32	Haze_lv3(1-0)—Haze_lv4(0-1)
	200.5	375.5	PM25_lv5(1-0)—PM25_lv6(0-1)		0.32	0.36	Haze_lv4(1-0)—Haze_lv5(0-1)
	375.5	+∞	PM25_lv6(1)		0.36	0.4	Haze_lv5(1-0)—Foggy(0-1)
					0.4	1	Foggy(1)

Fig. 3. Fuzzy negation parameters

Uncertain Transaction. Given a triplet vector (ti_k, im_k, pm_k), we apply feature extractors FE_y and fuzzy negation FU_z to generate the uncertain transaction $t_{tid} = (tid, ts, Y)$, where $tid = k$, $ts = ti_k$, and

$$Y = \{\cup_{z,y} FU_z(FE_y(im_k)), FU_z(pm_k)\} \tag{3}$$

3.3 Periodic Frequent-Pattern Mining from Uncertain Temporal Databases

Periodic-frequent patterns are patterns (e.g., a sequence of items) that 1) appear in a dataset with its frequency less than or equal to the predefined threshold (i.e., minsup) and 2) repeat themselves with a specific period in a given sequence.

For example, the pattern $b,e:[1.1, 6]$ expresses that if 2-pattern happens (i.e., b and e are items of the pattern), then within six periods, it will appear again with the probability (or sup) is 1.1.

In general, if $PA = \{pa_i\}_{i=1..L}$, where L is the number of mined patterns, is a set of mined patterns, the format of a mined pattern i^{th} is

$$pa_i = (\{item_j\}) : [sup, maxPer]. \tag{4}$$

Periodic-frequent pattern mining tries to discover all periodically occurring frequent patterns in a temporal database. In our research, we want to mine such periodic-frequent patterns from an uncertain temporal dataset, as mentioned before.

The main idea of the algorithm introduced by Uday et al. [19] is to build a new tree structure and a pattern-growth algorithm to effectively find all desired patterns in the database. This algorithm aims to mine a set of patterns whose format is as in Eq. (4). Readers can refer to [19] to have more details.

3.4 PM2.5 Estimation

We define the function EPM to estimate PM2.5 level pm_c at current time ti_c from image im_c as $pm_c \leftarrow EPM(im_c, EffC)$. We utilize the EfficientNet [22] model that pre-trained on the ImageNet dataset to build our embedding vector generator and classifier, since in [22], the authors did prove the advantage of the EfficientNet comparing to ResNet both in accuracy and latency. Figure 2: bottom-left sketches out our classification model where "backbone" is Efficient-NetB7 re-trained with all unfrozen layers on our dataset. We add three full connection layers as the end of the backbone with the output for six PM2.5 fuzzy levels.

3.5 Future PM2.5 Level Prediction

Algorithm 1 is built to predict the future PM2.5 ppm that might be measured within $[ti_c, ti_{c+\Delta t}]$ period, where Δt is the $maxPer$ extracted from found patterns. First, we use the deep-learning prediction model to detect the current

PM2.5 level pm_c (i.e., EfficientNetB7, in our case). Second, we extract features and create the uncertain transaction t_c by fuzzy-based merging high-semantic handcraft features (urban nature ratio, moving object ratio, haze degree) and the predicted PM2.5 level. Finally, Δt is assigned by $pa.maxPer$, and ppm is assigned by pm_c if $pa \in PA$ and all $item_i$ of this pa appears in t_c. Hence, we can say that within $[ti_c, ti_{c+\Delta t}]$ period, the air pollution might reach the pm_c value.

Algorithm 1. Predict future PM2.5 level using patterns and current PM2.5 level estimation

---

    Input: im_c, ti_c, PA
    Output: Predicted PM2.5 level ppm within $[ti_c, ti_{c+\Delta t}]$
$pm_c \leftarrow EPM(im_c, EffC)$ ▷ estimate PM2.5
Apply Equation (3) to generate the uncertain transaction of the triplet vector (ti_c, im_c, pm_c)
$t_c \leftarrow (tid = c, ts = ti_c, Y = \{\cup_{z,y} FU_z(FE_y(im_c)), FU_z(pm_c)\})$ where $y = 1..3$, $z = 1..3$
if $\exists pa \in PA : \forall pa.item_i \in t_c$ **then**
    $\Delta t \leftarrow pa.maxPer$
    return $(pm_c, \Delta t)$
end if
return (pm_c, NULL) ▷ $ppm \leftarrow NA$
                                                            ▷ $\Delta t \leftarrow 0$

---

4   Experimental Results

In this section, we describe datasets and evaluation metrics used to evaluate our method. We also report our results and compare them to other methods that use images as the source for predicting/estimating air pollution.

We use three different datasets to evaluate our methods: MNR-HCM [16], Tokyo-Japan [24], and VisionAir India [1]. Table 1 describes the detail of each dataset. The two first datasets were constructed by using first-view cameras (e.g., GoPro) and personal PM2.5 sensors. Hence, the image's contents reflect mostly the urban traffic with a narrow depth of view. The last dataset was formed by gathering images from fixed-position cameras to capture urban nature mainly (e.g., sky) to measure haze and smoke levels.

The accuracy of our method and others is judged by using $precision = tp/(tp + fp)$, where $tp = $ true positive and $fp = $ false positive.

We define the evaluation strategy to examine the productivity of the proposed method and to compare it with others as follows:

– Randomly pick out 100 images $\{im_c\}$.
– **State 1**: Estimate/Predict the PM2.5 level of each image by using the image-2-PM25 models (i.e., EfficientNetB7 in our case)

Table 1. Dataset information

Dataset	Number of images	Collection time	Periodic
VisionAir India	6087	2019/06/10–2019/07/09	15 min
MNR-HCM	13,292	2020/02/28–2020/03/06	3 min
Tokyo-Japan	119,928	2019/04/15–2019/04/25	3 s

– Create the pattern by fuzzy-based integration of high-semantic handcraft features and the PM2.5 level.
– Find the most similar pattern from the mined pattern database.
– Record the PM2.5 level and the maxPer of the found pattern if it satisfies the condition mentioned in Algorithm 1.
– **Stage 2**: Check within the $[t\_c, t\_c + maxPer]$ (at the same route/place of the image) is there any of the same PM2.5 level. If yes, we count the true positive; otherwise, false positive.

This test is repeated ten times and reports the average precision.

We do the same test with the method introduced in [11]. Table 2 shows the results from two different methods. The extra high-semantic handcraft features (e.g., Haze) could explain our method's better productivity than the old one. The second reason might be the first PM2.5 level estimation/prediction of the current image. The result of stage 1 and stage 2 also are impacted heavily by the volume of data, especially the density of images during a specific period.

Table 2. PM2.5 prediction evaluation (1)

Dataset	Stage 1 (Ours)	Stage 2 (Ours)	Stage 1 [11]	Stage 2 [11]
MNR-HCM	0.96	0.86	0.75	0.86
VisionAir India	0.95	0.86	0.84	0.69
Japan	0.75	0.90	0.63	0.84

We define the second evaluation strategy to see how significant improvement of our method compares to the traditional time-series image-based supervised practice in PM2.5 prediction. We select the LSTM-based model introduced in [6] as the competitor. The strategy is defined as follows:

– Randomly selects 100 consecutive images taken at the same route/image along the time.
– Apply Stage 1 and Stage 2 to predict the PM2.5 level and get the time where the PM2.5 level is measured, say $T\_predict$, for each image.
– A fixed-size time-window image sequence is used to predict the next PM2.5 level at the time $T\_predict$ by utilizing the LSTM-based model [6].
– Calculate the precision of two methods.

Table 3. PM2.5 prediction evaluation (2)

Dataset	Stage 2 (Ours)	LSTM-based prediction [6]
MNR-HCM	0.88	0.73
VisionAir India	0.87	0.82
Japan	0.92	0.84

Table 3 denotes the better result of our method compared to LSTM-based model. Furthermore, LSTM-based method needs to define the fixed-size time window to predict, while ours does not require that. That proves the strength of our approach.

Table 4 describes some interesting patterns mined from VisionAir, MNR-HCM, and Tokyo-Japan datasets. For VisionAir case, we can see how urban nature and urban traffic can impact on PM2.5 level. For MNR-HCM case, we can see a clear association between the number of vehicles traveling on streets, especially motorbike, with PM2.5 high level. For Tokyo-Japan, PM2.5 high-level associates with the number of vehicles running on big roads/streets. We can also see that haze and foggy correlate tightly with the number of vehicles and non- or less-green zones.

Table 4. Pattern Mining

Dataset	Patterns
MNR-HCM	R_high_motorcycle, Foggy, PM25_lv4 :[1195, 203]
	S_low_tree, Foggy, PM25_lv4, S_high_road, R_high_motorbike :[1000, 200]
	R_med_car, S_low_sky, PM25_lv4, R_med_motorbike :[1048, 202]
VisionAir	Haze_lv2, PM25_lv3, S_high_building :[140, 234]
	Foggy, S_med_sky, PM25_lv3 :[102, 309]
	S_low_sidewalk, PM25_lv3, S_med_sky S_low_tree :[61, 298]
Tokyo	S_low_road, S_high_building, R_high_car, PM25_lv2 :[7408, 16431]
	R_low_truck, S_high_building, PM25_lv2 :[6056, 16443]
Japan	S_low_tree, S_high_road, PM25_lv2 :[6679, 17747]

5   Conclusions and Future Works

We introduced a novel method to estimate and predict PM2.5 levels using images captured by personal or IoT devices. The backbone of the method is an uncertain transaction created by converting a set of high-semantic-level features extracted from an image and its relevant air pollution value into a set of approximate reasoning. Here, we pay attention to discovering the association between urban nature segments (e.g., tree, building, road, sky), moving objects (e.g., car, truck, pedestrian), haze degree, and air pollution (e.g., PM2.5). We applied transfer learning on a pre-trained model (e.g., EfficientNET) and re-trained this model with our image-PM2.5 dataset to construct our PM2.5 estimator. After estimating the PM2.5 level with the current image, we generate the image's uncertain transaction and search for the best match pattern from our pattern dataset. The found pattern will return the predicted PM2.5 in the short- and medium-term of time. By utilizing our methods, people can understand well the influences of urban traffic and urban greening on air pollution towards having an excellent strategy to adjust them all.

In the future, we will investigate other visual pollution features and expand image sources (e.g., CCTV, open street map, SNS, and dashcam) to discover more exciting patterns. We will also compare our method to other short- and medium-term prediction methods under accuracy and latency perspectives. We will deploy the system into a city to have better testing field evaluation and see how well the system can contribute to a smart city.

Acknowledgements. We thank Prof. Rage Uday Kiran, supervisor of Kazuki, for sharing his ideas and expertise on fuzzy sets, uncertain database creation, and mining patterns.

References

1. Visionair (2020). https://vision-air.github.io/
2. Awan, F.M., Minerva, R., Crespi, N.: Improving road traffic forecasting using air pollution and atmospheric data: experiments based on LSTM recurrent neural networks. Sensors **20**(13), 3749 (2020)
3. Bedregal, B.C.: On interval fuzzy negations. Fuzzy Sets Syst. **161**(17), 2290–2313 (2010)
4. Bo, Q., Yang, W., Rijal, N., Xie, Y., Feng, J., Zhang, J.: Particle pollution estimation from images using convolutional neural network and weather features. In: 2018 25th IEEE International Conference on Image Processing (ICIP), pp. 3433–3437. IEEE (2018)
5. Cao, F., Bao, Q.: A survey on image semantic segmentation methods with convolutional neural network. In: 2020 International Conference on Communications, Information System and Computer Engineering (CISCE), pp. 458–462. IEEE (2020)
6. Dao, M.-S., Zettsu, K., Rage, U.K.: IMAGE-2-AQI: aware of the surrounding air qualification by a few images. In: Fujita, H., Selamat, A., Lin, J.C.-W., Ali, M. (eds.) IEA/AIE 2021. LNCS (LNAI), vol. 12799, pp. 335–346. Springer, Cham (2021). https://doi.org/10.1007/978-3-030-79463-7_28
7. Ganji, A., Minet, L., Weichenthal, S., Hatzopoulou, M.: Predicting traffic-related air pollution using feature extraction from built environment images. Environ. Sci. Technol. **54**(17), 10688–10699 (2020)
8. Hien, T.T., Chi, N.D.T., Nguyen, N.T., Takenaka, N., Huy, D.H., et al.: Current status of fine particulate matter (pm2.5) in Vietnam's most populous city, Ho Chi Minh city. Aerosol Air Qual. Res. **19**(10), 2239–2251 (2019)
9. Junfei, Q., Zengzeng, H., Shengli, D.: Prediction of pm2.5 concentration based on weighted bagging and image contrast-sensitive features. Stochastic Environ. Res. Risk Assess. **34**(3–4), 561–573 (2020)

10. Ke, L., Tai, Y.W., Tang, C.K.: Deep occlusion-aware instance segmentation with overlapping bilayers. In: Proceedings of the IEEE/CVF Conference on Computer Vision and Pattern Recognition, pp. 4019–4028 (2021)
11. La, T.V., Dao, M.S., Kazuki, Tejima, R.K.U., Zettsu, K.: Improving the awareness of sustainable smart cities by analyzing lifelog images and IoT air pollution data. In: IEEE Big Data, pp. 3589–3594 (2021)
12. Liang, L., Gong, P.: Urban and air pollution: a multi-city study of long-term effects of urban landscape patterns on air quality trends. Sci. Rep. **10**(1), 1–13 (2020)
13. Liu, L., Liu, W., Zheng, Y., Ma, H., Zhang, C.: Third-eye: a mobilephone-enabled crowdsensing system for air quality monitoring. Proc. ACM Interactive Mob. Wearable Ubiquitous Technol. **2**(1), 1–26 (2018)
14. Ma, J., Li, K., Han, Y., Yang, J.: Image-based air pollution estimation using hybrid convolutional neural network. In: 2018 24th International Conference on Pattern Recognition (ICPR), pp. 471–476. IEEE (2018)
15. Mao, J., Phommasak, U., Watanabe, S., Shioya, H.: Detecting foggy images and estimating the haze degree factor. J. Comput. Sci. Syst. Biol. **7**, 226–228 (2014)
16. Nguyen-Tai, T.L., Nguyen, D.H., Nguyen, M.T., Nguyen, T.D., Dang, T.H., Dao, M.S.: MNR-HCM data: a personal lifelog and surrounding environment dataset in Ho-Chi-Minh city, Viet Nam. In: Proceedings of the 2020 on Intelligent Cross-Data Analysis and Retrieval Workshop, pp. 21–26 (2020)
17. Núñez-Alonso, D., Pérez-Arribas, L.V., Manzoor, S., Cáceres, J.O.: Statistical tools for air pollution assessment: multivariate and spatial analysis studies in the Madrid region. J. Anal. Methods Chem. **2019**, 9753927 (2019)
18. Pochwała, S., Anweiler, S., Deptuła, A., Gardecki, A., Lewandowski, P., Przysiężniuk, D.: Optimization of air pollution measurements with unmanned aerial vehicle low-cost sensor based on an inductive knowledge management method. Optim. Eng. **22**(3), 1783–1805 (2021). https://doi.org/10.1007/s11081-021-09668-2
19. Uday Kiran, R., Likhitha, P., Dao, M.-S., Zettsu, K., Zhang, J.: Discovering periodic-frequent patterns in uncertain temporal databases. In: Mantoro, T., Lee, M., Ayu, M.A., Wong, K.W., Hidayanto, A.N. (eds.) ICONIP 2021. CCIS, vol. 1516, pp. 710–718. Springer, Cham (2021). https://doi.org/10.1007/978-3-030-92307-5_83
20. Rijal, N., Gutta, R.T., Cao, T., Lin, J., Bo, Q., Zhang, J.: Ensemble of deep neural networks for estimating particulate matter from images. In: 2018 IEEE 3rd International Conference on Image, Vision and Computing (ICIVC), pp. 733–738. IEEE (2018)
21. Shan, Y., Wang, X., Wang, Z., Liang, L., Li, J., Sun, J.: The pattern and mechanism of air pollution in developed coastal areas of china: From the perspective of urban agglomeration. PLoS ONE **15**(9), 1–21 (2020)
22. Tan, M., Le, Q.: Efficientnet: rethinking model scaling for convolutional neural networks. In: International Conference on Machine Learning, pp. 6105–6114. PMLR (2019)
23. Wu, D., Gong, J., Liang, J., Sun, J., Zhang, G.: Analyzing the influence of urban street greening and street buildings on summertime air pollution based on street view image data. ISPRS Int. J. Geo Inf. **9**(9), 500 (2020)
24. Zhao, P., Dao, M.S., Nguyen, T., Nguyen, T.B., Duc-Tien, D.N., Gurrin, C.: Overview of mediaeval 2020 insights for wellbeing: multimodal personal health lifelog data analysis. In: Proceedings of the MediaEval 2020 Workshop, vol. 2882. CEUR-WS.org (2020)

Two-Stage Traffic Clustering Based on HNSW

Xu Zhang[1], Xinzheng Niu[1(✉)], Philippe Fournier-Viger[2], and Bing Wang[3]

[1] School of Computer Science and Engineering, University of Electronic Science
and Technology of China, Chengdu, Sichuan, China
2386100@qq.com
[2] Shenzhen University, Shenzhen, China
[3] Southwest Petroleum University, Chengdu, China

Abstract. Traffic flow clustering is a common task to analyze urban traffic using GPS data of urban vehicles. Existing density-based traffic flow clustering methods generally have two important problems, that is to not consider the characteristics of urban roads and not handle well different sizes of urban areas. In this paper, we propose a novel method, called TSST-HDBC (**T**wo-**S**tage **S**patial-**T**emporal **H**ierarchical **D**ensity-based **C**lustering), which solves the above problems using a two-stage clustering approach. In the first stage, the characteristics of the input trajectory are considered as a whole in terms of time and space, and an appropriate similarity function is selected. Then, the HNSW (**H**ierarchical **N**avigable **S**mall **W**orld) structure is used to set different search radii at each layer, and preliminary clustering results are obtained, which are recorded as sub-clusters. In the second stage, the sub-clusters are re-clustered, and the similarity measurement function is applied according to the road characteristics to obtain the final clustering result. Experimental results show that the proposed TSST-HDBC method can effectively solve two problems and improve the accuracy of traffic clustering in urban arears.

Keywords: Trajectory clustering · Urban traffic · Taxi GPS data

1 Introduction

In recent years, with the improvement of mobile positioning equipment and economic development, people tend to travel to more diversified locations, and a large amount of GPS trajectory data is being generated. Many studies were carried out on these traffic flows to perform various types of analyses, such as traffic clustering [1] and abnormal trajectory detection [2]. Clustering is most commonly applied to spatial-only trajectories, which can be roughly divided into two categories: partition-based and density-based [3].

This paper focus on density-based traffic clustering. Although there are many studies on trajectory clustering, most algorithms do not consider the features of road networks and are not suitable for practical application in urban traffic analysis. To preserve the locality of trajectories, sub-trajectory segmentation [4] and

© Springer Nature Switzerland AG 2022
H. Fujita et al. (Eds.): IEA/AIE 2022, LNAI 13343, pp. 609–620, 2022.
https://doi.org/10.1007/978-3-031-08530-7_51

grid-based segmentation [5] are widely used in the preprocessing stage. However, neither of those strategies directly considers the relationship between a trajectory and the road network, and the sub-trajectories cannot express road characteristics. For the calculation of similarity between trajectories, most studies take the spatial distance between two sub-trajectories as the similarity. However, apart from the spatial distance, other dimensions also have a great influence on the clustering result, such as the dissimilarity in the temporal dimension. In urban traffic flow analysis, the most widely used clustering models include K-Means [6], and DBSCAN [7]. However, applying most clustering methods require setting parameters manually, for example, K-Means requires that the user sets the number of clusters, while DBSCAN requires setting the Eps and MinPts parameters, but values for these parameters are generally not suitable for all urban roads.

Considering the shortcomings of previous studies, a novel two-stage trajectory clustering method based on the HNSW [8] (**H**ierarchical **N**avigable **S**mall **W**orld) structure is proposed in this paper. In the first stage, a hierarchical structure is built, and the trajectories are stored as vertices in the graph of each layer, so that clustering results of different granularity can be obtained. In the second stage, the imprecision caused by the fixed search radius is addressed. This is done by building a model for road geospatial features. The final clustering results indicate the main road commonly used in the urban area.

The remainder of the paper is organized as follows. Section 2 shows the related work. Definitions and problem statement are given in Sect. 3. Then, the proposed method and the details of its modules are explained in Sect. 4. Section 5 describes experimental results. Finally, we conclude the paper in Sect. 6.

2   Related Work

This section reviews the related work on density-based trajectory clustering and summarizes differences between the proposed method and prior work.

Clustering is the most commonly used method in urban traffic flow analysis, and density-based clustering is an active research topic. Birant and Kut [9] extended the DBSCAN algorithm to process spatio-temporal data. Their algorithm ST-DBSCAN has the ability to discover clusters based on non-spatial, spatial, and temporal values of a target. Agrawal et al. [10] proposed an algorithm named ST-OPTICS (Spatio-Temporal-Ordering Points to Identify Clustering Structure), which is an improvement of the OPTICS clustering algorithm [11]. ST-OPTICS can be adapted to cluster data of arbitrary density and shape, and overcome the difficulties of nested clustering. An improved density-based spatial clustering algorithm was proposed by Liu et al. [12] to better handle noise when measuring the similarity between different trajectories. Yang et al. [13] designed a trajectory clustering algorithm named TAD based on spatio-temporal density analysis of data. TAD considers the trajectory characteristics from the perspective of time and space, and constructs a new density function by synthesizing NMA theory, data distribution and residence time. To protect the personal privacy information contained in a trajectory, Zhao et al. [14] proposed a clustering privacy protection method that ensures differential privacy of a trajectory. The method adds Laplacian noise to the trajectory positions to protect against the continuous query attacks.

To sum up, in previous studies, most spatial-temporal clustering algorithms require users to set multiple parameters, which greatly affects the clustering results. In addition, choosing an exact number of clusters is also difficult for users [15]. For this reason, the TSST-HDBC (**T**wo-**S**tage **S**patial-**T**emporal **H**ierarchical **D**ensity-based **C**lustering) algorithm is proposed. Compared with prior work, it has the following advantages: (1) It measures similarity by taking into account spatial and temporal aspects, as well as road features. (2) The clustering process is divided into two stages. In the second stage, the influence of road features on clustering results is emphasized, which effectively solves the problem of inaccurate clustering caused by manual parameter settings.

3   Problem Formulation

Before describing the proposed algorithm, basic concepts and definitions are introduced in this section.

Definition 1. _Trajectory (TR):_ _In this paper, $TRS = \{TR_1, TR_2, ..., TR_n\}$, represents a Trajectory dataset. And TR_i is a sequence of ordered points, denoted as $TR_i = \{p_i^1, p_i^2, ..., p_i^j, ..., p_i^{maxLength}\}$, where the superscript j is used to denote the jth point of TR_i. Each point consists of GPS coordinates and a timestamp, where $p_i^j = (latitude, longitude, t)$._

Definition 2. _Sub-trajectory (STR):_ _A sub-trajectory is a sequence of ordered trajectory segments obtained by dividing a trajectory according to certain rules. It is denoted as $TR_i = \{STR_i^1, STR_i^2, ..., STR_i^j, ..., STR_i^n\}$. The sequence of STRs can cover TR_i, and there is no overlap among STRs._

Definition 3. _Distance between STRs_ _(STR_{dist}): $STR_{dist} = spatial_{dist} + temporal_{dist}$. The distance between STRs is used to measure the dissimilarity between STRs, where $spatial_{dist}$ represents the spatial distance and $temporal_{dist}$ refers to the temporal distance._

Definition 4. _Direction angle_ _α:α is calculated by the Law of Cosines. Each trajectory slice can be defined as $traj\_slice = \{p_i^j, p_i^{j+1}, ..., p_i^{j+k}\}$, for a trajectory point p_i^j, the direction angle can be calculated as follows:_

$$\alpha = arccos\frac{\|p_i^{j+1} - p_i^j\|^2 + \|p_i^j - p_i^{j-1}\|^2 - \|p_i^{j+1} - p_i^{j-1}\|^2}{2\|p_i^{j+1} - p_i^j\|\|p_i^j - p_i^{j-1}\|} \tag{1}$$

where $\|p_i^{j+1} - p_i^j\|$ is the Euclidean distance between p_i^{j+1} and p_i^j.

Definition 5. Trajectory Cluster (TRC): *A TRC consists of many subTRCs respecting certain rules, where subTRCs are the results of the first stage clustering. Each TRC corresponds to a road or a set of roads, which are frequently used roads, called main roads (*M-roads*).*

The purpose of this approach is to obtain *TRCs* with high spatial similarity, as well as high temporal similarity, so as to obtain *M-roads* in a certain area, called main roads. The clustering problem can be defined as follow.

Definition 6. Problem Statement: *Let there be a trajectory data set $TRS = \{TR_1, TR_2, ..., TR_n\}$, where each TR_i is divided into a sequence of sub-trajectories according to certain rules, $\{STR_i^1, STR_i^2, ..., STR_i^m\}$. The distance between STRs is determined by STR_{dist}. Through two-stage clustering, TRCs having a high similarity will be obtained, as well as M-roads of the region.*

In this paper, the *S_Dbw* [16] and Calinski-Harabasz (CH) score [17] are used to evaluate the accuracy of clustering results. These score functions are defined as follow.

$$S\_Dbw(c) = Scat(c) + Dens\_bw(c) \tag{2}$$

$$Scat(c) = \frac{1}{m} \sum_{i=1}^{m} \frac{\|\sigma(v_i)\|}{\|\sigma(S)\|} \tag{3}$$

$$Dens\_bw(c) = \frac{1}{m(m-1)} \sum_{i=1}^{m} [\sum_{j=1}^{m} \frac{density(u_{ij})}{max\{density(v_i), density(v_j)\}}] \tag{4}$$

In Eq. (3) and Eq. (4) v_i, v_j represent the center of cluster c_i and c_j respectively, m is the number of clusters, and $\|\sigma(v_i)\|$ is the Standard Deviation of the ith cluster c_i. In this paper, u_{ij} is the trajectory that is exactly halfway between v_i and v_j. $density(\cdot)$ is defined as $density(u) = \sum_{i=1}^{n} f(x, u)$,

$$f(x, u) = \begin{cases} 0 & if \quad \|x - u\| > stdev \\ 1 & otherwise \end{cases} \tag{5}$$

where n is the number of trajectories that belong to the cluster having u as center, and $stdev$ is given as $stdev = \frac{1}{m} \sqrt{\sum_{i=1}^{m} \|\sigma(u_i)\|}$.

$$CH = \frac{tr(B_c)/(k-1)}{tr(W_c)/(N-k)} \tag{6}$$

$$B_c = \sum_{i=1}^{M} n_i \|u_i - u_E\| \tag{7}$$

$$W_c = \sum_{i=1}^{M} \sum_{x \in C_i} \|x - u_i\| \tag{8}$$

where N is the number of trajectories, k is the number of trajectories that belong to a cluster c and M is the number of clusters. $tr(B_c)$ denotes the trace of variance matrix among all clusters, and $tr(W_c)$ denotes the trace of variance matrix of cluster c. In Eq. (7) and Eq. (8), u_i denotes the center trajectory of cluster c_i.

In general, a lower S_Dbw and a higher CH score indicate a more accurate clustering.

4   Method

This section proposes a clustering method named TSST-HDBC. An overview of the proposed method is given in Fig. 1. To solve the defined problems, trajectory data need to be preprocessed first. This includes map-matching and segmentation, which maximize the representation of road features by trajectories segments used in clustering. In the first stage of clustering, an improved clustering method based on proximity search is used, which considers both spatial and temporal dimensions to calculate similarity. To obtain the main roads of a region, the second stage is carried out. At this stage, sub-clusters conforming to the measurement standard are combined as one based on the geographical features of the road.

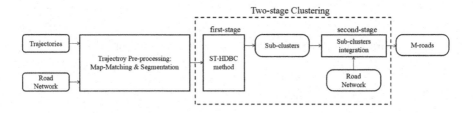

Fig. 1. Overview of the Two-stage clustering method

4.1   GPS Trajectory Pre-processing

Due to the response latency of GPS positioning devices, raw trajectories may not match well with urban roads, which adversely affects to the analysis of urban main roads and traffic flow. Therefore, map matching is firstly carried out. The Shortest-path algorithm and the urban road information provided by OSMnx [18] are used to calculate the points between GPS points with a large response time interval in raw trajectories, so as to obtain the trajectories that highly match with roads. This can effectively avoid the incomplete expression of urban road features caused by inaccurate trajectory data.

After matching, the trajectories are segmented. The principle is shown in Fig. 2. Each point will form a triangle with its neighbor points, as shown in the shaded part of Fig. 2. The direction angle α of each point is calculated using Eq.

(1). As map-matching may lead to dense GPS points, a window function with a size of k is set to judge whether a point is a split-point or not.

$$\delta_i = \frac{\Delta\alpha_i}{\sum_{j=i-k}^{i+k}\Delta\alpha_j} \qquad i = 0, 1, ..., n \qquad (9)$$

$$\Delta\alpha_i = \begin{cases} |\alpha_i - \alpha_{i-1}| & i = 1, ..., n \\ 0 & i = 0 \end{cases} \qquad (10)$$

Eq. (9) is used to calculate the change rate of the direction angle denoted as δ_i. If $\delta_i > 0.5$, we reckon that the direction angle of this point has a sudden change, which may represent a road corner or the intersection of two roads and this point is a split-point. In this way, we get sub-trajectories containing road features denoted as $TR_i = \{STR_i^1, STR_i^2, ..., STR_i^j, ..., STR_i^n\}$.

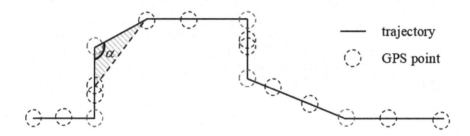

Fig. 2. The schematic diagram of trajectory segmentation

4.2   Two-Stage Clustering

A method based on the DBSCAN and the HNSW structure called FISHDBC was proposed in a previous study [8]. Inspired by the idea of hierarchical structure, we think that it can obtain clustering results of varied granularity through one clustering. But there is a problem: traditional clustering methods based on a similarity function cannot adaptively find an appropriate minimum search radius suitable for all urban regions. Sub-trajectories distributed on the same road may have very small distance similarity due to the long length of the road itself, but they all essentially reflect the usage of a same road. Hence, we propose TSST-HDBC, which is described in details in the rest of this section.

Spatial-Temporal HDBC. Each layer of this hierarchical structure can be regarded as a clustering process at a different scales. As shown in Fig. 3, upper layers reflect the traffic flow clustering among urban areas, while the lower layers reflect the detailed clusters in small regions. In each layer, trajectories are stored as vertices in an undirected graph, and edges with weights represent the STR_{dist} between trajectories.

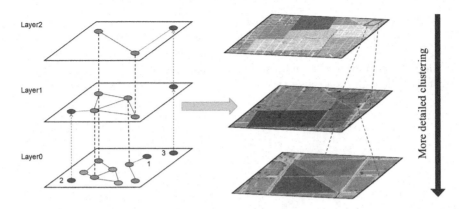

Fig. 3. Illustration of the ST-HDBC structure.

In this paper, we use $STR_{dist}(TR_i, TR_j)$ to refer to the distance between vertices in the graph, defined as follow:

$$STR_{dist}(TR_i, TR_j) = S_{dist}(TR_i, TR_j) + T_{dist}(TR_i, TR_j) \qquad (11)$$

For spatial similarity, because points of trajectories may be inconsistent, the DTW [19] method is used. The recursive formula of Eq. (12) is used to calculate the cumulative distance γ.

$$\gamma(i, j) = d(p_k^i, p_m^j) + min\{\gamma(i-1, j-1), \gamma(i-1, j), \gamma(i, j-1)\} \qquad (12)$$

$$d(p_k^i, p_l^j) = \|p_k^i - p_l^j\| \qquad i = 1, 2, ..., n; j = 1, 2, ..., m$$

where p_k^i is the ith point on TR_k, and p_m^j is the jth point on TR_l. After recursion, we set $S_{dist} = \gamma(i, j)$ which denotes the spatial distance between TR_k and TR_l.

For temporal similarity, the time span is used to assess the time distance between two trajectories. Equation (13) measures the difference of start time and end time between two trajectories. t_k^s represents the start time of TR_k and t_k^e represents the end time.

$$T_{dist} = |(t_k^s - t_m^s) + (t_k^e - t_m^e)| \qquad (13)$$

DBSCAN is applied at each layer from bottom to top to find trajectories within the search radius. When a new trajectory T_k is found, as shown in Fig. 3, there are three cases. Assuming the search radius of the bottom layer is r_0, after inserting it into the bottom layer as a new vertex, T_k will be connected with other vertices with STR_{dist} less than r_0, and the clusters of related vertices will be updated, which is the first case. If the new vertex cannot be assigned to a cluster, it will be passed up and get a larger search radius, called $r_1 (r_1 > r_0)$. The new vertex will be passed up until there is an existed vertex within the search radius (the second case) or it has reached the top layer (the third case). On the top layer, each vertex without other vertices in its search radius is temporarily considered as an outlier.

Sub-cluster Integration Based on Geographic Features. After the first stage of clustering, some trajectories on the same road are grouped into several clusters. Closer to the bottom of the hierarchy, the clustering results are more detailed which may be redundant. The goal of this stage is to merge clusters that cover the same road, like the green and red clusters shown in Fig. 4.

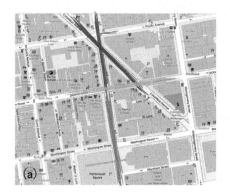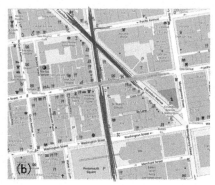

Fig. 4. The second stage combines clusters located on the same road (Color figure online)

We consider the similarity between two clusters from two aspects, distance and geographical features. For example, in Fig. 4(a), clusters in red and blue are close in spatial distance, but they have very different geographical features and belong to two different roads respectively. To accurately integrate clusters with the same geographical features without influencing the already correct correct results, we designed the discriminant rule as follow.

$$
\begin{cases}
|\bar{\alpha}_s - \bar{\alpha}_t| < \epsilon \\
|\theta - \frac{1}{2}(\bar{\alpha}_s + \bar{\alpha}_t)| < \epsilon \\
STR_{dist}(TR_s, TR_t) < r_{l+1}
\end{cases}
\tag{14}
$$

$$
\theta = arctan\frac{y_2 - y_1}{x_2 - x_1}, \quad \epsilon = \frac{\pi}{18}, \quad p_s^i = (x1, y1), \quad p_t^j = (x2, y2)
$$

where $\bar{\alpha}_s$ and $\bar{\alpha}_t$ respectively denotes the mean direction angle of $subTRC_s$ and $subTRC_t$. Assuming that TR_s and TR_t are centers of $subTRC_s$ and $subTRC_t$, p_s^i and p_t^j are GPS points randomly selected on TR_s and TR_t. θ represents the offset angle of the line connecting centers of two clusters. If the clustering result generated on the lth layer in the first stage, r_{l+1} represents the search radius of the upper layer.

5   Experiments

This section describes experiments conducted to evaluate the proposed TSST-HDBC algorithm. We first describe the experimental setup, including the two data sets used and the experimental parameters. The results obtained by the proposed algorithm are compared with FISHDBC [20] with different parameter settings. Finally, the results are discussed.

The goal is to demonstrate that the improved algorithm has better clustering effect on spatial-temporal data by evaluating TSST-HDBC and FISHDBC under different parameter settings. We conducted comparative experiments on two data sets and two different parameter settings.

Dataset. The first dataset (GeoLife[1]) contains trajectory data for 182 users, consisting of a series of points in chronological order, each containing longitude, latitude, and altitude. To test the effect of TSST-HDBC in a small area, we selected 524 trajectories recorded at different times near Tsinghua University. The second dataset (San Francisco cab[2]) contains 30 days of trajectory data from 500 taxis that include longitude and latitude. On this dataset, we tested TSST-HDBC over a large area. Therefore, we selected 2088 user trajectories for the experiment.

Parameter Setting. The experiments mainly involves two parameters. The first is the parameter $MinPts$ of DBSCAN. According to the suggestions of previous studies [21], we use a value of $Minpts = 10$. The second one is the parameter ef of the HNSW structure, which controls search quality. A small ef value saves running time but may produce inaccurate results and a large one will increase time cost but yield more accurate clustering results. We followed parameter settings from a prior study [20], and respectively set the value $ef = 20$ and $ef = 50$. Other parameters kept the default values of Malkov and Yashunin [8] in the HNSW structure.

Experimental Results. To illustrate the effectiveness of the TSST-HDBC algorithm, we compare the S_Dbw index and $Calinski - Harabasz(CH)$ index of the clustering results in experiments, which both reflect the cohesion and coupling of clusters. A smaller S_Dbw value and larger CH value indicate better performance of an algorithm. The results are shown in Table 1, we can see that the proposed algorithm has better performance than FISHDBC. In the Geolife dataset, we test the clustering effect of TSST-HDBC on traffic flow in a small area. When $ef = 50$, the algorithm shows a significant advantage, with the best results on both S_Dbw and CH. In the San Francisco Cab dataset, we test performance of proposed algorithm in a large area. For S_Dbw index, parameter ef seems to not significantly affect the performance of the algorithm, but TSST-HDBC still shows better clustering results than FISHDBC. For the CH index, when $ef = 50$, the algorithm has optimal results.

---

[1] https://www.microsoft.com/en-us/download/details.aspx?id=52367.

[2] http://crawdad.org/epfl/mobility/20090224/.

Table 1. Experimental evaluation indexes under different data sets (rows) and parameter ef (columns)

	Data Set	TSST-HDBC $(ef = 20)$	TSST-HDBC $(ef = 50)$	FISHDBC $(ef = 20)$	FISHDBC $(ef = 50)$
S_Dbw [16]	GeoLife	0.889	**0.232**	0.867	0.917
	San Francisco	**0.914**	0.93	0.957	0.952
CH [17]	GeoLife	247.763	**931.045**	82.285	113.766
	San Francisco	7.226	**19.376**	5.727	4.337

To intuitively see the clustering effect of TSST-HDBC, we selected the clustering result when $ef = 50$ and created a visualization, shown in Fig. 5. Through the proposed method, the main roads in a region are obtained, marked in red in the Fig. 5. This method can also obtain detailed clustering of main roads, as shown in Fig. 5. Different clusters are marked with different colors. It can be seen intuitively that the clustering result obtained by this method has high similarity within each cluster and no overlap between clusters.

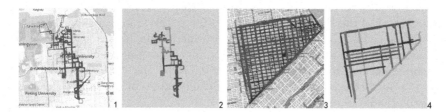

Fig. 5. The clustering result produced by TSST-HDBC when $ef = 50$ (Color figure online)

6   Conclusion

In this paper, a two-stage spatial-temporal clustering algorithm based on HNSW was proposed. The method preprocesses trajectory data by using the change rate of direction angle so that trajectory segments contain road features. The relationship between urban roads and trajectories is fully considered and a two-stage clustering method is designed. In the first stage, the similarity function is considered for both the spatial and temporal dimensions, and clustering results of different granularities are obtained. In the second stage, the angle similarity function is calculated by using the average direction angle in sub-clusters, and the sub-clusters are integrated to obtain the final clustering result. Experimental results show that TSST-HDBC achieves a highly accurate clustering results without increasing the running time, and the result reflects the main urban roads. In future work, we will focus on the clustering of urban traffic flow in different time periods, hoping to expand the model to obtain dynamic urban traffic flow clustering in both spatial and temporal aspects.

Acknowledgement. This research is sponsored by the Science and Technology Planning Project of Sichuan Province under grant No. 2021YFS0391.

References

1. Mustafa, H., Barrus, C., Leal, E., Gruenwald, L.: Gtraclus: a local trajectory clustering algorithm for gpus. In: 2021 IEEE 37th International Conference on Data Engineering Workshops (ICDEW), pp. 30–35 (2021)
2. Zhao, X., Su, J., Cai, J., Yang, H., Xi, T.: Vehicle anomalous trajectory detection algorithm based on road network partition. Applied Intelligence, pp. 1–19 (2021)
3. Wang, S., Bao, Z., Culpepper, J.S., Cong, G.: A survey on trajectory data management, analytics, and learning. ACM Comput. Surv. (CSUR) **54**(2), 1–36 (2021)
4. Lee, J.G., Han, J., Whang, K.Y.: Trajectory clustering: a partition-and-group framework. In: Proceedings of the 2007 ACM SIGMOD International Conference on Management of Data, pp. 593–604 (2007)
5. Wang, J., Yuan, Y., Ni, T., Ma, Y., Liu, M., Xu, G., Shen, W.: Anomalous trajectory detection and classification based on difference and intersection set distance. IEEE Trans. Veh. Technol. **69**(3), 2487–2500 (2020)
6. Krishna, K., Murty, M.N.: Genetic k-means algorithm. IEEE Trans. Syst. Man Cybern. Part B (Cybern.) **29**(3), 433–439 (1999)
7. Ester, M., Kriegel, H.P., Sander, J., Xu, X., et al.: A density-based algorithm for discovering clusters in large spatial databases with noise. In: KDD, vol. 96, pp. 226–231 (1996)
8. Malkov, Y.A., Yashunin, D.A.: Efficient and robust approximate nearest neighbor search using hierarchical navigable small world graphs. IEEE Trans. Pattern Anal. Mach. Intell. **42**(4), 824–836 (2018)
9. Birant, D., Kut, A.: St-dbscan: an algorithm for clustering spatial-temporal data. Data Knowl. Eng. **60**(1), 208–221 (2007)
10. Agrawal, K., Garg, S., Sharma, S., Patel, P.: Development and validation of optics based spatio-temporal clustering technique. Inf. Sci. **369**, 388–401 (2016)
11. Ankerst, M., Breunig, M.M., Kriegel, H.P., Sander, J.: Optics: ordering points to identify the clustering structure. ACM SIGMOD Rec. **28**(2), 49–60 (1999)
12. Li, H., Liu, J., Wu, K., Yang, Z., Liu, R.W., Xiong, N.: Spatio-temporal vessel trajectory clustering based on data mapping and density. IEEE Access **6**, 58939–58954 (2018)
13. Yang, Y., Cai, J., Yang, H., Zhang, J., Zhao, X.: Tad: a trajectory clustering algorithm based on spatial-temporal density analysis. Expert Syst. Appl. **139**, 112846 (2020)
14. Zhao, X., Pi, D., Chen, J.: Novel trajectory privacy-preserving method based on clustering using differential privacy. Expert Syst. Appl. **149**, 113241 (2020)
15. Ansari, M.Y., Ahmad, A., Khan, S.S., Bhushan, G., et al.: Spatiotemporal clustering: a review. Artif. Intell. Rev. **53**(4), 2381–2423 (2020)
16. Halkidi M, V.M.: Clustering validity assessment: finding the optimal partitioning of a data set. In: Proceedings 2001 IEEE International Conference on Data Mining, pp. 187–194 (2001)
17. T Caliński, J.H.: A dendrite method for cluster analysis. Commun. Stat. **3**(1), 1–27 (1974)
18. Boeing, G.: Osmnx: new methods for acquiring, constructing, analyzing, and visualizing complex street networks. Comput. Environ. Urban Syst. **65**, 126–139 (2017)

19. Berndt, D.J., Clifford, J.: Using dynamic time warping to find patterns in time series. In: KDD Workshop, vol. 10, Seattle, WA, USA, pp. 359–370 (1994)
20. Dell'Amico, M.: Fishdbc: Flexible, incremental, scalable, hierarchical density-based clustering for arbitrary data and distance (2019)
21. Schubert, E., Sander, J., Ester, M., Kriegel, H.P., Xu, X.: Dbscan revisited, revisited: Why and how you should (still) use dbscan. ACM Trans. Database Syst. **42**(3) (2017)

Explainable Online Lane Change Predictions on a Digital Twin with a Layer Normalized LSTM and Layer-wise Relevance Propagation

Christoph Wehner[1]($\boxtimes$)(iD), Francis Powlesland[2](iD), Bashar Altakrouri[2](iD), and Ute Schmid[1,3](iD)

[1] University of Bamberg, Bamberg, Germany
{christoph.wehner,ute.schmid}@uni-bamberg.de
[2] IBM Deutschland GmbH, Munich, Germany
francis.powlesland1@ibm.com,
[3] fortiss GmbH, Munich, Germany

Abstract. Artificial Intelligence and Digital Twins play an integral role in driving innovation in the domain of intelligent driving. Long short-term memory (LSTM) is a leading driver in the field of lane change prediction for manoeuvre anticipation. However, the decision-making process of such models is complex and non-transparent, hence reducing the trustworthiness of the smart solution. This work presents an innovative approach and a technical implementation for explaining lane change predictions of layer normalized LSTMs using Layer-wise Relevance Propagation (LRP). The core implementation includes consuming live data from a digital twin on a German highway, live predictions and explanations of lane changes by extending LRP to layer normalized LSTMs, and an interface for communicating and explaining the predictions to a human user. We aim to demonstrate faithful, understandable, and adaptable explanations of lane change prediction to increase the adoption and trustworthiness of AI systems that involve humans. Our research also emphases that explainability and state-of-the-art performance of ML models for manoeuvre anticipation go hand in hand without negatively affecting predictive effectiveness.

Keywords: XAI · Prototype · Digital Twin · Manoeuvre anticipation · Safety-critical AI

1 Introduction

Digital transformation trends such as Artificial Intelligence (AI), Digital Twins and internet of things plays an increasing and integral role in driving innovation

This research was co-funded by the Bavarian Ministry of Economic Affairs, Regional Development and Energy, project Dependable AI, IBM Deutschland GmbH, and IBM Research, and was carried out within the Center for AI jointly founded by IBM and fortiss.

H. Fujita et al. (Eds.): IEA/AIE 2022, LNAI 13343, pp. 621–632, 2022.
https://doi.org/10.1007/978-3-031-08530-7_52

and becoming ubiquitous in various domains and applications such as intelligent driving [16,23]. These trends enable smart systems with novel capabilities that were never possible without AI. However, with the increasing adoption of these complex AI driven systems, new challenges emerge, especially related to the acceptance and trustworthiness of these systems by human users. Hence, there are increasing voices that demand more transparent and explainable AI models and systems.

AI engines for predicting lane changes can be implemented using white-box models that come with the advantage of being explainable by default. Alternatively, black-box classifiers currently outperform white-box models on the lane change prediction task but struggle as they are not explainable [24]. For mission critical tasks that involve human, such as predicting lane changes, both explainability and performance are crucial and equally important.

This paper explores how the decisions making process of a complex black-box classifier can be made explicit and explained to a user. We present a state-of-the-art approach for lane change predictions that are explainable and a novel technical proof-of-concept implementation.

2   Related Work

Predicting lane changes for vehicles is an ongoing field of research. In their paper, Xing et al. [24] presented a general discussion and a survey of the latest technology trends around this topic.

The literature suggests various approaches to predict lane changing behaviour that mainly varying in the data and AI architecture used. Chen et al. [7] aimed to train an attention-based deep reinforcement agent based on visual data in a simulated environment that predicts lane changes. Another approach, by Tang et al. [22], uses tabular data to train an adaptive fuzzy neural network to predict if a lane change takes place soon. Furthermore, a popular machine learning architecture for this type of problem is the recurrent neural network, as it is optimized to deal with problems related to time sequence analyses [18].

With the increasing attention on trustworthiness and transparency of machine learning models and systems, we are seeing a focus in recent literature on explainable models for the lane change prediction task, to be able to explain the reasons behind a predicted vehicle lane change for a human user or road stakeholder. The main goal is to move away from predicting with black-box models and aiming to increase the performance of white-box models like expert systems and other explainable classifiers [10,11]. One paper by Dank et al. reformulated the prediction task based on tabular data to a regression problem [8].

While white-box models come with the advantage of being explainable [19], they are outperformed on the lane change prediction task by black-box classifiers [24]. Nonetheless, the latter are not explainable. For safety critical tasks, both explainability and performance are crucial and equally important.

AI systems cannot be implemented without reliable data resources. Advancement in the area of Internet of Things (IoT) and Digital Twins within the automotive area, especially around autonomous and intelligent driving, can be seen in recent literature and successfully deployed projects and systems [9,17,20]. Alongside the vehicles themselves, infrastructure, such as roads and highways have also undergone modernization in places, so that these elements also can relay their "state" back to operators in real time. A real example of this is the Providentia++ Digital Twin [2], which covers a section of Autobahn between Munich and Munich Airport. Here, the Providentia++ team decided to use cameras placed at regular points along the road, combined with visual recognition to identify vehicles. The setup is capable of relaying the position of every vehicle on the track, with a high level of accuracy and frequent update cycles.

This paper shows an approach to combining data from the Providentia++ Digital Twin with an explainable machine learning model to predict lane changes in real-time. The following section shall introduce this approach.

3 Approach

In this section, we present our suggested approach towards explainable lane change prediction supported by an extensible technical implementation.

3.1 Lane Change Predictions by a Layer Normalized LSTM

The lane change predictions are computed by a layer normalized long short-term memory proposed for this purpose by Patel et al. in 2018 [18]. This section shall introduce the prediction model and its input features regarding relevant perspectives for generating explanations of its predictions. Please consult Patel et al.'s paper [18] for further information and evaluation of the ML model.

First, a layer normalized LSTM [5] considers at each time step t a 1-dimensional array of vehicles $[v_t^i | \forall i \in [0,1,2,3,4,5,q]]$. The vehicles v^i with $i \in [0,1,2,3,4,5]$ are the closest existing neighbours of v^q. Each vehicle v^i at the time k is represented by the following array of features:

$$v_k^i = [v_{x_k}^i, v_{y_k}^i, \psi_k^i, x_k^i, y_k^i, n_l^i, n_r^i], \tag{1}$$

where x_k^i and y_k^i are the absolute world-fixed positions in meters, $v_{x_k}^i$, $v_{y_k}^i$ the respective velocities in meters per second, ψ_k^i is the heading angle of the vehicle in radiance and n_l^i, n_r^i the number of lanes to the left and right. Furthermore, at each time step, the layer normalized LSTM considers as an input the previous cell state $c_{k-0.5s}$ and the previous recurrent state $h_{k-0.5s}$ [13]. Formally, the layer normalized LSTM is defined in Eq. 2 [18].

$$(h_k, c_k) = lnLSTM([v_k^0; ...; v_k^5; v_k^q], h_{k-1}, c_{k-1}) \tag{2}$$

Its output at each time step is the cell state c_k and the recurrent state h_k. For each prediction $k \in [t_{-1.5s}, t_{-1.0s}, t_{-0.5s}, t]$ time steps are shown to the layer normalized LSTM layer. t is the time the prediction is generated. Therefore, the layer normalized LSTM observes four frames of a vehicle and its surroundings within 1.5 s, before it creates a prediction.

Layer normalization [5] is applied before the non-linearities of the LSTM to increase its robustness.

Layer normalization $\vartheta(\cdot)$ is defined as follows:

$$\vartheta(a) = f\left[\frac{g}{\sigma} \odot (a - \mu) + b\right] \quad \sigma = \sqrt{\frac{1}{H}\sum_{i=1}^{H}(a_i - \mu)^2} \quad \mu = \frac{1}{H}\sum_{i=1}^{H}a_i, \quad (3)$$

where μ is the mean of a. a is the activation vector along the feature axis before the non-linearities of the gated interactions inside an LSTM cell. H denotes the number of hidden units in a layer, σ is the standard deviation of a, g are the learned gain parameters, and b is a learned bias [5].

Ba et al. showed that layer normalization stabilizes the gradient [5]. This results in a more stable and faster convergence of the validation loss to an optimum at training time and increases classification performance at inference time.

The model's output represents if v^q changes to the left or right lane or stays on the same lane within the next 2.5 s. The labels are a one-hot encoding of the three classes.

3.2   Explanations of the Prediction Generated by LRP

Layer normalized LSTM's show state-of-the-art performance at the lane change prediction task [18,24]. However, their decision-making process is considered a black box, as it is too complex and complicated to be understood by a user.

We follow the increasing demand and research efforts to explain the decision-making process of a black-box classifier.

The core of our proposed prediction engine applies the Layer-wise Relevance Propagation (LRP) attribution method on the layer normalized LSTM. We aim to make the decision-making process of the lane change prediction explicit, by identifying which part of the input is relevant for the classification.

LRP assigns each input dimension of the layer normalized LSTM a relevance value. The relevance values represent how much each input dimension contributed to the prediction.

LRP starts at the output layer, where the relevance for each neuron is set to be the prediction function value of the class to be explained $f_c(x)$. Layer by layer, the relevance is completely redistributed, from higher layer neurons to lower layer neurons by employing layer-specific LRP rules, where neurons that contribute most to the higher layer receive the most relevance from it, as explained in [6].

Arras et al. propose a chaining of (1) the LRP-ϵ rule for the linear mappings, (2) the LRP-*all* rule for the gated interactions, and (3) the LRP accumulation rule to explain the interactions of a standard LSTM [3].

We extend their approach to layer normalized LSTM's by applying in addition the LRP-Ω rule to the model-specific interaction of a layer normalized LSTM. In particular, we propose the novel LRP-Ω rule to explain layer normalization.

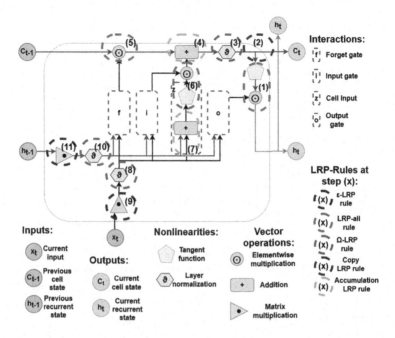

Fig. 1. Step by step chaining of the LPR rules for layer normalized LSTM's.

Figure 1 visualizes the LRP rule chaining for the layer normalized LSTM architecture.

LRP-all Rule for the Gated Interactions. The relevance flow of the gated interactions in step (1), (5), and (6) of Fig. 1 are retraced by the LRP-*all* rule. With the LRP-*all* rule, all relevance flows to the source units. However, the gate units receive no relevance, as they do not hold information themselves but control the information flow [3].

LRP Accumulation Rule. Accumulations are interactions appearing in step (4) and (7) on Fig. 1. At accumulations, the relevance is split proportional to the magnitude of each addend, as suggested by [4].

LRP-ε Rule for Linear Mappings. Linear mappings are the interactions depicted in step (2)[1], (9) and (11) of Fig. 1. As suggested by [3], the LRP-ε rule is used to retrace the relevance flow of the linear mapping. The linear mapping is equivalent to a dense layer with a linear activation function and a zero bias.

LRP-Ω Rule for Layer Normalization. Layer normalization requires a specific LRP rule. According to our knowledge, LRP for layer normalization is not yet explored by the literature. In principle, layer normalization is similar to batch normalization [15]. However, they differ in the normalization dimension. While

---

[1] Step (2) in Fig. 1 is called the copy LRP rule. The copy LRP rule is a particular case of the LRP-ε Rule, where one lower-level node and n upper-layer nodes exist, the weights are set to one, the bias is zero, and the activation function is linear.

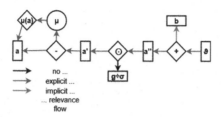

Fig. 2. Heuristic relevance decomposition of layer normalization in Eq. 3 by the LRP-Ω rule. A black arrow signalizes no relevance flows to the term according to the LRP-Ω rule. A red arrow signalizes relevance flows to the term, and the LRP-Ω rule explicitly calculates it. Finally, a blue arrow signalizes relevance flows to the term, but it is not explicitly calculated in the LRP-Ω rule as the term is a relevance sink. (Color figure online)

batch normalization normalizes over the whole batch, layer normalization normalizes over one instance [5,15]. We have explored and applied different LRP approaches for batch normalization to layer normalization, including the LRP identity rule [3], LRP-$|z|$ [14], LRP-ϵ [14], LRP fusion [12], and LRP heuristic rule [1].

While the previous approaches made tremendous progress on explaining batch normalization, none fully consider the mean's impact on the relevance. Thus, we propose the novel LRP-Ω rule for layer normalization. The LRP-Ω rule decomposes layer normalization into a series of summations and scalings. This is shown in Fig. 2. In particular, the LRP-Ω rule acknowledges the impact of the mean shift in Eq. 3 on the relevance flow. Furthermore, it propagates the relevance assigned to the mean further down to the input of the layer normalization. The LRP-Ω rule is formalized in Eq. 4.

$$R_{i \leftarrow j} = \left(z_i - \frac{z_i}{H} \right) \cdot \frac{g_i}{\sigma} \cdot \frac{R_j}{z_j} \tag{4}$$

R is the relevance signal from the input value z of the higher layer unit j to the input value of the layer normalization unit i. H is the length of the input to the layer normalization.

We outlined in this section how LRP calculates relevance values as explanations. The following section shall introduce how the relevance values are transformed into comprehensible explanations.

3.3   Comprehensible Explanations

LRP assigns relevance values to the 4×49 dimensions of the layer normalized LSTM's input. This is LRP's explanation of the model's prediction. The user of the maneuver anticipation system, i.e. the driver, will find 196 relevance values to be incomprehensible. Hence, we have adopted an aggregation approach that utilizes the adaptive nature of LRP in terms of dimensionality reduction. Due to LRPs' relevance conservation and redistribution property [6], relevance values of terminal units can be added together without invalidating them. The aggregated value represents the relevance of the aggregated units. Thus, features in the input space can be aggregated to meaningful virtual super-features, and their aggregated attribution values represents their relevance for the classification.

The aggregation of relevance values makes it possible to communicate comprehensible explanations to the user. To explain the lane change prediction of the layer normalized LSTM, the time dimensions are aggregated. Therefore, the relevance values of input features representing a vehicle are added together over every time step:

$$\forall i : R_{v^i} = \sum_{\forall k} R_{v_k^i} \tag{5}$$

Furthermore, the relevance values of the individual features of the vehicles are aggregated to the virtual and weighted super-features movement m and position p:

$$\forall i : R_{v^i} = [R_{m^i}, R_{p^i}];$$
$$\text{with } R_{m^i} = \frac{R_{v_x^i} + R_{v_v^i} + R_{\psi^i}}{3} \text{ and } R_{p^i} = \frac{R_{x_k^i} + R_{y_k^i} + R_{n_l^i} + R_{n_r^i}}{4} \tag{6}$$

Finally, the three most relevant super-features are communicated to the user via the demonstrator in real-time, as shown in Fig. 4. The three most relevant features are visualized via their name and logo to the driver. In addition, a color scheme describes their relative impact on the classification.

This section described how high dimensional explanations by LRP are reduced to make them comprehensible by the driver while steering a vehicle. Up next, the implementation details of the prototype and an evaluation shall be provided.

3.4 Prototype Architecture

This prototype has been designed as a distributed set of containers, and as such can be deployed on any Kubernetes cluster with minimal configuration. This approach was chosen in order to maximize resiliency and redundancy across the application, whilst also logically separating concerns, permitting independent horizontal scaling. Up next, the elements that describe this prototype are detailed.

Live Adaptor. The Live Adaptor takes the protobuf stream from the digital twin, decodes and enriches it, so it can be consumed by other parts of the application. This optional step improved the workflow for the rest of the application by propagating the data as JSON. It also checks each vehicle that comes through the digital twin and assigns a UUID. This was necessary since the digital twin itself only assigns vehicle ID's in the cycle 1–10,000, meaning that we lose vehicle uniqueness if we record data that contains over 10,000 vehicles directly from the digital twin. To address this, we looked at each original vehicle ID coming through the digital twin, and checked to see when the ID was last present, if the original vehicle ID has not been present for a period of time, we assume that the vehicle is new, and it is issued with a new UUID. To store UUIDs we use redis as the in-memory cache, preserving state across application restarts and failures. Vehicle IDs contain no identifying information about the vehicle itself.

Prediction Engine. The prediction engine is composed of the **Prediction Model** and the **Service Broker**, that together enable the user to consume live predictions on demand, in a scalable way. To realize this, we kept the prediction model in a python container that communicates with the service broker over a standard HTTP protocol. Instead of having the prediction model handle connections to the user, we created a "sessioning" platform in the service broker which listens for user requests to open a "prediction session" for a specific UUID. While the session is open, snapshots are repeatedly collected from the live adaptor and are then sent off for inference. The novel element here is that the service broker can handle many connections at once, enabling multiple users, and handles all internal state about user sessions. Because of this, the prediction model itself is stateless and can be scaled horizontally. Once a vehicle leaves the digital twin, the service broker will automatically terminate the session, running any garbage collection.

General Considerations. Because of the nature of this domain, specifically our data source being a live digital-twin, considerations were made across every facet of this project to make sure we utilised an event driven architecture. In practice, this meant heavily utilising technologies such as websockets for two-way communication between the system and the user, as well as using websockets to manage state across the system itself. Kubernetes was chosen as our platform as it allowed us to deploy highly customised containers with relative ease.

The architecture of the prototype is fully mapped. Up next, the explanations of the lane change predictions shall be evaluated and the prototype's GUI shall be discussed.

4   Evaluation and Discussion

This section evaluates the explanations provided by LRP in terms of their faithfulness to the layer normalized LSTM's behavior. Furthermore, the GUI of the prototype is presented and critically discussed. Please consult [18] for an in-depth evaluation and comparison of the layer normalized LSTM in contrast to other machine learning models for predicting lane changes.

4.1   Evaluation of the Explanations

A perturbation test is deployed to evaluate the explanation. The perturbation test is a behaviouristic approach to evaluate the faithfulness of an explanation. It asks if the explanations reflect the model's behaviour.

For the perturbation test, classifications and attributions, i.e. relevance values, of a representative amount of instances are calculated. Next, the instances

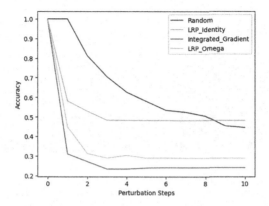

Fig. 3. Perturbation test on the layer normalized LSTM for the lane change prediction task. The perturbation test is conducted on 2315 instances. The instances are randomly drawn from a set, uniformly distributed over the labels.

are split into correct and wrong classified instances. For the correct classified instances, the most important super-feature is occluded. The occluded instances are classified, and the accuracy is measured. Again, the remaining most relevant super-feature is occluded, and the model's accuracy on the newly created instances is measured. The previous step may be repeated until there is no more feature to occlude. A faithful explanation method produces results that decrease the accuracy significantly more than randomly occluding features.

To set the faithfulness of LRP for layer normalized LSTM's into context, LRP is compared with the attribution method Integrated Gradients [21]. Furthermore, two versions of LRP are compared: LRP with the LRP identity rule [3] applied to layer normalization and LRP with the LRP-Ω rule for layer normalization.

The results of the perturbation test are depicted in Fig. 3.

LRP with the LRP identity rule applied to layer normalization performs worst in the perturbation test. The perturbation test converges towards 50% accuracy for this rule combination. After nine perturbation steps, it is outperformed by random occlusion. The rule captures the most relevant features accurately, but fails to distribute relevance to minor impactful parts of the input.

The LRP-Ω rule outperforms the LRP identity rule significantly. Heuristically redistributing the relevance from the layer's output to its input, while fully considering the impact of the mean on the relevance flow, increases the performance. The LRP-Ω rule allows capturing the impact of every part of the input accurately.

Integrated Gradients outperforms LRP for the layer normalized LSTM in terms of faithfulness. At first glance, this is surprising. Gradient-based attribution methods tend to not perform well on standard LSTM's [4] because the gradient of the sigmoid- and tanh- non-linearities of the LSTM cell is close to zero for activations outside the interval $[-4; 4]$ and respectively $[-2; 2]$. However, through the layer normalization, the inputs to the non-linearities are brought

Fig. 4. The "driver perspective", showing our selected vehicle (highlighted with "Q"), the surrounding vehicles, and their impact on the latest prediction.

closer to those intervals, stabilizing the gradient and leading to faithful explanations of gradient-based attribution methods.

On the Computational Expenses of LRP. We implemented Integrated Gradients and LRP in Tensorflow 2.4, running on a workstation with two Nvidia 2080TI, CUDA 11.1, 64 GB RAM, and an AMD Ryzen Threadripper 2920X. LRP computed the explanations on average 10.47 times faster than Integrated Gradients for 2335 randomly drawn instances. This is due to approximating integrals being computationally expensive.

LRP performs in terms of faithfulness comparably to Integrated Gradients while being significantly more computationally efficient. Thus, LRP is our method of choice for explaining the online lane changes predictions.

Next, the GUI, where the prediction and explanation by LRP are presented, is discussed.

4.2   Discussion of the Prototype's GUI

The visualization component is the user facing web application that shows the capabilities of the demonstrator (Fig. 4). This allows the user to "jump in" to a vehicle and get various insights as if they were driving the vehicle themselves. The user can see real-time stats such as the nearest neighbours, number of vehicles on the road, the next prediction and the reasons associated with it. The UI also instructs the prediction engine to start or stop predictions for a specific vehicle, rendering the output. In the explainability domain, our chosen approach here is to use a heatmap, where the neighbouring vehicles change color depending upon their actions and how much they impacted the latest prediction.

5   Conclusion

This paper showcased how to predict and explain lane changes given live data of a digital twin. For that reason, the layer normalized LSTM is outlined as a state-of-art prediction model. However, its decision-making process is too complicated and complex to be understood by a user. In safety-critical applications like lane change predictions, a user must understand the reasoning of the prediction engine. Thus, we developed LRP for layer normalized LSTM's to make the decision-making process of the layer normalized LSTM explicit. LRP shows performant results in terms of faithfulness to the models' behavior while being computationally lightweight. Thus, it is the method of choice for explaining the lane change predictions in real-time. Furthermore, this paper gave implementation insides on how to realize a scalable, high-performance prototype for making explainable lane change predictions. In addition, we presented the user interface and critically discussed it. Future work includes implementing multimodal communication strategies of the computed relevance values beyond heatmaps. And the communication strategies of the prototype shall be evaluated in a user study. Furthermore, it is an interesting open question on how to use the explanation of the model's prediction so that the user interactively improves the prediction model to make it more performant and trustful.

Attribution methods provide deep insights into a black box machine learning model's decision-making process. Let us use those insights to create more trustful and safe machine learning applications.

References

1. Alber, M., et al.: Innvestigate neural networks! J. Mach. Learn. Res. **20**(93), 1–8 (2019). http://jmlr.org/papers/v20/18-540.html
2. Krämmer, A., Christoph Schöller, D.G., Knoll, A.: Providentia - a large scale sensing system for the assistance of autonomous vehicles. In: Robotics: Science and Systems (RSS), Workshop on Scene and Situation Understanding for Autonomous Driving (2019). https://sites.google.com/view/uad2019/accepted-posters
3. Arras, L., et al.: Explaining and Interpreting LSTMs, pp. 211–238. International Publishing, Cham (2019). https://doi.org/10.1007/978-3-030-28954-6_11. https://doi.org/10.1007/978-3-030-28954-6_11
4. Arras, L., Montavon, G., Müller, K., Samek, W.: Explaining recurrent neural network predictions in sentiment analysis. CoRR (2017). http://arxiv.org/abs/1706.07206
5. Ba, J.L., Kiros, J.R., Hinton, G.E.: Layer normalization (2016)
6. Bach, S., Binder, A., Montavon, G., Klauschen, F., Müller, K.R., Samek, W.: On pixel-wise explanations for non-linear classifier decisions by layer-wise relevance propagation. PLOS ONE **10**(7), 1–46 (2015). https://doi.org/10.1371/journal.pone.0130140. https://doi.org/10.1371/journal.pone.0130140
7. Chen, Y., Dong, C., Palanisamy, P., Mudalige, P., Muelling, K., Dolan, J.M.: Attention-based hierarchical deep reinforcement learning for lane change behaviors in autonomous driving. In: 2019 IEEE/CVF Conference on Computer Vision and Pattern Recognition Workshops (CVPRW), pp. 1326–1334 (2019). https://doi.org/10.1109/CVPRW.2019.00172
8. Dang, H.Q., Fürnkranz, J., Biedermann, A., Hoepfl, M.: Time-to-lane-change prediction with deep learning. In: 2017 IEEE 20th International Conference on Intelligent Transportation Systems (ITSC), pp. 1–7 (2017). https://doi.org/10.1109/ITSC.2017.8317674
9. El Marai, O., Taleb, T., Song, J.: Roads infrastructure digital twin: a step toward smarter cities realization. IEEE Network **35**(2), 136–143 (2020)

632   C. Wehner et al.

10. Gallitz, O., De Candido, O., Botsch, M., Melz, R., Utschick, W.: Interpretable machine learning structure for an early prediction of lane changes. In: Farkaš, I., Masulli, P., Wermter, S. (eds.) ICANN 2020. LNCS, vol. 12396, pp. 337–349. Springer, Cham (2020). https://doi.org/10.1007/978-3-030-61609-0_27

11. Gallitz, O., De Candido, O., Botsch, M., Utschick, W.: Interpretable feature generation using deep neural networks and its application to lane change detection. In: 2019 IEEE Intelligent Transportation Systems Conference (ITSC), pp. 3405–3411. IEEE (2019)

12. Guillemot, M., Heusele, C., Korichi, R., Schnebert, S., Chen, L.: Breaking batch normalization for better explainability of deep neural networks through layer-wise relevance propagation. CoRR (2020). https://arxiv.org/abs/2002.11018

13. Hochreiter, S., Schmidhuber, J.: Long short-term memory. Neural Comput., 1735–80, December 1997. https://doi.org/10.1162/neco.1997.9.8.1735

14. Hui, L.Y.W., Binder, A.: BatchNorm decomposition for deep neural network interpretation. In: Rojas, I., Joya, G., Catala, A. (eds.) IWANN 2019. LNCS, vol. 11507, pp. 280–291. Springer, Cham (2019). https://doi.org/10.1007/978-3-030-20518-8_24

15. Ioffe, S., Szegedy, C.: Batch normalization: Accelerating deep network training by reducing internal covariate shift. CoRR (2015). http://arxiv.org/abs/1502.03167

16. Khan, M.Q., Lee, S.: A comprehensive survey of driving monitoring and assistance systems. Sensors **19**(11) (2019). https://doi.org/10.3390/s19112574. https://www.mdpi.com/1424-8220/19/11/2574

17. Kumar, S.A.P., Madhumathi, R., Chelliah, P.R., Tao, L., Wang, S.: A novel digital twin-centric approach for driver intention prediction and traffic congestion avoidance. J. Reliable Intell. Environ. **4**(4), 199–209 (2018). https://doi.org/10.1007/s40860-018-0069-y

18. Patel, S., Griffin, B., Kusano, K., Corso, J.J.: Predicting future lane changes of other highway vehicles using RNN-based deep models. CoRR (2018). http://arxiv.org/abs/1801.04340v1

19. Schwalbe, G., Finzel, B.: Xai method properties: a (meta-)study. ArXiv abs/2105.07190 (2021)

20. Steyn, W.J., Broekman, A.: Development of a digital twin of a local road network: a case study. J. Testing Eval. **51**(1) (2021)

21. Sundararajan, M., Taly, A., Yan, Q.: Axiomatic attribution for deep networks. CoRR (2017). http://arxiv.org/abs/1703.01365

22. Tang, J., Liu, F., Zhang, W., Ke, R., Zou, Y.: Lane-changes prediction based on adaptive fuzzy neural network. Expert Syst. Appl., 452–463 (2018). https://doi.org/10.1016/j.eswa.2017.09.025

23. Thevendran, H., Nagendran, A., Hydher, H., Bandara, A., Oruthota, U.: Deep learning and computer vision for IoT based intelligent driver assistant system. In: 2021 10th International Conference on Information and Automation for Sustainability (ICIAfS), pp. 340–345 (2021). https://doi.org/10.1109/ICIAfS52090.2021.9605823

24. Xing, Y., et al.: Driver lane change intention inference for intelligent vehicles: framework, survey, and challenges. IEEE Trans. Veh. Technol., 1, March 2019. https://doi.org/10.1109/TVT.2019.2903299

An Agenda on the Employment of AI Technologies in Port Areas: The TEBETS Project

Adorni Emanuele[1,2], Rozhok Anastasiia[1,2], Revetria Roberto[1(✉)], and Suchev Sergey[2]

[1] Università degli Studi di Genova, 16100 Genoa, Italy
Roberto.Revetria@unige.it
[2] Bauman Moscow State Technical Univesity, Moscow 105005, Russia

Abstract. This paper presents the agenda extending the project reported at the International Maritime Transport and Logistics Conference "Marlog 10" in 2021. Within the framework of the Smart Port model project proposed by the European Union institutions, we examined the control model of the port of Genoa and the digital twin representation of the Terminal San Giorgio. The application of digital technologies to intermodal systems like ports and harbors would give highly positive results in terms of efficiency and effectiveness. The resources employed were a Digital Twin realized with the Optimize software, a simulation developed through Anylogic, a tracking system operated by OpenGTS, and an AI platform developed through Bonsai. In addition, we have considered the libraries that, in our opinion, are necessary when designing a port simulation model. Our proposed model includes loading, unloading, storage planning, and other operations taking place at the port of Genoa. The article presents various agents which play a fundamental role in the simulation.

Keywords: Simulation · Anylogic · Microsoft Project Bonsai · Port safety

1 Introduction

Artificial Intelligence and simulation have the potential to develop mutually beneficial relationships, particularly in three key areas:

- synthetic data
- test bench
- learning environment

"Synthetic data" and "Test bench" areas refer to the fact that simulation models can be used as a source to generate an unlimited amount of data that are representative of the outputs of real systems and can also be used as a virtual environment to test the implications of incorporating AI into existing systems. "Learning environment" means how helpful information extracted from a simulation model is used. The simulation model can be used to perform experiments and learn from the virtual representation

© Springer Nature Switzerland AG 2022
H. Fujita et al. (Eds.): IEA/AIE 2022, LNAI 13343, pp. 633–647, 2022.
https://doi.org/10.1007/978-3-031-08530-7_53

of your real system in a risk-free way. These experiments facilitate decision-making and help analysis and strategy development. In this scenario, Bonsai was used as the Reinforced Learning system coupled with a Terminal San Giorgio (TSG) digital twin to implement the import and export procedures for the trailers.

This paper is an agenda extending the project presented at the International Maritime Transport and Logistics Conference "Marlog 10" in 2021. The work is based on data from the port area of Genoa. The city is one of the main cargo ports of Italy, which requires a high level of control in the area. The Technological Boost for Efficient port Terminal operations following safety-related events (TEBETS) project is described and presented as a successful example of the implementation of digital technologies in real-life scenarios.

The paper is structured in eight sections, where the first one is the introduction. In the second section we describe the state of the art of the simulation projects that have been used to develop the TEBETS project. The third section is used to give a brief description of Microsoft Project Bonsai, Anylogic and Optimize. These software have been used to support the modelling and simulation processes of the TEBETS project. The fourth section is dedicated to a brief description of the TEBETS project that continues with the description of its characteristics and of how the model for the simulation is built (fifth and sixth section). The seventh section is then presenting a discussion on how the TEBETS project was developed and which levels of autonomy of the designed AI.

2 Previous Works

Simulation engines have been seen as a powerful tool since the last decade of the previous century. Because of the location of our city, the implementation of such software in the maritime environment, to develop safer and more natural friendly infrastructures, has been, since the early 2000s, necessary. Bruzzone et al. [4] provided an interesting study on the risk analysis advantages of such technology. Risk related to neglected port operations and emergency events can be prevented with simulation models to develop effective port designs. The development of the project Maritime Environment for Simulation Analysis (MESA) by the authors is still nowadays active and effective within the harbor community.

With the improvement of the technology in the European Union, it started being necessary to build ports implemented with digital technologies. With the work of Agatić et al. [2], we got a clearer idea of the agenda of the EU institutions on how and why the Smart Port model will be needed in the next decade. Nowadays, the increase of processes and the amount of information that a port generates can be processed by digital technologies, for example, Artificial Intelligence (AI) and the Internet of Things (IoT). These paradigms allow better management of all the harbor's related processes, not only the movements inside the port and the loading/unloading activities but also the safety management, quality control, etc. Furthermore, the employment of digital technologies in planning the logistics of a harbor area supports the personnel in the decision-making process. In the case of a system made by several processes all interconnected, the System Dynamics paradigm is a helpful tool. It can be used to forecast how the system should change in the case of a delay of a cargo ship. This and other operative scenarios have been studied by our department [5] and helped to build the simulation model presented in this article.

New technologies are starting to become an essential part of the workers' workplaces. Devices can be implemented to be a tool to ensure the efficiency of the safety measures. The problems related to human error, the inefficiencies due to inadequate training, and costs can be solved by applying new technologies [16]. As Mladenov et al. [15] presented, augmented reality, as the aggregation of several technologies, demonstrated the right capability to improve workplace conditions. Several AR technologies can be employed in the production field [12], each with its advantages, and their implementation in the Internet of Things (IoT) framework allows to have virtual feedback on their application.

The employing of artificial intelligence and simulation models also improves the safety of port areas. By following modern studies on the mathematical modeling of the traffic flow of ships [21], it would be possible to increase the smoothness of the navigation of container ships. Nevertheless, this problem, always connected to the significant amount of data that has to be collected, is still not easy to resolve. Augmented reality has been studied as an effective means of providing workplace safety criteria [7]. The possibility of employing digital technologies is nowadays a factual reality. Equipment is much more available on the market than it was five years ago, the technology has grown, and the industry's vision is gradually pointing more to the Industry 4.0 framework. A technology developed by our department has been augmented reality equipment for overheating and mechanical risk prevention and fire risk and emergency management. Due to the successful results of the first studies, with the presentation of two cases, we can say that the employment of such technology in coordination with IoT technology would guarantee improvements in the transition to an autonomous system where human interaction is still present. Given the possibility of acquiring information and a human-friendly interface, the AR technology proposed, would also improve the safety standards in workplaces where they may not be always guaranteed. Within the next decade, autonomy will be a necessary step for industries. Nowadays, we already have examples of industries using fully automated plants [1, 17]. This idea has behind the well-being of the humans behind the processes. Depending on the complexity of the systems, it would be needed a higher or lower level of automation for risk analysis management, and the system should have, autonomously, the ability to perceive the situation, decide the best course of action, and intervene [11]. The field of application is infinite, but we can still not implement a suitable technology for every system. This is a very challenging point in risk analysis. The machine learning process works with artificial intelligence training through examples, which it is not always possible to retrieve especially in failure cases.

A previous study [8] has highlighted improvements in the employment of digital twin processes, with applications in several environments [10, 20]. The new digital technologies can find in ports and harbors excellent employment. For example, for freights delivering containers, it is necessary to implement a system that would support the logistic center of the port to have an estimation of the time of arrival and departure. For example, machine learning models can be implemented to estimate three-time intervals (container pick up, estimated delivery time, and real-time delivery of a container) [3]. The problem that we still have to face nowadays is that the amount of data to model and simulate is so huge that it is necessary to use several techniques simultaneously to predict the time interval and compute significant outputs. We think that recent technologies, like the study of sonar images [22], may aid the collection of certain types of data when the available equipment on land cannot collect such data. However, with the evolution of technology, we have several space-based technologies that can support gathering necessary data. As said before, this data can be fully exploited through data processing algorithms. So, the safety within the harbor and machine learning and artificial intelligence make it possible to implement safety systems to protect the marine habitat near the coastal areas [18]. However, we still do not have the technology to analyze the collected data correctly.

3   The AI Platform

Microsoft Project Bonsai is a resource handy for these types of simulations. It presents many benefits:

- no need of using data science to develop an AI
- it trains the AI with the instructions given by the trainers
- prepares the AI for real environments through simulation
- employs the AI to work independently and in cooperation with other people

The digital twin representation of the Terminal San Giorgio was possible with the implementation of Bonsai. The simulation through training allowed us to represent the real-world systems and develop a realistic environment for the Bonsai brains.

This resource was chosen because it is supported by many simulation software, among which AnyLogic, is the chosen software for the project. The AI is trained through simulations and deep reinforcement learning (DRL). The simulation of the TSG is possible to be developed with Bonsai because it respects the preferred characteristics of Bonsai (these can be read on the Microsoft Bonsai website) [14]. Anylogic is a powerful tool when we need to develop simulation models. Many application fields can be used, from healthcare management time studies [10] to the support of the management for a maritime terminal [9]. The employment of this software allows the creation of a digital twin and, consequently, allows the researchers to understand the best choice of parameters to minimize the time function and maximize efficiency. The proper data collection is based on the perfect implementation of such software. With the implementation of sensors and other technologies, such as the IoT concept, increasing the amount of data collected would be possible.

The MESA S.r.l. group employed Optimize simulation software based on thee Anylogic clouds to allow all the users to carry out simulation models online using only the available browsers. The cloud computing simulation tools supply the users with functions to carry out complex problems, visualize the results, and provide online analysis and simulations. Optimize simplifies the integration of templates into operational workflows and facilitates the creation of digital twins. Model sharing is immediate and always ready to use in a cloud environment without software implementation. In addition, Optimize simplifies the update process and management. The dashboard is interactive and customizable, with web-based animations and a shared results database. You can store and organize experiments and scenario data in the cloud, including different model versions, run parameters, and simulation results. The Cloud Scenario Comparison option allows you to visualize the results of different experiments to highlight critical metrics better.

4 TEBETS

The TEBETS project's objective was to design an intelligent surveillance system for the flow of people and vehicles in a port using interconnected sensors to share information with real-time operations management.

The data acquisition system is designed from actually installed utilities (surveillance cameras, sensors, etc.) and updated with new devices and systems capable of being linked with IT systems. Surveillance areas include routes of people within the terminal, security checkpoints, and tracking people and vehicles' positions. Sensors used for the collection of the data must be reliable. New technologies to harvest energy must be studied and applied to prevent running out of battery, minimize other battery-related problems [19], and endure the continuous connectivity and exchange of data with the control system [6, 13].

With the development of Digital Twin and what-if analysis systems, Artificial Intelligence components in the simulation model must be envisaged to enable testing and forecasting. Simulation modeling found new valuable assets in applying artificial intelligence for optimization and calibration. However, many parameters characterize Agent-based systems, and the development of all the permutations would require a lot of run times. Here machine learning and intelligent sampling can be employed effectively.

The studied scenario was very challenging and retributive from the artificial intelligence point of view. The port is a complex operational structure. Historically, there has always been a series of "control rooms" capable of coordinating the complex port operations carried out by a plurality of subjects with often conflicting roles operating in restricted spaces and with a high level of intensity. Over the years, technical and technological evolution has brought considerable changes to the organization of work and the level of integration between the systems and players involved. Suppose we want to define a "level" of automation and integration between systems and cyber-physical entities operating in the place. In that case, we can imagine a subdivision into six categories, from level "0" (no autonomy in the systems and all decisions are made by human operators) to level "5" (total autonomy in the decision-making and implementation process, even in the absence of humans). In the TEBETS project experimentation framework, the applicability of the level 3, 4, and 5 scenarios was explored, which led to the definition of a complete "exercise" involving the practical operation of the Terminal San Giorgio. Based on what was developed during the project, a Digital Twin (Optimize), a tracking system (OpenGTS), and an AI platform (Bonsai) were integrated (and three reports were developed).

5 Development of the Study

The TEBETS technology demonstrator was built using a modular architecture with three main components (Systems): the Terminal San Giorgio entity tracking system (GIS), the Port Operations Digital Twin (DT), and the Artificial Intelligence (AI) Services Platform.

The first step is determining the event that records the number of people in the designated area, the TSG. Such an event will be acquiring the badge number once the people enter the area. For the sake of the project, anonymity was kept. To collect data from each person, it was given an 8-character code, randomly generated and randomly given. Each subject would have had to enter the temporary ID (IDTemp) in their *Safety and Geolocalisation* app. The app then started recording the position of each individual at the time of entering the area through the turnstiles until the badge was not passed to exit the area. For the drivers, anonymity is not needed. In fact, at each gate, an interchange would happen. In this way, only the authorized personnel with the IDTemp will be inside the TSG. The position of the trucks is then registered by collecting the data on if there is a driver on board, latitude, and longitude. Finally, the localization of the goods is determined by the information on the arriving and departing goods (Table 1).

Table 1. Field, type, and attribute of each variable of the system

Field	Type	Attribute
IDOperator	int32	Primary key
Profile	Varchar	
IDTemp	Varchar	
LastCheckIn	TimeStamp	
TermNameIn	Varchar	
LastCheckOut	TimeStamp	
TermNameOut	Varchar	
StatusIn	Bit	

The simulation of the TSG has been represented through a Digital Twin model. As a result, the simulation would have been able to support every decisional phase by representing the operative processes and safety events.

5.1 Simulation Model

The TSG can be classified in three different areas:

- the operational area for the ships, where the ships interact with the landside through cranes. These would load and unload the containers to the storage area;
- the operative area for the trucks, where the containers are loaded on the trucks;
- the operative area for the trains, where the containers are loaded.

To be the most efficient possible, ports need the synergy of multiple activities, which have to be coordinated and work together.

Each ship has its mooring. The companies usually report the arrival time through Electronic Data Interchange (EDI) for big ships. The allocation of the mooring starts within a period of 3 weeks from the arrival of the ship and before the arrival of the containers for the ship. Several characteristics have to be considered: the technical data of the ships and the dock cranes (this is because not every crane can be used for every ship); the dimensions of the ship; and the extension of the crane. There is then an optimization target for the mooring that must be respected. In general, the total distance between the bank and the containers has to be minimal. In this way, we would have the maximum efficiency of the ship operations. So we understand how it is vital to avoid confusion and the automatic and optimized location of each mooring in case of the ships' delay.

After the mooring, there is the stowing. This concept is crucial for the planning of a ship. First, the shipping line must provide the port authorities with its plan for each port the ship will visit. The position of each container has to be specified for each ship. Third, the shipping line classifies the containers into categories: length or type of the container and the weight or weight class of a container. Depending on their category, containers are stored in specific areas of the ship. The final objective of this phase is to minimize the loading/unloading rounds at the arrival port.

After the two phases, there is the dock crane assignment. This choice usually depends on the length of the ship. In the case of ships coming from overseas, assigning from three to five cranes is usual. The constraints that the assignment of a crane has to respect are related to the technical data of the crane and the accessibility of the crane to the mooring. Because not every crane can be piloted to each mooring, the optimization of the assignment of the dock cranes can be summarized as the reduction of delay for each ship or the maximization of the performance of a ship, or the balanced employment of the cranes.

A crucial element is played in the port by storing the containers, whose number constantly increases with reduced available space. Usually, the storage area is divided into blocks, and the containers are stacked on different levels, depending on the port's resources (available straddle carrier). The storage area is generally divided between spaces for imported containers or containers to be exported, refrigerated containers, or dangerous materials. Statistically, containers stay in a storage area for not more than a week. Consequently, it is crucial to have an efficient stowing and stacking process for the containers. For example, given the reduced spaces in the storage area, containers at higher levels have to be removed first. Automated planning would also be able to collect missing data from the shipping line or lost data, reducing future problems. An example of a problem is the location of the containers on trains. To avoid accidents, the needed containers are pre-stowed to minimize errors. However, this operation requires time, and it is not economically efficient. To have an efficient stowing process, different scenarios can be developed in an unloading area. It is possible to think about different unloading ways depending on criteria like the unloading port, the type/length of the container, or its weight. An example is loading the heavier containers over the lighter ones thinking they will be loaded first on the ship to avoid stabilization problems. For exported containers, if the land transport is known, it is possible to group the containers separating them per medium of transport.

A container ship can leave a port only once it has been clarified that it is fit for sailing, with all the stability parameters within limits, especially the draught. Before leaving the port, a document is produced. This document is the stowing plan with the maximum capacity of the fuel tanks. By knowing these characteristics, it is possible to maintain the ship's stability at every moment. Plus, the stowing plan allows the destination port to know how the containers are positioned for better coordination of the unloading in terms of efficiency of crane movements (too many movements mean additional economic efforts) and containers' importance (to have the priority containers are placed on top to ensure the fastest unloading). In a few words, correct stowing would ensure the ship's safety, the cargo's protection, the most efficient employment of the available space on the ship, efficient loading and unloading of the containers, and the safety of the land personnel. To achieve these objectives, simulation tools play an essential role in increasing productivity and minimizing the cost of ports. Through simulation and analysis of the hypotheses, it is possible to describe the agents' behavior once they enter the port. For example, suppose we are simulating the boarding on a ferry. In that case, the entity representing the agent inside the simulation allows us to perceive how the simulation evolves (in this case, the agent becomes a vehicle which can be described through the dimension of the vehicle, its acceleration, and the speed). Depending on the arrival time of the ships at the terminal, the agent decides when to arrive at the terminal too. This information is modeled through a PDF modeled considering the element of the model.

5.2   Modeling and Simulation (M&S)

The employment of modeling and simulation processes in engineering is well known. It is possible to develop a simulation, starting from a model, to verify the satisfaction of the management and technical factors. The possibility to develop a simulation also allows to reduce the costs (not needing effective testing), increase the quality of the products and the systems (comprehending the constraint of the model), and study possible scenarios (to forecast eventual problems). However, a proper simulation can be developed exclusively if the model is correctly built for every domain. This is why professionals (as engineers or analysts) are needed.

In developing the project TEBETS, the simulation environment employed was Optimize. Through the implementation of Anylogic, it was possible to simulate discrete events, agent-based simulations, and system dynamics.

System dynamics is a well-known simulation approach developed in the second half of the XX century. It is possible to digitally represent something that we want to model (from people to several products). Any process is then modeled through flow diagrams between stocks (for example, raw materials to finished products).

A discrete-event simulation focuses on the simulation of the operations as a discrete sequence of events in time. Each event occurs at a particular instant in time and marks a state change in the system. The basic concept is to consider the system as a process. Then, the process is represented through a block diagram from a source block, which develops entities and injects them into a sink block, which removes the entity from the model.

The agent-based simulation is the most recent simulation approach, developed at the end of the XX century with the evolution of computers. The bottom-up approach allows describing the system as the sum of the agents which interact with their behaviors.

5.3 Optimize

Optimize has different types of libraries. These characteristics allow the program to plan and simulate many types of flow in very complex ways, for example, a traffic flow. In this case, the library considers the driving rules (from speed control to collision prevention) and the drivers' behaviors. We must remember how each car will be modeled as an agent, with its characteristics, physical parameters, and behaviors. With the possibility of developing 2D and 3D models, all this information allows for modeling an environment that would allow a better understanding of reality.

In our application in the scenario of a port area, the libraries that may come in handy are the Road Traffic Library, the Process Modelling Library, and the Material Handling Library. The first one can be employed to plan the transport of goods and the management of the traffic roads for eventual transport by truck. The second one allows modeling many operations in different frameworks accurately. This one allows us to analyze the flow and understand the processes' dynamics. It allows to model real-world systems as processes (operations that require queues, delays, and resources), entities of the flow (clients and products), and resources employed to influence the process.

In the case of Optimize, its Process Modelling Library allows to model business processes through a discrete event simulation. These models then show the operational flows as a sequence of discrete events. Finally, flow diagrams are used to analyze the information.

In the framework of the TSG, it is necessary to model the environment to evaluate the capacity the efficiency, locate bottlenecks for pedestrians, and plan eventual emergency routes. Optimize's Library can be extremely useful in our scenario. Modeling is an essential action to balance the stowing constraints and optimization of the production plan. In this case, the Handling Material Library comes in handy. Modeling the stowing processes and the flow of entities within the area can be quickly simulated.

6 The Model

The TSG is located near the access to the highway in the industrial part of the city of Genoa, spreading over an area of 206.000 m^2. Many are the operations carried out in the terminal: ferry transfers, docking of container ships, and handling maritime containers. The idea behind designing a terminal container is to facilitate and develop an efficient flow of containers between the landside and the seaside. To do this, the terminal has to be divided into areas. The TSG is divided into an area for container ships, for ferries, and one for external vehicles. As a consequence of the intermodal dependencies between the three areas, it is then necessary to have areas for the necessary tools for each area, for example, stowing areas for containers, stowing areas for trailers, etc. The model will consider these concepts, and for example, some areas will not be accessible to all agents (Fig. 1).

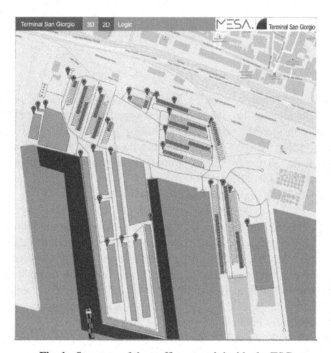

Fig. 1. Structure of the traffic network inside the TSG

Equipment and operating areas are two sources that differ by dimensions, capacity, loading and unloading operations, and type of service. The operation areas, being divided in two, are maximized by allowing the moving of materials and parallel operations. For the simulation, it was necessary to identify the stowing areas of the containers, the trailers, and the maneuvering areas of the vehicles. Once the containers are unloaded from the ship, they are stowed to be loaded again on a land transport or another ship. The stowing is developed on multiple levels, where the heavier containers are on the upper levels. Containers can be classified as outbound or inbound, depending on whether they are coming from a ship or waiting to be delivered. As we previously said, the efficiency of the loading/unloading operations depends on the resources available (such as the dock cranes). Within the terminal, it is necessary to develop a good traffic area to guarantee efficient movement between the different areas (such as a strategic position of the outbound containers near the nearest exit on the road to be delivered by truck/train).

Precise analysis and identification are necessary to avoid human factor-related problems. For example, different colors can be used to avoid confusion.

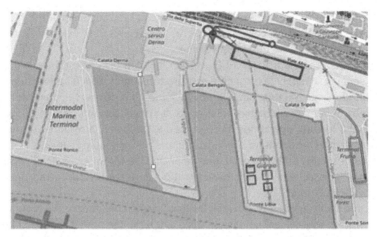

Fig. 2. Paths and areas reserved for specific vehicles (internal trucks, external trucks, and reach stackers carrying out several operations)

To optimize the movements within the terminal (Fig. 2), it was used:

- Yellow for areas or paths that the trucks can use inside the terminal;
- Green for areas where reach stackers and trucks can travel. The transfer area is the zone where containers are loaded on trucks;
- Purple for areas where reach stackers can move for stowing operations;
- Blue for external trucks to follow to arrive inside the terminal
- red will represent the network that internal trucks can follow

Studying and analyzing the characteristics of the model is a necessary operation to maximize the capacity and efficiency of a terminal. Think about the available space, the interface between seaside and landside, the interconnectivity between the terminal and other means of transport.

The high-end target is the movement of goods within the perimeter of the TSG area. The "shipping container" will represent the main agent of the simulation and other components in the model will be the different means of transport, storing, and sorting the shipping containers for different destinations. Thanks to the libraries, as mentioned earlier, it is possible to create a model and move the agent through it. Depending on the mean of transport, the model is divided into external entities (shipped containers, ferries, and trucks), and within the terminal, we have internal entities that are moved by protocols and transferred to external agents. Using this simulation makes it possible to identify the performance indicators and the bottlenecks associated with the model and better understand what can happen in the TSG. The simplified model represents transferring a shipping container to a terminal. We will consider the sources of the container (such as a ship, a ferry, or external vehicles) and the processes within the modelled TSG.

The container ship is an external agent who brings the container agent to our process. Because of the ship's capacity, many agents can be brought to the process. So the unloading operations are carried out by cranes. Through the Process Modelling Library, it is possible to coordinate the arrival and departure of the "container ship" entity, but it is an external factor, and the delays are not considered a result of external factors. Once the ship arrives at the dock, the processes start.

Through the process, it is possible to keep the ship docked until the unloading is finished (creating a "delay") for then departure to its next stop. The function *inject ()* controls the ship's arrival time, and *StopDelayForall ()* controls the departure time.

The ferry is an external agent similar to the container ship, but it differs in the unloading and loading operations. Ferries need a "tugboat," a resource in our model. Cranes are not employed, but we need internal trucks. The loading capacity is inferior to the container ship due to the reduced dimension and the transport for smaller distances. Nevertheless, the exact functions before controlling the arrival and departure times, *inject ()*, and *StopDelayForall ()*.

The truck agent represents the vehicles that bring a container to the terminal, leave the container, pick up the containers assigned and leave the port. The *inject ()* function controls this agent. The Traffic Control Library is used to model the traffic processes.

From the container ship, the containers are unloaded through cranes. Then the containers are stored in storage areas until the critical moment. Then, it will be taken by a reach stacker and loaded on the trucks at that point. Because we need the processes carried out by the dock cranes, the operations of loading and unloading for a container ship are much more complex than the ones for an external truck. The logic process is the following: the containers are unloaded by the cranes; they are then transferred to the storage area; the reach stacker is called and the container for the transfer; the truck takes the inners roads and moves to the parking lot; in the parking lot, the truck is loaded with the container, and it takes the road to go out. Keep in mind that when we use the system's resources, we have to ensure the return of these resources to have a correct simulation. For example, it may happen that the simulation of the trucks does not proceed appropriately due to traffic road problems (Fig. 3).

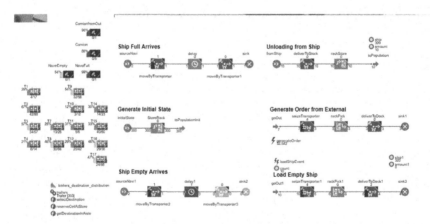

Fig. 3. Events within the TSG

7 Discussion

The TEBETS project was aimed at designing an autonomous surveillance system. By modeling the agents and the several flows (of involved vehicles and personnel), the system was designed to collect data in real-time through several sensors and share them with the workers through dedicated devices. This technology was necessary to recover and collect data for the safety of the Terminal San Giorgio designated area. Due to the high traffic, the study of vehicle flow has been necessary to prevent accidents that may involve risks for the personnel and delay operations. For loading and unloading operations, it was necessary to rely on machine learning algorithms to minimize the time function of loading and unloading operations while maximizing the efficiency of the resources available (dock cranes, reach stackers, and more if needed).

The digital technologies that nowadays are available allow, in theory, to elaborate the complicated simulation of inter-modal systems. However, the main problem that such simulations nowadays cannot predict is how to autonomously correctly respond to every type of risk (knowledge which would require not available data sometimes) and to predict the behavior of agents outside our simulation (the delay of the cargo ships, for example).

In particular, through the employment of a Digital Twin and a Machine Learning engine applied to a harbor scenario, all the functions were planned for the TEBETS project. The studied levels were:

- Level 3, cyber-physical entities support the port control room by autonomously reacting suggesting to the operators which actions to pursue as the answer to a specific event and receiving feedback on which one the operator chose
- Level 4, the AI suggests the actions to pursue in case of an event, and the AI takes control over the operator only if his actions may result not be correct.
- Level 5, where the AI has complete autonomy in the decision process and can take autonomous decisions also in the absence of a human.

8 Conclusions

This article was aimed to present the agenda related to the employment of artificial intelligence and simulation software for the logistic analysis of a harbor area. Developed the TEBETS project, we wanted to present here the features related to its purpose, its characteristics and its development. In this agenda we analyzed the employment of Microsoft Bonsai, Anylogic and Optimize. Especially the first two software are essential for the development of the model and the simulation process. The Terminal San Giorgio is presented as a complicate system where many means of transport are employed for the moving of goods within the port area. These goods arrive to the TSG by truck, by train and by ship. These goods have to be exported but first, they have to be stored and classified. The employment of AI and machine learning algorithms in such system has been proved (with different levels of autonomy of the AI) to be a reliable paradigm to ensure the safety of the personnel within the terminal and to properly develop a system for the moving of goods.

9   Future Projects

In the conclusion of our article, we should mention significant, critical disadvantages of the application of TEBETS at present. According to the law, the first relates to the fact that when an accident occurs in the port, there must be a person responsible for the accident that occurred because of automatic decision-making. Who, in such a case, should be held responsible? The developer of the TEBETS program or the safety engineer, the operator who operated the equipment, or the person who manages the program? Furthermore, this issue is essential since this simulation is required to ensure safety and save lives. Moreover, a very similar question applies to artificial intelligence. The second disadvantage is artificial intelligence's inability to make decisions in situations that were not predetermined or predicted in advance. For example, if the library loaded into the program did not contain a specific scenario, we cannot know how the machine will behave in this case. Again there is the question of safety and risking the lives of workers. In the case of an unforeseen situation, a person makes decisions based on experience, but not the machine. The artificial intelligence has no experience other than that downloaded into it by the manager.

Thus, TEBETS helps to improve logistics and port security, but further developments in legislation are needed to implement it in the real process.

Acknowledgment. The authors would like to acknowledge the University of Genoa and Silvestro Vespoli (University of Bergamo), developers of the TEBETS project.

References

1. ABB. The factory of the future is smart, connected and already here. https://new.abb.com/news/detail/2176/thefactory-of-the-future-is-smart-connected-and-already-here (2017)
2. Agatić, A., Jugović, T. P., Tijan, E., Kolanović, I.: European union policies and funding for smart port model implementation. In: NAŠE MORE 2021, 2nd International Conference of Maritime Science & Technology (2021)
3. Antamis, T., et al.: AI-supported forecasting of intermodal freight transportation delivery time. In: 2021 62nd International Scientific Conference on Information Technology and Management Science of Riga Technical University (ITMS) (2021)
4. Bruzzone, A.G., Mosca, R., Revetria, R., Rapallo, S.: Risk analysis in harbor environments using simulation. Saf. Sci. **35**, 75–86 (2000)
5. Caballini, C., Mosca, M., Revetria, R.: A system dynamics decision cockpit for a container terminal: the case of voltri terminal Europe. Int. J. Math. Comput. Simul. **3**, 55–64 (2009)
6. Chebrolu, K., Rmana, B., Mishra, N., Valiveti, P.K., Kumar, R.: Brimon: a sensor network system for railway bridge monitoring. In: Proceedings of 6th International Conference on Mobile Systems, Applications, and Services, pp. 2–14. ACM (2008)
7. Damiani, L., Revetria, R., Morra, E.: Safety in Industry 4.0: the multi-purpose applications of augmented reality in digital factories. Adv. Sci. Technol. Eng. Syst. J. **5**(2), 248–253 (2020)
8. Galli, G., Mosca, M., Revetria, R., Mosca, R.: Artificial intelligence for supporting maritime terminal management, safety and security. In: International Maritime Transport and Logistics Conference "Marlog 10", pp. 55–63 (2021)

9. Galli, G., Patrone, C., Battilani, C., Revetria, R.: Simulation and business process manage-ment notation in support of business process re-engineering. In: Transactions on Engineering Technologies, pp. 47–58 (2021)

10. Galli, G., Patrone, C., Bellam, A.C., Annapareddy, N.R., Revetria, R.: Improving process using digital twin: a methodology for the automatic creation of models. In: Proceedings of the World Congress on Engineering and Computer Science 2019 WCECS 2019 (2019)

11. Gamer, T., Kloepper, B., Hoernicke, M.: The way toward autonomy in industry - taxonomy, process framework, enablers, and implications. In: IECON 2019 - 45th Annual Conference of the IEEE Industrial Electronics Society, pp. 565–570 (2019)

12. Guizzi, G., Revetria, R., Rozhok, A.: Augmented reality and virtual reality: from the industrial field to other areas, augmented reality and virtual reality (2020)

13. Lee, R.G., Chen, K.C, Chiang, S.S., Lai, C.C., Liu, H.S., Wei, M.S.: A backup routing with wireless sensor network for bridge monitoring system. In: Annual Conference on Communication Networks and Services Research. IEEE Computer Society, pp. 157–161 (2006)

14. Microsoft Project Bonsai (2021). https://docs.microsoft.com/en-us/bonsai/product/

15. Mladenov, B., Damiani, L., Giribone, P., Revetria, R.: A short review of the SDKs and wearable devices to be used for AR application for industrial working environment. In: WCECS 2018, vol. 1 (2018)

16. Rozhok, A.P., Storozhenko, A.S., Valiaeva, A.V., Sushchev, S.P., Ugarov, A.N., Revetria, R.: Methods of monitoring the Ground-Climate-Pipeline system in sections with hazardous pro-cesses. In: IOP Conference Series: Earth and Environmental Science, Sino-Russian ASRTU Forum "Ecology and Environmental Sciences", vol. 864 (2020)

17. Ship Technology Global: ABB supports remotely operated passenger ferry trial in Finland (2018). https://www.shiptechnology.com/news/abb-supports-ferry-trial/

18. Soldi, G., et al.: Space-based global maritime surveillance. Part ii: artificial intelligence and data fusion techniques IEEE Aeros. Electron. Syst. Mag. **36**, 30–42 (2021)

19. Sudevalayam, S., Kulkarni, P.: Energy harvesting sensor nodes: survey and implications IEEE Commun. Surv. Tutor. **13**(3), 443–461 (2011)

20. Tantik, E., Anderl, R.: Integrated data model and structure for the asset administration shell in Industrie 4.0. In: 27th CIRP Design, vol. 60, pp. 86–91 (2017)

21. Vasyutina, A.A., Popov, V.V., Kondratyev, A.I., Boran-Keshishyan, A.L.: Improvement of the vessel traffic control system for accident-free electronic navigation in the port area. In: Journal of Physics: Conference Series, Volume 2061, International Conference on Actual Issues of Mechanical Engineering (AIME 2021) (2021)

22. Wawrzyniak, N., Stateczny, A.: MSIS image positioning in port areas with the aid of comparative navigation methods. Polish Maritime Res. **24**, 32–41 (2017)

Modelling and Solving the Green Share-a-Ride Problem

Elhem Elkout[1,3]([⊠]) [iD] and Olfa Belkahla Driss[2,3] [iD]

[1] Ecole Nationale des Sciences de l'Informatique ENSI, University of Manouba, Manouba,
Tunisia
elhem.elkout@ensi-uma.tn
[2] Ecole Supérieure de Commerce de Tunis, University of Manouba, Manouba, Tunisia
olfa.belkahla@esct.uma.tn
[3] LARIA LA Recherche en Intelligence Artificielle, ENSI, University of Manouba, Manouba,
Tunisia

Abstract. Transport occupies an important place in the economic life of modern
societies. Emerging challenges, such as increased road traffic congestion, rising
oil prices, and growing environmental concerns, have driven research towards
shared mobility systems. Indeed, road transport is one of the main contributors
to greenhouse gas emissions. We introduce, for the first time in this paper the
Green Share-a-Ride Problem (Green-SARP) which represents an extension of the
classical Share-a-Ride problem by considering a limited refueling infrastructure.
We present a linear mixed-integer mathematical programming formulation for the
Green-SARP where a vehicle can refuel at any Alternative Fuel Station (AFS) to
eliminate the risk of running out of fuel during a route. Experiments are performed
using the CPLEX solver on benchmarks from the literature and have proved the
effectiveness of the proposed model.

Keywords: Green transport · Share-a-Ride problem · Transportation · Modelling

1 Introduction

Today, the number of cars on the roads is one of the main causes of traffic congestion, affecting energy consumption and passenger mobility, particularly in urban areas.
Indeed, our lifestyles must be changed into new fashions that are more organized and
that preserve natural resources and fight pollution. One solution to this problem is to use
the Share-A-Ride (SARP) transport mode by Li et al. [1] in 2014. In existing transport
systems, passenger and parcel requests are always served separately due to their different
characteristics. In the urban context, parcels are generally small in size and weight, so
they can be transferred to passengers without significantly affecting private space and
comfort. These problems have been studied intensively in recent years as part of the
Share-A-Ride problems.

Shared Mobility Problems issues with the rideshare concept based on vehicle sharing
are rapidly becoming a trend in many urban cities around the world. Examples of these

© Springer Nature Switzerland AG 2022
H. Fujita et al. (Eds.): IEA/AIE 2022, LNAI 13343, pp. 648–658, 2022.
https://doi.org/10.1007/978-3-031-08530-7_54

issues are ridesharing, taxisharing, buspooling, carpooling, and multimodal carpooling. It is the new way to access transport services for those driving the sharing economy, where access rather than ownership is the new norm proposed by [2] in 2013.

The Share-a-Ride (SARP) problem, in which passengers and parcels are transported at the same time by the same taxi, was introduced in 2014 by Li et al. [1]. SARP is an extension of the Dial-A-Ride Problem (DARP) which was first introduced by Cordeau and Laporte in 2003 [3] and it consists in designing vehicle routes and schedules for a number of users who specify delivery requests as defined by pickup and drop-off points. SARP has been put in place to minimize traffic congestion by reducing the number of vehicles on the road as passengers and parcels are generally served separately. Indeed, SARP is used to find a set of taxi routes that maximizes the total profit of a taxi company that meets the demands of passengers and parcels. In addition, SARP represents a multi-criteria combinatorial optimization problem. It belongs to the class of NP-difficult problems of combinatorial optimization [1].

Given the importance of SARP, and given that reducing toxic gas emissions is one of the points of interest to researchers for environmental protection, there are a wide variety of problems related to green transport. This work addresses, for the first time, the Share-A-Ride issue in the context of green logistics; the problem is named Green Share-a-Ride Problem (Green-SARP) which represents an extension of the classical Share-a-Ride problem by considering a limited refueling infrastructure.

We focus on minimizing the emission of carbon dioxide (CO_2) due to its adverse effects on the environment in the field of transport. For this reason, we have introduced the green aspect into the SARP problem taking into account the refueling nodes (AFS). Vehicles can refuel from any alternative service station to eliminate the risk of running out of fuel along the way.

This paper is organized as follows. In Sect. 2, we provide a literature review on studies for the Share-a-Ride Problem and green transport with Alternative Fuel Stations. We give description of the Green-SARP and we present our model formulation in Sect. 3. Then, in Sect. 4, we develop the model formulation for the Green SARP and we present the experimental settings as well as the numerical results. Section 5 concludes the paper with a summary and directions for future research.

2   Related Work

This section presents a literature review on Share-a-Ride Problems and green transport with Alternative Fuel Station.

2.1   The Share a Ride Problem

In existing transport systems, passenger and parcel requests are still served separately due to their different characteristics. Every day there are a large number of empty journeys for the transport of passengers and parcels. This increases operating costs. In the urban context, parcels are generally small size and weight, so they can be transferred to passengers without significantly affecting private space and comfort of passengers. On the contrary, passengers are willing to share a taxi with parcels if comfort and security are guaranteed [4].

Li et al. [1] introduced for the first time, in 2014, the Share-A-Ride Problem (SARP) which simultaneously combines two types of transport of passengers and parcels, i.e., people and parcels are handled in an integrated manner by the same network of taxis, to minimize traffic congestion by reducing the number of vehicles on the road. The authors proposed a new mathematical model and a formal definition of SARP; they studied the static version and the dynamic version of the problem. In addition, they presented a heuristic to solve the dynamic SARP whose initial solution is built using a heuristic that aims to insert the unheeded applications into a route. Then they used neighborhood search to optimize the route.

Nguyen and al. in 2015 proposed an alternative SARP resolution approach [5]. The main objective of this article is to study a new hybrid transport model for the city of Tokyo, which combines the transport of passengers and parcels in the same taxi. In fact, the authors modelled a new formulation based on the model of Li et al. [1]. They proposed two heuristics to test the effectiveness and feasibility of the model.

In 2016, Li et al. [6] used the same mathematical formulation proposed in [1] but proposed a time slack strategy for route planning. They developed a heuristic called Adaptive Large Neighborhood Search (ALNS) to solve static SARP. In the proposed approach, an initial solution is constructed using a "basic greedy insertion heuristic" to randomly insert the selected applications to the vehicles. Then the ALNS algorithm with simulated annealing as a local search frame was used.

The same authors published another article in which they considered two stochastic variants of the Share-a-Ride problem: one with stochastic travel times and one with stochastic delivery locations. Both variants are formulated as a two-stage regression stochastic programming model. The goal is to maximize the expected profit from handling a group of passengers and parcels with a set of homogeneous vehicles. The authors' contribution integrates adaptive large neighborhood search heuristics and three sampling strategies for generating the scenario (fixed sample size sampling, sample average approximation, and sequential sampling procedure) [7].

Li et al. [1] formulated and solved SARP with two aspects: dynamic and static. They presented a new model of the public transport system in the urban area that ensures taxi sharing between passengers and parcels taking into account the speed window. In addition, they classified the speed windows according to the different zones and according to the levels of congestion during a day in the urban area. The authors used a heuristic to construct the initial solution. Then they used a local search technique to improve the solution. In addition, they developed two heuristics where the first aims to solve the case of direct transport without sharing of demands while the second aims to solve the case of dynamic shared transport.

The general ride-sharing problem (G-SARP), proposed by Vincent et al. in 2018 [8], generalizes the SARP by relaxing some limitations. In G-SARP, if the time window of requests and vehicle capacity are satisfied, the vehicles can serve as many requests as they can simultaneously allow. In other words, there is no priority of service between passengers and parcels, nor is there any consideration of the maximum travel time for passengers. This study proposes a simulated annealing algorithm to solve G-SARP, and a tabu search algorithm has been developed to test the effectiveness of simulated annealing. The new problem is called G-SARP-EVs [9].

2.2   Green Transport with Alternative Fuel Station

In recent years, there has been a trend in green transportation research. Many studies focus on minimizing the emissions of carbon dioxide from the transport sector. This gas is considered one of the main sources of greenhouse gases that contribute to the climate change. For example, in 2019, Yu et al. [10] proposed a bi-objective nonlinear mathematical model. The first objective is to minimize carbon emissions. The second objective is to maximize the average profit of the trip so that the interest of each driver can be satisfied. The authors have developed an exact three-step method to solve the Bi-objective green ride-sharing problem. The highlight of the exact method is to cut most of the non-Pareto optimal solutions and use a decomposition method. Indeed, a lot of work focuses, for example, on the introduction of electric vehicles, while others take into account more relevant constraints (for example, intermediate refueling stops). In the following, we focused on this type of problem. In fact, one of the most interesting problems that was studied and introduced in this regard is the Green-VRP (G-VRP) by Erdogan and Miller-Hooks in 2012 [11], in which refueling stops were incorporated. The authors proposed a mixed integer linear model to minimize the travel distance considering the AFS as well as the number of trips and their limitations.

NurMayke et al. in 2017 [12] investigated the problem of Green Vehicle Capacitive Routing (CGVRP), which is a development of the Green Vehicle Routing (GVRP) problem, characterized by the aim of harmonizing the environmental and economic costs of implementing efficient routes to meet all environmental concerns while meeting the customers' requirements. The goal of CGVRP is to minimize the total distance traveled by an alternative fuel powered vehicle (AFV). The authors formulated the CGVRP mathematical model and proposed a simulated annealing (SA) heuristic for its solution, where CGVRP is configured as a mixed integer linear program (MILP).

Andelmin and Bartolini in 2017 [13] proposed an exact algorithm to solve the Green Vehicle Routing Problem, where the G-VRP models the optimal routing of a fleet of alternative fuel vehicles to serve a set of customers. Vehicle fuel autonomy and possible en-route refueling stops are explicitly modeled and maximum duration restrictions are imposed on each vehicle route. The G-VRP was modeled as a set partition problem in which the columns represent feasible routes corresponding to simple circuits in a multigraph.

Recently, Masmoudi et al. in 2019 [14] proposed an efficient Adaptive Large Neighborhood Search (ALNS) algorithm for the G-DARP solution. The Green Dial-a-Ride Problem (G-DARP), which is an extension of the classic Dial-a-Ride problem, considering the limited range of vehicles combined with limited refueling infrastructure. In G-DARP, the vehicle can refuel from any alternative fuel station with partial refueling to eliminate the risk of running out of fuel in route. Computational experiments confirm that the proposed algorithm provides high-quality solutions in an economical and ecological way.

3   The Green Share-a-Ride Problem Formulation

3.1   Description

We present as follows a formal description of Green-SARP. We have a graph $G = (V, A)$. Let $1,..., 2\sigma$ be the origin and destination set of σ requests. Let Vp and Vf correspond respectively to the stops of the requests for passengers and for parcels. More precisely, $Vp = Vp,o \cup Vp,\text{d}$, where Vp,o is the set of passenger origins and Vp,d is the set of passenger destinations.

The same, $Vf = Vf,o \cup Vf,d$, where $Vf,0$ and Vf,d are the set of origins and destinations of passengers. Each request has a pair of collect and delivers nodes $\{i, i + \sigma\}$ and must be served. And let $F = 2\sigma + 1,..., 2\sigma + f$ is the set of AFS nodes. The nodes 0 and $2\sigma + f + 1$ correspond respectively to the origin depot and the destination depot. The set of nodes is $V = Vp \cup Vf \cup F \cup \{0, 2\sigma + f + 1\}$, while $A = (i, j): i, j \in V, i \neq j$ represents the set of arcs connecting each pair of knots.

An arc (i, j) in the set A, is associated with a non-negative distance d_{ij} and a travel time t_{ij}. In addition, it is assumed that the vehicles travel through each arc (i, j) with constant speeds. The quantity of the demand i is denoted by qi, where c $qi = 1$ for the parcel demand and $qi = 3$ for the passenger demand. Let K be the set of vehicles available at the depot.

In addition, we assume that the number of stops the vehicle can make to refuel is unlimited. At the time of refueling, it is assumed that the tank must be filled with a quantity Qf.

Regarding visits to the peaks, we assume that some peaks may have multiple visits, while others are only visited once. To allow this, we add a set of fictitious vertices f', $F' = \{2n + f + 2,..., 2n + f + f' + 1\}$ on the graph G, where each fictitious node corresponds to a potential visit to an AFS node or depot that serves as AFS. As a result, we get the graph $G' = (V', A')$, where $V' = V \cup F'$.

The time window to visit any AFS node is set to $[0, T]$, where T is the length of the planning horizon. In addition, we have a homogeneous fleet of vehicles $k \in K$ in terms of resource capacities to serve the σ demands.

3.2   Notations

The notations used to formulate the Green-SARP are defined as follows:

- m: Number of parcel requests
- n: Number of passenger requests
- σ: Total number of all requests ($\sigma = m + n$)
- Vf: Set of parcel stops
- Vp: Set of passenger stops
- Vf,o: All the origins of the parcels
- Vp,o: All the origins of the passengers
- Vf,d: Set of parcel destinations
- Vp,d: Set of passenger destinations
- F: set of AFS nodes

- q_i: Quantity of type c request at stop i
- t_{ij}: Travel time between stops i and j
- d_{ij}: Distance traveled between stops i and j
- [ei, li]: Stop time window i
- Qk: Taxi capacity k for demand
- Tk: Maximum duration of the taxi trip k
- α: Initial price charged for the delivery of a passenger
- β: Initial price charged for the delivery of a parcel
- $\gamma 1$: Price invoiced for the delivery of one passenger per kilometer
- $\gamma 2$: Price invoiced for the delivery of a parcel per kilometer
- $\gamma 3$: Average cost per kilometer for the delivery of requests
- $\gamma 4$: Discount factor in the event of exceeding the time limit for direct passenger delivery
- η: The maximum number of parcels served during the passenger service
- ωi: The maximum delivery time for request i
- R: Vehicle fuel consumption rate (gallons per mile)
- Qf: Vehicle fuel tank capacity

Decision variables:

- τ_j^k: The time when the vehicle k begins to serve the demand i
- r_i^K: The time when the request i stays in the taxi k.
- w_i^k: The taxi charge k after visiting the stop i
- Pi: The demand index i in the service sequence of a taxi
- X_{ij}^K: A binary decision variable, which is equal to 1 if the taxi k goes directly from stop i to j. Otherwise it is equal to 0.
- y_j: Fuel level variable specifying the level of fuel remaining in the tank on arrival at the top j. It is returned to Qf at each summit of supply station i and at the depot.

3.3 The Mathematical Model

The resulting mathematical model is as follows:

$$MAX \sum_{i \in Vp, 0} \sum_{j \in V} \sum_{k \in K} (\alpha + \gamma 1 d_{i,i+\sigma}) X_{ij}^K + \sum_{i \in Vf, 0} \sum_{j \in V} \sum_{k \in K} (\beta + \gamma 2 d_{i,i+\sigma}) X_{ij}^K - \gamma 3$$

$$\sum_{i \in V\prime} \sum_{j \in V\prime} \sum_{k \in K} d_{i,j} X_{ij}^K - \gamma 4 \sum_{i \in Vp, 0} \sum_{k \in K} (r_i^K / t_{i,i+\sigma} + 1) \tag{1}$$

$$\sum_{j \in V} \sum_{k \in K} X_{ij}^k \leq 1 \quad \forall i \in Vp, 0 \cup Vf, 0 \tag{2}$$

$$\sum_{j \in V\prime} X_{0i}^k = \sum_{i \in V\prime} X_{i,2\sigma+f+1}^k = 1 \quad \forall f \in F', k \in K \tag{3}$$

$$\sum_{i \in V} X_{0i}^k = \sum_{i \in V} X_{2\sigma+f+1,i}^k = 0 \quad \forall f \in F', \forall k \in K \tag{4}$$

$$\sum_{i\in V} X_{ij}^k = \sum_{i\in V} X_{i,j+\sigma}^k \quad \forall j \in Vp, 0 \cup Vf, 0; \forall k \in K \tag{5}$$

$$\sum_{i\in V\prime} X_{ij}^k = \sum_{i\in V\prime} X_{ji}^k \quad \forall j \in V'; \forall k \in K \tag{6}$$

$$\tau_j^k \ge \left(\tau_i^k + tij\right) X_{ij}^k \quad \forall i,j \in V'; \forall k \in K \tag{7}$$

$$w_j^k \ge \left(w_i^k + qj\right) X_{ij}^k \quad \forall i,j \in V; \forall k \in K \tag{8}$$

$$r_i^k = \tau_{i+\sigma}^k - \tau_i^k \forall i \in Vp, 0 \cup Vf, 0; \quad \forall k \in K \tag{9}$$

$$Tk \ge \tau_{2\sigma+1}^k - \tau_0^k \quad \forall k \in K \tag{10}$$

$$ei \le \tau_i^k \le li \quad \forall i \in V' \tag{11}$$

$$ti, \sigma + i \le r_i^k \le \omega i \quad k \in K; \forall i \in Vp, 0 \tag{12}$$

$$w_i^k \ge \max\{0, qi\}, \quad \forall i \in V; \forall k \in K \tag{13}$$

$$w_i^k \ge \max\{QK, QK + qi\}, \forall i \in V; \forall k \in K \tag{14}$$

$$M(\sum_{k\in K} X_{ij}^k - 1) + Pj - 1 \le Pi \forall i,j \in Vp \cup Vf; \forall k \in K \tag{15}$$

$$M(1 - \sum_{k\in K} X_{ij}^k) + Pj - 1 \ge Pi \forall i,j \in Vp \cup Vf; \forall k \in K \tag{16}$$

$$P_{j+s} - P_j - 1 \ge \eta \quad \forall i,j \in Vp, 0 \tag{17}$$

$$\tau_j^k - \tau_i^k \ge tij + M_{ij}^k\left(X_{ij}^k - 1\right) \quad \forall i,j \in Vp \cup Vf; k \in K \tag{18}$$

$$w_j^k - w_i^k \ge qi + W_{ij}^k\left(X_{ij}^k - 1\right) \quad \forall i,j \in Vp \cup Vf; k \in K \tag{19}$$

$$y_j \le y_i - R.d_{ij}X_{ij}^k + Qf\left(1 - X_{ij}^k\right) \quad \forall j \in V; \forall i \in V' \tag{20}$$

$$y_j \ge Qf \quad \forall j \in F' \tag{21}$$

$$y_j \ge \min\{R.d_{j0}, R(d_{j1} + d_{l0})\} \quad \forall j \in V; \forall l \in F' \tag{22}$$

$$X_{ij}^k \in \{0, 1\}; \tau_j^k, w_j^k \in R+; Pi \in [0, 2(m+n)] \tag{23}$$

Our proposed mathematical model is based on the model of Li et al. in 2014 [1] for the SARP and the model of Erdogan and Miller-Hooks in 2012 [11] for the green VRP. In this formulation, Eq. (1) models the objective function of the problem. Indeed, it serves to maximize the following four criteria:

- The profit obtained by the passengers.
- The profit obtained by the parcels.
- Reducing the cost of the distance traveled
- Reduced time when passengers stay in the taxi compared to drop delivery time.

The constraints (2) and (5) ensure that each request is served once by the same vehicle, while constraints (3) and (4) ensure that each vehicle starts and ends its route from depot. The constraint (6) indicates that each stop (except the origin and the destination of the taxi) must have a predecessor and a successor. The constraints (7), (8) and (9) respectively define the travel time, the taxi load and the travel time, while the constraint (10) limits the working hours of the taxi drivers. The constraint (11) presents the time window, while the constraint (12) reflects the fact that each passenger request must be processed within a given time frame and the passenger's origin is visited before its destination. The constraints (13) and (14) define the capacity of the vehicles, while the constraints (15) and (16) define the service sequence of requests. The constraint (17) guarantees that the passenger service has a higher priority, that is to say that we can insert η requests between the pickup and delivery point of a passenger with $\eta \in \{0, 1, 2...\}$. The constraint (20) for monitoring the fuel level. The constraint (21) resets the fuel level to Qf on arrival at the depot or at an AFS peak. Constraint (22) ensures that there will be enough remaining fuel to return to the depot directly or by means of an AFS from any customer location en-route. This constraint aims to ensure that vehicles will not be blocked. Finally, binary integrality is guaranteed through constraint (23).

4 Numerical Results

In this section, we present the details of the results obtained of the mathematical model defined in the previous section and solved using the CPLEX 12.10 solver, on an Asus personal computer with Intel Core i5-9300H at 2.4 GHz and 8 GB of RAM.

To test our mathematical model we implemented our CPLEX code through the instances generated by [15]. Readers can find more details in [15]. The number of requests is between 10 and 100 in these cases, with a number of vehicles varying between 2 and 12, the fuel tank capacity of 60 gallons and fuel consumption rate of 0.2 gallons per mile were set based on average values for biodiesel-powered AFVs [11].

The results are calculated using a small data set of instances. We note that, for example, the instance "a0_10_1"; "a0" refers to the type of family class in the benchmark instances of [15]. The second number "10" in the instance name corresponds to the number of request and the last number"1" shows the number of file.

Two fundamental factors influence the complexity of the Green-SARP: the number of taxis and the total number of requests. The experiments reported here aim to test the influence of these two factors. Table 1 lists the parameters of our dataset. Distances between nodes are calculated using Euclidean distance.

Table1. Parameters used for the Green SARP [1].

Parameters	Values
α	3.5
β	2.33
$\gamma 1$	2.70
$\gamma 2$	0.90
$\gamma 3$	0.80
$\gamma 4$	3.5
η	2.00
ωi $(i \leq n)$	$2t$ $i,\sigma + i$

To validate our mathematical model, we used the data set instances of [15] with one vehicle or two vehicles and 10 to 12 requests. These instances are then solved using CPLEX 12.10 solver.

Table 2. Computational results

Instance	Nb-reqt	Nb-AFS	Nb-Veh	Obj
A0–10-1	10	3	1	65.912
A0–10-1	10	3	2	102.42
A0–10-2	10	3	1	
A0–10-2	10	3	2	64.050
A0–12-3	12	4	2	--.--
A1–10-1	10	3	1	63.42
A1–10-1	10	3	2	103.95
A1–10-2	10	3	1	27.165
A1–10-2	10	3	2	34.95
A1–12-3	12	4	2	--.--
A2–10-1	10	3	1	73.15
A2–10-1	10	3	2	112.02
A2–10-2	10	3	1	70.12
A2–10-2	10	3	2	108.40
A2–12-3	12	4	2	--.--

Table 2 shows the results obtained. The "Nb-Veh" and "Nb-reqt" columns respectively present the number of vehicles and the number of requests on each instance. The "Nb-AFS" column provides the number of AFS; the "obj" column represents the value of the objective function. For each instance, we have defined the upper two thirds as passenger requests and the lower one third as parcel requests.

From Table 2 we concluded that the number of vehicles acts on the results found indeed for all the instances used, we found that the results of 2 vehicles are better than a single vehicle. For all the instances used we also noticed that most vehicles did not go through AFS and there are some that went to just a single AFS node.

From the same Table 2 we also observe that CPLEX is able to solve instances containing one or two vehicles and up to 10 requests (user and parcel). However, we can observe that CPLEX might not be able to find a feasible solution for all instances with three or four vehicles and having up to 60 and 100 requests, as well as some instances containing more than two vehicles.

Given the complexity of the problem, CPLEX could not find feasible solutions for a number of requests greater than 10. We propose in our future work solutions based on heuristics.

5   Conclusion and Future Directions

In this paper, we have presented a new extension of the share-a-ride problem (SARP), with a fleet of alternative fuel vehicles (AFV). According to the limited range, the AFV may visit some Alternative Fuel Stations (AFS) to be refueled with a refuel amount during its trip to meet all requests. The proposed variant is called the Green Share-A-Ride Problem (Green-SARP).

We have proposed a new mathematical formulation of the Green-SARP based on the formulation of Li et al. [1] for the SARP and the model of Erdogan and Miller-Hooks [11] for the green VRP. We tested our formulation with literature instances, and we found that the number of vehicles and requests are the only factors that act on our formulation as in the SARP.

In order to contribute to the resolution of Green-SARP, we will first improve our development with CPLEX in order to find good results for our new proposed mathematical formulation. Then another interesting line of research consists in proposing metaheuristics to solve this problem such as tabu search, variable neighborhood search, optimization by ant colony, etc.

References

1. Li, B., Krushinsky, D., Reijers, H.A., Woensel, T.V.: The share-a-ride problem: people and parcels sharing taxis. Eur. J. Oper. Res. 238(1), 31–40 (2014)
2. Ting, K.H., Lee, L.S., Pickl, S., Seow, H.-V.: Shared mobility problems: a systematic review on types, variants, characteristics, and solution approaches. Appl. Sci. 11, 7996 (2021). https://doi.org/10.3390/app11177996
3. Cordeau, J.F., Laporte, G.: A tabu search heuristic for the static multi-vehicle dial a ride problem. Transp. Res. Part B: Methodol. 37(6), 579–594 (2003)
4. Mourad, A., Puchinger, J., Chu, C.: A survey of models and algorithms for optimizing shared mobility. Transp. Res. Part B 123, 323–346 (2019)

5. Nguyen, N.Q., Dung, N.V., Do, P.T., Le, K.T., Nguyen, M.S., Mukai, N.: A practical dynamic share-a-ride problem with speed windows for Tokyo city. In: Proceedings of the Sixth International Symposium on Information and Communication Technology (2015)
6. Li, B., Krushinsky, D., Woensel, T.V., Reijers, H.A.: An adaptive large neighborhood search heuristic for the share-a-ride problem. Comput. Oper. Res. **66**, 170–180 (2016)
7. Li, B., Krushinsky, D., Woensel, T.V., Reijers, H.A.: The Share-a-Ride problem with stochastic travel times and stochastic delivery locations. Transp. Res. Part C **67**, 95–108 (2016)
8. Yu, V.F., Purwanti, S.S., PerwiraRedi, A.A.N., Chung-Cheng, L., Suprayogi, S., Jewpanya, P.: Simulated annealing heuristic for the general share-a-ride problem. Eng. Optim. **50**(7), 1178–1197 (2018). https://doi.org/10.1080/0305215X.2018.1437153
9. Li, Y.-T., Lu, C.-C.: Optimization model for the General Share-A-Ride problem with electric vehicles. In: Proceedings of the 24th International Symposium on Logistics (ISL 2019) (2019)
10. Yu, Y., Wu, Y., Wang, J.: Bi-objective green ride-sharing problem: model and exact method. Int. J. Prod. Econ. **208**, 472–482 (2019)
11. Erdogan, S., Miller-Hooks, E.: A green vehicle routing problem. Transp. Res. Part E: Logist. Transp. Rev. **48**(1), 100–114 (2012)
12. Mayke, N., Yu, V.F., Bachtiyar, C., Sukoyo: A simulated annealing heuristic for the capacitated green vehicle routing problem. Hindawi Math. Prob. Eng. **2019**, Article ID 2358258, 18 pages (2019)
13. Andelmin, J., Bartolini, E.: An exact algorithm for the green vehicle routing problem. Transp. Sci. **51**(4), 1288–1303 (2017)
14. Masmoudi, M.A., Hosny, M., Demir, E.: An adaptive large neighborhood search heuristic for the green Dial-a-Ride Problem. Published by ISTE Ltd and John Wiley & Sons, Inc. Solving Transport Problems: Towards Green Logistics (2019)
15. Masmoudi, M.A., Hosny, M., Demir, E., Genikomsakis, K.N., Cheikhrouhou, N.: the dial-a-ride problem with electric vehicles and battery swapping stations. Transp. Res. Part E: Logist. Transp. Rev. **118**, 392–420 (2018)

Machine Learning Techniques to Predict Real Time Thermal Comfort, Preference, Acceptability, and Sensation for Automation of HVAC Temperature

Yaa T. Acquaah[1]([✉]), Balakrishna Gokaraju[1], Raymond C. Tesiero III[2], and Kaushik Roy[3]

[1] Department of Computational Data Science and Engineering, North Carolina A & T State University, Greensboro, USA
ytacquaah@aggies.ncat.edu
[2] Department of Civil, Architectural and Environmental Engineering, North Carolina A & T State University, Greensboro, USA
[3] Department of Computer Science, North Carolina A & T State University, Greensboro, USA

Abstract. The control of Heating, Ventilation, and Air Conditioning (HVAC) system automatically is one of the progressive areas of research. The collective importance of the HVAC system is to maintain indoor thermal comfort while ensuring energy efficiency. This study explores the thermal comfort, acceptability, preference, and sensation of fifteen subjects from February to September 2021. Multiclass-multioutput Decision Tree, Extra Trees, K-Nearest Neighbors and Random Forest classification models were developed to predict the thermal comfort metrics, of subjects in a room based on gender, age, indoor temperature, humidity, carbon dioxide concentration, activity level and time series features. It is important to understand occupants' thermal comfort in real time to automatically control the environment. The best mean accuracy and mean squared error of 68% and 2.15 respectively was achieved by multiclass-multioutput Extra Tree classification model, when all the features were used in training and testing. Through this study, the feasibility of using machine learning techniques to predict thermal comfort, preference, acceptability, and sensation at the same time for HVAC control was established.

Keywords: Multiclass-multioutput · Thermal comfort · Thermal preference · Thermal acceptability · Thermal sensation

1 Introduction

To understand the thermal needs of individuals in a building it is imperative to know the robust interpretation of the occupant's thermal comfort state and behavior. Thermal comfort as defined by the American Society of Heating, Refrigerating and Air-Conditioning Engineers (ASHRAE Standard 55) is the condition of mind that expresses satisfaction with the thermal environment and is assessed by subjective evaluation [1]. Thermal

© Springer Nature Switzerland AG 2022
H. Fujita et al. (Eds.): IEA/AIE 2022, LNAI 13343, pp. 659–665, 2022.
https://doi.org/10.1007/978-3-031-08530-7_55

sensation is defined in ISO 7730 and ASHRAE standard 55 with a scale ranging from −3 (cold) to +3 (hot) [1, 2]. Thermal acceptability is a subjective metric that measures if an occupant is comfortable or not. Thermal preference identifies occupants' suitable temperature and provides them a means to control the room temperature. In the last decade, occupancy-based machine learning HVAC automation methodologies have been explored by many researchers to curtail high energy consumption [3–11]. However, the problem of building occupants' dissatisfaction persists. Knowing the number of people in a building or occupancy level alone does not depict comfortability and thus, not sufficient for HVAC control. There is therefore the need of expanded research to explore the thermal comfort, preference, acceptability, and sensation of occupants in a building for automation of the HVAC control. This is to maintain comfortable indoor temperatures without over heating or cooling when necessary. The study presents the development of four multiclass-multioutput classification models based on real world data collected in a laboratory experiment. The structure of this paper is organized as follows: Sect. 2 discusses the related work, Sect. 3 discusses the research methodology, Sect. 4 presents the results and discussion, and Sect. 5 concludes the paper.

2   Related Work

The predicted mean vote is not considered an optimal method for the prediction of the thermal comfort of individuals due to the restrictions of average response calculated from the data of a group of people [12]. Previous studies have predicted thermal comfort, preference, sensation, and acceptability separately leveraging myriad cutting edge machine learning techniques [13–24]. Ubiquitous features utilized in the training of these machine learning algorithms include skin temperature extraction, environmental and physiological parameters. Throughout the literature, there is consistent evidence of the successful performance of these machine learning algorithms in prediction of thermal comfort, preference, sensation, and acceptability separately. However, research has not been conducted to explore the possibility of predicting the four thermal comfort metrics (that is thermal comfort, preference, sensation, and acceptability) at the same time using machine learning techniques. In this paper, we present the development of four multiclass-multioutput classification models for the concurrent prediction of thermal comfort, preference, sensation, and acceptability.

The main contributions of this study are: (1) the data collection approach, (2) The multiclass-multioutput classification model development for the determination of thermal comfort, sensation, acceptability, and preference for a holistic understanding of an individual's thermal comfort need.

3   Methodology

In this section, we present the methodology used in this research (see Fig. 1). We designed and implemented both hardware and software for data collection. The data collection hardware consisted of carbon dioxide concentration sensor, temperature and humidity sensor interfaced with an Arduino Uno micro-controller connected to a desktop computer. The data collection software was a CoolTerm software and a Google Form survey, which resided on the desktop computer. The data collection is followed with data preparation before the data is fed into a multiclass-multioutput classification model (Scikit-learn MultiOutputClassifier API).

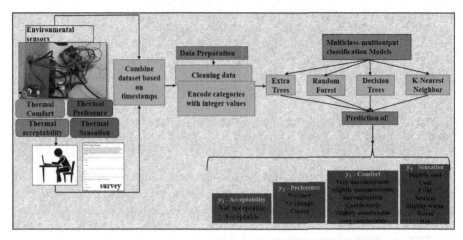

Fig. 1. Methodology Diagram for workflow.

3.1 Data Collection

The experiment was carried out in the Artificial intelligence and Visualization laboratory at North Carolina Agricultural and Technical State University from February to September of 2021. The cooling strategy of the laboratory is air conditioned. Fifteen subjects in the age range of nineteen years to thirty-four years, comprising nine males and six females were recruited to aid in the data collection. Environmental data (within the laboratory) was collected using carbon dioxide, temperature, and humidity sensors. The temperature and humidity sensors are calibrated with the Optrix PI 160 camera which interrogated the same area these sensors had been placed. Data were collected from these sensors interfaced with an Arduino Uno microcontroller and connected to a desktop computer, as shown in Fig. 2. Environmental data was streamed every minute with the help of the CoolTerm software at the same time as subjects completed a Google Form survey about their thermal sensation, comfort, acceptability, and preferences. The environmental data and the survey data were combined based on the timestamp for further analysis. One hundred and forty observations were utilized in this study.

3.2 The Multiclass-Multioutput Classification Model

The multiclass-multioutput classification is a classification task with more than one target variable and has target classes more than two. Scikit-learn API provides a Mulit Output Classifier class that helps to classify multi-output data.

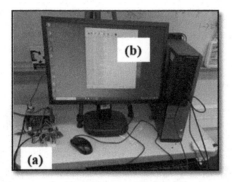

Fig. 2. The experiment setup showing (a) environmental sensors and (b) CoolTerm software on computer screen.

The multiclass-multioutput classification models were trained on gender, age, indoor temperature, humidity, carbon dioxide concentration, activity level and time series features. The target variables are presented in Fig. 3, indicating the classes in each target namely, thermal comfort, thermal preference, thermal acceptability, and thermal sensation. These are the subjective votes collected through the survey of the subjects. Due to the text nature of the target variables, they were encoded using label encoding to make them compatible with the machine learning algorithms. The dataset was split into 80% and 20% for training and testing respectively.

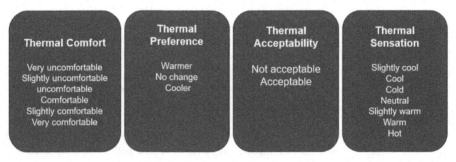

Fig. 3. Expected Output of model

4   Results and Discussion

The dataset was trained and tested on four Multiclass-multioutput models supported by the Scikit-learn API namely, Decision Tree, Extra Trees, K-Nearest Neighbors and Random Forest multioutput Classifiers. The results are presented in Table 1.

Table 1. Results of Multioutput Classifiers trained and tested on dataset

Multiclass-multioutput classifiers	Mean accuracy (%)	Mean squared error
Extra trees	68	2.15
Random forest	58	2.79
Decision trees	53	3.05
K-nearest neighbor	43	3.12

The best mean accuracy and mean squared error of 68% and 2.15 respectively was achieved by multiclass-multioutput Extra Tree classification model, when all the features were used in training and testing. The first eight important features for multiclass-multioutput Extra tree classification model were temperature, humidity, Carbon dioxide concentration, Age, Month, Day, Hour, and Minute.

5 Conclusion

In this paper, four new multiclass-multioutput classification models were trained and tested to predict the thermal comfort metrics of occupants in real time based on gender, age, indoor temperature, humidity, carbon dioxide concentration, activity level and time series features. Experimental results show that the proposed modelling technique for the prediction of thermal comfort, sensation, acceptability, and preference performs well in terms of mean accuracy and mean squared error. The best mean accuracy and mean squared error of 68% and 2.15 respectively was achieved by multiclass-multioutput Extra Tree classification model. The proposed model is expected to achieve automated control of built environments to improve human thermal comfort, health and for energy efficiency.

This paper forms part of an ongoing research. Future work will consider combining classes in the target variables to improve the models. Additionally, the dataset will also be expanded. Development of machine learning model to predict thermal preference of individuals (warmer, no change and cooler) based on thermal images and environmental parameters is also under consideration. We will develop secure machine learning models by validating against the white-box and black-box threat models for security of heterogeneous data.

Acknowledgements. This research was funded by a National Centers of Academic Excellence in Cybersecurity Grant (H98230–21-1–0326), which is part of the National Security Agency.

References

1. ANSI/ASHRAE Standard 55–2017, Thermal Environmental Conditions for Human Occupancy
2. ISO_7730, Ergonomics of the Thermal Environment -Analytical Determination

3. Acquaah, Y.T., Gokaraju, B., Tesiero, R.C., Monty, G.H.: Thermal imagery feature extraction techniques and the effects on machine learning models for smart HVAC efficiency in building energy. Remote Sens. **13**(19), 3847 (2021)
4. Acquaah, Y., Steele, J.B., Gokaraju, B., Tesiero, R., Monty, G.H.: Occupancy detection for smart HVAC efficiency in building energy: a deep learning neural network framework using thermal imagery. In: Proceedings of the 2020 Applied Imagery Pattern Recognition Workshop (AIPR), 13–15 October, pp. 1–6. IEEE (2020)
5. Kumar, S., Singh, J., Singh, O.: Ensemble-based extreme learning machine model for occupancy detection with ambient attributes. Int. J. Syst. Assur. Eng. Manag. **11**(2), 173–183 (2020). https://doi.org/10.1007/s13198-019-00935-1
6. Vela, A., Alvarado-Uribe, J., Davila, M., Hernandez-Gress, N., Ceballos, H.G.: Estimating occupancy levels in enclosed spaces using environmental variables: a fitness gym and living room as evaluation scenarios. Sensors (Basel) **20**, 6579 (2020)
7. Weber, M., Doblander, C., Mandl, P.: Towards the Detection of Building Occupancy with Synthetic Environmental Data (2020)
8. Wang, C., Jiang, J., Roth, T., Nguyen, C., Liu, Y., Lee, H.: Integrated sensor data processing for occupancy detection in residential buildings. Energy Build. **237**, 110810 (2021)
9. Beltran, A., Erickson, V., Cerpa, A.: ThermoSense: occupancy thermal based sensing for HVAC control, pp. 1–8 (2013)
10. Tyndall, A., Cardell-Oliver, R., Keating, A.: Occupancy estimation using a low-pixel count thermal imager. IEEE Sens. J. **16**, 3784–3791 (2016)
11. Sirmacek, B., Riveiro, M.: Occupancy prediction using low-cost and low-resolution heat sensors for smart offices. Sensors **20**, 5497 (2020)
12. Park, H., Park, D.Y.: Prediction of individual thermal comfort based on ensemble transfer learning method using wearable and environmental sensors. Build. Environ. **207**, 108492 (2022)
13. Xu, G., An, Q.: Prediction of human thermal sensation based on improved PMV model. In; Paper Presented at the IOP Conference Series: Earth and Environmental Science, vol. 680, no. 1 (2021)
14. Wu, Y., et al.: Individual thermal comfort prediction using classification tree model based on physiological parameters and thermal history in winter. Build. Simul. **14**(6), 1651–1665 (2021). https://doi.org/10.1007/s12273-020-0750-y
15. Hepokoski, M., Curran, A., Viola, T., Ockfen, A.: Thermal acceptability limits for wearable electronic devices. In: Paper Presented at the 37th Annual Semiconductor Thermal Measurement, Modeling and Management Symposium, SEMI-THERM - Proceedings, pp. 16–19 (2021)
16. Gao, N., Shao, W., Rahaman, M.S., Zhai, J., David, K., Salim, F.D.: Transfer learning for thermal comfort prediction in multiple cities. Build. Environ. **195**, 107725 (2021)
17. Chai, Q., Wang, H., Zhai, Y., Yang, L.: Using machine learning algorithms to predict occupants' thermal comfort in naturally ventilated residential buildings. Energy Build. **217**, 109937 (2020)
18. Luo, M., Xie, J., Yan, Y., Ke, Z., Yu, P., Wang, Z., Zhang, J.: Comparing machine learning algorithms in predicting thermal sensation using ASHRAE comfort database II. Energy Build. **210**, 109776 (2020)
19. Cosma, A.C., Simha, R.: Machine learning method for real-time non-invasive prediction of individual thermal preference in transient conditions. Build. Environ. **148**, 372–383 (2019)
20. Cosma, A.C., Simha, R.: Thermal comfort modeling in transient conditions using real-time local body temperature extraction with a thermographic camera. Build. Environ. **143**, 36–47 (2018)

21. Jin, L., Liu, T., Ma, J.: Modeling thermal sensation prediction using random forest classifier. In: Han, Q., McLoone, S., Peng, C., Zhang, B. (eds.) LSMS/ICSEE -2021. CCIS, vol. 1469, pp. 552–561. Springer, Singapore (2021). https://doi.org/10.1007/978-981-16-7213-2_53
22. Megri, A., Naqa, I.: Prediction of the thermal comfort indices using improved support vector machine classifiers and nonlinear kernel functions. Indoor Built Environ. **25**, 6–16 (2014)
23. Wang, Z., Yu, H., Luo, M., et al.: Predicting older people's thermal sensation in building environment through a machine learning approach: modelling, interpretation, and application. Build. Environ. **161**, 106231 (2019)
24. Zhong, C., Liu, T. Zhao, J.: Modeling the thermal prediction using the fuzzy rule classifier. In: Chinese Automation Congress (CAC 2019), 22–24 November 2019, pp. 3184–3188. IEEE, Hangzhou (2019)

Neural Networks

Serially Disentangled Learning
for Multi-Layered Neural Networks

Ryotaro Kamimura[1(✉)] and Ryozo Kitajima[2]

[1] Kumamoto Drone Technology and Development Foundation, Tokai University,
2880 Kamimatsuo Nishi-ku, Kumamoto 861-5289, Japan
ryo@keyaki.cc.u-tokai.ac.jp
[2] Tokyo Polytechnic University, 1583 Iiyama, Atsugi, Kanagawa 243-0297, Japan
r.kitajima@eng.t-kougei.ac.jp

Abstract. The present paper aims to propose a new type of learning method called "serially disentangled learning." This method is inspired by representation learning, but our focus is not on the production of the disentangled representation but on the computational procedures themselves to make it possible to produce the representation. We try to separate or disentangle several contradictory computational procedures and to force them to perform their tasks serially and independently, meaning that we try to resolve, though seemingly, the contradiction among computational procedures. This concept can be used to disentangle complex information control procedures, where information minimization and maximization procedures are forced to perform each procedure separately, independently, and serially. The method was applied to relations between the mission statements of companies and their profitability, where conventional learning methods could not show better generalization performance. The new method with de-hierarchical disentanglement by network compression produced compressed weights close to the point-biserial correlation coefficients between inputs and targets of the original data set. In addition, we could detect an input related to improved generalization.

1 Introduction

1.1 Entanglement and Disentanglement

The neural networks have been applied to many problems, showing better generalization performance. However, as the complexity of neural network architecture and data sets becomes greater, the necessity of identifying and disentangling the underlying important factors, hidden in complicated and entangled representations [1], has been increasing. Thus, there has been a surge in interest in the disentanglement of informational factors or components in neural networks, for example, architectural, de-hierarchical, de-biasing, and model disentanglement, for interpretation and improved generalization. First, in the architectural disentanglement the original network architecture is disentangled into several sub-architectures, corresponding to specific tasks, which can be easily interpreted [2]. Second, complicated models are simplified, compressed, distilled, and de-hierarchized into smaller and more interpretable ones [3]. For example, model compression has been applied to the de-hierarchical processing, where

H. Fujita et al. (Eds.): IEA/AIE 2022, LNAI 13343, pp. 669–681, 2022.
https://doi.org/10.1007/978-3-031-08530-7_56

multi-layered neural networks can be simplified and compressed into the simplest ones without hidden layers [4]. Recently, more attention has been paid to the biases in data sets. The presence of biases in data sets has been harmful to improving generalization performance. Then, there has been a number of de-biasing techniques from the data set [5,6], using the adversarial learning [7,8] and lowering the biased influences in learning [9].

However, hidden in those studies on the extraction of informational representations, there are the other types of complexity or entangled phenomena. This means that the optimizing and regularizing procedures needed to realize useful representations tend to be complicated, interwoven, and entangled with each other. For example, to produce disentangled representations, as mentioned above, many different kinds of regularization have been used to penalize the optimization processes. Naturally, regularization terms are contradictory to optimizers in their essence, and one of the serious problems lies in difficulty in compromising between them.

Considering this difficulty, the present paper aims to disconnect or disentangle several computational, regularization, and optimization procedures, because they are interwoven and entangled, making learning processes considerably complicated and hard. To make learning as efficient as possible, we need to disentangle those entangled computational procedures as much as possible. In this paper, we call this learning with disentangled or independently treated computational procedures "disentangled learning," in which all different computational procedures should be applied as independently as possible.

1.2 Information and Entanglement

The present paper tries to show one example of disentangled learning in particular, namely, disentanglement of information-theoretic procedures. In the information-theoretic methods, one of the main difficult problems is a contradiction between information maximization and minimization. This paper takes a storage view of information, and the information should represent how much information content can be stored in a neural network. Then, information minimization aims to distribute information in inputs into as many components as possible and, in particular, as evenly as possible. Information maximization aims to distribute information contained in inputs as unevenly as possible, focusing on some specific components. For different objectives, different learning strategies on information should be undertaken.

For example, by paying attention to interpretation or generalization, information should be differently controlled. If we try to interpret the meaning of components, we must increase the information on the components as much as possible [10]. On the other hand, if we try to improve generalization performance, too much information specific to some inputs or components can be harmful, corresponding to information minimization. This information minimization has been extensively applied in improving generalization performance, having a long tradition dating back to the preeminent and excellent works by Deco [11].

When we try to maximize more complicated measures of mutual information, whose use in neural networks has been initiated by the pioneer works on the maximum information preservation by Linsker [12], the properties of information maximization and minimization are actually mixed or entangled, and more strongly, those are contradictory to each other. For example, in mutual information, information maximization in terms of conditional entropy minimization is contradictory to information minimization in terms of entropy of averaged components. Those two types of information control are performed in two different levels, but still difficulty has existed in compromising between them in actual learning. Furthermore, in a more recent approach with multiple operations of mutual information [13], the problem of computational entanglement has become more serious, in particular, for practical implementation.

In addition, another example of co-existence of information maximization and minimization can be pointed out in terms of selectivity. As has been well known, the selectivity has played important roles in the neurosciences from the beginning [14, 15], with many experimental results accumulated [16]. The selectivity has also played important roles in neural networks in improving generalization [17]. In particular, the selectivity has been so far discussed in the field of convolutional neural networks (CNN). For example, the majority of interpretation methods have tried to show which components in a neural network are the most responsible for a specific input pattern or specific output, [18, 19]. However, there have been recent discussions on the importance of selectivity in neural networks, saying that the selectivity should be undermined and eliminated as much as possible, especially for improving generalization performance [20, 21]. This discussion also suggests that selectivity or selective information should be minimized and maximized, depending on different situations in neural networks.

1.3 Serially Disentangled Learning

Considering the necessity of simplification of complicated computational procedures, this paper tries to propose a new type of disentangled learning in which information minimization and maximization and error minimization are disentangled or independently and serially treated. Contrary to the simultaneous and entangled information minimization and maximization, we process these procedures independently and serially. For example, information is first minimized and then maximized, and this process of information minimization and maximization is repeated many times. In addition, error minimization is applied also independently of information minimization and maximization.

Focusing on the disentanglement of computational procedures in neural networks, the present paper tries to show the following three points on the disentangled learning, namely, how to disentangle information minimization and maximization and error minimization, interpretation by de-hierarchical disentanglement, and application to the detection of important features. First, we try to show how to disentangle information minimization and maximization and error minimization. The error minimization is applied also independently, which aims to assimilate informational effects. Then, by the de-hierarchical disentanglement, we try to simplify multi-layered neural networks into the simplest ones and to show how compressed weights can be changed by controlling the information. Finally, applied to the real data set, we try to show which features are responsible in linear or non-linear relations for improving generalization performance.

2   Theory and Computational Methods

2.1   Serial Disentanglement

The present paper tries to disentangle the complicated entanglement of information minimizer and maximizer and error minimizer (assimilator) into completely separated and independently operated procedures. Figure 1(a) shows the entanglement of three procedures, where they are processed in complicated and interwoven ways, which makes it hard to compromise among three types of processing.

Figure 1(b) shows that the three procedures are separated and serially operated. In this processing, the error minimizer is called "information assimilator," which is operated separately from the corresponding information minimizer and maximizer. For example, the information minimizer is applied, which can produce a minimized information state independently of the information assimilator. Then, the information obtained by the information minimizer, should be assimilated by the information assimilator independently of the information minimizer, which is usually an error minimizer.

The same process is applied to the information maximization, which is followed by the information assimilation. This cycle of information maximization and minimization with information assimilation can be repeated to assimilate the effects of information minimization and maximization sufficiently.

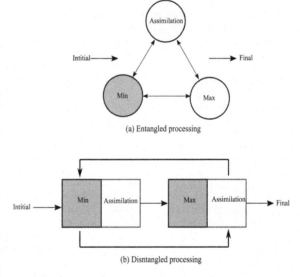

(a) Entangled processing

(b) Disntangled processing

Fig. 1. Conventional entangled processing (a) and serially disentangled processing (b).

2.2   Structural Information

In this paper, we use a new type of information measure, close to the entropy of information theory, but its meaning can be concretely interpreted in terms of the number of connection weights. Let us define the structural information in terms of the number of connection weights for the concept of information.

For simplicity, we focus on connection weights between the second and third layer $(2, 3)$, as shown in Fig. 2, and the absolute strength of weights are computed by

$$u_{jk}^{(2,3)} = \mid w_{jk}^{(2,3)} \mid \tag{1}$$

Then, we normalize this by its maximum value, which can be computed by

$$g_{jk}^{(2,3)} = \frac{u_{jk}^{(2,3)}}{\max_{j'k'} u_{j'k'}^{(2,3)}} \tag{2}$$

where the max operation is over all connection weights between the layer. In addition, we define the complementary one

$$\bar{g}_{jk}^{(2,3)} = 1 - \frac{u_{jk}^{(2,3)}}{\max_{j'k'} u_{j'k'}^{(2,3)}} \tag{3}$$

We call this absolute strength "potentiality," and "complementary potentiality," because they can be used to increase or decrease the structural information, defined below. By using this potentiality, structural information can be computed by

$$\bar{G}^{(2,3)} = \sum_{j=1}^{n_2} \sum_{k=1}^{n_3} \left[1 - \frac{u_{jk}^{(2,3)}}{\max_{j'k'} u_{j'k'}^{(2,3)}} \right] \tag{4}$$

When all potentialities become equal, naturally, the structural information becomes zero. On the other hand, when only one potentiality becomes one, while all the others are zero, the structural information becomes maximum. For simplicity, we suppose that at least one connection weight should be larger than zero.

In the serially disentangled learning in Fig. 2, in the first place, structural information is minimized (a1), where the strength of connection weights is forced to be larger, contrary to the assumption of the other regularization, and the strength of all connection weights is also forced to be the same. This large strength for information minimization is introduced to strengthen the forces of information minimization. Then, this effect by information minimization is assimilated in connection weights in Fig. 2(a2). When the assimilation ends, information maximization is applied in Fig. 2 (b1), and the number of connection weights is forced to be smaller, and the strength of the weights is also forced to be smaller. Then, the effect by information maximization should be assimilated in Fig. 2 (b2).

Connection weights are changed by multiplying them by the corresponding potentialities. For example, to increase the structural information and for the $n +$ 1th learning step, denoted by $(n + 1)$, we have

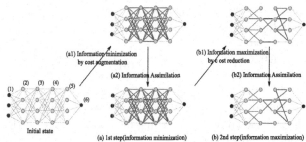

Fig. 2. Serially disentangled learning only for the first and second learning step.

$$w_{jk}^{(2,3)}(n + 1) = g_{jk}^{(2,3)}(n) \, w_{jk}^{(2,3)}(n) \tag{5}$$

For decreasing information, $\bar{g}$ should be used instead of g, and for assimilation, the usual weight update rules should be used. Finally, we should note the computational method, in which the actual potentialities should be modified to stabilize the learning.

For example, the complemental potentiality $\bar{g}$ can be changed into the corresponding modified one

$$\bar{h}_{jk}^{(2,3)} = \theta_1 \left[1 - \frac{u_{jk}^{(2,3)}}{\max_{j'k'} u_{j'k'}^{(2,3)}} + \theta_3 \right]^{\theta_2} \tag{6}$$

The parameter θ_3 is introduced to eliminate zero potentiality. Except for the parameter θ_1 to increase the strength of weight, all the other parameters should be much smaller ones.

2.3 De-Hierarchical Disentanglement

For interpreting multi-layered neural networks, we first compress them into the simplest ones, as shown in Fig. 3. We try here to trace all routes from inputs to the corresponding outputs by multiplying and summing all corresponding connection weights.

First, we compress connection weights from the first to the second layer, denoted by $(1, 2)$, and from the second to the third layer $(2, 3)$ for an initial condition and a subset of a data set.

Then, we have the compressed weights between the first and the third layer, denoted by $(1, 3)$.

$$w_{ik}^{(1,3)} = \sum_{j=1}^{n_2} w_{ij}^{(1,2)} w_{jk}^{(2,3)} \tag{7}$$

Those compressed weights are further combined with weights from the third to the fourth layer $(3,4)$, and we have the compressed weights between the first and the fourth layer $(1, 4)$.

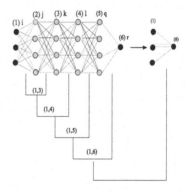

Fig. 3. De-hierarchical disentanglement to the simplest network.

$$w_{ik}^{(1,4)} = \sum_{k=1}^{n_3} w_{ik}^{(1,3)} w_{kl}^{(3,4)} \tag{8}$$

By repeating these processes, we have the compressed weights between the first and sixth layer, denoted by $w_{iq}^{(1,5)}$. Using those connection weights, we have the final and fully compressed weights $(1, 6)$.

$$w_{ir}^{(1,6)} = \sum_{q=1}^{n_6} w_{iq}^{(1,5)} w_{qr}^{(5,6)} \tag{9}$$

Considering all routes from the inputs to the outputs, the final connection weights should represent the overall characteristics of connection weights of the original multi-layered neural networks.

3   Results and Discussion

3.1   Experimental Outline

This experiment aimed to examine relations between the mission statements of companies and their profitability. We collected the mission statements of 300 companies, listed in the first section of the Tokyo Stock Exchange, which were summarized by five input variables, extracted by the natural language processing systems. For demonstrating the performance of disentangled learning, we used the very redundant ten hidden-layered neural networks, where each hidden layer had ten neurons.

The data set seemed to be very easy, but the conventional methods such as the linear regression and random forest could not improve generalization. In addition, simple information maximization and minimization also could not improve generalization. We repeated the process of information minimization and maximization ten times, because improved performance could not be seen even if we repeated them further. In the experiments, we tried to increase gener-

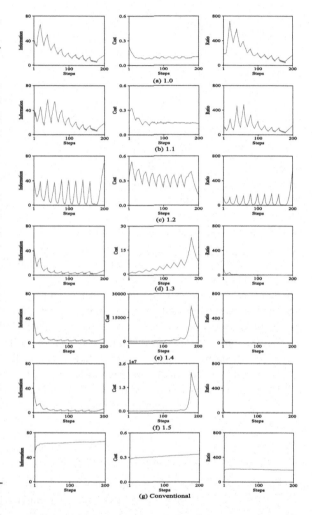

Fig. 4. Structural information (left), cost (middle), and ratio of information to its cost (right) as a function of the number of steps, when the parameter θ_1 was increased from 1.0 (a) to 1.5 (f), and by the conventional method (g) for the mission statement data set.

alization performance, focusing on information minimization, where information was first minimized and then maximized. One learning step contained the variable number of learning epochs for assimilating the information by information minimizer and maximizer. In this experiment, the number of learning epochs increased gradually when the number of learning steps increased. In the experiments, several learning parameters should be controlled, and among them, a parameter θ_1, which was changed from

1 to 1.5, to control the strength of weights or cost was important. When this parameter increased, the strength of weights or potentialities (cost) increased gradually to decrease the information content.

3.2   Structural Information and Cost

The experimental results showed that information minimization and maximization could be repeated, and when the parameter was appropriately chosen, information could be increased in spite of the forces to decrease information.

Figure 4 shows structural information, cost (strength of weights), and the ratio of information to its cost, when the number of learning steps increased to 200 and the parameter to increase the cost was increased from 1 (a) to 1.5 (f), and when the conventional method without the information control was used (g). When the parameter increased from 1 (a) to 1.1 (b), the cost in terms of the sum of all potentialities in the middle deceased gradually, though there was a slight increase and decrease. Correspondingly, the structural information (left) and the ratio decreased gradually with a more explicit increase and decrease. When the parameter was 1.2 in Fig. 4(c), the fluctuations of cost, structural information, and the ratio increased considerably. In the end, the structural information (left) and the ratio (right) increased sharply. As shown in Table 1, the highest generalization performance was obtained when the parameter was 1.2. Thus, this increase in information can be related to improved generalization. When the parameter increased further from 1.3 in Fig. 4 (d) to 1.5 (f), the cost (middle) increased considerably, and the structural information (left) and the ratio (right) were forced to be extremely small. Finally, when we used the conventional method without information control in Fig. 4(g), the information (left), cost (middle), and the ratio (right) did not show any changes, and remained almost the same for all learning steps.

The results showed that when the parameter was appropriately large, an increase and decrease in information could be clearly detected. Then, by this forced information control, the structural information increased sharply in spite of the forces to decrease the information.

3.3   Connection Weights and Potentialities

The results showed that symmetric properties by the smaller parameter values could be obtained. Then, with the parameter value producing the best generalization performance, the connection weights became more sparse. When the parameter was further increased, and information was further decreased, connection weights became randomly activated.

Figure 5 shows connection weights, when the parameter θ_1 increased from 1(a) to 1.5 (f). When the parameter θ_1 increased from 1 (a) to 1.1 (b), symmetric connection weights appeared, and by seeing the potentialities (2), we could see that a neuron in a precedent layer tended to respond strongly to a group of neurons in the corresponding subsequent layer. When the parameter increased to 1.2 in Fig. 5(c), the number of strong weights and potentialities became smaller. As was mentioned, the parameter value of 1.2 produced the best generalization performance. This smaller number of strong weights in addition to the regularity observed in the values of 1 (a) and

1.1 (b), contributed to the improvement of generalization. Then, when the parameter θ_1 increased from 1.3 (d) to 1.5 (g), connection weights tended to have the same absolute values, and in terms of potentiality, almost all potentialities became white, meaning that all potentialities became equally higher, reducing the information content. Finally, Fig. 5(g) shows connection weights and potentialities by the conventional methods without information control. As can be seen in the figure, connection weights and potentialities were almost random, without any regularity.

The results showed that, when the information was smaller, more symmetric weights and potentialities could be seen. Then, when the information was controlled to get the best generalization, more sparse weights and potentialities could be seen. However, when the information was further decreased, connection weights became random.

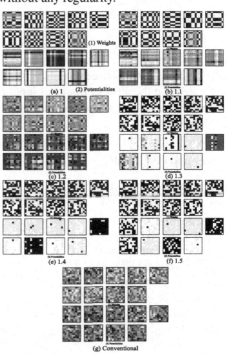

Fig. 5. Weights (1) and potentialities (2) for all hidden layers by the structural information when the parameter θ_1 was 1.0 (a) to 1.5 (f), and by the conventional method without information control (g) for the mission statement data set.

3.4 Compressed Weights

The compressed weights became closest to the correlation coefficients between inputs and targets of the original data set by the conventional method. By controlling information, gradually, different features could be seen, different from the correlation coefficients. When the parameter was appropriately increased, input No. 5 (abstract things), with the weakest correlation coefficient, tended to have the strongest relative importance.

Figure 6 shows compressed weights (left), relative compressed weights (middle), and the original correlation coefficients between inputs and targets (right). The relative compressed weights were obtained by dividing compressed weights (left) by the corresponding correlation coefficients (right). The conventional method in Fig. 6(g) produced the highest correlation with the original correlation coefficients. When the parameter θ_1, to increase the cost in terms of strength of weights, increased from 1 (a) to 1.3 (d), the compressed weights became slightly different from the original correlation (right). However, we could see that input No. 2 had the largest strength, corresponding to the original correlation coefficient (right).

One of the major characteristics could be seen in the relative collective weights in the middle. Input No. 5 had the largest strength relative to the correlation coefficients (middle). In particular, when the parameter θ_1 was 1.2 and 1.3 in Fig. 6(c) and (d), input No. 5 had the largest relative strength.

The experimental results showed that the compressed weights were close to the original correlation coefficients between inputs and targets for the conventional method, in which input No. 2 (organizations) had the largest importance. When the parameter increased gradually, the compressed weights became different from the correlation coefficients, and input No. 5 (abstract things) became the largest with respect to the corresponding value of correlation coefficients. This means that, to improve generalization, this input No.5 (abstract things) played an important role in addition to the correlation coefficients between inputs and targets, represented by input No. 2 (organizations). Because the abstract things of input No.5 represent the values and principles of companies, the results, obtained by this method, seem to be a plausible interpretation of the data set.

3.5 Summary of Results

The results showed that the information control could increase generalization at the expense of decreasing the correlation coefficients. However, we could still see that generalization performance could be considerably increased, even keeping relatively higher correlation coefficients.

Table 1 shows the summary of experimental results in terms of correlation coefficients and corresponding generalization accuracies. As can been in the table, the conventional method without the information control produced the highest correlations coefficient of 0.986. However, the highest accuracy of 0.602 was obtained when the parameter θ_1 was 1.2. The accuracy increased from 0.585 to 0.602, when the parameter θ_1 increased from 1.0 to 1.2. Then, the

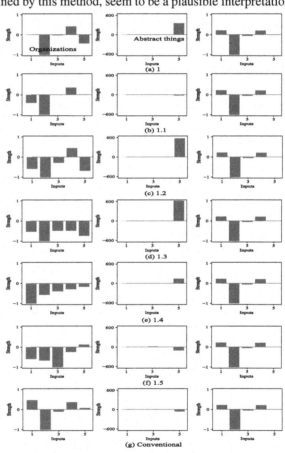

Fig. 6. The compressed weights (left), relative weights (middle), and the original correlation coefficients (right) when the parameter θ_1 increased from 1.0 (a) to 1.5 (f), and by the conventional method (g) for the mission statement data set.

accuracy decreased, and when the parameter θ_1 was the highest at 1.5, the generalization became lowest among the present methods. The random forest produced the second lowest correlation of 0.244, and the accuracy was also the second worst of 0.540. Finally, the logistic regression produced a higher correlation of 0.862, but the accuracy was the lowest at 0.511.

The results show that the new method could produce higher accuracies by controlling structural information and reducing the correlation coefficients. However, we could see that the method produced higher accuracies, even keeping higher correlation coefficients.

4   Conclusion

The present paper proposed a new type of disentangled learning, called "serially disentangled learning," where information minimizer and maximizer and error minimizer are separately applied, because those procedures are entangled in a complicated manner with each other, making it hard to compromise among them in the actual learning. The information minimization and maximization, followed independently by information assimilation, are repeated many times to reach the final state in which we can expect that information minimization and maximization are not so contradictory to each other.

The method was applied to real data representing relations between mission statements and profitability of companies. The experimental results showed that information could be decreased, minimizing and maximizing information. The de-hierarchical entanglement showed that the conventional method could extract the linear correlation coefficients between inputs and targets almost perfectly, in spite of much redundancy of ten-hidden-layered neural networks.

Table 1. Summary of experimental results on average correlation coefficients and generalization performance for the mission statement data set. The numbers in the method represent the values of the parameter θ_1 to increase the potentialities.

Method	Correlation	Accuracy
1.0	0.892	0.585
1.1	0.821	0.597
1.2	0.649	**0.602**
1.3	0.894	0.599
1.4	0.009	0.587
1.5	0.274	0.570
Conventional	**0.986**	0.585
Random forest	0.244	0.540
Logisitc	0.862	0.511

In addition, by controlling the information, generalization performance could be improved, and we could extract the important input responsible for the non-linear relations.

Though we do not know exactly how many times we should use these disentangled procedures, the present paper could show at least that the repeated application of serially disentangled learning can be used to compromise among many contradictory procedures in neural networks.

References

1. Bengio, Y., Courville, A., Vincent, P.: Representation learning: a review and new perspectives. IEEE Trans. Pattern Anal. Mach. Intell. **35**(8), 1798–1828 (2013)
2. Hu, J., et al.: Architecture disentanglement for deep neural networks. In: Proceedings of the IEEE/CVF International Conference on Computer Vision, pp. 672–681 (2021)
3. Hinton, G., Vinyals, O., Dean, J.: Distilling the knowledge in a neural network. arXiv preprint arXiv:1503.02531 (2015)
4. Kamimura, R.: Neural self-compressor: collective interpretation by compressing multi-layered neural networks into non-layered networks. Neurocomputing **323**, 12–36 (2019)
5. Cubuk, E.D., Zoph, B., Mane, D., Vasudevan, V., Le, Q.V.: Autoaugment: learning augmentation strategies from data. In: Proceedings of the IEEE/CVF Conference on Computer Vision and Pattern Recognition, pp. 113–123 (2019)
6. Gupta, A., Murali, A., Gandhi, D., Pinto, L.: Robot learning in homes: improving generalization and reducing dataset bias. arXiv preprint arXiv:1807.07049 (2018)
7. Kim, B., Kim, H., Kim, K., Kim, S., Kim, J.: Learning not to learn: training deep neural networks with biased data. In: Proceedings of the IEEE/CVF Conference on Computer Vision and Pattern Recognition, pp. 9012–9020 (2019)
8. Wang, T., Zhao, J., Yatskar, M., Chang, K.-W., Ordonez, V.: Balanced datasets are not enough: estimating and mitigating gender bias in deep image representations. In: Proceedings of the IEEE/CVF International Conference on Computer Vision, pp. 5310–5319 (2019)
9. Hendricks, L.A., Burns, K., Saenko, K., Darrell, T., Rohrbach, A.: Women also snowboard: overcoming bias in captioning models. In: Ferrari, V., Hebert, M., Sminchisescu, C., Weiss, Y. (eds.) ECCV 2018. LNCS, vol. 11207, pp. 793–811. Springer, Cham (2018). https://doi.org/10.1007/978-3-030-01219-9_47
10. Kamimura, R.: Som-based information maximization to improve and interpret multi-layered neural networks: from information reduction to information augmentation approach to create new information. Expert Syst. Appl. **125**, 397–411 (2019)
11. Deco, G., Obradovic, D.: An Information-Theoretic Approach to Neural Computing. Springer (2012). https://doi.org/10.1007/978-1-4612-4016-7
12. Linsker, R.: Self-organization in a perceptual network. Computer **21**(3), 105–117 (1988)
13. Tishby, N., Zaslavsky, N.: Deep learning and the information bottleneck principle. In: 2015 IEEE Information Theory Workshop (ITW), pp. 1–5. IEEE (2015)
14. Hubel, D.H., Wisel, T.N.: Receptive fields, binocular interaction and functional architecture in cat's visual cortex. J. Physiol. **160**, 106–154 (1962)
15. Bienenstock, E.L., Cooper, L.N., Munro, P.W.: Theory for the development of neuron selectivity. J. Neurosci. **2**, 32–48 (1982)
16. Schoups, A., Vogels, R., Qian, N., Orban, G.: Practising orientation identification improves orientation coding in v1 neurons. Nature **412**(6846), 549–553 (2001)
17. Ukita, J.: Causal importance of low-level feature selectivity for generalization in image recognition. Neural Netw. **125**, 185–193 (2020)
18. Nguyen, A., Yosinski, J., Clune, J.: Understanding neural networks via feature visualization: a survey. In: Samek, W., Montavon, G., Vedaldi, A., Hansen, L.K., Müller, K.-R. (eds.) Explainable AI: Interpreting, Explaining and Visualizing Deep Learning. LNCS (LNAI), vol. 11700, pp. 55–76. Springer, Cham (2019). https://doi.org/10.1007/978-3-030-28954-6_4
19. Montavon, G., Binder, A., Lapuschkin, S., Samek, W., Müller, K.-R.: Layer-wise relevance propagation: an overview. In: Samek, W., Montavon, G., Vedaldi, A., Hansen, L.K., Müller, K.-R. (eds.) Explainable AI: Interpreting, Explaining and Visualizing Deep Learning. LNCS (LNAI), vol. 11700, pp. 193–209. Springer, Cham (2019). https://doi.org/10.1007/978-3-030-28954-6_10

20. Morcos, A.S., Barrett, D.G., Rabinowitz, N.C., Botvinick, M.: On the importance of single directions for generalization. Stat **1050**, 15 (2018)
21. Leavitt, M.L., Morcos, A.: Selectivity considered harmful: evaluating the causal impact of class selectivity in DNNs. arXiv preprint arXiv:2003.01262 (2020)

Detecting Use Case Scenarios in Requirements Artifacts: A Deep Learning Approach

Munima Jahan[1]([✉]), Zahra Shakeri Hossein Abad[2], and Behrouz Far[1]

[1] Department of Electrical and Software Engineering, University of Calgary,
Calgary, Canada
{munima.jahan,far}@ucalgary.ca

[2] Department of Computer Science, University of Calgary, Calgary, Canada
zshakeri@ucalgary.ca

Abstract. Use case scenarios (UCS) written in natural languages like English are popular tools for requirements elicitation. They also play an important role in the model driven design process by being used as an initial input for many automated behavior modeling such as generating sequence diagrams and class diagrams. However, there is no unified approach used by the engineers to represent use cases and make it hard to identify use case statements from requirements artifacts without doing it manually by human experts, which is time consuming and tedious. In this paper, we propose a novel approach for automatically identifying use case scenarios within requirements documents written in English. We employ a machine learning based approach which is applicable to a wide range of specifications in various domains. We train and evaluate different state-of-the-art machine learning models including transformers over an independently labeled dataset to compare the performance of different models and select the best classifier based on the result. Our experimental result shows that ULMFiT outperforms other models with an accuracy of 95%. Moreover, we disclose our dataset as well as the source code to foster the discourse on the automatic use case detection within the research community.

Keywords: Use case scenario · Natural language processing · UML model · Machine learning · Deep learning · Transfer learning

1 Introduction

Requirements specification (RS) is the central element for any successful software development process. The software requirements specification (SRS) are predominantly written in natural language (NL) to facilitate better understanding, clarity and communication across stakeholders with varying backgrounds

This research is partially supported by NSERC (Natural Sciences and Engineering Research Council), Canada and AITF (Alberta Innovates Technology Futures).

© Springer Nature Switzerland AG 2022
H. Fujita et al. (Eds.): IEA/AIE 2022, LNAI 13343, pp. 682–694, 2022.
https://doi.org/10.1007/978-3-031-08530-7_57

and skills [22]. For business personnel, requirements must be simple to comprehend. Designers, on the other hand, demand precision and clarity. Unfortunately, most requirement notations do not offer both of these features. As a result, notations that machines can easily handle are difficult to interpret without expert knowledge. NL notations are readable by humans, but they allow too much room for interpretation and lack the technical people's rigor. The ideal notation would allow as many people as possible to be involved in the eliciting process to ensure a higher-quality system [2]. One of the most commonly used notations for requirements in the modeling world is use case scenarios (UCS). However, the semantics of use cases as defined in UML are extremely ambiguous [5]. Lack of clear separation between UCS and the problem domain descriptions can cause inconsistencies in requirements artifacts [23].

UCSs are often considered as the initial inputs for many design processes and behavior modelings, such as generating sequence diagrams or class diagrams [10]. However, collecting scenarios from textual SRS documents manually is time-consuming and tedious. Motivated by the need for automated extraction of UCS from NL requirements, we propose a machine learning based approach to distinguish the use case specifications from the rest of the content within a requirements document. To the best of our knowledge and belief, this is the first study that classifies UCS and non-UCS statements. The key contributions of our proposed method are: (1) we employ deep learning (DL) to devise a novel approach to classify UCS statements within requirements documents, (2) our DL model is based on generic features that can be applied to any SRS document irrespective of the problem domain or specific template or style used, (3) no external input or user intervention is required, (4) we empirically evaluate our approach using a dataset prepared from several industrial SRSs from various application domains.

Besides, we investigate different DL models to compare the performance on different settings to achieve the most reliable classifier. Finally, we make our dataset and the code available to the research community to facilitate replication.

The paper is organized as follows: Sect. 2 provides the background. Section 3 reviews the related works. Section 4 presents our proposed method. Section 5 discusses the experimental results. Finally, the conclusion of the paper and the possible future research directions are provided in Sect. 6.

2   Background

This section reviews a few concepts and terms related to our approach.

2.1   Deep Learning Methods

Deep Neural Networks (DNNs) have achieved significant results in different application domains like computer vision [11] and speech recognition [6]. Recently, it has become more common to use DNNs in natural language processing (NLP), where much of the work involves learning word representations through neural language models and then performing a composition over the learned word vectors for classification.

For text classification, DNNs combine feature extraction and classification. Recurrent neural networks (RNN) are artificial neural networks that are frequently utilized in NLP. RNNs identify the progressive features of the input and use patterns to forecast the following most probable outcome.

However, an RNN is a skewed model, with recent words being more significant than older ones. When leveraged to capture the semantics of a whole document, this may reduce efficiency. As a consequence, the Long Short Term Memory (LSTM) model was developed to address the RNN's shortcomings [7]. For some jobs, such as part-of-speech tagging, where we must assign a tag to each word in a phrase, it would be more efficient to use both past and future words. That is what the Bidirectional Neural Network (BNN) does; it consists of two LSTMs. One runs forward from left to right, and the other runs backward from right to left [20]. We have experimented with Dense, LSTM, and Bi-LSTM neural networks to compare the performance.

2.2 Transfer Learning

Transfer learning is a method of using a deep learning model trained on an extensive dataset to perform similar tasks on a new dataset. Word embeddings, pre-trained vector representations of words taken from large volumes of text data, have been used in different language-based tasks with promising results. Word2vec and Glove are two of the most often used context-independent neural embeddings. With notable advancements such as Universal Language Model Fine-Tuning (ULMFiT) [8], Embedding from Language Models (ELMO) [13], OpenAI's Generative Pre-trained Transformer (GPT) [14], and Google's BERT model [3], the year 2018 has been a watershed moment for many NLP tasks. BERT is the first unsupervised, deeply bidirectional language representation learned using only a plain text dataset. Masked LM and Next Sentence Prediction are two new prediction tasks in BERT. In a variety of downstream NLP tasks, the pre-trained BERT model outperformed ELMo and OpenAI GPT significantly [3].

Identifying use cases from requirement documents is a complicated task due to the lack of available labeled data and the inability of surface features to capture the subtle semantics in the text. To address this issue, we have used the pre-trained language model BERT and ULMFiT and fine-tune for a specific task by leveraging information from different transformer encoders. The process of transfer learning in our experiment can be visualized as in Fig. 1.

3   Related Work

Software requirements are typically written in natural language to make them understandable to the stakeholders. However, the use of NL increases the possibility of introducing several inherent issues such as ambiguity, inconsistency,

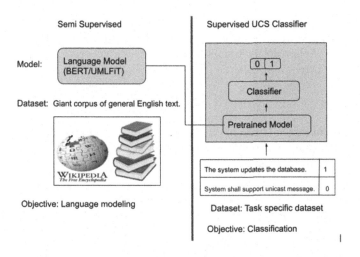

Fig. 1. The process of transfer learning.

incorrectness, and incompleteness in the specification. UML models are gaining popularity to mitigate the limitations of NL and are used in software engineering to understand and analyze the envisaged systems and business processes. However, building models from NL requirements is complex and time-consuming. Extensive research has been carried out on the automated and semi-automated generation of UML diagrams from NL requirements such as use case, class diagrams sequence diagrams [10]. However, to the best of our knowledge, no approach is available that classifies the UCS statements from requirement specifications. We are the first to propose a machine learning-based method for automatically distinguishing use case statements from other contents in the textual requirements specifications.

NLP is widely used in requirements verification and model generation [17,19]. LOLITA is an NLP-based system proposed in [19] that automatically generates object models from textual requirements. LIDA [12] is a semi-automatic tool that helps designers in creating class diagrams. Similar NLP techniques have been adopted in other works [9] for identifying concepts in NL requirements and constructing class diagrams based on those concepts. Automatic approaches to derive the sequence diagrams from UCSs using NLP and rule-based approach are carried out by several researchers [10,17,21].

ML-based approaches are also becoming popular in requirements engineering for model transformation and verification. An approach for automated demarcation of requirements in textual specifications using machine learning is proposed in [1]. A neural network-based approach to domain modeling relationships and patterns recognition is presented in [15].

UCSs have been used as the key input by many researchers for creating and transforming design models [17,21]. However, not having a unique standard semantics to define UCSs within the elicitation document made it hard to extract them automatically. No research has been addressed this issue and demands further attention. An automatic classifier for extracting the UCS statements from the SRS documents is crucial to facilitate the generation of future design models from NL requirements.

4   Method

Our method combines three main steps: 1) Dataset preparation, 2) Training different models, and 3) Applying the classifier and evaluation. Figure 1 presents an overview of our approach (Fig. 2).

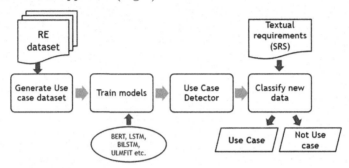

Fig. 2. Basic workflow of the proposed approach.

4.1   Dataset

To prepare the training data for predicting UCS, we analyzed 79 SRS documents from the PURE (PUblic REquirements dataset) [4] dataset. PURE is a dataset of 79 publicly available natural language requirements documents collected from the Web. The dataset includes 34,268 sentences and has been used by many researchers for NLP tasks typical in requirements engineering. After analyzing all the documents, we found a few use cases as part of the specification. We collected 582 sentences representing UCS statements from different SRS documents containing both basic snd alternative scenarios. We also gathered 738 sentences from random sections within the documents that are not part of the use cases. Labeled the first set of sentences as "use_case" and the second set of the sentences as "not_use_case. We finally tagged the sentences with binary labels, '1' for use_cases and '0' for the sentences that are not_use_case. Table 1 represents a few samples from our dataset.

Table 1. Sample data from UCS dataset.

SL	Text	Label
1	User selects an editor from list alt	1
2	The unique tag must be at most characters long	0
3	central trading system save the sell instruction	1

Before training our models, we applied standard text preprocessing steps to our dataset, including tokenization, sequencing, and padding. Additionally, we checked for any duplicate entries and removed them. After removing the duplicate entries, we downsized the not_use_case data to balance each class with 582 data points.

Finally, we considered 70% of the dataset as training data, 15% as validation data to measure the out-of-sample performance of the model during training, and 15% as test data to measure the out-of-sample performance after training.

4.2 Development of Predictive Models for Classifying Use Case Scenario

We considered the problem of distinguishing UCSs from nonUCS statements as a binary classification task. We trained different machine learning (ML) models, including pre-trained language models, and compared their performance with varying parameters.

It is well established that no single machine learning classifier consistently achieves the best classification performance and demonstrates different strengths depending on the specific task and dataset. We, therefore, experimented with two classical ML models (Naive Bayes, Random forest), three popular state-of-the-art deep learning (DL) models (Dense, LSTM, BiLSTM), and three transformers (BERT base model, DistilBERT, ULMFiT) to find the possible best classifier. We considered Naive Bayes and the Random Forest as our baseline ML models. Three different classical deep learning models used in this study are discussed in the subsequent sections.

Dense Architecture: We developed the dense architecture with 24 neurons for the dense hidden layer. The first layer is the embedding layer which maps each word to an N-dimensional vector of real numbers. The size of this vector is 16 in our case. We used average pooling and converted the layer to one dimension. Next, we used a dense layer with activation function 'relu' followed by a dropout layer to avoid overfitting and a final output layer with the sigmoid activation function. As there are only two classes (use case or not use case) to classify, we used only a single output neuron.

Long Short Term Memory (LSTM) layer architecture: For the LSTM model, we added an LSTM cell with 20 nodes. The return_sequences property is set to True to ensure that the LSTM cell returns all of the outputs from the unrolled LSTM cell through time.

Bi-directional Long Short Term Memory (BiLSTM) Model: Bi-LSTM is a special type of LSTM network that learns patterns from both directions, before and after a given token within a document. The Bi-LSTM back-propagates in both backward and forward directions in time. Due to this, the computational time is increased compared to LSTM. However, in most cases, Bi-LSTM results in better accuracy. To implement a Bi-directional LSTM architecture, we used the Bidirectional wrapper to LSTM.

Additionally, we used three different pre-trained language models (PLM).

BERT Base Model: BERT (Bidirectional Encoder Representations from Transformers) is a multi-layer bidirectional Transformer encoder. It is pre-trained on a large corpus of unlabelled text, including the entire Wikipedia and Book Corpus. Unlike traditional sequential or recurrent models, the attention architecture processes the whole input sequence simultaneously, enabling all input tokens to be processed in parallel. The layers of BERT are based on transformer architecture. The pre-trained BERT model can be fine-tuned with just one additional layer to obtain state-of-the-art results in a wide range of NLP tasks.

There are different techniques for fine-tuning the BERT model. We can either train the entire architecture or some layers while freezing others. It is also possible to freeze the entire architecture.

For our experiment, we froze all the layers of BERT during fine-tuning and appended a dense layer and a softmax layer to the architecture. Moreover, we used the BERT base model as the task-specific dataset we created is comparatively small in size. Additionally, we employed a lighter version of BERT called DistilBERT [16] to compare the performance of the models.

DistilBERT: DistilBERT is a smaller version of BERT developed and open-sourced by the team at HuggingFace. DistilBERT has 40% fewer parameters than bert-base-uncased, runs 60% faster while preserving over 95% of BERT's performances as measured on the GLUE language understanding benchmark [16]. The model is claimed to be a smaller, faster, cheaper, lighter, and distilled version of BERT that can achieve a sensible lower-bound on BERT's performances with the advantage of quicker training. That is why we considered DistilBERT as a better match for our experiment with limited data.

ULMFiT: ULMFiT stands for Universal Language Model Fine-tuning for Text Classification and is a transfer learning technique that involves creating a Language Model that is capable of predicting the next word in a sentence, based on unsupervised learning of the WikiText 103 corpus. The ULMFiT model uses multiple LSTM layers, with dropout applied to every layer [8]. We use this pre-trained AWD-LSTM based model and follow the recommended fine-tuning stages for UCS classification.

4.3   Implementation of the Predictive Models

We implemented our models using Python 3.6 language with Keras framework and Tensorflow in the back-end. We used Python Scikit-learn package for implementing Naive Bayes and Random Forest. For feature extraction, we tried both TfidfVectorizer and CountVectorizer. We used Keras's sequential model for the dense classifier where layers are added in a sequence. For LSTM and BiLSTM we also used Keras sequential models and added LSTM layers. The word vectors used within the neural networks were embedded using the Keras embedding layer. To add enough diversity and considering the data size we chose 500 as our embedding size. To implement BERT models, we used the pytorch transformer library containing the pre-trained BERT model and text tokenizer. In addition, we used the BERT_base model and the DistilBERT model. The Google Collaboratory tool is used as the implementation environment, a free research tool with a Tesla K80 GPU and 12G RAM. As an input, we tokenized each sentence with the BERT tokenizer. As mentioned in the previous section, we froze all the model layers before fine-tuning to prevent updating the model weights during fine-tuning. For the implementation of ULMFiT we use fastai libraries. The source code along with the dataset is available at "https://github.com/Jahan2021/UCS-Classification".

5   Results

This section presents the experimental results and discussion on the performance of different models.

5.1   Prediction Performance

To evaluate the performance of the different models we use the standard evaluation metrics precision, recall, accuracy and F1-score. We also consider the AUC score as part of our evaluation.

Table 2. Comparing the precision, recall and F_1 scores for different models.

	Models	Precision(%)	Recall(%)	F1 Score(%)	Accuracy(%)
ML models	Naive Bayes	87	87	87	87
	Random Forest	89	89	89	89
DL models	Dense	91	91	91	91
	LSTM	76	73	72	73
	Bi-LSTM	88	87	87	87
Transformers	ULMFiT	95	95	95	95
	BERT_base	91	90	90	90
	DistilBERT	94	94	94	94

From Table 2 we notice that all algorithms perform quite well in terms of the standard metrics, with the state-of-the-art algorithms having a small but significant advantage. Among the classical ML models, RF has better performance than NB, which is also proven in other research works [18]. In the case of the classical DL models, Dense architecture outperforms LSTM and BiLSTM. To have a better understanding of the performance of DL models, we include the learning curves for different models in Fig. 3 and Fig. 4.

The training and validation learning curve for LSTM clearly indicates a possibility of overfitting, which essentially affects the outcome for LSTM. From the eight predictive models developed and tested, LSTM performed the worst across all measures (average accuracy: 73%, F1: 72%, precision: 76%, recall: 73%). Which reflects the requirements of a large dataset for training an LSTM model.

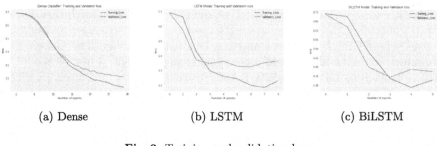

(a) Dense   (b) LSTM   (c) BiLSTM

Fig. 3. Training and validation loss

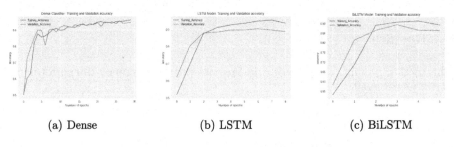

(a) Dense   (b) LSTM   (c) BiLSTM

Fig. 4. Training and validation accuracy

The Naive Bayes model produced the second-worst results with an average F1 of 87%. From the remaining six models, the top-three classifiers were ULMFiT, DistilBERT, and BERT_base model. Table 2 reports the average performances of these models. In addition to the evaluation results listed in Table 2, Fig. 5 confirms the superior performance of the transformers over other models. It is readily seen that these models significantly outperformed all classical ML and DL models with higher recall, F1, and AUC scores.

While comparing the three pre-trained models, ULMFiT gives us the best performance score with an accuracy of 95% (see: Table 2). However, in terms of AUC score in Table 3, BERT_base model and DistilBERT show better results.

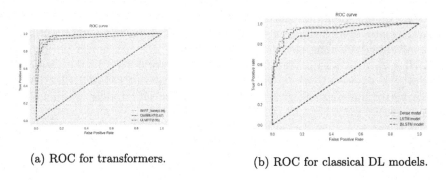

(a) ROC for transformers.   (b) ROC for classical DL models.

Fig. 5. Comparision between different models with respect to ROC curve.

Table 3. Comparison between the performance of different models.

Models	AUC score %	Epochs	Training Time(seconds)
ULMFiT	95	24	87
BERT_base	96	50	391
DistilBERT	99	4	74

5.2 Discussion

The key objective of this study was to develop machine learning models to accurately classify statements within requirements documents that represent UCS from the rest of the content. A few important conclusions can be reached from observing the experimental results. Firstly, the dataset is simple enough that even naive baseline approaches can achieve good results. Secondly, the state-of-the-art approaches do outperform the classical ML and DL classifiers. The accuracy of the BERT basic model is slightly lower than the distilled version of the model. Both BERT and ULMFiT have similar results. Potential reasons could be the simplicity of the dataset and the limited data size.

Another interesting observation worth mentioning here is the performance of the fully connected dense model that beats the BERT_base model with a 91% accuracy. This indicates that in the case of UCS classification, relevant terms in a sentence may have more discriminating ability than the sentence structure or the relationship between the words within a sentence. On the other hand, it is also possible that the data size is not large enough to learn the deep structure of the sentences.

Among the three pre-trained models, ULMFiT shows better performance in terms of accuracy than the two BERT models. However, the AUC score suggests further investigation of the binary classification threshold where the standard 0.5 may not be the best choice. We considered accuracy as a better performance metrics compared to AUC score as our dataset is balanced.

One final aspect worth examining is the training time spent on state-of-the-art algorithms. Table 3 shows the number of epochs spent on training each algorithm along with the execution time. It also displays the AUC scores for the transformers. We only included the metrics for the transformers as they appeared to be the predominant among all classifiers. The data shows that DistilBERT is the fastest with the lowest number of epochs which is expected based on the model's characteristics. It is worth noting that the number of epochs and execution time for ULMfit includes the training time for the language model and the fine-tuning of the classifier.

5.3 Threat to the Validity

The main concern regarding the validity of our approach is the dataset. We prepared the data from a few SRS documents in [4] that contain use cases. Thus the domain spaces for the application areas are not exhausted. Moreover, the sentence structure may not be generalized enough compared to the industrial-level SRS documents. However, use cases are usually written in simple sentences and present form. Therefore, to include different sentence structures within our training data, we collected random sentences from different parts of the SRS documents and labeled them as not_use_case. Besides, we applied our approach to entirely different SRS documents, and the result was promising.

6   Conclusion and Future Work

Use cases are commonly used as a notation for capturing functional requirements through scenarios and are considered as the starting point of object-oriented software development. The problem is that use cases are written as natural language documents, and there is no universal notation for use case representations.

In this paper, we propose a machine learning-based approach to distinguish use case statements from other contents within the requirements documents written in English. The key feature of our approach is that it applies to a wide range of requirements specifications without any external input from the users. The features that we use for learning are based on linguistic information applicable to diverse application domains. We empirically evaluated our method using industrial requirements specifications. We have applied different deep learning models to compare the performance of various classifiers. The results indicate that the pre-trained language models outperform classical ML and DL models, with ULMFiT showing the best average precision and recall of 95%. Additionally, we prepared a dataset related to the classification task and made it publicly available for future research.

We consider this study the starting point and demand further investigation and improvement. Moreover, with a small dataset, it is difficult to analyze the true behavior of the models effectively. Therefore, preparing a larger dataset and experimenting with hybrid models with different parameters are some of the aspects we consider as part of our future work.

References

1. Abualhaija, S., Arora, C., Sabetzadeh, M., Briand, L.C., Traynor, M.: Automated demarcation of requirements in textual specifications: a machine learning-based approach. Empir. Softw. Eng. **25**(6), 5454–5497 (2020). https://doi.org/10.1007/s10664-020-09864-1
2. Alexander, I.F., Maiden, N.: Scenarios, Stories, Use Cases: Through the Systems Development Life-Cycle. Wiley, New York (2005)
3. Devlin, J., Chang, M.W., Lee, K., Toutanova, K.: Bert: pre-training of deep bidirectional transformers for language understanding. arXiv preprint arXiv:1810.04805 (2018)

4. Ferrari, A., Spagnolo, G.O., Gnesi, S.: Pure: A dataset of public requirements documents. In: 2017 IEEE 25th International Requirements Engineering Conference (RE), pp. 502–505 (2017). https://doi.org/10.1109/RE.2017.29

5. Génova, G., Llorens, J., Metz, P., Prieto-Díaz, R., Astudillo, H.: Open issues in industrial use case modeling. In: Jardim Nunes, N., Selic, B., Rodrigues da Silva, A., Toval Alvarez, A. (eds.) UML 2004. LNCS, vol .3297, pp. 52–61. Springer, Heidelberg (2005). https://doi.org/10.1007/978-3-540-31797-5_6

6. Graves, A., Mohamed, A.R., Hinton, G.: Speech recognition with deep recurrent neural networks. In: 2013 IEEE International Conference on Acoustics, Speech and Signal Processing, pp. 6645–6649. IEEE (2013)

7. Hochreiter, S., Schmidhuber, J.: Long short-term memory. Neural Comput. 9(8), 1735–1780 (1997)

8. Howard, J., Ruder, S.: Fine-tuned language models for text classification. arXiv preprint arXiv:1801.06146, 1–7 (2018)

9. Ibrahim, M., Ahmad, R.: Class diagram extraction from textual requirements using natural language processing (NLP) techniques. In: 2010 Second International Conference on Computer Research and Development, pp. 200–204. IEEE (2010)

10. Jahan, M., Abad, Z.S.H., Far, B.: Generating sequence diagram from natural language requirements. In: 2021 IEEE 29th International Requirements Engineering Conference Workshops (REW), pp. 39–48. IEEE (2021)

11. Krizhevsky, A., Sutskever, I., Hinton, G.E.: Imagenet classification with deep convolutional neural networks. Adv. Neural. Inf. Process. Syst. 25, 1097–1105 (2012)

12. Overmyer, S.P., Benoit, L., Owen, R.: Conceptual modeling through linguistic analysis using Lida. In: Proceedings of the 23rd International Conference on Software Engineering. ICSE 2001. pp. 401–410. IEEE (2001)

13. Peters, M.E., et al.: Deep contextualized word representations. arXiv preprint arXiv:1802.05365 (2018)

14. Radford, A., Narasimhan, K., Salimans, T., Sutskever, I.: Improving language understanding by generative pre-training (2018)

15. Saini, R., Mussbacher, G., Guo, J.L., Kienzle, J.: A neural network based approach to domain modelling relationships and patterns recognition. In: 2020 IEEE Tenth International Model-Driven Requirements Engineering (MoDRE), pp. 78–82. IEEE (2020)

16. Sanh, V., Debut, L., Chaumond, J., Wolf, T.: Distilbert, a distilled version of bert: smaller, faster, cheaper and lighter (2020)

17. Sharma, R., Gulia, S., Biswas, K.: Automated generation of activity and sequence diagrams from natural language requirements. In: 2014 9th International Conference on Evaluation of Novel Approaches to Software Engineering (ENASE), pp. 1–9. IEEE (2014)

18. Singh, A., Halgamuge, M.N., Lakshmiganthan, R.: Impact of different data types on classifier performance of random forest, Naive Bayes, and k-nearest neighbors algorithms (2017)

19. Smith, M.H., Garigliano, R., Morgan, R.G.: Generation in the Lolita system: An engineering approach. In: Proceedings of the Seventh International Workshop on Natural Language Generation (1994)

20. Xiao, Y., Cho, K.: Efficient character-level document classification by combining convolution and recurrent layers. arXiv preprint arXiv:1602.00367 (2016)

21. Yue, T., Briand, L.C., Labiche, Y.: aToucan: an automated framework to derive UML analysis models from use case models. ACM Trans. Software Eng. Methodol. (TOSEM) 24(3), 1–52 (2015)

22. Zhao, L., et al.: Natural language processing (NLP) for requirements engineering: a systematic mapping study. arXiv preprint arXiv:2004.01099 (2020)
23. Śmiałek, M., Bojarski, J., Nowakowski, W., Ambroziewicz, A., Straszak, T.: Complementary use case scenario representations based on domain vocabularies, vol. 4735, pp. 544–558 (09 2007). https://doi.org/10.1007/978-3-540-75209-7_37

Hybrid Deep Neural Networks for Industrial Text Scoring

Sidharrth Nagappan(✉)🆔, Hui-Ngo Goh🆔, and Amy Hui-Lan Lim🆔

Faculty of Computing and Informatics, Multimedia University, Cyberjaya, Malaysia
sidharrth2002@gmail.com, {hngoh,amy.lim}@mmu.edu.my

Abstract. Academic scoring is mainly explored through the pedagogical fields of Automated Essay Scoring (AES) and Short Answer Scoring (SAS), but text scoring in other domains has received limited attention. This paper focuses on industrial text scoring, namely the processing and adherence checking of long annual reports based on regulatory requirements. To lay the foundations for non-academic scoring, a pioneering corpus of annual reports from companies is scraped, segmented into sections, and domain experts score relevant sections based on adherence. Subsequently, deep neural non-hierarchical attention-based LSTMs, hierarchical attention networks and longformer-based models are refined and evaluated. Since the longformer outperformed LSTM-based models, we embed it into a hybrid scoring framework that employs lexicon and named entity features, with rubric injection via word-level attention, culminating in a Kappa score of 0.9670 and 0.820 in both our corpora, respectively. Though scoring is fundamentally subjective, our proposed models show significant results when navigating thin rubric boundaries and handling adversarial responses. As our work proposes a novel industrial text scoring engine, we hope to validate our framework using more official documentation based on a broader range of regulatory practices.

Keywords: Natural Language Processing (NLP) · Deep Learning (DL) · Longformer

1 Introduction

Millions of students sit for standardised writing tests like IELTS every year; a teacher then reads and scores them individually. The subjective, time-consuming and bias-ridden process of manual scoring opened doors for the widely explored Automated Essay Scoring (AES), and Short Answer Scoring (SAS) frameworks, from the naive rule-based engine [10] to the now groundbreaking intervention of deep neural networks [6,12,14,18]. However, to the best of our knowledge, there has been no research exploring automated text scoring in non-academic domains, where voluminous text is also prevalent. While academic documents are scored based on linguistic proficiency and pedagogical competence, non-academic text is more verbose and is primarily evaluated on content alone.

© Springer Nature Switzerland AG 2022
H. Fujita et al. (Eds.): IEA/AIE 2022, LNAI 13343, pp. 695–706, 2022.
https://doi.org/10.1007/978-3-031-08530-7_58

One use case is the scoring of organisations' annual reports, based on thematic demands that require them to disclose relevant business activities and demonstrate compliance to norms set out by regulating bodies. Hence, an organisation should dedicate separate report sections for each regulatory practice, and descriptions should be comprehensive enough to comply with standards. To ensure the documentation is satisfactory, regulating officers read these long reports and assign an adherence score based on fine-grained rubrics. Since there are over 41,000 listed companies globally [4], one can ask whether a computer can algorithmically sift through all this text, comprehend it, identify parts of interest, and score in real-time with the backing of modern computational power. The need for automation stems from the same epiphany that once motivated [10] to automate the processing of student essays.

By migrating the document length aspect of AES and the content-scoring strategies of SAS, this research proposes a novel, extensible framework for industrial text scoring. To enable our research, a pioneering corpus of annual reports is aggregated, segmented into relevant sections, and subsequently scored by domain experts to validate performance. As a starting point, the chosen state-of-the-art models from AES are neural attention-based LSTM networks [6,14] and ensemble BERT [15]. For our work, we substitute BERT with the Longformer, a newer long document counterpart, to accommodate the parsing of lengthy reports while still leveraging the attentional superiority of transformers.

The major contributions of our research can be summarised as follows: (a) a novel deep hybrid framework that scores long corporate documents and scales to a wide range of document lengths by leveraging linear attention transformers and (b) a mechanism that allows domain experts to inject rubrics and domain-relevant lexicons in a hybrid manner to more closely influence neural scoring models. Our generalisable framework's source code is publicly available[1].

2   Related Work

Our research amalgamates principal strategies employed in both AES and SAS.

Automated scoring is a computer's ability to evaluate and score written prose without human interference [13]. A typical pipeline would go from a response input to feature extraction to a learning algorithm that assigns an ordinal score through classification or regression. Initial AES models required manual feature engineering before this was replaced by neural networks, particularly Recurrent Neural Networks (RNN) and transformers. They showed superior performance on the ASAP-AES benchmark student essay dataset.

A naive RNN that encompassed word embedding, convolutional, Long Short Term Memory (LSTM), pooling and fully connected layers (FCL) set an initial Kappa benchmark of 0.761 [14]. As opposed to this preliminary approach of modelling documents non-hierarchically via linear word sequences, [6] was of the view that documents constitute sentences, which in turn constitute words,

---

[1] https://github.com/sidharrth2002/text-scoring

so they proposed hierarchical attention networks that used separate convolutional and recurrent encoders at both the word and document level, setting a new benchmark of 0.764. With a tree-like structure, they attested the effectiveness of their architecture for long document modelling. Later research primarily modified these hierarchical and non-hierarchical LSTM-based models for more sophisticated tasks, such as multi-task trait scoring and coherence modelling. Subsequently, the intervention of transformers such as BERT in 2018 [5] outperformed former neural systems that used context-independent embedding schemas such as GLoVE [11]. However, given the quadratic complexity $O(n^2)$ of BERT, current research attempts to apply transformers for longer sequences, trailing on linear $O(n)$ attention transformers such as the Longformer [1] and BigBird [19]. Extensive pretraining on large corpora and transfer learning allowed researchers to utilise these modern transformers in smaller domains such as AES [9,15].

A key turning point was the fusion of handcrafted features with neural scoring models to influence the scoring mechanism more closely. A combination of BERT embedding outputs and a multimodal vector of essay-level features produced the current AES benchmark of 0.801 [15]. Such hybrid frameworks also began to unfold the black box [8] of neural scoring models, lending an explainability element and countervailing for smaller datasets [16].

While AES is multifaceted and holistic, SAS research looked at specific content scoring strategies that opened doors for the injection of rubrics and marking schemes into scoring models. [16] consolidated word-level attention for keyword matching with a standard neural pipeline and showed noticeable performance compensations in low-data settings on the ASAP-SAS short answer dataset. The keyword similarity method was more sophisticated than the boolean presence or absence of prompt-relevant lexicons [3,15], simultaneously locating relevant information in responses and allowing one to visualise the semantic relevance of a response to the key ideas it was expected to address.

3 Experimental Framework

3.1 Data Aggregation and Preprocessing

A collection of annual reports is aggregated from companies. Then, we break down the documents into chapters of interest and find the chapter relevant to each regulatory practice. After this, the text in the chapter is scored for adherence by domain experts. Finally, two corpora for separate regulatory practices are compiled, each spanning 2500 responses. The schema is outlined below in Table 1, consisting of the regulatory practice, the explanation written to prove adherence and the score.

The original dataset is first augmented with additional score 0 incoherent responses using NLPAug[2] sequence-to-sequence translation and random synonym replacement, in an effort to help the model distinguish between coherent and incoherent text. In addition, since each response is directly extracted from

---

[2] https://github.com/makcedward/nlpaug.

Table 1. Schema of corpus

Attribute	Description
practice	Regulatory practice that the response is written for. Each practice comes with a set of adherence rubrics
explanation	Response written to show evidence of compliance, detail the process and list outcomes (extracted from the annual report)
score	Score signifying relevance, comprehensiveness and overall rubric adherence

Portable Document Format (PDF) files, our preprocessing pipeline consists of standard whitespace removal, expansion of contractions, removal of non-textual characters and spelling correction. Next, a separate corpus is compiled for feature extraction, with additional steps taken such as stopword removal and lemmatization. Each corpus is then distributed into a stratified 60%–20%–20% train-test-validation split, in line with the norm in AES literature [6,12,14].

3.2 Modelling

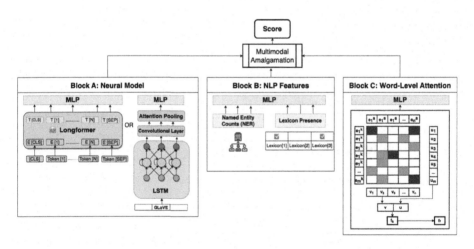

Fig. 1. Proposed Framework, consisting of a Block A neural model, Block B categorical and numerical features and Block C word-level attention for rubric injection

As seen in Fig. 1, our hybrid proposal consists of blocks A, B and C.

Block a (Neural Model). Block A can either be (i) Non-hierarchical LSTM, (ii) Hierarchical LSTM or (iii) Longformer. (i) and (ii) are traditional RNNs, and (iii) is a transformer. We aim to choose the best of the three architectures.

- **(i) Non-hierarchical LSTM** - As the naive baseline, this model is based on the architecture of [14]. It consists of word tokenization, 300-dimensional pre-trained GLoVE embeddings, a convolutional layer for n-gram level feature extraction, an LSTM layer, pooling of LSTM states via mean-over-time or attention, followed by a 32-cell FCL, and another FCL with either a discrete softmax (classification) or scalar sigmoid (regression) activation. We replicate the same attention layer implementation of [14], which involves the learning of an attention vector signifying the importance of each time step.
- **(ii) Hierarchical LSTM** - While non-hierarchical LSTMs model text in a linear sequence, hierarchical LSTMs first split the text into sentences, and then words, resulting in a 3-dimensional input vector of shape *(batch size, number of sentences, number of words)* [6]. The word and sentence encoders can either be convolutions or LSTMs. Hence, our experiments test permutations of different encoders at both levels.
- **(iii) Longformer** - To handle sequences of text longer than 512 words, we employ the Longformer [1]. By introducing a dilated sliding window and combining local and global attention, the Longformer can handle sequence lengths of up to 4096 words, pre-trained on autoregressive language modelling tasks. While the Longformer is not the current state-of-the-art, we contend that it stands on the balance between fast computations and performance compared to RoBERTA and BigBird [19]. Our longformer-based model consists of tokenization, Longformer block embeddings (finetuned), dropout and FCL.

Block B (NLP Features). Block B is a module that processes the categorical presence of domain-specific lexicons and the numerical count of selected named entity families. This is based on the understanding that more detailed responses will make mention of relevant n-grams and named entities in their description. We use the Term Frequency-Inverse Document Frequency (TF-IDF) ranking of n-grams in the lemmatized corpus along with Spacy's[3] rule-based PhraseMatcher to select and mark the 30 most meaningful lexicons, while using a Named Entity Recognition (NER) tagger to annotate the corpus before counting the frequency of each NER family (organisations, laws, persons) in each response. A Support Vector Classifier is used with Recursive Feature Elimination to identify handcrafted features that most impact the final score. For standardisation, the numerical features are transformed using a quantile normal distribution, and categorical features are one-hot encoded.

Block C (Rubric Word-Level Attention). Based on the work of [2,16], we compute the attentional similarity between expert-defined keywords (that one can expect a high-scoring response to use) and each response. The careful selection of key phrases allows the injection of scoring rubrics into the model.

For a response r with n words and key phrase k with m words, GLoVE word embedding sequences $\{e_1^r, e_2^r, e_3^r, ..., e_n^r\}$ and $\{e_1^k, e_2^k, e_3^k, ..., e_m^k\}$ are generated.

---

[3] https://spacy.io.

1. The dot product of the two sequences is computed.

$$z_{i,j} = e_i^k \cdot e_j^r \tag{1}$$

2. Softmax is computed over the rows and columns of the matrix to obtain α_i^k and α_j^r, where α_i^k intuitively signifies the attention that the word i in the key phrase pays to every word in r.

$$\{\alpha_i^k = softmax(z_{i,1}, ..., z_{i,n}), \ \alpha_j^r = softmax(z_{1,j}, ..., z_{m,j})\} \tag{2}$$

3. Attentional vectors are computed based on α_i^k and α_j^a using a weighted sum for both key phrase to response and response to key phrase.

$$\{u = \frac{1}{m}\sum\sum \alpha_{i,j}^k e_j^r, \ v = \frac{1}{n}\sum\sum \alpha_{j,i}^r e_i^k\} \tag{3}$$

4. A feature vector $f_k = [u; v]$ is output for k key elements before being concatenated into an overall word-level attention vector $f = [f_1, f_2, f_3, ...f_k]$.

Combining Module for Multimodal Amalgamation. The outputs of blocks B and C are combined with the logits output of the neural block either via an (a) attention sum [7] or (b) a Multi-layer Perceptron (MLP). The attention sum adds the neural outputs, categorical and numerical features and the word-level attention feature vector before the neural outputs query the result vector. For example, if $\mathbb{F}$ is the final feature vector, W is the weight matrix, x is the neural textual features, c represents categorical features, n represents numerical, and w represents attentional features, the final feature vector $\mathbb{F}$ and attention $\alpha_{i,j}$ are:

$$\mathbb{F} = \alpha_{x,x}W_x x + \alpha_{x,c}W_c c + \alpha_{x,n}W_n n + \alpha_{x,w}W_w w \tag{4}$$

$$\alpha_{i,j} = \frac{exp(LeakyReLU(a^T[W_i x_i || W_j x_j]))}{\sum_{k \in \{x,c,n,w\}} exp(LeakyReLU(a^T[W_i x_i || W_k x_k]))} \tag{5}$$

3.3   Evaluation Metric

The Hewlett Essay Scoring Competition used the Quadratic Weighted Kappa (QWK) as its principal evaluation metric because of its ability to account for both the frequency and severity of scoring errors. Given its wide adoption, our experimental setup also employs it. QWK is tabulated as follows:

$$W_{i,j} = \frac{(i-j)^2}{(N-1)^2} => k = 1 - \frac{\sum_{i,j} W_{i,j} O_{i,j}}{\sum_{i,j} W_{i,j} E_{i,j}} \tag{6}$$

W is a weight matrix, i is the ground truth score, j is the predicted score and N is the number of possible scores. O is a matrix denoting the number of responses that have a ground truth of i and prediction of j. E is the expected count, which accounts for random agreement. It is then normalised before k is computed.

4   Results and Discussion

4.1   Chosen Regulatory Practices

For framework validation, we choose regulatory practices X and Y. With a grading scale between 0 and 4, the descriptions and score distributions are shown in Table 2. The skewed distribution is satisfactory because it resembles the real world (most organisations make sure to meticulously check for adherence before submitting). Score 1 also does not exist as there were no provided rubrics.

Table 2. Description of regulatory practices and original score distribution

Prac.	Description	0	1	2	3	4
X	Practice X **requires all organisations to appoint directors to the board in a fair and objective manner, using independent assessors to make hiring decisions**; a perfect response would first acknowledge that the board made appointments, list the sources, describe the hiring process and the outcomes.	76	0	1259	513	368
Y	Practice Y **requires organisations to annually evaluate the board to ensure optimal performance**, so a perfect response would acknowledge evaluation, list the evaluating criteria, process and outcomes.	40	0	911	1188	367

4.2   Training

We use the Adam optimiser to minimise the loss and gradient clipping to avoid vanishing gradients due to long sequence lengths. All models are trained for a fixed number of epochs: 30 for LSTM-based and 4 for transformer-based and hybrid models. We freeze the neural block to prevent transformer weights overfitting for hybrid models, where several layers are trained from scratch. Finally, we run each configuration twice and take the best result. Transformer-based and hybrid models are implemented in Pytorch to accommodate the compatibility of Huggingface transformers [17], while LSTM-based models are written in Tensorflow 2.0. All models are trained on a Google Cloud Compute Engine, with a dedicated Tesla T4 GPU and 15 GB RAM.

4.3 Ablation Study

Critical hyperparameter decisions for the LSTM-based models were made using an ablation study to address the discrepancies in the literature, choosing between (a) classification or regression, (b) attention pooling or mean over time pooling to aggregate LSTM outputs and (c) unidirectional or bidirectional LSTMs. We find that treating scoring as a classification problem produces a better Kappa score and is practical because we use a uniform grading scale. For the LSTM layer, we find that unidirectional LSTMs are both more effective and train faster than their bidirectional counterparts. Attention pooling also proves more effective than mean-over-time [6,12].

4.4 Neural Models

Table 3. QWK Performance of different configurations of Block A neural models

Architecture	Practice X	Practice Y
Non-hierarchical LSTM-CNN-Attention	0.9500	0.7702
Hierarchical Unidirectional LSTM Word Encoder + Unidirectional LSTM Sentence Encoder	0.9459	0.7782
Hierarchical Bidirectional LSTM Word Encoder + Bidirectional LSTM Sentence Encoder	0.9381	0.762
Hierarchical Convolutional Word Encoder + Unidirectional LSTM Sentence Encoder	0.9307	0.7769
Longformer + Fully Connected Layer(s)	**0.9506**	**0.7948**

A substantial disparity can be seen between the range of Kappa scores for Practices X and Y in Table 3. As Practice X primarily consists of short responses, with an average length of 200 words, it is observable that models perform markedly better when handling shorter sequences, consistent with past results [15]. The non-hierarchical LSTM baseline [14] is significantly outperformed by both hierarchical networks [6,18] and the longformer [1], with the highest Kappa scores (X score 0.9506 with 0.06% increase and Y score 0.7948 with 3.2% increase from the baseline). Conceivably, Practice Y drastically benefits from the more sophisticated models because it spans verbose responses with an average length of 600 words and the fundamental use case of both hierarchical networks and the longformer is long document modelling. On the other hand, Practice X shows little improvement from the baseline, suggesting that model complexity is not proportional to performance for shorter documents.

During error analysis, we identify the best model's wrongly scored responses, as shown in Table 4. Scoring is a very subjective process, and there are thin boundaries in the rubrics that any human scorer or model needs to navigate. E1 is a testament to this hypothesis where the prediction is off by only a score of 1, and a second look at the response may lead a human scorer to re-evaluate their original decision.

Table 4. Examples of incorrectly scored responses with their respective ground truths (GT) and predictions (P)

	Response	GT	P
E1	An annual board evaluation is conducted for the fiscal period under review. The Chairman of the Board encourages the engagement of all directors via an open discussion during the assessment process of each directors' performance, and that it is crucial to the organisation's operational efficiency. The directors are assessed based on their general commitment, competency in discharging their duties and contribution to decision-making. In conclusion, the Board finds that the current composition is sufficient and appropriate.	3	4
E2	The Nomination Committee only met once during the current financial year under review and all the activities executed by the Committee were properly recorded.	0	2

For E2, however, we consider it an adversarial response that is designed to exploit the grading mechanism. While it does use prominent domain-relevant keywords such as "Nomination Committee", "financial year" and "review", the rubrics require the response to acknowledge that a review was conducted, but there is no such acknowledgement. Moreover, the word "review" is in the context of "financial year under review", where "review" is a semantic auxiliary referring to the period and not the evaluation process itself. Therefore, one can contend that the model needs to differentiate between the properly coherent, contextually meaningful responses and those that simply manage to use the right words.

4.5 Hybrid Models

Since the longformer performed most consistently and produced the highest Kappa Score among all runs, we chose it as Block A in our hybrid framework. Based on TF-IDF ranks, specialist consultation, rubric analysis and feature selection, we select lexicons, NER families and key phrases for both corpora, with a subset of features for Practice Y shown in Table 5.

Table 5. Lexicons, NER families and key phrases used in Practice Y hybrid frameworks

Categorical Lexicons to look for (Block B)	Numerical Counts (Block B)	Key phrases for word-level attention (Block C)
annual assessment, evaluation form, skill experience, independent, board composition, .., board effectiveness	ORG(Organisations), LAW(Documents), GPE(Geopolitical Entities), PER(Persons), EVENT(Occurrences), Token Length	undertake board evaluation, independent sources used, use questionairres, conduct interviews, based on criteria, tabled to board, board satisfied, satisfactory outcome, ..., formulate action plan

Table 6. QWK Performance of different configurations of hybrid frameworks

Run	Architecture	Practice X	Practice Y
1	Block A	0.961	0.802
2	Block A + B (Combined with MLP)	0.964	0.810
3	Block A + B + C (Combined with MLP)	**0.967**	**0.820**

We proposed two approaches for multimodal amalgamation and found that the MLP (0.802 Kappa) performed better than the attention-weighted sum (0.756 Kappa). To explain this phenomenon, we observe the loss for the attention sum, showing large fluctuations and failing to converge, even after lowering the learning rate. Hence, we uniformly set the combining model to use the MLP for hybrid runs 2 and 3. Furthermore, we balance out the training set through augmentation, resulting in increased test accuracy.

As seen in Table 6, the addition of Block B's lexicon and NER features empirically improves performance for both Practice X and Y. The same can be observed during the inclusion of Block C, although Practice Y benefits more discernibly. Perhaps, the keywords extracted for Practice X's Block C are suboptimal because Practice X has less detailed scoring rubrics, thinner boundaries, and, naturally, increased grader bias. Nonetheless, it leads us to question the actual contribution of the hybrid modules. In [16], Wang found that the impact of the word-level attention module decreased when more training data was added and suggested that hybrid additions are supplementary rather than ameliorating. Our additional finding is that the Run 3 hybrid model is more effective in flagging adversarial responses. For example, when passing E2 with its corresponding feature vector into Block A + B + C, it was now correctly classified as a 0. 18 of 20 other irrelevant responses were also correctly flagged as score 0, while the best neural model only flagged 8.

4.6   Visualising Word-Level Attention

The word-level attention module is indicative of rubric adherence, and its vectorial contribution can be visualised, such that each response generates a separate heatmap per key phrase. We initially used all the words in the response when calculating attentional similarity, but the existence of stopwords and multiple variations of the same word proved visually distracting. Hence, we removed stopwords, used the lemmatized version of responses and set a word-pair attention threshold of 0.8 when generating the heatmaps. Examples of attentional similarity between words in key phrases and responses are shown in Fig. 2.

We can observe that the heatmap correctly identifies pairs of words that are highly related. For instance, the words "review" and "evaluation", "board" and "committee", "satisfactory" and "satisfy" are correctly associated. Surprisingly, it also identified more complex relationships, such as associating "Audit Com-

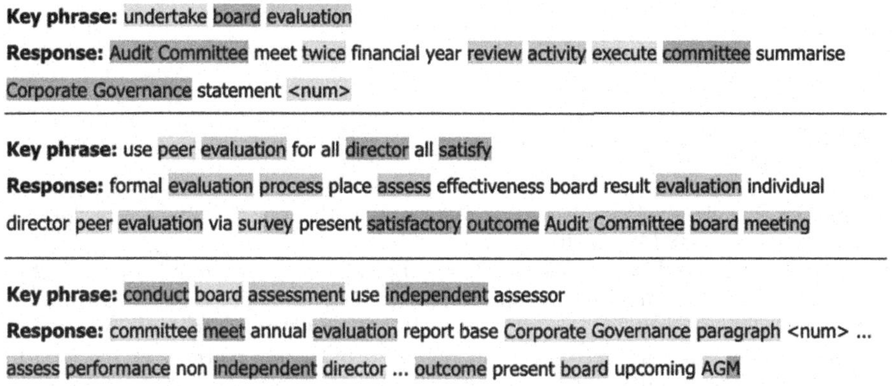

Key phrase: undertake board evaluation

Response: Audit Committee meet twice financial year review activity execute committee summarise Corporate Governance statement <num>

Key phrase: use peer evaluation for all director all satisfy

Response: formal evaluation process place assess effectiveness board result evaluation individual director peer evaluation via survey present satisfactory outcome Audit Committee board meeting

---

Key phrase: conduct board assessment use independent assessor

Response: committee meet annual evaluation report base Corporate Governance paragraph <num> ... assess performance non independent director ... outcome present board upcoming AGM

Fig. 2. Examples of attentional similarity heatmaps between key phrases and responses. Word pairs of the same colour display high attention.

mittee" and "board", where the committee was a subset of the board. "peer" and "survey" are also correctly associated with "evaluation".

However, since the basis of the attention module is GLoVE embeddings, words that co-occur frequently but possess no semantic meaning or deep contextual dependencies are wrongly highlighted, such as "outcome" and "directors" or "board" and "paragraph". As the embedding layer is finetuned, two words that appear within the vicinity of one another often enough, can be incorrectly associated. Furthermore, words used in different contexts also depict high attention, such as "independent" in "non-independent director" and "independent" in "independent assessors". Consequently, the attentional heatmap can, at times, become an inadequate representation of a response's actual rubric adherence.

The macroscopic deduction, as observed in Table 6, is that the word-level attention module is heavily dependent on the careful, experimental selection of key phrases since only Practice Y, which had comprehensive rubrics, benefitted markedly from its inclusion.

5 Conclusion

This paper proposed a novel industrial text scoring engine to automate the laborious adherence checking process of annual reports in regulatory bodies. We explored non-hierarchical and hierarchical LSTMs, the Longformer, hybrid models with lexical, NER and word-level attention features, with our best model achieving a Kappa of 0.9670. Our proposition is the ameliorating effect of hybrid models, both as a neural supplement and as a visual aid lending explainability.

Though the attention module is intuitive, we hope to improve performance by using embedding schemes that consider context instead of just acting on the word level. Furthermore, we hope to refine a feedback mechanism that allows one to identify exact shortcomings in each response based on the rubrics. Above all, our current corpus is limited to only two regulatory practices, so we hope to further validate our framework on more official documentation with regulatory rubrics, while also examining the nuances of the framework's performance on academic essays (ASAP-AES).

References

1. Beltagy, I., Peters, M.E., Cohan, A.: Longformer: the long-document transformer. CoRR abs/2004.05150 (2020)
2. Chen, Q., Zhu, X., Ling, Z., Wei, S., Jiang, H.: Enhancing and combining sequential and tree LSTM for natural language inference. CoRR abs/1609.06038 (2016)
3. Dasgupta, T., Naskar, A., Dey, L., Saha, R.: Augmenting textual qualitative features in deep convolution recurrent neural network for automatic essay scoring. In: NLP-TEA@ACL (2018)
4. De La Cruz, A., Medina, A., Tang, Y.: Owners of the world's listed companies. OECD Capital Market Series (2019)
5. Devlin, J., Chang, M., Lee, K., Toutanova, K.: BERT: pre-training of deep bidirectional transformers for language understanding. CoRR abs/1810.04805 (2018)
6. Dong, F., Zhang, Y., Yang, J.: Attention-based recurrent convolutional neural network for automatic essay scoring, pp. 153–162, August 2017
7. Gu, K., Budhkar, A.: A package for learning on tabular and text data with transformers. In: Proceedings of the Third Workshop on Multimodal Artificial Intelligence, pp. 69–73. Association for Computational Linguistics, June 2021
8. Kumar, V., Boulanger, D.: Explainable automated essay scoring: deep learning really has pedagogical value. Front. Educ. **5**, 186 (2020)
9. Mayfield, E., Black, A.W.: Should you fine-tune Bert for automated essay scoring? In: BEA (2020)
10. Page, E.B.: Project essay grade: Peg. J. Educ. Technol. (2003)
11. Pennington, J., Socher, R., Manning, C.: Glove: Global vectors for word representation, vol. 14, pp. 1532–1543 (2014)
12. Riordan, B., Horbach, A., Cahill, A., Zesch, T., Lee, C.M.: Investigating neural architectures for short answer scoring, pp. 159–168. Association for Computational Linguistics, September 2017
13. Shermis, M.D., Burstein, J.: Automated essay scoring: a cross-disciplinary perspective. In: Proceedings of the 2003 International Conference on Computational Linguistics, p. 13 (2003)
14. Taghipour, K., Ng, H.T.: A neural approach to automated essay scoring. In: Proceedings of the 2016 Conference on Empirical Methods in Natural Language Processing, pp. 1882–1891. Association for Computational Linguistics, November 2016
15. Uto, M., Xie, Y., Ueno, M.: Neural automated essay scoring incorporating handcrafted features. In: COLING (2020)
16. Wang, T., Inoue, N., Ouchi, H., Mizumoto, T., Inui, K.: Inject rubrics into short answer grading system. In: Proceedings of the 2nd Workshop on Deep Learning Approaches for Low-Resource NLP, pp. 175–182 (2019)
17. Wolf, T., et al.: Huggingface's transformers: State-of-the-art natural language processing. CoRR abs/1910.03771 (2019)
18. Yang, Z., Yang, D., Dyer, C., He, X., Smola, A., Hovy, E.: Hierarchical attention networks for document classification. pp. 1480–1489. Association for Computational Linguistics, June 2016
19. Zaheer, M., et al.: Big bird: transformers for longer sequences. CoRR abs/2007.14062 (2020)

Benchmarking Training Methodologies for Dense Neural Networks

Isaac Tonkin[✉][iD], Geoff Harris[iD], and Volodymyr Novykov[iD]

Centre for Data Analytics, Bond Business School, Bond University,
Robina, QLD 4227, Australia
itonkin@bond.edu.au

Abstract. Multi-Layer Perceptrons (MLP) trained using Back Propagation (BP) and Extreme Learning Machine (ELM) methodologies on highly non-linear, two-dimensional functions are compared and benchmarked. To ensure validity, identical numbers of trainable parameters were used for each approach. BP training combined with an MLP structure used many hidden layers, while ELM training can only be used on the Single Layer, Feed Forward (SLFF) neural network topology. For the same number of trainable parameters, ELM training was more efficient, using less time to train the network, while also being more effective in terms of the final value of the loss function.

Keywords: Extreme learning machine · Back Propagation · Neural networks · Non-linear function approximation

1 Introduction

The Multi-Layer Perceptron (MLP) is a classic Deep Neural Network topology that has been demonstrated, amongst many other AI actions, to be capable of learning any range of non-linear functions. Indeed, it has been cited many times that given enough nodes in the hidden layers, the MLP can approximate any universal function via Cybenko's theorem (Cybenko 1989). The MLP was a popular Neural Network in the 1980's but then gave way to the Support Vector Machine (SVM). An SVM can be trained very quickly and demonstrated superior predictive power over the MLP trained by the gradient Back Propagation (BP) methodology. However, with the successes in recent decades of Deep Learning (see, e.g., Emmert-Streib et al. (2020)), along with significant advances in computing hardware, research and commercial interest has returned to the uses, and thus training, of an MLP.

Another theoretical aspect of the MLP topology is that any number of hidden layers can be replaced by a single hidden layer. This was proved by Lippmann (1987) who was able to show that an MLP with a single hidden layer can, via the universal approximation theorem and with any compressing activation function, can learn any Borel measurable function to any desired non-zero error provided that the network has enough nodes in that hidden layer. Nonetheless it is also

© Springer Nature Switzerland AG 2022
H. Fujita et al. (Eds.): IEA/AIE 2022, LNAI 13343, pp. 707–713, 2022.
https://doi.org/10.1007/978-3-031-08530-7_59

well known (Mingard et al. 2019) that multiple hidden layered networks typically use a far fewer total number of hidden nodes than the single hidden layered MLP to reach the desired accuracy. Paradoxically the re-distribution of hidden nodes from a single hidden layer into many hidden layers is both empirically driven and often requires significantly more time to train despite the reduced number of network parameters.

The contribution to the literature by this research is an objective evaluation of both the efficiency and efficacy of training an MLP using BP versus a single hidden layer MLP (with the exact same number of trainable parameters) trained using the Extreme Learning Machine methodology popularized by Huang (see, for example, Huang et al. (2006)). This paper is laid out as follows. In the next section we provide a short review of the literature on academic attempts to compare BP to other training methodologies. Following this section is presented a set of well-known non-linear functions, often used for benchmarking algorithms. The results of both the efficacy and efficiency of the different training methods are then compared.

2   Literature Overview

The MLP has been demonstrated to have value in computationally complex applications such as the N-body gravitational problem (Breen et al. 2020) and the solution of stochastic differential equations (Namadchian and Ramezani, 2020). Breen et al. (2020) had significant success with the classic n-body problem, training a deep MLP (consisting of nearly 150,000 trainable parameters arranged into a ten hidden layer MLP) on a defined data subset of the chaotic three-body problem. This MLP was then used to accurately calculate this non-linear function in other regions of its phase space. The mathematical foundation for this approach is given by Hornik (1991), who showed that artificial neural networks (ANNs) can be used to approximate, with any given degree of precision, any continuous function that contains dependent and independent variables related through this function.

In addition, training an MLP to approximate a non-linear function has been investigated, showcasing one of the well-known uses of this ANN structure. These include Trenn (2008), Pinkus (1999) and Namadchian and Ramezani (2020). Such studies typically cover the effectiveness and efficiency sides of the problem, demonstrating how well a network can approximate a non-linear function together with indicated topology choices. Trenn (2008) is an example of the mathematical derivation of topology choices for a network. Huang et al. (2006) applied a different approach and created a SLFF neural network to approximate a non-linear function using an ELM methodology.

Despite attempts to compare the effectiveness and efficiency of an MLP function approximation to other types of neural networks (Zainuddin and Ong 2008), a thorough review of Scopus, Google Scholar and other academic search engines identified no other academic works comparing effectiveness and efficiency of the two distinct training methodologies, backpropagation, and ELM, in this task.

ELM-trained neural networks have seen recent success across a wide variety of applications (see e.g., Wang et al. (2021)), however, are yet to be applied directly to the non-linear function approximation task.

The present paper contributes to this literature by using a comparative approach, examining these claims by empirically testing the approximation capabilities of a subset of neural network architectures. Given independently chosen functions, the focus is on single-layer feed-forward neural networks and multi-layer perceptrons. Two distinct methodologies, ELM and TensorFlow- based BP, are used for the design, training, and testing of these fully connected MLP architectures. Additional comparison is drawn between the two main training methodologies and a naïve benchmark, namely, the k-nearest neighbours algorithm.

3   Choice of Non-linear Function and Empirical Data Generation

To enable a legitimately strong challenge, the first two optimization functions listed in the Wikipedia benchmark functions were chosen for this initial empirical study (the Rastrigin and Ackley functions). Despite being trivial for a human, these functions are extremely difficult for algorithms to locate the global optimum value. Importantly, the input dimensionality of this function is completely variable from 1 dimension through to as many dimensions as are required for testing purposes. The equations of the Rastrigin and Ackley functions are as follows:

$$f(x, y) = 20 + [x^2 - 10 \times \cos(2\pi x)] + [y^2 - 10 \times \cos(2\pi y)] \tag{1}$$

$$f(x, y) = -20e^{-0.2\sqrt{0.5(x^2+y^2)}} - e^{0.5(\cos(2\pi x)+\cos(2\pi y))} + e + 20 \tag{2}$$

These functions are highly non-linear in a way that represents a clear challenge to any Neural Network modelling approach. The functions are analytic and thus it is easy to generate high accuracy datasets with insignificant roundoff error. It is important to note that there was no white noise added to for the current study. This paper evaluates different training methodologies, both in terms of time taken to train and the precision that training can attain. The impact of a normally distributed error term is a complication for future study.

To this end only training and holdout datasets were generated. To ensure no turning points were directly mapped into the training sets, an evenly spaced grid on the plane $-5.12 < x < 5.12$ and $-5.12 < y < 5.12$ was chosen as the basis domain for both functions for the purposes of this study and digitized with 500 divisions in each dimension resulting in 250,000 training vectors. In the same manner, an evenly spaced grid was used with 250 divisions to create a set of 62,500 additional data vectors as the holdout validation data.

4   Neural Network Models and Training Methodologies

The training of an MLP, as well as the entire field of deep learning, relies heavily on BP methodology to update their initial random weights towards values that result in the Loss Function (typically the MSE for regression) being minimized. The BP algorithm does this by calculating this loss and then backwardly feeding the loss through the network using calculus to calculate adjustments to each weight based upon minimizing the entire Loss Function. The process is repeated until either the changes to each parameter (weight) are negligible or the MSE (Loss Function) is within acceptable bounds. This method dominates the academic literature in terms of the most widely used and accepted method for training the parameters within a Neural Network. It is common practice to utilize the efficiency of the now mature Keras libraries and utilities, incorporated with TensorFlow, to do the BP training. In this work we leverage that technology for training the MLP parameters.

The ELM training methodology is significantly easier to understand than the BP method for MLPs described above. It first randomly generates weights for the parameters between the input layer and the hidden layer. This subset of non-trainable parameters are then represented by a matrix, $\mathbf{W}$, and the bias nodes denoted $\mathbf{B}$. Using a non-linear activation function, such as $g(x) = tanh(x)$ and similarly denoting the matrix $\mathbf{V}$ for the final set of weights (the only trainable parameters in the ELM methodology) allows the mechanics of these MLP neural networks to be expressed in the notation of Linear Algebra succinctly as follows:

$$\mathbf{Y} = \mathbf{V} \times \mathbf{H} \tag{3}$$

where $\{\mathbf{X}, \mathbf{Y}\}$ are the input and output vectors of the independent and dependent variable observations respectively. Once expressed in this manner, and using the Universal Approximation Theorem (see, e.g., Huang and Chen (2007) and Hornik (1991)), the trainable parameters (the elements of just $\mathbf{V}$) can be solved in a single pass using the generalized Moore-Penrose inverse where $\mathbf{H} = g(\mathbf{W} \cdot \mathbf{X})$:

$$\mathbf{V} = \mathbf{Y} \times \mathbf{H}^{-1} \tag{4}$$

The salient point is that, unlike BP, the parameters of the matrices are calculated in a single pass with no iterative updating required thereafter. The use of this training method is gaining in popularity (Wang et al., 2021) due to its significantly reduced training time versus BP. In this work we use the ELM method for training the SLFF networks.

The mathematical algorithm for the ELM training method (given above) was implemented directly into Python using the NumPy library for all linear algebra calculations. The importance of benchmarking is noted by Nanda et al. (2016), and as such, the k-nearest neighbor (kNN) shallow technique was chosen as an appropriate naive benchmark. As is standard, we adopt just $k = 1$ as the naive "closest is good enough" baseline. Prediction systems unable to generate more accurate results than this are not worth any computational effort (Nanda et al. (2016)). Using this approach, the baseline metrics for the Rastrigin and Ackley functions were 0.142 and 0.0007, respectively.

5   Results and Discussion

The output for the ELM-trained SLFF and the BP-trained MLP are presented in Table 1 below. These results were obtained using the 64-core AMD Ryzen Threadripper 3990X.

Table 1. Summary of ELM and MLP results for the Rastrigin and Ackley function

The Rastrigin Function				The Ackley Function			
Time taken (secs)		MSE		Time taken (secs)		MSE	
nH	ELM	MLP	ELM	MLP	nH	ELM	MLP	ELM	MLP
681	10	604	6.030	41.95	681	18	467	0.0331	0.0060
2513	76	391	0.166	23.22	2513	93	567	0.0107	0.0012
5497	247	614	0.029	0.12	5497	348	827	0.0091	0.0006
9633	902	>2000	0.021	99.00	9633	1032	921	0.0090	0.0005
14921	1852	3500	0.017	207.00	14921	2508	1900	0.0090	0.0008

From Table 1 the ELM training method is far more efficient than the BP method (TensorFlow). Once the number of trainable parameters increases significantly, however, the ELM method increasingly takes more time. At this research stage this is thought to be due to the quadratic increase in the sizes of the Generalized Matrix inverse needing to be computed. Perhaps most interesting is that the ELM training is significantly and consistently more effective at learning the Rastrigin function than using BP in TensorFlow, albeit with human-supervised management of the training. The MSE drops faster with the ELM method and consistently gets to an order of magnitude more accurate than the BP-trained MLP for the same number of trainable parameters. In addition, the ELM methodology was able to outperform the kNN naive benchmark when the number of trainable parameters was increased. This outperformance justifies the additional computation required to train the network. Meanwhile, we find that the ELM method is slightly outperformed by the BP-trained MLP for the Ackley function. This may be due to the relative smoothness of this function compared to the Rastrigin function.

We also chose not to include additional features or layers in the MLP such as dropout. Dropout is a regularization technique used to help prevent overfitting. Such features were excluded to ensure that our comparative analysis was not biased. It is noted that overfitting is not a significant issue for these particular non-linear functions and certainly in the above results.

6   Conclusion and Future Research

The efficacy and efficiency of the ELM training method for learning highly non-linear functions has been demonstrated as superior to the Back Propagation training method for certain applications. It also highlights a situation where an MLP trained via BP underperforms a SLFF network trained via the ELM methodology. In contrast to the BP approach, which requires packages such as TensorFlow, a complete ELM framework was shown to be implementable in a few lines of Python/NumPy code. Further, mathematical libraries allow for ELMs to be trained on GPUs, which can significantly reduce training time.

Future research will investigate a variety of highly non-linear functions to determine if the finding of this study holds true for other highly non-linear analytic functions. Naturally, this would also include investigation of what factors limit the performance of MLPs. Where relevant, the inclusion of derivatives as inputs to networks approximating highly non-linear functions would also be an interesting addition. Further, examination of other such cases where BP underperforms the ELM methodology may provide insight for improving network design and more clearly defining optimal use-cases for different network types and training methodologies. We also suggest that the ELM training methodology be implemented into a greater number of supervised learning works so as to provide additional evidence of the efficacy and efficiency of various neural network training methodologies.

References

Breen, P., Foley, C., Boekholt, T., Zwart, S.: Newton versus the machine: solving the chaotic three-body problem using deep neural networks. Mon. Not. R. Astron. Soc. **494**(2), 2465–2470 (2020). https://doi.org/10.1093/MNRAS/STAA713

Cybenko, G.: Approximation by superpositions of a sigmoidal function. Math. Control Sig. Syst. **2**(4), 303–314 (1989). https://doi.org/10.1007/BF02551274

Emmert-Streib, F., Yang, Z., Feng, H., Tripathi, S., Dehmer, M.: An introductory review of deep learning for prediction models with big data. Front. Artif. Intell. **3** (2020). https://doi.org/10.3389/frai.2020.00004

Hornik, K.: Approximation capabilities of multilayer feedforward networks. Neural Netw. **4**(2), 251–257 (1991). ISSN 0893-6080, https://doi.org/10.1016/0893-6080(91)90009-T

Huang, G.B., Chen, L.: Convex incremental extreme learning machine. Neurocomputing **70**(16–18), 3056–3062 (2007). https://doi.org/10.1016/j.neucom.2007.02.009

Huang, G.B., Zhu, Q.Y., Siew, C.K.: Extreme learning machine: theory and applications. Neurocomputing **70**(1–3), 489–501 (2006). https://doi.org/10.1016/j.neucom.2005.12.126

Lippmann, R.: An introduction to computing with neural nets. IEEE ASSP Mag. **4**(2), 4–22 (1987). https://doi.org/10.1109/MASSP.1987.1165576

Mingard, C., Skalse, J., Valle-Pérez, G., Martínez-Rubio, D., Mikulik, V., Louis, A.A.: Neural networks are a priori biased towards Boolean functions with low entropy. arXiv preprint (2019). https://doi.org/10.48550/arxiv.1909.11522

Namadchian, A., Ramezani, M.: Analytical solution of stochastic differential equation by multilayer perceptron neural network approximation of Fokker-Planck equation. Numer. Methods Partial Differ. Equations **36**(3), 637–653 (2020). https://doi.org/10.1002/num.22445

Nanda, U., Rajput, S., Agrawal, H., Goel, A., Gurnani, M.: On context awareness and analysis of various classification algorithms. Adv. Intell. Syst. Comput. **381**, 175–181 (2016). https://doi.org/10.1007/978-81-322-2526-3_19

Pinkus, A.: Approximation theory of the MLP model in neural networks. Acta Numer. **8**, 143–195 (1999). https://doi.org/10.1017/S0962492900002919

Trenn, S.: Multilayer perceptrons: approximation order and necessary number of hidden units. IEEE Trans. Neural Netw. **19**(5), 836–844 (2008). https://doi.org/10.1109/TNN.2007.912306

Wang, J., Lu, S., Wang, S.-H., Zhang, Y.-D.: A review on extreme learning machine. Multimedia Tools Appl. 1–50 (2021). https://doi.org/10.1007/s11042-021-11007-7

Zainuddin, Z., Ong, P.: Function approximation using artificial neural networks. WSEAS Trans. Math. **7**(6), 333–338 (2008)

Proposing Novel High-Performance Compounds by Nested VAEs Trained Independently on Different Datasets

Yoshihiro Osakabe[(✉)] [iD] and Akinori Asahara [iD]

Hitachi, Ltd. Research and Development Group, Tokyo, Japan
`yoshihiro.osakabe.fj@hitachi.com`

Abstract. Materials informatics (MI), which uses artificial intelligence
and data analysis techniques to improve the efficiency of materials development, is attracting interests from industry. One of its main applications is the rapid development of novel high-performance compounds.
Recently, several deep generative models have been utilized to suggest
candidate compounds that are expected to satisfy the desired performance. However, it is practically difficult to obtain sufficient amount of
experimental data to train such models. Thus, the authors proposed a
deep generative model with nested two variational autoencoders (VAEs).
The outer VAE learns the structural features of compounds using large-
scale public data, while the inner VAE learns the relationship between
the latent variables of the outer VAE and the properties from small-scale
experimental data. To generate high performance compounds beyond
the range of the training data, the authors also proposed a loss function
that amplifies the correlation between a component of latent variables of
the inner VAE and material properties. The results indicated that this
loss function contributes to improve the probability of generating high-
performance candidates. Furthermore, as a result of verification test with
an actual customer in chemical industry, it was confirmed that the proposed method is effective in reducing the number of experiments to 1/4
compared to a conventional method.

Keywords: Variational Autoencoder · Deep generative model ·
Automatic compounds design · Materials informatics

1 Introduction

Technological innovation in the materials industry is important for a wide range
of related manufacturing industries, and materials informatics (MI) that utilizes
AI and data analytics technology has attracted much interest due to recent
advances. In particular, MI is expected to make it possible to find new materials
that meet the required performance in a shorter period of time than before by
utilizing past experimental data and numerical simulation data.

© Springer Nature Switzerland AG 2022
H. Fujita et al. (Eds.): IEA/AIE 2022, LNAI 13343, pp. 714–722, 2022.
https://doi.org/10.1007/978-3-031-08530-7_60

For example, in the development of organic materials, the virtual screening method is used to efficiently extract candidate compounds to be tested. In this method, a machine learning model trained with experimental data is used to screen a large number of chemical formulas by predicting their properties. If the prediction accuracy is high, unnecessary experiments can be avoided and the number of experiments can be reduced. However, there are two problems with this method.

The first problem is the extrapolability of the prediction. Machine learning models commonly used in virtual screening methods are interpolative, which can only make effective predictions within the range of the training data. However, the task of searching for new materials that outperform known materials is a matter of extrapolability. As a result, promising compounds that were initially expected to be discovered may be sifted out.

The second problem is the quality of the screening targets. When screening a list of compounds irrespective of their performance, it is difficult to shorten the time to discover new materials even if the prediction accuracy is high. Conventionally, the "structure generation process," which prepares compounds to be screened, has been the bottleneck. For example, BRICS [2] method, which generates structures by mechanically combining substructures of known compounds, is inefficient because it generates a large number of chemical structures that cannot exist in nature.

Recently, various deep generative models have been proposed to more directly generate compounds with high probability of performance improvement. Gómez-Bombarelli et al. proposed ChemicalVAE [4], which uses a variational autoencoder (VAE) to obtain a continuous representation from a chemical formula expression SMILES [11] given in string form. In general, the latent representation of a VAE is represented by a real-valued vector, which is called a latent vector. ChemicalVAE has a separate neural network that predicts the properties of compounds from the latent vectors of VAE, and the output of this predictor is also used for training the VAE. With this feature, the latent space of ChemicalVAE reflects both the similarity of chemical structures and the similarity of property values. In addition to being able to generate new chemical formulas by specifying arbitrary latent vectors, ChemicalVAE has been shown to be able to discover compounds with optimal properties by exploring the latent space. However, deep generative models such as VAE commonly require a huge amount of training data to achieve sufficient performance. Practically, it is difficult to collect a large amount of experimental data in which chemical formulas are paired with property values. In fact, it is often possible to accumulate only a hundred to a thousand data at most in practical cases, and a method that can be effective even with a small amount of experimental data.

Therefore, Osakabe et al. proposed a deep generative model, called Mat-VAE, which consists of two nested VAEs independently trained on different datasets [10]. The first (outer) VAE, which is trained on a huge open dataset of SMILES, is a universal generator of chemical structural formulae. The second (inner) VAE, which is trained on a small experimental dataset, learns the structure-property relation. This nested structure allows us to generate candidate compounds with high probability of having the desired property values. Their previous work indicates that the model can generate more than five times as many compounds with improved property compared to ChamicalVAE in the case of the experimental data is limited to 1000 records [10]. Their network structure and the learning method are designed to perform even with small amounts of experimental data. However, in the generation process, MatVAE are only searching the latent space for the neighborhood of the best performing compound among the known compounds. For further improvement in efficiency, more direct and extrapolative approaches are required to generate candidates with more desirable performance.

In this study, we propose a method that can generate high-performance compounds more directly than nearest neighbor search in latent space. Specifically, we introduce a loss function into MatVAE such that a given component of the latent vector is strongly correlated with a property value, thereby creating a linear dimension in the latent space with respect to the property value. If the ideal latent space can be obtained by learning, it is expected that the properties of the generated compound can be adjusted according to the bias to the correlated component.

2   Related Works: Deep Generative Models for Chemical Design

Deep neural networks (DNNs) have outperformed other methods in various fields on regression and classification tasks. DNN have been also utilized to generate a novel sample similar to the training data, i.e., a deep generative model (DGM). DGM assumes that the observed data x is generated from an unobserved latent variable z and aims to learn a transformation rule $p(x|z)$.

DGM has been extensively studied to directly obtain chemical structures that have desirable properties. Previously, methods for training a generative adversarial network (GAN) with the reinforcement learning (RL) framework have been reported [1,9]. The major difference between them is the representation of the compounds; REINVENT [9] and MolGAN [1] use text-based and graph-based representations, respectively. Another approach uses variational autoencoder (VAE) models; for example, JT-VAE using graph-based representation [6] and ChemicalVAE using text-based representation [4]. With VAEs, the chemical structure can be directly obtained by specifying one point in its latent space. Studies have shown that it is possible to optimize the property values by searching the latent space because of the continuity of the space. However, it is important to note that the previous studies required huge training datasets. ChemicalVAE and REINVENT were trained using supervised learning with 250,000 and 350,000 compound data extracted from the ZINC database [5], respectively. In most cases, when developing industrial chemical products, only a few hundred or thousand supervised training data, i.e., data with property values measured by experiments, are available for a single product family.

3 Proposal: MatVAE

3.1 Methodological Overview

In order to realize a DGM that generates chemical formulas as candidate compounds for actual experiments, VAE has to learn two objectives. The first is the characteristics of chemical structures that can actually exist. When the training data is a representation of a compound in string form, such as SMILES [11], the main task is to learn the pattern of elemental symbol sequences. The second is the relationship between the structural features of the compound and its properties. It is considered that trying to acquire these two learning objectives in a single network causes the unnecessary requirement of a large amount of experimental data for training. Thus, in the proposed method, VAEs are trained independently for each of these two learning objectives.

Figure 1 shows a schematic diagram of the proposed model MatVAE. As described earlier, it consists of two VAEs. The outer network (M^{out}) is a VAE that acquires structural features of compounds as latent expressions based on a large number of chemical formulas obtained from open data. Its encoder, M_{enc}^{out}, consists of several 1D convolutional layers and several fully connected layers.

In this study, SMILES [11] is used for representation of molecules. One SMILES representation becomes an array H of $d \times M \times N$ dimensions, where d is a batch size, M is a number of chemical symbols, and N is the max length of the SMILES string, respectively. Accordingly, M_{enc}^{out} takes H as its input, and outputs a P-dimensional latent vector, Z^{out}. The decoder M_{dec}^{out} consists of several fully connected layers and Gated Recurrent Unit (GRU) layers, which inversely convert the P-dimensional vectors into H.

The inner network (M^{in}) takes the $(P+Q)$-dimensional vector concatenated with Z^{out} and Q-dimensional vector representing experimental conditions such as the temperature, amount of compounding, and catalyst during synthesis. The encoder M_{enc}^{in} and the decoder M_{dec}^{in} both consist of several fully connected layers, and its latent variable Z^{in} is an R-dimensional vector where $R < P + Q$.

As shown in Fig. 1, the training process is divided into two steps, and M^{out} is trained first. The training data is a dataset of tens or hundreds of thousands of chemical formulae curated from open data, ZINC [5]. A common loss function L_{recon} for VAE [7] is applied so that M_{dec}^{out} outputs the same array H that is input to M_{enc}^{out}, and the weights and parameters of M^{out} are optimized.

Next, M^{in} is trained with about a hundred or a thousand experimental data. As same as the training of M^{out}, the normal VAE loss L_{recon} is used for training M^{in}. In this study, we propose an additional loss function L_{corr} such that a predetermined component of Z^{in} is strongly correlated with the property value as described in the next section.

3.2 Additional Loss Function L_{corr} to Amplify Correlation Between Latent Component and Property

In many cases of material development, the goal is to discover a new compound that outperforms the existing performance for a certain property. In other words,

Step 1: Training outer VAE M^{out} with a huge open data

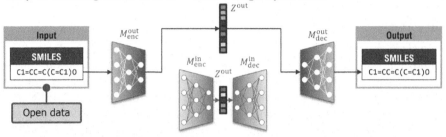

Step2: Training inner VAE M^{in} with a small experimental data

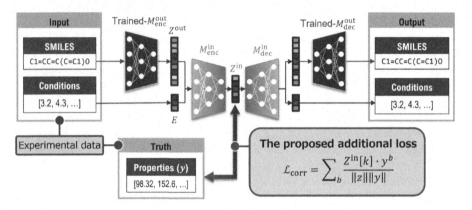

Fig. 1. Schematic diagram of the proposed model, MatVAE. The model consists of two VAEs, and they are trained independently in two steps.

Table 1. The details of the training datasets

Role	Contains	# of records	Conditions
open data	SMILES	500,000	None
experimental data	SMILES with SAS	1,000	$2.5 < S < 4.5$

we want to propose a compound that exists in the extrapolation region of the known data for the target property value. Therefore, we designed the loss function L_{corr} so that the predefined k-th component $Z^{in}[k]$ of Z^{in}, where k is an integer such that $0 \leq k < R$, has a strong correlation with property values. For instance, L_{corr} can be defined by following the cosine similarity between the b-th properties y^b and the $Z^{in}[k]$, such as

$$L_{corr} = \sum_b \frac{z[k] \cdot y^b}{\|z\|\|y\|}. \tag{1}$$

The correspondence between b and k, in other words, which property and components are correlated, is determined beforehand. The whole loss function L for M^{in} is given as $L = L_{recon} + \alpha L_{corr}$, where α is a hyperparameter that takes negative value because L_{corr} will becomes large if the training succeeds.

The combination of the trained M^{out} and M^{in} is used as a generative model for automatic chemical design. Due to the collaboration of the two networks, the latent space Z^{in} is a continuous representation in which the similarity of chemical structures and the similarity of properties are simultaneously reflected as distances. Furthermore, due to the effect of L_{corr}, $Z^{in}[k]$ is correlated with the objective property. Hence, if the ideal learning is achieved, $Z^{in}[k]$ can be used as a parameter for candidate generation. In other words, if we encode a chemical formula that has particularly high performance among known compounds, then add a certain bias to its $Z^{in}[k]$, and finally decode the latent vector Z^{in}, we can directly obtain a chemical formula for experimental candidates with high accuracy of performance improvement. This method would allow us to obtain the latent vector that gives improved compounds more directly than the conventional neighborhood sampling in the latent space.

4   Evaluation of the Proposed Loss, L_{corr}

To verify the effectiveness of the proposed loss function L_{corr}, the following experiments are conducted. In principle, the performance of the model should be evaluated through actual chemical experiments. However, in this study, we prioritize the investigation of the informatics aspect of the proposed method and conduct experiments by treating the Synthetic Accessibility Score (SAS) [3] as a substitute for actual properties. SAS, S, is an index value that quantifies the difficulty of synthesis as a real number between $S = 1$ (easy to fabricate) and $S = 1$ (difficult to fabricate), and can be calculated only from SMILES using the open source software, RDkit [8].

All the data used in the experiments were randomly selected data sets from an open database of commercial compounds called ZINC [5]. We prepared two types of the training data, one to be considered as open data and the other as experimental data. The details of the datasets are shown in Table 1.

To confirm the effect of the proposed loss function L_{corr}, we examined the correlation coefficients between each component of the latent variable $Z^{in}[k]$ and SAS. The correlation coefficients calculated for 1000 compounds included in the experimental dataset are shown in Fig. 2(a). It can be seen that $Z^{in}[0]$ has strongly correlated with SAS in the case where L_{corr} is applied compared with in the case where L_{corr} is not applied. For the rest of the components ($k > 0$), it can be also confirmed that the magnitude of the correlation coefficient does not affected by the usage of L_{corr}. In fact, as shown in Fig. 2(b) and (c), the linear relationship that SAS increases along with the magnitude of $Z^{in}[0]$ is caused by L_{corr}. Therefore, it is clear that L_{corr} is indeed functioning as intended.

Next, we investigated whether this $Z^{in}[0]$ can actually be utilized as a parameter for the generation of improved candidate compounds. We applied positive

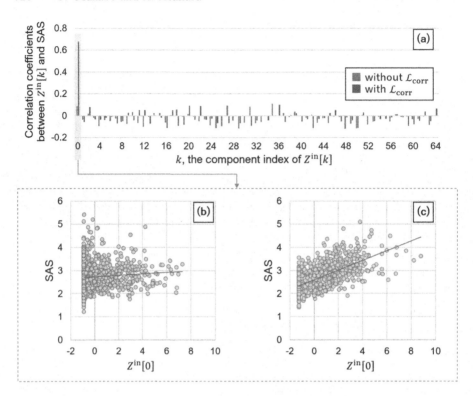

Fig. 2. The experimental results of verifying the effect of the loss function L_{corr}. (a) Correlation coefficients between z for SAS for each component of the latent variable. In the case of (b) without and (c) with L_{corr}, the scatter plots of $Z^{in}[k]$ and SAS.

biases to the latent vector $Z^{in}[0]$ obtained by encoding a known compounds, and then calculated the SAS of the decoded SMILES from the biased $Z^{in}[0]$. It is considered a success when the SAS of generated compounds becomes larger than the original SAS when a positive bias is added. Figure 3(a) shows a plot of such success rate after 1000 trials. It was demonstrated that the SAS of the generated candidate can be selectively modulated depending on the magnitude of the applied bias.

Finally, we investigated whether the proposed method can produce compounds that exceed the existing properties, which is the ultimate goal of material development. The prepared experimental data was filtered so that their SAS is in the range of $2.5 < S < 4.5$. Therefore, if the SAS of generated compounds exceeds this range, it can be understood that the proposed model could generate compounds that exist in the extrapolation region. Figure 3(b) shows the probability of generating compounds having $S > 4.5$ when 1000 compounds were generated by adding a positive bias to $Z^{in}[0]$. This task of proposing a compound with a larger SAS can be understood as a task of generating the structure of a more complex compound, which can only be successful if the relationship

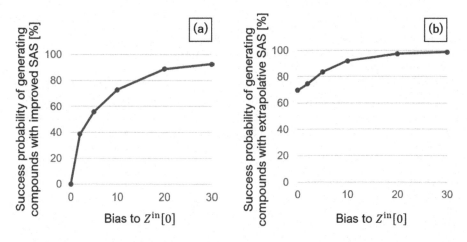

Fig. 3. Success rates of generating improved compounds by controlling the bias to $Z^{in}[0]$. (a) The probability of successfully generating compounds that exceed the SAS of the original compounds, relative to the magnitude of the applied bias. (b) The probability of successfully generating compounds that exceed the SAS of the learned compounds in training datasets, relative to the magnitude of the applied bias.

between structure and properties is properly acquired. The experimental results show that the model is successful in proposing compounds with extrapolatory SAS with high probability. In fact, the model outputs longer and more diverse SMILES to the extent that they satisfy the grammar rules.

These results shown in this section indicate that the proposed loss function L_{corr} works as expected to strongly correlate the predetermined latent components with the objective properties, and with the proposed model, the properties of the generated compounds can be controlled by adding bias to the correlated components. If the quality of the proposed compound is thus improved, the number of experiments required for the discovery of new materials can be reduced. Indeed, we applied the improved MatVAE and investigated its effectiveness on the past experimental data provided by the actual manufacturer of the chemical industry. Although the detailed results are not disclosed due to the confidential information of the target material data, the results show that the number of experiments can be reduced by a quarter compared to the conventional virtual screening method using the XGBoost (XGB) model.

5 Conclusion

In this study, we investigated a deep generative model to shorten the material development time by proposing compounds with high probability of performance improvement as experimental candidates. This study improved the deep generative model, called MatVAE, which has a nested structure of two VAEs trained with different data set and learning objectives, and verified its performance. In particular, the loss function L_{corr} was proposed to form the latent space strongly correlated with the target property, which enabled us to generate candidate compounds with desirable properties more directly than the neighbor search in the latent space. The validation using open data confirmed the validity of using a predefined component $z^{in}[k]$ of the latent vector, which is strongly correlated with the property values to be improved via L_{corr}, as a parameter for candidate generation. In fact, another verification using past experimental data also confirmed that the method can produce high performance compounds, and it was confirmed that the method is effective in reducing the number of experiments required for the discovery of high performance materials by a 1/4 compared to the virtual screening method based on XGB model.

References

1. De Cao, N., Kipf, T.: MolGAN: an implicit generative model for small molecular graphs (2018). http://arxiv.org/abs/1805.11973
2. Degen, J., Wegscheid-Gerlach, C., Zaliani, A., Rarey, M.: On the art of compiling and using 'drug-like' chemical fragment spaces. ChemMedChem 3(10), 1503–1507 (2008). https://doi.org/10.1002/cmdc.200800178, http://doi.wiley.com/10.1002/cmdc.200800178
3. Ertl, P., Schuffenhauer, A.: Estimation of synthetic accessibility score of drug-like molecules based on molecular complexity and fragment contributions. J. Cheminformatics 1(1), 8 (2009). https://doi.org/10.1186/1758-2946-1-8, https://jcheminf.biomedcentral.com/articles/10.1186/1758-2946-1-8
4. Gómez-Bombarelli, R., et al.: Automatic Chemical Design Using a Data-Driven Continuous Representation of Molecules. ACS Central Sci. 4(2), 268–276 (2018). https://doi.org/10.1021/acscentsci.7b00572, https://pubs.acs.org/doi/10.1021/acscentsci.7b00572
5. Irwin, J.J., Sterling, T., Mysinger, M.M., Bolstad, E.S., Coleman, R.G.: ZINC: a free tool to discover chemistry for biology. J. Chem. Inf. Model. 52(7), 1757–1768 (2012). https://doi.org/10.1021/ci3001277, https://pubs.acs.org/doi/10.1021/ci3001277
6. Jin, W., Barzilay, R., Jaakkola, T.: Junction tree variational autoencoder for molecular graph generation. In: 35th International Conference on Machine Learning, ICML 2018 5, pp. 3632–3648 (2018)
7. Kingma, D.P., Welling, M.: Auto-Encoding Variational Bayes, December 2013. https://arxiv.org/abs/1312.6114
8. Landrum, G.: Rdkit: Open-source cheminformatics (2013). https://rdkit.org
9. Olivecrona, M., Blaschke, T., Engkvist, O., Chen, H.: Molecular de-novo design through deep reinforcement learning. J. Cheminform. 9(1) (2017). https://doi.org/10.1186/s13321-017-0235-x
10. Osakabe, Y., Asahara, A.: Matvae: Independently trained nested variational autoencoder for generating chemical structural formula. In: The AAAI 2021 Spring Symposium on Combining Artificial Intelligence and Machine Learning with Physical Sciences, vol. 2964. AAAI (2021). http://ceur-ws.org/Vol-2964
11. Weininger, D.: SMILES, a Chemical Language and Information System: 1: Introduction to Methodology and Encoding Rules. J. Chem. Inf. Comput. Sci. 28(1), 31–36 (1988). https://doi.org/10.1021/ci00057a005

Clustering

Monotonic Constrained Clustering: A First Approach

Germán González-Almagro[1,2(✉)], Pablo Sánchez Bermejo[2],
Juan Luis Suarez[1,2], José-Ramón Cano[3], and Salvador García[1,2]

[1] DaSCI Andalusian Institute of Data Science and Computational Intelligence,
University of Granada, Granada, Spain
`germangalmagro@ugr.es`
[2] Department of Computer Science and Artificial Intelligence (DECSAI),
University of Granada, Granada, Spain
[3] Department of Computer Science, EPS of Linares, University of Jaén, Campus
Científico Tecnológico de Linares, Cinturón Sur S/N, 23700 Linares, Jaén, Spain

Abstract. Clustering has been proven to produce better results when applied to learning problems that fall under semi-supervised paradigms, where only incomplete or partial information about the dataset is available to perform clustering. Classic constrained clustering and recent monotonic clustering problems belong to the semi-supervised learning paradigm, although a combination of both has never been addressed. This study aims to prove that the fusion of the background knowledge leveraged by the two aforementioned semi-supervised clustering techniques results in improved performance. To do so, a hybrid objective function combining them is proposed and optimized by means of an expectation minimization scheme. The capabilities of the proposed method are tested in a wide variety of datasets with incremental levels of background knowledge and compared to purely monotonic clustering and purely constrained clustering methods belonging to the state-of-the-art. Bayesian statistical testing is used to validate the obtained results.

Keywords: Pairwise instance-level constraints · Monotonicity constraints · Expectation-minimization · Hybrid objective function

1 Introduction

Clustering constitutes a key research area in the unsupervised learning framework, where no information on how data should be handled is available. It can be viewed as the task of grouping instances from a dataset into groups (or clusters), with the aim to extract new information from them. Background knowledge can be incorporated into the classic clustering framework, thus reframing it into the semi-supervised learning paradigm, where partial or incomplete information about the dataset is given to the clustering technique.

Our work has been supported by the research projects PID2020-119478GB-I00, A-TIC-434-UGR20 and PREDOC_01648.

H. Fujita et al. (Eds.): IEA/AIE 2022, LNAI 13343, pp. 725–736, 2022.
https://doi.org/10.1007/978-3-031-08530-7_61

When this additional information is given in the form of constraints, the constrained clustering problem arises. Constraints can be understood in three main ways: cluster-level [2], pairwise instance-level (or simply pairwise) [6] and feature-level constrained clustering [14]. From among these types, this study focuses on pairwise constraints. These constraints tell us whether two specific instances of a dataset must be placed in the same or in different clusters, resulting in Must-link (ML) and Cannot-link (CL) constraints, respectively. Pairwise constraints can be enforced in two ways: hard [17] and soft [11] constraints. The former must necessarily be satisfied in the output partition of any algorithm making use of them, while the latter are interpreted as strong suggestions by the algorithm but can be only partially satisfied in the output partition.

Recently, a new type of background knowledge coming from the supervised learning paradigm has been integrated into the clustering process. Monotonic classification is a particular case of supervised learning where classes are a set of ordered categories and classification models must respect monotonicity constraints among instances based on their descriptive features. This means that, if an instance a has greater feature values than those of instance b, its assigned class must also be higher (later) in the ordering than that of b [13]. Monotonicity constraints are a type of background knowledge that can be used to produce more accurate predictive models [3]. The monotonicity information can be incorporated into the clustering problem by biasing the solution space towards partitions satisfying monotonicity to a variable extent.

This study addresses the combination of the two types of background knowledge described above: pairwise constraints and monotonicity constraints. To the best of our knowledge, there are no previous studies on this topic, as monotonic clustering has emerged very recently. In this study the viability of monotonic constrained clustering (MCC) is proved, and an expectation-minimization (EM) scheme is proposed to optimize a hybrid objective function combining both monotonicity and pairwise constraints. The proposed hybrid objective function is composed of a monotonic distance metric and a pairwise constraint-based penalty term.

The rest of this study is organized as follows: background knowledge concerning classic clustering, pairwise constrained clustering and monotonic clustering is presented in Sect. 2, whose content is later used in Sect. 3 to introduce the proposed MCC method. After presenting the experimental setup used to carry out our experiments in Sect. 4, Sects. 5 and 6 presented and analyzed the experimental results obtained with the proposed method. Lastly, our conclusions are discussed in Sect. 7.

2 Background

As already mentioned, partitional clustering is the task of grouping instances of a dataset into k clusters. A dataset $X = \{x_1, \cdots, x_n\}$ contains n instances, each one described by u features. The ith instance from X is noted as $x_i = (x_{[i,1]}, \cdots, x_{[i,u]})$. The goal of a clustering algorithm is to assign a class label

l_i to each instance in X. The result is a list of labels $L = [l_1, \cdots, l_n]$, with $l_i \in \{1, \cdots, k\} \ \forall i \in \{1, \cdots, n\}$, that effectively splits X into k non-overlapping clusters c_i to form a partition called $C = \{c_1, \cdots, c_K\}$. The label associated with a given cluster c_i can be accessed as $l(c_i)$. The cluster membership of every instance is determined by the similarity of the instance to the rest of instances in the same cluster, and the dissimilarity to instances in other clusters. Many types of distance measurements can be used to determine pairwise similarities [10]. This definition is extended to incorporate pairwise constraints and monotonicity constraints in Sects. 2.1 and 2.2 respectively.

2.1 Constrained Clustering

In real world applications, it is common to have some information about the analyzed datasets, even if this information is not given in the form of labels. In pairwise constrained clustering, a set of constraints is given to guide the clustering process. Constraints involve pairs of instances, indicating whether they must or must not belong to the same cluster; thus, two types of pairwise constraints can be formalized:

Must-link (ML) constraints $C_=(x_i, x_j)$: instances x_i and x_j from X must be placed in the same cluster.
- Cannot-link (CL) constraints $C_{\neq}(x_i, x_j)$: instances x_i and x_j from X cannot be assigned to the same cluster.

It is known that ML constraints are transitive, reflexive and symmetric, and therefore they constitute an equivalence relationship. This is not the case for CL constraints; however, they can be chained to deduce new ML constraints [17].

In CC (Constrained Clustering), the goal is to find a partition (clustering) of k clusters such that $C = \{c_1, \cdots, c_k\}$ of X ideally satisfying all constraints (in hard CC) or as many constraints as possible (in soft CC). The classic clustering requirements also have to be observed: it must be fulfilled that the sum of instances in each cluster c_i is equal to the number of instances in X, which has been defined as $n = |X| = \sum_{i=1}^{k} |c_i|$.

2.2 Monotonic Clustering

In [13] the monotonicity constraints are integrated into unsupervised learning; particularly, they are integrated into the clustering task, producing the monotonic clustering framework. To do so, the classic symmetrical notion of distance in pattern recognition is replaced with the asymmetrical notion of preference from the Multi Criteria Decision Aid (MCDA) paradigm. The preference of an instance over another evaluates the global advantages of the former over the latter with respect to some preference criteria. The notion of preference can be seen as a decomposition of a distance measure taking into account the sign of the differences. To cluster instances in monotonic clustering, the similarity between every pair of instances is evaluated in terms of preferences taking all the other alternatives into account. Having this in mind, two instances are similar if they are preferred to or by the same set of instances. To formalize these concepts, let us consider the weighted L_1 distance (for the maximization case and without loss of generality) as in Eq. 1, which can be simplified as in Eq. 2, with $w_d \in [0, 1]$ being the weight assigned to the dth feature.

$$L_1(x_i, x_j) = \sum_{d=1}^{u} w_d |x_{[i,d]} - x_{[j,d]}|. \tag{1}$$

$$L_1(x_i, x_j) = \sum_{d:x_{[i,d]}>x_{[j,d]}}^{u} w_d x_{[i,d]} - w_d x_{[j,d]} + \sum_{d:x_{[j,d]}>x_{[i,d]}}^{u} w_d x_{[j,d]} - w_d x_{[i,d]}. \tag{2}$$

Then, let us define the preference of x_i over x_j as in Eq. 3. To put this into words, $r(x_i, x_j)$ quantifies the sum of differences between x_i and x_j limited to the features in which x_i is better than x_j. Intuitively, the preference $r(x_i, x_j)$ indicates the cumulative quantified value of the advantage of x_i over x_j. Please note that, as it has already been mentioned, the preference is not symmetrical: $r(x_i, x_j) \neq r(x_j, x_i)$ in most cases.

$$r(x_i, x_j) = \sum_{d:x_{[i,d]}>x_{[j,d]}}^{u} w_d x_{[i,d]} - w_d x_{[j,d]}. \tag{3}$$

Finally, note that the weighted L_1 distance between two instances can always be expressed as in Eq. 4. This decomposition can be done the same way for any L_p distance.

$$L_1(x_i, x_j) = r(x_i, x_j) + r(x_j, x_i). \tag{4}$$

3 The Proposal: Monotonic Constrained Clustering

In this study, the combination of two different types of background knowledge (namely: instance-pairwise constraints and monotonicity constraints) is investigated. To do so, an Expectation-Minimization (EM) optimization scheme is used, along with a hybrid objective function which takes into account both pairwise constraints and monotonicity constraints. To this end, a distance measure designed on the basis of the definition of preference, and a pairwise constraint-based penalty term are combined to produce the already mentioned function. We named this approach Pairwise Constrained K-Means - Monotonic (PCKM-Mono).

The EM optimization scheme is widely used in the literature to approach clustering problems ranging from classic clustering problems to constrained clustering [16] and monotonic clustering [13]. Two steps build the EM optimization scheme: (1) in the Expectation step (E step), given a set of cluster representatives (centroids) $\{\mu_1, \cdots, \mu_K\}$, every instance x_i is assigned to the cluster c_j

that minimizes its contribution to the objective function, computed with respect to the cluster representatives; (2) in the Minimization step (M step), the cluster representatives $\{\mu_1, \cdots, \mu_K\}$ are reestimated for the current cluster assignment $\{c_1, \cdots, c_K\}$ to minimize the objective function. The EM optimization scheme iterates between these two steps until some convergence criteria are met. With this in mind, two elements need to be defined in order to apply the EM scheme to the constrained monotonic problem: the objective function and the centroid computation criteria.

Cost Function. The cost function of the proposed PCKM-Mono algorithm combines two main elements: a monotonic distance measure (proposed in [13]) and a pairwise constraint-based penalty term. Equation 5 defines the hybrid objective function optimized by PCKM-Mono, where $\mathbb{1}[\![\cdot]\!]$ is the indicator function (returns 1 if the predicate given as argument holds and 0 otherwise), and μ_k is the centroid associated with cluster k. The first term in Eq. 5 is a preference-based distance metric, while the other two terms refer to the cost of violating CL and ML constraints respectively (the penalty term). Please note that the first term of Eq. 5 produces completely stratified clusters when applied alone, which would produce perfectly monotonic partitions. However, this is not a desirable result in most real-world problems, as will be proved in Sect. 5.

$$
\begin{aligned}
J_{PCKMM} = \quad & \tfrac{1}{K} \sum_{k=1}^{K} \sum_{x_i \in c_k} |(r(x_i, \mu_k) - r(\mu_k, x_i))| \\
& + \sum_{(x_i, x_j) \in C_=} \mathbb{1}[\![l_i \neq l_j]\!] + \sum_{(x_i, x_j) \in C_{\neq}} \mathbb{1}[\![l_i = l_j]\!]
\end{aligned}
\tag{5}
$$

This cost function can be translated into an assignation rule as in Eq. 6, which can be intuitively interpreted as: assign each instance to its closest (preferred) cluster among those where it produces the least violated constraints.

$$
\begin{aligned}
x_i \in c_{h^*} \ \text{if} \ h^* = \quad & \mathbf{argmin}_h \left(|\sum_{j=1}^{u} (x_{[i,j]} - \mu_{[h,j]})| \right. \\
& \left. + \sum_{x_j : (x_i, x_j) \in C_=} \mathbb{1}[\![l(c_h) \neq l_j]\!] + \sum_{x_j : (x_i, x_j) \in C_{\neq}} \mathbb{1}[\![l(c_h) = l_j]\!] \right)
\end{aligned}
\tag{6}
$$

Centroid Update Rule. Regarding the computation of the centroid for every cluster after the E step, it is done by following its traditional form: every centroid is computed as the average of all instances belonging to the cluster it represents. This can be formalized as in Eq. 7.

$$
\mu_i = \frac{1}{|c_i|} \sum_{x_i \in c_i} x_i
\tag{7}
$$

The overall PCKM-Mono optimization procedure is summarized in Algorithm 1.

Algorithm 1. Pairwise Constrained K-Means - Monotonic (PCKM-Mono)

Input: Dataset X, constraint sets $C_=$ and $C_{\neq}$, the number of clusters K.
Output: Partition C of K non overlapping clusters.

[1] Initialize centroids $\{\mu_1, \cdots, \mu_K\}$ randomly
    do
        // Expectation Step
[2]   **for** $i \in \{i, \cdots, n\}$ **do**
[3]   Assign each instance x_i to cluster h^* following Eq. 6.
[4]   **end**
        // Minimization Step
[5]   **for** $i \in \{i, \cdots, K\}$ **do**
[6] $\mu_i = \frac{1}{|c_i|} \sum_{x_i \in c_i} x_i$
[7]   **end**
[8] **while** *not converged*;
[9] **return** C

---

4   Experimental Setup and Calibration

A total of 10 datasets are used in our experiments, and three constraint sets are generated for each one. These datasets can be found in the Keel-dataset repository[1] [15]. Since the Euclidean distance is used to measure pairwise distances in all compared algorithms, a standard normalization procedure is applied to all datasets. No other preprocessing step is performed on the datasets.

A combination of monotonic and non-monotonic datasets is employed to measure the performance of every compared method with respect to different quality measures. Constraints are generated following the method in [17]. Three constraint sets are generated for every datasets, namely: CS_{10}, CS_{15} and CS_{20}. Each constraint set is associated with a small percentage of the size of the dataset: 10%, 15% and 20%, respectively. The formula $(n_f(n_f-1))/2$ tells us how many artificial constraints will be created for each constraint set, with n_f being the fraction of the size of the dataset associated with each of these percentages. Table 1 displays a summary of all datasets and constraint sets used in our experiments. Datasets marked with $*$ are commonly used as monotonic classification benchmarks [7]; the rest of them are standard classification datasets.

4.1   Evaluation Method and Validation of Results

Given the hybrid nature of our proposal, different features of the obtained partitions results have to be inspected to assess their quality in terms of different measures. The Adjusted Rand Index (ARI) will be used to measure the overall degree of agreement between the obtained partitions and the ground truth [9]. The Rand Index measures the degree of agreement of two partitions C_1 and C_2 for the same given dataset X, with C_1 and C_2 viewed as collections of

---

[1] https://sci2s.ugr.es/keel/category.php?cat=clas.

Table 1. Datasets and constraint sets summary

Dataset	Instances	Classes	Features	CS_{10}		CS_{15}		CS_{20}	
				ML	CL	ML	CL	ML	CL
Balance*	625	3	4	832	1059	1799	2479	3332	4418
Car*	1728	4	6	7961	6745	18167	15244	32076	27264
LEV*	1000	5	4	1381	3569	3174	8001	5692	14208
MachineCPU	209	4	6	41	149	99	366	205	615
Newthyroid	215	3	5	112	98	275	221	489	414
Bostonhousing	506	4	13	284	941	686	2089	1266	3784
SWD*	1000	4	10	1566	3384	3674	7501	6583	13317
Iris	150	3	4	29	76	82	149	140	295
Wisconsin*	683	2	9	1273	1005	2834	2317	5146	4034
Wine	178	3	13	34	102	116	209	186	409

$n(n-1)/2$ pairwise decisions. This measure is corrected for chance to obtain the ARI. For more details on ARI see [9]. An ARI value of 1 indicates total agreement between C_1 and C_2, while -1 means total disagreement. The quality with respect to the monotonicity of the obtained partition can be measured with the Non-Monotonic Index (NMI), which measures the degree to which monotonicity constraints are violated. It is defined as the rate of violations of monotonicity divided by the total number of examples in a dataset [7]. Finally, the *Unsat* measure is used to evaluate the quality of the results from the point of view of constrained clustering. *Unsat* is computed as the rate of violated constraints in a given partition [8].

Bayesian statistical tests are used in order to validate the results (which will be presented in Sect. 5), instead of using the classic Null Hypothesis Statistical Tests (NHST), whose disadvantages are analyzed in [1], where a new statistical comparative framework is also proposed. The Bayesian version of the frequentist non-parametric sign test is used in this study. In the Bayesian sign test, the statistical distribution of a given parameter ρ is obtained according to the differences between two sets of results, assuming it is a Dirichlet distribution. To do so, the Bayesian sign test proceeds as follows: the number of times that $A - B < 0$, the number of times where there are no significant differences, and the number of times that $A - B > 0$, then the weights of the Dirichlet distribution are iteratively updated and finally sampled to obtain a large sample of the distribution. In order to identify cases where there are no significant differences, the region of practical equivalence (rope) $[r_{min}, r_{max}]$ is defined, so that $P(A \approx B) = P(\rho \in \text{rope})$. The result of this process is a set of triplets with the form described in Eq. 8. The rNPBST R package is employed to apply the test, whose documentation and guide can be found in [4].

$$[P(\rho < r_{min}) = P(A - B < 0), \quad P(\rho \in \text{rope})P(\rho > r_{max}) = P(A - B > 0)]. \tag{8}$$

4.2 Calibration

To demonstrate the capabilities of the proposed PCKM-Mono algorithm, it is compared against four other previous EM-style clustering algorithms, including a purely monotonic clustering algorithm, two purely constrained clustering algorithms (including the most recent one), and a classic clustering algorithm:

- P2Clust: The first approach to monotonic clustering. It modifies the distance measure used in the expectation step of the EM scheme to produce purely monotonic partitions. Monotonicity constraints are never violated in partitions produced by P2Clust [13]. It does not consider pairwise constraints, so it is purely monotonic.
- COP-Kmeans: COnstrained Partitional K-means constitutes the first approach to constrained clustering [17]. It is taken as the baseline comparison for any constrained clustering method. To integrate constraints into the clustering process, it modifies the assignment rule of instances to a cluster in such a way that no constraints can be violated. The algorithm halts when a dead-end is reached. It produces partitions which satisfy all constraints when it does not arrive at dead-ends. It a purely constrained clustering algorithm.
- Kmeans: The original Kmeans algorithm proposed in [12]. Neither pairwise constraints nor monotonicity constraints are considered in Kmeans.
- PCSKMeans: The Pairwise Constrained Sparse K-Means algorithm is an extension of the classic Sparse K-Means algorithm that integrates constraints by means of a weighted penalty term [16]. It constitutes the most recent EM-style approach to constrained clustering.

Regarding the parameter setup, all algorithms use an EM scheme to find a partition of the datasets, so many of their parameters are shared. The k parameter, which indicates the number of clusters of the output partition is always set to the number of classes for every dataset (in Table 1). The maximum number of iterations allowed before convergence is set to 100 in all cases. The convergence criterion is centroid shifting: the EM optimization procedure is considered to have converged when average centroid shifting is less than 10^{-4}. Random centroid initialization is used for all algorithms. The P2Clust algorithm allows us to parameterize the computation of its internal α coefficient; this parameter is set to 1.1. The sparsity level of the PCSKMeans algorithm is set to 1.1. All parameters have been set by following the guidelines of the authors, and PCKM-Mono parameters have been decided upon via preliminary experimentation. The final purpose of this work is to provide a fair comparison between algorithms, assessing their robustness in a common environment with multiple datasets.

5 Experimental Results

The experimental results obtained for all datasets and constraint sets are presented in this section. Since non-deterministic procedures are present in every compared method (such as the random initialization of centroids), the average results of 50 runs are presented in Tables 2, 3 and 4, aiming to mitigate the effects that stochastic procedures may cause. Please note that, in cases where the COP-Kmeans algorithm is not able to produce a partition, we assign that particular run the worst possible benchmark values. Cases where no result is reported are cases in which COP-Kmeans was never able to produce an output partition. Let us remember that ARI is a maximization external quality index, while NMI and Unsat are both for minimization.

By looking at the results, it seems obvious that the proposed algorithm, PCKM-Mono, is able to find a balance between constraint satisfaction and the monotonicity of the output partition. It is clear that P2Clust, which is a purely monotonic algorithm, always produces the best results with respect to NMI, and purely constrained clustering algorithms (COP-Kmeans and PCSKMeans) produce the best results with respect to Unsat. However, PCKM-Mono is able to produce the best average NMI results, while also achieving better NMI results than purely constrained clustering algorithms, and better Unsat results than purely monotonic clustering algorithms. This is indicative of the viability of the combination of pairwise and monotonic constraints to solve benchmark problems in both areas, as well as it provide; moreover, it provides evidence in favor of the proposed EM optimization scheme, which is simple but can be nonetheless suitable for this task.

Table 2. Results obtained by the five compared methods for the CS_{10} constraint set.

Dataset	ARI (↑)					NMI (↓)					Unsat (↓)				
	PCKM-Mono	P2Clust	COP-KMeans	KMeans	PCSKMeans	PCKM-Mono	P2Clust	COP-KMeans	KMeans	PCSKMeans	PCKM-Mono	P2Clust	COP-KMeans	KMeans	PCSKMeans
Balance	0.016	0.005	1.000	0.146	1.000	0.610	0.000	0.767	0.914	0.631	0.053	0.477	0.000	0.406	0.000
Car	0.825	0.036	1.000	0.112	0.993	0.058	0.000	0.057	0.128	0.164	0.008	0.500	0.000	0.458	0.000
LEV	1.000	−0.224	1.000	0.071	0.999	0.989	0.000	0.971	1.000	0.971	0.000	0.558	0.000	0.345	0.000
MachineCPU	0.156	0.159	0.987	0.224	0.196	0.143	0.000	0.258	0.364	0.258	0.034	0.385	0.000	0.339	0.011
Newthyroid	0.233	0.233	1.000	0.623	1.000	0.009	0.000	0.153	0.167	0.104	0.009	0.403	0.000	0.181	0.000
Bostonhousing	0.123	0.122	1.000	0.124	0.657	0.000	0.000	0.000	0.000	0.000	0.091	0.332	0.000	0.356	0.005
SWD	0.217	0.111	1.000	0.066	0.955	0.947	0.000	0.947	0.973	0.933	0.093	0.380	0.000	0.390	0.001
Iris	0.589	0.557	1.000	0.676	1.000	0.027	0.000	0.027	0.933	0.437	0.011	0.250	0.000	0.136	0.000
Wisconsin	1.000	0.857	1.000	0.849	1.000	0.009	0.000	0.009	0.764	0.009	0.000	0.070	0.000	0.074	0.000
Wine	0.297	0.242	1.000	0.843	1.000	0.000	0.000	0.000	0.208	0.000	0.088	0.350	0.000	0.074	0.000
Mean	0.446	0.210	0.999	0.373	0.880	0.279	0.000	0.319	0.545	0.351	0.039	0.371	0.000	0.276	0.002

Table 3. Results obtained by the five compared methods for the CS_{15} constraint set.

Dataset	ARI (↑)					NMI (↓)					Unsat (↓)				
	PCKM-Mono	P2Clust	COP-KMeans	KMeans	PCSKMeans	PCKM-Mono	P2Clust	COP-KMeans	KMeans	PCSKMeans	PCKM-Mono	P2Clust	COP-KMeans	KMeans	PCSKMeans
Balance	1.000	0.005	1.000	0.140	1.000	0.566	0.000	0.790	0.914	0.778	0.000	0.482	0.000	0.392	0.000
Car	0.999	0.029	1.000	0.113	1.000	0.057	0.000	0.057	0.120	0.131	0.000	0.501	0.000	0.461	0.000
LEV	0.934	−0.225	0.250	0.062	0.896	0.971	0.000	0.982	1.000	0.985	0.003	0.555	0.375	0.340	0.000
MachineCPU	1.000	0.159	1.000	0.216	0.803	0.212	0.000	0.258	0.349	0.258	0.000	0.344	0.000	0.377	0.000
Newthyroid	0.956	0.233	0.984	0.617	1.000	0.009	0.000	0.140	0.167	0.180	0.007	0.408	0.000	0.116	0.000
Bostonhousing	0.999	0.122	1.000	0.127	0.989	0.000	0.000	0.000	0.000	0.000	0.000	0.342	0.000	0.349	0.000
SWD	0.997	0.114	1.000	0.068	1.000	0.947	0.000	0.947	0.963	0.947	0.001	0.381	0.000	0.394	0.000
Iris	1.000	0.559	−0.306	0.855	1.000	0.027	0.000	0.662	0.933	0.027	0.000	0.176	0.650	0.170	0.000
Wisconsin	1.000	0.857	-	0.848	1.000	0.009	0.000	-	0.764	0.009	0.000	0.069	-	0.075	0.000
Wine	0.985	0.243	0.339	0.838	1.000	0.000	0.000	0.325	0.208	0.000	0.000	0.344	0.325	0.122	0.000
Mean	0.987	0.210	0.527	0.368	0.969	0.280	0.000	0.516	0.542	0.331	0.001	0.360	0.235	0.280	0.000

6   Statistical Analysis of Results

In contrast with NHST, it is possible to create illustrative graphical representations of the results of the Bayesian sign test. To do so, the obtained distribution

Table 4. Results obtained by the five compared methods for the CS_{20} constraint set.

Dataset	ARI (↑)					NMI (↓)					Unsat (↓)				
	PCKM-Mono	P2Clust	COP-KMeans	KMeans	PCSKMeans	PCKM-Mono	P2Clust	COP-KMeans	KMeans	PCSKMeans	PCKM-Mono	P2Clust	COP-KMeans	KMeans	PCSKMeans
Balance	**1.000**	0.004	**1.000**	0.136	**1.000**	0.509	**0.000**	0.831	0.914	0.825	**0.000**	0.479	**0.000**	0.407	**0.000**
Car	0.999	0.030	–	0.107	**1.000**	0.057	**0.000**	–	0.122	0.063	**0.000**	0.504	–	0.464	**0.000**
LEV	**0.997**	−0.229	–	0.078	0.012	0.971	**0.000**	–	0.996	0.973	**0.000**	0.551	–	0.343	0.001
MachineCPU	**0.987**	0.159	0.231	0.216	0.949	0.258	**0.000**	0.241	0.359	0.258	**0.000**	0.365	**0.000**	0.406	**0.000**
Newthyroid	0.923	0.233	0.750	0.622	**1.000**	0.009	**0.000**	0.259	0.167	0.165	0.007	0.383	0.125	0.195	**0.000**
Bostonhousing	1.000	0.122	–	0.128	**1.000**	0.000	**0.000**	–	0.000	0.000	0.000	0.342	–	0.340	**0.000**
SWD	0.998	0.113	–	0.066	**1.000**	0.947	**0.000**	–	0.963	1.000	0.001	0.385	–	0.397	**0.000**
Iris	**0.948**	0.557	0.850	0.681	0.810	0.018	**0.000**	0.100	0.913	0.048	0.003	0.202	0.075	0.156	**0.000**
Wisconsin	0.945	0.857	**1.000**	0.848	0.836	0.008	**0.000**	0.009	0.041	0.009	0.016	0.068	**0.000**	0.071	0.004
Wine	0.817	0.247	0.950	0.824	**0.965**	0.000	**0.000**	0.025	0.208	0.001	0.015	0.343	0.025	0.084	**0.000**
Mean	**0.961**	0.209	0.078	0.370	0.857	0.278	**0.000**	0.547	0.468	0.334	0.004	0.362	0.423	0.286	**0.000**

is sampled to obtain a set of triplets, which are interpreted as barycentric coordinates in an equilateral triangle, thus producing a cloud of points with varying density. This is known as a heatmap. Figure 1 shows heatmaps comparing the proposed method PCKM-Mono with the rest of the benchmarked methods for the three measures obtained: ARI, NMI and Unsat. The region of practical equivalence is set to $rope = [-0.02, 0.02]$ for ARI, and to $rope = [-0.01, 0.01]$ for NMI and Unsat, following the guidelines in [5]. The results produced by PCKM-Mono are always taken as B in 8, and the set of results obtained by the method it is compared against as A. Please note that, as ARI is a measure to maximize, a cloud of points located in the region of the map corresponding to MPCK-Means would indicate statistically significant differences between the two methods in favor of MPCK-Means. The contrary case is found for NMI and Unsat.

All heatmaps reinforce the conclusions obtained in the Experimental Results Sect. 5. It is clear that PCKM-Mono represents a statistically significant improvement over all other compared method with respect to ARI, with the comparison against PCSKMeans being the most disputed one. Heatmap 1d gives the advantage to PCKM-Mono, but not by a wide margin, indicating no significant differences in some cases and advantage of PCSKMeans in even fewer cases. Regarding the comparison with respect to NMI, and Unsat, conclusions remain unchanged. Heatmap 1e confirms the indisputable superiority of purely monotonic algorithms with respect to NMI. However, 1l reveals no statistically significant differences between PCKM-Mono and PCSKMeans with respect to Unsat, and 1j an advantage of PCKM-Mono over COP-Kmeans for the same measure. Having this in mind, it is reasonable to assert that, for the experiment carried out in this study, the proposed PCKM-Mono algorithm has the same or better capabilities than previous CC algorithms to incorporate constraints into the clustering process. Note that 1h indicates a general advantage of PCKM-Mono over PCSKmeans in terms of the NMI measure, with some exceptions in which both algorithms present no statistical differences. This is due to the influence of non-monotonic datasets, where PCSKMeans generally performs better than PCKM-Mono.

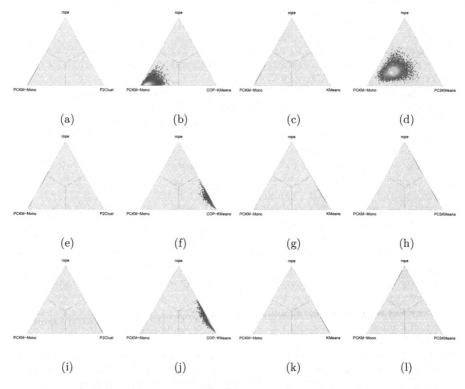

Fig. 1. Heatmaps 1a to 1l compare the proposed method PCKM-Mono to the other four compared methods: P2Clust, COP-Kmeans, Kmeans and PCSKMeans respectively from left to right in every row. The first row compares the results for the ARI measure, the second row does so for NMI and the third is for Unsat.

7 Conclusion

In this study, the first method addressing MCC is proposed: Pairwise Constrained K-Means - Monotonic (PCKM-Mono). An expectation-minimization scheme is used to locally optimize a hybrid objective function integrating a monotonic distance metric and a pairwise constraint-based penalty term. The experimental results obtained for a variety of datasets and their posterior statistical analysis confirm the viability of the proposed method when compared with purely monotonic and purely pairwise constrained clustering methods/techniques. Even if PCKM-Mono obtains results similar to those obtained by previous approaches for specific monotonicity and pairwise constraint satisfaction, there exists strong statistical evidence in favor of PCKM-Mono regarding general clustering quality measures.

References

1. Benavoli, A., Corani, G., Demšar, J., Zaffalon, M.: Time for a change: a tutorial for comparing multiple classifiers through Bayesian analysis. J. Mach. Learn. Res. **18**(1), 2653–2688 (2017)
2. Bradley, P.S., Bennett, K.P., Demiriz, A.: Constrained K-means clustering. Microsoft Research, Redmond, vol. 20 (2000)
3. Cano, J.R., Gutiérrez, P.A., Krawczyk, B., Woźniak, M., García, S.: Monotonic classification: an overview on algorithms, performance measures and data sets. Neurocomputing **341**, 168–182 (2019)
4. Carrasco, J., García, S., del Mar Rueda, M., Herrera, F.: rNPBST: an R Package covering non-parametric and Bayesian statistical tests. In: Martínez de Pisón, F.J., Urraca, R., Quintián, H., Corchado, E. (eds.) HAIS 2017. LNCS (LNAI), vol. 10334, pp. 281–292. Springer, Cham (2017). https://doi.org/10.1007/978-3-319-59650-1_24
5. Carrasco, J., García, S., Rueda, M., Das, S., Herrera, F.: Recent trends in the use of statistical tests for comparing swarm and evolutionary computing algorithms: practical guidelines and a critical review. Swarm Evol. Comput. **54**, 100665 (2020)
6. Davidson, I., Basu, S.: A survey of clustering with instance level constraints. ACM Trans. Knowl. Discov. Data **1**, 1–41 (2007)
7. González, S., Herrera, F., García, S.: Monotonic random forest with an ensemble pruning mechanism based on the degree of monotonicity. N. Gener. Comput. **33**(4), 367–388 (2015)
8. González-Almagro, G., Luengo, J., Cano, J.R., García, S.: DILS: constrained clustering through dual iterative local search. Comput. Oper. Res. **121**, 104979 (2020)
9. Hubert, L., Arabie, P.: Comparing partitions. J. classif. **2**(1), 193–218 (1985)
10. Jain, A.K., Murty, M.N., Flynn, P.J.: Data clustering: a review. ACM Comput. Surv. (CSUR) **31**(3), 264–323 (1999)
11. Law, M.H.C., Topchy, A., Jain, A.K.: Clustering with soft and group constraints. In: Fred, A., Caelli, T.M., Duin, R.P.W., Campilho, A.C., de Ridder, D. (eds.) SSPR /SPR 2004. LNCS, vol. 3138, pp. 662–670. Springer, Heidelberg (2004). https://doi.org/10.1007/978-3-540-27868-9_72
12. Lloyd, S.: Least squares quantization in PCM. IEEE Trans. Inf. Theory **28**(2), 129–137 (1982)
13. Rosenfeld, J., De Smet, Y., Debeir, O., Decaestecker, C.: Assessing partially ordered clustering in a multicriteria comparative context. Pattern Recogn. **114**, 107850 (2021)
14. Schmidt, J., Brandle, E.M., Kramer, S.: Clustering with attribute-level constraints. In: 2011 IEEE 11th International Conference on Data Mining, pp. 1206–1211. IEEE (2011)
15. Triguero, I., et al.: KEEL 3.0: an open source software for multi-stage analysis in data mining. Int. J. Comput. Intell. Syst. **10**(1), 1238–1249 (2017)
16. Vouros, A., Vasilaki, E.: A semi-supervised sparse K-means algorithm. Pattern Recogn. Lett. **142**, 65–71 (2021)
17. Wagstaff, K., Cardie, C., Rogers, S., Schrödl, S.: Constrained K-means clustering with background knowledge. In: Proceedings of the Eighteenth International Conference on Machine Learning, pp. 577–584. Morgan Kaufmann Publishers Inc. (2001)

Extractive Text Summarization on Large-scale Dataset Using K-Means Clustering

Ti-Hon Nguyen[✉] and Thanh-Nghi Do

Can-Tho University, Can-Tho, Vietnam
{nthon,dtnghi}@ctu.edu.vn

Abstract. Extractive text summarization is one of the most important tasks in natural language processing. In this work, we use K-Means clustering to create the clusters on the Vietnamese large-scale dataset, then use these clusters to extract the most relevant sentences on the single-document to produce the summary. At first, we collected the articles in the Vietnamese online newspapers, cleaned up and packaged them into the dataset, after that we applied our summarization model for the experimentation. The best F-Score of this model based on ROUGE-2 and ROUGE-L are 15.48% and 28.68%.

Keywords: Text summarization · K-Means · Clustering · Large-scale dataset

1 Introduction

Text summarization was first introduced in 1958 by H. P. Luhn [11], which is the task to produce a smaller version of a document or a set of documents. There are two kinds of summarization methods based on the output technique, the extractive method and abstract method. The extractive methods try to find out the most relevant sentences in the original text to produce the summary; on the other hand, the abstractive methods paraphrase the original text into the summary.

Nowadays, text summarization attracts a lot of research in natural language processing [3]. It caused the release of some English large-scale dataset for this kind of research such as GigaWord [5,15], CNN/Daily Mail [17]. However, there are only a few works based on Vietnamese [14] and it does not have many large-scale Vietnamese dataset [12] for evaluating in the summarizing research. Besides, most Vietnamese researchers are working on abstractive methods [9,14] and multi-documents summarization [13,14]. Therefore, in this work we built the Vietnamese large-scale dataset for text summarization research and focused on summarization on a single-document.

In the next section, we will discuss the summary methods. In the model section, we will introduce our summary model, the K-Means clustering model,

© Springer Nature Switzerland AG 2022
H. Fujita et al. (Eds.): IEA/AIE 2022, LNAI 13343, pp. 737–746, 2022.
https://doi.org/10.1007/978-3-031-08530-7_62

distance metrics, the evaluating method and the relationship between these topics with this work. In the experiment section, we will present the dataset and the evaluated results. The conclusion section will show the contribution, the advantages of our summary model and the future works.

2   Related Work

In spite of the large number of text summarization research [3], there is only a few research using clustering models for extractive text summarization. In summarizing the single-document, this kind of model uses the clustering algorithm to cluster the sentences of a document and choose some good sentence candidate to combine into the summary [1,18].

Other side, in multi-document summarization, the clusters of documents will be created first by applying the clustering algorithm on the corpus, then each in the corpus are clustered into sentence clusters, after that the best scoring sentences from sentence clusters are selected into the final summary [2,4].

In this work, we use a clustering algorithm to find the clusters of the documents in the training set and then, in the summary process, we use these cluster centroids to select the sentences in a document to produce its summary.

3   Model

3.1   Summary Model

In this work, we construct the extractive summarization model for summarizing the single-document. The model has two steps: the training step and the summarizing step. In the training step we use the K-Means clustering model to create the clusters of the documents in the training set. In the summarizing step, based on the trained K-Means model the summary model will choose the pertinent sentences in the document to produce the summary. To find the best number of clusters n for training the K-Means model, we choose n in the set of $\{50, 100, 200, 300, 500, 1000, 2000\}$. The details of the summary algorithm will be described in the **Algorithm 1**.

In **Algorithm 1**, $|C| = n$, vectors x of s are based on the Bag-of-word [6] model and the vocabulary is also built from on the training set.

3.2   K-Means

K-Means [7] is a simple and useful algorithms of clustering. In this summary model, we use Mini-batch K-Means [16] (Algorithm 2), a variant of K-Means which is improved for working on the large-scale dataset, to find the clusters of the articles in the training set. Then in the summary process, these clusters are used to select the sentence in the test article to construct its summary.

In **Algorithm 2**: C is the set of cluster centers, $c \in R^m$ is cluster center, $|C| = n$, X is the set of vectors x, $f(C, x)$ is the function that returns the nearest cluster center $c \in C$ to x by using Euclidean distance.

Algorithm 1. Summary function for a single-document d (the summarizing step)

1: Given: K-Means model from the training step $kmeans\_model$, document d, number of output sentence k
2: Initialize S is an empty set of sentences, X is an empty set of vectors, V is an empty set of vectors, $summary$ is a zero-length text
3: $S \leftarrow s$ sentences is splitted from d
4: **for** $s \in S$ **do**
5:     $X \leftarrow x$ vector of s
6: **end for**
7: $C \leftarrow c$ cluster centroid in $kmeans\_model$
8: **for** $x \in X$ **do**
9:     $c \leftarrow f(x, C)$ $\triangleright$ $f(x, C)$ is the function for finding closest $c \in C$ with x
10:     $V \leftarrow (x\_idx, x\_dis)$ $\triangleright$ x_idx is index of x in X, x_dis is distance of x and c
11: **end for**
12: sort V by increasing x_dis
13: **for** $i = 1$ to k **do**
14:     $summary = summary + S[V[i].x\_idx]+$ ". "
15: **end for**
16: **return** $summary$

3.3 Distance Metrics

Distance metrics are the standard for measuring how a vector is similar to another same dimension vector. In our work, Euclidean distance is used in the K-Means algorithm implementation and cosine distance is used to compute the distance of document sentences with the cluster's centroid of the trained K-Means model.

Cosine Distance. Cosine distance of two same dimension vectors $\vec{a} = [a_1, a_2, ..., a_n]$ and $\vec{b} = [b_1, b_2, ..., b_n]$ is determined by the Eq. (1)

$$cosine\_distance(\vec{a}, \vec{b}) = 1 - cos(\theta) = 1 - \frac{\vec{a} \cdot \vec{b}}{|\vec{a}||\vec{b}|} = 1 - \frac{\sum_{i=1}^{n} a_i b_i}{\sqrt{\sum_{i=1}^{n} a_i^2}\sqrt{\sum_{i=1}^{n} b_i^2}} \quad (1)$$

where n is the dimension of vector $\vec{a}$, vector $\vec{b}$ and θ is the angle formed by $\vec{a}$ and $\vec{b}$. Thus, if $\vec{a}$, $\vec{b}$ and $\vec{c}$ are the vectors of document A, B and C, then cosine distance of $(\vec{b}, \vec{a})$ is larger than cosine distance of $(\vec{c}, \vec{a})$ we can infer that document B is less similar to document A than document C similar to document A.

Euclidean Distance. Euclidean distance d of two same dimension vectors $\vec{a} = [a_1, a_2, ..., a_n]$ and $\vec{b} = [b_1, b_2, ..., b_n]$ is determined by the Eq. (2)

$$d(\vec{a}, \vec{b}) = \sqrt{\sum_{i=1}^{n}(a_i - b_i)^2} \quad (2)$$

Algorithm 2. Mini-batch K-Means [16].

1: Given: k, mini-batch size b, iterations t, data set X
2: Initialize each $c \in C$ with an x picked randomly from X
3: $v \leftarrow 0$
4: **for** $i = 1$ to t **do**
5:      $M \leftarrow b$ examples picked randomly from X
6:      **for** $x \in M$ **do**
7:          $\mathbf{d[x]} \leftarrow f(C, x)$ ▷ Cache the center nearest to x
8:      **end for**
9:      **for** $x \in M$ **do**
10:         $\mathbf{c} \leftarrow \mathbf{d[x]}$ ▷ Get cached center for this x
11:         $\mathbf{v[c]} \leftarrow \mathbf{v[c]} + 1$ ▷ Update per-center counts
12:         $\eta \leftarrow \frac{1}{\mathbf{v[c]}}$ ▷ Get per-center learning rate
13:         $\mathbf{c} \leftarrow (1 - \eta)\mathbf{c} + \eta\mathbf{x}$ ▷ Take gradient step
14:      **end for**
15: **end for**

---

where n is the vector dimension. Thus, if $\vec{a}$, $\vec{b}$ and $\vec{c}$ are the vectors of document A, B and C, then $d(\vec{b}, \vec{a}) > d(\vec{c}, \vec{a})$ we can infer that document B is less similar to document A than document C similar to document A.

3.4 ROUGE

ROUGE [10] is a set of metrics for evaluating the text summary automatically. It works by comparing the system summary, which is produced by the summarization model, against reference summaries, which are the provided summaries in the corpus for evaluating the summary models. In ROUGE, the "overlapping word" is a word that is visible in both system summary and reference summary.

In addition, the "overlapping word" is determined by n-gram in ROUGE-N, longest common subsequence in ROUGE-L, weighted longest common subsequence in ROUGE-W and skip-bigram co-occurrence statistics in ROUGE-S.

According to the author ROUGE-2, ROUGE-L, ROUGE-W, and ROUGE-S are good choices for evaluating the single-document summarization tasks. The formula of Recall, Precision and F measure in ROUGE metric is present in Eqs. 3, 4 and 5.

$$R = \frac{S_O}{S_R} \tag{3}$$

$$P = \frac{S_O}{S_S} \tag{4}$$

$$F = \frac{(\beta^2 + 1)PR}{\beta^2 P + R} \tag{5}$$

where R is the Recall, P is the Precision, F is the F measure, when $\beta = 1$ the F measure become F_1-score, S_O is the number of "overlapping words", S_R is the total words in the reference summary, S_S is the total words in the system summary.

4 Experiment

4.1 Dataset

The dataset **VNText** was built by collecting the articles from the Vietnamese online newspapers that based on the main idea of building the **CNN/Daily Mail** dataset[1] [8,17]. These articles were collected in HTML form, then were cleaned up by eliminating HTML tags, unrelated information like links and advertising. The final VNText dataset has 1, 101, 101 articles, every article included the title, subtitle and main content. The subtitle of each article will be used as the reference summary and its content will be used as the sample content.

Next, the VNText was splitted into three subsets, these are the train set with 880, 895 articles, the validation set with 110, 103 articles and the test set with 110, 103 articles, the detail of number of articles in these subsets is shown in Table 1. The train set is used for training the K-Means models, the validation set is used for turning parameters of the K-Means models and the summary model, the test set is used for evaluating and has the result in Sect. 5.

Table 1. Detail of the VNText

Information	Train	Test	Val	Test-Ref	Val-Ref
Articles	880,895.00	110,103.00	110,103.00	110,103.00	110,103.00
Sentences	18,738,333.00	2,342,296.00	2,333,519.00	155,573.00	155,496.00
Words	452,686,377.00	56,563,630.00	56,417,036.00	4,163,723.00	4,170,809.00
Words/Sent	24.16	24.15	24.18	26.76	26.82
Words/Art	513.89	513.73	512.40	37.82	37.88
Sents/Art	21.27	21.27	21.19	1.41	1.41

In Table 1, $Words/Sent$ is average number of words per sentence, $Words/Art$ is average number of words per article, $Sents/Art$ is average number of sentences per article and Val is the validation set.

4.2 Parameters

In the training step we have to choose the number of clusters n for K-Means clustering, which has been presented in the summary model session (3.1).

In the summary step, there are two parameters, the number of sentences in summary k and the *distance metric*, which is used to find the closest cluster centroid with the sentence vector (line 9-th in Algorithm 1). For eleluating, we choose k from 1 to 5, and based on the held-out we choose *cosine* as the *distance metric*.

---

[1] This dataset is popular used in text summary research.

4.3   The Evaluating Computer

We use one computer configuration for all tasks in this work. It consists of CPU ARM model Neoverse-N1 4 cores 2.8 GHz, single thread per core, RAM 24 GB, HDD read and write speed about 150.34 MB/sec.

5   Results

5.1   The Summary

Table 2 presents the average number of words in the summary, which is produced by the summary model. This result gives us a general view about the length of the output summary. In this table, k is the number of sentences in the summary, we use the value of k from 1 to 5 as mentioned in the parameters section, n is the number of clusters in the K-Means model, $w/sent$ is the average words per sentence in the summary, w/sum is the average words per summary.

Table 2. Average number of words in the summary

n	$k = 1$		$k = 2$		$k = 3$		$k = 4$		$k = 5$	
	$w/sent$	w/sum	$w/sent$	w/sum	$w/sent$	w/sum	$w/sent$	w/sum	$w/sent$	w/sum
50	40.60	40.60	40.75	80.52	40.81	120.01	40.81	158.77	40.77	196.50
100	40.35	40.35	40.47	79.97	40.50	119.11	40.53	157.68	40.53	195.37
200	39.81	39.81	39.98	79.00	40.03	117.72	40.07	155.88	40.07	193.14
300	39.59	39.59	39.83	78.70	39.89	117.30	39.92	155.30	39.91	192.37
500	38.80	38.80	39.16	77.38	39.31	115.59	39.36	153.13	39.39	189.86
1000	38.77	38.77	39.12	77.30	39.24	115.41	39.29	152.87	39.31	189.47
2000	36.55	36.55	37.39	73.88	37.78	111.09	38.02	147.91	38.16	183.94

In Table 2, the length of the summary is not a high difference based on n but has a down trend when n goes up. Besides, the number of words in the summary increases with the stable ratio when k goes up.

5.2   ROUGE Score

Tables 3, 4 and 5 show the F_1-score of the summary model based on ROUGE-1, ROUGE-2 and ROUGE-L with use variant n. These results are also visualized in **Fig.** 1.

In ROUGE-1, the highest F_1-score is **49.73%**, with $n = 50$ and $k = 1$, the F_1-score is slightly decrease with the increasing of n but go down significantly when k go up, it because of the length of the reference summary, which shown in Table 1.

The best F_1-score is **15.48%** based on ROUGE-2, and **28.68%** based on ROUGE-L when $n = 100, k = 1$ and remains a fluctuation trend when the n is changing. F_1-score also remains in the down direction when k goes up. This is an important result because according to ROUGE's author [10] when summarizing

Table 3. F_1-Score (%) of the summary model based on ROUGE-1

n	k = 1	k = 2	k = 3	k = 4	k = 5
50	**49.73**	44.46	36.97	31.31	27.22
100	49.72	44.54	37.14	31.47	27.35
200	49.01	44.68	37.60	32.00	27.85
300	48.94	44.76	37.66	32.07	27.94
500	48.05	44.90	38.05	32.48	28.32
1000	48.09	44.83	38.02	32.49	28.39
2000	44.54	44.48	38.66	33.26	29.06

Table 4. F_1-Score (%) of the summary model based on ROUGE-2

n	k = 1	k = 2	k = 3	k = 4	k = 5
50	15.47	14.28	12.37	10.88	9.76
100	**15.48**	14.32	12.46	10.96	9.82
200	15.31	14.45	12.79	11.37	10.24
300	15.34	14.52	12.81	11.39	10.27
500	15.07	14.58	13.00	11.61	10.50
1000	15.09	14.61	13.02	11.64	10.56
2000	14.04	14.27	12.95	11.66	10.59

Table 5. F_1-Score (%) of the summary model based on ROUGE-L

n	k = 1	k = 2	k = 3	k = 4	k = 5
50	28.66	27.20	24.43	21.98	19.99
100	**28.68**	27.26	24.52	22.07	20.06
200	28.42	27.30	24.65	22.23	20.23
300	28.39	27.32	24.67	22.27	20.28
500	27.92	27.36	24.81	22.43	20.44
1000	28.09	27.33	24.79	22.43	20.46
2000	26.18	27.12	25.07	22.78	20.81

the single-document, ROUGE-2 and ROUGE-L are better to use than ROUGE-1.

In short, our summary model get best F_1-score based on ROUGE-2 and ROUGE-L when $n = 100$ and $k = 1$.

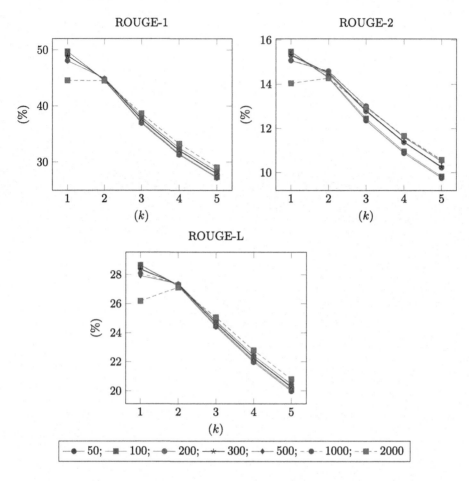

Fig. 1. F_1-Score of the summary model based on ROUGE

5.3   Time

Table 6 show the costing time of training K-Means model on the train set, the time to run summary model on the test set, the time of running ROUGE evaluating and the total time. The results show that the costing time is dependent on the summary step and high increase when n runs into higher. In addition, when $n = 100$ it took $12,258.51$ seconds for summarizing $110,103.00$ documents in the test set in 5 times ($k \in \{1, 2, 3, 4, 5\}$). It is about 22.27 ms for summarizing one document, this is an acceptable costing time.

Table 6. Time for evaluating (in seconds)

n	Train K-Means	Summary	Calculate rouge	Total seconds
50	2,209.73	8,080.84	2,239.95	12,530.52
100	2,220.00	12,258.51	2,236.03	16,714.53
200	2,241.51	20,437.45	2,209.66	24,888.62
300	2,263.66	28,739.38	2,194.12	33,197.15
500	2,318.06	45,207.08	2,161.51	49,686.65
1000	2,502.38	85,156.52	2,151.51	89,810.41
2000	2,996.58	169,041.61	1,937.72	173,975.92

6 Conclusion

The results of the experiment shows that the K-Means clustering model can be used as the base of an extractive summary model. Moreover, it also indicates that the VNText dataset can be used for evaluating text summarization by clustering. The main contribution of this work is to propose the summary model, which can work well with the large-scale dataset. The benefits of this model are: it is simple to implement, requires less resources for running and needs no more than seconds to create the summary of a document in the test set.

However, the vector Bag-of-words in this model is too long, so it takes a lot of time in training and summary steps. In the next work, we can improve the training time and the summary time by using shorter vector models for text presentation such as Word-to-vector, Glove or Fasttext. In addition, the extractive technique currently only based on the training set, in the next work we should consider the extractive techniques based on the context of the document, which is being summarized.

References

1. Agrawal, A., Gupta, U.: Extraction based approach for text summarization using K-means clustering. Int. J. Sci. Res. Publ. **4**(11), 1–4 (2014)
2. Akter, S., Asa, A.S., Uddin, M.P., Hossain, M.D., Roy, S.K., Afjal, M.I.: An extractive text summarization technique for Bengali document(s) using K-means clustering algorithm. In: 2017 IEEE International Conference on Imaging, Vision & Pattern Recognition (icIVPR), pp. 1–6. IEEE (2017)
3. Allahyari, M., et al.: Text summarization techniques: a brief survey. arXiv preprint arXiv:1707.02268 (2017)
4. Deshpande, A.R., Lobo, L.: Text summarization using clustering technique. Int. J. Eng. Trends Technol. **4**(8), 3348–3351 (2013)
5. Graff, D., Kong, J., Chen, K., Maeda, K.: English gigaword. Linguis. Data Consortium Philadelphia **4**(1), 34 (2003)
6. Harris, Z.S.: Distributional structure. Word **10**(2–3), 146–162 (1954)
7. Hartigan, J.A., Wong, M.A.: Algorithm as 136: a K-means clustering algorithm. J. Roy. Stat. Soc. Ser. C (Appl. Stat.) **28**(1), 100–108 (1979)

8. Hermann, K.M., et al.: Teaching machines to read and comprehend. In: Advances in Neural Information Processing Systems, pp. 1693–1701 (2015)

9. Le, H.T., Le, T.M.: An approach to abstractive text summarization. In: 2013 International Conference on Soft Computing and Pattern Recognition (SoCPaR), pp. 371–376. IEEE (2013)

10. Lin, C.Y.: ROUGE: a package for automatic evaluation of summaries. In: Text Summarization Branches Out, pp. 74–81 (2004)

11. Luhn, H.P.: The automatic creation of literature abstracts. IBM J. Res. Dev. **2**(2), 159–165 (1958)

12. Nguyen, V.H., Nguyen, T.C., Nguyen, M.T., Hoai, N.X.: VNDS: a Vietnamese dataset for summarization. In: 2019 6th NAFOSTED Conference on Information and Computer Science (NICS), pp. 375–380. IEEE (2019)

13. Nguyen-Hoang, T.A., Nguyen, K., Tran, Q.V.: TSGVi: a graph-based summarization system for Vietnamese documents. J. Ambient. Intell. Human. Comput. **3**(4), 305–313 (2012). https://doi.org/10.1007/s12652-012-0143-x

14. Quoc, H.T., Van Nguyen, K., Nguyen, N.L.T., Nguyen, A.G.T.: Monolingual versus multilingual bertology for Vietnamese extractive multi-document summarization. arXiv preprint arXiv:2108.13741 (2021)

15. Rush, A.M., Chopra, S., Weston, J.: A neural attention model for abstractive sentence summarization. In: Proceedings of the 2015 Conference on Empirical Methods in Natural Language Processing (2015). https://doi.org/10.18653/v1/d15-1044

16. Sculley, D.: Web-scale K-means clustering. In: Proceedings of the 19th International Conference on World Wide Web, pp. 1177–1178 (2010)

17. See, A., Liu, P.J., Manning, C.D.: Get to the point: summarization with pointer-generator networks. arXiv preprint arXiv:1704.04368 (2017)

18. Zhang, P.Y., Li, C.H.: Automatic text summarization based on sentences clustering and extraction. In: 2009 2nd IEEE International Conference on Computer Science and Information Technology, pp. 167–170. IEEE (2009)

Multi-Granular Large Scale Group Decision-Making Method with a New Consensus Measure Based on Clustering of Alternatives in Modifiable Scenarios

José Ramón Trillo⬤, Ignacio Javier Pérez⬤, Enrique Herrera-Viedma⬤,
Juan Antonio Morente-Molinera⬤, and Francisco Javier Cabrerizo$^{(\boxtimes)}$⬤

Department of Computer Science and Artificial Intelligence, Andalusian Research
Institute of Data Science and Computational Intelligence, DaSCI,
University of Granada, 18071 Granada, Spain
{jrtrillo,ijperez}@ugr.es, {viedma,jamoren,cabrerizo}@decsai.ugr.es

Abstract. Because a decision-making process in the real world involves
more people and it is more heterogeneous nowadays, this paper develops
an innovative large scale multi-granular decision-making method that
is intended to be used in situations where many experts are involved,
who do not have to provide their opinions about the alternatives at the
same time. The proposed method has three main characteristics. Firstly,
experts provide their judgments using the numerical labels set of their
choice. Secondly, experts can join the decision-making process in any
of its rounds. Thirdly, this method performs a clustering that sorts the
experts according to their judgments, obtaining how many experts belong
to the same group and the consensus among them. These two parameters
allow the weight of the group to be adjusted. Finally, the method uses
inter-group consensus measures to verify that the groups agree with the
decision made.

Keywords: Multi-granular · Large scale group decision-making ·
Modifiable scenarios · Clustering · Consensus

1 Introduction

New technologies using the Internet and social networks have globalised decision-making processes because they are accessible to everyone through their smart electronic devices. Traditionally, the decision-making process was based on a fixed and narrow set of experts expressing their ideas and opinions. These ideas and opinions could be shown and discussed during a debate or simply shown to the rest of the experts. To automate the group decision-making (GDM) process, GDM systems have been developed [2]. However, the use of the Internet and social networks have increased the number of experts that can belong to a group decision-making process since the set of experts does not need to travel to a specific location in person [1]. Furthermore, this increase has made information

© Springer Nature Switzerland AG 2022
H. Fujita et al. (Eds.): IEA/AIE 2022, LNAI 13343, pp. 747–758, 2022.
https://doi.org/10.1007/978-3-031-08530-7_63

management more difficult for three reasons. The first reason is the amount of information generated by the experts. The second reason is that experts do not provide information in the same way, and the rise in the number of experts increases the diversity of the information provided. Finally, the third reason is the complexity to gather a large number of people at the same time in the same place.

Initially, GDM is an area of research that aids experts in making decisions when they have to decide between a set of alternatives. Within this research area, there is another area called large scale decision-making (LSGDM) [9]. LSGDM methods, unlike other GDM systems, allow an undetermined significant number of users to provide their preferences and participate in the process of selecting alternatives [23].

During LSGDM processes, a large number of experts can choose from a set of alternatives [15]. In this process, it is desirable that the experts provide their opinions in the most possible natural way. Consequently, in this paper, experts will use a numerical labels set to compare two options belonging to the set of alternatives. However, it may be the case that an expert does not want to provide an opinion on a comparison between two alternatives. This is known as GDM systems with incomplete information [6]. Moreover, within LSGDM processes, it may be the case that experts do not use the same label set to provide their opinions. Within the LSGDM processes, there is a type of system called Multi-Granular LSGDM (MgLSGDM) methods [21]. They are processes where a large number of experts can contribute with their opinions on a set of alternatives using the numerical label set that best suits them.

The process of providing opinions can be done after a discussion process has taken place. However, in LSGDM methods, by using an indeterminate set of experts, some experts may provide their opinions when the process has started or may leave the process without completing it. To overcome this problem, the method proposed in this paper uses an open debate. It allows experts to provide their opinions even if the process has started or to leave the process even if it has not finished. Moreover, the open debate is an advantage because new experts who are familiar with the arguments that arise during the debate can be brought in. Moreover, in LSGDM methods, when using preference relations, the GDM system needs to manage a high amount of information. This amount of information can be difficult to manage and classify [17]. One solution to manage a large number of opinions is to group the experts under certain similarities and obtain a single preference relation matrix associated with each group [20]. In this way, the method works with all the information in a more organised way and makes consensus and ranking calculations more efficient and understandable. Finally, it is possible to analyse the consensus among experts without having to compare all experts with each other. A new consensus measure is proposed that allows comparing the distances in opinions between experts in the same group and then the differences between the groups. This allows for a more comprehensive analysis of the decision-making process.

Based on the above features, an MgLSGDM method with an open discussion is created to cluster experts according to their preferred alternative [8,11]. Additionally, this method determines the number of experts who have a preference for a particular alternative, which allows knowing how popular a preference can be beforehand without the need to perform the ranking. Moreover, the method proposed in this paper explains whether there is a majority consensus within each group, since experts have a favourite alternative. Furthermore, the method is adapted to the needs of the experts, as they can provide their opinions through a numerical label set. Finally, experts can participate at any time during the process [18].

This paper is divided into six sections. In Sect. 2, we describe the basic concepts of LSGDM. In Sect. 3, we explain the stages of the proposed method. In Sect. 4, we provide a case study in e-government to show how this new method works. In Sect. 5, we discuss the advantages of our method. Finally, in Sect. 6, conclusions are drawn.

2   Preliminaries

In this section, we introduce the basic knowledge needed to understand the LSGDM method developed in this paper.

An LSGDM procedure is characterised by an indeterminate number of experts who have to choose between a set of alternatives [19]. The set of experts is defined by a vector $E = \{e_1, \ldots, e_N\}$, $N \in \mathbb{N}$, and the set of alternatives from which the experts will have to choose is denoted by another the vector $A = \{A_1, \ldots, A_M\}$, $M \in \mathbb{N}$. Various ways of displaying experts' opinions and ratings can be found in the literature, such as the use of linguistic labels [13]. However, in this study, we will use different numerical label sets, which will be chosen by the expert [7]. Subsequently, the numerical label sets will be normalised to a single representation which will be in the interval $[0, 1]$. The normalised values represent the comparison between two alternatives, as discussed above. With these values, a preference relation can be created for each expert e_i, $i \in 1, \ldots, N$. The preference relation, denoted as p_{e_i}, $i = 1, \ldots, N$, is a matrix of dimensions $M \times M$ whose main diagonal is empty and, in the rest, it has values of the comparisons between two alternatives belonging to the set A. Assigning the values and normalising them, the preference relation is created for each expert as follows: $\mu_i : A \times A \to [0, 1] \cup \{-1\}$. A matrix is defined as $p_{e_i} = (P_i^{tj}, t \neq j = 1, \ldots, M)$, where $P_i^{tj} = \mu_i(A_t, A_j)$, $A_t, A_j \in A$.

The problems that are solved by an LSGDM process have different steps that allow users to reach a solution. Furthermore, this proposed method detects whether the difference between the experts is significant enough to have to repeat the process. The steps of which an LSGDM process is composed are:

- Input of opinions and assessments. In this first step, the experts have a discussion and provide their opinions on the alternatives [22]. As this is an open discussion, this process is carried out throughout the LSGDM process. Consequently, an expert can contribute his/her ideas at the beginning of the

LSGDM process or once the process has already started. Likewise, the expert can leave the process whenever he/she wants without the need for his/her preferences to be taken into account. The preference relation is used to show the ratings given by the experts.

- Consensus analysis. The assessments provided by the groups must be compared to verify whether the decision to be taken is consolidated by the participants. Consequently, this LSGDM process will perform a consensus analysis. The consensus analysis assesses the distance between the groups of experts and verifies that the differences are not significant enough to repeat the consensus process [16]. The consensus analysis obtains a numerical value, called the consensus value. To consider the consensus value as high enough, a value called consensus threshold, denoted as $\alpha \in [0, 1]$, must be defined. This consensus threshold is the minimum value that the consensus value must obtain for the LSGDM process to consider that the distance between the groups is not significant [12].
- Aggregating the information. In this step of the LSGDM process, all the preference relations expressing by the experts are joined together to form a single preference relation, called the collective preference relation and denoted by G [13]. Here, a weighted mean is used to aggregate the information given by the experts and by the groups of experts [18].
- Ranking of alternatives. When the consensus threshold value has been exceeded, the LSGDM process uses the G matrix to generate a ranking of alternatives. This ranking allows the alternatives to be ordered, from the most preferred one by the experts to the alternative they favour the least. To perform the ranking, we have to apply a method, for this paper we have chosen to apply the quantified-guided dominance degree (QGDD) [5].

Nowadays, there is a variety of works in the current literature that propose a solution, by using an LSGDM process, to a specific problem. For example, in [3], an LSGDM process with cooperative behaviours is created to solve a financial problem. In [10], an LSGDM process is developed based on a consensus model using the k-means algorithm to group experts. Finally, in [14], an LSGDM system employs probabilistic terms to use the imprecise information given by the users in the decision-making process.

3   A MgLSGDM Method with a New Consensus Measure Based on Clustering of Alternatives in Modifiable Scenarios

In this section, we describe the MgLSGDM method proposed in this paper, which is composed of the following steps:

- Establishing initial parameters and providing preferences. In this first step, the initial alternatives' and experts' sets are defined. In addition, the experts provide their opinions in the form of preference relations using an specific set of numerical labels.

- Standardisation of information. When the experts have already provided their opinions and preferences, the information is transformed in order to represent all the information using the same numerical label set.
- Cluster creation. Experts are ranked according to the alternative with the highest average score. Once the experts have been classified, the consensus among them is calculated and a group preference relation is obtained for each group of experts.
- Consensus calculation. Using the preference relations of each group, the process analyses the consensus of the groups. If the consensus obtains a value higher than the consensus threshold, it is considered that there is consensus among the groups and the method proceeds to the next step. However, if the consensus value obtained is lower than the consensus threshold, the process is repeated to try to increase the consensus.
- Aggregation of the groups' preference relations and obtaining the ranking of alternatives. In this last step, the groups' preference relations are aggregated to obtain the collective preference relation. Using it, the ranking of alternatives is calculated, reflecting the order of the alternatives chosen by the experts.

In the following subsections, these five steps are elaborated on in greater detail.

3.1 Establishing Initial Parameters and Providing Preferences

In this first step, the initial experts, defined by the vector $E = \{e_1, \ldots, e_N\}$, $N \in \mathbb{N}$, discuss the alternatives, denoted by the vector $A = \{A_1, \ldots, A_M\}$, $M \in \mathbb{N}$, and provide their opinions or ideas. However, any expert can start or end the process at any time. This implies that the experts do not have to start at the same time. If an expert drops out, his or her preference list is empty. However, in the case that a new expert is added, he/she will be denoted as e_{N+1} and then the number of experts in set is updated.

Initially, each expert expresses a preference relation represented as a matrix. This matrix, denoted by p_{e_i}, $i = 1, \ldots, N$, is an $M \times M$ matrix with an empty main diagonal. The rest of the positions of the matrix can be empty or filled by the numerical label set that the expert has previously chosen. Each numerical label set is denoted by the set $L_{e_i}^l$, $l \in \mathbb{N} \cup 0$. Consequently a matrix p_{e_i} is defined as follows:

$$
p_{e_i} = \begin{bmatrix} - & \cdots & P_i^{1n} \\ \vdots & \ddots & \vdots \\ P_i^{n1} & \cdots & - \end{bmatrix} \tag{1}
$$

3.2 Standardisation of Information

In this step, all the preference relations provided by the experts are taken and homogenised. To homogenise them, the following expression is used:

$$
NP_i^{tj} = \frac{P_i^{tj}}{\max(L_{e_i}^l)}, \quad t \neq j; \ t, j = 1, \ldots, M \wedge P_i^{tj} \neq -1 \tag{2}
$$

where $\max(L^l_{e_i})$ is the maximum value of the set $L^l_{e_i}$. When the values are normalised, each value is approximated to a label of the label set $L^0 = \{0.0, 0.1, \ldots, 1\}$. If the comparison is empty and does not belong to the main diagonal, it is replaced by the value -1. Once they have been assigned a label from the set $L^0 \cup \{-1\}$, in the next step, we proceed to include each expert into a single cluster.

3.3 Cluster Creation

In this step, experts are ranked according to their preferred alternative. To perform the classification, we must use the normalised p_{e_i} preference relation and apply the following expression:

$$classif(e_i) = \{t; \ \max(\sum_{\substack{j=1; \ t \neq j; \ NP^{tj}_i \neq -1}}^{M} NP^{tj}_i) : t = 1, \ldots, M\} \quad (3)$$

When the experts are classified, we proceed to calculate the preference relation for each group and the weight associated with each group. To calculate the preference relation for each group, it is necessary to define as num_{A_k} the number of experts who have as a preferred option A_k. Moreover, we define as nop^{tj}_k the number of experts belonging to the same group who have made the comparison between the alternatives A_t and A_j. Therefore, the preference relation associated with a group, denoted as p_{A_k}, is defined as follows:

$$p_{A_k} = \begin{bmatrix} - & \cdots & \frac{\sum_{i=1; \ NP^{1n}_i \neq -1}^{num_{A_k}} NP^{1n}_i}{nop^{1n}_k} \\ \vdots & \ddots & \vdots \\ \frac{\sum_{i=1; \ NP^{n1}_i \neq -1}^{num_{A_k}} NP^{n1}_i}{nop^{n1}_k} & \cdots & - \end{bmatrix} \quad (4)$$

To calculate the weights of each group, W_{A_k}, we define the number of experts belonging to the same group, denoted as nop_{A_k}, and the consensus value, denoted as $cons_{A_k}$, among the experts of the same group. This consensus value is calculated by using the Euclidean distance and comparing the preferences of the experts, provided that two experts have made the comparison. When the consensus has been computed, the weight for each group is calculated as follows:

$$w_{A_k} = \frac{cons_{A_k} + \frac{nop_{A_k}}{N}}{2} \quad (5)$$

Once they have been computed, they are normalised using the following expression:

$$W_{A_k} = \frac{w_{A_k}}{\sum_{o=1}^{M} w_{A_o}} \quad (6)$$

3.4   Consensus Calculation

For obtaining the consensus value between the experts' groups, first, it is necessary to calculate the distance between each of the groups. The expression that measures the distance between two preference relations is defined as follows:

$$Differ_{A_k}^{A_w} = 1 - \cfrac{\sqrt{\displaystyle\sum_{o=1}^{M} \sum_{h=1;\ h\neq o;\ P_{A_k}^{oh}\neq -1;\ P_{A_w}^{oh}\neq -1}^{M} (P_{A_k}^{oh} - P_{A_w}^{oh})^2}}{M \cdot (M-1)} \tag{7}$$

Using this expression, we can calculate the consensus value, denoted as *cons*, as follows:

$$cons = \cfrac{\displaystyle\sum_{k=1}^{M} \sum_{w>k}^{M} Differ_{A_k}^{A_w}}{\sum_{q=1}^{M-1} q} \tag{8}$$

The consensus value, *cons*, has to be greater than the consensus threshold, α. If it is lower, the process will be repeated. To prevent the an infinite number of rounds, a maximum number of rounds, denoted as $R \in \mathbb{N}$, is set. In the case of repeating R times, the ranking of alternatives will be obtained, regardless of the consensus value reached.

3.5   Aggregation of the Groups' Preference Relations and Obtaining the Ranking of Alternatives

In this step, the groups' preference relations are aggregated to obtain the collective preference relation and, using it, the ranking of alternatives is obtained. Initially, the collective preference relation, denoted as G, is generated. In order to obtain it, the weighted averaging operator is used. This operator uses the preference relation of each group and the calculated weight. With these two elements, the following collective preference relation is calculated:

$$G = \sum_{k=1}^{M} W_k \cdot p_{A_k} \tag{9}$$

When the collective preference relation has been computed, we proceed to generate the ranking of alternatives. To build the ranking of alternatives, it is necessary to use G and the QGDD operator [4], which is applied on the G as follows:

$$QGDD_o = \phi(G^{oh}), \quad h = 1, \ldots, M \wedge o \neq h \tag{10}$$

where ϕ is the mean operator and G^{oh} represents the value of the G at position (o, h). Once all the values provided by the QGDD operator are obtained, they are ordered in decreasing order, and the position obtained from the order is the position obtained by the alternative in the ranking.

4   A Case Study in E-government

In this section, we develop a case study based on an e-government problem to illustrate the application of the MgLSGDM method proposed in this paper. We define the set of initial citizens as $E = \{e_1, \ldots, e_{1000}\}$. The citizens must choose which measures their city should take to improve its quality of life among a set of possible alternatives. This set is defined as $A = \{A_1, A_2, A_3, A_4\}$, where A_1 is to improve road safety, A_2 is to increase green areas in the city, A_3 is to increase sanitary measures and A_4 is to improve access for people with reduced mobility. Each citizen can choose any numerical label set she/he wish. For instance, e_1 has chosen the label set $L_{e_1}^{11} = \{0, 1, \ldots, 10\}$ and e_{324} has chosen the label set $L_{e_{324}}^3 = \{0, 1, 2\}$. Once the initial discussion with the 1000 citizens has been finished, they proceed to express their preference relations. For example, the citizens e_1 and e_{324} provide the following preference relations:

$$
p_{e_1} = \begin{bmatrix} - & 5 & 0 & - \\ 5 & - & 2 & 4 \\ 10 & 8 & - & 7 \\ - & 6 & 3 & - \end{bmatrix} \qquad p_{e_{324}} = \begin{bmatrix} - & 1 & 0 & 2 \\ 1 & - & 0 & - \\ 2 & 2 & - & 1 \\ 0 & - & 1 & - \end{bmatrix}
$$

Once the process has been initialised, a new citizen, e_{1001}, appears and uses the label set $L_{e_{1001}}^6 = \{0, 1, \ldots, 5\}$. Next, the data is normalised, the gaps are filled in and the set is approximated by a chosen label set, in this case $L^0 = \{0, 0.1, \ldots, 1.0\}$. Therefore, we obtain for the citizens e_1, e_{324} and e_{1001}, the following preference relations:

$$
p_{e_1} = \begin{bmatrix} - & 0.5 & 0.0 & -1 \\ 0.5 & - & 0.2 & 0.4 \\ 1.0 & 0.8 & - & 0.7 \\ -1 & 0.6 & 0.3 & - \end{bmatrix} \quad p_{e_{324}} = \begin{bmatrix} - & 0.5 & 0.0 & 1.0 \\ 0.5 & - & 0.0 & -1 \\ 1.0 & 1.0 & - & 0.5 \\ 0.0 & -1 & 0.5 & - \end{bmatrix} \quad p_{e_{1001}} = \begin{bmatrix} - & -1 & 1.0 & 0.2 \\ -1 & - & 0.4 & 0.2 \\ 0.0 & 0.6 & - & 0.0 \\ 0.8 & 0.8 & 1.0 & - \end{bmatrix}
$$

As it can be seen, while citizens e_1 and e_{324} prefer alternative A_3, citizen e_{1001} prefers alternative A_4. The citizens are then ranked according to their preferred alternative. For this purpose, as explained in the previous section, the average of each alternative is calculated and the one with the highest average is selected. Moreover, the consensus among the citizens belonging to the same group is calculated. Furthermore, using (5), the weight of each group is calculated. When we have calculated the weight for each group, we proceed to normalise the weights using (6). Finally, we obtain the weights for each group, which are shown in Table 1.

Table 1. Distribution of citizens and the consensus among them and weight associated with each group

Group	Nº Citizens	Consensus	Weight
Group A_1	220	0.9156	0.2400
Group A_2	175	0.9325	0.2344
Group A_3	305	0.9586	0.2670
Group A_4	301	0.9226	0.2586

In this paper, citizens have the same initial weight. According to it, we obtain the preference relations for each group:

$$p_{A_1} = \begin{bmatrix} - & 0.8 & 0.6 & 0.7 \\ 0.2 & - & 0.4 & 0.5 \\ 0.4 & 0.6 & - & 0.8 \\ 0.3 & 0.5 & 0.2 & - \end{bmatrix} \quad p_{A_2} = \begin{bmatrix} - & 0.3 & 0.3 & 0.5 \\ 0.7 & - & 0.6 & 0.8 \\ 0.7 & 0.4 & - & 0.9 \\ 0.5 & 0.2 & 0.1 & - \end{bmatrix}$$

$$p_{A_3} = \begin{bmatrix} - & 0.4 & 0.0 & 0.5 \\ 0.6 & - & 0.0 & 0.5 \\ 1.0 & 1.0 & - & 1.0 \\ 0.5 & 0.5 & 0.0 & - \end{bmatrix} \quad p_{A_4} = \begin{bmatrix} - & 0.5 & 0.2 & 0.1 \\ 0.5 & - & 0.1 & 0.2 \\ 0.8 & 0.9 & - & 0.4 \\ 0.9 & 0.8 & 0.6 & - \end{bmatrix}$$

Once we have obtained the preference relations and the weights, we proceed to compute the consensus between the groups. In this case study, a value of $\alpha = 0.8$ has been chosen. Using (8), we obtain that the consensus value is equal to 0.90. Since the it is higher than α, we proceed to compute the collective preference relation, which is equal to:

$$G = \begin{bmatrix} - & 0.4984 & 0.2660 & 0.4446 \\ 0.5016 & - & 0.2625 & 0.4927 \\ 0.7340 & 0.7375 & - & 0.7734 \\ 0.5554 & 0.5073 & 0.2266 & - \end{bmatrix}$$

After calculating the collective preference relation G, we proceed to compute the ranking of alternatives. The ranking of alternatives is determined by the value obtained by the QGDD, the higher the value of the QGDD, the higher the alternative will be in the ranking. The results obtained are the following:

$$QGDD = \{0.4030,\ 0.4189,\ 0.7483,\ 0.4298\}$$
$$\text{Ranking} = \{A_3,\ A_4,\ A_2,\ A_1\}$$

As it can be seen, alternative A_3 is the alternative that has obtained the highest value in the QGDD. Consequently, the alternative A_3, which refer to increase sanitary measures, is the one chosen by citizens.

5   Discussion

This paper has introduced an innovative MgLSGDM method that includes an open debate and can be applied to real situations. It presents the following advantages:

- Open debate. The proposed method includes an open debate. It means that any expert can enter to the process once it has started or left it without completing the process. Moreover, being in an MgLSGDM process, the appearance of a new participant or one dropping out may not be significant for the ranking of alternatives since the number of experts in an MgLSGDM process is very large.
- Consensus based on groups of similarly minded experts. The method calculates the consensus based on the previously formed groups. This is an optimisation in the comparison of preference relations since experts do not have to be compared with each other.
- Allowing incomplete preference relation. In the MgLSGDM process, experts do not have to make all comparisons with each other. This situation leads to experts giving their opinions only on the alternatives they are familiar with. Furthermore, experts can use the numerical label set that best suits their judgments.
- First approximation of the ranking of alternatives. The clustering applied to this MgLSGDM process allows to know how convinced the experts are about the choice made and the number of experts who prefer an alternative before ranking them.

Finally, it can be stated that this method can be used in a process where an expert can enter the process even if it is already started, which improves some MgLSGDM methods that use a scenario where the number of experts is fixed. Moreover, the proposed method allows the experts not to have to compare all the alternatives with each other, which implies that it takes into account that not all the experts have to compare all the alternatives and using the same numerical label set. This process proposed in this paper, by using clustering, also allows knowing beforehand the number of experts that prefer an alternative before performing the ranking.

6   Conclusions

This paper have introduced a novel MgLSGDM method that uses clustering to classify experts according to their favourite preference. This clustering allows measuring the consensus between experts belonging to the same group. Consequently, it make possible to know the number of experts who have a particular alternative as their choice and to know the consensus among the experts, i.e. it is possible to know how important the alternative is in each group.

In the current literature, in most of the decision-making processes, the experts start and end the process and must give their opinions on all alternatives. However, in the real world, experts do not start to provide their opinions on a topic at the same time and experts do not have knowledge on all alternatives. Nevertheless, the method described in this paper allows experts to select which options they want to compare and start the process when they want. Furthermore, the method is adapted to the needs of the experts because experts do not express themselves in the same way. Finally, it allows the experts to be classified according to their preferred alternative and allows them to obtain information before the overall consensus of all groups is reached.

Acknowledgments. This work was supported by FEDER/Junta de Andalucía-Consejería de Transformación Económica, Industria, Conocimiento y Universidades / Proyecto B-TIC-590-UGR20, by the Andalusian Government through the project P20_00673, and by the project PID2019-103880RB-I00 funded by MCIN / AEI / 10.13039/501100011033.

References

1. Alonso, S., Pérez, I.J., Cabrerizo, F.J., Herrera-Viedma, E.: A linguistic consensus model for web 2.0 communities. Applied Soft Computing **13**(1), 149–157 (2013)
2. Cabrerizo, F.J., Trillo, J.R., Morente-Molinera, J.A., Alonso, S., Herrera-Viedma, E.: A granular consensus model based on intuitionistic reciprocal preference relations and minimum adjustment for multi-criteria group decision making. In: 19th World Congress of the International Fuzzy Systems Association (IFSA), 12th Conference of the European Society for Fuzzy Logic and Technology (EUSFLAT), and 11th International Summer School on Aggregation Operators (AGOP). pp. 298–305. Atlantis Press (2021)
3. Chao, X., Kou, G., Peng, Y., Viedma, E.H.: Large-scale group decision-making with non-cooperative behaviors and heterogeneous preferences: an application in financial inclusion. Eur. J. Oper. Res. **288**(1), 271–293 (2021)
4. Herrera, F., Herrera-Viedma, E., Chiclana, F.: A study of the origin and uses of the ordered weighted geometric operator in multicriteria decision making. Int. J. Intell. Syst. **18**(6), 689–707 (2003)
5. Herrera, F., Herrera-Viedma, E., Verdegay, J.L.: Direct approach processes in group decision making using linguistic OWA operators. Fuzzy Sets Syst. **79**(2), 175–190 (1996)
6. Kou, G., Peng, Y., Chao, X., Herrera-Viedma, E., Alsaadi, F.E.: A geometrical method for consensus building in gdm with incomplete heterogeneous preference information. Appl. Soft Comput. **105**, 107224 (2021)
7. Li, C.C., Dong, Y., Xu, Y., Chiclana, F., Herrera-Viedma, E., Herrera, F.: An overview on managing additive consistency of reciprocal preference relations for consistency-driven decision making and fusion: Taxonomy and future directions. Inf. Fusion **52**, 143–156 (2019)
8. Li, G., Kou, G., Peng, Y.: Heterogeneous large-scale group decision making using fuzzy cluster analysis and its application to emergency response plan selection. IEEE Transactions on Systems, Man, and Cybernetics: Systems (2021)
9. Liu, B., Zhou, Q., Ding, R.X., Palomares, I., Herrera, F.: Large-scale group decision making model based on social network analysis: Trust relationship-based conflict detection and elimination. Eur. J. Oper. Res. **275**(2), 737–754 (2019)

10. Liu, Q., Wu, H., Xu, Z.: Consensus model based on probability k-means clustering algorithm for large scale group decision making. Int. J. Mach. Learn. Cybern. **12**(6), 1609–1626 (2021)
11. Lu, Y., Xu, Y., Herrera-Viedma, E., Han, Y.: Consensus of large-scale group decision making in social network: the minimum cost model based on robust optimization. Inf. Sci. **547**, 910–930 (2021)
12. Morente-Molinera, J.A., Ríos-Aguilar, S., González-Crespo, R., Herrera-Viedma, E.: Dealing with group decision-making environments that have a high amount of alternatives using card-sorting techniques. Expert Syst. with Appl. **127**, 187–198 (2019)
13. Morente-Molinera, J.A., Wu, X., Morfeq, A., Al-Hmouz, R., Herrera-Viedma, E.: A novel multi-criteria group decision-making method for heterogeneous and dynamic contexts using multi-granular fuzzy linguistic modelling and consensus measures. Inf. Fusion **53**, 240–250 (2020)
14. Song, Y., Li, G.: A large-scale group decision-making with incomplete multi-granular probabilistic linguistic term sets and its application in sustainable supplier selection. J. Oper. Res. Soc. **70**(5), 827–841 (2019)
15. Tang, M., Liao, H., Xu, J., Streimikiene, D., Zheng, X.: Adaptive consensus reaching process with hybrid strategies for large-scale group decision making. Eur. J. Oper. Res. **282**(3), 957–971 (2020)
16. Tomlin, D.: Consensus decision-making: performance of heuristics and mental models. Evolution and Human Behavior (2021)
17. Trillo, J.R., Fernández, A., Herrera, F.: Hfer: Promoting explainability in fuzzy systems via hierarchical fuzzy exception rules. In: 2020 IEEE International Conference on Fuzzy Systems (FUZZ-IEEE). pp. 1–8. IEEE (2020)
18. Trillo, J.R., Herrera-Viedma, E., Cabrerizo, F.J., Morente-Molinera, J.A.: A multi-criteria group decision making procedure based on a multi-granular linguistic approach for changeable scenarios. In: International Conference on Industrial, Engineering and Other Applications of Applied Intelligent Systems. pp. 284–295. Springer (2021)
19. Xu, X., Yin, X., Chen, X.: A large-group emergency risk decision method based on data mining of public attribute preferences. Knowl.-Based Syst. **163**, 495–509 (2019)
20. Xu, Y., Gong, Z., Wei, G., Guo, W., Herrera-Viedma, E.: Information consistent degree-based clustering method for large-scale group decision-making with linear uncertainty distributions information. International Journal of Intelligent Systems (2021)
21. Yao, S., Gu, M.: An influence network-based consensus model for large-scale group decision making with linguistic information. Int. J. Comput. Intell. Syst. **15**(1), 1–17 (2022)
22. Zhang, Z., Kou, X., Yu, W., Gao, Y.: Consistency improvement for fuzzy preference relations with self-confidence: An application in two-sided matching decision making. Journal of the Operational Research Society pp. 1–14 (2020)
23. Zhang, Z., Yu, W., Martínez, L., Gao, Y.: Managing multigranular unbalanced hesitant fuzzy linguistic information in multiattribute large-scale group decision making: A linguistic distribution-based approach. IEEE Trans. Fuzzy Syst. **28**(11), 2875–2889 (2019)

Optimal User Categorization from a Hierarchical Clustering Tree for Recommendation

Wei Song[(⊠)] [iD] and Siqi Liu

School of Information Science and Technology, North China University of Technology,
Beijing 100144, China
songwei@ncut.edu.cn

Abstract. Recommender system (RS) and clustering are two main types of data mining techniques that have wide applications. An RS helps users to acquire useful online resources efficiently and effectively, whereas clustering groups similar objects together and separates dissimilar objects as much as possible. Recently, in an increasing number of studies, similar users/items are grouped before recommendation to improve the recommendation quality. Following this routine, we group similar users from a binary cluster tree formed using hierarchical clustering, for which no user-determined number of clusters is required. To extract optimal clusters, we incorporate the formation order of clusters, and propose two cluster quality measures based on lifetime and variance. The first measure favors stable clusters and the second measure tends to group users whose ratings have low variance. Extensive tests on public datasets demonstrate the superiority of the proposed method.

Keywords: Recommender system · Agglomerative hierarchical clustering · Order difference · Lifetime with order · Variance with order

1 Introduction

As an effective tool for information filtering, the recommender system (RS) [12] is receiving an increasing amount of attention. Various methods for the RS have been proposed, such as collaborative filtering [9], content-based recommendation [11], matrix factorization (MF)-based methods [8], and sequential pattern-based algorithms [10].

Clustering is a data mining technique that has been used in RSs for pre-processing [4]. The advantage of hierarchical clustering over partition-based clustering is that it is not necessary to specify the number of clusters. Thus, using hierarchical clustering can improve the stability of the recommendation results [6].

Recently, Neto et al. proposed two measures for extracting user partitions from hierarchical clustering results for an RS [6]. Specifically, one is based on the notion of cluster lifetime, which is used to extract stable clusters, and the other is based on the sparsity of interactions within a cluster and is used for minimizing sparsity within clusters. Since ensuring a minimum number of users in each cluster, this method of

© Springer Nature Switzerland AG 2022
H. Fujita et al. (Eds.): IEA/AIE 2022, LNAI 13343, pp. 759–770, 2022.
https://doi.org/10.1007/978-3-031-08530-7_64

user partition based on hierarchical clustering results is more generic for improving recommendation accuracy.

In this paper, we also study the partition extraction problem from hierarchical clustering for an RS by improving the measures proposed in [6]. We do not consider the sparsity-based measure because we prove that it is problematic. When using this measure, child clusters are definitely selected over the parent cluster. The main motivation of our work is to incorporate the formation order difference between child/parent clusters in the binary tree that represents the hierarchical clusters. Using the cluster formation order, users can be categorized into fine groups to achieve high recommender quality. The main contributions of this paper are as follows.

First, we improve the lifetime-based measure [6] by incorporating the formation order difference between child/parent clusters.

Second, using the order difference, we propose a new cluster quality measure based on the variance of users' ratings.

Third, we report the experimental results and compare them with the recommendation results using the lifetime-based measure [6].

2 Clustering Extraction from a Clustering Tree

2.1 Agglomerative Hierarchical Clustering

A fundamental task of data mining is clustering groups of data objects into subsets in such a manner that similar objects are grouped together and dissimilar objects are grouped separately [3].

There are two main categories of clustering algorithms: partitional and hierarchical. Partitional clustering algorithms are clustering methods used to categorize data objects into multiple groups based on their similarity. Such methods typically require that the number of clusters is pre-set by the user. Hierarchical clustering methods construct clusters by recursively partitioning the objects in either a top-down (divisive) or bottom-up (agglomerative) manner. As in [6], agglomerative hierarchical clustering (AHC) is used in our paper. For many applications, AHC is superior to partitional clustering because it is not necessary to set the number of clusters in advance.

2.2 Framework for the Optimal Selection of Clusters

Let $D = \{d_1, d_2, ..., d_n\}$ be a dataset that contains n data objects. Generally, the result of hierarchically clustering of these n objects is represented as a binary cluster tree with $(2 \times n - 1)$ nodes. For this tree, the root node contains all the data objects.

The framework for optimal selection of clusters (FOSC) is a globally optimal solution for extracting clusters from different levels of the hierarchical cluster tree [2]. The aim of FOSC is to form a partition $\mathcal{P}$ of D by extracting a collection of k clusters $\{C_{i_1}, C_{i_2}, ..., C_{i_k}\}$ from the binary cluster tree. To determine the final partition $\mathcal{P}$, the objective function $J(\mathcal{P})$ needs to be optimized using

$$Maximize\ J(\mathbf{P}) = \sum_{j=1}^{k} S(C_{i_j}), \tag{1}$$

subject to

$$\begin{cases} C_{i_1} \cup C_{i_2} \cup ... \cup C_{i_k} = \mathbf{D}, \ C_{i_j} \in \mathbf{P}, \ 1 \le j \le k \\ C_{i_m} \cap C_{i_n} = \emptyset, \qquad \text{for any } m \ne n \end{cases} \quad (2)$$

It should be noted that C_1 represents the root node; hence, it is not considered in the optimal partition. We can see from Eqs. 1 and 2 that each data object belongs to one and only one cluster in $\mathcal{P}$. In [6], the problem determined by Eq. 1 is solved recursively and equivalently using

$$J^*_{C_n} = \begin{cases} S(C_n), & \text{if } C_n \text{ is a leaf cluster} \\ max\{S(C_n), J^*_{C^l_n} + J^*_{C^r_n}\}, & \text{otherwise} \end{cases}, \quad (3)$$

where $J^*_{C_n}$ is the optimal value of the objective function corresponding to $\mathcal{P}$ that can be extracted from the subtree rooted at C_n; C^l_n and C^r_n are the left and right child nodes of a non-leaf node C_n of the cluster tree, respectively; and $S(C_n)$ is a cluster quality measure. We can see from Eq. 3 that if the quality of C_n is better than the aggregated value of its two sub-clusters, C_n is retained; otherwise, C_n is discarded.

We illustrate the main idea of FOSC determined by Eq. 3 using an example with seven data objects. By performing Ward's AHC [6], a binary cluster tree is constructed as shown in Fig. 1.

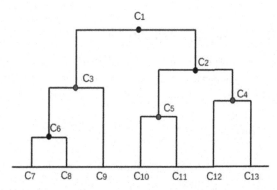

Fig. 1. Example cluster tree.

We denote the 13 nodes in the cluster tree by $C_1, C_2, ..., C_{13}$, where C_1 represents the root cluster and C_7 to C_{13} represent the seven leaf clusters. Clearly, selecting a large number of leaf nodes as clustering results is of little help in solving many practical problems. Thus, we set all the values of the cluster quality measures of C_7 to C_{13} to $-\infty$. According to Eq. 3, their aggregated values are all $-\infty$. Similarly, it is also meaningless to select the root cluster. Thus, we also set $S(C_1)$ to $-\infty$. The values of $S(C_n)$ (denoted by S) and $J^*_{C_n}$ (denoted by J) of each internal cluster are shown in Table 1.

We can see from Fig. 1 and Table 1 that $\{C_3, C_4, C_5\}$ is an optimal solution of FOSC in this example. It should be noted that the values of $S(C_n)$ in Table 1 are only for illustrating how to calculate the optimal value of the objective function. For practical

Table 1. Values of the cluster measures and aggregated quantities of each non-leaf cluster

Cluster	S	J
C_1	$-\infty$	6
C_2	2	3
C_3	3	3
C_4	2	2
C_5	1	1
C_6	1	1

problems, $S(C_n)$ can be any popular quality measure for Ward's AHC, such as distance, density, or lifetime.

3 Proposed Method

As in [6], the main aim of this paper is to categorize users initially, and then recommend items to the target user using information within the cluster to which the user belongs. To achieve this, the most important factor is the cluster quality measure, that is, $S(C_n)$ in Eq. 3.

3.1 Problem of the Sparsity-Based Quality Measure

As proposed in [6], sparsity with a minimum cluster size (SparMCS) is a measure that favors clusters that are as dense and large as possible when the size of the cluster is no lower than the minimum size threshold. SparMCS is formally defined as

$$SparMCS(C_n) = \begin{cases} -\infty, & \text{if } |C_n| < M_{CS} \\ |C_n| \times den(C_n), & \text{otherwise} \end{cases}, \tag{4}$$

where M_{CS} is the minimum cluster size threshold, $|C_n|$ is the number of users in C_n, and $den(C_n)$ is the density of C_n defined as

$$den(C_n) = \frac{Inter(C_n)}{|C_n| \times Items(C_n)}, \tag{5}$$

where $Inter(C_n)$ is the number of interactions by users that belong to C_n and $Items(C_n)$ is the set of items with which at least one user in C_n has had an interaction. Substituting Eq. 5 into Eq. 4 yields

$$SparMCS(C_n) = \begin{cases} -\infty, & \text{if } |C_n| < M_{CS} \\ \frac{Inter(C_n)}{Items(C_n)}, & \text{otherwise} \end{cases}. \tag{6}$$

We can prove that SparMCS always favors child nodes without considering M_{CS}. Let C_n^l and C_n^r be left and right child clusters of C_n, $Inter(C_n^l)$ and $Inter(C_n^r)$ be the number of interactions by users that belong to C_n^l and C_n^r, and $Items(C_n^l)$ and $Items(C_n^r)$ be the number of items interacted with by users in C_n^l and C_n^r, respectively. Suppose $Items(C_n^l) \geq Items(C_n^r)$. Then, considering Eqs. 3 and 6,

$$SparMCS(C_n) - (J_{C_n^l}^* + J_{C_n^r}^*)$$

$$= \frac{Inter(C_n)}{Items(C_n)} - (\frac{Inter(C_n^l)}{Items(C_n^l)} + \frac{Inter(C_n^r)}{Items(C_n^r)})$$

$$= \frac{Inter(C_n^l) + Inter(C_n^r)}{Items(C_n^l) + Items(C_n^r)} - (\frac{Inter(C_n^l)}{Items(C_n^l)} + \frac{Inter(C_n^r)}{Items(C_n^r)}).$$

$$< \frac{Inter(C_n^l) + Inter(C_n^r)}{Items(C_n^l)} - (\frac{Inter(C_n^l)}{Items(C_n^l)} + \frac{Inter(C_n^r)}{Items(C_n^r)})$$

$$= \frac{Inter(C_n^r) \times (Items(C_n^r) - Items(C_n^l))}{Items(C_n^l) \times Items(C_n^r)} \leq 0 \qquad (7)$$

Similarly, we can prove that $SparMCS(C_n) < (J_{C_n^l}^* + J_{C_n^r}^*)$ when $Items(C_n^l) \leq Items(C_n^r)$. Thus, in our method, the cluster quality based on sparsity is not considered.

3.2 Consideration of the Cluster's Order

Lifetime is a measure of cluster quality with the premise that good clusters survive long after they are formed [2]. The lifetime of a given cluster is typically defined as the length of the dendrogram scale along the hierarchical levels in which that cluster exists, and the dendrogram scale can be calculated using any similarity measure, such as single linkage, average linkage, or complete linkage. As we explained in Sect. 3.1 SparMCS is not an appropriate cluster measure. We improve the lifetime using the minimum cluster size (LftMCS) proposed in [6], which is formally defined as

$$LftMCS(C_n) = \begin{cases} -\infty, & \text{if } |C_n| < M_{CS} \\ |C_n| \times Lft(C_n), & \text{otherwise} \end{cases}, \qquad (8)$$

where $Lft(C_n)$ is the lifetime of C_n.

As we can see from Eq. 8, LftMCS is a good measure for considering users within a cluster. In this paper, we improve this measure by considering factors between clusters. Specifically, we incorporate the formation order difference between the measuring cluster and its parent cluster, and we define this measure as *lifetime with order* (LftOrd).

Definition 1. Let C_m and C_n be two clusters, where C_m is the parent cluster of C_n in the cluster tree. Lifetime with order of C_n is defined as

$$LftOd(C_n) = \begin{cases} -\infty, & \text{if } |C_n| < M_{CS} \\ |C_n| \times Lft(C_n) \times OdDf(C_n), & \text{otherwise} \end{cases}, \qquad (9)$$

where $OdDf(C_n)$ is the order difference between C_m and C_n, and is defined as

$$OdDf(C_n) = n - m. \qquad (10)$$

We can see from Eq. 9 that $Lft(C_n)$ measures the period of users that belong to C_n, whereas $OdDf(C_n)$ measures the period between the formation of C_n and the formation of its parent cluster. Thus, from Definition 1, both survival factors within a cluster and between child and parent clusters are considered; that is, the higher the difference between the order of C_n and the order of its parent cluster, the higher the value of $LftOd(C_n)$; and the longer C_n survives, the higher the value of $LftOd(C_n)$.

3.3 Variance-Based Quality Measure

The LftOrd is a general measure that can be used to determine the quality of a cluster provided similarity can be calculated. Considering that the recommended items in many recommendation tasks have ratings, we propose a rating-based clustering quality measurement.

Our main idea is to use the variance of ratings to measure the similarity between users. Furthermore, we also incorporate the order difference between a cluster and its parent cluster. We define this measure as *variance with order* (VarOrd).

Definition 2. Let C_m and C_n be two clusters, where C_m is the parent cluster of C_n in the cluster tree. Variance with order of C_n is defined as

$$VarOrd(C_n) = \begin{cases} -\infty, & \text{if } |C_n| < M_{CS} \\ |C_n| \times \frac{OdDf(C_n)}{Var(C_n)}, & \text{otherwise} \end{cases}, \tag{11}$$

where $Var(C_n)$ is the variance of all ratings within C_n, and is defined as

$$Var(C_n) = \frac{1}{|R_n|} \times \sum_{r \in R_n} (r - \overline{R_n})^2, \tag{12}$$

where R_n is the set of all ratings in C_n, $|R_n|$ is the number of ratings in R_n, and $\overline{R_n}$ is the average rating of all ratings in R_n.

We can see from Definition 2 that the measure of VarOrd favors clusters that are larger and have a higher order difference than their parent cluster, and lower variance of ratings within those clusters.

3.4 General Recommendation Roadmap

Data for an RS are typically represented by a rating matrix (RM). Let $\mathcal{U} = \{u_1, u_2, ..., u_M\}$ be a set of users and $\mathcal{I} = \{i_1, i_2, ..., i_N\}$ be a set of items. The associated RM R is an $M \times N$ matrix. Each entry $r_{j,k}$ of R corresponds to user u_j's ($1 \leq j \leq M$) preference for item i_k ($1 \leq k \leq N$). If $r_{j,k} \neq 0$, then u_j has rated i_k; otherwise, the user has not rated this item.

Similar to the methodology in [6], we use the two proposed cluster quality measures to group users before recommendation. The general routine of an RS based on AHC is as follows.

Step 1. Generate a binary cluster tree by performing Ward's AHC algorithm;

Step 2. Compute an objective function of either LftOrd or VarOrd;

Step 3. Extract a partition of disjoint clusters of users using Eq. 3;

Step 4. Recommend items to the target users only considering information within the clusters to which they belong.

It should be noted that when the quality of a cluster is the same as that of its child clusters, we prefer the parent cluster if it is not the root node, because it is simpler than its child clusters.

4 Experimental Results

To test the two proposed cluster quality measures, we used three recommendation methods: User-KNN [5], BPR-MF [7], and Most Popular [1], in Step 4 of the proposed algorithm. User-KNN [5] recommends items based on users that are similar to the target users. BPR-MF [7] uses Bayesian personalized ranking to optimize matrix factorization, and recommends items that have the optimized low-rank user-factor and item-factor matrices. Most Popular [1] is a non-personalized algorithm that recommends items on the basis of their popularity within the cluster to which they belong.

For each of the three methods, we compared the different effects using LftMCS [6], LftOrd, and VarOrd.

4.1 Datasets and Parameter Settings

We used three datasets, MovieLens 100K[1], FilmTrust[2], and Last.FM[3], for performance evaluation. Table 2 presents the characteristics of the datasets used in the experiments. All experimental results were averaged values of 5-fold cross-validation.

Table 2. Characteristics of the datasets

Dataset	Users	Items	Interactions
MovieLens 100K	943	1,682	100,000
FilmTrust	1,508	2,071	35,497
Last.FM	1,892	17,632	92,800

We used Ward's AHC with cosine similarity to construct the hierarchical clustering tree. For User-KNN [5], we set the number of users to 30. For BPR-MF [7], we set the number of factors to 10. When calculating LftMCS [6], we used the average distance, and set M_{CS} to 50.

---

[1] https://grouplens.org/datasets/movielens/.

[2] https://dataverse.harvard.edu/dataset.xhtml?persistentId=doi:10.7910/DVN/AKVGJ9.

[3] https://grouplens.org/datasets/hetrec-2011/.

4.2 Evaluation Metrics

Precision is one of the most popular evaluation metrics used in an RS, and is defined as

$$P = \frac{|ES \cap AS|}{|ES|}, \tag{13}$$

where ES is the set of items recommended to a user and AS is the set of items selected by that user. Precision measures how many of the recommended items were actually selected by the user.

The precision measure reflects the overall accuracy of all recommendations and does not take ordering into account. Sometimes, we want to know the precision up to the kth recommended item. This can be measured using the precision at cutoff k, which is defined as

$$P@k = \frac{1}{|U|} \times \sum_{u \in U} \frac{AS(u)@k}{k}, \tag{14}$$

where $AS(u)@k$ is the set of items selected by user u up to the kth recommended item. We can see from Eq. 14 that the higher the value of $P@k$, the more accurate the recommendation results up to the kth recommended item.

In addition to accuracy, it is important that the correct recommended items are listed in front of the recommendation list. Thus, we also used average precision in our experiments. If we need to recommend N items to user u, the average precision AP@N is defined as

$$AP(u)@N = \frac{1}{|AS(u)|} \times \sum_{k=1}^{N} P(u)@k \times r(k, u), \tag{15}$$

where $AS(u)$ is the set of items selected by u, $|AS(u)|$ is the number of items in $AS(u)$, $P(u)@k$ is the precision at cutoff k for u, and $r(k, u)$ is an indicator that represents whether the kth item is relevant to u ($r(k, u) = 1$) or not ($r(k, u) = 0$).

AP applies to a single user, whereas mean average precision MAP@N averages the AP across all users, and is defined as

$$MAP@N = \frac{1}{|U|} \times \sum_{u \in U} AP(u)@N, \tag{16}$$

where $|U|$ is the number of users. We can see from Eq. 16 that the higher the value of $MAP@N$, the higher the likelihood that the recommended items in front of the recommendation list are selected by the users.

Commonly used in information retrieval, normalized discounted cumulative gain (NDCG) is another metric for measuring ranking quality. The NDCG at cutoff N is defined as

$$NDCG@N = \frac{1}{|U|} \times \sum_{u \in U} NDCG(u)@N, \tag{17}$$

where $NDCG(u)@N$ is the NDCG of user u at cutoff N, which is defined as

$$NDCG(u)@N = Z_N \times \sum_{k=1}^{N} \frac{2^{r(k,u)} - 1}{log_2(k+1)}, \qquad (18)$$

where Z_N is a normalizer used to guarantee that the perfect recommendation score is 1.

The NDCG measures the performance of an RS based on the relevance of the recommended items that varies from 0.0 to 1.0, where 1.0 represents the ideal ranking of the items. We can see from Eq. 17 that the higher the value of $NDCG@N$, the higher the chances that the correct recommender items ranked higher are in front of the recommendation list.

4.3 Comparison Results

We tested the performance of the two proposed cluster quality measures using the three recommender methods.

Results of the User-KNN Method. Tables 3, 4 and 5 show the comparison results on the three datasets using User-KNN [5]. For each dataset, the underlined value in each column represents the optimal value of the measure.

Table 3. Comparison on MovieLens 100K for user-KNN

	P@10	MAP@10	NDCG@10
LftMCS	0.3126	0.5186	0.6371
LftOrd	0.3112	0.5187	0.6378
VarOrd	0.3053	0.5152	0.6335

Table 4. Comparison on FilmTrust for user-KNN

	P@10	MAP@10	NDCG@10
LftMCS	0.3370	0.5559	0.6532
LftOrd	0.3389	0.5599	0.6545
VarOrd	0.3390	0.5596	0.6544

For the User-KNN method, LftOrd performed best, and the performance of VarOrd was comparable with that of LftMCS. This is mainly because the quality of User-KNN-based recommendation depends on similar users within extracted clusters. Incorporating the order difference between child/parent clusters further optimized the quality of the cluster. Thus, the recommendation quality improved compared with the original LftMCS. By contrast, similar users are not considered in VarOrd. Thus, the variance-based measure is not well-suited for the User-KNN recommender method.

Table 5. Comparison on Last.FM for User-KNN

	P@10	MAP@10	NDCG@10
LftMCS	0.1484	0.3626	0.4731
LftOrd	0.1489	0.3632	0.4718
VarOrd	0.1486	0.3612	0.4697

Results of the BPR-MF Method. Tables 6, 7 and 8 show the comparison results on the three datasets using BPR-MF [7].

Table 6. Comparison on MovieLens 100K for BPR-MF

	P@10	MAP@10	NDCG@10
LftMCS	0.2465	0.4504	0.5669
LftOrd	0.2539	0.4582	0.5755
VarOrd	0.2645	0.4621	0.5827

Table 7. Comparison on FilmTrust for BPR-MF

	P@10	MAP@10	NDCG@10
LftMCS	0.3412	0.5369	0.6382
LftOrd	0.3308	0.5375	0.6321
VarOrd	0.3322	0.5424	0.6353

Table 8. Comparison on Last.FM for BPR-MF

	P@10	MAP@10	NDCG@10
LftMCS	0.0933	0.2450	0.3209
LftOrd	0.0966	0.2500	0.3295
VarOrd	0.1038	0.2638	0.3490

For the BPR-MF method, both LftOrd and VarOrd performed better than LftMCS, on average, and VarOrd performed best. These results show that the variance of ratings helped to form appropriate factors for MF. Thus, recommendation quality improved.

Results of the Most Popular Method. Tables 9, 10 and 11 show the comparison results on the three datasets using Most Popular [1].

Table 9. Comparison on MovieLens 100K for most popular

	P@10	MAP@10	NDCG@10
LftMCS	0.2541	0.4566	0.5729
LftOrd	0.2648	0.4693	0.5863
VarOrd	<u>0.2762</u>	<u>0.4811</u>	<u>0.5980</u>

Table 10. Comparison on FilmTrust for most popular

	P@10	MAP@10	NDCG@10
LftMCS	<u>0.3455</u>	0.5542	0.6511
LftOrd	0.3432	<u>0.5606</u>	<u>0.6536</u>
VarOrd	0.3430	0.5604	0.6532

Table 11. Comparison on Last.FM for most popular

	P@10	MAP@10	NDCG@10
LftMCS	0.0799	0.1986	0.2771
LftOrd	0.0821	0.1880	0.2701
VarOrd	<u>0.0882</u>	<u>0.2001</u>	<u>0.2841</u>

Similar to the results for the BPR-MF method, VarOrd performed best, followed by LftOrd. These results show that without considering similarity or other factors, the two proposed measures also improved the recommendation quality.

Discussion. From the experimental results on three datasets using three recommendation methods, we can determine the following two insights.

Using the order difference between child/parent clusters, the two proposed measures outperformed the LftMCS measure, in most cases.

For the two proposed measures, LftOrd was suitable for the User-KNN method; whereas VarOrd was suitable for methods without neighbor users.

5 Conclusions

We studied the problem of clustering users for an RS. Because it is not necessary to specify the number of clusters, we chose AHC. To extract optimal clusters from the hierarchical clustering tree, we considered the formation order of clusters. Using this order, we proposed two clustering quality measures. One is a stable lifetime-based measure that considers the order difference between a cluster and its parent cluster. The other is based on the variance of all ratings within a cluster. Extensive experiments on publicly available datasets demonstrated that the two proposed measures not only improved the recommendation accuracy but also ranked the relevant items in front of the recommendation list.

In future work, we will attempt to generate a binary cluster tree using AHC algorithms other than Ward's algorithm. Furthermore, designing new object functions to extract users' groups will be investigated in future studies.

Acknowledgments. This work was partially supported by the National Natural Science Foundation of China (61977001), and Great Wall Scholar Program (CIT&TCD20190305).

References

1. Adomavicius, G., Bockstedt, J., Curley, S., Zhang, J.: Understanding effects of personalized vs. aggregate ratings on user preferences. In: Proceedings of the Joint Workshop on Interfaces and Human Decision Making for Recommender Systems, pp. 14–21 (2016)
2. Campello, R.J.G.B., Moulavi, D., Zimek, A., Sander, J.: A framework for semi-supervised and unsupervised optimal extraction of clusters from hierarchies. Data Min. Knowl. Discov. **27**(3), 344–371 (2013). https://doi.org/10.1007/s10618-013-0311-4
3. Chao, G., Sun, S., Bi, J.: A survey on multiview clustering. IEEE Trans. Artif. Intell. **2**(2), 146–168 (2021)
4. Das, J., Majumder, S., Mali, K.: Clustering techniques to improve scalability and accuracy of recommender systems. Int. J. Uncertain. Fuzziness Knowl. Based Syst. **29**(4), 621–651 (2021)
5. Koren, Y.: Factor in the neighbors: scalable and accurate collaborative filtering. ACM Trans. Knowl. Discovery Data **4**(1), 1–24 (2010)
6. Neto, F.S.A., Costa, A.F.D., Manzato, M.G., Campello, R.J.G.B.: Pre-processing approaches for collaborative filtering based on hierarchical clustering. Inf. Sci. **534**, 172–191 (2020)
7. Rendle, S., Freudenthaler, C., Gantner, Z., Schmidt-Thieme, L.: BPR: Bayesian personalized ranking from implicit feedback. In: Proceedings of the 25th Conference on Uncertainty in Artificial Intelligence, pp.452–461 (2009)
8. Song, W., Li, X.: A Non-negative matrix factorization for recommender systems based on dynamic bias. In: Torra, V., Narukawa, Y., Pasi, G., Viviani, M. (eds.) MDAI 2019. LNCS (LNAI), vol. 11676, pp. 151–163. Springer, Cham (2019). https://doi.org/10.1007/978-3-030-26773-5_14
9. Song, W., Liu, S.: Collaborative filtering based on clustering and simulated annealing. In: Proceedings of the 3rd International Conference on Big Data Engineering, pp.76–81 (2021)
10. Song, W., Yang, K.: Personalized recommendation based on weighted sequence similarity. In: Wen, Z., Li, T. (eds.) Practical Applications of Intelligent Systems. AISC, vol. 279, pp. 657–666. Springer, Heidelberg (2014). https://doi.org/10.1007/978-3-642-54927-4_62
11. Trinh, T., Wu, D., Wang, R., Huang, J.Z.: An effective content-based event recommendation model. Multimedia Tools Appl. **80**(11), 16599–16618 (2020). https://doi.org/10.1007/s11042-020-08884-9
12. Zheng, Y.: Utility-based multi-criteria recommender systems. In: Proceedings of the 34th ACM/SIGAPP Symposium on Applied Computing, pp. 2529–2531 (2019)

Classification

A Preliminary Approach for using Metric Learning in Monotonic Classification

Juan Luis Suárez$^{(\boxtimes)}$ ⓘ, Germán González-Almagro ⓘ, Salvador García ⓘ, and Francisco Herrera ⓘ

Department of Computer Science and Artificial Intelligence,
Andalusian Research Institute in Data Science and Computational Intelligence,
DaSCI, University of Granada, 18071 Granada, Spain
jlsuarezdiaz@ugr.es, {salvagl,herrera}@decsai.ugr.es

Abstract. The purpose of this paper is to introduce, for the first time, a distance metric learning algorithm for monotonic classification. Monotonic classification is performed on labeled data in which both input and output data have an order relation. It addresses the problem of predicting labels in such a way that, if the data to be predicted is greater or lower than any training sample, its prediction should be greater or lower as well. On a different note, the use of distance metric learning algorithms enables significant performance improvements in distance-based classifiers, such as the nearest neighbors classifier. Several distance-based classifiers that are able to respect the monotonicity constraints of the datasets have been proposed. However, the development of distance metric learning algorithms remains a challenge, since, when these algorithms transform the space they can negatively modify the monotonic constraints. In our work, we propose a new methodology for learning distances that does not corrupt these constraints and allows for the reduction of the non-monotonicity of the dataset as well. The conducted experimental analysis also shows that the learned distances allow to improve the performance of the analyzed classifiers.

Keywords: Distance metric learning · Monotonic classification · Nearest neighbors · Triplet loss · M-Matrix

1 Introduction

Monotonic constraints are frequent in some real-world problems. They arise in prediction problems in which the variables to be predicted are of ordinal nature, and their order depends on how the input data are ordered. For example, when we talk about house pricing, it is to be expected that in two houses in the same area and with similar features, the bigger one will have a higher price. Or when

Our work has been supported by the research projects PID2020-119478GB-I00 and A-TIC-434-UGR20, and by a research scholarship (FPU18/05989), given to the author Juan Luis Suárez by the Spanish Ministry of Science, Innovation and Universities.

we talk about students' grades, if student A has has better grades than B during the course, it is to be expected that A will have also a higher final grade. This problems are known as monotonic classification problems [4]. The study of these problems is of interest in areas such as credit risk modeling [5] and lecturer evaluation [3].

When tackling this type of problem we are not only interested in models having high accuracy. It is also very important that the predictions obtained respect as much as possible the monotonic constraints existing in the data. And it should also be borne in mind that the cost of misclassifying a prediction should be higher the further the prediction is from its true value. Therefore, it is necessary to use classifiers that are able to handle these monotonic constraints and take them into account when making their predictions.

Similarity-based learning methods have been applied to monotonic classification problems with success. This type of learning is inspired by the human ability to recognize objects by their resemblance to other previously seen objects. This same idea can be extended to fulfill the monotonicity constraints, simply by restricting the similar objects or instances to those that comply with these constraints. The well-known nearest neighbors rule for classification [6] has been extended following this idea, so that the nearest neighbors are filtered in order to meet the monotonic constraints [8]. Recently, a new proposal restated the previous idea using a fuzzy approach [9], in order to gain robustness against possible noise in the monotonicity constraints.

All of the above algorithms require a distance metric to work. Normally, standard distances such as Euclidean distance are used. However, using a distance metric that better fits the data can help improve the performance of these classifiers. Distance metric learning [15] performs this task. Distance metric learning has been successfully applied in ordinal problems without monotonic constraints [11,14]. However, its application when monotonicity constraints are available adds a significant difficulty. Distance metrics have the ability to transform the space, and, although this could allow us to transform the data in order to reduce the instances that may break the monotonicity of the dataset, the reality is that it is difficult to ensure that no new false monotonic constraints are introduced in the process, thus worsening the quality of the data.

In our work we propose a new distance metric learning algorithm for monotonic classification. This algorithm is designed to transform the input space so that no new monotonic constraints can be introduced, overcoming the previous problem. For this purpose, we rely on monotonic matrices and M-matrices [2], which have very special properties for defining distance metrics that are well suited for monotonic datasets, as we will see later on this paper.

The paper is organized as follows. Section 2 describes the current state of distance metric learning and monotonic classification, from the similarity-based learning point of view. Section 3 shows our proposal of distance metric learning for monotonic classification. Section 4 describes the experiments conducted to evaluate the performance of our algorithm, and the results obtained. Finally, Section 5 ends with the concluding remarks and future work.

2   Background

In this section we will discuss the main problems we have addressed in this paper: distance metric learning, monotonic classification and how similarity-based methods are employed to address monotonic classification nowadays.

2.1   Distance Metric Learning

Distance metric learning [15] arises with the purpose of improving the similarity-based (or, equivalently, distance-based) learning methods such as the k-nearest neighbors classifier, or k-NN. For this purpose, distance metric learning methods aim at learning distances that facilitate the detection of hidden properties in the data that a standard distance, such as the Euclidean distance, would not be able to discover. Here, we will refer as *distance* to any map $d\colon \mathcal{X} \times \mathcal{X} \to \mathbb{R}$, where $\mathcal{X}$ is a non-empty set, satisfying the following conditions:

1. Coincidence: $d(x, y) = 0 \iff x = y$, for every $x, y \in \mathcal{X}$.
2. Symmetry: $d(x, y) = d(y, x)$, for every $x, y \in \mathcal{X}$.
3. Triangle inequality: $d(x, z) \leq d(x, y) + d(y, z)$, for every $x, y, z \in \mathcal{X}$.

The linear distance metric learning is the most common approach to learning distances among numerical data. It consists of learning Mahalanobis distances, which are parameterized by positive semidefinite matrices. Given a positive semidefinite matrix $M \in \mathcal{M}_d(\mathbb{R})_0^+$, and $x, y \in \mathbb{R}^d$, the Mahalanobis distance between x and y defined by M is given as

$$d_M(x, y) = \sqrt{(x - y)^T M (x - y)}.$$

Since every positive semidefinite matrix M can be decomposed as $M = L^T L$ with $L \in \mathcal{M}_d(\mathbb{R})$, and therefore, $d_M(x, y) = \|L(x - y)\|_2$ [15], learning a Mahalanobis distance is equivalent to learning a linear map L, and then measuring the Euclidean distance after applying that linear map. Thus, the linear distance metric learning approach comes down to learning a positive semidefinite matrix (also called metric matrix) M or a linear map matrix L. Both approaches are equivalent. Learning M usually facilitates convexity during the optimization, while learning L facilitates other tasks, such as dimensionality reduction [7].

2.2   Monotonic Classification

Monotonic classification [4] involves the development of models that respect the properties of monotonic datasets. Given a dataset $D = \{(x_1, y_1), \ldots, (x_N, y_N)\}$, with $\mathcal{X} = \{x_1, \ldots, x_N\} \subset \mathbb{R}^d$ and $y_1, \ldots, y_N \in \{1, \ldots, C\}$, D is a *monotonic* dataset if $\mathcal{X}$ is endowed with the *product order* relation $\leq$ in $\mathbb{R}^d$, the labels are also ordered with the usual order of the natural numbers, and for every pair of comparable samples in $\mathcal{X}$, the greater value in $\mathcal{X}$ has also the grater label, that is, for every $x_i, x_j \in X$, $x_i \leq x_j \iff y_i \leq y_j$. This current definition considers that all the input attributes are directly related to the output label, but it can be extended to consider attributes that are inversely related to it or even no related at all [4].

It is important to remark that, in real scenarios, due to the subjective nature of the labeling process or to measurement errors, some datasets may not be fully monotonic and there may be several pairs of instances for which the monotonicity is broken. In any case, the goal of *monotonic classification* is to provide algorithms that, when predicting new labels, are able to respect the monotonicity constraints of the datasets, and that are also robust against monotonicity clashes that may arise when the dataset is not fully monotonic.

Similarity-based learning is highly related to ordinal classification problems. Typically, it is to be expected that if two samples are close their labels will also be close, and the farther apart the samples are the more different their labels will be as well. If our data also have monotonic constraints, it is necessary to take additional caution, since we want the values predicted by the classifier to satisfy these constraints as much as possible. To this end, two proposals extend the traditional k-NN to take into account the monotonic constraints. The *monotonic k-nearest neighbors* classifier (Mon-k-NN) modifies the neighbor selection rule of the k-NN so that only the neighbors whose label is in a range that meets the constraints are considered for voting [8]. The *monotonic fuzzy k-nearest neighbors* classifier (Mon-F-k-NN) [9] relies on Mon-k-NN and the fuzzy k-NN [10] to obtain membership probabilities from the monotonic nearest neighbors that are robust against monotonicity violations. Two variants of these algorithms can be considered. The *in-range* variants consider from the beginning only nearest neighbors that comply with the constraints. The *out-range* variants consider initially any type of neighbor. Then, these neighbors are filtered, in Mon-k-NN, or used to guide the membership probabilities, in Mon-F-k-NN.

3   Algorithm Description

In this section we will describe our distance metric learning proposal for monotonic classification. First, we will introduce the concepts needed to apply the algorithm. Then, we will describe the algorithm and finally we will show its optimization procedure. We named this approach *Large Margin Monotonic Metric Learning* (LM^3L).

3.1   Preliminary Definitions

As mentioned above, one of the problems of learning a distance by means of a linear transformation is that this transformation disturbs the monotonic constraints and therefore some new constraints that are not necessarily true can be added. This may happen if we pick a distance defined by a generic $L \in \mathcal{M}_d(\mathbb{R})$. Consider, for example, the extreme case of a matrix L defining a $90°$ rotation in $\mathbb{R}^2$. If such a matrix transforms the dataset, all the monotonic constraints of the original dataset are lost, and furthermore, all those pairs of instances that were not comparable become false monotonic constraints with this rotation. However, this can be avoided by restricting ourselves to the appropriate subset of matrices, as the one we will define below.

Definition 1. *A linear transformation or square matrix $L \in \mathcal{M}_d(\mathbb{R})$ is said to be* monotone *[2] if for any real vector $x \in \mathbb{R}^d$, we have that*

$$Lx \geq 0 \implies x \geq 0,$$

where $0 \in \mathbb{R}^d$ is the vector with zeros in all its entries and $\geq$ is the product order in $\mathbb{R}^d$.

Observe that, if L is monotone, if we have two samples $x_i, x_j \in \mathcal{X}$ so that $Lx_i \geq Lx_j$, then $L(x_i - x_j) \geq 0$ and therefore $x_i \geq x_j$. This means that any pair of samples that meets a monotonicity constraint after applying L was also meeting the constraint before applying the transformation. So, when L is monotone, no new monotonic constraints will be added after the dataset is transformed. However, the reciprocal is not true, that is, if $x_i \geq x_j$, it does not necessarily follow that $Lx_i \geq Lx_j$. Consequently, some monotonic constraints may be lost in this transformation. This will allow the algorithms that use this type of matrices to select the constraints that may be more relevant in the dataset, without ever adding new incorrect monotonicity constraints after applying the transformation.

Monotone matrices are tough to use in optimization settings, since they cannot be adequately parameterized for this purpose. However, there is a subset of monotone matrices with much better properties for use in differential optimization. We describe them below.

Definition 2. *A linear transformation or square matrix $L \in \mathcal{M}_d(\mathbb{R})$ is an M-Matrix [2] if it can be expressed as $L = sI - B$, where I is the identity matrix of dimension d, $B \in \mathcal{M}_d(\mathbb{R})$ is a positive matrix, and $s \in \mathbb{R}$ verifies that $s \geq \rho(B)$, where $\rho(B)$ is the spectral radius of the matrix B.*

M-matrices are monotone [2] and, since they depend on the real value s and the positive matrix B, they can be used easily and efficiently to optimize a differentiable objective function.

3.2   Objective Function and Optimization

Once we have established the linear applications that allow us to control the monotonicity of the dataset, we are going to define the function to be optimized. Since monotonicity is already implicitly controlled by the linear application, the objective function will focus on evaluating a goodness-of-classification metric. This metric should take into account the ordinal nature of the dataset, in the sense that the farther is the real label from the predicted label the higher the penalty for that prediction must be.

Inspired by the large margin proposals for distance metric learning in other classification tasks [11,18] we propose a triplet-based objective function that considers, for each anchor sample x_i in the dataset, a positive sample x_j and a negative sample x_l so that $y_i \leq y_j < y_l$ or $y_i \geq y_j > y_l$. The distance from x_i to x_j is then minimized while the distance from x_i to x_l is maximized. The objective function and the constrained optimization problem are defined as follows:

$$\min_{L \in \mathcal{M}_d(\mathbb{R})} f(L) = \sum_{x_i \in \mathcal{X}} \sum_{\substack{x_j, x_l \in \mathcal{U}(x_i) \\ y_i \leq y_j < y_l \\ \text{or} \\ y_i \geq y_j > y_l}} \left[\|L(x_i - x_j)\|^2 - \|L(x_i - x_l)\|^2 + \lambda \right]_+$$

$$\text{s. t. } : L = sI - B$$
$$B_{ij} \geq 0, (i, j = 1, \ldots, d)$$
$$s \geq \rho(B).$$

(1)

In the previous optimization problem, $[z]_+ = \max\{z, 0\}$, λ represents a margin constant so that the distance from the negative sample to the anchor sample should not be lower than the distance from the positive sample to the anchor sample plus that margin constant, and, for each $x_i \in \mathcal{X}$, $\mathcal{U}(x_i)$ is a neighborhood containing the K nearest neighbors to x_i for the Euclidean distance. This neighborhood is calculated before the optimization process and it allows for filtering the instances that are initially more distant, giving a local character to the method and reducing the computational cost.

While the constraints in the optimization problem from Eq. 1 ensure that no new monotonic constraints can be added when the dataset is transformed, the objective function pulls data from nearby classes closer together while pushing data from more distant classes further apart, so that a transformed dataset that minimizes Eq. 1 has optimal ordinality and monotonicity properties to be learned by a similarity-based classifier.

To optimize Eq. 1, we propose a stochastic projected gradient descent method. Since L is fully parameterized by s and B the optimization problem can be rewritten as

$$\min_{\substack{s \in \mathbb{R}, B \in \mathcal{M}_d(\mathbb{R}) \\ s \geq \rho(B) \\ B_{ij} \geq 0 \; \forall i,j}} f(L) = \sum_{x_i \in \mathcal{X}} \sum_{\substack{x_j, x_l \in \mathcal{U}(x_i) \\ y_i \leq y_j < y_l \\ \text{or} \\ y_i \geq y_j > y_l}} \left[\|(sI - B)(x_i - x_j)\|^2 \right.$$

(2)

$$\left. - \|(sI - B)(x_i - x_l)\|^2 + \lambda \right]_+ .$$

At each gradient epoch we can update the pair (s, B) using the partial derivatives. Since we follow the stochastic gradient approach, at each step we will pick a random instance $x_i \in \mathcal{X}$ and use the i-th component of the gradient function to update s and B. Thus, we obtain the following update rules [12]:

$$s_{new} = s_{old} - \eta \sum_{x_j, x_l \in \mathcal{A}(x_i)} d_{ij}^T [2sI - (B + B^T)] d_{ij} - d_{il}^T [2sI - (B + B^T)] d_{il},$$
(3)

$$B_{new} = B_{old} - \eta \sum_{x_j, x_l \in \mathcal{A}(x_i)} 2(B - sI)(O_{ij} - O_{il}),$$
(4)

where η is a pre-established learning rate, $d_{ij} = x_i - x_j$, $O_{ij} = d_{ij} d_{ij}^T$, and $\mathcal{A}(x_i)$ is the set of active (positive, negative) 2-tuples associated with the anchor sample x_i and $L = sI - B$, that is,

$$\mathcal{A}_L(x_i) = \{(x_j, x_l) : x_j, x_l \in \mathcal{U}(x_i), [(y_i \leq y_j < y_l) \text{ or } (y_i \geq y_j > y_l)] \text{ and }$$
$$\|L(x_i - x_j)\|^2 - \|L(x_i - x_l)\|^2 + \lambda > 0\}.$$

Since the above update rules do not ensure that s and B meet the constraints to which they are subject, it is necessary to project them into the constrained set. Therefore, after applying the update rules, we convert the negative entries of B to zero and if S is lower than $\rho(B)$ we equal it to that value:

$$\pi(B) = (\tilde{B}_{ij}), \text{ where } \tilde{B}_{ij} = \max\{B_{ij}, 0\}, \text{ for each } i, j = 1, \ldots, d.$$
(5)
$$\pi(s) = \max\{s, \rho(B)\}.$$
(6)

This concludes the optimization process of LM^3L. In short, at each epoch the samples $x_i \in \mathcal{X}$ are chosen randomly. With each of the samples, s and B are updated using the rules from Eqs. 3 and 4 and then projected into valid values with Eqs. 5 and 6. The process is repeated until a maximum of epochs is reached or the algorithm converges. With the final values of s and B, the obtained distance is retrieved by means of the linear transformation $L = sI - B$.

4   Experiments

In this section we describe the experiments we have developed with our algorithm and the results we have obtained.

4.1   Experimental Framework

We have evaluated the distance learned by LM^3L using several distance-based classifiers. All of them are variants of the nearest neighbors classifier, namely: the original (majority-vote) k-NN, the median-vote k-NN, the monotonic k-NN (both in-range and out-range versions) and the monotonic fuzzy k-NN (both in-range and out-range versions). The first k-NN is the usual approach when dealing with non-ordinal classification problems, the median-vote k-NN is the natural adaptation for the k-NN to unconstrained ordinal regression, without taking into account any monotonic constraints. The rest of the k-NN versions are the monotonic nearest neighbors proposals discussed in Section 2.

The purpose of these experiments is to detect whether the distance learned by LM^3L allows for the improvement of these versions of k-NN in two different aspects: in the performance when classifying new data and in the fulfillment of the monotonic constraints. To this end, we will compare all the k-NN versions using both the Euclidean distance and the distance learned by LM^3L.

We have used a number of neighbors of $k = 9$ in all the k-NN versions. The distances used by each of the classifiers will be evaluated using a stratified 5-fold cross validation, that is, a cross validation that preserves the original class proportions in each fold. We have used 10 different monotonic datasets from different sources [1,17]. These datasets are numerical and have no missing values. All the features have monotonic constraints, but some of them are inverse. These attributes have been sign-switched prior to the execution of the experiments. A *min-max* normalization to the interval $[0, 1]$ has also been applied. It is important to note that, due to the nature of the data, some datasets are not fully monotonic and have pairs of samples that violate the monotonicity constraints. The datasets, their dimensions and their monotonicity properties are shown in Table 1.

Table 1. Datasets used in the experiments.

Dataset	# Samples	# Features	# Classes	Attribute directions	Non-Monotonic pairs / Comparable pairs (%)
autoMPG8	392	7	5	(-,-,-,-,+,+,+)	0.044 / 36.14
car	1728	6	4	All direct (+)	0.246 / 39.67
ERA	1000	4	9	All direct (+)	3.349 / 16.77
ESL	482	4	7	All direct (+)	0.585 / 59.81
LEV	1000	4	5	All direct (+)	1.330 / 24.08
machineCPU	209	6	5	(-,+,+,+,+,+)	1.196 / 46.24
pima	768	8	2	All direct (+)	0.151 / 6.576
SWD	1000	10	4	All direct (+)	0.949 / 12.62
balance	625	4	3	(-,-,+,+)	0.000 / 25.64
boston-housing	506	13	5	(-,+,-,+,-,+,-,+,-,-,+,-)	0.299 / 14.60

4.2 Metrics and Results

To evaluate the classification performance of the distances with each classifier, we have used two metrics [11]: the *mean absolute error* (MAE), which penalyzes the classification error according to the distances between the labels, and the *concordance index* (C-INDEX), which measures the ratio between the number of ordered pairs in both true labels and predictions and the number of all comparable pairs.

The parameters proposed for LM^3L in the execution of these experiments are a fixed neighborhood size of 50 for the anchor samples, a maximum number of 300 optimization epochs, a neighborhood margin λ of 0.1 and an adaptive learning rate η. This rate starts in 10^{-6} and, at each epoch, if the objective function improves it is increased by 1%, otherwise it is halved, following the adaptive approach in [18]. The choice of these default parameters follows the guidelines of the algorithms on which this method is inspired. The code of LM^3L used for these experiments is available in pyDML [13], an open-source Python Library with a wide range of distance metric learning algorithms.

Table 2 shows the results of the classification performance. This table also includes, for each combination of distance and classifier, its average ranking over all the combinations of distance and classifiers (AVG RANK [ALL]) and its average ranking within the distances that use the same classifier (AVG RANK [IN]).

Table 2. MAE and C-INDEX of the distance and classifiers on each dataset.

Dataset	k-NN		Med-k-NN		Mon-k-NN (IR)		Mon-k-NN (OR)		Mon-F-k-NN (IR)		Mon-F-k-NN (OR)	
	Euclidean	LM^3L	Euclidean	LM^3L	Euclidean	LM^3L	Euclidean	LM^3L	Euclidean	LM^3L	Euclidean	LM^3L
C-INDEX												
autoMPG8	0.918457	0.920312	0.920898	**0.925040**	0.920753	**0.925040**	0.919799	**0.925040**	0.919230	0.911326	0.918868	0.912257
car	0.963044	0.972001	0.962430	**0.974398**	0.971015	**0.974398**	0.970817	**0.974398**	0.964425	0.736149	0.942335	0.718687
ERA	0.692304	0.683062	0.696339	0.692771	0.675817	0.662343	0.675817	0.662343	**0.702484**	0.698284	0.700877	0.701683
ESL	0.911469	0.914401	0.917153	0.918642	0.912686	0.901638	0.911558	0.902055	0.917847	0.915939	**0.918671**	0.916644
LEV	0.803299	0.816627	0.804052	0.819050	0.768287	0.736397	0.768397	0.730397	0.818089	0.815131	**0.822355**	0.821664
machineCPU	0.857039	0.854280	0.860909	0.867868	**0.880012**	0.877510	0.873755	0.874879	0.863247	0.869615	0.855029	0.870145
pima	0.705144	0.712881	0.705144	0.712881	0.706292	0.706473	0.709292	0.706473	**0.719280**	0.667439	0.685984	0.665552
SWD	0.742335	0.743421	0.749917	0.752555	0.685411	0.674434	0.689080	0.674434	0.759628	0.757493	**0.761008**	0.754048
balance	0.903139	0.898828	0.933961	0.940843	0.946904	0.948984	0.946882	0.948657	0.948366	**0.951353**	0.944890	0.944437
boston-housing	0.780491	0.848367	0.795450	**0.853051**	0.803741	**0.853051**	0.799271	**0.853051**	0.819694	0.810555	0.773578	0.810751
AVG SCORE	0.827672	0.836418	0.834625	**0.845710**	0.827092	0.826027	0.826456	0.825773	0.843229	0.813328	0.832360	0.811587
AVG RANK [ALL]	8.863636	6.409091	6.954545	**3.772727**	6.818182	6.454545	7.363636	6.636364	4.090909	7.090909	6.363636	7.181818
AVG RANK [IN]	1.666667	**1.333333**	1.833333	**1.166667**	1.416667	1.583333	**1.416667**	1.583333	**1.250000**	1.750000	**1.333333**	1.666667
MAE												
autoMPG8	0.342740	0.322085	0.332608	**0.306889**	0.334720	**0.306889**	0.337281	**0.306889**	0.329528	0.359831	0.331758	0.360091
car	0.052658	0.050339	0.054392	0.049193	**0.030085**	0.049193	0.031826	0.049193	0.056113	0.256990	0.090264	0.270272
ERA	1.411089	1.465980	1.300229	1.310030	1.527210	1.556053	1.527210	1.556053	**1.291049**	1.300199	1.293127	1.292118
ESL	0.331710	0.327648	0.321226	**0.317207**	0.352483	0.396309	0.356799	0.396309	0.342489	0.346530	0.335982	0.338046
LEV	0.438115	0.415123	0.421073	0.405062	0.496170	0.544144	0.496170	0.544144	0.398021	0.407012	0.384050	0.389021
machineCPU	0.613086	0.612891	0.574255	0.560191	**0.489040**	0.513241	0.508696	0.527438	0.549397	0.527777	0.572187	0.543366
pima	0.247441	0.247432	0.247441	0.247432	0.244826	0.251328	**0.240905**	0.251328	0.244835	0.259138	0.247449	0.260453
SWD	0.479069	0.482130	0.448037	0.450157	0.503911	0.536098	0.499966	0.536098	0.442002	0.447038	**0.440997**	0.450057
balance	0.148918	0.158031	0.147267	0.134311	0.087984	**0.083351**	0.083351	0.091236	0.092091	0.092927	0.129795	0.129666
boston-housing	0.588659	0.422635	0.549184	**0.406773**	0.499192	**0.406773**	0.517014	**0.406773**	0.672612	0.473747	0.570193	0.477609
AVG SCORE	0.465349	0.450429	0.439571	**0.418725**	0.456099	0.464801	0.459922	0.466546	0.441904	0.447119	0.439580	0.451070
AVG RANK [ALL]	8.590909	6.590909	6.409091	**4.318182**	6.000000	7.227273	6.454545	7.590909	5.363636	6.636364	5.818182	7.000000
AVG RANK [IN]	1.750000	**1.250000**	1.833333	**1.166667**	**1.166667**	1.833333	**1.166667**	1.833333	**1.250000**	1.750000	**1.333333**	1.666667

Finally, to evaluate the fulfillment of the monotonic constraints we rely on the *non monotonicity index* (NMI). This metric is a normalized measure of how many samples do not fulfill a monotonic constraint. This can be used to evaluate both the monotonicity of the transformed training dataset after applying LM^3L and the monotonicity of the predicted samples with respect the training dataset. For a training set $\mathcal{X}$ and a labeled point $(x, y) \in \mathbb{R}^d \times \{1, \ldots, C\}$ we define

$$NClash_{\mathcal{X}}(x) = |\{x_i \in \mathcal{X} \colon (x_i < x \text{ and } y_i > y) \text{ or } (x_i > x \text{ and } y_i < y)\}|.$$

Table 3. Results of the monotonicity analysis on each dataset. The (C.I.) column title states that the metrics do not depend on the classifier used, only on the distance.

Metric	NMI-TRAIN (C.I.)		CP-TRAIN (C.I.)		NMI-TEST k-NN		Med-k-NN		Mon-k-NN (IR)	
Classifier										
Distance	Euclidean	LM^3L	Euclidean	LM^3L	Euclidean	LM^3L	Euclidean	LM^3L	Euclidean	LM^3L
autoMPG8	0.002558	**0.000000**	**0.401125**	0.000539	0.001534	**0.000000**	0.001390	**0.000000**	0.000834	**0.000000**
car	0.000058	**0.000000**	**0.143672**	0.003914	0.000126	**0.000000**	0.000116	**0.000000**	0.000028	**0.000000**
ERA	0.033560	0.021917	**0.167776**	0.075397	0.026396	0.018346	0.025813	0.017834	0.027845	0.020012
ESL	0.009530	**0.003307**	**0.703749**	0.348972	0.006029	0.002217	0.005926	0.002212	0.007266	0.002425
LEV	0.013335	**0.005879**	**0.240869**	0.048001	0.008462	0.003953	0.008401	0.003911	0.010639	0.005150
machineCPU	0.013873	**0.003957**	**0.497229**	0.289424	0.011322	0.002963	0.011395	0.002906	0.006639	0.001817
pima	0.001640	**0.000045**	**0.073187**	0.001316	0.000896	**0.000015**	0.000896	0.000015	0.000396	0.000002
SWD	0.009505	**0.004307**	**0.126258**	0.008387	0.005609	0.003078	0.005274	0.003100	0.007336	0.003595
balance	**0.000000**	**0.000000**	**0.256326**	0.209876	0.000051	0.000041	0.000038	0.000016	**0.000000**	**0.000000**
boston-housing	0.002775	**0.000000**	**0.149025**	0.000000	0.001934	**0.000000**	0.001731	**0.000000**	0.001443	**0.000000**
AVG SCORE	0.008683	**0.003941**	**0.275922**	0.098583	0.006236	0.003061	0.006098	0.002999	0.006243	0.003300
AVG RANK	1.954545	**1.045455**	**1.000000**	2.000000	10.681818	4.636364	9.772727	4.090909	9.500000	4.318182

Metric	NMI-TEST Mon-k-NN (OR)		Mon-F-k-NN (IR)		Mon-F-k-NN (OR)		CP-TEST (C.I.)	
Classifier								
Distance	Euclidean	LM^3L	Euclidean	LM^3L	Euclidean	LM^3L	Euclidean	LM^3L
autoMPG8	0.000762	**0.000000**	0.000747	**0.000000**	0.001657	**0.000000**	**0.400190**	0.000507
car	0.000028	**0.000000**	0.000085	0.000005	0.000222	4.18e-07	**0.143342**	0.003887
ERA	0.027845	0.020012	0.024477	0.017358	0.024352	**0.017335**	**0.167337**	0.075719
ESL	0.007138	0.002404	0.006103	0.001991	0.006096	**0.001914**	**0.703130**	0.351790
LEV	0.010639	0.005150	0.007547	0.003831	0.007555	**0.003815**	**0.240689**	0.047908
machineCPU	0.006346	0.001614	0.008060	**0.001277**	0.011200	0.002779	**0.489219**	0.286028
pima	0.000411	**0.000002**	0.000231	**0.000000**	0.000744	0.000012	**0.073161**	0.001432
SWD	0.006880	0.003595	0.005230	0.002951	0.005212	**0.002945**	**0.125906**	0.008363
balance	**0.000000**	**0.000000**	**0.000000**	**0.000000**	0.000003	**0.000000**	**0.256840**	0.211008
boston-housing	0.001438	**0.000000**	0.000441	**0.000000**	0.001719	**0.000000**	**0.147432**	0.000000
AVG SCORE	0.006149	0.003278	0.005292	**0.002741**	0.005876	0.002880	**0.274725**	0.098664
AVG RANK	8.954545	4.045455	7.545455	**2.545455**	9.181818	2.727273	**1.000000**	2.000000

Then, the NMI of the labeled dataset $\mathcal{X}$ with respect to the labeled dataset $\mathcal{Y}$ is defined as

$$NMI(\mathcal{X}, \mathcal{Y}) = \frac{1}{|\mathcal{X}||\mathcal{Y}| - |\mathcal{X} \cap \mathcal{Y}|} \sum_{x \in \mathcal{Y}} NClash_{\mathcal{X}}(x).$$

We can use the NMI in different ways. If we want to measure the monotonicity of the original training set $\mathcal{X}$, we can use $NMI(\mathcal{X}, \mathcal{X})$. If we want to measure the monotonicity of the training set after being transformed by a linear map $L \in \mathcal{M}_d(\mathbb{R})$, we can use $NMI(L\mathcal{X}, L\mathcal{X})$. Finally, if we want to measure the monotonicity of a set of test samples and their predictions, $\mathcal{X}_t$, with respect to the training set, we can use $NMI(L\mathcal{X}, \mathcal{X}_t)$. Table 3 shows the results regarding the fulfillment of the monotonic constraints. In this table, the metric NMI-TRAIN represents the NMI fot the training sets, for the Euclidean distance (that is, with no transformations applied) and for the transformed dataset using the distance learned by LM^3L. The NMI-TEST metric represents each of the NMIs of the training sets for each distance, with respect to the sets of predicted values by each of the classifiers for the test set. We also show the total of comparable pairs in the training set (CP-TRAIN), and between the training and test sets (CP-TEST), for each of the distances.

4.3   Analysis of Results

Looking at the results of Tables 2 and 3, we can draw the following conclusions. First, we can see that our algorithm considerably improves the median-vote k-NN (Med-k-NN) in terms of MAE and C-INDEX, being in fact the combination of LM^3L and Med-k-NN the one that obtains the best results in the experiments. However, when LM^3L is combined with the monotonic classifiers the results are not so good, often getting worse than the results obtained by the Euclidean distance with these classifiers. Therefore, in terms of classification performance we see that the distance learned by LM^3L is able to achieve significantly better results than the Euclidean distance, although this improvement is achieved only using the more traditional k-NN and Med-k-NN classifiers. This may be due to the fact that the monotonic classifiers already optimize the constraint aspect a lot and the combination with LM^3L becomes counterproductive.

Finally, according to the monotonicity results, we can see that the transformation that our algorithm learns considerably reduces the number of non-monotonic pairs of samples after transforming the training set. If we observe the monotonicity of the predicted samples with respect to the training set, we see again that the transformation learned by LM^3L allows to reduce the number of predictions that violate a monotonic constraint, for all the classifiers. Here we can clearly highlight the ability of our method with regards to not introducing new wrong monotonic constraints when transforming the dataset, thanks to the use of the M-matrices in the optimization process. It should be noted that the reduction of the NMI is at the cost of also reducing the number of comparable instances in the dataset, as can also be seen in Table 3, where the number of comparable pairs is always higher in the untransformed dataset. In any case, this reduction can help us to detect instances that due to noise or lack of accuracy were wrongly linked in monotonic constraints.

5   Conclusion

In this paper we have developed a new distance metric learning algorithm for monotonic classification, that, for the first time, exploits the potential of linear transformations to reduce the non-monotonicity of the dataset, thanks to the use of the M-matrices. In addition, the distances learned allow us to improve the classification performance of the classifiers analyzed.

As a future work, we plan to go deeper in the control of the monotonic constraints, so that the algorithm does not arbitrarily filter the constraints but it is possible to determine which constraints are more important in the dataset and make them immutable when transforming it. Another scope of study consists in the extension of our algorithm to deal with data of a highly non-linear nature. For this purpose, tools such as kernel functions [16] of deep metric learning [19] may be of interest.

References

1. Ben-David, A.: Automatic generation of symbolic multiattribute ordinal knowledge-based DSSS: methodology and applications. Decis. Sci. **23**(6), 1357–1372 (1992)
2. Berman, A., Plemmons, R.J.: Nonnegative matrices in the mathematical sciences. SIAM (1994)
3. Cano, J.R., Aljohani, N.R., Abbasi, R.A., Alowidbi, J.S., Garcia, S.: Prototype selection to improve monotonic nearest neighbor. Eng. Appl. Artif. Intell. **60**, 128–135 (2017)
4. Cano, J.R., Gutiérrez, P.A., Krawczyk, B., Woźniak, M., García, S.: Monotonic classification: an overview on algorithms, performance measures and data sets. Neurocomputing **341**, 168–182 (2019)
5. Chen, C.C., Li, S.T.: Credit rating with a monotonicity-constrained support vector machine model. Expert Syst. Appl. **41**(16), 7235–7247 (2014)
6. Cover, T.M., Hart, P.E., et al.: Nearest neighbor pattern classification. IEEE Trans. Inf. Theory **13**(1), 21–27 (1967)
7. Cunningham, J.P., Ghahramani, Z.: Linear dimensionality reduction: survey, insights, and generalizations. J. Mach. Learn. Res. **16**(1), 2859–2900 (2015)
8. Duivesteijn, Wouter, Feelders, Ad.: Nearest neighbour classification with monotonicity constraints. In: Daelemans, Walter, Goethals, Bart, Morik, Katharina (eds.) ECML PKDD 2008. LNCS (LNAI), vol. 5211, pp. 301–316. Springer, Heidelberg (2008). https://doi.org/10.1007/978-3-540-87479-9_38
9. González, S., García, S., Li, S.T., John, R., Herrera, F.: Fuzzy k-nearest neighbors with monotonicity constraints: moving towards the robustness of monotonic noise. Neurocomputing **439**, 106–121 (2021)
10. Keller, J.M., Gray, M.R., Givens, J.A.: A fuzzy k-nearest neighbor algorithm. IEEE Trans. Syst. Man Cybern. **4**, 580–585 (1985)
11. Nguyen, B., Morell, C., De Baets, B.: Distance metric learning for ordinal classification based on triplet constraints. Knowl.-Based Syst. **142**, 17–28 (2018)
12. Petersen, K.B., Pedersen, M.S., et al.: The matrix cookbook. Tech. Univ. Den. **7**(15), 510 (2008)
13. Suárez, J.L., García, S., Herrera, F.: pydml: a Python library for distance metric learning. J. Mach. Learn. Res. **21**(96), 1–7 (2020)
14. Suárez, J.L., García, S., Herrera, F.: Ordinal regression with explainable distance metric learning based on ordered sequences. Mach. Learn. **110**(10), 2729–2762 (2021). https://doi.org/10.1007/s10994-021-06010-w
15. Suárez, J.L., García, S., Herrera, F.: A tutorial on distance metric learning: mathematical foundations, algorithms, experimental analysis, prospects and challenges. Neurocomputing **425**, 300–322 (2021)
16. Torresani, L., Lee, K.C.: Large margin component analysis. In: Advances in Neural Information Processing Systems 19, p. 1385 (2007)
17. Triguero, I., et al.: Keel 3.0: an open source software for multi-stage analysis in data mining (2017)
18. Weinberger, K.Q., Saul, L.K.: Distance metric learning for large margin nearest neighbor classification. J. Mach. Learn. Res. **10**(Feb), 207–244 (2009)
19. Yi, D., Lei, Z., Liao, S., Li, S.Z.: Deep metric learning for person re-identification. In: 2014 22nd International Conference on Pattern Recognition, pp. 34–39. IEEE (2014)

Deep Learning Architectures Extended from Transfer Learning for Classification of Rice Leaf Diseases

Hai Thanh Nguyen[1]([✉]), Quyen Thuc Quach[2], Chi Le Hoang Tran[3], and Huong Hoang Luong[4]

[1] Can Tho University, Can Tho, Vietnam
nthai.cit@ctu.edu.vn
[2] Soc Son High School, Kien Giang, Vietnam
c3socson.kiengiang@moet.edu.vn
[3] FPT Polytechnic, Can Tho, Vietnam
chithl@fe.edu.vn
[4] FPT University, Can Tho, Vietnam

Abstract. Rice is one of the world's five main food crops. The problem helps farmers identify diseases on rice leaves early and develop a plan to prevent diseases in time; at the same time, helping them reduce damage and increase crop yields is of great interest to the agricultural sector. However, with the cultivation on a large scale, the detection of rice diseases by experience or manual form is still limited. In recent years, the application of Deep Learning techniques to detect disease identification in rice through images has yielded many superior results compared to traditional methods. This study has leveraged and extended transfer learning convolutional neural network architectures including DenseNet-121, VGG-16, MobileNet-V2, and ResNet-50 to identify the four most common rice leaves diseases in the Mekong Delta, Vietnam, such as bacterial leaf blight, tungro, blast, and brown spot, and obtained better performances compared to the original architectures with accuracies of 0.9930, 0.9703, 0.9740, and 0.9770, respectively.

Keywords: Transfer learning · Rice leaf · Rice diseases · Deep learning

1 Introduction

Plant diseases are a significant threat to food quality. Especially for rice is the staple food of Asians in general and Vietnam in particular. Bacterial leaf blight, tungro, blast, and brown spot (as illustrated in Fig. 1) are common diseases in rice. When the disease arises, it spreads very quickly on a large scale and can cause loss of productivity if not prevented in time. Therefore, identifying and detecting plant diseases based on computer vision is increasingly popular in agriculture. Moreover, applying deep learning technology to identify diseases on rice plants through images has become a research issue of great interest.

© Springer Nature Switzerland AG 2022
H. Fujita et al. (Eds.): IEA/AIE 2022, LNAI 13343, pp. 785–796, 2022.
https://doi.org/10.1007/978-3-031-08530-7_66

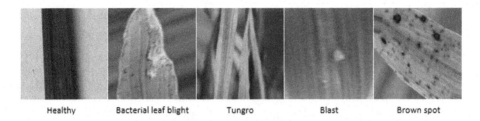

Healthy   Bacterial leaf blight   Tungro   Blast   Brown spot

Fig. 1. Disease symptoms of rice leaves.

In this study, we propose an approach based on the famous Convolutional Neural Network (CNN) architecture to identify diseases in rice plants through images of rice leaves[1]. Besides, we compare the results to prove that our model is more efficient than other architectures. Our contributions include:

- We present a deep learning-based approach for identifying rice leaf diseases using transfer learning techniques. Moreover, we have added two neural layers to the original famous considered architectures such as DenseNet-121 [1], VGG-16 [2], MobileNet-V2 [3,4], ResNet-50 [5] architectures, and Support Vector Machine (SVM) algorithm. From obtained results, this extension gives an improvement in classification performance. In addition, DenseNet-121 has exhibited the best result in both the original architecture and the extended architecture compared to other techniques.
- Data augmentation method has also revealed the efficiency in improving rice leaves disease classification performance. The dataset used for training with the proposed model includes 6,000 rice images (one type of rice with normal development and four types of rice diseases: Bacterial leaf blight, tungro, blast, and brown spot).

The rest of this article is as follows. First, in Sect. 2, we introduce some related work. Then, Sect. 3 presents and illustrates the proposed method and overall architecture. Next, the experimental results are presented in Sect. 4. Finally, the conclusions are presented in Sect. 5.

2   Related Work

There have been many successful applications of deep learning models in many areas of computer vision to identify plant diseases [6–9]. For example, in research [10], the authors proposed a model using SIFT feature extraction and SVM classification to process images on leaves of rice plants to detect and recognize four types of rice leaves, including zebra blast, rice blast, leaf rollers, and brown-backed hoppers. In another study [11], the author applied BLASTRec operating on an Android operating system based on two classification algorithms, Naive

---

[1] https://cezannec.github.io/Convolutional_Neural_Networks/.

Bayes and Decision Tree. Experimental results showed that the classification of the two algorithms has an accuracy of over 90% to support farmers in preventing rice blast disease. In addition, Jun Liu and Xuewei Wang [12] used a semi-supervised learning method on a small dataset to detect plant diseases based on deep learning. In addition, the study [13] evaluated different image processing techniques to increase recognition ability. The authors also presented a system for detecting and classifying rice leaf diseases, including Bacterial leaf blight, Brown spot, and Leaf smut on their dataset.

The appearance of the CNN network has promoted the development of computer vision [14,15], different architectures of CNN have been studied more and more to find optimization methods [16,17]. For example, the authors in [18] used Information and Communication Technology (ICT) tools, applied traditional image processing techniques combined with CNN networks to build and train a model. for image analysis of datasets collected for different classes for classification and detection of rice leaf diseases. In [19], the authors proposed a Faster R-CNN model to detect rice leaf diseases in real-time. At the same time, the authors also compared and evaluated the results with other studies; The results showed that their model had the highest accuracy. Besides, DenseNet neural network was also applied to predict COVID-19 by CT images [20] with 2,482 images and achieved 95% accuracy.

3   Method

This study has deployed transfer learning and data augmentation techniques with the DenseNet-121 model to identify diseases on rice leaves and compare them with VGG-16, MobileNet-V2, ResNet-50, and SVM architectures with details exhibited in Fig. 2.

The dataset includes images of diseases on rice leaves (bacterial leaf blight, tungro, blast, and brown spot) and healthy rice leaves. Data were collected from some sources[2,3,4] and from some rice fields in Kien Giang province, Vietnam. We have labeled images with the diseases of rice leaves and healthy rice leaves from several rice fields in Kien Giang province, Vietnam. Some images which contain many diseases are removed to ensure each image only reveals one disease or health. Some common diseases on rice leaves (as exampled in Fig. 1) with specific manifestations are as follows:

- Bacterial leaf blight: The disease causes gray, yellow-brown, or red-brown burning spots along the edges of the leaves. The disease spreads very quickly after the rice blooms.
- Tungro: The disease first appears as a small pale yellow spot, round or more often oval. Spots spread rapidly and extend yellow stripes toward the tips of leaves. Finally, yellow stripes spread out into orange-yellow streaks. The

---

[2] https://www.kaggle.com/bahribahri/riceleaf.
[3] https://archive.ics.uci.edu/ml/datasets/Rice+Leaf+Diseases.
[4] https://archive-beta.ics.uci.edu/ml/datasets?name=rice.

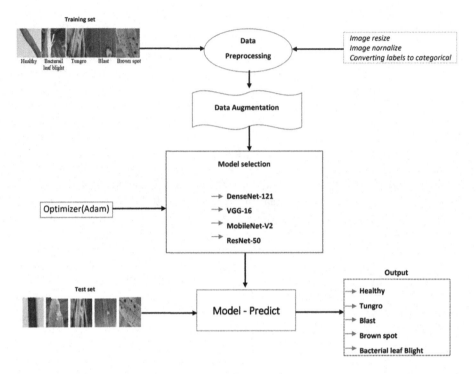

Fig. 2. Overview architecture of disease identification on rice leaves.

severe disease causes the leaves to turn wholly orange-yellow, burning the leaves. Yellow leaves disease appears at tillering stage of rice.

– Blast: At first as small green dots or faint oil stains, then turn to light gray; the lesions are extensive, rhombus-shaped, thick, light brown, sometimes with a yellow halo; the middle part of the disease is gray-brown. Blast disease is highly destructive to rice in temperate and subtropical regions, rice is flattened, and cotton neck is broken.

– Brown spot: Due to the fungus Helminthosporium oryzae and the fungus Curvularia lunata; Initially, the disease is only as small as a light brown needle tip; then, it gradually expands to a slightly oval shape, almost like a sesame seed, brown, dark brown on both sides of the disease, surrounded by a tiny yellow halo, reducing rice yield.

We adjust the image size to 224 × 224 as input for considered architectures. Each type of rice disease image initially consisted of 200 images (see details in Table 1). We have divided the dataset into the training set, validation set, and test set. To increase the efficiency of applying deep learning architecture, we perform a data augmentation technique on the training set to increase the size of the dataset. To increase the size of the dataset, we use the data augmentation technique from an original image to create six images by rotating, flipping, and zooming, as exhibited in Table 2.

The model selection is performed based on the classification performance on the validation set. Finally, the best model is saved to evaluate the test set during the learning phase.

We have deployed the DenseNet-121, MobileNet-V2, VGG-16, and ResNet-50 architectures for comparison. Moreover, we have added two neural layers to the original, such as transfer learning architectures, to improve the classification tasks. Besides, we also compare the performance of some famous transfer learning algorithms such as MobileNet-V2, VGG-16, ResNet-50, and SVM.

Table 1. The number of samples of sets dataset.

	Healthy	Bacterial leaf blight	Tungro	Blast	Brown spot	Total
Training and validation images after data augmentation	1,200	1,200	1,200	1,200	1,200	6,000
Training rate in Training and validation set	80:20	80:20	80:20	80:20	80:20	80:20
Test set	200	200	200	200	200	1,000

Table 2. The considered augmentation techniques and their values in the experiments.

Data augmentation methods	Range
rotation_range	45
width_shift_range	0.1
height_shift_range	0.1
zoom_range	−0.2

4   Results

This section presents comparison results between the original famous CNN architectures and the extended architectures. Moreover, some results reveal the effectiveness of the data augmentation technique in the image classification of rice leaf diseases.

We evaluate the model's performance with the Accuracy (ACC), Matthews Correlation Coefficient (MCC), Area Under the Curve (AUC), Precision, Recall, and F1 Score metrics. Hyper-parameters used for all architectures include a *learning rate* of 10^{-4}, a *batch size* of 16 running to 100 *epochs*. The best performance is stored and used for model selection during the learning through many epochs.

For comparison with classical machine learning techniques, we run the training with the SVM algorithm and then evaluate the test set with four kernels, including linear, poly, Radial basis function (RBF), and sigmoid. The results with SVM show that the classification measures are much lower than our four CNN architectures.

Table 3. Experimental results on four original architectures on the test set.

	ACC	MCC	AUC	Precision	Recall	F1-Score
DenseNet-121	0.8620	0.8325	0.9712	0.8801	0.8620	0.8613
VGG-16	0.6810	0.6026	0.8917	0.6790	0.6810	0.6800
MobileNet-V2	0.7900	0.7406	0.9501	0.7909	0.7900	0.7905
ResNet-50	0.7100	0.6483	0.8996	0.7402	0.7100	0.7248
SVM (kernel:"linear")	0.2286	0.0360	0.4807	0.2192	0.2286	0.2199
SVM (kernel:"poly")	0.2317	0.0400	0.4680	0.2276	0.2317	0.2272
SVM (kernel:"RBF")	0.2444	0.0558	0.5000	0.2462	0.4444	0.2421
SVM (kernel:"sigmoid")	0.2254	0.0500	0.5000	0.2406	0.2254	0.1518

Table 3 shows that the original DenseNet-121 architecture results are significantly higher than the other original architectures. Therefore, we experiment with extending the DenseNet-121 architecture to improve the results by adding two neural layers of 256 neurons and 128 neurons, respectively.

Figure 3 illustrates the performances of DenseNet-121 architecture with various variances. The accuracy has gradually increased from 0.8620 (original architecture) to 0.9141 (extended architecture) and reached the highest value of 0.9930 (when combining both extended architecture and data augmentation technique).

This proves that our proposed model has a clear and comprehensive effect, significantly improving the results of disease classification on rice leaves. Therefore, we continue to train the proposed model with the remaining architectures (VGG-16, MobileNet-V2, ResNet-50) to compare the results with the original architectures. The results are detailed in Table 4.

Table 4. Detailed comparison of experimental results on four extended architectures with data augmentation technique on test set.

	ACC	MCC	AUC	Precision	Recall	F1-Score
DenseNet-121	0.9930	0.9913	0.9975	0.9931	0.9930	0.9830
VGG-16	0.9703	0.9700	0.9950	0.9700	0.9700	0.9701
MobileNet-V2	0.9740	0.9675	0.9980	0.9741	0.9740	0.9740
ResNet-50	0.9770	0.9713	0.9978	0.9770	0.9770	0.9769

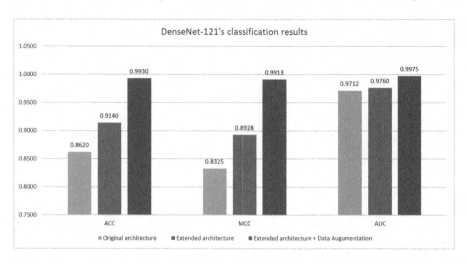

Fig. 3. The chart compares the results of DenseNet-121 between the original architecture and the proposed architecture.

DenseNet-121 architecture with the proposed model gives the best results, so we use this architecture to conduct the classification and get the results as the Table 5.

Table 5. Classification results on various classes with different scores.

	ACC	Precision	Recall	F1-Score
Bacterial leaf blight	0.9600	0.9700	0.9600	0.9600
Tungro	0.9950	0.9900	0.9900	0.9900
Blast	0.9700	0.9900	1.0000	1.0000
Brown spot	0.9950	0.9900	0.9900	0.9900

Loss is the distance between the actual label vector and the predicted label vector; the more the predicted model deviates from the actual value, the larger the Loss and vice versa. If the prediction is closer to the actual value, the Loss will gradually decrease to 0. We proceed to calculate the Loss by Formula 1.

$$Loss = -\sum_{i=1}^{n}\sum_{j=1}^{m} y_{i,j} \log_e(p_{j,j}) \tag{1}$$

where n is the number of data points while m is the number of classes, $y_{i,j}$ is the actual label of the data ($y_{i,j} = 1$ if data point i belongs to class j, otherwise $y_{i,j} = 0$), and $p_{i,j}$ is the predicted label ($p_{i,j}$ represents the probability that the model predicts data point i of class j).

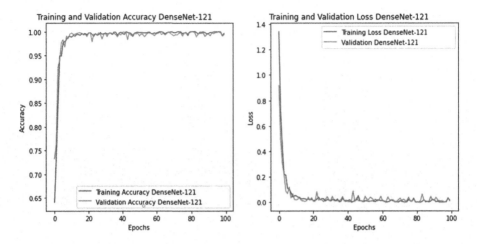

Fig. 4. Training and validation of DenseNet-121 architecture.

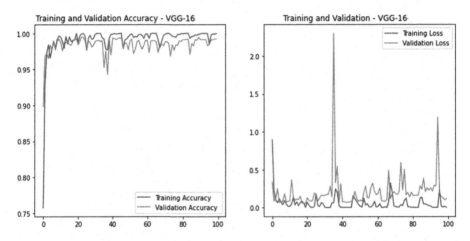

Fig. 5. Training and validation of VGG-16 architecture.

The charts in Figs. 4, 5, 6, 7, represent the results of four architectures through 100 epochs. Figure 8 shows the Confusion matrix of the four architectures, and we conclude that the DenseNet-121 architecture tends to be more stable than the VGG-16, MobileNet-V2, ResNet-50 architectures. As the epoch increases, the accuracy gradually increases, and at the same time, the error decreases.

We use the same dataset for training and one dataset of tests on different network architectures. We use the data augmentation technique to create large datasets for training with deep learning problems that give much higher results than the original datasets. Besides, we also use the method of extended architecture for training. Experimental results show that the accuracy is higher than

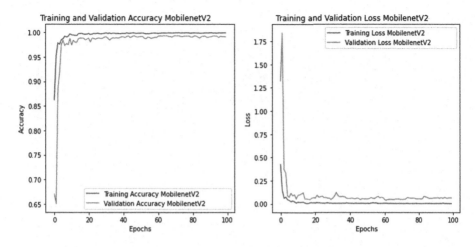

Fig. 6. Training and validation of MobileNet-V2 architecture.

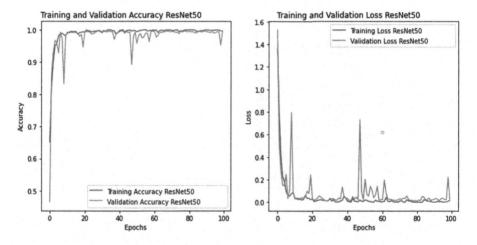

Fig. 7. Training and validation of ResNet-50 architecture.

the original architectures. Combining extended architecture and the data augmentation technique gives the best and highest results (0.9930) compared to the original and extended architecture. In four architectures, all of them are trained on the same dataset and the same test set, but the network architecture DenseNet-121 (0.9930) gives higher results and runs more stable than the other three architectures such as VGG-16 (0.9703), MobileNet-V2 (0.9740), and ResNet-50 (0.9770).

In this study, we have selected, adjusted the hyper-parameters in the network architecture, applied the transfer learning method on large datasets in the problem of rice disease identification through images; DenseNet-121 network architecture for high accuracy results in a good performance, stability.

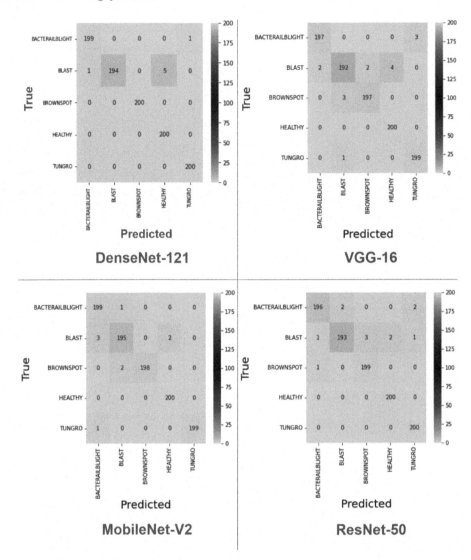

Fig. 8. Confusion matrix of four extended architectures including DenseNet-121, VGG-16, MobileNet-V2, ResNet-50.

5   Conclusion

This study has demonstrated that using transfer learning and data augmentation techniques in training rice leaves disease recognition models achieves very high accuracy on architectures: DenseNet-121 (0.9930), VGG-16 (0.9703), MobileNet-V2 (0.9740), and ResNet-50 (0.9770). Our model shows the influence of the dataset size to build the model on the system's accuracy. Furthermore, the model works well on large datasets, reducing overfitting. Especially for the

DenseNet-121 architecture, which adopts the join structure, can effectively save the parameters and reduce the computation, thus saving the bandwidth and storage cost; simultaneously, the DenseNet algorithm has powerful anti-overfitting performance. Besides, the experimental results show that the DenseNet architecture is well used for disease identification in rice. In addition, our study provides an analysis and outlook on the future trends of plant disease and pest detection based on deep learning. This technology has excellent development potential and application value because of the early recognition of plant diseases. It is very beneficial for preventing and controlling plant diseases and pests. From there, it is possible to prevent their spread and development.

References

1. Huang, G., Liu, Z., Maaten, L.V.D., Weinberger, K.Q.: Densely connected convolutional networks. In: 2017 IEEE Conference on Computer Vision and Pattern Recognition (CVPR). IEEE Publishing, July 2017. https://doi.org/10.1109/cvpr.2017.243
2. Simonyan, K., Zisserman, A.: Very deep convolutional networks for large-scale image recognition (2015)
3. Nawaz, M.A., et al.: Plant disease detection using internet of thing (IoT). Int. J. Adv. Comput. Sci. Appl. **11**(1) (2020). https://doi.org/10.14569/ijacsa.2020.0110162
4. Sandler, M., Howard, A., Zhu, M., Zhmoginov, A., Chen, L.C.: MobileNetV2: inverted residuals and linear bottlenecks. In: 2018 IEEE/CVF Conference on Computer Vision and Pattern Recognition. IEEE Publishing, June 2018. https://doi.org/10.1109/cvpr.2018.00474
5. He, K., Zhang, X., Ren, S., Sun, J.: Deep residual learning for image recognition. In: 2016 IEEE Conference on Computer Vision and Pattern Recognition (CVPR). IEEE Publishing, June 2016. https://doi.org/10.1109/cvpr.2016.90
6. Dhingra, G., Kumar, V., Joshi, H.D.: A novel computer vision based neutrosophic approach for leaf disease identification and classification. Measurement **135**, 782–794 (2019). https://doi.org/10.1016/j.measurement.2018.12.027
7. Nanehkaran, Y.A., Zhang, D., Chen, J., Tian, Y., Al-Nabhan, N.: Recognition of plant leaf diseases based on computer vision. J. Ambient Intell. Hum. Comput. (2020). https://doi.org/10.1007/s12652-020-02505-x
8. Loey, M., ElSawy, A., Afify, M.: Deep learning in plant diseases detection for agricultural crops. Int. J. Serv. Sci. Manage. Eng. Technol. **11**(2), 41–58 (2020). https://doi.org/10.4018/ijssmet.2020040103
9. Roy, A.M., Bhaduri, J.: A deep learning enabled multi-class plant disease detection model based on computer vision. AI **2**(3), 413–428 (2021). https://doi.org/10.3390/ai2030026
10. Nguyen, N.T., Bui, T.T.P., Le, H.N., Ngo, N.T.: Rice pests and diseases identification using sift feature. Version B Vietnam J. Sci. Technol. **61**(8) (2019). https://b.vjst.vn/index.php/ban_b/article/view/158
11. Thu, T.N.M., Nguyen Thi Thanh Lan, N.H.M.: Content-based recommendation system to support farmers in blast prevention. Can Tho Univ. J. Sci. Inf. Technol. **2017**, 164 (2017). https://doi.org/10.22144/ctu.jsi.2017.022
12. Liu, J., Wang, X.: Plant diseases and pests detection based on deep learning: a review. Plant Methods **17**(1) (2021). https://doi.org/10.1186/s13007-021-00722-9

13. Prajapati, H.B., Shah, J.P., Dabhi, V.K.: Detection and classification of rice plant diseases. Intell. Dec. Technol. **11**(3), 357–373 (2017). https://doi.org/10.3233/IDT-170301

14. Wang, W., Yang, Y.: Development of convolutional neural network and its application in image classification: a survey. Opt. Eng. **58**(04), 1 (2019). https://doi.org/10.1117/1.oe.58.4.040901

15. Yoo, H.J.: Deep convolution neural networks in computer vision: a review. IEIE Trans. Smart Process. Comput. **4**(1), 35–43 (2015). https://doi.org/10.5573/ieiespc.2015.4.1.035

16. Jmour, N., Zayen, S., Abdelkrim, A.: Convolutional neural networks for image classification. In: 2018 International Conference on Advanced Systems and Electric Technologies (IC_ASET). IEEE Publishing, March 2018. https://doi.org/10.1109/aset.2018.8379889

17. Sultana, F., Sufian, A., Dutta, P.: Advancements in image classification using convolutional neural network. In: 2018 Fourth International Conference on Research in Computational Intelligence and Communication Networks (ICRCICN). IEEE Publishing, November 2018. https://doi.org/10.1109/icrcicn.2018.8718718

18. Fan, F., Roy, T., Roy, K.: Classification and detection rice leaf diseases using information and communication technology ICT tools. Int. J. Adv. Eng. Res. Sci. **7**(6), 460–470 (2020). https://doi.org/10.22161/ijaers.76.56

19. Bari, B.S., et al.: A real-time approach of diagnosing rice leaf disease using deep learning-based faster r-CNN framework. PeerJ Comput. Sci. **7**, e432 (2021). https://doi.org/10.7717/peerj-cs.432

20. Hasan, N., Bao, Y., Shawon, A., Huang, Y.: DenseNet convolutional neural networks application for predicting COVID-19 using CT image. SN Comput. Sci. **2**(5) (2021). https://doi.org/10.1007/s42979-021-00782-7

Height Estimation for Abrasive Grain of Synthetic Diamonds on Microscope Images by Conditional Adversarial Networks

Joe Brinton[✉], Shota Oki, Xin Yang, and Maiko Shigeno

University of Tsukuba, 1-1-1 Tennodai, Tsukuba, Japan
s202020441@s.tsukuba.ac.jp

Abstract. In this paper, we show an example of a 2-class classification of single-view synthetic diamond images based on their height with pix2pix, an image-to-image translation method using conditional adversarial networks. We first gradationally color each diamond image based on each pixel's height (grad-color image), then we train the pix2pix model to convert a raw image into a grad-color image. In the testing phase, we use the trained model to estimate grad-color images of raw image inputs whose height is unknown, then classify each image as normal or abnormal based on the generated grad-color image. Results show a possibility of implementing this method to real world cases where images can only be taken from a single angle to be practicable.

Keywords: pix2pix · Generative Adversarial Networks · Deep learning

1 Introduction

One of the popular industrial applications of synthetic diamond is cutting tools and polishing tools. When such tools are used for high-precision processing, the quality of such will affect the quality of the processed product, thus it is critical to maintain its quality at a high standard. Degradation of such tools is mainly caused by abnormally sharp synthetic diamond particles. If abnormally sharp synthetic diamonds can be detected from single-view images, the quality of such tools can be improved. We interpret "abnormally sharp" as "abnormally tall" in this paper and consider a method to effectively detect such tall synthetic diamonds from a single-view image of them (see Fig. 1). In other words, given an "anomaly threshold", we want to determine whether the maximum height of a grain of synthetic diamond is higher than this threshold or not from a single-view image of it. Each diamond image has a size of 768×768 as shown in Fig. 1. The images were taken by a laser microscope and provided together with data of each pixel's height (see Fig. 2). Since acquiring such data consumes time, our task is to detect whether a grain of synthetic diamond has an abnormal height only from a single-view image like Fig. 1.

© Springer Nature Switzerland AG 2022
H. Fujita et al. (Eds.): IEA/AIE 2022, LNAI 13343, pp. 797–804, 2022.
https://doi.org/10.1007/978-3-031-08530-7_67

	A	B	C	D	E	F	G	H
1	37760	37665	37770	37990	37870	37955	38300	382
2	37635	37935	38110	38145	37800	37865	38160	379
3	37620	37880	38915	32620	38170	37815	37915	378
4	37675	37685	37655	37925	38210	37870	37810	379
5	37720	37700	37510	37700	37960	38260	38140	379
6	37685	37615	37690	37580	37955	38500	38410	379
7	37765	37645	37600	37685	37840	38300	37995	378
8	37715	37630	37730	37780	37745	37805	37645	376
9	37720	37740	37735	37670	37605	37630	37565	376
10	37680	37645	37560	37705	37695	37500	37620	376
11	37795	37690	37495	37715	37560	37740	37515	376
12	38225	37735	37540	37650	37690	37590	37640	376
13	38650	38025	37690	37595	37690	37635	37730	376

Fig. 1. A sample image of a grain of synthetic diamond

Fig. 2. An example of height data corresponding to Fig. 1

There have been many works that researched methods to estimate three-dimensional features from two-dimensional images. Hara et al. [1] have proposed a method to calculate the height of a sphere-shaped object by overlaying an image taken from vertically above the object to an image created by applying distortion correction to one taken from a diagonal angle and using actual optical displacement values acquired from this overlaid image. Wu et al. [2] have proposed a method to estimate a 3D model of a human face from an single-view image by applying their training method based on an auto-encoder to facial images taken from various angles which showed superior accuracy compared to other methods. Nie el al. [3] developed a method that allows indoor scene understanding and object reconstruction simultaneously from a single image, and their method outperformed other existing methods in indoor layout estimation, 3D object detection, and mesh reconstruction. In these works, images taken from different angles are used in the process of estimating three-dimensional features. However, our situation limits us from acquiring images from different angles. Thus, this method mentioned above cannot be applied in this case.

In this paper, we use pix2pix, an image-to-image translation method using conditional adversarial networks, which was proposed by Isola et al. [4], to tackle this problem. There are other works that use pix2pix to resolve different problems in the image processing field such as [5–7], and this paper is one of these application cases.

2   Method

2.1   Overview of pix2pix

As mentioned above, pix2pix is a network proposed by Isola et al. [4] which learns a translation from one image to another. Its structure is based on Conditional GAN (CGAN) proposed by Mirza and Osindero [8], which is a conditional version of GAN [9]. The method of pix2pix enables converting an aerial photograph to a map of the same location, a doodle of a cat to a realistic image of it, etc. Since colorized images can be estimated from edge maps using pix2pix, we employ this method expecting it to be capable of coloring images according to height values. The structure of pix2pix is similar to that of CGAN - conditional image is used instead of conditional vector (see Fig. 3). The discriminator uses pairs of real or generated images and condition images as training data, and learns the correspondence between them. In our application - the model to learn a translation from a diamond image to a colored image in accordance with its height (grad-color image) - they each correspond to conditional image and real image respectively.

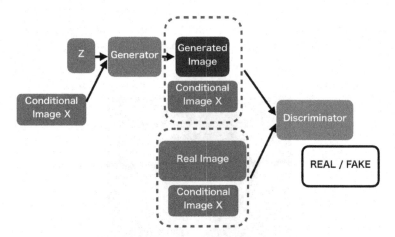

Fig. 3. Illustration of pix2pix (Color figure online)

2.2 Proposed Method

In this section, we explain the method to classify abrasive grain of synthetic diamonds depending on their height. We first prepare gradationally colored images for each abrasive grain of synthetic diamond as training data. We then use pairs of a raw image of a grain of synthetic diamond and its grad-color image to train the pix2pix model to learn the translation from a raw image to a colored image. By doing so, we can estimate a grad-color image from a diamond image whose height is unknown, and determine whether its height is higher than a given anomaly threshold.

In grad-color images, each pixel is colored blue if its corresponding height is below the anomaly threshold, and the higher it becomes than the threshold the more (resp. less) green (resp. blue) it becomes such as Fig. 4. The color of each pixel is defined by RGB decimal code $c = (x_R, x_G, x_B)$ where each component varying from 0 to 255. Since our focus is to differentiate pixels whose corresponding height is higher than the anomaly threshold (abnormal pixels) from those that are lower than it (normal pixels), x_R for red is constantly set to 0, and values depending on the height will be allocated to x_B for blue and x_G for green for each pixel. The height of the anomaly threshold is given by z, and we set $c = (0, 0, 255)$ for its color.

We define t_{anom}, t_{color} as parameters. t_{anom} represents the anomaly threshold height ratio to the set of maximum heights of each diamond, H. For instance, if $t_{anom} = 0.75$, the anomaly threshold is equal to the third quartile of H. t_{color} sets the height ratio of the border z of blue and green. If $t_{color} = 0.5$, the border of blue and green is at the height of the second quartile. The reason t_{color} is defined aside from t_{anom} is because the whole data set may become biased depending on the value of t_{anom}. If such bias appears, when we want to differentiate pixels above the height defined by t_{anom} from the ones below it by coloring them with different colors, the model may not be able to learn the data effectively since there will be fewer data with the color used for pixels above z; in our case, green. Thus, we set the value of t_{color} lower than t_{anom} to increase the amount of color used for pixels above z; in our case, to obtain grad-color images with more green area.

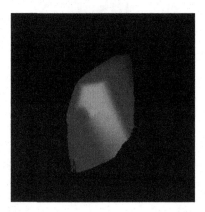

Fig. 4. An example of a gradationally colored image for a grain of synthetic diamond displayed in Fig. 1 (Color figure online)

To assign color values for the height values between z and $\frac{3}{2}z$ where blue and green are mixed together, we scale the height by $m = \lceil \frac{z}{255 \times 2} \rceil$. For height h, its color is defined by

$$
c = (x_R, x_G, x_B) = \begin{cases} (0,\ 0,\ 0) & h \leq b \\ (0,\ 0, \lfloor \frac{h}{m} \rfloor) & b < h \leq z \\ (0, \lfloor \frac{h}{m} \rfloor - 510, 765 - \lfloor \frac{h}{m} \rfloor) & z < h \leq \frac{3}{2}z \\ (0, 255, 0) & \frac{3}{2}z < h \end{cases}
$$

where b is the height of the base. If the height is less than b, its pixel is regarded as the base of a grain of synthetic diamonds and is colored by black. When $h \leq z$, we have $\lfloor \frac{h}{m} \rfloor \leq 255$ and the pixel will be colored blue. If $z < h \leq \frac{3}{2}z$, it will be a color with blue and green mixed together, and finally, if $\frac{3}{2}z < h$, this pixel will be colored completely green.

We use another parameter g_r in the classification phase. In this phase, we scan each pixel of the generated image for each diamond, and define pixels with $\lambda \times x_B \geq x_G$ as normal pixels, and $\lambda \times x_B < x_G$ as abnormal pixels where

$$\lambda = \frac{t_{anom} - t_{color}}{1 - t_{anom}}.$$

Then, if the ratio of abnormal pixels to the total number of pixels on the surface of the grain of synthetic diamond is higher than g_r, such diamond will be estimated to be abnormal.

3   Experiments

3.1   Settings

For our model, we used a pix2pix code provided by [10] in Python. LeakyReLU and ReLU are used as activation functions in the convolution phase and up-convolution phase of the generator respectively. Dropout, epochs, batch size, and patch size are 0.5, 100, 64, and 5 respectively. The ratio of train data and test data is 9 : 1 (900 for training and 100 for testing) since we need enough data for the model to learn the features each grain of synthetic diamond has. Since the anomaly threshold is set to the third quartile of H, $t_{anom} = 0.75$.

3.2   Results

Here, we show the estimation results using pix2pix. Fig. 5 shows examples of the obtained images. The top image shows the raw image of a grain of synthetic diamond, the middle shows the generated grad-color image, and the bottom shows the real grad-color image. Figures 6 and 7 show examples of grad-color images with different t_{color} values applied to the image of synthetic diamond showed in Fig. 1. Lowering the t_{color} value allows the model to generate grad-color images with more green areas, increasing the accuracy of them when compared to real grad-color images.

Table 1 shows results with different t_{color} values. We see an increase in Accuracy and Recall and a decrease in Precision as g_r decreases. When $t_{color} = 0.50, 0.35$, Precision and Recall were 0. This could be because the grad-color images became less likely to be classified as abnormal since $\lambda > 1$ is multiplied to x_B. In addition, there were not enough green areas in the estimated grad-color images at these t_{color} values to be classified as abnormal. At $t_{color} = 0.1$, Accuracy was over 0.8 and Recall reached its peak 0.750 at $g_r = 0.005$.

While in most cases the estimated grad-color image was accurate, in some cases they differed greatly from the real grad-color image (see Fig. 8). The generated image shows green areas on the base below the synthetic diamond. A common aspect these images share is that the base below the synthetic diamond is brighter than most other images, even similar to the brightness of the surface of the diamond. Due to this factor, the model may assume the base part

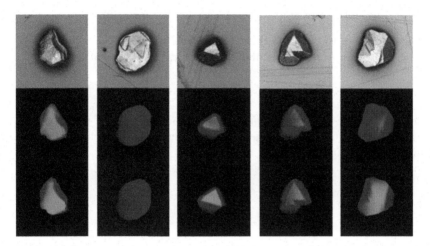

Fig. 5. pix2pix estimation examples (Color figure online)

as the surface of the diamond. Thus, it will be difficult to accurately estimate a grad-color image when some images have different levels of overall brightness compared to most others. In our case, the ratio of such failure images was 4%, and the brightness of the base part of all failure cases was high. This problem can be avoided by re-obtaining the image such that the brightness is equal to most other images. However, if this is not possible, we need to for instance extract the contour of the synthetic diamond and adjust the brightness of the area outside the contour. This way the model will be able to differentiate the base part and the surface of the diamond.

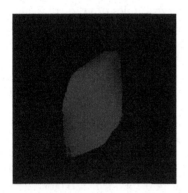

Fig. 6. t_{color} = 0.75 (Color figure online)

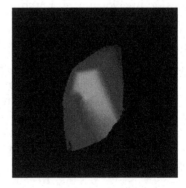

Fig. 7. t_{color} = 0.10 (Color figure online)

Table 1. Estimate result

t_{color}	g_r	Accuracy	Precision	Recall
0.75	0.05	0.590	0.260	0.632
	0.01	0.190	0.190	1.000
	0.005	0.190	0.190	1.000
	0.001	0.190	0.190	1.000
0.50	0.05	0.740	0.000	0.000
	0.01	0.740	0.000	0.000
	0.005	0.740	0.000	0.000
	0.001	0.740	0.000	0.000
0.35	0.05	0.760	0.000	0.000
	0.01	0.760	0.000	0.000
	0.005	0.760	0.000	0.000
	0.001	0.760	0.000	0.000
0.10	0.05	0.850	0.882	0.530
	0.01	0.810	0.655	0.679
	0.005	0.810	0.636	0.750
	0.001	0.770	0.568	0.750

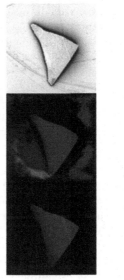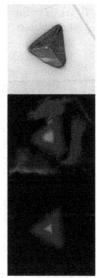

Fig. 8. Failure cases (Color figure online)

4 Conclusion

In this paper, we showed an example of applying pix2pix to a 2-class classification problem using single-view images only, and showed a possibility of implementing this method to real world cases where images can only be taken from a single angle to be practicable. While results of when t_{color} in grad-color images were not low enough showed little effectiveness, by lowering this value well enough and adjusting the classification condition, results showed effectiveness of this method on classifying objects by their height despite its information is limited. We also showed examples of inaccurate estimations of grad-color images. Since in these cases the overall brightness of the raw image differed from most other images, it can be said that little to no difference in the brightness between images is ideal for accurate estimation through pix2pix.

Acknowledgement. This research is supported in part by the JSPS Grant-in-Aid for Scientific Research (B) under Grant No. 20H02382. We would like to specially thank READ Co., Ltd. for preparing and providing all of the data used in this paper and for taking the time to discuss with us.

References

1. Hara, Y., et al.: Detection of height of minute balls based on triangulation using image distortion correction. In: The 24th Japan Institute of Electronics Packaging Research Presentation Conference (2010)

2. Wu, S., Rupprecht, C., Vedaldi, A.: Unsupervised learning of probably symmetric deformable 3D objects from images in the wild. arXiv:1911.11130 (2020)
3. Yinyu, N., Xiaoguang, H., Shihui, G., Zheng, Y., Chang, J., Zhang, J.J.: Total3DUnderstanding: joint layout, object pose and mesh reconstruction for indoor scenes from a single image. In: Proceedings of the IEEE/CVF Conference on Computer Vision and Pattern Recognition, pp. 55–64 (2020)
4. Isola, P., Zhu, J.-Y., Zhou, T., Efros, A.A.: Image-to-Image translation with conditional adversarial networks. In: Proceedings of the IEEE Conference on Computer Vision and Pattern Recognition, pp. 1125–1134 (2017)
5. Yanyun, Q., Yizi, C., Jingying, H., Xie, Y.: Enhanced pix2pix dehazing network. In: Proceedings of the IEEE/CVF Conference on Computer Vision and Pattern Recognition (CVPR), pp. 8160–8168 (2019)
6. Christovam, L.E., Shimabukuro, M.H., Galo, M.L.B.T., Honkavaara, E.: Pix2pix conditional generative adversarial network with MLP loss function for cloud removal in a cropland time series. Remote Sens. **14**(1), 144 (2022)
7. Liu, X., Meng, X., Wang, Y., Yin, Y., Yang, X.: Known-plaintext cryptanalysis for a computational-ghost-imaging cryptosystem via the pix2pix generative adversarial network. Opt. Express **29**, 43860–43874 (2021)
8. Mirza, M., Osindero, S.: Conditional generative adversarial nets. arXiv:1411.1784 (2014)
9. Goodfellow et al.: Generative adversarial networks. In: Advances in Neural Information Processing Systems 27 (2014)
10. Tommy, pix2pix-keras-byt. https://github.com/tommyfms2/pix2pix-keras-byt (2017)

Pattern Mining and Tsetlin Machines

Fast Weighted Sequential Pattern Mining

Zhenqiang Ye[1], Ziyang Li[1], Weibin Guo[1], Wensheng Gan[2,3(✉)],
Shicheng Wan[1], and Jiahui Chen[1]

[1] Guangdong University of Technology, Guangzhou 510006, China
[2] Jinan University, Guangzhou 510632, China
wsgan001@gmail.com
[3] Pazhou Lab, Guangzhou 510330, China

Abstract. In the real world, ordered sequence data is commonly seen, and sequence analysis plays an important role in a wide range of real applications, such as market basket analysis. The weight concept helps to find more interesting sequences, whereas they may be treated as meaningless patterns in sequential pattern mining. Therefore, how to effectively discover these high weighted sequences from a quantitative sequential database is an urgent task. Based on the remaining weight concept, we propose a novel algorithm called Fast Weighted Sequential Pattern Mining (FWSPM) by utilizing an upper-bound called the remaining sequence maximum weight. Based on this upper-bound, an effective pruning strategy is designed to reduce the search space and save memory cost. Experimental results on both real and synthetic datasets show that the designed FWSPM algorithm is more efficient than the existing algorithms, and also has good scalability on large-scale datasets.

Keywords: Pattern mining · Sequence · Weight · Upper-bound

1 Introduction

Discovering and analyzing interesting patterns (i.e., itemsets, rules, and sequences) from various types of data (e.g., transaction data, sequence, graph, stream data, etc.) is a vital and challenging task in the data mining domain [7,9,17,21]. Frequency is usually an important yardstick in data mining, and there are many kinds of mining algorithms adopting the frequency metric, such as association rule mining [2] and sequential pattern mining [15,22]. Traditional sequential pattern mining (namely SPM for short) [8,11] seeks to discover patterns in a specific order in large databases. Until now, it is well-known that different SPM technologies have been applied in a variety of fields to help people make better decisions, including market basket analysis [24], web-log mining [8], energy reduction in smart-homes [23], bioinformatics analysis [25], etc. However, finding out these interesting sequential patterns from large databases is not an easy task [1,7]. A fundamental limitation of SPM is the generation of a huge number of redundant and unpromising candidates. In order to solve this challenge,

H. Fujita et al. (Eds.): IEA/AIE 2022, LNAI 13343, pp. 807–818, 2022.
https://doi.org/10.1007/978-3-031-08530-7_68

researchers proposed a well-recognized anti-monotone concept called the downward closure property [4]. It meas that a frequent pattern cannot consist of any infrequent sub-pattern. Nonetheless, since SPM ignores the relative importance of each pattern [14,19], therefore some more useful and meaningful sequential patterns could not be discovered fully. In order to solve this limitation of SPM, Yun *et al.* [26] first formulated a new research task called weighted sequential pattern mining (WSPM) and proposed an algorithm called WSpan. In WSPM, different weights are assigned to the importance of distinct items. However, in WSPM the downward closure property in SPM is broken [12,20,26]. Therefore, it is a critical issue to develop a suitable algorithm for fast mining weighted sequential patterns.

To solve the above problems, in this paper, we propose an effective WSPM algorithm with a new upper-bound and a projection-based technique. Our novel algorithm is called **F**ast **W**eighted **S**equential **P**attern Mining (**FWSPM**). The major contributions of this paper are summarized as follows.

1. We propose an upper-bound called remaining sequence maximum weight (abbreviated as RSMW) to measure the weighted support of sub-sequence in the mining process. Besides, by utilizing a projection technique, the updated version of RSMW can speed up the mining process efficiently.
2. Based on RSMW, an effective pruning strategy is adopted to prune a number of unpromising sub-sequences in the recursive mining process. Therefore, the efficiency of finding weighted sequential patterns can be improved visibly.
3. The descending order is utilized in each sequence to facilitate the procedure of discovering the remaining maximum weight itemsets.
4. Extensive experiments on both real and synthetic datasets are tested to compare FWSPM with the state-of-the-art algorithms. Experimental results show that FWSPM performs better than other algorithms in terms of execution time and memory.

2   Related Work

Sequential pattern mining (SPM) [11] aims at discovering the interesting patterns with order. Many Apriori-based mining algorithms [2–4] were proposed to discover association rules from a transaction database with different constraints. However, all these algorithms do not consider the support metric with weight concept together and may lead to unreasonable results. Therefore, some studies consider the different significance of patterns in various applications, such as utility-driven SPM [10,13,27], weighted SPM (abbreviated as WSPM) [12,16,26], and other constraint-based SPM. In order to generate fewer candidates and reduce computation usage, researchers have proposed some upper-bound models. For example, Yun *et al.* [26] adopted the maximum weight (*MaxW*) from all items in database as an approximate value for all patterns in the WSpan algorithm. Since the average weight of any pattern will be no higher than the *MaxW*, this upper-bound value is always larger than or equal to the actual weighted support of its extension pattern. By this way, the $MaxW$-based model can hold the downward closure property during the mining procedure and reduce the search space. Lan *et al.* [16] proposed the IUA algorithm with a new upper-bound called MWU. The IUA algorithm can further tighten the upper-bound value when compare with WSpan. Since then IUA has better performance than WSpan. However, there is still room to improve the upper-bound model. Therefore, it motivated us to design a more efficient algorithm for the WSPM task.

3   Preliminaries

Firstly, there is a sequence database shown in Table 1, in which each sequence includes two characteristics, sequence identification (SID), and sequence itself. There are eight items in this example, respectively denoted as A to H. Moreover, assume the weight value of each item as $\{A: 0.1, B: 0.28, C: 0.38, D: 0.49, E: 0.51, F: 0.72, G: 0.82, H: 0.9\}$. A set of terms related to the problem of weighted sequential pattern mining (WSPM) are then defined as follows.

Table 1. A set of sequences

SID	Sequence
S_1	$\langle B\ A\ \ C\ B \rangle$
S_2	$\langle D\ E\ C\ H\ G \rangle$
S_3	$\langle A\ \ C\ F\ (DE)\ F \rangle$
S_4	$\langle (FG)\ H \rangle$
S_5	$\langle C\ (DA)\ C\ E\ F \rangle$

Definition 1 (Itemset's weight). We assume the weight value range of an item i, denoted as w_i, is from 0 to 1. The weight value of an itemset X, w_X, is $w_x = \frac{\sum_{i \in X}^{|X|} w_X}{l_X}$, where l_X is the number of items in X.

Cai *et al.* [6] first used the itemset weight to define weighted significance for fuzzy association rule mining. For example, in Table 1, the weight values of the two items C and D are 0.38 and 0.49, and the weight value of itemset $\{CD\}$ is the summation of C and D divided by the number of items in itemset $\{CD\}$. Thus, $w_{\{CD\}} = (0.38 + 0.49)/2 = 0.435$.

Definition 2 (Pattern's weight). Since each sequence consists of one or more itemsets, previous studies [18, 26] estimated the weight of the pattern by calculating the average itemset weight of that pattern for the same biasness issue. The weight value of a pattern $w_{\mathcal{P}}$ is $w_{\mathcal{P}} = \frac{\sum_{X \in \mathcal{P}}^{|\mathcal{P}|} w_X}{|\mathcal{P}|}$, where $|\mathcal{P}|$ and w_X are the number of itemsets in $\mathcal{P}$ and the weight value of the itemset X in $\mathcal{P}$, respectively.

In Table 1, taking the third sequence $\langle ACF(DE)F \rangle$ as an example. There are five itemsets $\{A, C, F, (DE), F\}$, and the weights of these five items are 0.1, 0.38, 0.72, $(0.49 + 0.51)/2 = 0.5$, and 0.72, respectively. Thus, we have $w_{\langle ACF(DE)F \rangle}$ $= (0.1 + 0.38 + 0.72 + 0.5 + 0.72)/5 = 0.484$.

In WSPM, there are two main methods to calculate the weighted support (*wsup*) of a pattern: one is the absolute weighted support [26], and the other is the relative weighted support [16]. The relative weighted support of a pattern is defined as the total sequence maximum weight (*tsmw*) in the database divided by the absolute weighted support with this pattern [16]. In this paper, we adopt the concept of relative weighted support. In order to calculate the *tsmw*, we introduce some definitions as follows.

Definition 3 (The *tsmw* in sequence database). The maximum weight value of a sequence $\mathcal{S}$, denoted as $MWU_{\mathcal{S}}$, is the maximum weight among all itemsets in $\mathcal{S}$. In the concepts of relative weighted support of a pattern [16], the *tsmw* is the summation of the $MWU_{\mathcal{S}}$ in a sequence database. That is: $tsmw = \sum_{\mathcal{S} \subseteq \mathcal{D}} MWU_{\mathcal{S}}$.

When considering $\mathcal{S} = \langle ACF(DE)F \rangle$ in Table 1, the weight value of F is maximum, thus the $MWU_{\mathcal{S}}$ of the third sequence is 0.72. The $MWU_{\mathcal{S}}$ of the five sequences in Table 1 are 0.38, 0.9, 0.72, 0.9, and 0.72, respectively. Therefore, the *tsmw* in Table 1 is equal to $0.38 + 0.9 + 0.72 + 0.9 + 0.72 = 3.62$.

Definition 4 (The weighted support of a pattern). The weighted support value of a pattern $\mathcal{P}$, denoted as *wsup*, is the summation of the weight values of $\mathcal{P}$ in the sequences containing $\mathcal{P}$ in $\mathcal{D}$ divided by the *tsmw*. That is: $wsup_{\mathcal{P}} = \frac{\sum_{\mathcal{P} \subseteq \mathcal{S}_y \wedge \mathcal{S}_y \subseteq \mathcal{D}} w_{\mathcal{P}}}{tsmw}$.

We take the pattern $\langle DC \rangle$ in Table 1 as an example. The weight of it is equal to $(0.49 + 0.38)/2 = 0.435$ in accordance with the third definition, and it appears in two sequences, $\mathcal{S}_2$ and $\mathcal{S}_5$. And then the total sequence maximum weight is $tsmw = 3.62$. Thus the *wsup* of $\langle DC \rangle$ can be calculated as $(2 \times 0.435)/3.62$, which is 24.03%.

Definition 5. Let the minimum weighted support threshold be *minWS*, then a sequential pattern $\mathcal{P}$ can be regarded as a weighted sequential pattern (abbreviated as *WS*) if and only if $wsup_{\mathcal{P}} \geq minWS$.

It should be noted that the downward closure property could not be maintained during the mining process in weighted sequential pattern mining [26]. It is clear that one of the key challenges in WSPM is how to satisfy the downward closure property. To efficiently address this challenge, in this paper, we propose a new concept namely remaining sequence maximum weight (abbreviated as *RSMW*), which can keep the downward closure property in the mining process. Meanwhile, the *RSMW* can tighten the upper bound w.r.t. weighted support value more efficiently and then speed up the computation efficiency in finding weighted sequential patterns. The related definitions of the proposed *RSMW* concepts are given as follows.

Definition 6 (Remaining maximum weight). The remaining maximum weight of an itemset i, denoted as rmw_i, is the maximum weight among all itemsets after i in a sequence $\mathcal{S}$. If an itemset is the last itemset in one sequence, then its remaining maximum weight is zero because there are no remaining items.

For example, in Table 1, the item G appears in two sequences, Seq_2 and Seq_4, respectively. Moreover, it is the last itemset in Seq_2 thus $rsmw_G$ in Seq_2 is 0 but since there is an itemset $\langle H \rangle$ left. Therefore, the $rsmw_G$ in Seq_4 is 0.9. The possible extension itemsets of $\mathcal{P}$, denoted as $PET_\mathcal{P}$, is the number of remaining itemsets of $\mathcal{P}$ in the sequence $\mathcal{S}$. We consider $\langle D \rangle$ in Table 1 as an example. In $\mathcal{S}_2$, the four itemsets are $\langle E \rangle$, $\langle C \rangle$, $\langle H \rangle$ and $\langle G \rangle$, respectively. Therefore, the $PET_\mathcal{D}$ in $\mathcal{S}_2$ is 4. We can easily observe that the $PET_\mathcal{D}$ in $\mathcal{S}_3$ and $\mathcal{S}_5$ are 1 and 3, respectively.

Definition 7 (Residual utility of a pattern). The residual utility of a pattern, denoted as RUP, is defined below. There are two situations in the method of calculating the RUP of a pattern. The first case is that the weight of the current pattern is equal to or even larger than the rmw of it. In this case, we use the first formula. Otherwise, we adopt the second formula. Note that $w_\mathcal{P} \times length_\mathcal{P}$ calculates the current total weight of pattern $\mathcal{P}$.

$$RUP_\mathcal{P} = \begin{cases} \frac{w_\mathcal{P} \times length_\mathcal{P} + rmw_\mathcal{P}}{length_\mathcal{P} + 1}, & w_\mathcal{P} \geq rmw_\mathcal{P} \\ \frac{w_\mathcal{P} \times length_\mathcal{P} + PET_\mathcal{P} \times rmw_\mathcal{P}}{length_\mathcal{P} + PET_\mathcal{P}}, & w_\mathcal{P} < rmw_\mathcal{P} \end{cases}$$

When considering C in Table 1, there are two different situations in $\mathcal{S}_1$ and $\mathcal{S}_2$. In $\mathcal{S}_1$, the rmw_C is 0.28. Since the rmw_C is less than w_C, the RUP_C in $\mathcal{S}_1$ can be calculated as $(0.38 \times 1 + 0.28)/(1 + 1) = 0.33$. In $\mathcal{S}_2$, the rmw_C is 0.9 and the PET_C in $\mathcal{S}_2$ is 2, thus the RUP_C is $(0.38 + 2 \times 0.9)/(1 + 2) = 0.727$.

Definition 8 (Upper-bound of a pattern $\mathcal{P}$). The upper-bound of a pattern $\mathcal{P}$, denoted as $swub_\mathcal{P}$, is the summation of $RUP_\mathcal{P}$ over the $tsmw$. That is $swub_\mathcal{P}$
$= \frac{\sum_{\mathcal{P} \subseteq S_y \wedge S_y \subseteq \mathcal{D}} RUP_\mathcal{P}}{tsmw}$.

4   Proposed FWSPM Algorithm

In this section, the FWSPM algorithm is proposed to effectively address the problem of discovering the weighted sequential patterns in a sequence database. We propose a new upper-bound $RSMW$ and an efficient pruning strategy to speed up execution. The improved upper-bound model is first described below.

Our proposed FWSPM algorithm uses the pattern-growth approach, which begins with an empty sequence and then extends it with new items from the sequence database. As mentioned in the SPAM algorithm [5], the extension of a sequential pattern happens in two separate ways: *I-Step* and *S-Step*. *I-Step* has an extension of the itemset with a new item, while *S-Step* adds a new single-item itemset with the pattern. The new upper-bound model, $RSMW$, is proposed here to enhance the traditional weight upper-bound model [26]. The downward closed property can be kept and the procedure for tightening upper-bound values can be gradually executed during the WSPM process with the *RSMW* model. It is observed that the remaining maximal weight in a sequence can keep the downward closure property in the mining process. To illustrate the completeness of the new proposed upper-bound model, two lemmas are given to prove that the mining processes of FWSPM would not leave out any interesting weighted sequential patterns. Details are presented in the following.

Lemma 1. Based on the *RSMW* model, the upper-bound value of the current pattern $\mathcal{P}$ is larger than or equal to any of its extensions.

Lemma 2. The upper-bound value of a pattern $\mathcal{P}$ would not be less than its weighted support value.

The main procedure of FWSPM is described in Algorithm 1.

Algorithm 1: The FWSPM algorithm

Input: (1) $\mathcal{D}$: a sequence database; (2) *ItemsWeight*: weights of the items within weight range; (3) *minWS*: the minimum weighted-support threshold.

Output: *WS*: the complete set of weighted frequent pattern.

1 scan original $\mathcal{D}$ once to construct the transformed $\mathcal{D}$;
2 scan the *ItemsWeight* then give the items assignment respectively and then obtain all 1-length pattern;
3 sort the items in each itemset which in the $\mathcal{S}$;
4 initialize map $RMInSid(sid, index, rmw)$;
5 initialize map $numLeftItemsetInSid(sid, index, numLeftItemset)$;
6 call **procedure** $findRMAndPEI(\mathcal{D}, RMInSid, numLeftItemsetInSid)$;
7 initialize sets: *WFUB* and *WS*;
8 set $r = 1$;
9 calculate the $swub_r$ and $wsup_r$ of each r-pattern;
10 **for** each r-patterns i with order **do**
11   **if** $swub_i(i) \geq minWS$ **then**
12   add the i to $WFUB_r$;
13   **end**
14   **if** $wsup_i \geq minWS$ **then**
15   add the i to WS_r;
16   **end**
17 **end**
18 **if** $WFUB_r.size() \geq 1$ **then**
19   **for** each pattern $\mathcal{P} \in WFUB_r$ **do**
20   construct the projected database for $\mathcal{P}$, denoted as $\mathcal{D}'$;
21   call **recursive**$(\mathcal{P}, \mathcal{D}', r, minWS)$;
22   **end**
23 **end**
24 **return** *WS*

Algorithm 2: The findRMAndPEI procedure

Input: (1) $\mathcal{D}$; (2) $RMInSid(sid,\ index,\ numLeftItemset)$; (3)
$numLeftItemsetInSid(sid,\ index,\ numLeftItemset)$.

```
 1  for each S sid in D do
 2  │   initialize map RMOfCurrentItemset(index, rmw);
 3  │   initialize map numLeftOfCurrentItemset(index, numLeftItemset);
 4  │   set LeftItemset = 0;
 5  │   for each itemset I in S from backward do
 6  │   │   update the remaining maximum weight rmw;
 7  │   │   RMOfCurrentItemset.put(index, rmw);
 8  │   │   numLeftOfCurrentItemset.put(index, LeftItemset);
 9  │   │   LeftItemset++;
10  │   end
11  │   RMInSid.put(sid, RMOfCurrentItemset);
12  │   numLeftItemsetInSid.put(sid, numLeftOfCurrentItemset);
13  end
```

Algorithm Details: First, FWSPM scans the sequence database and sorts all items in the itemset in descending order. FWSPM then calls the procedure *findR-MAndPEI*. It traverses each sequence $\mathcal{S}$ in $\mathcal{D}$ by backwards. During this process, FWSPM records the remaining maximum weight and the number of left itemsets of each itemset. After that, FWSPM saves all the weighted frequent upper-bound r-patterns (r is initialized as 1) in the set $WFUB_r$. And the weighted sequential r-patterns are saved into the set WS_r in the meanwhile. During this process, if $WFUB_r$ is not empty, then it builds the projected database for $\mathcal{P}$ in $WFUB_r$. After that, FWSPM recursively calls the *recursive* procedure which uses the depth-first search for the rest mining processes.

Discussion: Based on Lemmas 1 and 2, FWSPM has the downward closure property to discover all the weighted sequential patterns in a sequence database. Based on theoretical analysis, the new proposed $RSMW$ model can be used to effectively tighten the upper-bound values and then reduce the search space of algorithm. For example, according to the traditional upper-bound model in IUA [16], the maximum weight values of sequences in Table 1 are 0.38, 0.9, 0.72, 0.9, and 0.72, respectively. Then these values are regarded as the upper-bound of any sub-sequence in each sequence in the sequence database. Taking the item C as an example, the item C appears in four sequences $\langle \mathcal{S}_1, \mathcal{S}_2, \mathcal{S}_3, \mathcal{S}_5 \rangle$, and the maximum weight of these four sequences are 0.38, 0.9, 0.72, and 0.72, respectively. Then the upper-bound of weight value of C is calculated as $(0.38 + 0.9 + 0.72 + 0.72)/3.62$, which is 75.14%. Based on our proposed model, the upper-bound value of the item C can be further tightened. First, the remaining maximum weights in the sequence are 0.28, 0.9, 0.72, and 0.72, respectively. According to the above definitions, we can get this upper-bound value of C as 64%. Through this example, we can see that the value 64% obtained by the new proposed model is obviously less than 75.14% obtained by the IUA algorithm. Therefore, the new proposed *RSMW* model can effectively tighten the upper-bounds compared with the existing upper-bounds [16].

5   Experimental Evaluation

In this section, we provide a comparative performance analysis of our proposed FWSPM algorithm based on experimental results. All the experiments were performed in RMD R5-5600X 6 cores 3.70 GHz CPU with 16 Gigabytes of RAM and Window 10 with a 64-bit operating system. The code of IUA and FWSPM was written in Java 11.0 language and executed in IntelliJ IDEA. We have conducted extensive experiments on both real and synthetic datasets to evaluate the performance of FWSPM. For efficiency elevation, we considered three aspects: execution time, memory consumption, and scalability.

Dataset Description. We tested five datasets which can be downloaded from the SPMF open-source data mining library[1]. **Leviathan** is a conversion of the novel Leviathan by Thomas Hobbes (1651) as a sequence dataset (each word is an item). **FIFA** is a click stream data from the website of FIFA World Cup 98. **Yoochoose**[2] is a real ecommerce dataset. **slen_10** is a synthetic dataset which is generated by the sequence dataset generator in SPMF. The details of these datasets are shown in Table 2. Moreover, for each dataset, a corresponding weight-table was also given in which a weight value in the range from 0 to 1 was randomly assigned to an item. Additionally, *NoS* represents the number of sequences, *UI* is the number of unite items, *ALS* is the average length of sequences, and *ALI* is the average length of itemsets.

Table 2. Information of tested datasets

	Leviathan	FIFA	Yoochoose	slen_10
NoS	5834	20450	234300	48467
UI	9025	2990	16004	75476
tsmw	5561.330	19543.510	130742.757	33744.498
ALS	33.81	36.24	1.54	2.84
ALI	1.00	1.00	1.98	8.00

Execution Time Analysis. The "runtime" indicates the average running time of each variant of the FWSPM algorithm, by varying thresholds. The abscissa presents the minimum weighted support threshold and the ordinate means the running time of the FWSPM algorithm. From Fig. 1, it can be easily observed that FWSPM has better performance in weighted sequential pattern mining.

---

[1] http://www.philippe-fournier-viger.com/spmf/.
[2] https://recsys.acm.org/recsys15/challenge/.

Moreover, the growth ratio of the running time of FWSPM is placid. The main reason is that FWSPM adopts the *RSMW* model and can further tighten the upper-bound value of every pattern. Therefore, FWSPM reduces the search space and speeds up the mining processes.

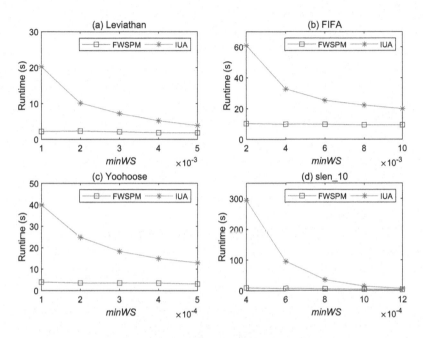

Fig. 1. Running time under different thresholds.

Memory Analysis. Next, we analyze the memory cost performance of the proposed algorithm. The abscissa presents the minimum weighted support threshold and the ordinate means the consumption of memory. Note that in all datasets, we use the Java API to count the maximal memory consumption for fair comparison. In Fig. 2, the missing value means the algorithm is out of memory. From Fig. 2 we can observe that FWSPM needs more memory to find out the all weighted sequential patterns. The less requirement for memory in IUA could be attributed to remove the unpromising items beforehand. Moreover, since the FWSPM algorithm adopts the pseudo projected method, it needs more memory in the mining process. Surprisingly, the cost of memory in *Yoochoose* the Fig. 2(c) has a better performance than the IUA algorithm when the threshold is from 0.01% to 0.05%. From Table 2 we can know that in *Yoochoose* the average length of sequence and itemset are 1.54 and 1.98, respectively. Since the average length of sequence in *Yoochoose* is low, the search space is small, and the consumption of memory can be reduced.

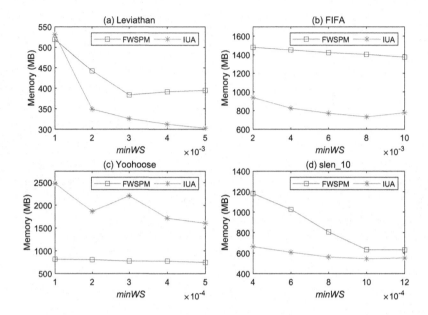

Fig. 2. Memory cost under different thresholds.

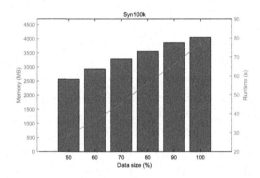

Fig. 3. Scalability test under various size of Syn100k ($minWS = 0.2\%$).

Scalability Test. To evaluate the scalability of the proposed FWSPM algorithm, we conducted it on *Syn100k*. This synthetic dataset has the following properties: 100000 sequences; total of 9999 unique items; the average number of itemsets per sequence is 6.20; the average itemset length is 4.32; and *tsmw* is 47069.58. Scalability test has been performed by selecting 50%, 60%, 70%, 80%, 90%, and 100% sequences from the whole dataset. We set the minimum weighted support threshold at 0.2% of *tsmw*, so the wsup value will be changed according to different dataset sizes. Figure 3 shows the required time and memory consumption for different sizes of *Syn100k*. The running time always increases linearly with the size of the dataset. From Fig. 3 we could also find that the consumption of memory is almost the same with a certain control range. Therefore, we can conclude that FWSPM has good scalability on large-scale datasets.

6   Conclusion

This paper introduced an efficient framework for mining weighted sequential patterns named FWSPM. It utilizes the projection technique to reduce the processed database size gradually, so it has superior performance. Moreover, it proposed the *RMW* measure, as an improved upper-bound, to prune the explosive growth of search space without compromising on completeness. Analysis of required memory for different datasets and threshold levels showed that the proposed FWSPM algorithm is a better choice for the WSPM task, when dealing with larger-scale datasets or lower thresholds since the consumption of memory is critical and this usually tends to increase exponentially in existing algorithms. Besides, in small-size datasets or large threshold levels, both the FWSPM and IUA algorithms performed with somewhat equivalent performance.

Acknowledgment. This research was supported in part by the National Natural Science Foundation of China (Grant Nos. 61902079 and 62002136), Guangzhou Basic and Applied Basic Research Foundation (Grant Nos. 202102020928 and 202102020277), and the Young Scholar Program of Pazhou Lab (Grant No. PZL2021KF0023).

References

1. Agrawal, R., Imielinski, T., Swami, A.: Database mining: a performance perspective. IEEE Trans. Knowl. Data Eng. **5**(6), 914–925 (1993)
2. Agrawal, R., Imieliński, T., Swami, A.: Mining association rules between sets of items in large databases. In: Proceedings of the ACM SIGMOD International Conference on Management of Data, pp. 207–216. ACM (1993)
3. Agrawal, R., Srikant, R.: Mining sequential patterns. In: Proceedings of the 7-th International Conference on Data Engineering, pp. 3–14. IEEE (1995)
4. Agrawal, R., Srikant, R., et al.: Fast algorithms for mining association rules. In: Proceedings of the 20-th International Conference on Very Large Data Bases, pp. 487–499 (1994)
5. Ayres, J., Flannick, J., Gehrke, J., Yiu, T.: Sequential pattern mining using a bitmap representation. In: Proceedings of the 8-th ACM SIGKDD International Conference on Knowledge Discovery and Data Mining, pp. 429–435 (2002)
6. Cai, C.H., Fu, A.W.C., Cheng, C.H., Kwong, W.W.: Mining association rules with weighted items. In: Proceedings of the International Database Engineering and Applications Symposium, pp. 68–77. IEEE (1998)
7. Chen, M.S., Han, J., Yu, P.S.: Data mining: an overview from a database perspective. IEEE Trans. Knowl. Data Eng. **8**(6), 866–883 (1996)
8. Fournier-Viger, P., Lin, J.C.W., Kiran, R.U., Koh, Y.S., Thomas, R.: A survey of sequential pattern mining. Data Sci. Pattern Recogn. **1**(1), 54–77 (2017)
9. Gan, W., Lin, J.C.W., Chao, H.C., Zhan, J.: Data mining in distributed environment: a survey. Wiley Interdisc. Rev.-Data Min. Knowl. Discov. **7**(6), e1216 (2017)
10. Gan, W., Lin, J.C.W., Fournier-Viger, P., Chao, H.C., Tseng, V.S., Yu, P.S.: A survey of utility-oriented pattern mining. IEEE Trans. Knowl. Data Eng. **33**(4), 1306–1327 (2021)

11. Gan, W., Lin, J.C.W., Fournier-Viger, P., Chao, H.C., Yu, P.S.: A survey of parallel sequential pattern mining. ACM Trans. Knowl. Discov. Data **13**(3), 1–34 (2019)
12. Gan, W., Lin, J.C.-W., Fournier-Viger, P., Chao, H.-C., Zhan, J., Zhang, J.: Exploiting highly qualified pattern with frequency and weight occupancy. Knowl. Inf. Syst. **56**(1), 165–196 (2017). https://doi.org/10.1007/s10115-017-1103-8
13. Gan, W., Lin, J.C.W., Zhang, J., Chao, H.C., Fujita, H., Yu, P.S.: ProUM: Projection-based utility mining on sequence data. Inf. Sci. **513**, 222–240 (2020)
14. Gan, W., Lin, J.C.W., Zhang, J., Fournier-Viger, P., Chao, H.C., Yu, P.S.: Fast utility mining on sequence data. IEEE Trans. Cybern. **51**(2), 487–500 (2021)
15. Han, J., Pei, J., Yin, Y., Mao, R.: Mining frequent patterns without candidate generation: a frequent-pattern tree approach. Data Min. Knowl. Disc. **8**(1), 53–87 (2004)
16. Lan, G.-C., Hong, T.-P., Lee, H.-Y.: An efficient approach for finding weighted sequential patterns from sequence databases. Appl. Intell. **41**(2), 439–452 (2014). https://doi.org/10.1007/s10489-014-0530-4
17. Lim, A.H., Lee, C.S.: Processing online analytics with classification and association rule mining. Knowl.-Based Syst. **23**(3), 248–255 (2010)
18. Lin, J.C.W., Gan, W., Fournier-Viger, P., Hong, T.P.: RWFIM: recent weighted-frequent itemsets mining. Eng. Appl. Artif. Intell. **45**, 18–32 (2015)
19. Lin, J.C.W., Gan, W., Fournier-Viger, P., Hong, T.P., Chao, H.C.: Mining weighted frequent itemsets without candidate generation in uncertain databases. Int. J. Inf. Technol. Dec. Mak. **16**(06), 1549–1579 (2017)
20. Lin, J.C.-W., Gan, W., Fournier-Viger, P., Hong, T.-P., Tseng, V.S.: Weighted frequent itemset mining over uncertain databases. Appl. Intell. **44**(1), 232–250 (2015). https://doi.org/10.1007/s10489-015-0703-9
21. Mannila, H., Toivonen, H.: Levelwise search and borders of theories in knowledge discovery. Data Min. Knowl. Disc. **1**(3), 241–258 (1997)
22. Pei, J., Han, J., Mortazavi-Asl, B., Wang, J., Pinto, H., Chen, Q., Dayal, U., Hsu, M.C.: Mining sequential patterns by pattern-growth: the prefixspan approach. IEEE Trans. Knowl. Data Eng. **16**(11), 1424–1440 (2004)
23. Schweizer, D., Zehnder, M., Wache, H., Witschel, H.F., Zanatta, D., Rodriguez, M.: Using consumer behavior data to reduce energy consumption in smart homes: Applying machine learning to save energy without lowering comfort of inhabitants. In: Proceedings of the 14-th International Conference on Machine Learning and Applications, pp. 1123–1129. IEEE (2015)
24. Srikant, R., Agrawal, R.: Mining sequential patterns: generalizations and performance improvements. In: Apers, P., Bouzeghoub, M., Gardarin, G. (eds.) EDBT 1996. LNCS, vol. 1057, pp. 1–17. Springer, Heidelberg (1996). https://doi.org/10.1007/BFb0014140
25. Wang, J., Han, J., Li, C.: Frequent closed sequence mining without candidate maintenance. IEEE Trans. Knowl. Data Eng. **19**(8), 1042–1056 (2007)
26. Yun, U., Leggett, J.J.: WSpan: Weighted sequential pattern mining in large sequence databases. In: Proceedings of the 3rd International Conference Intelligent Systems, pp. 512–517. IEEE (2006)
27. Zhang, C., Du, Z., Gan, W., Yu, P.S.: TKUS: mining top-k high utility sequential patterns. Inf. Sci. **570**, 342–359 (2021)

Parallel High Utility Itemset Mining

Gaojuan Fan[1], Huaiyuan Xiao[1,2], Chongsheng Zhang[1],
George Almpanidis[1(✉)], Philippe Fournier-Viger[3], and Hamido Fujita[4]

[1] School of Computer and Information Engineering,
Henan University, Kaifeng, China
cszhang@ieee.org, almpanidis@acm.org
[2] Bank of Zhengzhou Co., Ltd., Zhengzhou, China
[3] College of Computer Science and Software Engineering, Shenzhen University,
Shenzhen, China
philfv@szu.edu.cn
[4] Faculty of Software and Information Science, Iwate Prefectural University,
Takizawa, Japan
hfujita-799@acm.org

Abstract. Association rule mining is a popular data mining task for
finding relationships between values from the itemsets that co-occur frequently in a transactional database. Association rule mining has many
applications but the "support-confidence" framework it depends on is
inadequate for many cases. In recent years, a generalised task called high
utility itemset mining (HUIM) has gained much popularity; it aims at
discovering itemsets that yield a high revenue as measured by a utility
function. However, when facing large data volumes, the running time of
state-of-the-art HUIM algorithms often grows exponentially. In this work,
we investigate parallel HUIM algorithms (PHUIM) and adapt two state-
of-the-art sequential HUIM algorithms for parallel processing based on
the Apache Spark in-memory data processing platform. Extensive experiments on several benchmark and synthetic datasets show that the proposed methods improve considerably the efficiency of the baseline HUIM
algorithms.

Keywords: High-utility itemset · Apache spark · Parallel · d^2HUP ·
EFIM

1 Introduction

Association rule mining was first proposed by R. Agrawal in 1993 [1]. It aims at
discovering meaningful patterns and rules hidden in data [12]. These extracted
patterns and rules can be presented in the form of frequent itemsets or association rules and those extracted by various algorithms such as Apriori [2], FP-
growth [7], Eclat [18], and negFIN [3]. A traditional application of association
rule mining is the analysis of customer transactions in order to identify popular
itemsets (sets of items purchased together).

A drawback of association rule mining is that it only considers the co-
occurrence frequency of items while ignoring the practical utility and profits

© Springer Nature Switzerland AG 2022
H. Fujita et al. (Eds.): IEA/AIE 2022, LNAI 13343, pp. 819–830, 2022.
https://doi.org/10.1007/978-3-031-08530-7_69

of the itemsets. For instance, supermarkets could sell eggs and milk at very low prices to attract customers, and hence these items may co-appear frequently in transactions. However, the co-occurrences of eggs and milk do not necessarily yield much revenue to supermarkets. High utility itemset mining (HUIM) aims at finding the items that not only occur in the same transactions but also account for the most total revenue, i.e. it aims at identifying the frequent itemsets that are the most profitable [4]. HUIM has applications in various domains: in market-basket data analysis, bioinformatics, proof process learning [14], website click-stream data analysis, mobile data analysis, etc. [4].

In the past few years, HUIM has attracted a large amount of research effort, and many efficient algorithms have been developed [19]. Representative algorithms are UP-Growth [17], FHM [8], FHM+ [5], d^2HUP [10], EFIM [20], etc. Searching for high utility itemsets (HUIs) requires a minimum utility threshold to be set externally, then the HUIM algorithm will need to identify all the itemsets that have a utility (e.g. profit) higher than this threshold. All the HUIM algorithms have the same input and output, but they rely on different data structures, various search strategies and optimisation techniques to reduce the search space.

Despite such research efforts, HUIM algorithms often have a very long running time and scale poorly to large databases. To cope with the scalability issue, some researchers have developed approximate HUIM algorithms [13], [15]. However, although these algorithms can be faster than traditional algorithms, the results are generally incomplete. A promising solution for increasing the scalability of HUIM is to develop parallel HUIM (PHUIM) algorithms based on big data processing platforms, yet little effort has been directed towards this direction. PHUI-Growth [9] was the first PHUIM algorithm designed based on the Apache Hadoop platform but suffers from the limitations of Hadoop, i.e. it performs many costly map-reduce iterations. A promising direction is to use the Spark platform for its significantly faster in-memory processing speed than Hadoop MapReduce. P-FHM+ algorithm [16] is a parallel variant of the FHM+ algorithm using Spark. P-FHM+ is up to about 6 times faster than the original FHM+ algorithm. Nevertheless, some sequential algorithms such as EFIM and d^2HUP are up to 100 times faster than FHM [20].

Therefore, there is still a strong need to design efficient PHUIM algorithms. In this paper, based on two of the most efficient HUIM algorithms (EFIM and d^2HUP), we design PHUIM algorithms based on Spark to make them parallelable. Our proposed PHUIM framework first divides the search space by routing itemsets (sub-trees) having the same prefix head to different spark partitions. Next, using the same global threshold for high utility itemsets, it extracts high utility itemsets on each partition in parallel. Finally, it combines the results from different partitions and obtains the final high utility itemsets. Our proposal has theoretical guarantees on the correctness and completeness of the final set of HUIs derived by our designed PHUIM algorithms. Also, through experiments on seven real-world datasets and four large-scale synthetic datasets, we demonstrate that our proposed methods can significantly speed up the corresponding sequential HUIM algorithms while still obtaining the same HUIs.

2   Problem Definition

The problem of HUIM is defined as follows. Consider a finite set of distinct items $I = \{i_1, i_2, \ldots, i_n\}$ that are sold in a retail store. A transaction database D is a set of customer transactions $D = \{T_1, T_2, \ldots T_m\}$. A transaction $T_j \in D$ is a set of purchased items, where j is a unique identifier. For each item i appearing in a transaction T_i, a purchase quantity is associated to the item, denoted as $q(i, T)$, which is a positive integer. Moreover, each item i in a database D has a unit profit value, denoted as $p(i)$, which is a positive integer.

HUIM aims at discovering all itemsets that yield a high profit (have a high utility). An itemset X is a set of items $X \subseteq I$. For an itemset X, let $g(X)$ be the set of transactions containing X, that is $g(X) = \{T_j | X \subseteq T_j \in D\}$. The utility (profit) of an itemset X in a transaction T_j is defined as $u(X, T_j) = \sum_{i \in X} p(i) \times q(i, T_j)$. The utility (profit) of an itemset X in a database D is defined as $u(X) = \sum_{T_j \in g(X)} u(X, T_j)$. The utility of a database D is defined as $u(D) = \sum_{T_j \in D} u(T_j, T_j)$.

For example, consider a set of items $I = \{a, b, c\}$, representing some products sold in a store such as apple (a), bread (b), and cake (c). Let the unit profit of these items be defined as $p(a) = 1$, $p(b) = 2$, and $p(c) = 3$. Assume that there are two transactions $D = \{T_1, T_2\}$. The first transaction is $T_1 = \{a, b, c\}$ with $q(a, T_1) = 10$, $q(b, T_1) = 20$ and $q(c, T_1) = 30$. The second transaction is $T_2 = \{b, c\}$ with $q(b, T_2) = 3$ and $q(c, T_2) = 6$. Consider an itemset $X = \{b, c\}$. The utility of X in transaction T_1 is $u(X, T_1) = 2 \times 20 + 3 \times 30 = 130$. The utility of itemset X in transaction T_2 is $u(X, T_2) = 2 \times 3 + 3 \times 6 = 24$. Hence, the utility of X in the database D is $u(X) = u(X, T_1) + u(X, T_2) = 130 + 24 = 154$. The utility of D is $u(D) = 164$.

The goal of high utility itemset mining is to find all the high utility itemsets in a database D [17]. A high utility itemset is an itemset that has a relative utility that is no less than a user-defined threshold called $minUtil$ (minimum utility), chosen by the user. The relative utility $r(X, D)$ of an itemset X in a database D is its utility divided by the utility of the database, that is $r(X, D) = u(X)/u(D)$.

For example, the relative utility of $X = \{b, c\}$ in the previous example is $r(X) = 154/164$. If $minUtil = 0.9$, then X is a high utility itemset because $r(X) = 154/164 \geq 0.9 = minUtil$.

HUIM is notably a hard problem because the search space grows exponentially with respect to the number of items (there are $2^{|I|} - 1$ possible itemsets) and the utility measure is not monotonic (i.e. the utility of an itemset may be greater, equal, or less than that of its supersets) [11,17]. Hence, we need parallel HUIM (PHUIM) algorithms to speed up the computation efficiency. In this work, we utilise the Spark in-memory processing platform to design two efficient PHUIM algorithms, which are described below.

3   Methodology

We first introduce the proposed framework for PHUIM and then show how it is instantiated to parallelise two of the most efficient HUIM algorithms, namely d^2HUP and EFIM.

3.1   Overall Framework

Most HUIM algorithms utilise a depth-first search [10,17,20]. The proposed PHUIM framework consists of dividing the data into partitions such that item-sets starting by a same item (prefix head) can be explored using the same Spark partition. Since the items are ordered and each sub-tree has a unique prefix head, we can separately derive the HUIM on each sub-tree (partition) using the same global minimum utility threshold, then combine the HUIs extracted on different partitions.

Motivating Example. Our proposed PHUIM framework is illustrated in Fig. 1. Given items a-g, we route the sub-trees having different prefix heads to different partitions. For instance, node 3 is the partition that contains the sub-tree with prefix head c. On this partition, we compute the utility values of itemsets {c}, {c,a}, {c,b} and {c,b,a} respectively, and only keep the itemsets having utility values above the global utility threshold.

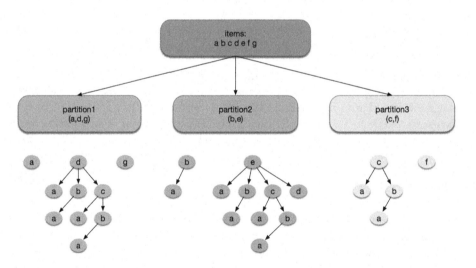

Fig. 1. The framework of our parallel HUIM algorithms.

Correctness Guarantees. Since all the items are ordered in advance according to a total order, e.g., {g, f, e, d, c, b, a}, therefore, if {c, b, a} appears on the third partition (with prefix head c) shown in Fig. 1, then cases like {b, a, c} will not appear in the second partition/sub-tree having prefix head b.

This ensures that all the itemsets starting with the same prefix head do not appear on the other partitions and sub-trees. Let the global HUI threshold be *t*. Since each sub-tree starting with a unique prefix head will be assigned to different partitions, we can use the same threshold to discover HUIs on each partition in parallel. The final HUI results from each partition will not overlap, for each sub-tree/partition has a unique prefix head. However, we note that this is only the first-level parallelisation for HUIM, because on each partition, the learning process for HUIs is still sequential. Actually, it is often challenging to make tree-based learning algorithms parallelisable. Nevertheless, our first-level parallelisation for HUIM (PHUIM) can lead to very significant efficiency improvements, as we will demonstrate in the experiments.

Load Balancing Strategies for PHUIM. It is worth noting that in the process of high-utility itemset mining, items are generally ordered by ascending Transaction Weighted Utility [11], which is the sum of the transaction utilities of all the transactions containing a specific itemset. The lower rank an item has, the deeper the corresponding prefix sub-tree is. The time cost of a high-utility itemset mining algorithm mainly depends on the number of recursions in the mining process. The deeper the sub-trees, the more recursions will happen. To achieve load balancing in the process of parallel high-utility itemset mining, three partition strategies are designed.

1. Partitioning Strategy 1: Randomly routing the sub-trees to different partitions. For instance, for sub-trees with prefix heads of a, b, c, d, e, f, g, we perform a modular operation between the node number and the partition ID to route the corresponding sub-tree. With this strategy, the numbers of sub-trees assigned to different partitions are roughly the same.
2. Partitioning Strategy 2: Balancing the difficulty of sub-trees on different partitions. For example, first route sub-trees having a and g as prefix heads to the first partition, then route sub-trees with b and f as prefix heads to the second partition, sub-trees with c and e as prefix heads to the third partition, and so on.
3. Partitioning Strategy 3: Routing the most difficult sub-trees to independent partitions. Since the lower-ranked item in {a, b, c, d, e, f, g} will have deeper prefix sub-trees, we route the sub-tree with prefix head {g} to a separate partition, then {e, f} to another partition and {a, b, c, d} are divided into another different partition, i.e. each partition has twice as many items as the previous partition.

3.2   Parallel d²HUP

The proposed framework is applied to parallelise the d²HUP algorithm [10]. The original algorithm preprocesses transactions into a linear structure, Chain of Accurate Utility Lists (CAUL), which is neither tree-based or graph-based, but simply consists of linear lists. CAUL maintains the original utility information for each enumerated pattern in a way that enables the computation of utility and an estimation of tight utility upper bounds efficiently. Our parallel d^2HUP algorithm first splits its CAUL structure according to its first-level sub-trees, then routes each sub-tree to different partitions on Spark. Then HUIs are derived from each partition on its corresponding sub-tree in parallel, using the same global minimum utility threshold. The final HUIs are finally collected by the master node from different partitions.

3.3 Parallel EFIM

The EFIM algorithm is also parallelised. EFIM contains two steps: (1) using the TWU (Transaction Weight Utility), unpromising generated candidates are filtered out and (2) the final HUIs are computed from the remaining candidates. Our parallel EFIM algorithm executes the first step on the master node, then routes the sub-trees of the remaining candidates (prefix-tree) to different partitions on Spark. Then, HUIs are derived on each partition in parallel, using similar strategies as parallel d^2HUP.

4   Experiments

4.1   Experiments Design

To evaluate the performance of the proposed algorithms, we have designed various experiments. We consider three factors that have a crucial influence on the performance of Spark-based high-utility itemset mining algorithms:

1. **Datasets.** To compare with existing methods in the literature, we use datasets obtained from the SPMF data mining library [6] as well as large-scale synthetic datasets from [19]. The list of the datasets that we use is shown in Table 1.
2. **Minimum utility threshold.** The minimum utility threshold is a key parameter for HUIM. In our experiments, the minimum utility thresholds for the synthetic datasets are selected in the same way as in the literature [19], so that the results can be verified in a consistent way.
3. **Spark Parameters.** We test parallel granularity parameters that are important in Spark-based data processing: the load balancing partitioning strategies, the number of partitions, and the number of executors and workers.

The major experiments were carried out on a server equipped with 4 Intel Xeon E7-4850v3 CPUs and total RAM of 256 GB, running RedHat Enterprise Linux operating system.

4.2   Experiment Results

In this subsection, we first evaluate the overall efficiency of the two proposed PHUIM algorithms, then check the influence of specific parameters on the algorithms respectively, including $minUtil$, the number of partitions and the partition strategies, and the number of executors and workers.

Table 1. Dataset descriptions.

Dataset	# of Transactions	# of Items	Max Trans. Length	File Size (MB)
chess	3,196	75	37	0.6
pumsb	49,046	2,113	74	25.1
BMS	59,602	497	267	1.3
connect	67,557	129	43	16.1
accidents	340,183	468	51	63.1
kosarak	990,002	41,270	2,498	53.3
chainstore	1,112,949	46,086	170	79.2
dh26 [19]	100,000	496,844	50	22.3
dh27 [19]	100,000	499,978	100	43.8
dh29 [19]	100,000	922,631	50	22.6
dh30 [19]	100,000	993,474	100	44.4

Table 2. Overall running time of serial and parallel HUIM algorithms.

Dataset	minUtil	d^2HUP-serial (s)	d^2HUP-parallel (s)	EFIM-serial (s)	EFIM-parallel (s)
chess	350,000	451	151	4	17
pumsb	12,500,000	1,673	608	37	36
BMS	2,230,000	4,340	2,884	2	5
connect	16,000,000	218	98	2	4
accidents	17,500,000	2,384	1,042	22	21
kosarak	1,000,000	2,254	765	444	38
chainstore	2,000,000	142	118	1,370	67
dh26	513	529	170	8,813	929
dh27	953	655	323	1,5070	2,123
dh29	510	822	572	3,439	821
dh30	953	3,324	470	4,9931	4,223

4.2.1 Overall Efficiency Comparison Between Parallel and Serial HUIM Algorithms

In Table 2, we compare the running time of the parallel and serial HUIM algorithms. We can see that the parallel d^2HUP and EFIM algorithms provide a great speed improvement over their serial counterparts when run using the same *minUtil* threshold value. In particular, for large-scale datasets and datasets with a large number of items, the improvements are very significant.

We verify that serial EFIM is very efficient due to its design, and its efficiency is remarkably better than d^2HUP for datasets with relatively small number of items, such as chess, BMS, and connect, but it is found that the parallel EFIM algorithm only provides little or no improvement over its serial counterpart. For datasets with a huge amount of items EFIM's efficiency drops remarkably.

Table 3. Overall running time of serial and parallel HUIM algorithms for different minUtils.

Dataset	minUtil	EFIM-serial (s)	EFIM-parallel (s)	speedup
chainstore	2,000,000	1,370	67	20.4x
chainstore	3,000,000	376	43	8.7x
chainstore	4,000,000	134	15	8.9x
kosarak	1,000,000	444	38	11.7x
kosarak	1,500,000	182	23	7.9x
Datasets	minUtil	d^2HUP-serial (s)	d^2HUP-parallel (s)	speedup
connect	15,000,000	1,251	967	1.3x
connect	16,000,000	218	98	2.2x

Parallel EFIM algorithm is often more than 9 times faster than serial EFIM in such situations, yet it fails by a large margin compared to d^2HUP.

For d^2HUP, we can see that the parallel d^2HUP algorithm consistently outperforms serial d^2HUP in all cases, and the speedup is often more than 2x (between 1.20 and 7.07 times faster).

Due to time and memory constrains, we mainly focused on measuring running time for default *minUtil* thresholds that also exist in the literature, as shown in Table 2. Nevertheless, we report on some limited *minUtil* scalability results in Table 3. For the datasets chainstore and kosarak, we find that the performance gain of parallel EFIM compared to serial EFIM is still significant for different values of *minUtil*. For the connect dataset, parallel d^2HUP is faster than serial d^2HUP, but the speedup gain seems to decrease monotonically with *minUtil* and the two variants have comparable running times (speedup ratio decreases to 1.3 when *minUtil* = 15000000). It must be noted that these findings are not conclusive and that connect is a dataset where d^2HUP generally underperforms [19].

4.2.2 Partitioning Strategies
We test the three different routing/partition strategies for load balancing in parallel high-utility itemset mining. The experimental results are shown in Table 4. We see that the first and second partitioning strategies are generally considerably better than the third one.

4.2.3 The Impact of the Number of Partitions on the Performance
A partition in Spark is a logical split of the data. To test the influence of the number of partitions on the performance of algorithms, we set the number of partitions to be 1, 2, 4, 8, and 16 in our experiments. Figure 2 shows the impact of the number of partitions on the parallel d^2HUP and EFIM algorithms. Overall, as the number of partitions increases, the running time of the algorithms drops monotonically.

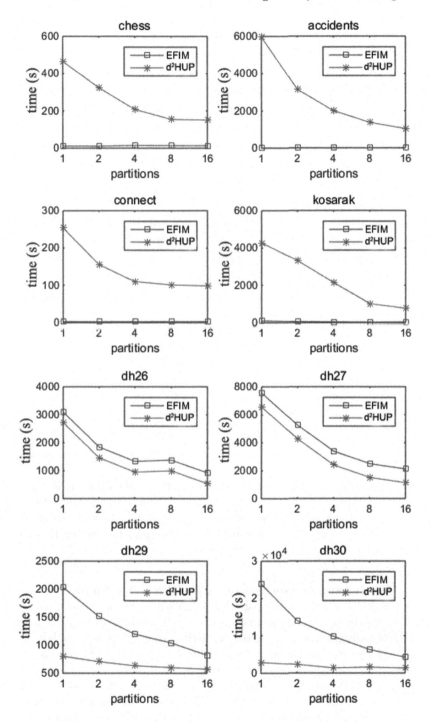

Fig. 2. The impact of the number of partitions on the performance of parallel HUIM algorithms on real-world and synthetic datasets.

Table 4. Effect of different routing strategies on the running time of parallel d^{2}HUP and EFIM algorithms.

Dataset	Strategy					
	d^{2}HUP-1 (s)	d^{2}HUP-2 (s)	d^{2}HUP-3 (s)	EFIM-1 (s)	EFIM-2 (s)	EFIM-3 (s)
pumsb	608	531	1,234	36	15	21
BMS	2,884	2,828	4,529	5	11	11
accidents	1,042	1,221	2,283	21	22	21
kosarak	765	598	1,460	38	45	80
dh26	170	153	683	929	956	3,852
dh27	323	283	1,209	2,123	1,263	8,767
dh29	572	591	723	821	804	2,000
dh30	470	554	2,673	4,223	3,978	22,009

Table 5. The impact of the number of executors on the running time of parallel d^{2}HUP and EFIM algorithms.

Dataset	Algorithm	# of executors			
		4 (s)	8 (s)	12 (s)	16 (s)
dh26	d^{2}HUP	233	229	170	121
dh29	d^{2}HUP	543	439	448	440
dh26	EFIM	1,412	1,661	1,423	1,106
dh29	EFIM	1,234	1,066	1,000	843

4.2.4   The Impact of Executors on Algorithm Performance

Executors are worker node processes responsible for running individual tasks within a given Spark job. As seen in Table 5, with the increase of the number of executors, the running time of parallel HUIM algorithms decreases to a certain extent.

4.2.5   The Impact of Workers on Algorithm Performance

Workers (slaves) are running Spark instances where executors live to execute tasks. We conducted experiments on a cluster of 32 desktop PCs (each with a 8 GB RAM and an Intel i5 processor) to verify the effect of the number of work nodes in the Spark cluster on the algorithms' performance. Shown in Table 6, we see that increasing the number of workers can improve the running speed of the algorithm. We note that, there is also the communication cost between the slave and master nodes. Therefore, for short overall running time, the influence of the communication cost can be greater, as can be observed over EFIM.

Table 6. The impact of the number of workers on the running time of parallel d²HUP and EFIM algorithms.

Dataset	Algorithm	# of workers					
		1 (s)	2 (s)	4 (s)	8 (s)	16 (s)	32 (s)
chess	d²HUP	305	233	150	138	148	156
pumsb	d²HUP	1,229	634	526	496	709	733
kosarak	d²HUP	2,958	1,575	1,255	1,277	1,201	1,027
chainstore	d²HUP	428	476	414	342	310	210
chess	EFIM	19	22	23	32	28	24
pumsb	EFIM	47	43	33	35	36	38
kosarak	EFIM	145	89	78	54	68	97
chainstore	EFIM	213	144	94	75	82	97

5   Conclusions

In this work, we propose parallel d²HUP and parallel EFIM algorithms for efficient high utility itemset mining using the Spark in-memory processing platform. We study the influence of the related parameters on the performance and run large-scale experiments to validate the efficiency of the algorithms under different situations. We demonstrate that our proposed algorithms can remarkably reduce the running time of the original sequential HUIM algorithms by a large margin.

In future work, we will consider applying the proposed framework to other variations of high utility itemset mining algorithms. Moreover, additional optimisations will be considered and larger-scale experiments involving different minimum utility thresholds will be done. It would also be interesting to compare the parallel d²HUP and EFIM with the parallel PHUI-Growth and P-FHM+ algorithms.

References

1. Agrawal, R., Imieliński, T., Swami, A.: Mining association rules between sets of items in large databases. In: Proceedings of the 1993 ACM SIGMOD International Conference on Management of Data, pp. 207–216 (1993)
2. Agrawal, R., Mannila, H., Srikant, R., Toivonen, H., Verkamo, A.I., et al.: Fast discovery of association rules. Adv. Knowl. Discov. Data Min. **12**(1), 307–328 (1996)
3. Aryabarzan, N., Minaei-Bidgoli, B., Teshnehlab, M.: negFIN: an efficient algorithm for fast mining frequent itemsets. Expert Syst. Appl. **105**, 129–143 (2018)
4. Fournier-Viger, P., Chun-Wei Lin, J., Truong-Chi, T., Nkambou, R.: A survey of high utility itemset mining. In: Fournier-Viger, P., Lin, J.C.-W., Nkambou, R., Vo, B., Tseng, V.S. (eds.) High-Utility Pattern Mining. SBD, vol. 51, pp. 1–45. Springer, Cham (2019). https://doi.org/10.1007/978-3-030-04921-8_1

5. Fournier-Viger, P., Lin, J.C.-W., Duong, Q.-H., Dam, T.-L.: FHM+: faster high-utility itemset mining using length upper-bound reduction. In: Fujita, H., Ali, M., Selamat, A., Sasaki, J., Kurematsu, M. (eds.) IEA/AIE 2016. LNCS (LNAI), vol. 9799, pp. 115–127. Springer, Cham (2016). https://doi.org/10.1007/978-3-319-42007-3_11

6. Fournier-Viger, P., et al.: The SPMF open-source data mining library version 2. In: Berendt, B., et al. (eds.) ECML PKDD 2016. LNCS (LNAI), vol. 9853, pp. 36–40. Springer, Cham (2016). https://doi.org/10.1007/978-3-319-46131-1_8

7. Han, J., Pei, J., Yin, Y., Mao, R.: Mining frequent patterns without candidate generation: a frequent-pattern tree approach. Data Min. Knowl. Disc. 8(1), 53–87 (2004)

8. Krishnamoorthy, S.: Pruning strategies for mining high utility itemsets. Expert Syst. Appl. 42(5), 2371–2381 (2015)

9. Lin, Y.C., Wu, C.-W., Tseng, V.S.: Mining high utility itemsets in big data. In: Cao, T., Lim, E.-P., Zhou, Z.-H., Ho, T.-B., Cheung, D., Motoda, H. (eds.) PAKDD 2015. LNCS (LNAI), vol. 9078, pp. 649–661. Springer, Cham (2015). https://doi.org/10.1007/978-3-319-18032-8_51

10. Liu, J., Wang, K., Fung, B.C.: Direct discovery of high utility itemsets without candidate generation. In: 2012 IEEE 12th International Conference on Data Mining, pp. 984–989 (2012)

11. Liu, M., Qu, J.: Mining high utility itemsets without candidate generation. In: Proceedings of the 21st ACM International Conference on Information and Knowledge Management, pp. 55–64 (2012)

12. Luna, J.M., Fournier-Viger, P., Ventura, S.: Frequent itemset mining: a 25 years review. Wiley Interdisc. Rev. Data Min. Knowl. Discov. 9(6), e1329 (2019)

13. Nawaz, M.S., Fournier-Viger, P., Yun, U., Wu, Y., Song, W.: Mining high utility itemsets with hill climbing and simulated annealing. ACM Trans. Manage. Inf. Syst. (TMIS) 13(1), 1–22 (2021)

14. Nawaz, M.S., Fournier-Viger, P., Zhang, J.: Proof learning in PVS with utility pattern mining. IEEE Access 8, 119806–119818 (2020)

15. Pramanik, S., Goswami, A.: Discovery of closed high utility itemsets using a fast nature-inspired ant colony algorithm. Appl. Intell. 52, 8839–8855 (2021). https://doi.org/10.1007/s10489-021-02922-1

16. Sethi, K.K., Ramesh, D., Edla, D.R.: P-fhm+: parallel high utility itemset mining algorithm for big data processing. Procedia Comput. Sci. 132, 918–927 (2018)

17. Tseng, V.S., Wu, C.W., Shie, B.E., Yu, P.S.: Up-growth: an efficient algorithm for high utility itemset mining. In: Proceedings of the 16th ACM SIGKDD International Conference on Knowledge Discovery and Data Mining, pp. 253–262 (2010)

18. Zaki, M.J.: Hierarchical parallel algorithms for association mining. In: Advances in Distributed and Parallel Knowledge Discovery, pp. 339–376 (2000)

19. Zhang, C., Almpanidis, G., Wang, W., Liu, C.: An empirical evaluation of high utility itemset mining algorithms. Expert Syst. Appl. 101, 91–115 (2018)

20. Zida, S., Fournier-Viger, P., Lin, J.C.W., Wu, C.W., Tseng, V.S.: Efim: a fast and memory efficient algorithm for high-utility itemset mining. Knowl. Inf. Syst. 51(2), 595–625 (2017)

Towards Efficient Discovery of Stable Periodic Patterns in Big Columnar Temporal Databases

Hong N. Dao⬤, Penugonda Ravikumar⬤, P. Likitha⬤,
Bathala Venus Vikranth Raj⬤, R. Uday Kiran$^{(\boxtimes)}$⬤, Yutaka Watanobe⬤,
and Incheon Paik⬤

The University of Aizu, Aizuwakamatsu, Fukushima, Japan
{udayrage,watanobe,paik}@u-aizu.ac.jp

Abstract. Extracting stable periodic-frequent patterns in very large temporal databases is a key task in big data analytics. Existing studies have mainly concentrated on discovering these patterns only in row temporal databases, and completely ignored the existence of these patterns in columnar databases, which are widely becoming popular for storing big data. In this paper we propose an efficient algorithm, Stable Periodic-frequent Pattern-Equivalence CLass Transformation (SPP-ECLAT), to find the desired patterns in a columnar temporal database. Empirical results demonstrate that the SPP-ECLAT algorithm is much faster and consumes significantly less memory than the state-of-the-art SPP-growth algorithm on sparse and dense databases.

Keywords: Columnar databases · Periodic patterns · Pattern mining

1 Introduction

Databases are broadly classified into two types based on the layout of data recording on a storage device, namely *row databases* and *columnar databases*[1]. Row databases store data as records, maintaining the complete data associated with a record in a storage device next to each other. These databases are primarily based on ACID[2] properties and are designed to read and write rows fast. MySQL and Postgres are two examples of horizontal databases. Columnar databases organize data into fields and store the complete data corresponding with a field in the same storage device. These databases are primarily based on

---

[1] Row and columnar databases are also referred to as horizontal and vertical databases, respectively.

[2] ACID stands for Atomicity, Consistency, Isolation, and Duration.

H. N. Dao, P. Ravikumar, P. Likitha and B. V. V. Raj—These authors equally contributed to 98% of this paper.

Y. Watanobe and I. Paik—These authors have equally contributed to 2% of this paper.

© Springer Nature Switzerland AG 2022
H. Fujita et al. (Eds.): IEA/AIE 2022, LNAI 13343, pp. 831–843, 2022.
https://doi.org/10.1007/978-3-031-08530-7_70

BASE[3] properties and are designed to be efficient when reading and computing on columns. Snowflake and BigQuery are two examples of columnar databases. Both row and columnar databases have their respective advantages and disadvantages. So, the user and/or application requirements determine the appropriate database layout. Generally, row databases are better suited to online transaction processing (OLTP), whereas columnar databases are better suited to online analytical processing (OLAP). Since the primary objective of OLAP is to uncover meaningful information in data, this paper makes an effort to discover stable periodic-frequent patterns in a columnar database.

Periodic-frequent pattern mining is a useful and essential big data analytical technique. It involves identifying all patterns that satisfy the *minimum support* (*minSup*) and *maximum periodicity* (*maxPer*) constraints which are specified by user. *MinSup* measure constraints the minimum number of transactions in a database where a pattern must appear. *MaxPer* measure constraints the maximum time interval within which a pattern must reappear. Periodic-frequent pattern has been used for the analysis of market-basket analysis, which involves finding the sets of items purchased by the customers periodically. Consider a following example:

$$\{Bread, Butter\} \; [support = 25\%, periodicity = 2\,h].$$

This pattern provides information that 25% of the customers have purchased the items 'Bread' and 'Butter' at least once every two hours. Such information may be helpful to the managers of a supermarket for inventory management and product placement. Periodic-frequent pattern mining was extended to find fuzzy periodic-frequent patterns [1], partial periodic patterns [2], and high utility periodic patterns [3]. However, this technique has the major disadvantage of being overly strict. The reason is that a pattern is discarded if there exist only one period that exceeds *maxPer*. For example, a pattern that shows a customer purchases bread every day would be discarded if the customer skipped only one day.

To deal with the above problem, some studies was proposed a model to find partial periodic patterns [4] by relaxing the *maxPer* constraint, i.e., some of the periods of a pattern is greater than the user-specified *maxPer* value. Unfortunately, this [4] model is accepting some of the patterns which are having very lengthy periods. For example, purchasing bread can be considered periodic even if a customer purchases it on multiple days but without purchasing it again for a month. The length of periods for some patterns can vary significantly in a real-world database so that traditional models for discovering periodic-frequent patterns are insufficient.

A new class of periodic-frequent patterns named stable periodic-frequent patterns were introduced by Philippe et al. [5], whose recurrence deviation in the database is within the user-specified threshold value. These patterns overcome the above mentioned limitations of the periodic-frequent patterns. Furthermore, a pattern-growth algorithm, called Stable Periodic-Frequent Pattern-growth (SPP-growth), was proposed to discover desired patterns in a temporal database. However, there exist two limitations as follows:

---

[3] BASE stands for Basically Available, Soft state, and Eventually consistent.

- SPP-growth is designed to discover stable periodic-frequent patterns only in row databases. So, desired patterns in columnar databases cannot be found by this algorithm. In addition, this would be costly to transform a big columnar database into a row database.
- In the SPP-growth algorithm, huge memory is required to complete a tree-based recursive mining process, and runtime consumption is also very high.

To that end, a novel algorithm, named Stable Periodic-frequent Pattern-Equivalence CLass Transformation (SPP-ECLAT), is proposed in this paper to discover stable periodic-frequent patterns in a columnar temporal database. The proposed algorithm is shown to be efficient in terms of both memory and runtime.

The remainder of the paper is organized as follows. The related work is presented in Sect. 2. The model of stable periodic-frequent pattern is provided in detail in Sect. 3. The SPP-ECLAT algorithm is then described in Sect. 4. Section 5 shows the evaluation results and discussions. Finally, conclusions and future research directions are given in Sect. 6.

2   Related Work

Tanbeer et al. [6] described a novel pattern-growth algorithm to discover periodic-frequent patterns in a transactional database. Amphawan et al. [7] have identified the most frequent patterns as candidate patterns and generated the Top-k periodic-frequent patterns with the help of the best-first search strategy. Uday et al. [8] have designed a novel concept named *local periodicity* to prune the non-periodic patterns locally. Authors have discarded the patterns whose *local periodicity* is less than the user-specified *maxPer* value. As a result, most of the non-periodic patterns tid-lists were not completely built, resulting in a decrease in the computational time of the proposed algorithm. Anirudh et al. [9] have designed an approach to reduce the memory consumption of the pattern growth approach. In general, PF-trees have maintained the transaction identifiers (tid) in a particular node named as tail-node. However, in real-world applications, it is highly impractical to maintain the complete tid-list. Hence the authors have designed a new strategy named periodic summaries to be maintained at the tail-node to reduce the memory consumption while generating periodic-frequent patterns. Ravi et al. [10] have introduced PF- ECLAT, to find periodic-frequent patterns in a columnar databases. All the algorithms mentioned above will discard a non-periodic pattern if any of the period or local period exceeds the value of the *maxPer* constraint. Kiran et al. [11] have classified patterns as partial periodic and full periodic patterns. In real-world applications, some patterns will occur only at a particular point of time named partial periodic patterns. However, this algorithm cannot be applied to mine stable periodic-frequent patterns. The proposed algorithm has calculated the *period − support* measure as a count of the periods of the patterns whose value is less than the user-specified *maxPer*. Unfortunately, it completely ignores the periods' deviation from the *maxPer*.

A new interestingness measure called lability was exploited by Philippe et al. [5] to determine the interestigness of stable periodic-frequent patterns in transactional databases. Authors had described a new strategy named lability. The lability of a pattern is the cumulative sum of the difference between each period length and *maxper*. A novel measure *maxLa* was also used to assess the stability of a pattern's periodic behavior in a database. Ruimeng et al. [12] discussed a model to find stable periodic-frequent patterns in uncertain databases. Fournier-Viger et al. [13] utilized the concept of top-K mining to generate the stable periodic-frequent patterns. It has to be noted that all of these algorithms find the patterns in only row databases. In this paper, we have devised an algorithm to find the patterns in columnar databases.

3   Model of Stable Periodic-Frequent Patterns

Assume that we have a **pattern** (or an itemset) Y, $Y \subseteq I$, where I is the set of items. Denoted k-**pattern** as the pattern that has k items, $k \geq 1$. $t_k = (ts, X)$ is a **transaction** with X being the pattern and ts being timestamp, $ts \in \mathbb{R}^+$. A set of transactions constitute a **temporal database** denoted by TDB over I, i.e., $TDB = \{t_1, \cdots, t_d\}$, $d = |TDB|$. Let ts^Y be the timestamp of pattern Y, $Y \subseteq X$, which occurs in transaction t_i (or t_i contains Y), $t_i = (ts, X)$, $i \geq 1$. Denoted TS^Y as a set of timestamps $\{ts_j^Y, \cdots, ts_i^Y\}$, j, $k \in [1, d]$ and $j \leq i$. TS^Y can be consider as an **ordered set of timestamps** of pattern Y.

Example 1. Assume that we have a set of items $I = \{p, q, r, s, t, u\}$. Table 1 shows a row temporal database. Table 2 shows a columnar temporal database which is converted from above row database. In Table 3 we show for each item the temporal occurrences over the whole database. The set of items 'r' and 'q', i.e., $\{r, q\}$ is a pattern. This pattern will be represented as 'rq' for brevity. This pattern is denoted as 2-pattern because it contains two items. The occurrences of pattern 'rq' are at the timestamps of 1, 3, 6, 8, 9, and 10. Therefore, we have a list of timestamps containing 'rq', i.e., $TS^{rq} = \{1, 3, 6, 8, 9, 10\}$.

Definition 1 (*The support of Y*). *Denoted $sup(Y)$ the **support** of Y which is the number of transactions containing Y in TDB. That is, $sup(Y) = |TS^Y|$.*

Example 2. The *support* of 'rq', i.e., $sup(rq) = |TS^{rq}| = 6$.

Definition 2 (*Frequentpattern Y*). *The pattern Y is a **frequent pattern** if $sup(Y) \geq minSup$, where minSup is a minimum support value indicated by user.*

Example 3. Suppose $minSup = 5$, then rq is a frequent pattern because of $sup(rq) \geq minSup$.

Definition 3. (*Periodicity of Y*). *Denoted ts_m^Y and ts_n^X, $j \leq m < n \leq k$ the two consecutive timestamps in TS^Y. The time difference between ts_n^Y and ts_m^Y is given by a **period** of Y, denoted by p_z^Y. That is, $p_z^Y = ts_n^Y - ts_m^Y$. Denoted $P^Y = (p_1^Y, p_2^Y, \cdots, p_n^Y)$ the set of all periods for pattern Y. The **periodicity** of Y, denoted by $per(Y) = maximum(p_1^Y, p_2^Y, \cdots, p_n^Y)$.*

Table 1. Row database

ts	Items	ts	Items
1	qrs	6	pqr
2	pq	7	stu
3	pqrs	8	qrs
4	tu	9	pqrs
5	prs	10	pqrs

Table 2. Columnar database

ts	p	q	c	s	e	u	ts	p	q	r	s	t	u
1	0	1	1	1	0	0	6	1	1	1	0	0	0
2	1	1	0	0	0	0	7	0	0	0	1	1	1
3	1	1	1	1	0	0	8	0	1	1	1	0	0
4	0	0	0	0	1	1	9	1	1	1	1	0	0
5	1	0	1	1	0	0	10	1	1	1	1	0	0

Table 3. Timestamp list of an item

Item	TS-list
p	2,3,5,6,9,10
q	1,2,3,6,8,9,10
r	1,3,5,6,8,9,10
s	1,3,5,7,8,9,10
t	4,7
u	4,7

Example 4. All periods of the pattern 'rq' are : $p_1^{rq} = 1 \ (= 1 - ts_{initial})$, $p_2^{rq} = 2 \ (= 3 - 1)$, $p_3^{rq} = 3 \ (= 6 - 3)$, $p_4^{rq} = 2 \ (= 8 - 6)$, $p_5^{rq} = 1 \ (= 9 - 8)$, $p_6^{rq} = 1 \ (= 10 - 9)$, and $p_8^{rq} = 0 \ (= ts_{final} - 10)$, where first transaction time stamp is denoted by $ts_{initial} = 0$ and the last transaction's time stamp is denoted by, $ts_{final} = |TDB| = 10$. The *periodicity* of rq, i.e., $per(rq) = maximum(1, 2, 3, 2, 1, 1, 1, 0) = 3$.

Definition 4 *(Periodic-frequent pattern Y).* *The frequent pattern Y be considered as* **periodic-frequent pattern** *if* $per(Y) \leq maxPer$, *here maxPer is maximum periodicity value which is specified by user.*

Example 5. Let the user-specified $maxPer = 3$, in this case the frequent pattern 'rq' is called as a periodic-frequent pattern as $per(rq) \leq maxPer$.

Definition 5 *(Lability of an itemset).* *Denoted* ts_{i+1}^Y *and* ts_i^Y, $i \in [0, sup(Y)]$ *two consecutive time stamps where Y occurs in TDB. We call i-th lability of Y denoted by* $la(Y,i) = max(0, la(Y, i-1) + p_i^Y - maxPer)$, *where* $la(Y, -1) = 0$. *For simplicity, the following short form is used*

$$la(Y, i) = max(0, la(Y, i-1) + ts_{i+1}^Y - ts_i^Y - maxPer)$$

The following is a list of periods which represent the lability of an itemset $Y: la(Y) = \{la(Y, 0), la(Y, 1), \cdots, la(Y, sup(Y))\}$, *and* $|la(Y)| = |per(Y)| = sup(Y) + 1$.

Example 6. Given an item p. If $maxPer = 2$, the parameters for calculating its lability are $la(p, 0) = max(0, la(p, -1) + p_0^p - maxPer) = max(0, 0 + 2 - 2) = 0$, $la(p, 1) = 0$, $la(p, 2) = 0$, $la(p, 3) = 0$, $la(p, 4) = 1$, $la(p, 5) = 0$, and $la(p, 6) = 0$. Therefore, the lability of p is $la(p) = \{0, 0, 0, 0, 1, 0, 0\}$.

Based on Definition 5, the periodic pattern can be considered as stable (lability is zero) if all its periods are less than or equal to $maxPer$. The lability of a period of a pattern will increase when a period of a pattern larger than $maxPer$, and these exceeding values are accumulated using the measure of lability. The value of lability will be reduced when periods of a pattern no more than $maxPer$. Therefore, according to the periodic characteristic of a pattern, its *lability* will vary over time, and each value exceeding $maxPer$ is accumulated. A periodic behavior is considered stable when lability value is low while a high value means an unstable one. So stable pattern can be found using this measure given a limit on the maximum lability.

Definition 6 *(Stable periodic-frequent pattern).* *For a pattern* Y, *denote* $la(Y)$ *the set of all i-th lability. The stability of the pattern is defined by* $maxla(Y)$ $= max(la(Y))$. *Pattern* Y *is a SPP if* $sup(Y) \geq minSup$ *and* $maxla(Y) \leq$ $maxLa$.

Example 7. Given the above example, if the user specified $minSup = 4$, $maxPer = 2$, and $maxLa = 1$, the complete set of SPPs are p: (6,1), ps: (4,0), psr: (4,0), pq: (5,0), pqr: (4,0), pr: (5,0), q: (7,1), qs: (5,0), qsr: (5,0), qr: (6,0), r: (7,0), rs:(6,0) and s: (7,0), where each SPP Y is annotated with Y: $(sup(Y),$ $maxLa(Y))$.

Be noted that if $maxLa = 0$, SPPs are the traditional PFPs. Therefore, the PFPs is a special case of SPPs.

Definition 7 *(Problem definition).* *Considering a temporal database (TDB) with* minimum support *(minSup),* maximum periodicity *(maxPer), and* maximum lability *(maxLa) constraints. The purpose of this task is discovering the complete set of stable periodic-frequent patterns that have support higher or equal to minSup and lability lower or equal to maxLa constraints.*

4   Our Mining Algorithm: SPP-ECLAT

This section shows the process of mining the stable periodic-frequent patterns in two steps using SPP-ECLAT algorithm and our algorithm works in two steps. First, SPP-ECLAT algorithm utilizes the Depth-First Search (DFS) strategy on the itemset lattice. Second, this algorithm employs the *downward closure property* (see Property 1) of stable periodic-frequent patterns to minimize the huge search space of the lattice effectively.

Property 1. If A is a stable periodic-frequent pattern, then $\forall A \subset B$ and $A \neq \emptyset$, A is also a stable periodic-frequent pattern.

4.1   Mining 1-Stable Periodic-Frequent Patterns

This part focuses on discovering 1-patterns by SPP-list. The detailed steps are shown in Algorithm 1, which works on a row database shown in Table 1. Let $minSup = 5$ and $maxPer = 2$ and $maxLa = 1$.

The 1-patterns are first generated by reading the whole database transactions at once. Then, the row database is converted to the columnar database. After reading the 1st transaction, "1 : qrs", with $ts_{cur} = 1$ inserts the items q, r and s, in the SPP-list. We have the timestamps of these items is 1 $(= ts_{cur})$. Similarly, ML and TS_l contents were updated to 0 and 1, respectively (lines 7 and 8 in

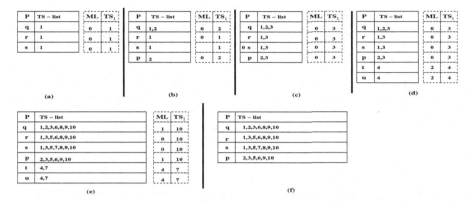

Fig. 1. SPP-list generation process. (a) content of the list after reading the 1st transaction, (b) after reading the 2nd one, (c) after reading the 3rd one, (d) after reading the 4th one, (e) Final content after reading the whole database, and (f) The complete list of 1-stable periodic-frequent patterns

Algorithm 1). Figure 1(a) shows the generated SPP-list from the 1st transaction. After reading the 2nd one, "2 : pq", with $ts_{cur} = 2$ inserts the new items p into the SPP-list by adding 2 $(= ts_{cur})$ in their TS-list. At the same instant, the ML and TS_l contents were updated to 0 and 2, respectively. Besides 2 $(= ts_{cur})$ was added to the TS-list of existing items q with ML and TS_l contents were updated to 0 and 2, respectively (lines 10 and 13 in Algorithm 1). The SPP-list which is generated after reading the 2nd one is shown in Fig. 1(b). After reading the 3rd one, "3 : $pqrs$", updates the TS-list, ML and TS_l values of p, q, r, and s in the SPP-list. Figure 1(c) shows the SPP-list which is generated after reading the 3rd one. After reading the 4th one, "4 : tu" with $ts_{cur} = 4$, inserts the new items e and u into the SPP-list by adding 4 $(= ts_{cur})$ in their TS-list. Simultaneously, the ML and TS_l values as 2 and 4. Figure 1(d) shows the SPP-list which is generated after reading the 4th. We repeat the whole process for the remaining transactions. Figure 1(e) depicts the final SPP-list which is generated after scanning the whole database. The pattern t and u are pruned (using the Property 1) from the SPP-list as its *support* value is no more than the $minSup$ value and ML value is greater than $maxLa$ (lines 15 to 20 in Algorithm 1). The complete list of patterns available in the SPP-list are considered as 1-stable periodic-frequent patterns. Those patterns are sorted in descending order in terms of their *support* values. Figure 1(f) shows the final SPP-list.

4.2 Finding All Interesting Patterns from SPP-ECLAT

The detailed procedure for finding stable periodic-frequent patterns is shown in Algorithm 2. Given the newly generated SPP-list, the procedure of this algorithm is carried out as follows. Initially we choose the pattern q, as this is the initial pattern in the SPP-list (line 2 in Algorithm 2). Figure 2(a) shows a record of its *support* and *lability*. Since q is a stable periodic-frequent pattern, we move

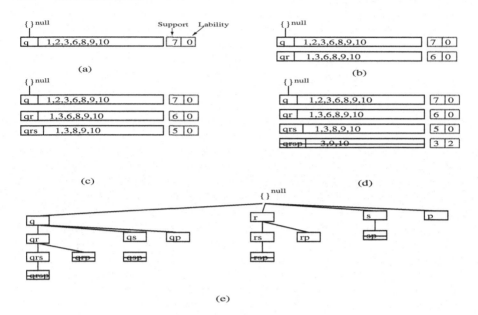

Fig. 2. The complete process of discovering stable periodic-frequent patterns using SPP-ECLAT algorithm

to its child node qr. TS-list of qr is generated by performing intersection of TS-lists of q and r, i.e., $TS^{qr} = TS^q \cap TS^r$ (lines 2 and 3 in Algorithm 2). This *support* and *lability* of qr are recorded, as shown in Fig. 2(b). We check whether qr is a stable periodic-frequent pattern or unstable periodic frequent pattern (line 4 in Algorithm 2). Since qr is stable periodic-frequent pattern we move it to its child node qrs. Next, TS-list will be generated by performing the intersection of TS-lists of qr and s, i.e., $TS^{qrs} = TS^{qr} \cap TS^s$. Figure 2(c) shows a record of *support* and *lability* of qrs. Then qrs is identified as a stable periodic-frequent pattern. Later, we shift to its child node $qrsp$. We produce its TS-list by performing intersection of TS-lists of qrs and p, i.e., $TS^{qrsp} = TS^{qrs} \cap TS^p$. Because a *support* of $qrsp$ is less than $minSup$ and lability is greater than $maxla$, the pattern $qrsp$ will be remove from the stable periodic-frequent patterns list as shown in Fig. 2(d). We repeat the process to find all stable periodic-frequent patterns for remaining nodes in the tree. Figure 2(e) shows the final list of generated stable periodic-frequent patterns. Since we can reduces the search space and the computational cost effectively our proposed approach is efficient.

5   Experimental Results

This section evaluates the performance of the SPP-ECLAT against the state-of-the-art algorithm named SPP-growth [5]. It shows that the SPP-ECLAT algorithm is more efficient in memory consumption and runtime than SPP-growth.

Algorithm 1. StablePeriodicFrequentItems(Temporal database (TDB), minimum support $(minSup)$, maximum periodicity $(maxPer)$, maximum Lability $(maxla)$:

1: Definition: $SPP\text{-}list = (Y, TS\text{-}list(Y))$ is a dictionary with the temporal occurrence information of a pattern in a TDB; TS_l is a temporary variable of list type to store the *timestamp* of the final occurrence of a pattern; la and ML are temporary variable of list type to store the *lability* and the *Maximum Lability* of a pattern; *last* is a term for the final timestamp; *support* is a temporary varibale of list type to store the *support* of a pattern.
2: Initate $ts_{cur} = 0$
3: **for** each transaction $t_{cur} \in TDB$ **do**
4:     Set $ts_{cur} = t_{cur}.ts$;
5:     **for** each item $j \in t_{cur}.Y$ **do**
6:         **if** j does not exit in SPP-list **then**
7:             SPP-list is updated by inserting j and corresponding timestamp value
8:             $la[j] = max(0, ts_{cur} - maxPer)$. Set $ML[j] = la[j]$
9:         **else**
10:            Add j's timestamp in the SPP-list.
11:            $la[j] = max(0, la[j] + ts_{cur} - TS_l[j] - maxPer$
12:            $ML[j] = max(la[j], ML[j])$
13:            Update $TS_l[j] = ts_{cur}$.
14:     $last = ts_{cur}$
15: **for** each item j in SPP-list **do**
16:     $la[j] = max(0, la[j] + last - TS_l[j] - maxPer)$
17:     $ML[j] = max(la[j], ML[j])$
18:     $s[j] = length(TS\text{-}list[j])$
19:     **if** $s[j] < minSup$ and $ML[j] > maxla$ **then**
20:         Prune j from SPP-list
21: After the pruning the final list of patterns available in the SPP-list is sorted in ascending order or descending order of the corresponding pattern's *support*. Initiate pi as Null. Call SPP-ECLAT(SPP-List, pi).

Algorithm 2. SPP-ECLAT(SPP-List, pi)

1: **for** each item j in SPP-List **do**
2:     Set $Y = j \cup pi$ and $TS^Y = TS^j \cap TS^{pi}$;
3:     Calculate *support* and *Lability* of X;
4:     **if** $sup(TS^Y) \geq minSup$ and $la(TS^Y) \leq maxla$ **then**
5:         Add j to pi and Y is considered as stable periodic-frequent pattern;
6:         $SPP\text{-}ECLAT(SPP\text{-}list[j+1:], pi)$;

The algorithms, SPP-growth and SPP-ECLAT, were developed in Python 3.7 and executed on a machine containing two AMD EPIC 7542 cpus and 600 GB RAM. The operating system of this machine is Ubuntu Server OS 20.04. The experiments have been conducted on real-world (T10I4D100K, Retail, and Mushroom) databases. The complete statistics of the databases is shown in the Table 4.

In this experiment, we have fixed the values of $minSup$, $maxPer$ for all the three databases. Subsequently, we have evaluated the performance of both the algorithms by varying the $maxLa$ parameter for all the three databases. Figure 3(a)–3(c) shows the number of stable periodic-frequent patterns generated in T10I4D100K, Retail, and Mushroom databases at different $maxLa$ values. After careful observation of the mentioned graphs, we can conclude that raises in $maxLa$ positively affect the total count of the number of stable periodic-frequent patterns. With an increase in the $maxLa$ threshold, most of the patterns have become stable periodic-frequent patterns.

Table 4. Statistics of the databases used

S. No	Database	Type	Nature	Transaction Length (in count)			Database Size (in count)
				min.	avg.	max.	
1	T10I4D100K	Synthetic	Sparse	1	10	29	1,00,000
2	Retail	Real	Sparse	2	12	77	88,162
3	Mushroom	Real	Dense	23	23	23	8,124

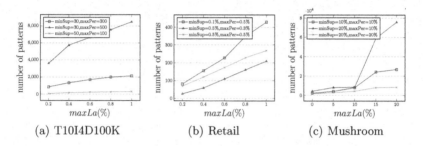

(a) T10I4D100K   (b) Retail   (c) Mushroom

Fig. 3. Number of stable periodic-frequent patterns generated in various datasets

We have varied the values of $minSup$, $maxPer$, and $maxLa$ parameters and shown the runtime requirements of SPP-growth and SPP-ECLAT algorithms in Fig. 4. Specifically, for Retail database, the runtime requirement of both algorithms are shown in Fig. 4(a)–4(c), respectively. For T10I4D100K database, the runtime requirement of both algorithms are shown in Fig. 4(d)–4(f), respectively. For Mushroom database, the runtime requirement of both algorithms are shown in Fig. 4(g)–4(i), respectively. After careful observation of the mentioned graphs, we can conclude that raises in the value of the $maxLa$ parameter shows the raising trend in the graphs. However, we can conclude that SPP-ECLAT algorithms always consumes relatively less runtime than the SPP-growth algorithm.

We have varied the values of $minSup$, $maxPer$, and $maxLa$ parameters and shown the memory consumption of SPP-growth and SPP-ECLAT algorithms in Fig. 5. Specifically, for Retail database, the memory consumption of both algorithms are shown in Fig. 5(a)–5(c), respectively. For T10I4D100K database, the runtime requirement of both algorithms are shown in Fig. 5(d)–Fig. 5(f), respectively. For Mushroom database, the runtime requirement of both algorithms are

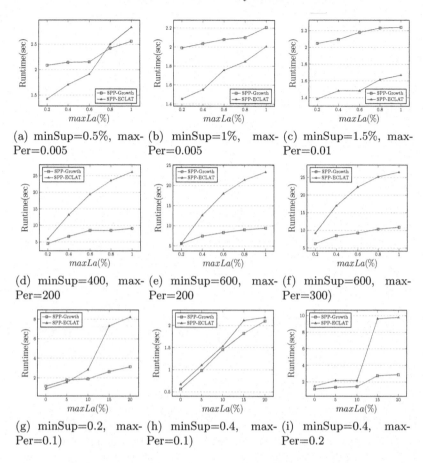

(a) minSup=0.5%, max-Per=0.005

(b) minSup=1%, max-Per=0.005

(c) minSup=1.5%, max-Per=0.01

(d) minSup=400, max-Per=200

(e) minSup=600, max-Per=200

(f) minSup=600, max-Per=300)

(g) minSup=0.2, max-Per=0.1)

(h) minSup=0.4, max-Per=0.1)

(i) minSup=0.4, max-Per=0.2

Fig. 4. Runtime comparison of the two algorithms

shown in Fig. 5(g)–5(i), respectively. After careful observation of the mentioned graphs, we can conclude that raises in the value of the $maxLa$ parameter shows the raising trend in the graphs. However, we can conclude that SPP-ECLAT algorithms always consumes relatively less memory than the SPP-growth algorithm.

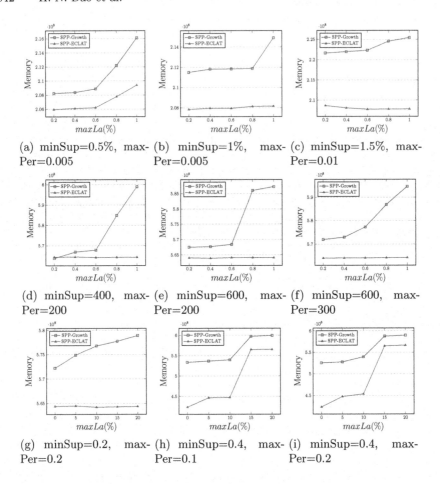

Fig. 5. Memory comparison of the two algorithms

6   Conclusions and Future Work

This paper proposes an efficient algorithm called stable periodic-frequent pattern-equivalence class transformation to discover stable periodic-frequent patterns from columnar temporal databases. The experiment was carryout with different real-world databases. Experimental results show that the proposed algorithm consumes less memory and can generate interesting patterns much faster than the state-of-the-art algorithm. We want to work on parallel algorithms to discover stable periodic-frequent patterns in vast temporal databases as part of future work.

References

1. Kiran, R.U., et al.: Discovering fuzzy periodic-frequent patterns in quantitative temporal databases. In: 2020 (FUZZ-IEEE), pp. 1–8 (2020)
2. Han, J., Dong, G., Yin, Y.: Efficient mining of partial periodic patterns in time series database. In: ICDE, pp. 106–115 (1999)
3. Fournier-Viger, P., Lin, J.C., Duong, Q., Dam, T.: PHM: mining periodic high-utility itemsets. In: ICDM, pp. 64–79 (2016)
4. Kiran, R.U., Venkatesh, J.N., Fournier-Viger, P., Toyoda, M., Reddy, P.K., Kitsuregawa, M.: Discovering periodic patterns in non-uniform temporal databases. In: PAKDD (2017)
5. Fournier-Viger, P., Yang, P., Lin, J.C., Kiran, R.U.: Discovering stable periodic-frequent patterns in transactional data. In: IEA/AIE, pp. 230–244 (2019)
6. Tanbeer, S.K., Ahmed, C.F., Jeong, B.-S., Lee, Y.-K.: Discovering periodic-frequent patterns in transactional databases. In: PAKDD, pp. 242–253 (2009)
7. Amphawan, K., Lenca, P., Surarerks, A.: Mining top-k periodic-frequent pattern from transactional databases without support threshold. In: Advances in Information Technology, pp. 18–29 (2009)
8. Kiran, R.U., Kitsuregawa, M.: Novel techniques to reduce search space in periodic-frequent pattern mining. In: DASFAA, pp. 377–391 (2014)
9. Anirudh, A., Kiran, R.U., Reddy, P.K., Kitsuregawa, M.: Memory efficient mining of periodic-frequent patterns in transactional databases. In: IEEE Symposium Series on Computational Intelligence 2016, pp. 1–8 (2016)
10. Penugonda, R., Palla, L., Rage, U.K., Watanobe, Y., Zettsu, K.: Towards efficient discovery of periodic-frequent patterns in columnar temporal databases. In: IEA/AIE. Springer International Publishing 2021, pp. 28–40 (2021)
11. Kiran, R.U., Shang, H., Toyoda, M., Kitsuregawa, M.: Discovering partial periodic itemsets in temporal databases. In: SSDBM, pp. 30:1–30:6 (2017)
12. He, R., Chen, J., Du, C., Duan, Y.: Stable periodic frequent itemset mining on uncertain datasets. In. IEEE CCET 2021, pp. 263–267 (2021)
13. Fournier Viger, P., Wang, Y., Yang, P., Lin, C.-W., Yun, U., Rage, U.: Tspin: mining top-k stable periodic patterns. Applied Intelligence, 02 2021

Cyclostationary Random Number Sequences for the Tsetlin Machine

Svein Anders Tunheim[1]([✉]), Rohan Kumar Yadav[1], Lei Jiao[1], Rishad Shafik[2], and Ole-Christoffer Granmo[1]

[1] Centre for Artificial Intelligence Research (CAIR), University of Agder, Grimstad, Norway
{svein.a.tunheim,rohan.k.yadav,lei.jiao,ole.granmo}@uia.no
[2] Microsystems Group, School of Engineering, Newcastle University, Newcastle upon Tyne, UK
rishad.shafik@newcastle.ac.uk

Abstract. The Tsetlin Machine (TM) constitutes an emerging machine learning algorithm that has shown competitive performance on several benchmarks. The underlying concept of the TM is propositional logic determined by a group of finite state machines that learns patterns. Thus, TM-based systems naturally lend themselves to low-power operation when implemented in hardware for micro-edge Internet-of-Things applications. An important aspect of the learning phase of TMs is stochasticity. For low-power integrated circuit implementations the random number generation must be carried out efficiently. In this paper, we explore the application of pre-generated cyclostationary random number sequences for TMs. Through experiments on two machine learning problems, i.e., Binary Iris and Noisy XOR, we demonstrate that the accuracy is on par with standard TM. We show that through exploratory simulations the required length of the sequences that meets the conflicting tradeoffs can be suitably identified. Furthermore, the TMs achieve robust performance against reduced resolution of the random numbers. Finally, we show that maximum-length sequences implemented by linear feedback shift registers are suitable for generating the required random numbers.

Keywords: Machine learning · Tsetlin machine · Cyclostationary random number sequences · Linear feedback shift registers

1 Introduction

The Tsetlin Machine (TM) is a novel machine learning (ML) algorithm that was introduced in 2018 [7]. TMs are based on propositional logic, leading to primarily Boolean operations. Such operations are in contrast to Deep Neural Networks (DNNs) where complex arithmetic is needed for Multiply-Accumulate Units (MACs). Furthermore, as propositional logic forms the basis of the algorithm, TMs have promising interpretability prospects. The unique features of TM make it suitable for low-energy hardware acceleration on Field Programmable

© Springer Nature Switzerland AG 2022
H. Fujita et al. (Eds.): IEA/AIE 2022, LNAI 13343, pp. 844–856, 2022.
https://doi.org/10.1007/978-3-031-08530-7_71

Gate Arrays (FPGA) and Integrated Circuits (ICs). Through compact implementation accelerated online training can be enabled for edge-nodes in Internet-of-Things (IoT) systems. To date TMs have been tested on tabular data, images, regression, natural language and speech, [3], and have shown competitive performance on several benchmarks in terms of accuracy, memory footprint and learning speed. For example, the convolutional TM (CTM) has obtained a peak test accuracy of 99.4% on MNIST, 96.31% on Kuzushiji-MNIST and 91.5% on Fashion-MNIST [8].

A key aspect of TMs is randomized choices during feedback based training. Effective randomization is crucial for avoiding deadlocks and overfitting using the training data. In the original software TM (Python/Cython) implementation [6], a randomization function is used to create a pseudo random number sequence [5]. The range of the numbers varies from 0 to RAND_MAX, where RAND_MAX is guaranteed to be at least 32767 (two bytes).

Python has a more advanced random sequence generator based on the Mersenne Twister [12]. This module produces 53-bit precision floating point numbers and has a period of $2^{19937} - 1$. It is also threadsafe and has been extensively tested.

Random sequence generation in hardware is, however, non-trivial. For low-complexity hardware implementation, the random sequence generation must be carried out efficiently. Storage of pre-generated random numbers in on-chip RAM/ROM is an alternative. However, this is only effective if relatively short sequences can be used. In a highly parallelized TM system, one will also need concurrent access to several independent random numbers, thus requiring many such sequences. Multi-word read capabilities from on-chip memory can reduce the number of sequences needed, but these require significant hardware resource allocation. Another alternative for random number generation is amplification of noise from an analog circuit module followed by analog-to-digital conversion. As true noise is the basis, it is possible to achieve very good stochastic properties [13]. However, mixed-signal design of the random number generator can lead to validation complexity as well as uncertainty due to the approximate nature of analog signals.

The concept of digital Pseudo Random Bit Sequences (PRBSs), typically implemented by Linear Feedback Shift Registers (LFSR) [9,14] is widely used for random number generation. LFSRs are suitable for hardware solutions as these can be implemented in digital and validation-friendly IC design flow. However, LFSRs usually require high level of switching depending on their sizes when the randomization process is active, which can consume non-negligible amount of energy.

In this paper, we aim to study an alternative randomization process suitable for hardware TM implementation. Core to this process is *whether a cyclostationary sequence of pre-generated random numbers could be applied as the source for random numbers in TMs*. Our main *contributions* are as follows:

- We evaluate if pre-generated cyclostationary sequences - with sequence address incremented for each lookup - can be applied for TM training and achieve accuracy similar to C/Python implementations.
- We study the impact of the number of elements in the sequence and therefore the tradeoffs between complexity and learning efficiency of the implementations.
- We evaluate the impact of the resolution (number of decimals) of the numbers in the sequence.

The remainder of this paper is organized as follows. Section 2 describes the TM architecture. Section 3 explores how cyclostationary sequences of random numbers can be applied to the TM. Section 4 details our experiment results together with discussions before we conclude in the final section.

2   Review of the Tsetlin Machine

A TM consists of several teams of Tsetlin Automata (TAs) that operates on literals, i.e., Boolean inputs and their complements. Each team of TAs forms a discriminative conjunctive clause by including or excluding literals as shown in Fig. 1(a). There are m clauses, c_j, where $j = 1, \ldots, m$, and m is an even integer. Half of the clauses, typically the odd numbered ones, are defined as positive, and the other half, the even numbered ones, are defined as negative, see Fig. 1(b). The outputs of the two groups of clauses are assembled in a majority voting unit to decide for the final classification, as shown in Fig. 1(c).

The TAs employed in a TM are of two-action type, i.e., a Tsetlin Automaton will either *include* or *exclude* a literal from a conjunctive clause. This is achieved during the learning process. Figure 2 shows the structure of a single TA with $2N$ states. Action 2 (*include*) is employed if the TA is in one of the states from $N+1$ to $2N$, while the states 1 to N result in Action 1 (*exclude*).

The input to the TM is a feature vector $X = [x_1, x_2, \cdots, x_o]$ consisting of o propositional Boolean variables, $x_u \in \{0,1\}^o$, $u = 1, \ldots, o$. The negation of the variables are appended to the input forming a new input vector $\mathbf{L}$ with in total $2o$ literals: $[x_1, \neg x_1, x_2, \neg x_2, \ldots, x_o, \neg x_o]$. The output of a single clause, c_j is given by:

$$c_j = \bigwedge_{k \in I_j} l_k \tag{1}$$

Here l_k is the literal with index k, and k belongs to $I_j \subseteq \{1, \ldots, 2o\}$. I_j denotes the set of indexes of all the TAs that select action "include" in c_j.

In a basic two-class TM, classification is given by

$$\hat{y} = \begin{cases} 1 & \text{if } v \geq 0 \\ 0 & \text{if } v < 0 \end{cases} \tag{2}$$

where the output sum, v, is defined in Eq. 3.

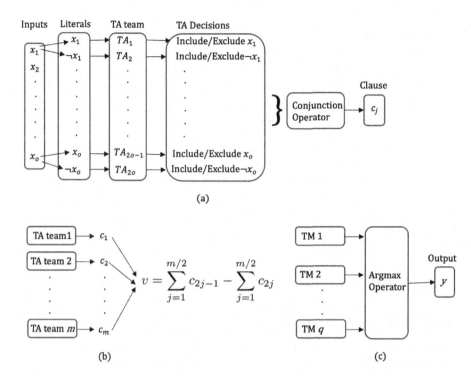

Fig. 1. (a) A TA team forms the clause c_j (b) A two-class TM with m clauses. (c) A q-class TM [7].

$$v = \sum_{j=1}^{m/2} c_{2j-1} - \sum_{j=1}^{m/2} c_{2j} \qquad (3)$$

A multiclass TM is constituted by several TMs, one for each class, 1 to q. As shown in Fig. 1(c), the final decision is made by an argmax operator that classifies the input data according to the highest vote sum [7].

Learning takes place in the TM through a novel finite state learning automata game [7]. It coordinates the collective of TAs and leverages resource-allocation and frequent pattern mining principles. Feedback mechanisms are employed and gives each TA either a *reward* or *penalty*, depending on the training input, the individual clause outputs and the TM output sum, as shown in Fig. 2. If a reward is applied, the TA will move deeper, i.e. towards state 1 or $2N$ depending on the action. With penalty the TA will move towards the center and will eventually jump to the other side of the action. In addition to the number of clauses, m, the hyperparameters T (Threshold value) and s determine the stochastic learning characteristics. As explained in [7], T decides the clause update probability. A higher T increases the robustness of learning by allocating more clauses to learn each sub-pattern [1], while greater values of s stimulates a TA team to include

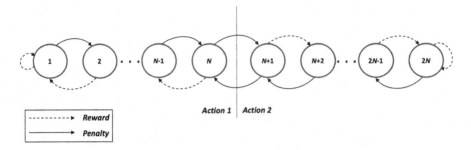

Fig. 2. A Tsetlin automaton for two-action environments [18].

more literals in the clause [7]. Optimum setting of m, T and s is dependent on the specific ML problem.

3 Proposed Random Number Generator in TM

In this section, we first give an overview of where stochasticity is employed in the TM algorithm. We then describe our proposed solution for applying the required random numbers to the TM. Thereafter, we elaborate how this solution can be implemented by LFSRs.

3.1 Randomization in Tsetlin Machine

Stochasticity is required by TM during training to avoid deadlocks and overfitting. Furthermore, the randomness helps the TM to allocate pattern recognition resources in an efficient way [7]. Stochasticity is applied during training at several different stages of the algorithm:

- Clause selection for update, based on the hyperparameter T, the class sum v, the clause type (positive/negative), and literal value l_k [7].
- TA update, based on the s hyperparameter [7].
- Patch selection for CTM [8]: A clause may output 1 for several patches during the convolution. TM randomly picks a patch among the ones that made the clause evaluate to 1 and trains the clause accordingly.
- For multiclass TM/CTM [1,7]: In addition to training the *target class*, TM randomly selects a different class, the *negative target*, to train against, for a given training example.
- Epoch level: A TM version [15] features randomly (in a user-specified percentage) dropping of the clauses per training epoch. This produces more distinct and well-structured patterns that improve the performance and the learning robustness. During inference all literals are utilized. In addition, for each training epoch, the data should be randomly reshuffled [7], which increases robustness when operating on new input data.

– In the "arbitrarily deterministic TM" [2], the update of a TA only occurs every d'th time with a probability of 0.5. Here, d is an integer hyperparameter.

The random numbers generated for TMs should follow uniform probability distribution, according to [7], with range $[0, 1)$. TMs have been implemented in various programming languages such as Python, C and CUDA utilizing these languages' random number generators, and a huge amount of random numbers are generated for training. However, those approaches are not appropriate for low-power hardware platforms due to their complexity. In addition, the generation of random numbers can be energy and computational expensive. We therefore study below the possibility to simplify the generating process, and evaluate its impact on the performance of TM.

3.2 Employing Pre-generated Random Number Sequences for TM

To simplify the procedure of generating random numbers, we utilize a *pre-generated sequence* of random numbers in the range $[0, 1)$ with uniform probability distribution, instead of generating a random number every time. The sequence is *cyclostationary*. That is, it repeats itself when the last element has been read. This implies that the statistical properties of the sequence vary *cyclically* with time.

The numpy.random.uniform Python routine [10] was adopted to generate the numbers in the sequence. Sequences of various lengths were generated and different number of decimals of the random numbers were tested. For our experiments, we used a single cyclostationary sequence for all the random number accesses needed in the TM algorithm. The TM Python/Cython code was modified the following way:

– The *rand()* function was replaced by a lookup to a pre-generated sequence of random numbers, addressed by an index counter.
– For each sequence access the index counter was incremented.
– When reaching the last number in the sequence, the index counter was reset, thus re-accessing the sequence from the beginning.

3.3 Generating Cyclostationary Random Number Sequences with LFSRs

In addition to the pre-generated sequences, to accommodate an effective implementation of random number generators in hardware, we adopt and study the potential of *Linear Feedback Shift Registers* (LFSRs) [14] for TM. An LFSR consists of D-flip-flops connected in series, as a shift register, with feedback from select D flip-flop outputs, so-called "taps". The taps are XOR'ed and fed to the input of the shift register. Figure 3 shows a 7-bit LFSR implementation where the taps 7 and 6 are XORed and fed to the input of the first D-flip-flop. We can describe this by the corresponding feedback polynomial $x^7 + x^6 + 1$.

By proper selection of taps, one can implement maximum length sequences (MLS) with the following properties [14]:

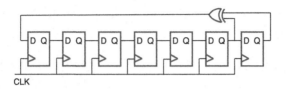

Fig. 3. A 7-bit LFSR.

- The period is $2^N - 1$, where N is the length of the shift register.
- In each period of an MLS, the number of 1s is always one more than the number of 0s. This is called the *balance property*.
- A *run* is a group of consecutively following 1s or 0s. Among the runs of 1s and 0s in each period of an MLS, one half of the runs of each kind are of length one, one-fourth of length two, one-eight of length three, and so on as long as these fractions represent meaningful number of runs. This is called the *run property*.
- The *autocorrelation property* of an MLS is periodic and binary-valued [14], as shown in Eq. (4). Higher N implies lower autocorrelation, which in general is a desired property of random number sequences.

$$R(n) = \begin{cases} 1 & n = 0 \\ -\frac{1}{N} & 0 < n < N \end{cases} \tag{4}$$

For the NoisyXOR problem, to be detailed in Subsect. 4.1, we evaluated three different LFSRs based on the following MLS feedback polynomials [11]:

- $x^{18} + x^7 + 1$
- $x^{16} + x^{12} + x^3 + x^1 + 1$
- $x^{12} + x^6 + x^4 + x^1 + 1$

The above sequences have lengths of 262143, 65535 and 4095 respectively and were chosen based on the results from the general experiments, in Sect. 4. We used the open source LFSR Python software in [4] to generate the LFSR random numbers by reading out the internal register state per clock period. Furthermore, with this software we also verified the general MLS characteristics of these sequences, as described earlier in this section. The random numbers generated by the LFSRs were scaled to the range [0,1), and the resolution tested were floored to 2, 3 and 4 decimals. This corresponds to binary number representations of about 7, 10 and 14 bits respectively. For illustration purposes, Fig. 4 shows an excerpt from the sequence generated by the 18 bit LFSR. The numbers here are scaled to the range [0, 1), and the sequence elements from 60000 to 60100 are shown.

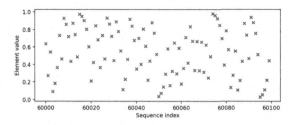

Fig. 4. Excerpt from sequence generated by the 18-bit LFSR.

4 Experimental Results and Discussion

To study the impact of different types of sequences on the performance of TM, experiments on two different ML problems were performed: The *NoisyXOR* and the *Binary Iris* datasets [6,7].

4.1 The Noisy XOR Dataset with Non-informative Features

The artificial dataset of Noisy XOR with non-informative features was one of the original datasets [6] used for testing the vanilla TM [7]. The dataset consists of 10,000 examples, and there are twelve Boolean inputs and one Boolean output. Ten of the inputs are completely random while two inputs follow the XOR-relation. Of the dataset, 50% is adopted for test and the other 50% for training. A high level of noise has been introduced in the training dataset by inverting 40% of the outputs. The motivation for this was to examine robustness of the TM [7]. The dataset is intended to uncover "blind zones" caused by XOR-like relations [7].

For comparison, the configuration of TM in this experiment is kept exactly the same as in [7]. The architecture is configured with 20 clauses and an s-value of 3.9, the threshold T value of 15. Each TA is allocated 100 states. We ran the TM algorithm for 200 epochs for each training experiment; the training data was reshuffled for each epoch. We performed 100 iterations of each experiment. Figure 5 shows the mean test accuracy for different sequence lengths and different resolution of the pre-generated random numbers from these experiments.

In the original experiments reported in [7], the average of the test accuracy was 99.3%. The 5%-percentile, 95%-percentile, Min and Max accuracy were 95.9%, 100.0%, 91.6% and 100% respectively. Applying the pre-generated cyclostationary random number sequence for the NoisyXOR case, we observe that sequences with more than about 50 k random numbers achieve mean test accuracy on par with the original results in [7], as shown in Fig. 5. For example, for a sequence of 100 k random numbers with 3 decimals, we obtained an average test accuracy of 99.5%, 5%-percentile of 97.7%, 95%-percentile of 100%, minimum accuracy of 93.1%, and maximum accuracy of 100%.

When it comes to resolution, applying a number with only 1 decimal significantly degrades the accuracy, independently of the sequence length.

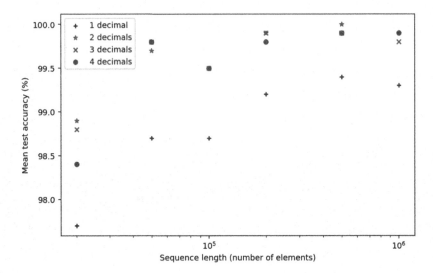

Fig. 5. Average accuracy for NoisyXOR test data versus sequence length and for different random number resolutions.

However, there are only minor accuracy differences between 2 and more decimals for sequence lengths greater than 50 k elements. For sequence lengths of 100 k and 200 k, we also tested rounding down to the closest decimal of the chosen resolution, i.e. the numbers were "floored", as this reflects a close-to-hardware representation. In this case, comparing to the normal rounding showed negligible difference.

As the sequence needs to be long to achieve a comparable accuracy, i.e., over 50 k numbers, storing it in on-chip RAM/ROM is not attractive. A sequence of 50 k numbers with, e.g., 10 bits representation, would need 0.5Mbit storage. Furthermore, with highly parallel operation of a TM-based system it is desirable to operate several such sequences in parallel, requiring more hardware resources. For this reason, LFSRs are considered as better candidates for random number generators in TM-based hardware systems. It should be noted, however, that increased register lengths will result in higher power consumption due to the digital switching activity. Therefore, TM training circuitry will only be enabled during training and will be switched off to save power during inference mode. In an IC, clock and power gating can be employed to enable the LFSRs. Clause and TA updating can be implemented with a high degree of parallelism, and several LFSRs can operate concurrently. In this case, the different LFSRs should be seeded differently, i.e., their start conditions should be different.

In Table 1, it is shown that LFSR registers, with lengths from 16 and upwards, provide random numbers that do not degrade the mean test accuracy. The 12 bit LFSR corresponds to a sequence length of only 4095, and the accuracy degradation for this is notable and as expected.

Table 1. Mean accuracy for NoisyXOR test data versus sequence length for different number resolutions. MLS/LFSR-based number sequences. The random numbers are floored.

	2 decimals	3 decimals	4 decimals
18 bit LFSR	99.9	99.9	99.9
16 bit LFSR	99.9	99.9	99.8
12 bit LFSR	88.7	85.8	82.0

For the results in Table 1, it should be noted that the sequences also included the startup transition part. More specifically, the sequences' probability distribution may not be precisely uniform. To test this further, we applied only 1/4 of the sequence for the 18 bit LFSR as the source for random numbers. In this case, we found that the mean accuracy was also high (99.9%). The reason for this is most likely the MLS's balance and run properties as shown in Sect. 3. Moreover, the quarter sequence also contains more than 50 k elements.

4.2   The Binary Iris Dataset

The Iris dataset is classical [16] and consists of only 150 examples. It was one of the datasets used when evaluating the vanilla TM performance [7]. Each example has four inputs and three possible outputs (classes). In [7], the dataset was converted into Boolean features the following way: Each input value was represented by a 4 bit number, i.e. the input sequence was 16 bits in total. This new dataset was denoted The Binary Iris Dataset. For training we used 80% of the dataset, and we randomly generated 100 different training and test partitions (ensembles). We adopted 300 clauses per class. The s-value was 3.0, the threshold T was 10 and each TA had 100 states. We ran the TM for 500 epochs for each data partition.

In Fig. 6, it is shown the mean accuracy for Binary Iris test data for different sequence lengths and number resolutions. The original mean test data accuracy was 95.0% [7]. As we can see from Fig. 6, even very short sequences provide good accuracy. For 11 elements and below we observed significant accuracy degradation. Similar to the NoisyXOR dataset, the resolution does not have a huge impact. However, with only one decimal there is a notable degradation. Thus, the chosen settings for the Binary Iris dataset are very robust with respect to the random number sequence.

The IC reported in [17] implements a TM-based classifier for the Binary Iris case. It the first reported chip based on the TM, and achieves 62.7Top/J during inference and 34.6Top/J during training. The chip employs one 8-bit LFSR per TA, and each LFSR is seeded differently. The accuracy of the IC was somewhat degraded compared with the original software version. The IC shows approximately 92.5% for the test accuracy. Our simulations show that even with very short sequence lengths, the TM operates robustly for this dataset. So the 8-bit LFSR with a sequence length of 255 should be sufficient. However, other IC

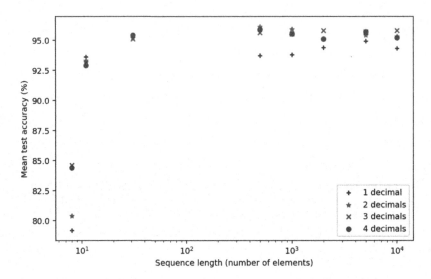

Fig. 6. Mean accuracy for Binary Iris test data for different sequence lengths and number resolutions.

implementation choices affect the test accuracy. Most important is the number of clauses applied per class, which for the IC in [17] was significantly less than for the results reported in [7].

The number of times a sequence is applied during one epoch scales approximately inversely proportional with the sequence length as expected. The number of sequence accesses depends on the amount of training examples and the general TM configuration (number of clauses, number of classes, T, and s). For the NoisyXOR case, a 50 k element sequence is applied approximately 13.9 times during one epoch, while the Binary Iris only requires 0.9 times.

Comparing the results from these two data sets, we can conclude that cyclostationary sequences can be used by TM for training purpose. For NoisyXOR, the sequence should contain more than 50 k elements, and the resolution of the random numbers should be minimum two decimals. For Binary Iris, the required length is significantly reduced. This indicates that the required sequence length is to a large degree dependent on the ML problem and the nature of the dataset. This shows that for any hardware implementations, if the application domain and the ML problem is known, the random sequence generator can be designed according to the nature of the problem. The approach may be tested in software implementations first, and then deployed accordingly in the hardware, which can best balance the complexity, performance, and power consumption. More specifically, one can modify the TM code by e.g., exchanging the random function calls with lookup to a pre-generated number sequence, and perform direct simulations of the effect of the sequence length and resolution.

5   Conclusions

Based on the empirical results, we conclude that pre-generated cyclostationary sequences of randomly generated numbers can be employed for TM training with test accuracy on par with vanilla TM. The required lengths of the sequences depend heavily on the ML problem. Using sequences with just sufficient lengths can reduce the complexity of the implementation of TM significantly. For hardware implementations, applying cyclostationary sequences from LFSRs is an attractive and robust solution for the random number sequence generation.

References

1. Abeyrathna, K.D., Granmo, O.C., Goodwin, M.: Extending the Tsetlin machine with integer-weighted clauses for increased interpretability. IEEE Access **9**, 8233–8248 (2021). https://ieeexplore.ieee.org/document/9316190
2. Abeyrathna, K.D., et al.: A Multi-Step Finite-State Automaton for Arbitrarily Deterministic Tsetlin Machine Learning. Expert Systems, John Wiley & Sons Ltd., Hoboken (2021). https://doi.org/10.1111/exsy.12836
3. Abeyrathna, K.D., et al.: Massively parallel and asynchronous tsetlin machine architecture supporting almost constant-time scaling. In: Meila, M., Zhang, T. (eds.) Proceedings of the 38th International Conference on Machine Learning. Proceedings of Machine Learning Research, vol. 139, pp. 10–20. PMLR, 18–24 July 2021. https://proceedings.mlr.press/v139/abeyrathna21a.html
4. Bajaj, N.: Nikeshbajaj/linear_feedback_shift_register: 1.0.6, April 2021. https://doi.org/10.5281/zenodo.4726667
5. C library function - rand(): https://www.tutorialspoint.com/c_standard_library/c_function_rand.htm
6. Granmo, O.C.: Github TM repo. https://github.com/cair/TsetlinMachine
7. Granmo, O.C.: The Tsetlin machine–a game theoretic bandit driven approach to optimal pattern recognition with propositional logic. arXiv e-prints (2018). https://arxiv.org/abs/1804.01508
8. Granmo, O.C., Glimsdal, S., Jiao, L., Goodwin, M., Omlin, C.W., Berge, G.T.: The convolutional Tsetlin machine. arXiv e-prints (2019). https://arxiv.org/abs/1905.09688
9. Haykin, S.: Communication Systems, 4th edn. John Wiley & Sons, Inc., Hoboken (2001)
10. Numpy library. https://www.numpy.org
11. Partow, A.: https://www.partow.net/programming/polynomials/index.html
12. Python 3.10.0 doc.: http://docs.python.org/3/library/random.html
13. Rahman, T., Shafik, R., Granmo, O.C., Yakovlev, A.: Resilient biomedical systems design under noise using logic based machine learning. Front. Control Eng. (in press, 2022). Adaptive, Robust and Fault Tolerant Control. https://www.frontiersin.org/articles/10.3389/fcteg.2021.778118/abstract
14. Sarwate, D., Pursley, M.: Crosscorrelation properties of pseudorandom and related sequences. Proc. IEEE **68**(5), 593–619 (1980)
15. Sharma, J., Yadav, R., Granmo, O.C., Jiao, L.: Human interpretable AI: Enhancing Tsetlin machine stochasticity with drop clause. arXiv (2021). https://arxiv.org/abs/2105.14506

16. UCI Machine Learning Repo. https://archive.ics.uci.edu/ml/datasets/iris
17. Wheeldon, A., Shafik, R., Rahman, T., Lei, J., Yakovlev, A., Granmo, O.C.: Learning automata based energy-efficient AI hardware design for IoT applications: learning automata based AI hardware. Royal Society Publishing (2020). https://doi.org/10.1098/rsta.2019.0593
18. Zhang, X., Jiao, L., Granmo, O.C., Goodwin, M.: On the convergence of tsetlin machines for the identity-and not operators. IEEE Trans. Pattern Anal. Mach. Intell. (2021). https://doi.org/10.1109/TPAMI.2021.3085591

Logics and Ontologies

Evolution of Prioritized $\mathcal{EL}$ Ontologies

Rim Mohamed[1($\boxtimes$)], Zied Loukil[1], Faiez Gargouri[1], and Zied Bouraoui[2]

[1] University of Sfax, MIRACL Laboratory, ISIMS, Sfax, Tunisia
rymmohammed2@gmail.com, zied.loukil@isims.usf.tn, faiez.gargouri@usf.tn
[2] CRIL CNRS & Univ Artois, Lens, France
bouraoui@cril.fr

Abstract. In this paper, we place ourselves in the context of the evolution of $\mathcal{EL}$ ontologies when a new piece of information that can be uncertain is available. The $\mathcal{EL}$ is a tractable family of lightweight description logics that underlay the *OWL2 EL* profile. To take into consideration the weights attached to the axioms reflecting their reliability or credibility, we propose an extension of the $\mathcal{EL}$ family within the possibility theory setting. This theory provides a natural framework to deal with the ordinal scale. We show how to induce a prioritized $\mathcal{EL}$ ontology in case we start with standard ontology,i.e., axioms without weights. The explored method exploits conflict statistical analysis between axioms to induce a vector space. Finally, we propose polynomial syntactic algorithms for the evolution process while preserving the consistency of the ontology.

Keywords: $\mathcal{EL}$ Ontology · Evolution of knowledge bases · Inconsistency and uncertainty management

1 Introduction

Description logic (DLs for short) is the most knowledge representation formalism in the context of the Semantic Web [17], it offers a powerful framework for reasoning about static knowledge in various domain areas such as ontology-based data access [18] and information and data integration [10]. The DLs contains two major elements: The terminological box (called also TBox) includes concepts, relationships, and constraints on a given domain. The assertional box (called also ABox) contains data. Recently, the *OWL2-EL* profile, which is a subset of *OWL2* gained a lot of attention in many applications areas, such as economics [14] and medicine [1]. In this profile, the basic reasoning problems can be done in polynomial w.r.t the size of the ontology [13]. The *OWL2-EL* provides powerful class constructors to express the very large biomedical ontology SNOMED CT[1] and gene ontology (GO)[2]. This profile is based on a family of lightweight DLs, called $\mathcal{EL}$ [3], which guarantees the tractability of the reasoning process, especially for concept classification.

---

[1] https://bioportal.bioontology.org/ontologies/SNOMEDCT.
[2] http://geneontology.org/.

© Springer Nature Switzerland AG 2022
H. Fujita et al. (Eds.): IEA/AIE 2022, LNAI 13343, pp. 859–870, 2022.
https://doi.org/10.1007/978-3-031-08530-7_72

Nowadays, the medical ontologies are involved over time (i.e., they are not static) [16]. For example, in the case of Alzheimer's illness, some studies proved that the symptoms of Alzheimer's are increased memory loss confusion, inability to learn new things, and difficulty with language. However, other studies showed that this illness starts with disorientation to time and place, and poor or decreased judgment. At each time a new study defines new symptoms. This leads us to revise the old one. The revision process consists of inserting some input information into the ontology while preserving its consistency.

The evolution process of DL consists of incorporating a new ontology into an existing ontology while taking into consideration the change that can occur. In general, the new ontology to incorporate is represented by a set of axioms that should satisfy some defined properties [8]. In case when the ontology interacts negatively with the old one, i.e., the ontology or a part of it becomes unsatisfiable, therefore, the input ontology cannot be simply added to the old one. For such reason, some changes need to be made to avoid the unsatisfiability of the ontology, e.g., handling the axioms that conflict with the ontology. This problem is similar to the belief revision problem in propositional logic [5], where the old belief is revised by adding new formulas to a knowledge base that can be sure or uncertain. This problem is also defined as the knowledge change and specified by the well-known AGM postulates [2]. Several works have been proposed to revise the DL ontology (e.g. [8,9]). In particular several model-based revision approaches have been proposed (e.g. [7,12,15]). This revision operator for a Dls is restricted to one part, either the ABox or TBox. However, in [16] it operates on the general KB.

In real applications, information are coming from several conflicting sources which lead to obtain a prioritized knowledge base [13]. In [4], a Prioritized Removed Sets Revision (PRSR) is proposed to revise DL-Lite ontology at the assertional level, it consists of removing from the knowledge base the information that contradicts the input while providing a minimal change in the ontology. In this paper, we study the syntactic evolution of $\mathcal{EL}$ when new information that can be uncertain is available. We explore some cases of evolution depending on the type of the input, i.e., when the input is consistent or not with the old ontology. Or whether the input can be inferred or not from the ontology and to what extent. When the evolution process starts with a flat ontology, we define a new method that exploits conflict statistical analysis between facts to induce a prioritized ontology. Finally, We show that our syntactic evolution is performed in polynomial time.

2   The Syntax and Semantics of $\mathcal{EL}$s

In this section, we briefly recall the syntax and semantics of the $\mathcal{EL}$ DL, the fragment underlying *OWL2-EL*.

Syntax. The syntax of $\mathcal{EL}$ is defined upon the three pairwise disjoint sets N_C, N_R, N_I, where N_C denotes a set of atomic concepts, N_R denotes a set of atomic roles and N_I denotes a set of individuals. The $\mathcal{EL}$ concept expressions are built according to the following syntax:

Table 1. Syntax and semantics of concept and role expressions.

	Syntax	Semantics
Atomic concept	A	$A^{\mathcal{I}} \subseteq \Delta^{\mathcal{I}}$
Atomic role	r	$r^{\mathcal{I}} \subseteq \Delta^{\mathcal{I}} \times \Delta^{\mathcal{I}}$
Individual	a	$a^{\mathcal{I}} \subseteq \Delta^{\mathcal{I}}$
Top	$\top$	$\Delta^{\mathcal{I}}$
Bottom	$\perp$	$\emptyset$
Conjunction	$C \sqcap D$	$C^{\mathcal{I}} \cap D^{\mathcal{I}}$
Existential restriction	$\exists r.C$	$\{x \in \Delta^{\mathcal{I}} \mid \exists y \in \Delta^{\mathcal{I}} \text{ tel que } (x, y) \in r^{\mathcal{I}} \text{ et } y \in C^{\mathcal{I}}\}$
Nominal	$\{a\}$	$\{a^{\mathcal{I}}\}$
Role chain	$r \circ s$	$\{< x, y > \mid \exists z \in \Delta^{\mathcal{I}} \text{ tel que } < x, z > \in r^{\mathcal{I}} \text{ et } < z, y > \in s^{\mathcal{I}}\}$

$$C, D \;\rightarrow\; \top \mid A \mid C \sqcap D \mid \exists r.C$$

where $A \in N_C$, $r \in N_R$.

An $\mathcal{EL}$ ontology (or knowledge base) consists of a set of general concept inclusion (GCI) axioms of the form $C \sqsubseteq D$, meaning that C is more specific than D or simply C is subsumed by D, a set of equivalence axioms of the form $C \equiv D$, which is the abbreviation of the two general concept inclusions $C \sqsubseteq D$ and $D \sqsubseteq C$, a set of concept assertions of the form $C(a)$, and a set of role assertions of the form $r(a, b)$. For more details, see for instance [3,11].

Semantics. The semantics is given in terms of interpretations $\mathcal{I} = (\Delta^{\mathcal{I}}, \cdot^{\mathcal{I}})$ which consist of a non-empty interpretation domain $\Delta^{\mathcal{I}}$ and an interpretation function $\cdot^{\mathcal{I}}$ that maps each individual $a^{\mathcal{I}} \in N_I$ to an element $a^{\mathcal{I}} \in \Delta^{\mathcal{I}}$, each concept $A \in N_C$ to a subset $A^{\mathcal{I}} \subseteq \Delta^{\mathcal{I}}$ and each role $r \in N_R$ to a subset $r^{\mathcal{I}} \subseteq \Delta^{\mathcal{I}} \times \Delta^{\mathcal{I}}$. Furthermore, the function $\cdot^{\mathcal{I}}$ is extended in a straightforward way for concept and role expressions as depicted in Table 1.

An interpretation $\mathcal{I}$ is said to be a model of (or satisfies) a GCI (resp. role inclusion, role composition) axiom, denoted by $\mathcal{I} \models C \sqsubseteq D$ (resp. $\mathcal{I} \models r \sqsubseteq s$, $\mathcal{I} \models r_1 \circ r_2 \sqsubseteq s$), if $C^{\mathcal{I}} \subseteq D^{\mathcal{I}}$ (resp. $r^{\mathcal{I}} \subseteq s^{\mathcal{I}}$, $(r_1 \circ r_2)^{\mathcal{I}} \subseteq s^{\mathcal{I}}$). Similarly, $\mathcal{I}$ satisfies a concept (resp. role) assertions, denoted $\mathcal{I} \models C(a)$ (resp. $\mathcal{I} \models r(a, b)$), if $a^{\mathcal{I}} \in C^{\mathcal{I}}$ (resp. $(a^{\mathcal{I}}, b^{\mathcal{I}}) \in r^{\mathcal{I}}$). An interpretation $\mathcal{I}$ is a model of an ontology $\mathcal{O}$ if it satisfies all the axioms of $\mathcal{O}$. An ontology is said to be consistent if it has a model. Otherwise, it is inconsistent. An axiom ϕ is entailed by an ontology, denoted by $\mathcal{O} \models \phi$, if ϕ is satisfied by every model of $\mathcal{O}$. We say that C is subsumed by D w.r.t an ontology $\mathcal{O}$ iff $\mathcal{O} \models C \sqsubseteq D$. Similarly, we say that a is an instance of C w.r.t $\mathcal{O}$ iff $\mathcal{O} \models C(a)$. A concept C is said to be unsatisfiable w.r.t. $\mathcal{O}$ iff $\mathcal{O} \models C \sqsubseteq \perp$, otherwise C is said to be satisfiable.

Section 3 provides the syntax and semantics of prioritized $\mathcal{EL}$ ontology. Then it provides some tractable algorithms for reasoning under such ontology. Section 3 shows how to induce a prioritized ontology in the case where the evolution process starts with a flat ontology, i.e., all its axioms are certain. Finally, Sect. 4 provides the syntactical evolution of prioritized $\mathcal{EL}$ ontology.

3   Prioritized $\mathcal{EL}$ ontology

The evolution process is the task of incorporating some information into the ontology while preserving its consistency. Considering the $\mathcal{EL}$ ontology, denoted by $\mathcal{O}$. Let (ϕ, w) be the new input where ϕ is an $\mathcal{EL}$ axiom and w is the weight of ϕ reflecting its priority (credibility). Depending on the input type, one can distinguish two scenarios, the first is where the input is consistent with the ontology. In this case, two situations hold, the first one, if the input is fully reliable, namely $w = 1$, then the evolution process consists in simply adding the information to the ontology. In the second situation, if the input ϕ is uncertain, namely $w < 1$, then the evolution process should ensure that ϕ will be inferred from the ontology with its prescribed weight w after revision. In the second scenario where the input ϕ is inconsistent with the ontology, then we need to repair the ontology and add ϕ with its prescribed weight w. To handle qualitative uncertainty of input information, we use prioritized $\mathcal{EL}$ ontologies within possibility theory [13].

Section 3.1 provides the syntax and semantics of such framework. Sections 4 provide evolution process of prioritized ontology when a new information (ϕ, w) is available

3.1   Syntax and Semantics of Prioritized $\mathcal{EL}$ Ontology

Syntax. A possibilistic $\mathcal{EL}$ ontology, denoted by $\mathcal{O}_\pi$ is a set of possibilistic axioms of the form (ϕ_i, w_i), where ϕ is an $\mathcal{EL}$ axiom and $w \in]0, 1]$ its certainty degree. Note that the higher the degree w, the more certain is the formula. Note that the axioms with w_i's equal to '0' are not explicitly represented in the ontology. Moreover, when all the degrees are equal to 1, $\mathcal{O}_\pi$ coincides with a standard $\mathcal{EL}$ ontology $\mathcal{O}$.

Definition 1. *Let $\mathcal{O}_\pi$ be a possibilistic $\mathcal{EL}$ ontology. We call the w-cut (resp. strict w-cut), denoted by $\mathcal{O}_{\pi \geq w}$ (resp. $\mathcal{O}_{\pi > w}$), the sub-base $\mathcal{O}_\pi$ contains the set of axioms having degree greater or equal (resp. strictly greater) than w.*

When the ontology is inconsistent, in that case, we assign a degree of inconsistency as follows:

Definition 2. *The inconsistency degree of $\mathcal{O}_\pi$ is syntactically defined as follows: $Inc(\mathcal{O}_\pi) = max\{w : \mathcal{O}_{\pi \geq w}$ is inconsistent$\}$.*

Semantics. The semantics of $\mathcal{O}_\pi$ ontology is defined by the possibility distribution. Let $\mathcal{L}$ be $\mathcal{EL}$ description language and $\Omega = \{\mathcal{I}_1, ...\mathcal{I}_n\}$ be a universe of discourse consisting of a set of $\mathcal{EL}$ interpretations. A possibility distribution is the main block of possibility theory, denoted by π and it is a mapping from Ω to the unit interval $[0.1]$. The possibility distribution $\pi(\mathcal{I})$ represents the degree of compatibility of $\mathcal{I}$ with the available knowledge. More specifically, when $\pi(\mathcal{I}) = 0$ this means that the interpretation $\mathcal{I}$ is impossible. Otherwise, $\pi(\mathcal{I}) = 1$ means that it is totally possible (i.e. fully consistent with available knowledge).

Definition 3. *The possibility distribution associated with $\mathcal{O}_\pi$ is obtained as follows:*

$$\forall \mathcal{I} \in \Omega, \pi(\mathcal{I}) = \begin{cases} 1 & \text{if } \forall(\phi_i, w_i) \in \mathcal{O}_\pi, \mathcal{I} \models \phi_i \\ 1 - max\{w_i : (\phi_i, w_i) \in \mathcal{O}_\pi, \mathcal{I} \not\models \phi\} & \text{otherwise.} \end{cases}$$

In the classical description logic, when the axioms are completely certain, then the ontology can be semantically described in terms of models and counter-models. Therefore, the ontology is consistent or inconsistent. This is not the case in a possibilistic setting, since its semantics in terms of possibility distributions attribute to the countermodels a degree of compatibility with the available knowledge.

An interpretation $\mathcal{I}$ is a model of $\mathcal{O}_\pi$ if it satisfies all the axioms of the ontology. In this case $\pi(\mathcal{I}) = 1$, which means that the possibility distribution $\pi_{\mathcal{O}_\pi}$ is normalized. Otherwise, the distribution is called sub-normalized. Given the possibility distribution π, two measures are defined, the possibility degree $\Pi_\pi(\phi) = max_{\mathcal{I} \in \Omega}\{\pi(\mathcal{I}) : \mathcal{I} \models \phi\}$ evaluates the extent to which ϕ is consistent with the available information encoded by π. The necessity measure, $N_\pi(\phi) = 1 - max_{\mathcal{I} \in \Omega}\{\pi(\mathcal{I}) : \mathcal{I} \not\models \phi\}$ evaluates to what extent ϕ is certainly entailed from the available knowledge encoded by π. Syntactically, an axiom ϕ_i has w_i as its certainty degree, means that $N(\phi_i) \geq w_i$.

In the following, we study reasoning in possibilistic $\mathcal{EL}$ ontology by given algorithms that compute the possibilistic entailment. The first step consists of transforming the ontology into the normal form using the normalization rules in Table 2, and then to compute the possibilistic entailment, we use the inference rules in Table 3. [13]. The possibilistic entailment using the inference rules in Table 3 is done in polynomial time.

3.2   Inducing Prioritized $\mathcal{EL}$ Ontology

Considering the flat $\mathcal{EL}$ ontology $\mathcal{O}$, i.e. an $\mathcal{EL}$ ontology where all the axioms are certain $(\phi_i, 1)$. In the following, we propose a new method for introducing the priority relation between axioms using the notion of conflict matrix when the new information ϕ is inconsistent with $\mathcal{O}$.

Definition 4. *Let $\mathcal{M}$ be a matrix that contains the set of ontological axioms. The conflict matrix $\mathcal{M}$ presents the conflict relations between axioms. If the*

Table 2. Possibilistic normalization rules.

(PNR_0) $\quad \frac{(C_1 \sqcap \top \sqcap C_2 \sqsubseteq D, \alpha)}{(C_1 \sqcap C_2 \sqsubseteq D, \alpha)}$

(PNR_1) $\quad \frac{(C_1 \sqcap \bot \sqcap C_2 \sqsubseteq D, \alpha)}{(\bot \sqsubseteq D, \alpha)}$:

(PNR_2) $\quad \frac{(C \sqsubseteq D_1 \sqcap D_2, \alpha)}{(C \sqsubseteq D_1, \alpha) \quad (C \sqsubseteq D_2, \alpha)}$

(PNR_3) $\quad \frac{(\exists r.C \sqsubseteq D, \alpha)}{(C \sqsubseteq A, 1) \quad (\exists r.A \sqsubseteq D, \alpha)}$: $C \notin N_c, A$ is a new concept

(PNR_4) $\quad \frac{(C \sqsubseteq D, \alpha)}{(C \sqsubseteq A, 1) \quad (A \sqsubseteq D, \alpha)}$: $C, D \notin N_c \cup \{\bot, \top\}, A$ is a new concept

(PNR_5) $\quad \frac{(B \sqsubseteq \exists r.C, \alpha)}{(B \sqsubseteq \exists r.A, 1) \quad (A \sqsubseteq C, \alpha)}$: $C \notin N_c, A$ is a new concept

(PNR_6) $\quad \frac{(C_1 \sqcap C \sqcap C_2 \sqsubseteq D, \alpha)}{(C \sqsubseteq A, 1) \quad (C_1 \sqcap A \sqcap C_2 \sqsubseteq D, \alpha)}$: $C \notin N_c \cup \{\bot, \top\}, A$ is a new concept

Table 3. Possibilistic inference rules.

(PIR_0) $\quad \frac{}{(A \sqsubseteq A, 1)}$: $A \in N_c \cup \{\bot, \top\}$

(PIR_1) $\quad \frac{}{(C \sqsubseteq \top, 1)}$: $C \sqsubseteq \top$

(PIR_2) $\quad \frac{}{(r, 1)}$: $r \in \mathcal{O}$

(PIR_3) $\quad \frac{(C \sqsubseteq D', \alpha_1) \quad (D' \sqsubseteq D, \alpha_2)}{(C \sqsubseteq D, \min(\alpha_1, \alpha_2))}$

(PIR_4) $\quad \frac{(A \sqsubseteq B_1 ... A \sqsubseteq B_n, \alpha_1) \quad (B_1 \sqcap ... \sqcap B_n \sqsubseteq B, \alpha_2)}{(A \sqsubseteq B, \min(\alpha_1, \alpha_2))}$: $A, B, B_i \in N_c \cup \{\bot, \top\}$

(PIR_5) $\quad \frac{(A \sqsubseteq \exists r.B, \alpha_1) \quad (B \sqsubseteq C, \alpha_2)}{(A \sqsubseteq \exists r.C, \min(\alpha_1, \alpha_2))}$: $A, B, C \in N_c \cup \{\bot, \top\}, C \neq \bot$

(PIR_6) $\quad \frac{(A \sqsubseteq \exists r.B, \alpha) \quad (B \sqsubseteq \bot, 1)}{(A \sqsubseteq \bot, \alpha)}$: $A, B \in N_c \cup \{\bot, \top\}$

(PIR_7) $\quad \frac{(A \sqsubseteq \exists r.B, \alpha_1) \quad (r \sqsubseteq s, \alpha_2)}{(A \sqsubseteq \exists s.B, \min(\alpha_1, \alpha_2))}$

(PIR_8) $\quad \frac{(r_1 \sqsubseteq r_2, \alpha_1) \quad (r_2 \sqsubseteq r_3, \alpha_2)}{(r_1 \sqsubseteq r_3, \min(\alpha_1, \alpha_2))}$

(PIR_9) $\quad \frac{(A \sqsubseteq \exists r_1.B, \alpha_1) \quad (B \sqsubseteq \exists r_2.A_2, \alpha_2) \quad (r_1 \circ r_2 \sqsubseteq s, \alpha_3)}{(A_1 \sqsubseteq \exists s.A_2, \min(\alpha_1, \alpha_2, \alpha_3))}$: $A_1, r_1, A_2 \in N_c \cup \{\bot, \top\}$

cell $M_{ij} = 0$, then there is no conflict between the ith axiom and j^{th} axiom. Otherwise, i.e., $M_{ij} = 1$, then the axioms are in conflict. More formally:

$$\forall ax_i, ax_j \in M, \text{ if } M_{ax_i, ax_j} = \begin{cases} 1 & \text{then } ax_i, ax_j \text{ are in conflict} \\ 0 & \text{then } ax_i, ax_j \text{ are consistent} \end{cases}$$

Example 1. Considering the following $\mathcal{EL}$ ontology.
$ax_1 = $ Amnesia $\sqsubseteq$ Alzheimer, $ax_2 = $ Amnesia $\sqsubseteq$ BrainCancer
$ax_3 = $ Aphasia $\sqsubseteq$ Alzheimer, $ax_4 = $ Apraxie $\sqsubseteq$ Paralysis
$ax_5 = $ Apraxie $\sqcap$ Alzheimer $\sqsubseteq \bot$, $ax_6 = $ Apraxie $\sqsubseteq$ Alzheimer
$ax_7 = \exists$InabilityToLearn.Language $\sqsubseteq$ Agnosia, $ax_8 = $ Agnosia $\sqsubseteq$ BrainCancer
$ax_9 = $ BrainCancer $\sqcap$ Alzheimer $\sqsubseteq \bot$, $ax_{10} = $ Apraxie $\sqcap$ Paralysis $\sqsubseteq \bot$

The conflict matrix of Example 1 is defined as follows (Table 4):

Table 4. Conflict matrix of Example 1

	ax_1	ax_2	ax_3	ax_4	ax_5	ax_6	ax_7	ax_8	ax_9	ax_{10}
ax_1	0	0	0	0	0	0	0	0	0	0
ax_2	0	0	0	0	0	0	0	0	1	0
ax_3	0	0	0	0	0	0	0	0	0	0
ax_4	0	0	0	0	0	0	0	0	0	0
ax_5	0	0	0	0	0	1	0	0	0	0
ax_6	0	0	0	0	1	0	0	0	0	1
ax_7	0	0	0	0	0	0	0	0	0	0
ax_8	0	0	0	0	0	0	0	0	0	0
ax_9	0	1	0	0	0	0	0	0	0	0
ax_{10}	0	0	0	0	0	1	0	0	0	0

The conflict matrix is square and sparse. Their diagonal values are '0' since there is no conflict between the same axioms. In real applications, the conflict matrix is difficult to manipulate since it contains huge data (several columns). For such reason, we need to reduce the dimensions of the matrix. In this paper, we use a technique of dimensionality reduction that preserves much information when it transforms the data from high dimensions space to low ones.

Multidimensional Scaling of Conflict Matrix. Multidimensional scaling [6] is a projection or a mapping from a high-dimensional space into a low-dimensional space. The MDS takes as input a square matrix that represents the conflict relationships between axioms that are in p dimensional space and transforms these data to a space with dimensions $m < p$. The obtained points are presented by a matrix D having the Euclidean distances between each pair of points. Based on the Euclidean distance in matrix D, we stratify the ontology from the least important to the most important.

The prioritized ontology is defined as follows (Table 5):

Table 5. Euclidean distance between axioms

	ax_1	ax_2	ax_3	ax_4	ax_5	ax_6	ax_7	ax_8	ax_9	ax_{10}
ax_1	0.000	0.100	0.000	0.000	0.100	0.199	0.000	0.00	0.100	0.100
ax_2	0.100	0.000	0.100	0.100	0.144	0.176	0.100	0.100000	0.200	0.162156
ax_3	0.000	0.100	0.000	0.000	0.100	0.199	0.000	0.000	0.100	0.100
ax_4	0.000	0.100	0.000	0.000	0.100	0.199	0.000	0.000	0.100	0.100
ax_5	0.100	0.162	0.100	0.100	0.000	0.299	0.100	0.100	0.170	0.043
ax_6	0.199	0.176	0.199	0.199	0.299	0.000	0.199	0.199	0.265	0.299
ax_7	0.000	0.100	0.000	0.000	0.100	0.199	0.000	0.000	0.100	0.100
ax_8	0.000	0.100	0.000	0.000	0.100	0.199	0.000	0.000	0.100	0.100
ax_9	0.100	0.200	0.100	0.100	0.116	0.250	0.100	0.100	0.000	0.117070
ax_{10}	0.100	0.162	0.100	0.100	0.049	0.299	0.100	0.100	0.117	0.000

Definition 5. *A prioritized $\mathcal{EL}$ ontology, denoted by $\mathcal{O}_\pi$ is defined as follows: $\mathcal{O}_\pi = S_1 \cup S_2, ..., \cup S_n$. Where S_1 contains lest reliable axioms and S_n, the most important ones.*

In the following section, we study evolution at the syntactic levels of the prioritized $\mathcal{EL}$ ontology.

4   Syntactic Evolution of $\mathcal{EL}$ Ontology

Revision at the syntactic level consists of adding a new input (ϕ, w) to the ontology $\mathcal{O}_\pi$ while preserving its consistency. Regarding the new input information, we first study revision when the input information is inconsistent with the ontology and then when the new information is consistent with $\mathcal{O}_\pi$.

4.1   Revision with Inconsistent Input

Let $\mathcal{O}_\pi$ be the prioritized ontology and (ϕ, w) be inconsistent input information. Two situations hold depending on the input information (ϕ, w). The first is when the information is inhibited by higher priority axioms that contradict it. The second is when (ϕ, w) is not inhibited. Adding the information with its prescribed weight to $\mathcal{O}_\pi$ while preserving its consistency is obtained using the following algorithm.

The algorithm takes as input the prioritized ontology $\mathcal{O}_\pi$ and the input (ϕ, w). In the first step, it adds the new information (ϕ) to the prioritized ontology $\mathcal{O}_\pi$ with the highest possible priority (namely $w = 1$). At the second step, it computes the inconsistency of the augmented ontology $w_{inc} = Inc(\mathcal{O}'_\pi)$, with $\mathcal{O}'_\pi = \mathcal{O}_\pi \cup \{(\phi, 1)\}$, then it remove every axiom has a priority level less or equal to w_{inc}. Now, let $\mathcal{O}_{\pi 1}$ be the obtained ontology that is consistent. As a final step, the algorithm adds ϕ with its prescribed level to the obtained coherent ontology $\mathcal{O}_{\pi 1}$ and adjusts the weights. Note that these steps guarantee the consistency of the obtained ontology before adding the information (ϕ, w) with its prescribed weight.

Example 2. Considering the following prioritized $\mathcal{EL}$ ontology $\mathcal{O}_\pi = \{$(Amnesia $\sqsubseteq$ Alzheimer, 0.7), (Amnesia $\sqsubseteq$ BrainCancer, 0.3), (Apraxia $\sqsubseteq$ Paralysis, 0.4), Apraxia $\sqsubseteq$ Alzheimer, 0.7 $\}$. Consider the first the input $(Alzheimer \sqcap Amnesia \sqsubseteq \perp, 0.2)$ and $(Alzheimer \sqcap Amnesia \sqsubseteq \perp, 0.9)$. Then the two input are inhibited (resp. not inhibited) respectively with the axioms having higher priority that contradicts it. We have $Inc(\mathcal{O}_\pi \cup (Alzheimer \sqcap Amnesia \sqsubseteq \perp, 0.1)) = 0.4$. In the first case $\mathcal{O}'_\pi = \{(Alzheimer \sqcap Amnesia \sqsubseteq \perp, 0.2), $(Apraxia $\sqsubseteq$ Alzheimer, 0.7)$\}$ and in the second case we have $\mathcal{O}'_\pi = \{(Alzheimer \sqcap Amnesia \sqsubseteq \perp, 0.9), $(Apraxia $\sqsubseteq$ Alzheimer, 0.7)$\}$.

Algorithm 1. Revision with Inconsistent Input

Data: $\mathcal{O}_\pi$, (ϕ, w)
Result: consistent ontology $\mathcal{O}_\pi''$
$\mathcal{O}_\pi' \leftarrow \emptyset$
$\mathcal{O}_\pi'' \leftarrow \emptyset$
   for i=1 to n **do**
      $\mathcal{O}_\pi' \leftarrow \mathcal{O}_\pi \cup (\phi, 1)$
      $Inc(\mathcal{O}_\pi') = \max\{w : \mathcal{O}'_{\pi \geq w} \text{is inconsistent}\}$
      $\mathcal{O}_{\pi 1} \leftarrow \mathcal{O}'_{\pi \geq Inc(\mathcal{O}_\pi')}$
      $\mathcal{O}_\pi'' \leftarrow \mathcal{O}_{\pi 1} \cup (\phi, w)$
   end for
   return $\mathcal{O}_\pi''$

The following proposition studies the computational complexity of Algorithm 2

Proposition 1. *The computation of the revised ontology can be done in polynomial time.*

In the following section, we study the syntactic revision when the input is consistent with the original ontology.

4.2 Revision with Consistent Input

Let (ϕ, w) be the consistent input information (namely $\Pi(\phi) = 1$). Two situations should take into consideration during the process of adding ϕ to the prioritized ontology $\mathcal{O}_\pi$. The first one is when the input is inferred from the ontology with a weight $w \leq 1$. The second is when the input cannot be inferred from $\mathcal{O}_\pi$. The revision, in the latter, is simply an expansion of the original ontology with (ϕ, w), namely $\mathcal{O}_\pi' = \mathcal{O}_\pi \cup \{(\phi, w)\}$.

Now, When the input is inferred from the ontology (namely $\mathcal{O}_\pi \models (\phi)$), and depending on the necessity measure of ϕ (i.e., $N(\phi, w) = w_b$) and the prescribe necessity measure $N'(\phi, w) = w$. two situations hold. The first one when necessity measure is greater than the prescribe necessity measure (namely $(w_b > w)$) means that the input is inferred from the prioritized ontology with necessity w_b greater than its prescribed weight w. The second is when $w_b < w$ means that the necessity degree of the inferred axiom is less than its prescribed weight. To determine to what extent ϕ is entailed from the ontology $\mathcal{O}_\pi$, we add first the assumption that ϕ is false by the following expression $\{(A \sqsubseteq B, 1), (C \sqsubseteq B, 1)\}$ if $\phi = A \sqcap C \sqsubseteq \bot$, with A, B and C are new concepts that are not in $\mathcal{O}_\pi$. Therefore, The revised ontology is obtained using the following algorithm.

Algorithm 2. Revision with Consistent Input

Data: $\mathcal{O}_\pi$, The assumption (ϕ, w) false
Result: a consistent ontology $\mathcal{O}_{\pi2}$
$\mathcal{O}'_\pi \leftarrow \emptyset$
$\quad \mathcal{S} \leftarrow \emptyset$
$\quad$ **for** i=1 to n **do**
$\qquad \mathcal{O}'_\pi \leftarrow \mathcal{O}_\pi \cup (\phi, 1)$
$\qquad w_b \leftarrow Inc(\mathcal{O}'_\pi)$
$\qquad$ **if** $w > w_b$ **then**
$\qquad\quad \mathcal{O}_{\pi1} \leftarrow \mathcal{O}_\pi \cup (\phi, w)$
$\qquad$ **else**
$\qquad\quad$ **while** $(\mathcal{S} \geq \mathcal{O}_{\pi \geq w}), (\mathcal{S} \leq \mathcal{O}_{\pi \leq w_b}), (\mathcal{S} \models \phi)$ **do**
$\qquad\qquad \mathcal{S} \leftarrow (\phi, w)$
$\qquad\quad$ **end while**
$\qquad$ **end if**
$\quad$ **end for**

The algorithm takes as input the assumption ϕ false and the prioritized ontology $\mathcal{O}_\pi$. In the first step, it adds the assumption that ϕ is false with the highest priority level namely $(w = 1)$. Second, it computes the inconsistency of the obtained ontology $\mathcal{O}'_\pi$ which is equal to w_b. There exists two cases, the first, if w is greater than w_b, then the revised ontology is $\mathcal{O}'_\pi = \mathcal{O}_\pi \cup \{(\phi, w)\}$. Second, when w_b is less than w, two solutions hold, either shift down the degree of axioms having priority between w and w_b to w or select the set $\mathcal{S} \in \mathcal{O}_\pi$ of axioms having a priority level between w and w_b and imply ϕ and assigning to them the degree w. The two solutions lead to inferring ϕ with its prescribed level. But, the second one ensures a minimal change of the ontology because it only change the weights of axioms responsible for inferring ϕ.

The following proposition gives the syntax definition the obtained ontology $\mathcal{O}'_\pi$.

Proposition 2. *Let $\mathcal{O}_\pi$ be the prioritized $\mathcal{EL}$ ontology. Let (ϕ, w) be the uncertain input. Let $\mathcal{O}'_\pi$ be the augmented ontology by the assumption that ϕ false. The degree of inconsistency of $\mathcal{O}'_\pi$ is $w_{inc} = Inc(\mathcal{O}'_\pi)$. The revised $\mathcal{O}'_\pi = \{(\phi, w)\} \cup \{(\phi_o, w_o) : (\phi_o, w_o) \in \mathcal{O}_\pi \text{ and } w > w_{inc}\} \cup \{(\phi_o, w_o) : (\phi, w) \in \mathcal{O}_\pi \text{ and } w_o < w\} \cup \{(\phi_o, w_o) : (\phi_o, w_o) \in \mathcal{O}_\pi \text{ and } w \leq w_o \leq w_{inc}\}$.*

Proposition 2 ensures that the degrees of axioms in $\mathcal{O}_\pi$ between w and w_{inc} should be minimised to w. However, we can improve the results using the second case, i.e., identifying the set of axioms $\mathcal{S}$ in $\mathcal{O}_\pi$ that implies ϕ.

Proposition 3. *Let $\mathcal{O}_\pi$ be the prioritized $\mathcal{EL}$ ontology. Let (ϕ, w) be the uncertain input. Considering that $\mathcal{O}_1$ be the augmented possibility ontology obtained by add the assumption ϕ false to $\mathcal{O}_\pi$. Let $w_{inc} = Inc(\mathcal{O}_1)$. The revised prioritized $\mathcal{EL}$ ontology, denoted $\mathcal{O}'_\pi$ is defined as: $\mathcal{O}'_\pi = \{(\phi, w)\} \cup \{\mathcal{O}_\pi \setminus \mathcal{S}\} \cup \{(\phi_o, w_o) : (\phi_o, w_o) \in \mathcal{S} \text{ and } w_o > w_{inc}\} \cup \{(\phi, w) : (\phi, w_{inc}) \in \mathcal{S} \text{ and } w_{inc} = w_o\}$.*

Example 3. Considering the following prioritized $\mathcal{EL}$ ontology $\mathcal{O}_\pi = \{$(Amnesia $\sqsubseteq$ Alzheimer, 0.7), (Amnesia $\sqsubseteq$ BrainCancer), 0.3), (Alzheimer $\sqsubseteq$ MentalIllness, 0.4)$\}$. Consider the first the input (Amnesia $\sqsubseteq$ MentalIllness, 0.9) and (Amnesia $\sqsubseteq$ MentalIllness, 0.2) we have $Inc(\mathcal{O}'_\pi) = 0.4$ since $(\mathcal{O}'_\pi) = (\mathcal{O}_\pi \cup \{$(Amnesia $\sqcap$ MentalIllness $\sqsubseteq \perp$, 1)$\}$. If the input is (Amnesia $\sqsubseteq$ MentalIllness, 0.9) then $\mathcal{O}''_\pi = \{$(Amnesia $\sqsubseteq$ MentalIllness, 0.9), (Amnesia $\sqsubseteq$ Alzheimer, 0.7), (Amnesia $\sqsubseteq$ BrainCancer), 0.3), Alzheimer $\sqsubseteq$ MentalIllness, 0.4)$\}$. Now the second case, if the input is (Amnesia $\sqsubseteq$ MentalIllness, 0.2) then $\mathcal{O}''_\pi = \{$(Amnesia $\sqsubseteq$ MentalIllness, 0.2), (Amnesia $\sqsubseteq$ Alzheimer, 0.7), (Amnesia $\sqsubseteq$ BrainCancer), 0.2), Alzheimer $\sqsubseteq$ MentalIllness, 0.2)$\}$.

5 Conclusion

In this paper, we studied the syntactical evolution of prioritized $\mathcal{EL}$ ontologies when a new piece of information that can be sure or uncertain is available. We showed how to induce prioritized ontology based on the Euclidean distances. Finally, we propose a polynomial syntactic algorithm for the evolution process while preserving the consistency of the ontology.

Acknowledgments. This work was supported by ANR CHAIRE IA BE4musIA and FEI INS2I 2022 EMILIE.

References

1. Achich, N., Ghorbel, F., Hamdi, F., Metais, E., Gargouri, F.: Certain and uncertain temporal data representation and reasoning in OWL 2. Int. J. Semant. Web Inf. Syst. (2021)
2. Alchourrón, C.E., Gärdenfors, P., Makinson, D.: On the logic of theory change: partial meet contraction and revision functions. J. Symb. Logic **50**(2), 510–530 (1985)
3. Baader, F., Brandt, S., Lutz, C.: Pushing the el envelope further (2008)
4. Benferhat, S., Bouraoui, Z., Papini, O., Würbel, E.: Prioritized assertional-based removed sets revision of DL-lite belief bases. Ann. Math. Artif. Intell. **79**(1–3), 45–75 (2017)
5. Benferhat, S., Dubois, D., Prade, H., Williams, M.A.: A practical approach to revising prioritized knowledge bases. Stud. Logica. **70**(1), 105–130 (2002)
6. Cox, M.A., Cox, T.F.: Multidimensional scaling. In: Cox, M.A., Cox, T.F. (eds.) Handbook of Data Visualization, pp. 315–347. Springer, Heidelberg (2008). https://doi.org/10.1007/978-3-540-33037-0_14
7. De Giacomo, G., Lenzerini, M., Poggi, A., Rosati, R.: On the approximation of instance level update and erasure in description logics. In: AAAI, pp. 403–408 (2007)
8. Flouris, G., Plexousakis, D., Antoniou, G.: Generalizing the AGM postulates: preliminary results and applications. In: NMR, pp. 171–179. Citeseer (2004)
9. Flouris, G., Plexousakis, D., Antoniou, G.: On applying the AGM theory to DLs and OWL. In: Gil, Y., Motta, E., Benjamins, V.R., Musen, M.A. (eds.) ISWC 2005. LNCS, vol. 3729, pp. 216–231. Springer, Heidelberg (2005). https://doi.org/10.1007/11574620_18

10. Goodhue, D.L., Wybo, M.D., Kirsch, L.J.: The impact of data integration on the costs and benefits of information systems. MIS Q. 293–311 (1992)
11. Kazakov, Y., Krötzsch, M., Simancik, F.: Practical reasoning with nominals in the el family of description logics. In: KR (2012)
12. Liu, H., Lutz, C., Milicic, M., Wolter, F.: Updating description logic aboxes. KR **6**, 46–56 (2006)
13. Mohamed, R., Loukil, Z., Bouraoui, Z.: Qualitative-based possibilistic el ontology. In: PRIMA 2018 (2018)
14. Prieto-Gonzalez, D., Castilla-Rodriguez, I., Gonzalez, E., Couce, M.L.: Automated generation of decision-tree models for the economic assessment of interventions for rare diseases using the radios ontology. J. Biomed. Inform. **110**, 103563 (2020)
15. Qi, G., Du, J.: Model-based revision operators for terminologies in description logics. In: Twenty-First International Joint Conference on Artificial Intelligence (2009)
16. Wang, Z., Wang, K., Topor, R.: A new approach to knowledge base revision in DL-lite. In: Proceedings of the AAAI Conference on Artificial Intelligence, vol. 24 (2010)
17. Wu, C., Potdar, V., Chang, E.: Latent semantic analysis – the dynamics of semantics web services discovery. In: Dillon, T.S., Chang, E., Meersman, R., Sycara, K. (eds.) Advances in Web Semantics I. LNCS, vol. 4891, pp. 346–373. Springer, Heidelberg (2008). https://doi.org/10.1007/978-3-540-89784-2_14
18. Xiao, G., et al.: Ontology-based data access: a survey. In: International Joint Conferences on Artificial Intelligence (2018)

A Comparison of Resource Data Framework and Inductive Logic Programing for Ontology Development

Durgesh Nandini[✉]

University of Bamberg, Bamberg, Germany
durgesh.nandini@uni-bamberg.de

Abstract. This study compares the expressive power of Resource Data Framework (RDF) and Inductive Logic Programing (ILP). While RDF and RDF Schema do not possess any rule language, ILP is a logic programing language that is fit for inferring facts. The research aims to identify and acknowledge the differences between RDF and ILP in terms of how much expressive power they hold within themselves and how efficient they infer knowledge when employed in ontologies. The paper represents ongoing work to compare RDF and ILP.

Keywords: Inductive logic programing · Ontology · Resource Data Framework (RDF)

1 Introduction

Traditionally, Resource Description Framework (RDF) [1] and Resource Description Framework Schema (RDFS) [2] have been used to build the semantic web, and Prolog [10] have been used to deduce inferences over triple knowledge of the RDF. Ontologies have been identified as a key technology for resolving semantic heterogeneity by providing common terms as well as formal specifications of their intended meaning in some logic.

Presently, RDF Triple [3] is widely used as a rule language for querying and inferring RDF clauses and predicates, providing full support for resources and their namespaces, models represented with sets of RDF triples, reification, and RDF data transformation. The language is intended to be used with a Horn-based inference engine.

The automatic identification of complex mappings between ontologies is an exciting and relevant challenge for the Inductive Logic Program (ILP) community. Developing linguistically data-compliant rules for entity extraction is usually an intensive and time-consuming process for any ontology engineer due to the complexity of the mapping paradigms involved. The use of ILP as a paradigm for addressing the problem is a natural fit as the goal is to learn complex logical rules based on instances and background knowledge [4] where prolog becomes an adequate language for managing graphs of RDF Triples.

© Springer Nature Switzerland AG 2022
H. Fujita et al. (Eds.): IEA/AIE 2022, LNAI 13343, pp. 871–876, 2022.
https://doi.org/10.1007/978-3-031-08530-7_73

RDF and RDF Schema lack the means for representing axioms and rules, which are still necessary to build applications, and different approaches originating from different motivations and requirements have been proposed. Meanwhile, the development of prolog has led to inference over data models represented in RDF, though mostly in an ad-hoc manner.

The paper represents the ideas of ongoing work to compare RDF and ILP. The research aims to identify and acknowledge the differences between RDF and ILP in terms of their expressivity and efficiency in inferring knowledge when employed in ontologies.

The rest of the paper is organised as follows: a brief overview and simple examples of RDF, Prolog and ILP have been provided in Sect. 2 of the paper, Sect. 3.1 and Sect. 3.2 illustrates the ideas to implement and compare notions of RDF and ILP using a case study of transport domain ontology, and then the Sect. 3.3 details the parameters and tools considered for comparison, Sect. 4 discusses the is the conclusions and future works. Additionally, Sect. 4 also states how the research can be applied in other scenarios.

2   An Overview: RDF, Prolog and ILP

In this section, RDF, Prolog and ILP concepts are introduced. The background knowledge for the three models has been introduced, and their examples are shown. The sections also describe the differences between the three.

2.1   RDF

The RDF data model is based on expressing knowledge in the form of triples of Subject, Predicate, and Object. Resources and literals are the two data types used in RDF, where resources are Universal Resource Identifiers (URI), and literals are individual atomic units.

The RDF triple is read as: subject has an attribute predicate with value object, where subject and predicate are resources and object can be either a resource or a literal.

A resource appearing as an object can also be the subject of another RDF triple. A set of RDF triples forms an RDF graph. A schema in RDF is added using RDFS; the schema will behave as a metamodel and impart hierarchies for defining concepts and relationships in an RDF

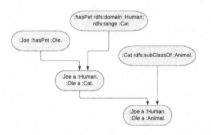

Fig. 1. A Simple RDF Graph

graph. A simple RDF graph is shown in Fig. 1. The graph asserts that Ole is Joe's cat by the relationship hasPet and that Joe is a Human and Ole is a Cat. since Ole is a Cat and Cat is a subclass of Animal; therefore Cat is also an

Animal. Additionally, Since Human can have a pet cat, and Joe is a Human and Ole is a Cat; hence this RDF graph is consistent.

Conceptual models in RDF and RDF Schema are represented using axioms. The axioms rdfs:subClassOf and rdfs:subPropertyOf are used to organise classes and properties into hierarchies. The axioms rdfs:domain and rdfs:range specify the attachment of properties to classes. In practicality, the domain and range axioms are classified as complex to be used because the conjunctive semantics of multiple occurrences of rdfs:domain or rdfs:range means that a property may be attached to an intersection of one or more classes, and not a union (disjunctive semantics). This poses some problems whenever a property has to be attached to several classes [5].

Several knowledge representation languages have been developed to work with RDF, OWL is one of the most extensively used one. OWL provides the advantage of extending the existing RDF Schema and introducing axioms and properties. The additional axioms can be equality axioms such as sameClassAs (denotes two classes are equivalent), samePropertyAs (denotes two properties are equivalent); and property axioms such as inverseOf, TransitiveProperty, to define inverse and transitive properties, respectively. The sets of axioms allow the modelling of numerous frequently needed constraints.

2.2 Prolog

Prolog is a logic programing language that relies on facts, rules and goals. The end goal, also known as query, derives its solution by analysing the relationship between the three. Prolog facts contain entities and their relationship. Prolog rules are clauses expressed using variables and/or facts. A Prolog rule usually consists of a head, followed by a neck, followed by a body.

A simple Prolog database with facts, queries and their interpretations is shown in Fig. 2. Figure 2.1 shows how facts will be represented in a database and how they will be interpreted. Figure 2.2 shows simple queries that can be run on the constructed facts and the output obtained by executing the queries.

Database		Queries on the Database	
Fact	The fact can be interpreted as:	Query and Output	Interpretation
friends(Alice, Bob).	Alice is friends with Bob.	?painter(Alice). true.	Since the database contains this fact, the output turns out to be true.
painter(Alice).	Alice is a painter.		
natural_number(16).	16 is a natural number.	?natutal_number(7). false.	Based on only the database knowledge, the output will be false, as there are no rules defining a natural number and only numeral 16 is explicitly stated as a natural number.

Fig. 2.1 Facts in a database   Fig. 2.2 Executing queries in a database

Fig. 2. Executing queries in a database

Prolog atoms are the equivalent of RDF resources. The facts in Prolog constitute the database of the system. By specifying queries or end goals, a user searches the database, and if the facts or rules are either already present in the database or can be derived/implied using the facts and rules in it, then the output/solution of the query is affirmative; otherwise, the solution is regarded as negative. Certain vital features make Prolog a strong logic programing language and thus differentiate it from RDF. The key features are Unification, Recursion and Backtracking. Unification allows to have the same structure from the terms, recursion allows reuse of the programing construct with multiple atoms/variables, and backtracking allows tracing and satisfying a previous task when a task has failed.

Prolog is an excellent tool for representing and manipulating data written in formal and natural languages. Its safe semantics and automatic memory management make it a prime candidate for logic programming. Efficient pattern matching due to recursion is enabled, and list handling is coherent.

2.3   Inductive Logic Programing

A subfield of symbolic artificial intelligence, Inductive logic programming (ILP), derives a hypothesised logic program entailing all positive and no negative examples from an encoding of known and given background knowledge. The background knowledge is given as a theory, usually as Horn clauses, and positive and negative examples are conjunction of unnegated and negated ground literals, respectively.

A correct hypothesis is a logic proposition satisfying the requirements of necessity, sufficiency, weak consistency and strong consistency.

3   Implementation

This work presents an idea to compare the expressive power of RDF and ILP and their efficiency in deriving solutions. An ontology will be developed using ILP using Prolog to demonstrate the comparison. For the research purpose, a transport domain ontology has been selected.

The following sections describe the transport domain ontology and the steps that would be used to construct the ontology using ILP.

3.1   Transport Ontology

In [6] the authors present a domain ontology for Transportation Systems. The authors have developed an ontology for a semantics-aware transportation system from the perspective of a traveller user, capable of answering general competence queries like the nearest bus stop to a particular place, the nearest parking slots available, and similar queries.

This ontology has been chosen because the authors have studied the transportation system of some of the biggest cities of the world and have developed a vocabulary that can be reused to any city with minor modifications. The alignment of the vocabulary of ontology with DOLCE [8], and the compliance with standard metadata protocols due to the use of MOD [7] makes the ontology an excellent fit to work with. The ontology comprises 46 classes, 42 distinct data properties, and eight distinct object properties.

3.2   Transport Ontology with ILP

Background theory and positive examples will be used for developing the transport ontology. The following will form the basis for the engineering of the transport ontology: The data properties and object properties will be used as facts. Some axioms will be needed to be specifically specified, for example, the rdfs:subClassOf, while the rest can be derived using the rules, for example, rdfs:subPropertyOf. Then, rules will be defined, stating that a class has to be made up of certain data property facts, and two or more classes hold relation and cardinality if certain object property rules are satisfied.

3.3   Analogical Analysis of RDF and ILP

After the development of the transport ontology, we want to compare the Transport Ontology built using RDF and ILP in terms of the following:

- Expressive Power of the logic programing languages
- Time taken for ontology development
- Manual labour is required for the development of ontology in terms of human involvement with respect to the whole process
- Efforts required for formulating queries/end goals
- Time taken for correct execution of end goal
- Ability to deduce facts without human intervention

While evaluating some of the parameters mentioned above would be quite straightforward, the others, such as the efforts required for formulating queries and end goals, time is taken for correct execution of end goal, and the ability to deduce facts without human intervention, will require an advanced procedure for a true unbiased judgement. For a much more civilised judgement, a survey will be conducted among ILP and RDF users (ranging from novice to sophisticated RDF and ILP users), and their viewpoints will be taken into consideration.

4   Conclusion and Future Work

The paper revolves around using inductive logic programing to construct ontology for the machine-readable data format to investigate the trade-off between RDF and ILP in terms of expressivity, expressibility and efficiency for learning and reasoning of knowledge. To test the trade-offs, a transport ontology has been chosen. The paper summarises the basic concepts of RDF, Prolog and ILP and illustrates their expressive power.

The concepts discussed in work can be applied to systems dealing with taxonomies and hierarchies. One such system is the Dare2Del [9]. The Dare2Del is a file assistance system that helps users in the decision-making process of deleting or hiding an irrelevant file. The files in a system are usually arranged in hierarchies that resemble hierarchies and taxonomy, just as classes and properties are arranged in ontologies. The ideas represented in the paper can be applied in scenarios where files are randomly placed with no categorisation or in no hierarchy (e.g. all files are on the Desktop), then the ideas in the paper can be applied to automatically generate hierarchies of files and folders and then assist the user in deletion/removal of non-essential files.

Acknowledgments. The research presented here has been carried out within the Dare2Del project under the DFG priority program Intentional Forgetting (SPP 1921). Dare2Del is a joint project of Cognitive Systems, University of Bamberg and the Chair for Work and Organisational Psychology, University of Erlangen.

References

1. Miller, E.: An introduction to the resource description framework. Bull. Am. Soc. Inf. Sci. Technol. **25**(1), 15–19 (1998)
2. Brickley, D., Guha, R.V., Layman, A.: Resource description framework (RDF) schema specification (1999)
3. Sintek, M., Decker, S.: TRIPLE-An RDF Query, Inference, and Transformation Language. In INAP, pp. 47–56, October 2001
4. Stuckenschmidt, H., Predoiu, L., Meilicke, C.: Learning complex ontology alignments-a challenge for ILP research. Inductive Logic Programming, 105 (2008)
5. Omelayenko, B.: Engaging prolog with RDF. In: Proceedings of the Workshop on Ontologies and Distributed Systems. IJCAI (2003)
6. Nandini, D., Shahi, G.K.: An ontology for transportation system. Kalpa Publications Comput. **10**, 32–37 (2019)
7. Dutta, B., Nandini, D., Shahi, G.K.: MOD: metadata for ontology description and publication. In: International Conference on Dublin Core and Metadata Applications, pp. 1–9, September 2015
8. Gangemi, A., Guarino, N., Masolo, C., Oltramari, A., Schneider, L.: Sweetening ontologies with DOLCE. In: Gómez-Pérez, A., Benjamins, V.R. (eds.) EKAW 2002. LNCS (LNAI), vol. 2473, pp. 166–181. Springer, Heidelberg (2002). https://doi.org/10.1007/3-540-45810-7_18
9. Niessen, C., Schmid, U.: Dare2Del In Internal and external IF-Empirical studies and development of an assist system for IF of digital information [Scholarly project]
10. Bratko, I.: Prolog programming for artificial intelligence. Pearson education (2001)

MDNCaching: A Strategy to Generate Quality Negatives for Knowledge Graph Embedding

Tiroshan Madushanka[1,2,3]($\boxtimes$)(iD) and Ryutaro Ichise[1,2]

[1] SOKENDAI (The Graduate University for Advanced Studies), 2-1-2 Hitotsubashi, Chiyoda-ku, Tokyo, Japan
[2] National Institute of Informatics, 2-1-2 Hitotsubashi, Chiyoda-ku, Tokyo, Japan
{tiroshan,ichise}@nii.ac.jp
[3] University of Kelaniya, Kelaniya, Sri Lanka
tiroshanm@kln.ac.lk

Abstract. Knowledge graph embedding (KGE) has become an integral part of AI as it enables knowledge construction and exploring missing information. KGE encodes the *entities* and *relations* (elements) in a knowledge graph into a low-dimensional vector space. The conventional KGE models are trained using positive and negative examples by discriminating the positives from the negatives. However, the existing knowledge graphs contain only the positives. Hence, it is required to generate negatives to train KGE models. This remains a key challenge in KGE due to various reasons. Among them, the quality of the negatives is a critical factor for KGE models to produce accurate embeddings of the observed facts. Therefore, researchers have introduced various strategies such as Bernoulli negative sampling to generate quality negatives that are hard to distinguish from positives. However, fixed negative sampling strategies are suffering from vanishing gradients and false negatives. Later, the dynamic negative sampling techniques were introduced to overcome the vanishing gradient, but the false negatives still remain as a challenge to the research community. The present research introduces a new strategy called MDNCaching (Matrix Decomposed Negative Caching), which generates negatives considering the dynamics of the embedding space while exploring the quality negatives with large similarity scores. Matrix decomposition is used to eliminate false negatives, and hence, the MDNCaching ensures the quality of the generated negatives. The performance of MDNCaching was compared with the existing state-of-art negative sampling strategies, and the results reflect that the proposed negative sampling strategy can produce a notable improvement in existing KGE models.

Keywords: Negative sampling · Knowledge graph embedding · Matrix decomposition

© Springer Nature Switzerland AG 2022
H. Fujita et al. (Eds.): IEA/AIE 2022, LNAI 13343, pp. 877–888, 2022.
https://doi.org/10.1007/978-3-031-08530-7_74

1   Introduction

A Knowledge Graph (KG) is a structured representation of facts, textual data in the form of $(head, relation, tail)$ known as a triplet, e.g., (*Shakespeare, isAuthorOf, Hamlet*). Knowledge graphs are constructed using the knowledge bases such as Freebase, DBpedia, WordNet, and YAGO. KGs have been utilized in many real-world applications, such as question answering, recommendation systems, and information retrieval. Although the knowledge bases contain vast volumes of facts, the KGs are often incomplete as they are created based on the available facts or the ground truth, which are often dynamic and evolving. For example, when considering the people's birthplaces, 71% and 66% are not found in Freebase and DBpedia, respectively. Therefore, it is worth having methods to complete the KGs automatically by adding the missing knowledge or the facts.

Recent research have shown that Machine Learning (ML) methods can be effectively used to complete knowledge graphs. However, applying ML methods is still a challenging task due to various facts such as high dimensionality. As a solution, knowledge graph embedding methods have been introduced by past research. Knowledge graph embedding (KGE) which maps *entities* and *relations* into a low dimensional vector space while preserving their semantic meaning. Moreover, KGE overcomes the difficulties in manipulating textual data in knowledge graphs, such as sparseness and computational cost [9]. Modern KGE strategies have shown promising results in knowledge acquisition tasks such as link prediction, triplet classification, and knowledge graph completion. Typically, KGE models accelerate training ML algorithms by extending the motivation of ranking the observed instances (positives) higher than the unobserved instances (negatives). However, the knowledge bases contain only positive examples. Hence, it is necessary to explore strategies to generate quality negatives that are hard to distinguish from positives as they have high similarity but are negatives. For instance, considering the positive (*Shakespeare, isAuthorOf, Hamlet*), we say that the generated negative (*Shakespeare, isAuthorOf, The Widow's Tears*) is a quality negative as it is hard to distinguish from positives instead the negative (*Shakespeare, isAuthorOf, London*). Generation of quality negatives enhances the KG embeddings, which is always challenging. Therefore, negative sampling becomes indispensable in knowledge representation learning as the KGE model's performance heavily relies on negative selection.

Most of the state-of-the-art strategies in generating negatives consider corrupting positives randomly (e.g., [2,14]), based on closed world assumption, or exploiting the KG structure when generating quality negatives (e.g., [1,17]). However, these strategies suffer from false negatives as they do not guarantee that the generated ones are always relevant, i.e., generating latent positives as negatives. As KGE models are sensitive to inputs, false negatives usually fool the model, losing the semantics of *entities* and *relations*. Furthermore, strategies that randomly corrupt positives suffer from vanishing gradients as they tend to generate triplets with zero gradients during the training phase.

To overcome the stated challenges, the present work proposes a negative sampling strategy that explores negatives, considering the dynamic distribution of embedding space and reducing false negatives by adopting matrix decomposition. We first trained the latent relation model that uses positives to utilize the matrix decomposition. Then we predict the latent relations and refer them with negative candidates generation. We utilize a caching technique to manage negative triplets with large similarity scores. To overcome the vanishing gradients problem, we up-date negative candidates concerning the changes to the embedding space with KGE model training. Furthermore, we propose a selection criterion that ensures "exploration and exploitation" that balances exploring all possible quality negative candidates and sampling a fixed number of negatives close to the positives.

The remainder of this paper is organized as follows. Section 2 discusses related work with knowledge graph embedding and negative sampling. In Sect. 3, we propose a new strategy in generating quality negatives with large similarity scores considering the dynamic distribution of embedding space while eliminating false negatives adopting matrix decomposition. In Sect. 4, we present an experimental study in which we compare our proposed negative sampling strategy with baseline results of benchmark datasets and analyze results with state-of-the-art. In Sect. 5, we conclude this paper.

2   Related Work

Various research work has been conducted in Negative Sampling and Knowledge Graph Embedding. KGE maps knowledge graph elements, i.e., *entities* and *relations*, into low dimensional continuous vector space to use numerical representation when carrying out knowledge acquisition tasks. Commonly, three mainstream KGE models are found: translational distance-based models, semantic matching-based models, and neural network approaches. Translational distance-based models represent the distance of projected KG elements (e.g., [2,6,14]). Using matrix decomposition, semantic matching-based models represent latent semantics organized in vectorized *entities* and *relations* (e.g., [8,11,16]). In addition, Neural network approaches have also gained attention in recent research work that utilizes the potential of neural networks and variants (e.g., [4,5]). Typically, KGE models learn knowledge representation by discriminating positives from negatives made by corrupting positives. However, the quality of negatives affects training and performances of knowledge representation downstream tasks. In abstract, knowledge graph embedding work has focused on providing a better representation for connection between *entities* and *relations* of the knowledge graph, while negative sampling strategies have focused on boosting the underlying embedding model.

2.1   Negative Sampling

Among the existing negative sampling strategies, Uniform negative sampling is a widely used strategy due to its simplicity and efficiency. For example, TransE [2],

ComplEx [11] and DistMult [16] use Uniform negative sampling to generate negatives. Uniform negative sampling randomly corrupts positives by replacing *head* or *tail* entities, and it reflects that generated negatives do not contribute to knowledge representation learning in most cases and generate false negatives. The Bernoulli negative sampling strategy was proposed to overcome this limitation by considering relation cardinality (i.e., 1-N, N-N, and N-1) to reduce false negatives [14]. However, both Uniform and Bernoulli sampling strategies are fixed sampling schemes. They both suffer from vanishing gradients as they generate triplets with zero gradients during the training phase [17]. Hence, Generative adversarial networks (GAN) based negative sampling strategies IGAN [12] and KBGAN [3] were introduced to generate negatives with large similarity scores considering the dynamic distribution of embeddings. The GAN-based strategies adversarially train the discriminator to produce quality negatives concerning a pre-trained KGE model as the generator. In KBGAN, the generator generates a candidate set of uniformly sampled negatives, i.e., $Neg = (\bar{h}, r, \bar{t})$, selects one with the highest probability from set Neg, and then feeds to the discriminator that minimizes marginal loss between positives and negatives to improve the final embedding. However, GAN-based strategies suffer from high variance in REINFORCE gradient [15], and the generator introduces additional parameters. Both KBGAN and IGAN require pre-trained, which adds extra costs. Recently, Structure Aware Negative Sampling (SANS) [1] strategy was introduced with a different perspective that utilizes available graph structure by selecting negatives considering the neighborhood. Since SANS explores potential negatives within a k-hop neighborhood, SANS also increases the possibility of generating false negatives. NSCaching [17] was proposed to overcome the challenges in generating quality negatives by introducing a cache that maintains negative triplets with large similarity scores and updating the cache using importance sampling. Despite this, NSCaching may produce false negatives with a high possibility as the latent positives also reflect large similarity scores.

3   MDNCaching

Although the literature shows diverse KGE models, generating quality negatives remains as a fundamental challenge in KGE. The present research introduces a new strategy called MDNCaching, which generates negatives considering the dynamics of the embedding space while exploring the quality negatives with large similarity scores. The novel strategy addresses existing challenges; 1). reducing false negatives in the generated candidates, and 2). generation of quality negatives with large similarity scores. To this end, we introduce a novel strategy that combines the dynamic updates of the embedding space to overcome the challenge of generating quality negatives with large similarity scores, avoiding the vanishing gradients problem. More precisely, the Matrix Decomposition technique is utilized to model the latent relations when avoiding false negatives that enhance the quality of the generated negatives. Before delving into the details of the proposed strategy, it is worth knowing the idea of matrix decomposition, a critical component of the proposed strategy.

3.1 Matrix Decomposition

The matrix decomposition (MD) technique utilizes matrix multiplication to generate latent features. Collaborative filtering is the typical application of MD to identify ratings between *item* and *user* entities [7]. Referring to collaborative filtering context, let U be a set of *Users*, D be a set of *items*, and R be a *rating* matrix between U and D, i.e., $R = R_{|U| \times |D|}$, including all product ratings given by users. With matrix decomposition, the goal is to generate latent rating features, given that the input of two matrices, P that represents the association between a *user* and features, and Q that represents the association between an *item* and features, i.e., $R \approx P \times Q^\top$ [7].

Even though the matrix decomposition techniques are utilized with KGE (e.g., [8]), to the best of our knowledge, the MD technique is yet under utilized in the negative sampling strategies. Considering the benefits of MD techniques in modeling hidden semantics, we apply a matrix decomposition technique to model the latent *relations* to predict potential false negatives. In our model, let h be a set of *Heads*, t be a set of *Tails*, and R be a *relation* matrix between h and t, i.e., $R = R_{|h| \times |t|}$, includes all relations between entities. Our goal is to generate latent relations referring to the matrix decomposition model such that $R \approx H \times T^\top$ where H represents the association between a *head* and features, and T represents the association between a *tail* and the features.

3.2 The Proposed Strategy

This section describes the proposed negative sampling strategy MDNCaching. Recall the stated challenges in negative sampling 1). reduce false negatives that fool the KGE model to lose the semantics of the KG, and 2). adopt dynamics of the embedding space when generating quality negatives with large similarity scores to avoid the vanishing gradient. The proposed strategy enhances the KGE by generating quality negatives with large similarity scores while reducing the possible false negatives in the sampling space.

The proposed MDNCaching is a dynamic distribution-based negative sampling strategy that integrates a matrix decomposition technique and utilizes the dynamics of the knowledge graph embedding to address the stated challenges. We integrate the matrix decomposition technique to eliminate false negatives by predicting latent *relations*. The reduction of false negatives decreases the possible discrimination on latent positives and enhances the KGE. We consider frequent updates to the embedding space to overcome the issue of generating quality negatives with large similarity scores. Furthermore, we utilize a caching technique that maintains negatives with large similarity scores for each positive in the training set S. Two caches are separately maintained as head-cache $\mathcal{H}$ that maintains candidates for *head* corruption and indexes negatives with *tail* and *relation* (t, r), while tail-cache $\mathcal{T}$ maintains candidates for *tail* corruption and indexes negatives with *head* and *relation* (h, r). We uniformly sample a negative from the cache efficiently without introducing any bias. The lazy update

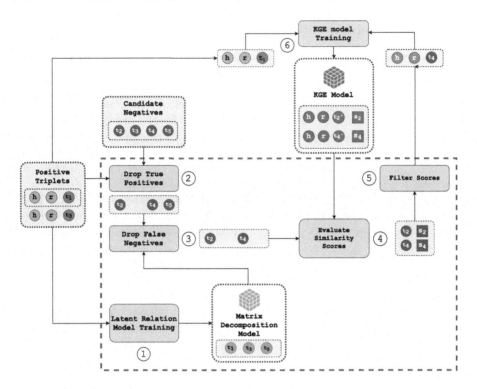

Fig. 1. The framework for the proposed negative sampling strategy MDNCaching.

technique in updating caches refreshes the cache after $\mathcal{N}$ number of epochs later rather than immediate.

Our framework for the proposed negative sampling strategy is depicted in Fig. 1, illustrating steps in generating a quality negative with *tail* corruption scenario. The proposed MDNCaching consists of six critical steps in generating quality negatives and executing the KGE task. In step 1, MDNCaching performs the **latent relation model training**. Detection and elimination of false negatives are critical tasks in the proposed negative sampling strategy. In order to predict latent positives, the matrix decomposition model is trained concerning the observed KG elements. In step 2, MDNCaching **drops true positives** from the candidate negatives. The candidate negatives are initiated as *entity* space $\mathcal{E}$ except the given positive elements. This technique ensures that the proposed strategy explores candidate negatives best. For example, given the positive (h, r, t_1), MDNCaching initializes the candidate negatives as $\{t_2, t_3, t_4, t_5\}$ where $\mathcal{E} = \{h, t_1, t_2, t_3, t_4, t_5\}$. However, the candidate negatives may comprise true positives since KG consists of 1-N, N-N and N-1 relations. Therefore, it is essential to drop true positives from the candidate negatives (e.g., given positive (h, r, t_3), candidate $\{t_3\}$ is removed from candidate negatives resulting $\{t_2, t_4, t_5\}$ as candidates). In step 3, MDNCaching **drops false negatives** from the can-

---

Algorithm 1: KGE model training with negatives from MDNCaching.

Input: Knowledge graph $\mathcal{G}$, and the matrix decomposition model f_{md}
Output: Knowledge graph embeddings

1 Initialize *head-features* embeddings $\forall h \in \mathcal{E}$, and *tail-features* embeddings $\forall t \in \mathcal{E}$ and train f_{md} for a certain number of epochs.

2 Initialize caches; head-cache $\mathcal{H}$, tail-cache $\mathcal{T}$ using Algorithm 2.

3 Initialize knowledge graph embeddings $\forall e \in \mathcal{E}, \forall r \in \mathcal{R}$.

4 **Loop**

5   **foreach** $(h, r, t) \in \mathcal{G}$ **do**

6   Index $\mathcal{H}$ by (t, r), i.e., $\mathcal{H}_{(t,r)}$ and $\mathcal{T}$ by (h, r), i.e., $\mathcal{T}_{(h,r)}$.

7   Uniformly sample $\bar{h} \in \mathcal{H}_{(t,r)}$ and $\bar{t} \in \mathcal{T}_{(h,r)}$.

8   Select negative $(\bar{h}, r, \bar{t})$ either as $(\bar{h}, r, t)$ or $(h, r, \bar{t})$ considering the relation cardinality of r.

9   Update knowledge graph embeddings discriminating (h, r, t) against $(\bar{h}, r, \bar{t})$.

10   **end foreach**

11   update cache $\mathcal{H}$ and $\mathcal{T}$ using Algorithm 2;

12 **end**

---

didate negatives utilizing the trained MD model at step 1. It is essential to identify false negatives before the score filtration since they also contain large similarity scores. Therefore, the proposed strategy predicts latent relations to exclude false negatives from the candidate negatives (e.g., given the (h, r) pair, $\{t_1, t_3, t_5\}$ are predicted, and the $\{t_5\}$ is removed from the candidate negatives, which is the latent). In step 4, the proposed strategy **evaluates similarity scores** for the candidate negatives, referring to the baseline scoring function. Since MDNCaching drops true positives and false negatives, the candidate negatives consist of potential negatives. In step 5, the quality negatives are filtered considering the similarity score, i.e., **filter scores**, and negatives with large similarity scores are selected (e.g., given s_2 and s_4 are similarity scores for (h, r, t_2), (h, r, t_4) respectively, and given $s_4 \geq filtration\ threshold$, we update the candidate negatives as $\{t_4\}$). In step 6, the proposed strategy performs the **KGE model training** by discriminating provided positives (e.g., (h, r, t_1)) against generated negatives (e.g., (h, r, t_4)).

3.3 Integration of MDNCaching with KGE Framework

Figure 1 describes the general framework of negative sample generation. However, we utilize a caching technique to manage generated negatives effectively in MDNCaching. Therefore, we describe the integration of MDNCaching with the typical KGE framework and the utilization of the caching technique in Algorithm 1. First, the matrix decomposition model training is performed. Then, the caches are initialized. Generally, we generate a triplet as a candidate for negatives by replacing either *head* or *tail*. The generated negatives are stored in two separate caches , i.e., head-cache $\mathcal{H}$ (indexed by (t, r)) and tail-cache $\mathcal{T}$ (indexed

---

Algorithm 2: MDNCaching cache update considering the dynamic distribution of embedding space.

Input: Baseline scoring function f, knowledge graph $\mathcal{G}$, matrix decomposition model f_{md}

Output: head-cache $\mathcal{H}$ and tail-cache $\mathcal{T}$

1 **foreach** $(h,r,t) \in \mathcal{G}$ **do**

2   │   Initialize negative candidates for $\mathcal{H}_{(t,r)}$ and $\mathcal{T}_{(h,r)}$.

3   │   Remove true positives from $\mathcal{H}_{(t,r)}$ and $\mathcal{T}_{(h,r)}$.

4   │   Predict latent relations from f_{md}, $\forall \bar{h} \in \mathcal{H}_{(t,r)}$, $\forall \bar{t} \in \mathcal{T}_{(h,r)}$ and drop false negatives from $\mathcal{H}_{(t,r)}$ and $\mathcal{T}_{(h,r)}$.

5   │   Evaluate similarity scores $\forall \bar{h} \in \mathcal{H}_{(t,r)}$ and $\forall \bar{t} \in \mathcal{T}_{(h,r)}$ considering the baseline scoring function f, i.e., $f(\bar{h}, r, t)$ and $f(h, r, \bar{t})$ respectively.

6   │   Select candidates with large similarity scores from $\mathcal{H}_{(t,r)}$ and $\mathcal{T}_{(h,r)}$.

7 **end foreach**

---

by (h, r)). Next, KGEs are initialized for KG elements, and KGE model training is performed iteratively for a certain number of epochs. When a positive triplet is received, the head-cache $\mathcal{H}$ and the tail-cache $\mathcal{T}$ are indexed. Then, a candidate negative triplet is generated referring to $\mathcal{H}_{(t,r)}$ and $\mathcal{T}_{(h,r)}$. Since the caches maintain quality negatives with large similarity scores, selecting any candidate from $\mathcal{H}_{(t,r)}$ or $\mathcal{T}_{(h,r)}$ avoids the vanishing gradient problem with high probability. Then, it performs the typical embedding update task referring to the baseline KGE model. Finally, caches are updated, adopting the changes to the KGE space, and strategy refers the Algorithm 2 to populate quality negatives in the head-cache $\mathcal{H}$ and the tail-cache $\mathcal{T}$. Algorithm 2 describes the process of generating quality negatives with large scores following the previously described steps 2–5 iteratively for each element in the KG.

In summary, the proposed MDNCaching strategy introduces an additional step to train a matrix decomposition model before KGE model training, and it introduces a caching technique to manage generated candidate negatives effectively. With flexibility in integrating any translational distance-based or semantic matching-based model, MDNCaching enables robustness in training models from scratch with fewer parameters than previous dynamic negative sampling work IGAN [12], and KBGAN [3]. The generator in GAN approaches tends to generate correct facts that are considered as *positives* instead of *negatives*, and in contrast to GAN approaches, the proposed MDNCaching strategy considers latent relations to eliminate plausible positive facts from the negative candidates by utilizing the matrix decomposition technique. Besides, MDNCaching extends the idea of caching candidate negative that proposed in NSCaching [17]. The proposed strategy *explores* the candidate space to the best at step 2 and *exploits* the candidates by carefully managing caches at step 5. In addition to that, the exploration of negatives with large similarity scores effectively impacts embedding training.

Table 1. Scoring functions for triple (h, r, t), and parameters. $diag(r)$ constructs the diagonal matrix with r.

Model	Scoring function	Definition	Parameters
Translational distance-based	TransE [2]	$\|h + r - t\|_i$	$h, r, t \in \mathbb{R}^n$
	TransD [6]	$\left\| h + w_r w_h^\top h + r - (t + w_r w_t^\top t) \right\|_i$	$h, r, t, w_h, w_t, w_r \in \mathbb{R}^n$
Semantic matching-based	DistMult [16]	$h \cdot diag(r) \cdot t^\top$	$h, r, t \in \mathbb{R}^n$
	ComplEx [11]	$Re(h \cdot diag(r) \cdot t^\top)$	$h, r, t \in \mathbb{C}^n$

Table 2. Statistics of the datasets used with experiments

Dataset	#entity	#relation	#train	#valid	#test
WN18RR	93,003	11	86,835	3,034	3,134
FB15K237	14,541	237	272,115	17,535	20,466

4 Experiments

We evaluated the proposed negative sampling strategy, i.e., MDNCaching, on the link prediction in KGs and compared results with the state-of-the-art negative sampling strategies. In this case, the task was to predict the missing *head* (h) or *tail* (t) entity for a positive triplet (h, r, t) and evaluate the rank of the *head* and *tail* entities among all predicted entities. We evaluated the results for link prediction with TransE [2], TransD [6], DistMult [16], and ComplEx [11] baseline KGE models, and definitions are described in Table 1.

4.1 Experimental Setup

Datasets. The experiments were conducted on two popular benchmark datasets WN18RR [13] and FB15K237 [10]. These datasets were constructed by removing inverse-duplicate relations from previous WN18 and FB15K datasets respectively. The experiments were carried out in these two variants as they were more challenging and realistic than originals. The statistics of the data sets are described in Table 2.

Performance Measurement. We consider the "Filtered" setting with performance evaluation so that valid entities outscoring the target are not considered mistakes. Hence they are skipped when computing the rank. We evaluate results based on the following metrics,

1. *Mean Rank (MR)* is the average of the obtained ranks; $MR = \frac{1}{|Q|} \sum_{q \in Q} q$. The smaller value of MR tends to infer better results. However, since MR is susceptible to outliers, the Mean Reciprocal Rank is widely used.

2. *Mean Reciprocal Rank (MRR)* is the average of the inverse of the obtained ranks; $MRR = \frac{1}{|Q|} \sum_{q \in Q} \frac{1}{q}$. The higher value of MRR tends to infer better results.
3. *Hit@K* is the ratio of predictions for which the rank is equal or lesser than a threshold k; $Hits@K = \frac{|\{q \in Q : q \leq K\}|}{|Q|}$. The higher value of Hits@K tends to infer better results.

Optimization and Implementation. A knowledge graph model was optimized by minimizing the objective function with Adam optimizer, and first, we tuned hyper-parameters referring to Bernoulli sampling strategy based on MRR. We conducted the evaluation for 1000 epochs and presented the best result for MRR. We started our experiments within the following ranges for hyper-parameters: embedding dimension $d \in \{50, 100, 250, 1000\}$, learning rate $\eta \in \{0.0005, 0.005, 0.05, 0.5\}$, margin value $\gamma \in \{1, 2, 3, 4, 5\}$ and optimized for best performance.

Results. We compare results with state-of-the-art negative sampling strategies concerning the reported performance comparison in NSCaching [17] work for Bernoulli, KBGAN, NSCaching concerning the training from scratch. Also, we directly consider the reported performance in SANS [1]. The performance comparison on link prediction is summarized in Table 3. When comparing results on translational distance-based, it is evident that the proposed negative sampling strategy gains substantial improvement for both datasets, i.e., WN18RR and FB15K237. When evaluating results for semantic matching-based KGE models, we observe that the proposed strategy outperforms the state-of-the-art negative sampling strategies. One can observe that MDNCaching consistently achieves better results with ComplEx than the state-of-the-art negative sampling strategies for both datasets with substantial improvements (i.e., Hits@10 by 4.40% and 10.91% for WN18RR and FB15K237 respectively). Although some results are competitive, experimental results reflect that MDNCaching enhances link prediction tasks against the state-of-the-art negative sampling strategies. For instance, when considering the Hits@10, we can witness 7.83% and 10.91% improvement with TransE and ComplEx respectively for the FB15K237 dataset while we observe 4.64% and 4.40% improvement with DistMult and ComplEx respectively for the WN18RR dataset. The results evidence that the proposed negative sampling strategy effectively enhances the KGE by generating quality negatives. The substantial improvements in MRR and Hits@10 reflect that the MDNCache successfully overcomes the stated challenges with negative generation.

Table 3. Comparison of state-of-the-art negative sampling strategies on WN18RR and FB15K237 datasets. Note that results of MR and results for TransD and ComplEx embedding models for SANS [1] is not available as the original did not include. We consider SANS with the random walk configuration.

Score function	Negative sampling strategy	WN18RR			FB15K237		
		MRR	MR	Hits@10	MRR	MR	Hits@10
TransE	Bernoulli [17]	0.1784	3924	45.09	0.2556	197	41.89
	KBGAN [17]	0.1808	5356	43.24	0.2926	722	46.59
	SANS [1]	0.2317	–	**53.41**	0.2981	–	48.50
	NSCaching [17]	0.2002	4472	47.83	0.2993	**186**	47.64
	MDNCaching	**0.2390**	3054	53.20	**0.3330**	200	**52.30**
TransD	Bernoulli	0.1901	3555	46.41	0.2451	**188**	42.89
	KBGAN	0.1875	4083	46.41	0.2465	825	44.4
	SANS	–	–	–	–	–	–
	NSCaching	**0.2013**	**3104**	**48.39**	**0.2863**	189	**47.85**
	MDNCaching	0.1737	4477	48.36	0.2683	354	46.43
DistMult	Bernoulli	0.3964	7420	45.25	0.2491	280	42.03
	KBGAN	0.2039	11351	29.52	0.2272	276	39.91
	SANS	0.4071	–	49.09	0.2021	–	41.46
	NSCaching	**0.4128**	7708	45.45	**0.2834**	**273**	**45.56**
	MDNCaching	0.3921	**2946**	**51.37**	0.2694	403	44.10
ComplEx	Bernoulli	0.4431	**4693**	51.77	0.2596	238	43.54
	KBGAN	0.3180	7528	35.51	0.1910	881	32.07
	SANS	–	–	–	–	–	–
	NSCaching	0.4463	5365	50.89	0.3021	**221**	48.05
	MDNCaching	**0.4729**	5312	**54.05**	**0.3594**	415	**53.29**

5   Conclusion

The present research proposed MDNCaching, which is an enhanced negative sampling strategy for KGE, addressing the problem of false negatives by reducing latent positives predicting through matrix decomposition. The proposed strategy effectively manages separate caches for *head* and *tail* candidates that contain quality negatives with large similarity scores, adopting the dynamic changes in the embedding space. Experimentally, we evaluated the MDNCaching on two datasets and four scoring functions covering translational-distance and semantic matching models. Experimental results reflect a substantial enhancement with TransE, DistMult, and ComplEx KGE models. Notably, the ComplEx KGE model with MDNCaching improves both datasets considerably. When carefully balanced the exploration and exploitation, MDNCaching requires considerable memory as it explores possible candidates, and utilization of memory handling will proceed as future works. Also, possible enhancements with latent relation prediction will continue for our future works.

References

1. Ahrabian, K., Feizi, A., Salehi, Y., Hamilton, W.L., Bose, A.J.: Structure aware negative sampling in knowledge graphs. In: Proceedings of Conference on Empirical Methods in Natural Language Processing, pp. 6093–6101 (2020)
2. Bordes, A., Usunier, N., García-Durán, A., Weston, J., Yakhnenko, O.: Translating embeddings for modeling multi-relational data. In: Proceedings of International Conference on Neural Information Processing Systems, vol. 26, pp. 2787–2795 (2013)
3. Cai, L., Wang, W.Y.: KBGAN: adversarial learning for knowledge graph embeddings. In: Proceedings of Conference of the North American Chapter of the Association for Computational Linguistics, pp. 1470–1480 (2018)
4. Dettmers, T., Minervini, P., Stenetorp, P., Riedel, S.: Convolutional 2D knowledge graph embeddings. In: Proceedings of AAAI Conference on Artificial Intelligence, vol. 32, pp. 1811–1818 (2018)
5. Dong, X., et al.: Knowledge vault: a web-scale approach to probabilistic knowledge fusion. In: Proceedings of ACM SIGKDD International Conference on Knowledge Discovery and Data Mining, pp. 601–610 (2014)
6. Ji, G., He, S., Xu, L., Liu, K., Zhao, J.: Knowledge graph embedding via dynamic mapping matrix. In: Proceedings of International Joint Conference on Natural Language Processing, pp. 687–696 (2015)
7. Koren, Y., Bell, R., Volinsky, C.: Matrix factorization techniques for recommender systems. Computer **42**, 30–37 (2009)
8. Nickel, M., Tresp, V., Kriegel, H.P.: A three-way model for collective learning on multi-relational data. In: Proceedings of International Conference on Machine Learning, pp. 809–816 (2011)
9. Shan, Y., Bu, C., Liu, X., Ji, S., Li, L.: Confidence-aware negative sampling method for noisy knowledge graph embedding. In: Proceedings of IEEE International Conference on Big Knowledge, pp. 33–40 (2018)
10. Toutanova, K., Chen, D.: Observed versus latent features for knowledge base and text inference. In: Proceedings of Workshop on Continuous Vector Space Models and their Compositionality, pp. 57–66 (2015)
11. Trouillon, T., Welbl, J., Riedel, S., Gaussier, E., Bouchard, G.: Complex embeddings for simple link prediction. In: Proceedings of International Conference on Machine Learning, pp. 2071–2080 (2016)
12. Wang, P., Li, S., Pan, R.: Incorporating GAN for negative sampling in knowledge representation learning. In: Proceedings of AAAI Conference on Artificial Intelligence, vol. 32, pp. 2005–2012 (2018)
13. Wang, Y., Ruffinelli, D., Gemulla, R., Broscheit, S., Meilicke, C.: On evaluating embedding models for knowledge base completion. In: Proceedings of Workshop on Representation Learning for NLP, pp. 104–112 (2019)
14. Wang, Z., Zhang, J., Feng, J., Chen, Z.: Knowledge graph embedding by translating on hyperplanes. In: Proceedings of AAAI Conference on Artificial Intelligence, pp. 1112–1119 (2014)
15. Williams, R.J.: Simple statistical gradient-following algorithms for connectionist reinforcement learning. Mach. Learn. **8**, 229–256 (1992)
16. Yang, B., Yih, W., He, X., Gao, J., Deng, L.: Embedding entities and relations for learning and inference in knowledge bases. In: Proceedings of International Conference on Learning Representations (2015)
17. Zhang, Y., Yao, Q., Shao, Y., Chen, L.: NSCaching: simple and efficient negative sampling for knowledge graph embedding. In: Proceedings of IEEE International Conference on Data Engineering, pp. 614–625 (2019)

Robotics, Games and Consumer Applications

Application of a Limit Theorem to the Construction of Japanese Crossword Puzzles

Volodymyr Novykov⬤, Geoff Harris$^{(\boxtimes)}$⬤, and Isaac Tonkin⬤

Centre for Data Analytics, Bond Business School, Bond University,
Robina, QLD 4227, Australia
gharris@bond.edu.au

Abstract. The generation of crossword puzzles is known to be NP-Complete. Optimization over structural characteristics is known to extend this into the NP-Hard regime. This paper discusses the application of a limit theorem to assist in optimizing the NP-Hard aspects of crossword puzzle generation of Japanese lexicons in particular. It is shown that the similarity of artificially enumerated lexicons to Japanese kana-based lexicons is greater than to English language lexicons. This greater similarity is exploited in the derivation of expressions for both the expected value of the crossword and the use of the central limit theorem to determine an empirical estimate on the upper limit of the associated optimization problem. Initial empirical outcomes attest to the expected efficacy of the central limit theorem application to the final expressions.

Keywords: Limit theorem · Crossword puzzle · NP-Hard optimization

1 Introduction

Solving a crossword under external optimization criteria remains a significant and unsolved testbed problem. We examine the subset of these problems represented as constructions of theme-based crossword puzzles. The reason for the choice of Japanese-based lexicon over English-based lexicons is the more uniform distribution of syllables in the Japanese language as opposed to the random multi-modal distribution of letters among English language words.

We prove a limit theorem for the distribution of scores of a Japanese crossword and apply this to finding the optimum solution. Similar types of NP-Hard word optimization problems such as the Crozzle, as well as more general word puzzles, have been studied in the works of Forster et al. (1992), Harris et al. (1993), Gower and Wilkerson (1996) and Agarwal and Joshi (2020).

Our rationale for employing a limit theorem is that it provides estimates of the number of solutions and the expected highest score achievable when given a partially completed crossword. To determine the number of solutions and maximum solution via a brute-force search is computationally infeasible for large

© Springer Nature Switzerland AG 2022
H. Fujita et al. (Eds.): IEA/AIE 2022, LNAI 13343, pp. 891–897, 2022.
https://doi.org/10.1007/978-3-031-08530-7_75

lexicons. However, these estimates via the limit theorem may prove to be crucial in deciding whether pursuing a solution path emanating from the given partially completed crossword, is likely to lead to the optimum crossword solution.

The remainder of this paper is as follows. In Sect. 2, the application of the method to the crossword puzzle is outlined. Here, artificially enumerated lexicon (AEL) datasets are used, which Harris and Forster (1993) exploited in automating the solution generation process in similar word games. In Sect. 3, the results of this work are presented and discussed. The paper then concludes with an outline of current and potential research efforts.

2   Description of Crossword Puzzle Construction and the Application of the Limit Theorem

Construction of these themed crossword puzzles, also known as Crozzles in some countries (see e.g., Harris and Forster (1992), Harris and Forster (1993) and Binkley and Kuhn (1997)), involves the placing of words from a specified lexicon (usually with a limited number of words) into an initially blank grid. On completion, the unfilled cells are filled with a black background as per normal convention. Scoring regimes for the purposes of optimizing the objective function are arbitrary.

Denote by $\mathcal{A}$ our alphabet of letters, $\mathcal{S}$ the set of feasible solutions of the crossword puzzle, $\mathcal{L} = \{w|w = word\}$ the set of words in our lexicon, $\mathcal{G}$ our $r \times c$ grid where our words are placed, $\underline{\alpha} = (\alpha_{i,j})_{i \in \{1,2,...,r\}, j \in \{1,2,...,c\}}$ and the $r \times c$ matrix of real numbers $\in (-\frac{1}{2}, \frac{1}{2}) \equiv I$. Typically, a word is written as a concatenation of letters in our alphabet $\mathcal{A}$, e.g. $w = dog$, and we adopt the convention that the k-th suffix of the word denotes the letter occurring in the k-th position, i.e. $w_1 = d$, $w_2 = o$, $w_3 = g$. Furthermore, to identify the start and end of a word, we employ the symbol ω, and to identify the adjacent positions of the word when it is placed in the grid, we employ the symbol ζ. These ancilliary symbols help us to check whether a placement of words from our lexicon in the grid $\mathcal{G}$ is valid. Additionally, to identify whether a letter is placed as part of a vertically placed word, we employ the prime symbol, for example d' if w' is the word dog placed vertically in the grid. Thus, we arrive at an augmented alphabet $\overline{\mathcal{A}} = \mathcal{A} \cup \mathcal{A}' \cup \{\zeta, \omega\}$.

We have a multidimensional integral involving the rc complex variables $z_{i,j}$ and we adopt as a contour of integration with respect to $z_{i,j}$ the circle of radius $\exp(-\sigma_{i,j})$, parametrized as $z_{i,j} = \exp(-\sigma_{i,j})e(\alpha_{i,j})$, where $\sigma_{i,j}$ is a positive real number and $\alpha_{i,j} \in I$. Consequently, our generating function in respect of a word $w = w_1 w_2 \ldots w_\ell$ is given by

$$G_w(\underline{\alpha}; \underline{\sigma}) = 1 + \sum_{(i,j) \in \mathcal{G}} G_w^{(i,j)}(\underline{\alpha}; \underline{\sigma}) + \sum_{(i,j) \in \mathcal{G}} G_{w'}^{(i,j)}(\underline{\alpha}, \underline{\sigma}), \tag{1}$$

where

$$G_w^{(i,j)}(\underline{\alpha}; \underline{\sigma}) = e^{-R_w^{(i,j)}(\underline{\sigma})} e(S_w^{(i,j)}(\underline{\alpha})),$$

$$S_w^{(i,j)}(\underline{\alpha}) = \alpha_{i,j-1}\omega + \alpha_{i,j}w_1 + \alpha_{i,j+1}w_2 + \ldots + \alpha_{i,j+\ell-1}w_\ell + \alpha_{i,j+\ell}\omega$$

$$+\alpha_{i-1,j}\zeta + \alpha_{i-1,j+1}\zeta + \ldots + \alpha_{i-1,j+\ell-1}\zeta$$
$$+\alpha_{i+1,j}\zeta + \alpha_{i+1,j+1}\zeta + \ldots + \alpha_{i+1,j+\ell-1}\zeta,$$

$$R_w^{(i,j)}(\underline{\sigma}) = \sigma_{i,j-1} + \sigma_{i,j} + \sigma_{i,j+1} + \ldots + \sigma_{i,j+\ell-1} + \sigma_{i,j+\ell}$$

$$+\sigma_{i-1,j} + \sigma_{i-1,j+1} + \ldots + \sigma_{i-1,j+\ell-1}$$
$$+\sigma_{i+1,j} + \sigma_{i+1,j+1} + \ldots + \sigma_{i+1,j+\ell-1}$$

and

$$G_{w'}^{(i,j)}(\underline{\alpha};\underline{\sigma})$$

is defined analogously for the placement of the word w in a vertical manner.

Our convention is that if any (i,j) grid reference is off the grid, i.e. either $i \notin \{1, 2, \ldots r\}$ or $j \notin \{1, 2, \ldots, c\}$, then it is ignored, i.e. we do not place anything there.

To facilitate a numerical result, we let $L = |\mathcal{L}|$, the number of words in our lexicon, and ascribe to the k-th letter in $\overline{\mathcal{A}}$ the integer value $1 + (L + 1)^{(k-1)}$. For example, if we enumerate our augmented alphabet as

$$a, b, c, \ldots, z, a', b', c', \ldots, \zeta, \omega,$$

then k is an integer in the set $\{1, 2, \ldots, 54\}$ and the letters a, b, c and a' are ascribed the values 1, $L+2$, $1+(L+1)^2$ and $1+(L+1)^{26}$ respectively. Ascribing the letters in our augmented alphabet to positive integers in this way ensures that we can identify invalidly placed words in the final crossword puzzle configuration.

An exact formula for the number of solutions to our crossword puzzle is

$$1 + |\mathcal{S}| = h(\underline{\sigma}) \prod_{w \in \mathcal{L}} g_w(\underline{\sigma}) \int_{\underline{\alpha} \in I^{rc}} \left\{ \prod_{w \in \mathcal{L}} r_w(\underline{\alpha};\underline{\sigma}) \right\} \times \phi(\underline{\alpha};\underline{\sigma})d\underline{\alpha}, \qquad (2)$$

where the validation function Φ is given by

$$\Phi(\underline{\alpha};\underline{\sigma}) = \prod_{(i,j) \in \mathcal{G}} \Phi^{(i,j)}(\alpha_{i,j};\sigma_{i,j}) \qquad (3)$$

and where

$$\Phi^{(i,j)}(\alpha_{i,j};\sigma_{i,j}) = 1 + \sum_{a \in \mathcal{A}} e^{\sigma_{i,j}}e(-\alpha_{i,j}a) + \sum_{a \in \mathcal{A}} e^{\sigma_{i,j}}e(-\alpha_{i,j}a') \qquad (4)$$

$$+\sum_{k=0}^{4} \sum_{(a,a') \in \Delta(\mathcal{A} \times \mathcal{A}')} e^{(k+2)\sigma_{i,j}}e(-\alpha_{i,j}(a + a' + k\zeta)),$$

$$g_w(\underline{\sigma}) = 1 + \sum_{(i,j)} e^{-R_w^{(i,j)}(\underline{\sigma})} + \sum_{(i,j)} e^{-R_{w'}^{(i,j)}(\underline{\sigma})} \tag{5}$$

$$r_w(\underline{\alpha};\underline{\sigma}) = G_w(\underline{\alpha};\underline{\sigma})/g_w(\underline{\sigma}), \tag{6}$$

$$h_{i,j}(\underline{\sigma}) = 1 + 2\sum_{a \in \mathcal{A}} e^{\sigma_{i,j}} + \sum_{k=0}^{4}\sum_{a \in \mathcal{A}} e^{(k+2)\sigma_{i,j}} \tag{7}$$

$$h(\underline{\sigma}) = \prod_{(i,j)\in\mathcal{G}} h_{i,j}(\sigma_{i,j}), \tag{8}$$

and

$$\phi(\underline{\alpha};\underline{\sigma}) = \Phi(\underline{\alpha};\underline{\sigma})/h(\underline{\sigma}). \tag{9}$$

Here the summation over $(a, a') \in \Delta(\mathcal{A} \times \mathcal{A}')$ indicates summation over the pairs of corresponding letters, $(a, a'), (b, b'), \ldots, (z, z')$.

Numerical computation of the integral in (2) is intractable and this motivates finding an asymptotic formula for the multi-dimensional integral which, under suitable conditions on the lexicon, provides a good approximation to our exact formula.

Taking the logarithms of $r_w(\underline{\alpha};\underline{\sigma})$ and $\phi(\underline{\alpha};\underline{\sigma})$, and expanding as power series in $\underline{\alpha}$ we obtain

$$\log\phi(\underline{\alpha};\underline{\sigma}) + \sum_w \log r_w(\underline{\alpha};\underline{\sigma}) = A_0 + 2\pi i\, A_1\underline{\alpha} + \frac{1}{2}(2\pi i)^2\,\underline{\alpha}^\top A_2\underline{\alpha} + O(\|\underline{\alpha}\|^3). \tag{10}$$

We find that A_0 vanishes and if by judicious choices of $\sigma_{i,j}$, for $(i,j) \in \mathcal{G}$, we can make the coefficient A_1 vanish, then we might hope that (2) can be estimated as

$$1 + |\mathcal{S}| = h(\underline{\sigma}) \prod_{w\in\mathcal{L}} g_w(\underline{\sigma}) \int_{\underline{\alpha}\in I^{rc}} \exp\left\{-2\pi^2\underline{\alpha}^T A_2\underline{\alpha}\right\}d\underline{\alpha} \times \left(1 + o(1)\right) \tag{11}$$

$$= h(\underline{\sigma}) \prod_{w\in\mathcal{L}} g_w(\underline{\sigma})\frac{1}{\sqrt{(2\pi)^{rc}|A_2|}} \times \left(1 + o(1)\right).$$

This motivates our main theorem.

Theorem 1. *Let A_1 and A_2 be given in (10) and let $\underline{\sigma}$ be the solution to the equation $A_1 = 0$. Then as the size of the lexicon becomes larger,*

$$1 + |\mathcal{S}| = h(\underline{\sigma}) \prod_{w\in\mathcal{L}} g_w(\underline{\sigma})\frac{1}{\sqrt{(2\pi)^{rc}|A_2|}} \times \left(1 + o(1)\right). \tag{12}$$

The proof of this theorem is too detailed for this paper, but follows in a logical form from (1), (5) and (6) we have

Furthermore, other relevant crossword puzzle statistics can be estimated in a similar manner, provided the estimations are valid. To prove validity, we rely on certain assumptions regarding the distributions of letters and words in the lexicon, which are satisfied for AELs.

The use of an artificially constructed character-based lexicon was introduced into NP-Hard word puzzles (see Harris and Forster (1993)) as the properties of such lexicons are logically dictated. Such purpose-built lexicons permit insight into the structural characteristics of these combinatorial NP-C word problems. A brute force implementation was developed to both count the number of solutions and to record the optimal value. The results for one example are shown in Figs. 1 and 2. Figure 1 shows the empirical frequencies plotted over the fitted normal distribution, whereas Fig. 2 shows frequencies that are smoothed using a symmetric 11-point moving average.

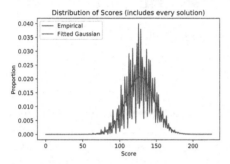
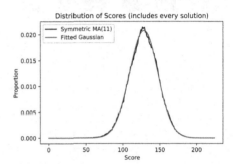

Fig. 1. Frequencies of crossword puzzle scores.

Fig. 2. Moving averages of frequencies of crossword puzzle scores.

3   Results and Discussion

A brute force implementation was used to generate a sample of multi-lingual crossword puzzles. The themed lexicons were English words from the Grade 2 Japanese Middle School textbook called New Horizon. The Japanese-themed lexicon consisted of the kana for the same words. Note that kanji, as provided in New Horizon, were used for the clues. Each lexicon comprised around 130 words. A 12 × 12 grid was chosen as the puzzle size and the implementation allowed to run for five minutes on each lexicon. The scoring was set arbitrarily at ten points per word and one point for each interlocking character. The reason for this weighting was to attempt to insert as many words as possible into the grid. As can be seen in Fig. 3, the highest scoring solution found by the implementation using the English lexicon in the time limit had 23 words. However, use of the Japanese lexicon, and the same implementation (modified for the different symbols) was able to insert 53 words. Interestingly, and perhaps not surprising, the density of unfilled cells (the black squares) was 32% for the Japanese crossword puzzle, but 47% for the English-based puzzle.

Fig. 3. Examples crossword puzzle solutions

From a structural perspective, it can be seen that the Japanese lexicon consisted of words of shorter syllabic length than the number of characters in the corresponding themed English lexicon. This by itself does not explain the difference in the density. It is therefore argued that the distribution of the syllables in the Japanese lexicon was more uniform, allowing a higher probability of finding interlocking syllables than the distribution of characters in the English lexicon, resulting in the fewer words and fewer interlocking positions shown in Fig. 3.

These results demonstrate how these themed lexicons can be used in an educational setting in a bilingual English/Japanese class. The addition of these clues, while manually demonstrated in Fig. 4, could be easily automated allowing educators a potentially useful tool in the classroom.

Fig. 4. Sample puzzle with Japanese language clues

4   Conclusion and Future Research

This paper examined the NP-Hard problem of constructing themed Japanese crossword puzzles subjected to an arbitrary objective scoring function. The derivation of a limit theorem applicable to such combinatorial problems has been presented and an expression derived for the number of solutions based upon assumptions relating to the distribution of characters in the supplied lexicon. Some results from a brute force implementation to generate such puzzles automatically were presented and a graph showing the distribution of the values of the objective function has shown a remarkable fit to a normal distribution. Future research will involve modification of this expression to determine the likely highest scoring solution. Integration of an algorithm for this modification into the brute force generation may enable a branch-and-bound style trimming of the search space to enable the location of the highest scoring solution in real time.

References

Agarwal, C., Joshi, R.K.: Automation strategies for unconstrained crossword puzzle generation. arXiv preprint (2020). https://doi.org/10.48550/arxiv.2007.04663

Binkley, D., Kuhn, B.: Crozzle: an NP-complete problem, pp. 30–34 (1997). https://doi.org/10.1145/331697.331705

Forster, J., Harris, G., Smith, P.: The Crozzle - a problem for automation. In: Proceedings of the 1992 ACM/SIGAPP Symposium on Applied Computing: Technological Challenges of the 1990s, pp. 110–115 (1992)

Gower, M., Wilkerson, R.: R-by-C Crozzle: an NP-hard problem. In: Proceedings of the 1996 ACM Symposium on Applied Computing, pp. 73–76 (1996)

Harris, G., Forster, J.: Additional papers: On the number of solutions, S(k, n), to a class of crossword puzzle. Comput. J. (1992)

Harris, G., Forster, J.: Automation of the Crozzle. Aust. Comput. J. 25(2), 41–48 (1993)

Harris, G., Forster, J., Rankin, R.: Basic blocks in unconstrained crossword puzzles, pp. 257–262 (1993). https://doi.org/10.1145/162754.162892

Non Immersive Virtual Laboratory Applied to Robotics Arms

Daniela A. Bastidas[1(✉)], Luis F. Recalde[2], Patricia N. Constante[1],
Victor H. Andaluz[1,2(✉)], Dayana E. Gallegos[1], and José Varela-Aldás[2]

[1] Universidad de las Fuerzas Armadas ESPE, Sangolqui, Ecuador
{dabastidas2,pnconstante,vhandaluz1,degallegos1}@espe.edu.ec
[2] SISAu Research Group, Universidad Tecnológica Indoamérica, Ambato, Ecuador
{fernandorecalde,victorandaluz,josevarela}@uti.edu.ec

Abstract. This article presents a non immersive virtual laboratory to emulate the behavior of Mitsubishi Melfa RV 2SDB robotic arm, allowing students and users to acquire skills and experience related to real robot, augmenting the access and learning of robotics in Universidad de las Fuerzas Armadas (ESPE). It was developed using the mathematical model of the robotic arm, thus defining the parameters for the virtual recreation. The environment, interaction and behavior of robotic arm was developed in a graphic engine (Unity3D) to emulate learning tasks such in a robotics's laboratory. In the virtual system, 4 inputs were development for the movement of the robot arm, further to program the robot a user interface was created where the user selects the trajectory such as point to point, line, arc or circle. Finally the hypothesis of the industrial robotic learning process is validated through the level of knowledge acquired after using the system.

Keywords: Virtual learning · Robot arm · Non-immersive reality

1 Introduction

Since the last years robotics have been a field of huge interest, producing relevant works in the research and educational field [2,4,10] due robotics is a interdisciplinary field that grows faster and it is necessary to merge the knowledge of different areas, given that the users spend a great number of hours doing laboratory exercises to achieve effective learning of new skills and experience [5,7]. The exercises mean physical experiments with real robots, some of the problems are the infrastructure and equipment that are really expensive and insufficient to be used by many users at the same time. [3].

Due the great advance of new technologies such tele-operation, internet and virtual environments is possible to emulate the real process and the physical infrastructure, so the creation of virtual laboratories could be a effective solution to increase the accessibility of this kind of systems [8], e.g.; The simulation of

© Springer Nature Switzerland AG 2022
H. Fujita et al. (Eds.): IEA/AIE 2022, LNAI 13343, pp. 898–906, 2022.
https://doi.org/10.1007/978-3-031-08530-7_76

simple robots to educational equipment using Unity 3D [9] and finally developing virtual environments to the industrial field [6].

In this context it is important to have a space in which the users development exercises, acquire new skills and a self-evaluate knowledge.

With all the above mentioned this paper presents the implementation of a non-immersive virtual laboratory in UNITY 3D that emulates the movements generated by the Mitsubishi Melfa 2SDB industrial robot, in order to develop a space for experimentation and testing where students and professionals can reinforce and acquire new skills in the field of industrial robotics. An interactive user interface was developed to manipulate the robot's movements, as well as to program different trajectories such as points, lines, arcs or circles. Finally checking the movements with the real robot through the CR1DA - 700 controller with TCP/IP communication and the user's learning process is validated by measuring the knowledge acquired after using the virtual system.

The remainder of this paper is structured as follows, Sect. 2 presents formulation of the problem. Section 3 shows the development process of the virtual laboratory, Sect. 4 presents the analysis of results and Sect. 5 concludes of the work.

2   Problem Structure

About 3 million industrial robots operate worldwide, approximately 435000 robot units are shipped to industries in 2021 according to Word Robotics 2021 report. [1].

For manufacturers who have these types of robots, a limitation exist in the search for qualified personnel, since the training process involves stoppages and costs in production, increased operating times and possible occupational hazards. Becoming an important point where education centers project their students to obtain such knowledge and skills. Although the university has laboratories suitable for this type of work, the constant advance of robotics in the industrial field generates an increase in the number of students and users in this area.

Both the laboratory space and the robots are not sufficient for optimal learning, since it requires practice time, given that class hours are not appropriate, In view of the above, the institutions require a high investment for the maintenance and development of these systems. Virtual reality allows simulations where one or more users, whether students or professionals, can understand and practice different processes through a virtual environment. Furthermore test new methods in order to save resources, eliminating delays in production and risks that endanger the safety of personnel or losses in the company or institution.

3   Development

3.1   Mathematical Model

This section presents the mathematical model of Mitsubishi Melfa RV 2SDB robotic arm shown in Fig. 1, where the position of the end-effector is given by the vector by $\boldsymbol{\eta} \in \mathbf{R}^n$ with $n = 3$ and the control space is defined $\mathbf{q} \in \mathbf{R}^m$ considering $m = 6$ such the number of actuators in the system. The instantaneous kinematic model of the manipulator gives the variations of point of interest $\dot{\boldsymbol{\eta}} = \frac{\partial f(\mathbf{q})}{\partial \mathbf{q}} \dot{\mathbf{q}}$ respect frame $< \mathcal{R} >$, considering $\dot{\boldsymbol{\eta}} = \begin{bmatrix} \dot{\eta}_x & \dot{\eta}_y & \dot{\eta}_z \end{bmatrix}$ is the derivative of the position and $\dot{\mathbf{q}}$ is the control vector defined by velocities of actuators.

$$\dot{\boldsymbol{\eta}}(t) = \mathbf{J}(\mathbf{q})\dot{\mathbf{q}}(t) \tag{1}$$

The movement of the robot arm is defined in (1), where $\mathbf{J}(\mathbf{q}) \in \mathbf{R}^{nxm}$ is the jacobian matrix that allows a linear mapping between control and operational space $\dot{\mathbf{q}} \to \dot{\boldsymbol{\eta}}$, it was used to emulate the behavior of the system in the virtual laboratory considering the forward and inverse kinematics.

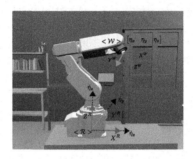

Fig. 1. Mitsubishi Melfa RV 2SDB robotic arm.

3.2   Control Scheme

This section presents the control law, the trajectory will be generated by the user and the control algorithm will guide the robot arms to the desired task. Due the manipulator has more degrees of freedom in the control than operational space $m > n$, it is possible to use the redundancy of the system to avoid singularities kinematic configurations.

With all the above the control law is defined in (2), where $\mathbf{q_0}$ is a arbitrary vector to avoid singularities during the execution of the tasks, $(\mathbf{I} - \mathbf{J}^{\#}\mathbf{J})$ considers the projection of the null space $\mathbf{q_0}$. The vector of control errors is defined in $\tilde{\boldsymbol{\eta}} = \boldsymbol{\eta}_d - \boldsymbol{\eta}$, where $\boldsymbol{\eta}_d = \begin{bmatrix} \eta_{x_d} & \eta_{y_d} & \eta_{z_d} \end{bmatrix}$ is the vector of desired positions of the end-effector, $\boldsymbol{\eta}_d = \begin{bmatrix} \eta_x & \eta_y & \eta_z \end{bmatrix}$ is the end-effector position, $\mathbf{K1}, \mathbf{K_2}, \mathbf{K_3}$ and $\mathbf{K_4}$ are constant matrices that modified the control actions in a smooth or aggressive way and finally $\mathbf{J}^{\#} = \mathbf{J}^{\intercal}(\mathbf{JJ}^{\intercal})^{-1}$ is the minimal norm solution by right side, guaranteeing the small number of movements of the manipulator.

$$\begin{aligned} \dot{\mathbf{q}}_c = \mathbf{J}^{\#}(\dot{\boldsymbol{\eta}}_d + \mathbf{K_2}\tanh(\mathbf{K_2}^{-1}\mathbf{K_1}\tilde{\boldsymbol{\eta}})) \\ + (\mathbf{I}_{6x6} - \mathbf{J}^{\#}\mathbf{J})\mathbf{K_3}\tanh(\mathbf{K_3}^{-1}\mathbf{K_4}\mathbf{q_0}) \end{aligned} \tag{2}$$

3.3 Environment Development

The virtual environment is shown in Fig. 2, the objects entered are 3D models in ".fbx" format, these are exported to the Unity video game engine and finally post-processing elements are added to the scene for higher image quality. The virtual classroom was designed in Sketch Up software and the Mitsubishi Melfa RV-2SDB robot model was obtained from the official Mitsubishi electric website.

The parent-child hierarchy allows the independence of each joint of the robot, by creating "GameObject" according to the positions set by the Denavit-Hartenberg (DH) parameters.

Fig. 2. Mitsubishi Melfa RV-2SDB user interface

3.4 User Interaction with the Virtual Environment

The user interface (UI) comprises a visual support that provides the operator's actions in the simulator, Fig. 2 shows the foreground view, where the user has the following elements: (A) Main camera intended for the front view of the user as a close-up.; (B) Secondary camera intended for the top view of the robot; (C) Secondary camera intended for the right side view of the robot ; (D) Forward or inverse kinematic panel of the robot, the states change panel that modifies the robot's joints or the end effector position. (E) Trajectory panel where the user generates the coordinates to simulate the movement of the robot and project on the physical arm; (F) Initial command system, the user has access to the home position of the robotic arm, connect the Wii Motion Plus Control (WMP), the instructions of the simulator operation and finally the start menu and exit the game.

The non-immersive system aims to communicate the virtual world through of internal elements (Fig. 3A. Sliders and Fig. 3B. Gizmo) and external elements (Fig. 3C. Keyboard and in Fig. 3D. WMP made as a "teach pendal") the use of this device was obtained from a study that guaranteed portability, communication via bluetooth and compatibility with the game Engine, provides interaction with the simulator, to increased user comfort. This work used external library, that allows the measurements from accelerometer data and generate the desired movements to the robot.

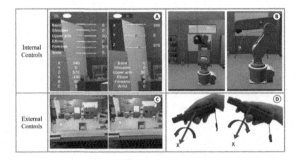

Fig. 3. Simulator input types.

3.5   System Structure

The System structure is divided into two main stages, the first one is the implementation of the non-immersive virtual reality system and the second one is the communication with the controller of the RV - 2SDB robotic arm, so that it can execute the movements of the real robot from the data obtained from the simulator, in Fig. 4 the general operation is represented.

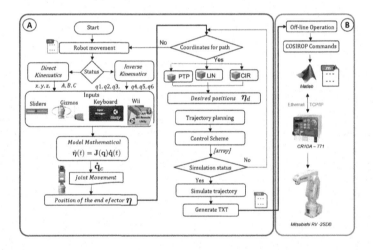

Fig. 4. Operation system flow chart.

The first part is shown in Fig. 4A. that presents the scripting programmed C# in Unity 3D. The simulator must interpret the states of the robot: (1) Direct Kinematic State (DK) modifies the joint angles of the robot to find the position and orientation of the end effector.; (2) Inverse kinematic State (IK) which sends the position data of the end-effector $\begin{bmatrix} \eta_x & \eta_y & \eta_z \end{bmatrix}$ of the robot, configuring the inverse control parameters in closed loop to find the joints of the system.; (3) Simulation state happens when the user has the trajectories as shown in the

Fig. 5 , these values are $\begin{bmatrix} \eta_x & \eta_y & \eta_z \end{bmatrix}$ which are registered to be projected by the system controller 2.

The programming of the Mitsubishi robot is handled from instructions that contain the information about the type of movement to be performed from an initial point $\begin{bmatrix} \eta_x & \eta_y & \eta_z \end{bmatrix}$ to a desired position $\begin{bmatrix} \eta_{x_d} & \eta_{y_d} & \eta_{z_d} \end{bmatrix}$. The trajectories used are: (1) PTP the shortest trajectory between 2 points starting point to end point (η_0, η_d); (2) LIN is the continuous motion from η_0 to η_d, the trajectory was implemented through the vector equation of the given line $\eta_d = (\eta_0, \vec{v})$;(3) CIR the trajectory is described from initial point, midpoint and end point (η_0, η_m ,η_d), through the general equation and the parametric equation of the circumference. These elements will appear as instances when the user wants to project them, as shown in Fig. 5.

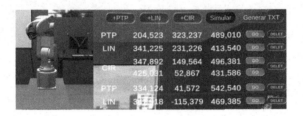

Fig. 5. Panel of trajectories, coordinates desired by the user.

The second stage Fig. 4B comprises the reading of the trajectory simulate and projected to the physical robot. To do this, the program is compiled in the Matlab software that receives such data, the communication and established through the TCP/IP protocol to the CR1DA - 700 controller.

4   Analysis of Results

This section shows the tests performed to analyze the correct operation of the virtual laboratory based on the precision between the Mitsubishi Melfa RV-2SDB and the developed system. In addition for the validation of learning hypotheses and acquisition of new skills of the users, the level of knowledge was measured before and after using the system.

4.1   Accuracy Virtual Laboratory

This work analyzes the precision with the following procedure: (1) Perform trajectories such circles in the simulator; (2) Perform the same trajectories using the "Teach Pendant" of the RV-2SDB robotic arm; (3) Compare the results generated by the virtual laboratory and the controller of the RV 2SDB robotic arm controller.

Circle Movements Accuracy. A 200 [mm] diameter arc was generated, considering the desired position over the plane $< \mathcal{Y} - \mathcal{Z} >$. The trajectory of the arc was sent to the real laboratory in order to see the performance of the system, it is shown in Fig. 6A. It is necessary to use the arc trajectory in the virtual manipulator, in order to see the behaviour under this kind of trajectories, it is shown in Fig. 6B.

Fig. 6. A. Movement of real robotic arm with arc with 200 [mm] of diameter. B. Movement of real robotic arm with arc trajectory 200 [mm] of diameter.

The number of experiments is determined by the statistical test t-Student that validates the precision hypothesis of the circular trajectory and determines if the two groups differ from each other, for which the sample size is small. Due same trajectories were executed in the real and virtual system, so this work compares the desired states of both systems and analyzed the precision using the Mean squared error (MSE), they are shown in Table 1.

The Table 1 shows the results, hence the systems have similar states during the control process, guaranteeing a good performance of the virtual laboratory.

Table 1. Mean squared error between both systems using Arc trajectory

Experiment	MSE	Experiment	MSE
1	0.50	4	0.13
2	0.24	5	0.06
3	0.40		

4.2   User Improvement Using the Virtual Laboratory

To proof the contribution of the system, this work uses a statistical method t-student, where the variables are the level of knowledge of robotics and the development of new skills in this kind of system. This work developed a test to measure the variables and it was applied to 2 groups of persons, one group doesn't receive the virtual laboratory and the other one receives it.

The groups are conformed by 10 persons and the scores are shown in Table 2, this work considers two hypothesis, H_0 The use of the system doesn't contribute in the learning process and development of new skill and H_1 the use of the system contribute in the learning process and learn new skills.

Table 2. User's scores

$N°$	Scores without the system	Scores with the system
1	3	8
2	2	7
3	4	8
4	3	8
5	5	7
6	5	7
7	4	8
8	5	8
9	4	8
10	2	8

After apply the t-student method, the results are the following, the critical value is $t_{vc} = \pm 2.10$ this value would be find in tables of t-student distribution with two degrees of freedom, finally the value of our work is $t = -10.09$

To reject or accept the hypothesis, the following comparison is verified considering:

$$t > t_{vc} \quad refuse\ H_0$$
$$t < t_{vc} \quad accepted\ H_0$$

. According to the analysis carried out, the statistical test value is higher, therefore, the null hypothesis H_0 is rejected , and the alternative hypothesis H_1 is accepted. The results show that the use of the non-immersive virtual laboratory system allowed to contribute to the industrial robotics learning process in users.

5   Conclusions

A robotic arm was implemented in the non-immersive virtual laboratory with the ability to generate trajectories for the user and to contribute to learning through virtual practices. The structure of the virtual-physical system was made through TCP/IP communication between the computer and the CRD1-700 controller. The present work is intended to contribute to the educational system, where the student is trained as qualified personnel showing skills in the handling of robotic arms, and in educational centers to reduce economic resources through virtual laboratories.

Acknowledgements. The authors would like to thank the Universidad de las Fuerzas Armadas ESPE; Universidad Tecnológica Indoamérica; SISAu Research Group, and the Research Group ARSI, for the support for the development of this work.

References

1. IFR presents World Robotics 2021 reports - International Federation of Robotics. https://ifr.org/ifr-press-releases/news/robot-sales-rise-again
2. Andaluz, V.H., et al.: Unity3D-MatLab simulator in real time for robotics applications. In: De Paolis, L.T., Mongelli, A. (eds.) AVR 2016. LNCS, vol. 9768, pp. 246–263. Springer, Cham (2016). https://doi.org/10.1007/978-3-319-40621-3_19
3. Candelas, F.A., Puente, S.T., Torres, F., Segarra, V., Navarrete, J.: Flexible system for simulating and tele-operating robots through the internet. J. Rob. Syst. **22**(3), 157–166 (3 2005). https://doi.org/10.1002/ROB.20056, https://onlinelibrary.wiley.com/doi/full/10.1002/rob.20056
4. Hilera, J.R., Otón, S., Martínez, J.: Aplicación de la Realidad Virtual en la enseñanza a través de Internet. Cuadernos de Documentación Multimedia **8**, 25–35 (2018). https://revistas.ucm.es/index.php/CDMU/article/view/59110
5. Kutia, V., Ruchel, F.L., Chrapek, K.: Simulation and programming of an industrial robot based on augmented reality. In: Modeling, Control and Information Technologies: Proceedings of International Scientific and Practical Conference, no. 3, pp. 184–186 (2019). https://doi.org/10.31713/MCIT.2019.59, https://itconfdoc.nuwm.edu.ua/index.php/ITConf/article/view/67
6. Lipton, J.I., Fay, A.J., Rus, D.: Baxter's homunculus: virtual reality spaces for teleoperation in manufacturing. IEEE Rob. Autom. Lett. **3**(1), 179–186 (2018). https://doi.org/10.1109/LRA.2017.2737046
7. Murhij, Y., Serebrenny, V.: An application to simulate and control industrial robot in virtual reality environment integrated with IR stereo camera sensor. IFAC-PapersOnLine **52**(25), 203–207 (2019). https://doi.org/10.1016/J.IFACOL.2019.12.473
8. Potkonjak, V., Vukobratović, M., Jovanović, K., Medenica, M.: Virtual mechatronic/robotic laboratory - a step further in distance learning. Comput. Educ. **55**(2), 465–475 (2010). https://doi.org/10.1016/J.COMPEDU.2010.02.010
9. Theofanidis, M., Sayed, S.I., Lioulemes, A., Makedon, F.: VARM: using virtual reality to program robotic manipulators. In: ACM International Conference Proceeding Series Part F128530, pp. 215–221 (2017). https://doi.org/10.1145/3056540.3056541
10. Whitney, D., Rosen, E., Ullman, D., Phillips, E., Tellex, S.: ROS reality: a virtual reality framework using consumer-grade hardware for ROS-enabled robots. In: IEEE International Conference on Intelligent Robots and Systems, pp. 5018–5025 (2018). https://doi.org/10.1109/IROS.2018.8593513

An Improved Subject-Independent Stress Detection Model Applied to Consumer-grade Wearable Devices

Van-Tu Ninh[1(✉)], Manh-Duy Nguyen[1(✉)], Sinéad Smyth[1], Minh-Triet Tran[2], Graham Healy[1], Binh T. Nguyen[2], and Cathal Gurrin[1]

[1] Dublin City University, Dublin, Ireland
tu.ninhvan@adaptcentre.ie
[2] VNU-HCM, University of Science, Ho Chi Minh City, Vietnam

Abstract. Stress is a complex issue with wide ranging physical and psychological impacts on human daily performance. Specifically, acute stress detection is becoming a valuable application in contextual human understanding. Two common approaches to training a stress detection model are subject-dependent and subject-independent training method. Although subject-dependent training method is proven to be the most accurate approach to build stress detection models, subject-independent one is a more practical and cost-efficient method, as it facilitates the deployment of stress level detection and management systems in consumer-grade wearable devices without requiring additional data for training from end-users. To improve the performance of subject-independent stress detection models, in this paper, we introduce a stress-related bio-signal processing pipeline with a simple neural network architecture using statistical features extracted from multimodal contextual sensing sources including Electrodermal Activity (EDA), Blood Volume Pulse (BVP), and Skin Temperature (ST) captured from a consumer-grade wearable device. Using our proposed model architecture, we compare the accuracy of stress detection models that use measures from each individual signal source with the one employing the fusion of multiple sensor sources. Extensive experiments on the publicly available WESAD dataset demonstrate that our proposed model outperforms conventional methods as well as providing 1.63% higher mean accuracy score compared to the state-of-the-art model while maintaining a low standard deviation. Our experiments also show that combining features from multiple sources produces more accurate predictions than using only one sensor source individually.

Keywords: Affective computing · Stress detection model · Human context · Multimodal sensing

---

V.-T. Ninh and M.-D. Nguyen—Contributed equally to this research.

H. Fujita et al. (Eds.): IEA/AIE 2022, LNAI 13343, pp. 907–919, 2022.
https://doi.org/10.1007/978-3-031-08530-7_77

1   Introduction

The development of sensor technology in recent years has lead to the availability of both consumer-grade and medical-grade wearable devices which has facilitated research into personal sensing with applications in self-quantification, lifelogging, and healthcare [11]. This has also resulted in the creation of many large multimodal personal datasets [11] comprising of different data types (e.g., passive visual capture, mobile device context, physiological data) [8] that enables research community to develop intelligent systems to track individual's health and gain more insights of an individual's personal data such as daily-life event segmentation [9], activities of daily-living identification as an indicator in health tracking systems [10], etc. Although multiple data sources are recorded in multimodal personal datasets [12], only the combination of visual and related metadata including semantic locations, daily-life activities, date and time are employed extensively in research [10,19] while others has not yet been exploited. Typically, physiological signals are usually ignored due to the limited amount of research conducted using this type of data as well as the limitations of recording devices in terms of the granularity signal measurement. Since consumer-grade wearable devices for health tracking increasingly allow the capture of real-time physiological signals (e.g. Empatica E4 wristband, Fitbit sensors, Garmin watches), researchers are now able to gather multi-sensor-source datasets as input for the study of developing automatic emotion recognition system and automatic stress detection models. Despite the possibilities, three commonly-known challenges mentioned by Gjoreski et al. [6] result in a limited amount of research in this field to-date.

There are two conventional approaches to building stress detection models, which correspond to two different training methods: subject-independent models and subject-dependent models. The hypothesis of the subject-dependent stress detection model is that the physiological response to stress stimuli is different for each individual and the stress monitoring systems need to adapt to the stress pattern individually [23]. Therefore, stress detection models are likely to perform more accurately when they are trained with each individual's data instead of using external data from various people. Nkurikiyeyezu et al. found that this hypothesis holds true by comparing the accuracy of both subject-dependent and subject-independent stress detection models trained on high-resolution EDA and ECG signals using SWELL [16] and WESAD [22] datasets [21]. Hence, most research to date has concentrated on the application of subject-dependent stress detection models while ignoring subject-independent ones despite their practicality and cost-efficiency for consumer-grade application scenarios. In order to address these issues, we investigate the usage of physiological data recorded from consumer-grade wearable devices for automatic stress detection and propose a new model that improves the accuracy of subject-independent stress detection. In summary, we present three main contributions of this paper:

1. Through extensive experiments, we prove that fusing multiple sensor sources of the consumer-grade wearable device enhances the accuracy of stress detection models compared to using each signal individually.

2. We propose a bio-signal processing pipeline with a novel training method for the subject-independent models that learn stress/non-stress patterns of EDA, BVP, ST, and their fusion.
3. Our proposed model outperforms traditional Machine Learning methods as well as being 1.63% more accurate than the state-of-the-art model on the same experiment dataset.

2   Related Work

Various human contextual data sources, such as Heart rate (HR), Heart Rate Variability (HRV), and Electrodermal Activity (EDA) are found to be discriminative signals for stress level measurement [1]. Using such multi-modal contextual signals, Nkurikiyeyezu et al., Schmidt et al., and Siirtola manage to build high-accuracy subject-dependent stress detection models [21,22,25]. Results of these works show that subject-dependent models outperformed subject-independent ones and the gap between the performance of the two models is huge.

In 2018, Schmidt et al. released a public multimodal dataset named WESAD which captured both high-resolution and low-resolution physiological contextual signals of 15 participants under different conditions [22]. They also provided preliminary work on their dataset by training a subject-independent stress detection model. In a binary classification task, they achieved an average accuracy score of 88.33% (0.25) using Random Forest classifiers trained on combinations of low-resolution sensor signals (EDA, BVP, TEMP). However, as the number of stress and non-stress samples in the WESAD dataset is unequal, this accuracy score cannot reflect the stress detection capability of the model completely as the model can achieve a high accuracy score by predicting the value of the majority class for all predictions.

Siirtola continued to evaluate the performance of subject-independent stress detection models using the same features as in the preliminary work of Schmidt et al. but with another appropriate evaluation metric and different window size [24]. The best model using Linear Discriminant Analysis (LDA) trained on three signals which include Skin Temperature (ST), BVP, and HR achieves the highest average balanced accuracy score of 87.4% (10.4). This result is high for subject-independent stress detection model. However, they suggested that the significant variation in recognition accuracy between study subjects can be alleviated by building subject-dependent stress detection instead.

In 2019, Nkurikiyeyezu et al. compared the performance of these two models using high-resolution EDA and HRV signals from both WESAD [22] and SWELL [16] dataset [21]. The accuracy score was chosen as the appropriate evaluation metric for their works since they down-sampled the dataset randomly to balance the number of samples in both classes. They provided evidence that the subject-dependent stress detection model outperformed the subject-independent one. They also proposed a hybrid calibrated model to improve the performance of the subject-independent model from 42.5% ± 19.9% to 95.2% ± 0.5% by including a small number of samples of the unseen subject ($n = 100$) [21]. However, their proposed hybrid calibrated model is not an enhanced version of the subject-independent model but a different way for subject-dependent model training as it requires a small amount of stress/non-stress annotated samples from the targeted user. Additionally, their work was only limited to the use of high-resolution signals, which are usually recorded using laboratory devices only, without analysing the performance of their models on low-resolution signals captured from consumer-grade wearable devices. More work is needed to analyse the possibility of improving the performance of the subject-independent stress detection model trained on low-resolution physiological signals.

3   Stress Detection Dataset

To improve the accuracy of the subject-independent stress detection model, we conducted experiments on the benchmarking dataset that is used extensively in related works [13,21,24]. The benchmarking dataset named WESAD [22] consists of four different types of low-resolution physiological data collected from 15 participants under two different study protocols in a laboratory environment. The low-resolution physiological signals including accelerometer (ACC), skin temperature (ST), Blood Volume Pulse (BVP), and Electrodermal Activity (EDA) are recorded using the Empatica E4 medical-grade wearable sensor, which facilitates real-time physiological data acquisition regardless of user context. Among the four signals, only three bio-signals which are related directly to the response of acute stress includes EDA, BVP, and ST. In our work, we concentrate on analysing the use of low-resolution EDA, BVP, and ST signals, which are recorded with a sampling rate 4 Hz, 64 Hz, 4 Hz respectively, to improve the stress prediction accuracy of a subject-independent model. Each study protocol in the dataset comprises of amusement, stress, meditation, and baseline conditions in different orders for each participant. However, only the amusement, stress, and baseline conditions are used to build and evaluate stress detection models [22].

Details of these three affective conditions are as follows:

1. **Baseline Condition**: This condition lasts for 20 min which aims to capture the neutral state of the participant. The participant is asked to sit or stand at a table with neutral reading material.
2. **Amusement Condition**: The participant watches a set of eleven funny video clips. A short neutral time period of five seconds is presented between the video clips. The total length for this condition is 392 s.
3. **Stress Condition**: The participant is exposed to the Trier Social Stress Test (TSST), where they are required to provide a five-minute speech on their strengths and weaknesses in front of a panel of three human resource specialists. Finally, the participant counts down from 2023 in decrements of 17, and is requested to start over if they make a mistake. The total length of this condition is about 10 min.

The total duration of the study protocol is about two hours, which is considered to be long enough to capture sufficient physiological data to train a stress detection model. Since previous works on this dataset employ study-protocol as the ground-truth of both train and test data [20–22,24], we also use the same ground-truth construction method as in previous works for consistent comparison of the results. In detail, the baseline and amusement condition are classified into non-stress class while the stress condition is considered as the stress one.

4 Experiments Description

In this research, we employ the bio-signals of the WESAD dataset, EDA, BVP, and ST signals, that can be recorded from separate sensors integrated on a low-cost consumer-grade wearable device to predict the stress pattern of an individual. We also propose a bio-signal processing pipeline for each signal individually before extracting statistical features, which is described in 4.1. Several statistical features identified in other researcher's findings are extracted from these signals, and then concatenated together to build our prediction models. We conduct many experiments with different training approaches to evaluate the effectiveness of combining features of multiple sensors in subject-independent models.

4.1 Bio-signal Processing and Statistical Feature Extraction of EDA, BVP, and ST

For both EDA and BVP, we extract statistical features using NeuroKit2 package[1] [17] and HRV-analysis library[2] for each 60-s segment. The window shift used in our experiment is 0.25 s. The values of the window size and window shift are the same as in the original paper of WESAD dataset for consistency when comparing the prediction results of the models [22]. As the physiological signals vary from person to person, we employ feature normalisation method to reduce the difference people's physiological responses. In addition, since the signals recorded using consumer-grade wearable device such as EDA, BVP, etc. contain many types of noise, we utilise different signal processing techniques to remove noises, baseline drifts, and outliers in the raw signal. These steps are combined together to clean the raw signal before extracting statistical feature, which is considered to be a bio-signal processing pipeline to improve the quality of the extracted feature.

For the EDA, the raw signal in each 60-s segment is firstly pre-processed to remove motion artifacts using the wavelet-based adaptive denoising procedure as described in [2]. The signal is then filtered by a fourth-order Butterworth low-pass filter with cut off frequency of 0.5 Hz to remove line noise. The min-max normalization is then applied to the cleaned signal to remove the inter-individual difference before it is inputted into the NeuroKit2 package for Skin Conductance

---

[1] https://github.com/neuropsychology/NeuroKit.
[2] https://github.com/Aura-healthcare/hrv-analysis.

Response (SCR) and Skin Conductance Level (SCL) decomposition using the cvxEDA method [7]. Other characteristics of SCR including SCR Peaks, SCR Onsets, and SCR Amplitude are also extracted. Finally, the statistical EDA features from three related works [3,21,22] are computed, which result in a 36-dimensional vector.

For the BVP, we firstly clean the raw signal in each window segment by removing the outlier values over the 98th and below the 2th percentile using winsorisation method as in [5] and removing the baseline drift using Butterworth high-pass filter with cut-off frequency of 0.5 Hz as in [14]. We then apply min-max normalization to the cleaned signal to minimise the physiological signal difference between individuals before following the previous research [20] to employ the Elgandi processing pipleine [4] for the photoplethysmogram (PPG) signal clearning [18] and the systolic peaks detection. The systolic peaks are used to compute a list of RR-intervals, which are then pre-processed using the hrv-analysis package to remove outliers and ectopic beats [27] as well as interpolating missing values. The cleaned RR-intervals are used to compute the NN-intervals, which are main items to compute time-domain, frequency-domain, geometrical, and Poincare-plot features. For frequency-domain HRV features, we employ the same parameters of low (LF: 0.04–0.15 Hz) and high (HF: 0.15–0.4 Hz) frequency bands as in [22]. The range of very-low frequency band used in our work is the same as in HRV-analysis package (0.003–0.04 Hz). In summary, we inherit most of the HRV features from [21,22] and combine them into a 30-dimensional vector.

For the ST, the statistical features are extracted on the raw 60-second segment signal as in [22]. The fusion of statistical features from three signal-sources is a 72-dimensional vector. The detail of extracted features is shown in Table 1.

4.2   Stress Detection Model Training Methodology

In this research, we build different classifiers to detect the stress condition of each participant in the WESAD dataset. Two conventional machine learning classifiers which are widely applied in this field – Random Forest (RF) and Support Vector Machine (SVM) – are applied as baseline models using our proposed feature extraction pipeline. The feature used for these machine learning models is either a feature vector combined from three signals (dimension of 72) or a feature vector of each signal only (dimension of 30 for BVP, 36 for EDA, and 6 for ST). Additionally, we introduce a neural network (NN) architecture that captures not only the local detail of EDA, BVP, and ST separately but also the fusion of these signals. The neural network model, as depicted in Fig. 1, contains three distinct embedding modules for each signal and a concatenating layer to learn the joint encoded features. The model is then added with three different classification layers for three branches which aims to optimise the performance of embedding stages of EDA, BVP, and ST signals prior to the concatenating step. The overall loss used to train the NN model is the sum of losses of all branches. In the testing phase, the NN model makes a prediction based on the average of all branches in order to gather the detail of each signal and their combined

Table 1. List of extracted features. Abbreviations: $\#$ = number of, $\sum$ = sum of, STD = standard deviation, RMS = Root Mean Square.

	Feature	Description
EDA	μ_{EDA}, σ_{EDA}, $\min_{EDA}$, $\max_{EDA}$	Mean, STD, min, max of the EDA
	∂_{EDA}	Slope of the EDA
	range_{EDA}, range_{SCR}	Dynamic range of EDA and SCR
	μ_{SCL}, σ_{SCL}	Mean, STD of the SCL
	$\text{corr}(SCL, t)$	Correlation btw SCL and time
	$\#_{Peak}$	# identified SCR peaks
	$\sum_{SCR}^{Amp}$, $\sum_{SCR}^{t}$	$\sum$ SCR startle magnitudes and response durations
	$\int_{SCR}$	Area under the identified SCRs
	μ_{SCR}, σ_{SCR}, $\max_{SCR}$, $\min_{SCR}$	Mean, STD, min, max of the SCR
	$\mu_{\nabla_{SCR}}$, $\sigma_{\nabla_{SCR}}$, $\mu_{\nabla(\nabla_{SCR})}$, $\sigma_{\nabla(\nabla_{SCR})}$	Mean and STD of the 1st and second derivative of the SCR
	μ_{Peak}, σ_{Peak}, $\max_{Peak}$, $\min_{Peak}$	Mean, STD, min, max of SCR Peaks
	kurtosis(SCR), skewness(SCR)	Kurtosis and skewness of SCR
	μ_{Onset}, σ_{Onset}, $\max_{Onset}$, $\min_{Onset}$	Mean, STD, min, max of SCR Onsets
	$\text{ALSC} = \sum_{n=2}^{N} \sqrt{1 + (r[n] - r[n-1])^2}$	Arc length of the SCR
	$\text{INSC} = \sum_{n=1}^{N} |r[n]|$	Integral of the SCR
	$\text{APSC} = \frac{1}{N} \sum_{n=1}^{N} r[n]^2$	Normalized average power of the SCR
	$\text{RMSC} = \sqrt{\frac{1}{N} \sum_{n=1}^{N} r[n]^2}$	Normalized RMS of the SCR
BVP	μ_{HR}, σ_{HR}, μ_{HRV}, σ_{HRV}	Mean and STD of Heart Rate and HRV
	kurtosis(HRV), skewness(HRV)	Kurtosis and Skewness of HRV
	f_{HRV}^{VLF}, f_{HRV}^{LF}, f_{HRV}^{HF}	Very low (VLF), Low (LF), and High (HF) frequency band in the HRV power spectrum
	f_{HRV}^{LFNorm}, f_{HRV}^{HFNorm} $f_{HRV}^{LF/HF}$	Normalized LF and HF band power Ratio of HRV LF and HRV HF
	$\sum\limits_{x \in \{\text{VLF, LF, HF}\}} f$	$\sum$ of the freq. components in VLF-HF
	NN50, pNN50, NN20 pNN20	# and percentage of HRV intervals differing more than 50 ms and 20ms
	HTI	HRV Triangular index
	rms_{HRV}	RMS of the HRV
	SD1, SD2	Short and long-term poincare plot descriptor of HRV
	RMSSD, SDSD	RMS and STD of all interval of differences between adjacent RR intervals
	SDSD_RMSSD	Ratio of SDSD over RMSSD
	RELATIVE_RR (μ, median, σ, RMSSD, kurtosis, skewness)	Mean, median, STD, RMSSD, kurtosis, and skewness of the relative RR.
ST	μ_{ST}, σ_{ST}, $\min_{ST}$, $\max_{ST}$	Mean, STD, min, max of ST
	range_{ST}, ∂_{ST}	Range and slope of ST

information. We also integrate batch normalization and dropout techniques to make the model converge faster and to address any over-fitting concerns.

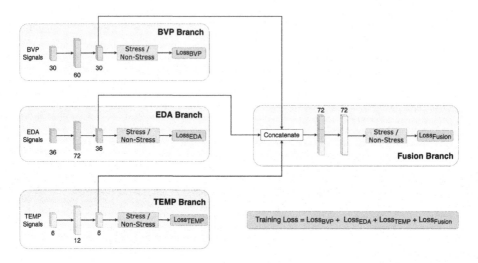

Fig. 1. The structure of our proposed neural network model. The numbers in the figure indicate the dimension of the input feature.

We train two models (RF and SVM) using the Leave-One-Subject-Out (LOSO) scheme as in [22]. For the NN model, the Leave-One-Subject-Out (LOSO) scheme is also employed, however, the data is split into train set (80%) and validation set (20%) using the Stratified Shuffle Split to deal with the imbalanced nature of the ground-truth distribution in the dataset. Additionally, the effectiveness of combining feature from multiple signals in stress detection is also considered. For each scenario described above, we run four trials either using features of each signal individually or using the fusion features from multiple sources. In the first three trials, the NN model is trained and tested only with its respective branch as illustrated in Fig. 1. By conducting these runs, we assess the effectiveness of using each signal in the stress prediction problem and evaluate if fusing signals can produce better prediction results.

4.3   Experimental Configuration

In our experiments, we set 250 trees for the RF model with enabled out-of-bag, bootstrap samples, max depth of 8, min sample splits of 2, and min sample leaves of 4. The radial basis function kernel was used in the SVM model with regularization parameter of 10. The remaining parameters of both RF and SVM model are kept default as in sklearn library[3]. Both models are set up to take the imbalance of the dataset into account by enabling balanced weights for each class when training. These are configurations that achieve the highest accuracy score after we conduct extensive experiments. The NN model is trained with an Adam optimiser [15] with a learning rate of 0.003 while the dropout level and the batch size are set at 10% and 2048 accordingly. Regarding the evaluation metrics, we

---

[3] https://scikit-learn.org/stable/.

Table 2. Comparison of the mean accuracy score between different subject-independent stress detection models in previous works and ours using biosensor signals (window size = 60 s, window shift = 0.25 s).

Sensor combinations	Schmidt [22]		Lam [13]	Proposed		
	RF	LDA	StressNAS	SVM	RF	NN
EDA+BVP+ST	88.33	86.46	92.87	92.71	91.53	**94.50**
EDA	76.29	78.08	**79.24**	76.32	75.53	77.57
BVP	84.18	85.83	81.16	86.95	87.39	**89.94**
ST	67.82	69.24	71.46	69.21	69.23	**74.07**

report both balanced accuracy and accuracy scores due to the imbalance between number of samples in the two classes in the dataset. Based on the analysis of Straube et al., balanced accuracy (BA) is both an appropriate choice and an intuitive metric to evaluate prediction results of a binary classification problem when dealing with an imbalanced dataset [26].

5 Results

In Table 2 and Table 3, we show the effect of combining statistical features from three sources of signal and compare the results of our proposed stress detection model with previous works. As can be seen from Table 2 and Table 3, all models of ours obtain higher evaluation scores when using the combining features from multiple sources of signals than using only each of them individually. According to Table 2, compared to the original work in [22], employing our proposed bio-signal processing pipeline before feature extraction step increases the mean accuracy score of conventional Machine Learning model (RF model) around 3.2%. Our proposed NN model improves the performance of state-of-the-art (SOTA) subject-independent stress detection model proposed by Lam et al. [13] around 1.63% using the same number of bio-signals. Conventional Machine Learning models also achieve competitive accuracy scores compared to the SOTA model when using the fusion feature of three sensor sources with appropriate signal processing before feature extraction. The mean accuracy scores of our SVM and RF models using the combined features as input are 92.71% and 91.53% with standard deviation of ±7.90% and ±7.24% respectively while the one of our NN model is 94.50% (±5.64%). The improvement of the NN model compared to conventional Machine Learning models comes from the difference in the final optimisation function of the NN model that takes the optimisation function of each stress detection branch trained on each signal individually into account. These results indicate that our proposed NN model not only increases the prediction accuracy of the subject-independent stress detection model in average, but it also does not result in a large difference of accuracy score between each subject's model.

Table 3. Comparison of the mean balanced accuracy score between different subject-independent stress detection models in previous work and ours using biosensor signals (window size = 120 s, window shift = 0.25 s).

Sensor combinations	Siirtola [24]			Proposed		
	RF	LDA	QDA	SVM	RF	NN
EDA+BVP+ST	81.00	78.80	81.60	93.36	93.09	**94.16**
EDA	**78.30**	73.50	69.70	71.34	70.12	77.52
BVP	81.40	81.20	67.90	87.57	**90.92**	88.28
ST	66.90	75.20	68.30	71.07	71.13	**77.97**

In terms of imbalanced-data insensitive evaluation metrics, we report the balanced accuracy scores of our models and compare them with corresponding related work [24]. For consistency in comparison the balanced accuracy score with [24], we use the window size and window shift of 120 s and 0.25 s respectively. According to the results in Table 3, our proposed NN model outperforms conventional Machine Learning approaches reported in [24]. In detail, the balanced accuracy score of our NN models is 94.16% ($\pm$6.90%), which is higher than the QDA model of [24] around 12.56% using the same number of bio-signals. In addition, our RF model achieves higher balanced score than the one in [24] approximately 12.09%, which proves that our bio-signal processing pipeline is efficient and necessary for feature extraction step to enhance subject-independent stress detection model. The balanced accuracy score of our SVM and RF model trained with features combined from different sensor sources are 93.36% ($\pm$9.14%) and 93.03 ($\pm$10.19%) respectively.

To facilitate for other future research to compare results with ours, we also report both balanced accuracy and accuracy scores of our models with different settings of window size and window shift that are not reported in Table 2 and Table 3. For window size of 120s, the accuracy scores of our SVM, RF, and NN models using combined feature are 95.23 ($\pm$5.32), 94.57 ($\pm$6.42), and 95.26 ($\pm$4.68) correspondingly For window size of 60s and window shift of 0.25 s, the balanced accuracy scores of our SVM, RF, and NN models are 90.85 ($\pm$9.99), 90.32 ($\pm$11.21), and 92.66 ($\pm$9.06) correspondingly.

6   Conclusion

In this paper, we build a model that uses a neural network (NN) architecture for subject-independent stress detection using three types of bio-signals that can be captured from a consumer-grade low-cost device. The model contains four different NN modules where each of the three modules learns the embedding features from each bio-signal individually while the remaining one learns the joint embedded feature of the three modules by concatenating the latent-space representation of each signal. The proposed model is evaluated against Random

Forest and Support Vector Machine models on the WESAD dataset using balanced accuracy and accuracy score. Our experiments show that using statistical features from multiple sensor sources can produce more accurate stress prediction results. Additionally, our experiments also show that our proposed NN model outperforms conventional Machine Learning approaches for subject-independent model training for both evaluation metrics. In detail, our NN model achieves higher evaluation scores than the SOTA model and conventional Machine Learning models while maintaining a low standard deviation score. We believe that our findings could help promote and guide future efforts in improving subject-independent stress detection models, in order to facilitate the integration of stress detection and management system into consumer-grade low-cost wearable devices in a practical and cost-efficient manner.

Acknowledgments. This publication is funded as part of Dublin City University's Research Committee and research grants from Science Foundation Ireland under grant numbers SFI/13/RC/2106, SFI/13/RC/2106_P2, and 18/CRT/6223.

References

1. Can, Y.S., Chalabianloo, N., Ekiz, D., Ersoy, C.: Continuous stress detection using wearable sensors in real life: algorithmic programming contest case study. Sensors (Basel, Switzerland) **19** (2019)
2. Chen, W.V., Jaques, N., Taylor, S., Sano, A., Fedor, S., Picard, R.W.: Wavelet-based motion artifact removal for electrodermal activity. In: 2015 37th Annual International Conference of the IEEE Engineering in Medicine and Biology Society (EMBC), pp. 6223–6226 (2015)
3. Choi, J., Ahmed, B., Gutierrez-Osuna, R.: Development and evaluation of an ambulatory stress monitor based on wearable sensors. IEEE Trans. Inf. Technol. Biomed. **16**(2), 279–286 (2011)
4. Elgendi, M., Norton, I., Brearley, M., Abbott, D., Schuurmans, D.: Systolic peak detection in acceleration photoplethysmograms measured from emergency responders in tropical conditions. PLoS One **8**(10), e76585 (2013)
5. Gjoreski, M.: Continuos Stress Monitoring using a Wrist Device and a Smartphone. Ph.D. thesis, Jožef Stefan International Postgraduate School, Ljubljana, Slovenia (09 2016)
6. Gjoreski, M., Luštrek, M., Gams, M., Gjoreski, H.: Monitoring stress with a wrist device using context. J. Biomed. Inf. **73**, 159–170 (2017). https://doi.org/10.1016/j.jbi.2017.08.006, https://www.sciencedirect.com/science/article/pii/S1532046417301855
7. Greco, A., Valenza, G., Lanata, A., Scilingo, E.P., Citi, L.: cvxeda: a convex optimization approach to electrodermal activity processing. IEEE Trans. Biomed. Eng. **63**(4), 797–804 (2016). https://doi.org/10.1109/TBME.2015.2474131
8. Gurrin, C., Albatal, R., Joho, H., Ishii, K.: A privacy by design approach to lifelogging. In: O'Hara, K., Nguyen, C., Haynes, P. (eds.) Digital Enlightenment Yearbook 2014, pp. 49–73. IOS Press, The Netherlands (2014)
9. Gurrin, C., Joho, H., Hopfgartner, F., Zhou, L., Albatal, R.: Overview of ntcir-12 lifelog task. In: NTCIR (2016)

10. Gurrin, C., et al.: Overview of the ntcir-14 lifelog-3 task. In: Proceedings of the 14th NTCIR Conference on Evaluation of Information Access Technologies (2019)
11. Gurrin, C., Smeaton, A., Doherty, A.: Lifelogging: personal big data. Found. Trends Inf. Retr. **8**, 1–125 (2014)
12. Gurrin, C., Joho, H., Hopfgartner, F., Zhou, L., Albatal, R., Healy, G., Nguyen, D.-T.D.: Experiments in lifelog organisation and retrieval at NTCIR. In: Sakai, T., Oard, D.W., Kando, N. (eds.) Evaluating Information Retrieval and Access Tasks. TIRS, vol. 43, pp. 187–203. Springer, Singapore (2021). https://doi.org/10.1007/978-981-15-5554-1_13
13. Huynh, L., Nguyen, T., Nguyen, T., Pirttikangas, S., Siirtola, P.: StressNAS: affect state and stress detection using neural architecture search, pp. 121–125. Association for Computing Machinery, New York (2021). https://doi.org/10.1145/3460418.3479320
14. Kher, R.: Signal processing techniques for removing noise from ECG signals. J. Biomed. Eng. Res. **1**, 1 (2019)
15. Kingma, D.P., Ba, J.: Adam: a method for stochastic optimization. In: Bengio, Y., LeCun, Y. (eds.) 3rd International Conference on Learning Representations, ICLR 2015, San Diego, CA, USA, 7–9 May 2015, Conference Track Proceedings (2015). http://arxiv.org/abs/1412.6980
16. Koldijk, S., Sappelli, M., Verberne, S., Neerincx, M.A., Kraaij, W.: The swell knowledge work dataset for stress and user modeling research. In: Proceedings of the 16th International Conference on Multimodal Interaction, ICMI 2014, pp. 291–298. Association for Computing Machinery, New York (2014). https://doi.org/10.1145/2663204.2663257
17. Makowski, D., et al.: Neurokit2: a python toolbox for neurophysiological signal processing. Behav. Res. Methods **53**, 1–8 (2021)
18. Nabian, M., Yin, Y., Wormwood, J., Quigley, K.S., Barrett, L.F., Ostadabbas, S.: An open-source feature extraction tool for the analysis of peripheral physiological data. IEEE J. Transl. Eng. Health Med. **6**, 1–11 (2018)
19. Ninh, V.T., et al.: Overview of imageclef lifelog 2020: lifelog moment retrieval and sport performance lifelog. In: CLEF (2020)
20. Ninh, V.T., Smyth, S., Tran, M.T., Gurrin, C.: Analysing the performance of stress detection models on consumer-grade wearable devices. In: SoMeT (2021)
21. Nkurikiyeyezu, K., Yokokubo, A., Lopez, G.: Effect of person-specific biometrics in improving generic stress predictive models. Sens. Mater. **32**, 703–722 (2020). https://doi.org/10.18494/SAM.2020.2650
22. Schmidt, P., Reiss, A., Duerichen, R., Marberger, C., Van Laerhoven, K.: Introducing wesad, a multimodal dataset for wearable stress and affect detection. In: Proceedings of the 20th ACM International Conference on Multimodal Interaction, pp. 400–408 (2018)
23. Schmidt, P., Reiss, A., Dürichen, R., Laerhoven, K.V.: Wearable affect and stress recognition: a review. ArXiv abs/1811.08854 (2018)
24. Siirtola, P.: Continuous stress detection using the sensors of commercial smartwatch. In: Adjunct Proceedings of the 2019 ACM International Joint Conference on Pervasive and Ubiquitous Computing and Proceedings of the 2019 ACM International Symposium on Wearable Computers (2019)
25. Siirtola, P., Röning, J.: Comparison of regression and classification models for user-independent and personal stress detection. Sensors (Basel, Switzerland) **20** (2020)

26. Straube, S., Krell, M.M.: How to evaluate an agent's behavior to infrequent events?-reliable performance estimation insensitive to class distribution. Front. Comput. Neurosci. **8**, 43 (2014)
27. M.arked, V.K.: Correction of the heart rate variability signal for ectopics and missing beats. In: Malik, M., Camm, A.J. (eds.) Heart Rate Variability (1995)

WDTourism: A Personalized Tourism Recommendation System Based on Semantic Web

Kaiyu Dai[1], Pengfei Ji[1], Xiaorui Zuo[2], and Daixin Dai[3](✉)

[1] Software School of Fudan University, Shanghai, China
[2] School of Economics, Fudan University, Shanghai, China
[3] College of Architecture and Urban Planning, Tongji University, Shanghai, China
daidaixin@tongji.edu.cn

Abstract. With the application of modern intelligent technologies, the smart tourism which aims to meet the personalized needs of tourists and provide high satisfaction service developed rapidly. In this paper, a personalized recommendation system for smart tourism named WDTourism was developed based on Semantic Web technology. The ontologies for scenic spots and their associations are constructed, including the information of over three hundred scenic spots in Taiwan. User portraits are calculated through three kinds of interactive behaviors between users and the system. The prototype system realized route recommendation based on spatio-temporal information and user-defined route planning. The feedbacks of the system users illustrated the effectiveness of the system.

Keywords: Smart tourism · Personalized recommendation · Semantic web · Ontology

1 Introduction

With the development and application of artificial intelligence technology, many scenic spots in China began to seek transformation and upgrading from "digital scenic spots" to "smart scenic spots". It is clearly stated in the 14th Five-year Tourism Development Plan issued by The State Council of China on December 22, 2021: "We will accelerate the development of smart tourism featuring digitalization, networking and intelligence, deepen the 'Internet plus tourism' model, and expand the application of new technologies" [1]. The main reason for developing smart tourism lies in the fact that with the deep application of Internet technology in various fields, problems such as information overload and resource confusion are becoming more and more serious. Although there are already some travel websites that provide various kinds of travel information, there are still the following major problems: Scenic spot search is mainly based on keyword retrieval, and also lack of user portrait analysis, then it is difficult to fully analyze the needs and intentions of users; The lack of semantic information about sites and their relationships makes it difficult to implement customized tour routes planning.

© Springer Nature Switzerland AG 2022
H. Fujita et al. (Eds.): IEA/AIE 2022, LNAI 13343, pp. 920–934, 2022.
https://doi.org/10.1007/978-3-031-08530-7_78

Semantic Web leverages the technologies of web and artificial intelligence, and provides promising solution to the above problems. Ontology is the key technology of Semantic Web, aiming to provide a normative description of shared concepts [2–5]. It can model the domain concepts and their properties and relationships, thus add the semantic information which support reasoning and knowledge sharing in a computer understandable way.

This paper used Semantic Web related technologies to realize a personalized tourism recommendation system named WDTourism, which is named after a research project called Wangdao. The highlight of the system is the personalized tourism planning based on user portraits and spatio-temporal relations. The rest of the paper is organized as follows: The second section gives a literature review of the field; The third section describes the overall architecture of WDTourism; The forth section discusses the realization of core functions in detail; The fifth section gives the realization and application effect of the prototype system, and the sixth section concludes the whole paper.

2   Related Works

Some research works have applied Semantic Web to the field of smart tourism. Roopa et al. constructed the ontology on tourism domain and user profiles to assist users in the planning of tourism routes [6], but the ontology model and algorithm are relatively simple, and the flexibility of personalized recommendation is limited. Olawande et al. constructed a comprehensive tourism ontology, including not only the related concepts and attributes of the scenic spots but also the tourism-related fields, such as transportation, food, festivals, destinations, etc. [7]. It has complete assistance to users in the planning of travel routes, but the user-specific information in the system is insufficient and the customization is not well supported. K.R. Ananthapadmanaban designed a comprehensive Tamilnadu Tourism Ontology to analyze user preferences and achieve personalized recommendations [8]. However, most of the user ontology requires manual filling, and the usability and user-friendliness are poor. David and others realized a set of personalized travel activity recommendation system by constructing an ontology of tourism activities, users, geography and other fields, combining with context-based and coordinating filtering recommendation algorithms [9]. The system is comprehensive and realizes personalized travel recommendation based on Semantic Web, but it can just recommend a single scenic spot, but not a tourism route consisted of several related spots. Am.A et al. presented a context-aware tourism recommendation system, which extracts a user's preferences by performing se- mantic clustering as well as sentiment analysis on their comments and reviews [10]. Feng, L Built the knowledge graph of Mount Tai scenic spot and user model, and calculated the similarity between the characteristic vectors of each attribute in Mount Tai and the attribute eigenvector of users, so as to decide whether to recommend Mount Tai scenic spot to users or not [11].

The above studies mainly focus on the matching between users and scenic spots, ignoring the association between scenic spots, such as similarities of scenic spots, etc. which makes it difficult to recommend routes with multiple scenic spots combination. In addition, spatiotemporal information is not considered. Furthermore, practical applications in this area of research are lacking. Navío-Marco et al. pointed out academic research in these disciplines is still in its infancy [12]. Lohvynenko. C., and D.

Nedbal gave the survey on the use of Semantic Web technology by Austrian the websites of regional tourism organizations. They found none of the vocabularies developed specifically for tourism were found in the examined websites [13].

In this paper, the above important factors in tourism planning are fully studied, and a personalized recommendation system for tourism routes, which supports semantic query of space and time, is implemented.

3   Overall System Architecture

The system architecture of WDTourism is shown in Fig. 1. The system is mainly divided into five modules. The main functions and design concerns of each module are described as follows:

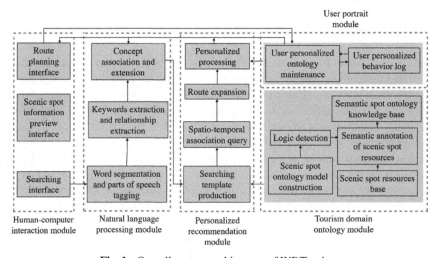

Fig. 1. Overall system architecture of WDTourism

Human-Computer Interaction Module: This module is used to realize the interaction between the user and the system, which is mainly divided into three interfaces of search, scenic spot information preview, and route planning. The searching interface provides not only navigation, but also natural language for the user to input search request, and then pass it to the pre-processing module. Scenic spot information preview interface enables users to preview the introduction of different scenic spots, comment information, etc. The route planning interface recommends travel routes to users and supports users to plan travel routes on the timeline. The interactive information, such as the time used by users to preview scenic spot information and the query statements entered by users, will be analyzed by the user personalized ontology module to maintain and update the user personalized preference model, or user portraits.

Natural Language Preprocessing Module: This module analyzes the user's retrieval request and extracts the user's intention, including word segmentation and parts of speech tagging, keyword and relation extraction, concept association expansion, with the usage of Stanford CoreNLP [14], which is a popular natural language processing software package. After extracting important verbs and nouns as keywords, Named Entity Recognition (NER) related technology is adopted to identify the corresponding concepts or instances of these words in the scenic spots ontology knowledge base. The corresponding semantic information can be obtained through SPASQL query. Finally, the concept is extended according to the user personalized ontology, and the results are passed to the intelligent retrieval layer for processing.

Personalized Recommendation Module: This module is the core of the whole system, which consists of four sub-modules: retrieval template generation, spatio-temporal associated query, route extension and personalized processing. The retrieval template generates a set of retrieval templates provided by the system that correspond to the results analyzed by the natural language preprocessing module. If the user's request contains spatial and temporal correlation information, the spatio-temporal associative query module will execute corresponding algorithms. Route expansion function utilizes associated attributes of scenic spots, such as route matching, to expand a single attraction to tourism routes. The personalized processing function filters and sorts retrieval results according to the user preference model and returns the results.

Tourism domain Ontology Module: This module is the data basis of the whole system, which mainly includes two tasks: The first is to construct the ontology model and fill scenic spots instances. Protege [15] is used to construct the tourism domain ontology, and Pellet [16] is utilized for consistency detection. The second task is to construct ontology knowledge base of scenic spots, which is done by crawling, extracting and analyzing the related information resources on the Internet, then carrying out semantic annotation under semi-supervision.

User Portrait Module: This module obtains user behavior information from human-computer interaction module, and constantly updates user preference model, which is applied to the concept association extension in the natural language preprocessing module and the personalized recommendation module described above.

4 Design and Implementation of Core Functions

4.1 The Construction of Scenic Spots Ontology

A seven-step domain ontology construction method [17] proposed by Stanford University was adopted, and the ontology is iteratively built and refined under the guidance of domain experts.

Step1: Determine the Scope and Construction Objectives of Domain Ontology.
The construction of the scenic spots ontology aims to clearly describe the specific properties like the type, activity, and location etc. of the scenic spot.

Step2: Consider Reusing Existing Ontologies. In the process of designing, the existing tourism ontology such as Harmonis-Ontology [18], Mondeca-Tourism-Ontology [19], OTA-Specification [20] was considered, and some common ontologies such as Schema.org and Microdata is reused.

Step3: Identify the Terminology for the Domain Ontology. WDTourism ontology focuses on describing scenic spots and identifies the following core concepts: accompanies (Accompanies), activities (Activity), ages (AgeGroup), scenic spot types (Attraction), destinations(Destination), and travel intentions(Purpose).

Step4: Define Classes and Hierarchies of Classes. Based on the terminologies identified in step3, and referring to Chinese Classified Thesaurus [21] and large tourism websites such as Ctrip for scenic spot classification, the classes and main hierarchical structure is shown in Fig. 2. A scenic spot can belong to multiple types. For example, Alishan scenic spot can belong to both Natural_Scenery and Local_Culture.

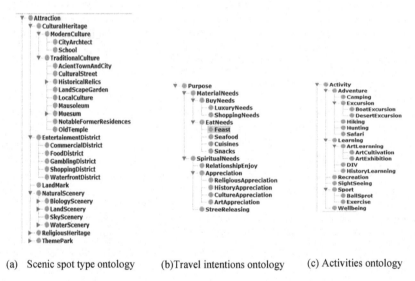

(a) Scenic spot type ontology   (b)Travel intentions ontology   (c) Activities ontology

Fig. 2. Some important ontology definitions

Step 5: Define the Related Properties of the Class and Their Relationships. The properties include two types: ObjectProperty and DataProperty. ObjectProperty is used to establish the relationship between objects, and DataProperty is used to establish the relationship between object and primitive values. For example, isNear = {Attraction; Attraction} is an ObjectProperty, indicating tow attractions is near to each other; hasStar = {Attraction; xsd:Int} is a DataProperty indicating the star rating of scenic spot.

Step 6: Define Constraints on Properties. The constraints exist between different properties. For example, there are relationship of InverseOf between locatedIn and hasAttr. HasPurpose and hasActivity do not have an intersection. The constraints also exist in properties themselves. For instances, the domain and the range of the property is Near are both Attraction, while the domain of property has Star is Attraction and its range is xsd:Int. The property is Near is Symmetry and is Similar is Transitive. These relationships and constraints play an important role in the consistency detection of ontology model and inference of knowledge.

Step 7: Create Instances of the Class and Assign the Attributes of the Instances.
The paper adopts semi-supervised method to generate ontology instances. The main flow chart is as Fig. 4 (Fig. 3):

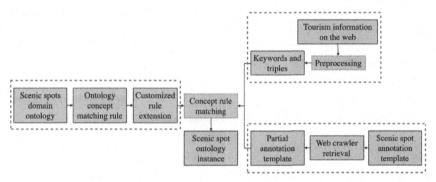

Fig. 3. Web information extraction module architecture flow chart

On one hand, the matching rules are specified manually according to the connotation of domain model. Then the paper adds customized rule extensions to improve the accuracy and recall rate of the system, which includes positive rules and negative rules. The positive rule means that if a scenic spot contains some instance concepts, it must also contain other instance concepts, such as the rule "If Attraction attr contains Camping and Hiking, then attr has Category MoutainArea". The negative rule means that if a scenic spot contains some instance concepts, it must not contain other instance concepts, such as the rule "If Attraction attr contains DesertExcursion or Adventure, then attr !suitForOldPeople". Finally, the matching rules are all expressed in the form of the main-predicate triad, because the basic RDF of OWL is essentially a triple description, such as {attraction, location, location}, {attraction, offer, event}, etc.

On the other hand, the information of some scenic spots, such as longitude and latitude, suggested travel time, can be retrieved directly from large tourism websites, such as Ctrip and Mafengwo, so some partial annotation templates can be easily generated through web crawler and templates. For massive unstructured text information, the system first preprocesses and extracts the triplets, and then identifies the instance ontology based on semantic matching. In the pre-processing stage, the system uses Stanford Deterministic Coreference Resolution System for coreference resolution, and Stanford CoreNLP for word segmentation and part of-of-speech tagging. Finally, HowNet thesaurus is called to complete the expansion of hypernym. Keyword triplet refers to the key information representation composed of <driver, driver relation, subject>, and its extraction needs dependency grammar analysis in natural language processing. This system by calling Stanford Neural Network Dependency Parser get all verbs nouns dependency grammar, and then select the words with driven relationship of nn, nsubj, conj, dep, dobj, assmod to be the candidate for semantic matching. Finally, the word similarity calculation method based on HowNet is used to match these words with the ontology vocabulary. If the matching degree exceeds the threshold, the words will be taken as the ontology instances. According to the conclusion of relevant research, the threshold value of this system is set as 0.75.

Finally, the consistency test and manual correction are carried out to form the final ontology instance of scenic spots. At this point, the ontology required by the system is completed. Taking MaoKongLanChe (Maokong Tea Garden) as an example, the instance ontology model is shown in Fig. 5, where "best For Loved Ones" means that the scenic spot is suitable for visiting with lovers, and "has Activity Nightwatch" means that the scenic spot provides night sightseeing activities. "Has Purpose Stress Releasing" means that visiting the attraction can relieve stress.

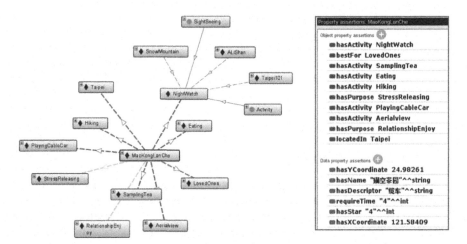

Fig. 4. Scenic spot instance ontology model of Maokong Tea Garden

4.2 Semantic Modeling of Scenic Spot Relationships

The semantic modeling of the relationship between scenic spots is key to the recommendation of tourism routes. This paper established three kinds of scenic spot relationships of adjacency, similarity, and route matching. It is relatively simple to measure adjacency by distances. Besides, if scenic spots A and B often appear together in the tourist guide information, it can be considered that they have the route matching relationship. Determination of the similarity relationships is given below.

This paper adopts the SDRPCM concept similarity calculation model proposed in literature [21], which is as follows:

$$Similar(A, B) = \theta_1 \times Relevant^{\text{hiearrchy}}(A, B)$$
$$+\theta_2 \times Relevant^{\text{anc}}(A, B)$$
$$+\theta_3 \times PathWeight(A, B) \qquad (1)$$

A and B represent the concept in ontology model respectively. Similar (A, B) represents the similarity between them, which is determined by three influence factors: conceptual hierarchy, overlap degree of upper relationship and weight of relationship path[22], which is represented as $Relevant^{\text{hiearrchy}}(A, B)$, $Relevant^{\text{anc}}(A, B)$ and $PathWeight(A, B)$ respectively. $\theta_1, \theta_2, \theta_3$ represent the weights of these impact factors. This paper uses empirical value concluded in literature [21], $\theta_1 = 0.2, \theta_2 = 0.3, \theta_3 = 0.5$, and proposes Calculation formula of similarity degree:

$$Sim(X, Y) = \delta_1 \times Sim_{\text{acc}}(X, Y)$$
$$+ \delta_2 \times Sim_{\text{act}}(X, Y) + \delta_3 \times Sim_{\text{attr}}(X, Y)$$
$$+ \delta_4 \times Sim_{\text{pur}}(X, Y) + \delta_5 \times Sim_{\text{des}}(X, Y) \qquad (2)$$

where, X and Y represent two scenic spots, and $Sim_{acc}(X,Y)$, $Sim_{act}(X,Y)$, $Sim_{ulur}(X,Y)$, $Sim_{pur}(X,Y)$ and $Sim_{des}(X,Y)$ represent the similarity in aspects of companionship, activities, types of scenic spots, tourist intention and destination. Taking the activity similarity as an example, the calculation algorithm of $Sim_{act}(X, Y)$ is given below, assuming that Actx and Acty represent the activity set provided by X and Y:

Algorithm 1.   The calculation of $Sim_{\text{act}}(X,Y)$

---

initial sim_val with 0
SimSet={Similar(α, β), for every $\alpha \in Act_x, \beta \in Act_y$}
do{
    sim is the element in SimSet with largest similar
    value as val with concept a b
    if($a \in Act_x, b \in Act_y$)
        sim_val+=val; $Act_x -= a$; $Act_y -= b$
    endif
    SimSet-=sim
}while($Act_x \neq \emptyset$ & $Act_y \neq \emptyset$)
$Sim_{acr}(X,Y) = \frac{sim\_val}{Min(|Act_x|,||)}$

The algorithm finds the activity concept pair with the highest similarity of the two scenic spots, calculates the similarities and average value, thus obtains the value of $Sim_{act}(X, Y)$. The calculation of the similarity of scenic spot type and tourist intention etc. are similar. Set weights $\delta_1, \delta_2, \delta_3, \delta_4, \delta_5$ as 0.1, 0.2, 0.4, 0.2, 0.1 respectively, according to the importance of each similarity. From formula (2), the similarity of the two scenic spots was obtained. If the value is greater than 0.7, the two scenic spots are considered to be similar scenic spots, and assign is Simuliar property to them.

4.3 Spatio-Temporal Associative Query Algorithm

The system of WDTourism takes into account spatio-temporal information, which is very important in smart tourism.

Time-related query is relatively simple. The third-party map service [23] can be invoked to query the traffic time between two scenic spots. Within the planned travel time, the three relationships mentioned above, such as similarity of scenic spots, proximity of scenic spots and route matching, are considered comprehensively, and the best extended scenic spots are obtained by combining the user's interest degree.

As to spatial-related queries, this paper describes the geographical location of scenic spots based on GeoRSS specification [24], and uses JTS Topology Suite [25] framework to realize the index query of geographical location. JTS provides an API for developing two-dimensional spatial applications, through a spatial index implementation of a quadtree. On this basis, this paper constructs a Jena model to enhance the quadtree, so that it can efficiently respond to spatial queries. The algorithm is described as follows:

Algorithm 2.   Spatial tourism route extension algorithm

Step 1: Parse out the locations Start, End from natural language text.

Step 2: Use Google Map Service to find the path information from Start to End and extract the locations sequence $Destinations = \{Start, M_1, M_2, ..., M_n, End\}$.

Step3: Each location in the sequence is taken as the center of the circle, and a circle is drawn with DISTANCE (determined by the given distance value in query) as the radius. All scenic spots falling within the circle are obtained to form a sub-graph of ontological scenic spots instance.

Step4: earch in the Sub-graph and get scenic spots collections Attrs that meet the requirements. For each scenic spot in Attrs, the route is extended according to the scenic spot association.

Step5: Output all Routes extended from Attrs, reorder and sort according to user preference model

---

4.4 User Personalization Modeling Based on Semantics

The ontology-based user model focuses on measuring the user's interest preference, and the implicit user preference can be obtained through three kinds of interaction between the user and the system: retrieval request, browsing scenic spot information, and adding scenic spot to planning route. Each behavior indicates the user's interest increment on a certain concept. The quantity relationship between the defined interest increment and user behavior is shown in Table 1. In the table, "Browsing scenic spot information" uses Logistic function model to describe the changing regular of users' interest in scenic spots with the elapsed time of browsing.

Table 1. Interest increment and user behavior.

Interaction	Interest incremental ΔI
Retrieval request (Query)	0.1
Browsing scenic spot information (Read)	$Logistic(T_p(A))$
Add to route planning (Visit)	0.5

In the Table 1,

$$Logistic(T_p(A)) = \frac{1}{1 + e^{-(a+b \times \mathrm{Tp(A)})}} \tag{3}$$

$$T_p(A) = \frac{Time(A)}{word(A)^u} \tag{4}$$

In formula (3), a and b are parameters to be estimated. We took $a = 7.356$, $b = 0.979$, which is estimated by maximum likelihood method from literature [26]. In formula (4), A indicates preview scenic spot; $Time(A)$ represents the total time spent in preview; $Word(A)$ represents the total word count of scenic spot information, and u is the adjustment parameter, which is used to eliminate the effect of too much content resulting in too long preview time on the results of interest measurement [27].

Algorithm 3.   The update algorithm for user's interest value

---

```
update(Con, ΔI )
    if Con is equal to "Thing", reaching the root of the tree
    return
    else if
    I(Con) = I(Con)+ ΔI ;
    Sup = GetParent(Con);
    ΔI = Similar(Sup,Con)* ΔI ;
    update(Sup, ΔI );
    endif
end
```

Because of the hierarchical relationship between ontology concepts, users' interest in a concept also implies that they have a certain interest in the related concepts. $I(Ci)$ is defined as the user's interest value (0–1) for a concept. If $I(Ci)$ is updated with value ΔI, initial Con represents the concept updated because of user behavior, Sup represents its epistatic concept, then the algorithm is described as follows:

Con represents the concept to be updated in the extension process. If Con is equal to Thing, it indicates that the top of the concept tree is reached, then the value is returned directly. Otherwise, increase concept interest value and find the parent concept of Con. Calculate the increment of interest of the parent concept iteratively until the root node is reached. Finally, the degree of interest should be normalized to the range of 0–1. Now the user's interest in scenic spots can calculated. Define the user's personalized preference vector as UCI and the feature description vector of scenic spot as ACI, as shown in formula (5). the calculation rules are as follows: If scenic spot A contains the concept Ci, (Ci) = 1; If scenic spot A does not contain Ci, (Ci) = 0.

$$UCI(A) = (I(C_1), I(C_2), \ldots, I(C_n))$$
$$ACI(A) = ((C_1), (C_2), \ldots, (C_n))$$

(5)

Pearson similarity calculation formula was used to calculate the similarity between $UCI(A)$ and $ACI(A)$, and the result was the user's interest in the scenic spot. Based on the calculation results, the sequence of scenic spots returned by the search can be sorted, and the scenic spots with high interest can be directly recommended to the user, while the calculation of route interest can be regarded as the sum average of scenic spots interest. So far, this system has realized the personalized recommendation of scenic spots and routes for users.

5   System Implementation and Results

WDTourism prototype system was implemented based on the information of 343 scenic spots in Taiwan. The front-end user interface is implemented using Angular framework and D3 Javascript API, and the back-end is implemented using the Java EE framework. Protégé was used to establish and maintain OWL ontologies, and Jena framework was used to program operations on ontologies. The user interface is mainly divided into three modules: retrieval module, route planning module and detailed information browsing module.

The interface of retrieval module is shown in Fig. 5. The module provides semantic retrieval function, and the results are marked with flags on Google Map. Users can click to query scenic spots and traffic information, and click "Learn more" to enter the detailed information preview module.

Fig. 5. Retrieval module interface

The route planning module provides personalized route planning and customization, supports spatial-temporal route recommendation and user-defined route planning, and displays the results with the force guide chart provided by D3. Figure 6 is a route planning based on space. Users input "From Taipei to Kaohsiung, want to visit the aboriginal culture of Taiwan on the way". The system will recommend a variety of routes for users to choose according to the spatial relationship of scenic spots.

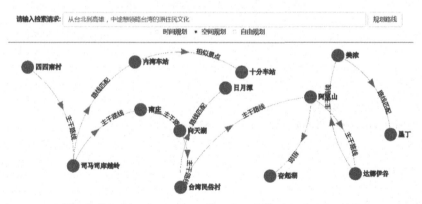

Fig. 6. Displays the query results of spatially associated routes

In order to verify the effectiveness of the similarity calculation of scenic spots and route matching algorithm, the calculated results were compared with the expert judgment results. The experimental results are shown in Table 2, which shows that the similarity between system judgment and expert judgment is as high as 95.57%.

Table 2. Similar scenic spot analysis data.

SpotA	SpotB	Similarity %	Determined by system	Determined by experts
Taipei 101	Gaoxiong85	93.62	Similar	Similar
Ye Liu	Kending	71.26	Similar	Dissimilar
Tailu ge	Ali Hill	86.12	Similar	Similar
Riyue Lake	Tailuge	45.12	Dissimilar	Dissimilar
Qixing Lake	Riyue Lake	67.22	Dissimilar	Dissimilar
Xingtian Temple	Palace Museum	11.34	Dissimilar	Dissimilar

The correctness of the functions of the system is illustrated by comparative experiment. Firstly, a total of 100 scenic spot retrieval statements were collected from people from different fields. For these retrieval statements, the scenic spots are selected from the scenic spots knowledge base in accordance with WDToufism system retrieval, expert recommendation and user's own choice respectively. The accuracy, recall rate and F-measure of the three groups of results were calculated. Comparative analysis was conducted, and the results are shown in Table 3, which shows that the accuracy of the system is close to 90%, and the recall rate is not less than 75%. It indicates that WDTourism system recommendation effect is pretty good.

Table 3. Comparative experiment results

The experimental type	Precision rate%	Recall rate %	F-Measure ($\beta = 1$)
WDTourism recommendation VS. experts recommendation	89.45	75.95	0.755 6
experts recommendation VS User's choice	91.75	81.77	0.805 4
WDTourism recommendation VS User's choice	93.35	70.83	6

6   Conclusion

The semantic web based tourism recommendation system constructed in this paper has the following main innovations and advantages compared with the traditional systems: The retrieval system provided by our system is ontology-based, which realizes the semantic annotation of the resources of the scenic spot. Besides, it utilizes the related technologies of natural language processing, thus enhances the expression ability of the user search, and can capture the user's implied intention. It excavates the relationship between different scenic spot, combined with the spatial-temporal modeling, thus realizing the dynamic customization of the tourism route, supporting the constraints of time and space. The system also implements ontology-based user portraits modeling, calculates the user preference dynamically by captures and analyzes the information generated in the process of user's interacting with the system. Combining the above functions, WDTourism system implements the intelligent personalized tourism recommendation.

Some aspects of the system are worth further studying and improving. For instance, semantic annotation of scenic spots still requires a lot of human involvement, and deep learning algorithms can be considered for automatic tagging. Besides, the system has limitations in processing users' natural language input, and complex input makes it difficult to accurately analyze user's intentions.

References

1. The State Council of the People's Republic of China. Tourism Development Plan of the 14th Five-year Plan (2021)
2. Neehes, R.F., Finin, R., Gruber, T., et al.: Enabling technology for knowledge sharing. AI Maga. **12**, 36–56 (1991)
3. Borst, W.N.: Construction of Engineering Ontologies for Knowledge Sharing and Reuse. PhD thesis, University of Twente, Enschede (1997)
4. Studer, R., Benjamins, V.R., Fensel, D.: Knowledge engineering, principles and methods. Data Knowl. Eng. **25**(1–2), 161–197 (1998)
5. Gruber, T.R.: Towards principles for the design of ontologies used for knowledge sharing. Int. J. Hum Comput Stud. **43**, 907–928 (1995)
6. Jakkilinki, R., Georgievski, M., Sharda, N.: Connecting destinations with ontology based e-tourism planner. In: ENTER 2007: 14th annual conference of IFITT, the International Federation for IT & Travel and Tourism. Ljubljana, Slovenia, 24–26 January 2007 (2007)
7. Daramola, O., Adigun, M., Ayo, C.: Building an ontology-based framework for tourism recommendation services. Inf. Commun. Technol. Tour. **2009**, 135–147 (2009)
8. K.R.Ananthapadmanaban, Dr.S.K.Srivatsa. Personalization of user Profile: Creating user Profile Ontology for Tamilnadu Tourism .International Journal of Computer Applications.Volume 23– No.8, June 2011 (42–47)
9. Moreno, A., Valls, A., Isern, D., et al.: Sigtur/e-destination: ontology-based personalized recommendation of tourism and leisure activities. Eng. Appl. Artif. Intell. **26**(1), 633–651 (2013)
10. Am, A., Vn, A., Js, B.: Tourism recommendation system based on semantic clustering and sentiment analysis. Expert Syst. Appl. **167**, 114324 (2020)
11. Feng, L.: Design of tourism intelligent recommendation model of mount tai scenic area based on knowledge graph. In: 2020 International Conference on E-Commerce and Internet Technology (ECIT) (2020)
12. Navío-Marco, J., Ruiz-Gómez, L.M., Sevilla-Sevilla, C.: Progress in information technology and tourism management: 30 years on and 20 years after the internet - Revisiting Buhalis & Law's landmark study about eTourism. Tour. Manag. **69**, 460–470 (2018)
13. Lohvynenko, C., Nedbal, D.: Usage of semantic web in Austrian regional tourism organizations. In: Acosta, M., Cudré-Mauroux, P., Maleshkova, M., Pellegrini, T., Sack, H., Sure-Vetter, Y. (eds.) SEMANTiCS 2019. LNCS, vol. 11702, pp. 3–18. Springer, Cham (2019). https://doi.org/10.1007/978-3-030-33220-4_1
14. Stanford. Stanford CoreNLP [EB/OL]. Accessed 01 Apr 2022. http://stanfordnlp.github.io/CoreNLP/

15. Stanford. Stanford Protégé[EB/OL]. Accessed 01 Apr 2022. http://protege.stanford.edu
16. Sirin E.Pellet in Java[EB/OL]. Accessed 01 Apr 2022. https://github.com/stardog-union/pellet
17. Noy, N.F., McGuinness, D.L.: Ontology Development 101: A Guide to Creating Your First Ontology (2001)
18. Dell'Erba, M., et al.: Harmonise: a solution for data interoperability. In: Towards the Knowledge Society: eCommerce, eBusiness and eGovernment eCommerce and Tourism Research Laboratory, ITC-Irst, Italy (2003)
19. Mondeca.Mondeca Tourism Ontology [EB/OL]. Accessed 01 Apr 2022. http://www.mondeca.com
20. OpenTravel Alliance. OTASpecification[EB/OL]. Accessed 01 Apr 2022. http://www.opentravel.org/Specifications/Default.aspx
21. Editorial Board of Chinese Library Law. Chinese Classified Thesaurus, 2nd edn. Beijing Library Press, Beijing (2005)
22. Li, Z., Du, J.: A computing model of concept semantic similarity based on tourism domain ontology. In: Processsdings of the 31st Chinese Control Conference, Hefei, China, pp. 3863–3868 (2012)
23. Google.Google Map API[EB/OL]. Accessed 01 Apr 2022. https://developers.google.com/maps/
24. OGC. GeoRSS[EB/OL]. Accessed 01 Apr 2022. http://www.georss.org
25. GeoConnections.JTS Topology Suite[EB/OL]. http://www.vividsolutions.com/jts
26. Pan, J.: Research on User Modeling Technology and Application Based on Semantics. Shanghai university, Diss (2010)
27. Li, Y.: Personalized User Modeling Technology and Application Based on Ontology in Intelligent Retrieval. National University of Defense Technology, Diss (2002)

Author Index

Abad, Zahra Shakeri Hossein   682
Acquaah, Yaa T.   659
Aldaej, Abdulaziz   112
Almpanidis, George   819
Altakrouri, Bashar   621
Amayri, Manar   431
Anastasiia, Rozhok   633
Andaluz, Victor H.   898
Anh, Bui Thi Mai   226, 240
Asahara, Akinori   714
Avtar, Ram   470

Baouya, Abdelhakim   555
Bastidas, Daniela A.   898
Bechikh, Slim   112
Belkahla Driss, Olfa   416
Bensalem, Saddek   555
Bermejo, Pablo Sánchez   725
Bhavsar, Mansi   505
Billami, Mokhtar Boumedyen   327
Bodó, Zalán   528
Bokolo, Biodoumoye George   516
Bortolaso, Christophe   211, 327
Bouguila, Nizar   431
Bouraoui, Zied   859
Brinton, Joe   797

Cabrerizo, Francisco Javier   747
Cai, Gaoyan   15
Cai, Hongming   51
Cano, José-Ramón   725
Cao, Hai-Nam   351
Cao, Tuan-Dung   351
Chai, Yuan   15
Chao, Guoqing   124
Chen, Chun-Hao   460
Chen, Jiahui   807
Chen, Sheng-Shan   101
Chen, Yu-Chun   77
Cheng, Ming-Shien   77
Chhabra, Vipul   470
Chiclana, F.   409
Constante, Patricia N.   898
Cristani, Matteo   315

Dai, Daixin   920
Dai, Kaiyu   920
Dai, Shengqi   199
Dang, Thai   303
Dao, Hong N.   831
Dao, Minh-Son   597
Dejournett, Rodney   452
del Moral, M. J.   409
Derras, Mustapha   211, 327
Deuse, Jochen   3
Diallo, Azise Oumar   397
Ding, Meirong   15
Dinh, Tran Thi   240
Do, Duc-Thai   351
Do, Nguyet Quang   497
Do, Thanh-Nghi   737
Doniec, Arnaud   397
Driss, Olfa Belkahla   648
Duc, Le Minh   226
Dumitru, Vlad Andrei   543

Elkout, Elhem   648
Emanuele, Adorni   633
Excell, Ying   29

Fan, Gaojuan   819
Fan, Wentao   431
Far, Behrouz   682
Fontanili, Franck   211
Fournier-Viger, Philippe   609, 819
Fujita, Hamido   275, 819

Gallegos, Dayana E.   898
Gan, Wensheng   15, 807
García, Salvador   725, 773
Gargouri, Faiez   859
Goh, Hui-Ngo   382, 695
Gokaraju, Balakrishna   505, 659
Gong, Chuanyang   339
González-Almagro, Germán   725, 773
Gosset, Camille   327
Granmo, Ole-Christoffer   844
Guo, Jiaxun   431
Guo, Weibin   807
Guo, Yike   184

Gurrin, Cathal   907
Gwyn, Tony   443, 452

Ha, Thanh-Le   303
Hamami, Mohd Ghazali Mohd   253
Hameed, Shilan S.   485
Harris, Geoff   707, 891
Healy, Graham   907
Hernandez V., Jeffrey J.   452
Herrera, Francisco   773
Herrera-Viedma, Enrique   409, 747
Ho, Ian H. J.   382
Hoang, Van-Dung   160
Honda, Kosuke   275
Hou, Guoxin   51
Hsu, Ping-Yu   77
Huang, Yaohui   124
Huynh, Anh T.   160
Hwang, Dosam   571

Ichise, Ryutaro   877
Ihara, Koya   63
Ismail, Zool H.   253

Jahan, Munima   682
Ji, Pengfei   920
Jiang, Weiliang   137
Jiao, Lei   844
Johnson, David   443

Kamimura, Ryotaro   669
Kato, Shohei   63
Kelly, John   505
Kim, Bong-Min   287
Kiran, R. Uday   470, 831
Kitajima, Ryozo   669
Krejcar, Ondrej   485, 497
Kurematsu, Masaki   275

Lafourcade, Mathieu   327
Lamine, Elyes   211
Latiff, Liza Abdul   485
Le, Trong T.   160
Le, Tung   303
Lee, Hou-Tsan   101
Li, Jingwen   184
Li, Zhenglong   199
Li, Ziyang   807
Likitha, P.   831
Lim, Amy Hui-Lan   695

Lim, Kok Cheng   497
Lin, Jerry Chun-Wei   460
Link, Sebastian   29
Liu, Siqi   759
Liu, Yue   184
Liu, Yuxi   89, 147
Liu, Zhipeng   505
Louati, Ali   112
Louati, Hassen   112
Loukil, Zied   859
Lozenguez, Guillaume   397
Luong, Huong Hoang   785

Maaloul, Alia   416
Madushanka, Tiroshan   877
Mandiau, René   397
Mester, Attila   528
Minh, Nguyen Le   303
Mohamed, Rim   859
Morente-Molinera, Juan Antonio   747

Nagappan, Sidharrth   695
Nandini, Durgesh   871
Nannuri, Udayasri   452
Nguyen, Bach Hoang Tien   370
Nguyen, Binh T.   583, 907
Nguyen, Cuong T. V.   583
Nguyen, Dung Manh   370
Nguyen, Hai Thanh   785
Nguyen, Ha-Thanh   363
Nguyen, Hien D.   160
Nguyen, Kim Anh   303
Nguyen, Le-Minh   363
Nguyen, Manh-Duy   907
Nguyen, Ngoc Thanh   571
Nguyen, Phuong   303
Nguyen, Thao   583
Nguyen, Ti-Hon   737
Nguyen, Trang Thi Thu   370
Nguyen, Trung T.   583
Ninh, Van-Tu   907
Niu, Xinzheng   609
Nouri, Houssem Eddine   416
Novykov, Volodymyr   707, 891

Oki, Shota   797
Oladimeji, Damilola   516
Osakabe, Yoshihiro   714
Ouchani, Samir   555

Pai, Tun-Wen   101
Paik, Incheon   831
Park, Seong-Bae   287
Pérez, Ignacio Javier   747
Phan, Huyen Trang   571
Pingaud, Hervé   211
Powlesland, Francis   621
Pu, Yangyi   42

Qian, Quan   184
Qin, Shaowen   89, 147
Qingge, Letu   443
Quach, Quyen Thuc   785

Raj, Bathala Venus Vikranth   831
Ratcliffe, Chandlor   516
Ravikumar, Penugonda   831
Razak, Shukor A.   485
Recalde, Luis F.   898
Reddy, P. Krishna   470
Roberto, Revetria   633
Roy, Kaushik   443, 452, 505, 659

Said, Lamjed Ben   112
Sakuma, Takuto   63
Satoh, Ken   363
Scannapieco, Simone   315
Schmid, Ute   621
Schwenken, Jörn   3
Selamat, Ali   485, 497
Sergey, Suchev   633
Shafik, Rishad   844
Shandiz, Amin Honarmandi   265
Shen, Bingqing   51
Shigeno, Maiko   797
Smyth, Sinéad   907
Song, Wei   759
Song, Young-In   351
Suárez, Juan Luis   725, 773
Sun, Yan   51

Tam, Vincent W. L.   199
Tan, Yi-Fei   382
Tapia, J. M.   409
Tejima, Kazuki   597
Tesiero III, Raymond C.   659
Thai-Nghe, Nguyen   173
Tomazzoli, Claudio   315
Tonkin, Isaac   707, 891
Tóth, László   265

Tran, Anh M.   583
Tran, Chi Le Hoang   785
Tran, Khanh   303
Tran, Minh-Triet   907
Tran, Viet-Trung   351
Trang, Nguyen Thi Thu   240
Trifa, Zied   416
Trillo, José Ramón   747
Trung, Bui Quoc   226
Tunheim, Svein Anders   844

Van-Binh, Nguyen   173
Varela-Aldás, José   898
Vu, Sang   160

Wan, Shicheng   807
Wang, Bing   609
Wang, Chao-Hung   101
Wang, Chongyu   51
Wang, Mingzhao   137
Wang, Zhijin   124
Watanabe, Minami   63
Watanobe, Yutaka   831
Wehner, Christoph   621
Wei, Zhihua   339
West, Nikolai   3
Wotawa, Franz   543
Wu, Xing   184

Xiao, Huaiyuan   819
Xiao, Juan   470
Xie, Juanying   137
Xu, Ni   77

Yadav, Rohan Kumar   844
Yan, Jingwen   124
Yan, Zhijie   51
Yang, Xin   797
Yang, Yi-Chen   460
Ye, Zhenqiang   807
Yeung, L. K.   199
Yu, Hongyang   42
Yuan, Xiaohong   452

Zeng, Biqing   15
Zettsu, Koji   597
Zhang, Chongsheng   819
Zhang, Hang   15
Zhang, Liwen   211

Zhang, Peisong 124
Zhang, Xu 609
Zhang, Zhenhao 147
Zheng, Cong-Han 77

Zhou, Bing 516
Zhu, Min 51
Zhu, Ping 339
Zuo, Xiaorui 920